Functionalized Nanomaterials I

Functionalized Nanomaterials I

Fabrications

Edited By
Vineet Kumar
Praveen Guleria
Nandita Dasgupta
Shivendu Ranjan

CRC Press
Taylor & Francis Group
Boca Raton London New York

CRC Press is an imprint of the
Taylor & Francis Group, an **informa** business

First edition published 2021
by CRC Press
6000 Broken Sound Parkway NW, Suite 300, Boca Raton, FL 33487-2742

and by CRC Press
2 Park Square, Milton Park, Abingdon, Oxon, OX14 4RN

ISBN: 978-0-8153-7041-3 (hbk)
ISBN: 978-1-351-02162-3 (ebk)

Typeset in Times
by Deanta Global Publishing Services, Chennai, India

Contents

Preface

The demand for affordable, environment-friendly, and sustainable materials has encouraged scientists in different fields to draw inspiration from nature in developing materials with unique properties, such as miniaturization, hierarchical organization, and adaptability. Together with the exceptional properties of nanomaterials, the field of functionalized nanomaterials has taken a huge leap over the past century. On one hand, the sophistication of hierarchical structures endows biological systems with multi-functionality. Simultaneously, synthetic control on the creation of nanomaterials permits/warrants the design of materials with specific functionalities. Biological templates are architectures that behave as containers such as viral capsids. Specifically, these biological containers can function as carriers for DNA assays, immunoassays, drugs, and catalysts, and can also be used in novel material synthesis. In recent times, researchers have also used biological macromolecular assemblies as templates for the construction of novel functional nanomaterials. The human biological system is made up of a nanoscale self-assembly of biological molecules. Functionalized nanomaterials have been applied for the fabrication of organs-on-chips and smart robotic devices. Thus, by varying the type of functionalization by chemistry, nanomaterials can be intentionally designed for several distinct applications.

This book contains nineteen chapters describing the basic principles, trends, and challenges in the synthesis of functionalized nanomaterials. **Chapters 1 and 2** are introductory chapters that provide a basic understanding of the strategies available for the functionalization of nanomaterials and their characterization. **Chapters 3–5** discuss the functionalization of metallic nanomaterials. **Chapter 6** describes the formation of nanofilms as and/or using functionalized nanomaterials. **Chapters 7–9** provide information regarding the functionalization of carbon-based nanomaterials and quantum dots. Currently, carbon-based nanomaterials and quantum dots are commonly used nanomaterials for industrial applications. **Chapters 10–13** describe the functionalization of polymeric nanomaterials including biological nanomaterials that are very useful for drug delivery and tissue engineering applications. **Chapters 14–17** focus over the functionalization of silica, clay, and other inorganic nanocomposites that are useful for tissue engineering, material chemistry, and environmental applications. **Chapter 18** discusses the use of microorganisms to functionalize nanomaterials for various applications. **Chapter 19** discusses the application of nanomaterials for drug delivery, enzyme immobilization, and molecular targeting applications. Collectively, this book addresses critical issues in the functionalization of nanomaterials, such as the way in which chemical functionalization, green synthesis, and bio-inspired synthesis have evolved, and the key benefits of functionalized nanomaterials over conventional synthetic approaches. It also summarizes the chemistry behind functionalized nanomaterials.

This book will be beneficial to help undergraduate and graduate students understand the detailed concept of functionalized nanomaterials and the strategies of nanomaterial fabrication. Doctoral and post-doctoral scholars will benefit from the use of this book through learning the basics of techniques, recent advancements, challenges, and opportunities in this field. This book will provide critical and comparative data for nanotechnologists and thus it will be beneficial for scientists and researchers working in this field. It will also be beneficial for academicians by providing the basics of the fabrication of functionalized nanomaterials, as many universities throughout the world include nano-biotechnology as a subject that cannot be completed without bio-inspired nanomaterials. Additionally, this book will also be of value to industry personnel, journalists, policy makers, and the common public, as it helps understand functionalized nanomaterials in detail and in depth.

Vineet Kumar
Praveen Guleria
Shivendu Ranjan
Nandita Dasgupta

Editors

Vineet Kumar is currently working as an assistant professor in the Department of Biotechnology, Lovely Professional University (LPU), Jalandhar, Punjab, India. Previously, he was assistant professor at DAV University, Jalandhar, Punjab, India, and UGC-Dr DSK post-doctoral fellow (2013–2016) at the Department of Chemistry and Centre for Advanced Studies in Chemistry (CAS), Panjab University, Chandigarh, U.T., India. He has worked in different areas of biotechnology and nanotechnology in various institutes and universities, namely, CSIR-Institute of Microbial Technology, Chandigarh, U.T., India; CSIR-Institute of Himalayan Bioresource Technology, Palampur, H.P. India; and Himachal Pradesh University, Shimla, H.P. India. His areas of interest include green synthesis of nanoparticles, nanotoxicity testing of nanoparticles, and application of nanoparticles in drug delivery, food technology, sensing, dye degradation, and catalysis. He has published many articles in these areas featuring in peer-reviewed journals. He is also serving as an editorial board member and reviewer for international peer-reviewed journals. He has received various awards, such as senior research fellowship, best poster award, post-doctoral fellowship, etc. He is currently in the final stages of editing two books each for CRC Press/ Taylor & Francis Group, Taylor & Francis, and Springer-Nature.

Praveen Guleria is presently working as assistant professor in the Department of Biotechnology at DAV University, Jalandhar, Punjab, India. She has worked in the areas of plant biotechnology, plant metabolic engineering, and plant stress biology at CSIR-Institute of Himalayan Bioresource Technology, Palampur, H.P. India. Her research interests include plant stress biology, plant small RNA biology, plant epigenomics, and nanotoxicity. She has published several research articles in various peer-reviewed journals. She is also serving as an editorial board member and reviewer for certain international peer-reviewed journals. She has been awarded the SERB-Start Up Grant by DST, GOI. She has also been awarded the prestigious "Bharat Gaurav Award" in 2016 by the India International Friendship Society, New Delhi. Furthermore, she has received various awards such as CSIR/ICMR-Junior Research Fellowship, CSIR-Senior Research Fellowship, and state-level merit scholarship awards. She is currently editing a book on nanotoxicity for Springer-Nature.

Nandita Dasgupta has vast working experience on micro/nanoscience and is currently serving in VIT University, Vellore, Tamil Nadu, India. She has been involved with various research institutes and industries, including CSIR-Central Food Technological Research Institute, Mysore, India, and Uttar Pradesh Drugs and Pharmaceutical Co. Ltd., Lucknow, India. Her areas of interest include the fabrication of micro/nanomaterials and their applications in different fields, especially medical, food, environmental, agricultural, biomedical, etc. She has published many books with Springer and has contracted a few with Springer, Elsevier, and CRC Press. She has also published many scientific articles in international peer-reviewed journals and is serving as an editorial board member and referee for reputed international peer-reviewed journals. She has received the Elsevier certificate for "Outstanding Contribution" in reviewing from Elsevier, The Netherlands. She has also been nominated for the advisory panel for Elsevier, The Netherlands. She is the associated editor in *Environmental Chemistry Letters* – a Springer journal of 3.59 impact factor. She has received several national and international awards and recognitions from different organizations.

Shivendu Ranjan is currently the director of the Centre for Technological Innovations and Industrial Research (CTIIR), at South Asian Institute for Advanced Research and Development (SAIARD), Kolkata (an institute certified by the Ministry of Micro, Small, and Medium Enterprises, Govt. of India). He is also working as guest/visiting faculty at the National Institute of Pharmaceutical Education and Research-R (NIPER-R),

Lucknow, Ministry of Chemicals and Fertilizers, Govt. of India. Previously, he worked as scientist at the DST-Centre for Policy Research, Lucknow, supported by the Ministry of Science and Technology, Govt of India. He is also serving as a senior research associate (visiting) at the Faculty of Engineering and Built Environment, University of Johannesburg, South Africa. He has a lot of expertise in micro/nanotechnology and is currently working in VIT University, Vellore, Tamil Nadu, India. His area of research is multidisciplinary, and includes micro/nanobiotechnology, nano-toxicology, environmental nanotechnology, nanomedicine, and nanoemulsions. He has published many scientific articles in international peer-reviewed journals. He has recently published five edited books with Springer and has contracted three books with Elsevier and four with CRC Press. All of these books cover vast areas of applied micro/nanotechnology. He has vast editorial experience. He is serving as associate editor in *Environmental Chemistry Letters* (Springer Journal with 3.59 impact factor); moreover, he is serving in the editorial panel of *Biotechnology and Biotechnological Equipment* (Taylor and Francis with 1.05 impact factor). He is also executive editor and expert board panel in several other journals. He has been recently nominated as Elsevier advisory panel, The Netherlands. He has received several awards and honors from different national and international organizations.

List of Contributors

S. Avinash
Department of Science and Humanities
MLR Institute of Technology
Hyderabad, India
and
School of Chemistry
Zhejiang University
Hangzhou, China

Kalpana Awasthi
Department of Physics
K.N. Govt. P.G. College Gyanpur
Bhadohi, India

Ayaz Baayramov
Institute of Physics Azerbaijan Academy of Sciences
Baku, Azerbaijan
and
Department of Physics
University of the Free State
Bloemfontein, South Africa

Wissam Bessa
UER Procédés Energétiques
Ecole Militaire Polytechnique
Algiers, Algeria

Aman Bhalla
Department of Chemistry and Center for Advanced Studies in Chemistry
Panjab University
Chandigarh, India

K. Byrappa
Mangalore University
Mangalore, India

Giuseppe Cavallaro
Dipartimento di Fisica e Chimica
Università degli Studi di Palermo
Viale delle Scienze
Palermo, Italy

A. Chander
Department of Chemistry
Zhejiang University
Hangzhou, China

M.A. Singara Charya
Department of Microbiology
Kakatiya University
Warangal, India

Ganga Ram Chaudhary
Department of Chemistry and Center for Advanced Studies in Chemistry
Panjab University
Chandigarh, India

Savita Chaudhary
Department of Chemistry and Centre of Advanced Studies in Chemistry
Panjab University
Chandigarh, India

Pooja Chauhan
Department of Chemistry and Centre of Advanced Studies in Chemistry
Panjab University
Chandigarh, India

Biswajit Choudhury
Institute of Advanced Study in Science and Technology
Paschim Boragaon, Vigyan Path
Assam, India

Subhasree Roy Choudhury
Institute of Nano Science and Technology
Habitat Centre, Phase-10
Punjab, India

Merideth A. Cooper
Department of Chemical and Biomolecular Engineering
The Ohio State University
Columbus, OH

Mehdi Derradji
UER Procédés Energétiques
Ecole Militaire Polytechnique
Algiers, Algeria

Nitesh Dhiman
Nanomaterials Toxicology Group
CSIR-Indian Institute of Toxicology Research (CSIR-IITR)
Lucknow, India
and
Academy of Scientific and Innovative Research (AcSIR)
Ghaziabad, India

Benjamin Ezekoye
Department of Physics and Astronomy
University of Nigeria
Nsukka, Nigeria

Veronica A. Ezekoye
School of General Studies
University of Nigeria
Nsukka, Nigeria

T. Guping
School of Chemistry
Zhejiang University
Hangzhou, China

M. Hazwan Hussin
Materials Technology Research Group (MaTReC)
Universiti Sains Malaysia
Penang, Malaysia

Neha Jindal
Department of Chemistry
DAV College Bathinda
Punjab, India

J. M. Kalita
Department of Physics
Cotton University
Guwahati, India

Rohini Kanwar
Department of Chemistry and Centre for Advanced Studies in Chemistry
Panjab University
Chandigarh, India

Shamsher S. Kanwar
Department of Biotechnology
Himachal Pradesh University
Summer Hill, India

Aditya K. Kar
Nanomaterials Toxicology Group
CSIR-Indian Institute of Toxicology Research (CSIR-IITR)
Lucknow, India
and
Academy of Scientific and Innovative Research (AcSIR)
Ghaziabad, India

Surajit Karmakar
Institute of Nano Science and Technology
Habitat Centre, Phase-10
Punjab, India

R. Karthik
Department of Electronics and Communication Engineering
MLR Institute of Technology
Hyderabad, India

Babita Kaundal
Institute of Nano Science and Technology
Habitat Centre, Phase-10
Punjab, India

Navneet Kaur
Department of Chemistry and Center for Advanced Studies in Chemistry
Panjab University
Chandigarh, India

Garima Khanna
Institute of Nano Science and Technology
Habitat Centre, Phase-10
Punjab, India

Gun-Do Kim
Department of Microbiology
Pukyong National University
Busan, Republic of Korea

G. Krishna
Department of Microbiology
Kakatiya University

Kishore Kumar
Department of Biotechnology
Himachal Pradesh University
Summer Hill, India

Giuseppe Lazzara
Dipartimento di Fisica e Chimica
Università degli Studi di Palermo
Palermo, Italy

S. S. Maitra
Jawaharlal Nehru University
New Delhi, India

Kiranmai Mandava
Department of Pharmaceutical Chemistry
Bharat Institute of Technology
Hyderabad, India

Oussama Mehelli
UER Procédés Energétiques
Ecole Militaire Polytechnique
Algiers, Algeria

S. K. Mehta
Department of Chemistry and Centre for Advanced Studies in Chemistry
Panjab University
Chandigarh, India

Stefana Milioto
Dipartimento di Fisica e Chimica
Università degli Studi di Palermo
Palermo, Italy

Ho Soon Min
Centre for Green Chemistry and Applied Chemistry
INTI International University
Negeri Sembilan, Malaysia

Ayan Mukherjee
RRS College
Patliputra University
Patna, India

B. Murali
School of Chemistry
University of Hyderabad
Hyderabad, TS, India

S. Neha
Chulalongkorn University
Bangkok, Thailand

Filippo Parisi
Dipartimento di Fisica e Chimica
Università degli Studi di Palermo
Palermo, Italy

Trilok Kumar Pathak
Teerthanker Mahaveer University
Moradabad, India

Satyakam Patnaik
Nanomaterials Toxicology Group
CSIR-Indian Institute of Toxicology Research (CSIR-IITR)
Lucknow, India
and
Academy of Scientific and Innovative Research (AcSIR)
Ghaziabad, India

Chander Prakash
M.L.S.M. College
Sunder Nagar, India

Maheshkumar Prakash Patil
Research Institute for Basic Sciences
Pukyong National University
Busan, Republic of Korea

V. Pranitha
Department of Botany
Kakatiya University
Warangal, India

M. Priyanka
Department of Science and Humanities
MLR Institute of Technology
Hyderabad, India

Mahaveer P. Purohit
Water Analysis Laboratory
CSIR-Indian Institute of Toxicology Research (CSIR-IITR)
Lucknow, India
and
Academy of Scientific and Innovative Research (AcSIR)
Ghaziabad, India

Uma Rajeswari B.
Department of Pharmaceutical Chemistry
Bharat Institute of Technology
Hyderabad, India

S. N. Rath
Department of Biomedical Engineering
Indian Institute of Technology
Hyderabad, India

Debabrata Saha
Department of Electrical Engineering
Indian Institute of Technology
Kanpur, India

R. Santhosh
School of Chemistry
University of Hyderabad
Hyderabad, India

M. P. Sarma
Department of Physics
Cotton University
Guwahati, India

Luciana Sciascia
Dipartimento di Scienze della Terra e del Mare
Università degli Studi di Palermo
Palermo, Italy

Abhishek Sharma
Department of Biotechnology
Himachal Pradesh University
Summer Hill, India

Ashok Kumar Sharma
School of Physics
Shri Mata Vaishno Devi University
Kakryal, India

Shweta Sharma
Directorate of Mushroom Research (DMR)
Chambaghat
Solan, India

Tanvi Sharma
Department of Biotechnology and Bioinformatics
Jaypee University of Information Technology
Waknaghat, India

Amrita Singh
Water Analysis Laboratory
CSIR-Indian Institute of Toxicology Research (CSIR-IITR)
Lucknow, India
and
Academy of Scientific and Innovative Research (AcSIR)
Ghaziabad, India

Kulvinder Singh
Department of Chemistry
Maharaja Agrasen University
Baddi, India

Vikas Kumar Singh
Department of Chemistry
Banaras Hindu University
Varanasi, India

A. H. Sneharani
Department of Studies in Biochemistry
Mangalore University
Chikka Aluvara, India

Anup K. Srivastava
Institute of Nano Science and Technology
Habitat Centre, Phase-10
Punjab, India

O. N. Srivastava
Department of Physics
Banaras Hindu University
Varanasi, India

Jingyao Sun
Department of Chemical and Biomolecular Engineering
Ohio State University
Columbus, Ohio
and
College of Mechanical and Electrical Engineering
Beijing University of Chemical Technology
Beijing, China

Reeja Sundaram
Forest College and Research Institute
Dulapally, India

Mahe Talat
Department of Physics
Banaras Hindu University
Varanasi, India

Ahmed Fouzi Tarchoun
UER Procédés Energétiques
Ecole Militaire Polytechnique
Algiers, Algeria

Djalal Trache
UER Procédés Energétiques
Ecole Militaire Polytechnique
Algiers, Algeria

K. Vinay
Ton Duc Thang University
Ho Chi Minh City, Vietnam

Daming Wu
College of Mechanical and Electrical Engineering
Beijing University of Chemical Technology
Beijing, China

Zhaogang Yang
Department of Chemical and Biomolecular Engineering
Ohio State University
Columbus, Ohio
and
Department of Radiation Oncology
University of Texas Southwestern Medical Center
Dallas, Texas

Manesh A Yewale
Department of Physics
Shivaji University
Kolhapur, India

Jing Zhu
Division of Pharmaceutics
Ohio State University
Columbus, Ohio

1

Comprehensive Array of Ample Analytical Strategies for Characterization of Nanomaterials

Nitesh Dhiman, Amrita Singh, Aditya K. Kar, Mahaveer P. Purohit, and Satyakam Patnaik

CONTENTS

1.1 Background

According to the Nanotechnology Characterization Laboratory (NCL) at the National Cancer Institute, National Institute of Health (NIH), Maryland, "nanotechnology refers to research and technology development at the atomic, molecular and macromolecular scale, leading to the skilled and directive study of structures and devices with length scales in the 1 to 100 nanometers (nm) range" (ongMcNeil Scott, 2005). This narrow scale range provides varied opportunities in diverse fields such as surface science (Boles et al., 2016), material sciences (Burda et al., 2003), electronics (Ma et al., 2015), environment science (Mueller and Nowack, 2008), catalysts (Prieto et al., 2012), sensors development (Kuang et al., 2011), energy storage (Wang et al., 2007), and drug delivery and diagnostic agents (Purohit et al., 2017), among many other traits.

However, this upsurge is certainly due to their ease of preparation and subsequent functionalization, which can be easily manipulated by using different ligands of choice. Unlike bulk materials whose material or physical and chemical properties can be assessed, such as molecular structure, chemical composition, melting point, boiling point, vapor pressure, flash point, pH, dispersion, and water octanol partition coefficient in a classical way, only some of the above features can be determined in the case of nanomaterials. Yet, specific properties akin to nanomaterials, such as size/size distribution, composition, porosity, pore size, pore volume, surface area, shape, wettability, zeta potential, adsorption isotherm, aggregation, and distribution of conjugated moieties and impurities of nanomaterials, require standardized protocols and sophisticated identification techniques (Lin et al., 2014). In this regard numerous methods comprising spectroscopic microscopy, X-ray diffraction, light scattering, circular dichroism, magnetic resonance, and zeta potential measurement, beside thermal techniques, centrifugation, chromatography, and electrophoresis, have been developed (Sapsford et al., 2011).

This chapter presents an overview of the principles, applications, merits, and boundaries of various physicochemical techniques that are used to characterize nanomaterials. Although a great deal of literature is already available on the subject, so far research has focused on metrology, stereology, size, and morphology, while focusing little attention on important features such as thermodynamics, surface area, pore volume, surface potential, etc. Therefore, in the present chapter, a sum-up of advanced monitoring techniques is discussed. Initially, we will discuss a summary of atypical macroscopic features which directly correlate to the behavior of nanomaterials. Moreover, the different characterization techniques based on their different features are portrayed in the following sections.

1.2 Overview of Physiochemical Characteristics of Nanomaterials

As discussed in the above section, nanomaterials have distinct physiochemical properties as compared to their bulk materials counterparts. These characteristics offered many advantages in medical applications, including enhanced efficacy and reduced side effects of drugs, often providing better prevention and treatment modalities. Also noteworthy is their applicability in other fields such as the optical, electrical, mechanical, and thermal industries, where they improve thermal or electrical properties (Lin et al., 2014). In the current research paradigm, many researchers are working on the concept of engineered nanostructures with a variety of sizes, morphologies, symmetries, and unorthodox applications to accelerate their potential use. Thus, practical approaches to reliable characterizations of nanomaterials are incumbent for quality assurance, and for the innocuous and rational development of advanced nano-products. The following subsections describe these unique features to nanomaterials.

1.3 Size

The novel properties of nanomaterials such as electronic (Nayak et al., 2017), optical (Kelly et al., 2003), magnetic (Dhiman, Fatima, et al., 2017), thermodynamic (bond length, melting point) (Shuai, 2016), and chemical reactivity (Rao et al., 2002; Robel, Kuno, and Kamat, 2007) are size dependent (Jortner and Rao, 2002). Innate to nanomaterials is the high surface to volume ratio, where an abundance of surface atoms are present vis-à-vis the total number of atoms and, consequently, when particle size decreases, this ratio reciprocally increases. Surface energy is the fundamental factor in the nucleation and growth of nanomaterials (Molleman and Hiemstra, 2018) and provides an explanation of the thermodynamic stability of the system (Elzey, 2010). Therefore, nanomaterials often tend to form agglomerates or aggregates to attain thermodynamic stability. The American Society for Testing and Materials (ASTM) has explained aggregates and agglomerates on the basis of forces that hold particles together. The former are a group of particles bound together by strong interactions or bonds that are not broken apart easily, while, in the latter, particles are bound with weak forces that can be easily disrupted (Walter, 2013). Size has an influence on the practical use of materials as it controls the kinetics of internalization, bio-distribution, cellular membrane deformity, and carrier loading efficiency in nanomaterial-based drug delivery systems. It is well documented that smaller nanomaterials escape the body's innate clearance mechanisms more efficiently and, hence, can remain in the blood for longer periods (Robertson et al., 2016). Various reports validated that nanoparticles (NPs) with an average size of 40 to 50 nm can recruit and bind to receptors for successful membrane wrapping in biological interactions (Behzadi et al., 2017). However, beyond the size of 50 nm, due to larger surface area several receptors can compete to wrap around the NP surface leading to delay in their cellular uptake process, while the smaller-size NPs (< 6 nm) can dodge renal clearance (Longmire, Choyke, and Kobayashi, 2008). NPs with a diameter larger than 200 nm are prone to accumulate in the spleen and liver (Albanese, Tang, and Chan, 2012). Within a confined geometric shape, a nanomaterial's dimensions are a strong determinant for its biological fate as explained by the fact that the rate of uptake and intracellular concentration in certain mammalian cells was maximized for the spherical gold NPs, silica NPs, single-walled carbon nanotubes, and quantum dots of 50 nm diameter (Albanese, Tang, and Chan, 2012).

There are innumerable techniques employed to access NPs size, including field flow fractionation (FFF) (Dulog and Schauer, 1996), nanoparticle tracking analysis (NTA) (Filipe, Hawe, and Jiskoot, 2010), UV-visible absorbance (Haiss et al., 2007), nuclear magnetic resonance (NMR) (Jores, Mehnert, and Mäder, 2003), transmittance electron microscope (TEM) (Dhiman, Singh, et al., 2017), dynamic light scattering (DLS) (Verma et al., 2016), inductively coupled plasma emission spectrometry (ICP-ES) (Pace et al., 2011), X-ray diffraction (XRD) (Pradeep, Priyadharsini, and Chandrasekaran, 2008), and fluorescence spectrometry (Parang et al., 2012).

1.3.1 Morphology

The primary NPs may be spheres, rods, tubes, squares, cubes, and prisms, or come in a variety of irregular shapes. Mock et al., 2002 have found that subtle changes in specific geometrical shapes give distinct spectral responses, causing a shift in the individual particle spectrum (Mock et al., 2002). For example, the effects of NPs with different shapes and their impact on the bond energy in a polyamide/$KTa_{0.5}Nb_{0.5}O_3$ composite were investigated using molecular dynamic simulations. The study concluded that, at the same doping volume fraction, a sphere nanocluster has higher bond energy compared to the NPs shaped differently (Lin et al., 2015). Thus, the differences in morphology also dictate the variation in the surface chemistry of nanomaterials. In the same context, it was observed that nanoscale CdSe aerogel formed from rods and branched particles offers greater surface area compared to hyper-branched NPs (Yu and Brock, 2008). Interestingly, the variations in the morphology of a particular NP elicit differential cellular interactions and, consequently, exhibit considerable impact on toxicity (Peng et al., 2011). Needless to point out, the relative reactivity, sturdiness, and other properties are equally dependent on size, shape, and structure (Khan, Saeed, and Khan, 2017). Through-focus scanning optical microscopy (TSOM) (Attota et al., 2010), TEM (Chen and Ho, 2008), SEM (Liu et al., 2005), atomic force microscopy (AFM) (De Falco et al., 2015), etc., are the leading analytical tools to assess the morphologies of nanomaterials.

1.3.2 Surface Properties

The properties of nanomaterials can be tuned by using protecting agents over the surface, which also dictates their chemical behavior. Not only does the large surface area to volume ratio of nanomaterials offer an enormous interface between a particle and its environment, but it also determines how the two communicate and influence each other (Biener et al., 2009). For drug targeting applications, surface properties are playing a primordial driver towards streamlining the synthesis of nanostructures to ensure the highest possible delivery efficiency or interaction with the target receptor (Desai, 2012). Surface functional density also dictates the cellular fate of nanomaterials. Needless to say, several studies have confirmed the critical role of NPs surface charge in eliciting favorable biological responses. As the cell membrane possesses a negative potential barrier, the cellular uptake of NPs is driven by electrostatic attractions (Behzadi et al., 2017), and hence positively charged NPs enter the cells at a faster rate than the neutral or negative ones. However, in the case of larger NPs (4 to 20 nm), the surface charge alters the lipid bilayers (Wang et al., 2008), forcing them to undergo reconstruction. As such, attachment of negatively charged NPs to lipid bilayer induces local gelation, whereas the attachment of positively charged NPs induces fluidity upon binding. Furthermore, the administration of charged NPs into a biological system also induces a dynamic phenomenon of corona formation made up of multiple proteins around the NPs and dictates long-term fate, metabolism, clearance, and immune response (Srivastav et al., 2019; Albanese, Tang, and Chan, 2012). Surface analytical techniques, such as X-ray photoelectron spectroscopy (Zhang et al., 2004), time-of-flight secondary ion mass spectrometry (TOF-SIMS) (Ghule et al., 2006), secondary neutral mass spectrometry (Laser-SNMS) (Haase et al., 2011), and diffusion-ordered nuclear magnetic resonance spectroscopy (DOSY-NMR) (Gun'ko and Turov, 2013; Zhu et al., 2013; Mansfield et al., 2014) are common methods used for surface coating analysis of NPs.

Contrarily, the negatively charged DNA can tie up with opposite-charged NPs for gene delivery or gene labeling (Trigueros et al., 2019). The neutral-charged NPs have longer blood half-life than the others. Positively charged NPs are cleared more quickly from the blood and cause several complications such as hemolysis and platelet aggregation (Sharma, Madhunapantula, and Robertson, 2012). In principle, there are many established methods for the surface characterization of nanostructured materials. The surface properties can be studied using electron microscopy techniques (Dhiman, Fatima, et al., 2017), AFM (Baer et al., 2013), Auger electron spectroscopy (Korin, Froumin, and Cohen, 2017), X-ray photoelectron spectroscopy (Techane, Gamble, and Castner, 2011), infrared (IR) spectroscopy, and zeta potential (Cheng et al., 2006; Kiefer, Grabow, Kurland, and Müller, 2015).

1.3.3 Composition and Purity

The wide applicability of NPs in cutting-edge sectors requires robust impurities profiling and compositional analysis. For example, TiO_2 exist naturally in three distinct crystallographic phases: anatase, brookite, and rutile. These forms of TiO_2 crystal lattice have found many individual technological applications. For example, anatase structure owing to their high photoactivity was preferred for excellent photocatalytic performance and solar energy conversion. (Li et al., 2005).

Through numerous studies it has become apparent that structure and composition play a pivotal role that dictates the fate of NPs. While within a bio-system, the NPs safety, biocompatibility, stability, bioavailability, encapsulation efficiency, active substance release, and dispersibility of materials are governed by their composition, in the physical domain of materials, composition and purity act upon parameters such as optical tunability, size, morphology, surface characteristics, shell thickness, etc. Taking advantage of the above features, researchers have designed and developed numerous compositions to achieve the desired endpoints such as nano-shells which are prepared by dielectric core material covered by thin metals, especially gold (Ahmadi and Arami, 2014), and nanocapsules made up by a liquid/solid core in which active ingredients are placed into a cavity (Kothamasu et al., 2012). Similarly, bimetallic and trimetallic

NPs were synthesized to improve their catalytic (Zhang and Toshima, 2013), electro-catalytic (Zhang et al., 2004), and electronic structure (Thomas et al., 2008). The same can be said for the magnetic particles with characteristics of magnetic resonance imaging, which further modified in multifunctional NPs for different applications, including drug delivery and magnetic resonance imaging (Jain et al., 2008), and dual-Color fluorescence imaging (Sun, Sun, et al., 2016).

Advanced techniques such as XPS (Mittal et al., 2016), energy dispersive X-ray spectroscopy (EDS) (Dhiman, Fatima, et al., 2017), mass spectrometry (MS) (Zordan, Pennington, and Johnston, 2010), ICP-MS (Hu et al., 2018), atomic absorption spectroscopy (AAS) (Liu et al., 2012), and XRD (Mansfield et al., 2014) are usually employed in assessing the purity and compositions of NPs.

1.3.4 Stability

Stability is one of the key components of the NPs features which helps them endure the different environmental interactions, resulting in their safety and efficacy (Wu, Zhang, and Watanabe, 2011). For practical applications, it is a pre-requisite for synthesized nanomaterials that they retain their phase after being introduced into the natural environment. The possible changes in their behavior might pose a risk often related to surface imbalance and sample quality (Labille and Brant, 2010; Zhang et al., 2009). Since the synthesis of monodisperse NPs is essential to enable a smooth translation (Robertson et al., 2016), the homogeneity of NPs can be achieved by a number of methods, including ultrasonic bath, high-speed stirring, ultrasonic disrupter, high-pressure homogenizer, and magnetron splitting system (Hwang et al., 2008). A freeze-drying technique is also employed to stabilize NPs that were synthesized through the emulsification-diffusion method (Choi et al., 2004). In another study, the stability of the metal oxide NPs was enhanced through the addition of natural organic matter, without which the NPs tends to aggregate in neutral water (Zhang et al., 2009). Just as the aggregation behavior of NPs opens a new dimension of application, Sun et al., have reported the salt-induced aggregation of gold NPs in biological media and used these aggregated NPs for photoacoustic imaging and photothermal cancer therapy (Sun et al., 2016). The general rule of thumb regarding the stability of NPs is that the higher the zeta potential (–30mV to +30mV), the better the stability. It was also noticed that organic coating onto NPs enhances the melting temperature, drastically altering their thermodynamic and thermal properties (Liang et al., 2004). To understand the stability of NPs, UV-Vis spectrometry, SEM, TEM (Ristau et al., 2009), DLS (Kaasalainen et al., 2017), zeta potential (Masoudipour et al., 2017), TGA (Ziegler-Borowska, Chełminiak, and Kaczmarek, 2015), TGA-MS, and TGA-FTIR (Chrissafis and Bikiaris, 2011) instrumentation and techniques are quite useful. A sketch to explain to physiochemical features of nanomaterials is represented in Figure 1.1.

1.4 Techniques for Physicochemical Characterization of NPs

A number of promising tools are available to predict the persisting or induced properties of nanomaterials. Microscopic techniques are used to create images related to the size, shape, and distribution of nanomaterials, whereas spectroscopy deals with the interaction of particles with electromagnetic waves that are appropriate to make inferences about particle concentration, size, shape, composition, and surface properties. Some miscellaneous yet important methods such as XRD, the Brunauer–Emmett–Teller method, centrifugation, and thermal analysis have also been recognized in the characterization of NPs. A flow chart of different techniques to emphasize the features of nanomaterials is shown in Figure 1.2.

1.4.1 Microscopic Techniques

Microscopic characterization allows visual inspection of surface coating and elemental composition analysis when paired up with a secondary detector (Mansfield et al., 2014). The main strategies

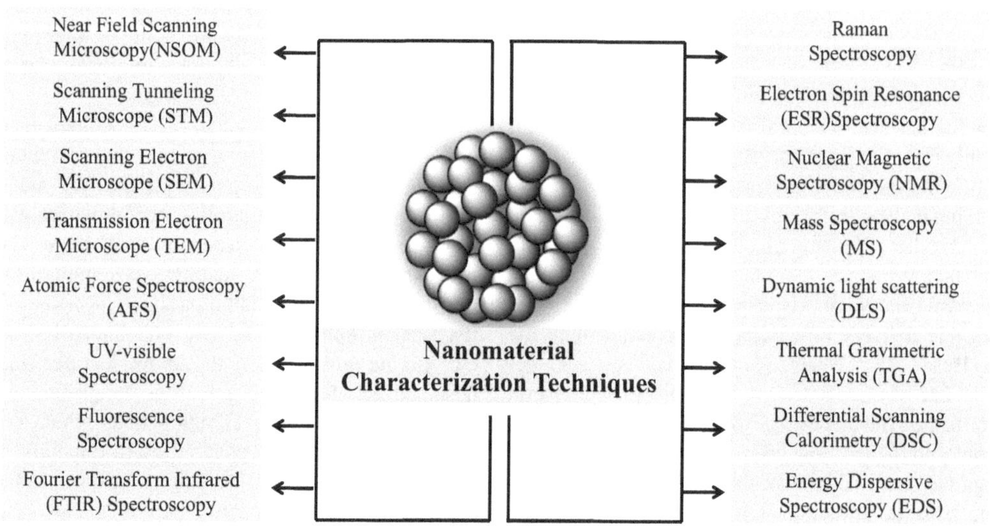

FIGURE 1.1 Characteristic features of nanomaterials in construct to bulk materials.

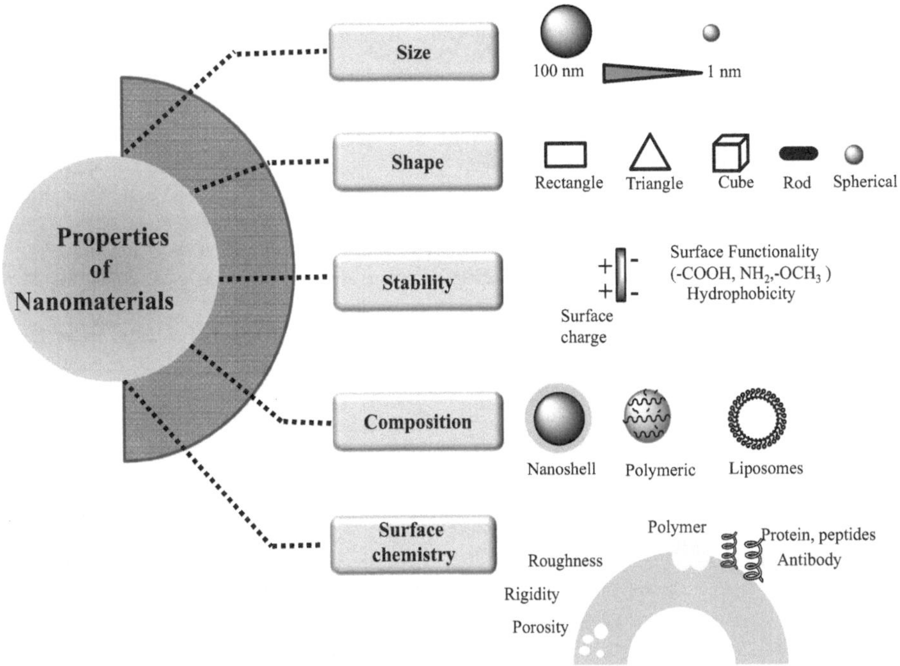

FIGURE 1.2 Common microscopic, spectroscopic, and thermal methods for nanomaterials characterization.

include optical microscopy, electron microscopy, and scanning probe microscopy. The latest reported articles explaining the characterization of nanomaterials are summarized in Table 1.1.

1.4.1.1 Near-Field Scanning Optical Microscopy (NSOM)

NSOM is a powerful method to characterize topography and fluorescence with a resolution down to the nanometer scale. The fundamental principle describing NSOM relies on the fact that the excitation laser is focused through a pinhole (sub-wavelength) diameter aperture and illuminates a sample that is placed within its near field, at a distance much shorter than that of the wavelength of the light. The sample is scanned at a distance below that of the aperture, while the optical resolution of transmitted or reflected light is governed only by the diameter of the aperture. The minimum resolution (d) for the optical element is defined by its aperture size, and expressed by the Rayleigh criterion (1.1)

$$d = 0.61\,\lambda / Na \tag{1.1}$$

where λ is wavelength excitation laser in vacuum, and Na denotes numerical aperture. (Oshikane et al., 2007) have observed that NSOM is applicable to 20 nm lateral resolution and 2–5 nm vertical resolution. (Teetsov and Vanden Bout, 2000) have used NSOM to identify the surface and fluorescence features of pristine films of stiff-chain polyfluorene. In another report, NSOM is used to probe single NPs, dimer, trimer, and small nanoaggregates of gold NPs. Small gold aggregates, including trimer, showed strong optical confinement of two-photon-induced photoluminescence with reference to that of single gold NPs (Hossain et al., 2014).

1.4.1.2 Scanning Electron Microscopy (SEM)

The obvious challenge in nanoscience and nanotechnology is the chemical mapping of the synthesized nanomaterials in a precise nanometer scale resolution using non-invasive techniques. The SEM technique uses a finely focused beam of electrons to produce images of solid specimens by scanning the sample surface. The signals derived from electron-sample interactions reveal information about the surface topography, chemical composition, crystallinity, and spatial orientation of materials. A resolution down to 1 nm scale can be achieved by SEM. The basic components of SEM are an electron source, namely Gun, electron lenses, a sample stage, and a detector. The beam of electrons having high kinetic energy is incident onto the solid sample and generates a variety of signals. These signals produce secondary electrons (which produce SEM images), characteristic X-rays (used for component analysis and continuum X-rays), backscattered electrons (BSE), diffracted backscattered electrons (EBSD, which are used to determine crystal lattice structures and stereo orientations of minerals), visible light (cathode luminescence or CL), and heat energy. Secondary electrons and BSE play a crucial role in sample imaging (Goldstein, 2012; Vernon-Parry, 2000). The quantum of secondary electrons produced with a particular energy beam of electrons depends on the atomic number of the element having a direct correlation with the image contrast. As no volume is lost during the sample analysis, the SEM technique is considered to be "non-destructive". While the limitation of SEM is that it cannot achieve atomic resolution, still, approximately 1 cm5 microns in width can be scanned to provide resolution between 1 and 20 nm and, in some cases, the SEM technique can achieve a resolution below 1 nm (magnification ranging from 20× to approximately 30,000×).

TABLE 1.1
The Latest Reported Articles to Explain the Characterization of Nanomaterials

S. No.	Techniques	Parameters	References
1	Near Field Optical Microscope (NSOM)	Nanoimaging	Kusch P.K et al., 2018; Tumkur T. et al., 2018
2	Scanning Electron Microscopy (SEM)	Surface morphology	Mukhopadhyay P. et al., 2018; Moradi G. et al., 2018
3	Transmission Electron Microscopy (TEM)	Size	Hara S. et al., 2018; Verma N.K. et al., 2016
4	Scanning Tunneling Microscopy (STM)	Nucleation density and apparent size	Motin A.M. et al., 2018; Naydenov B. et al., 2018
5	Atomic Force Microscopy (AFM)	Quantitatively determine the surface concentration	Morga M. et al., 2018; Mourdikoudis S. et al., 2018
6	UV-Vis Spectrometry	Absorbance	Maity P.et al., 2018; Sivasankar P. et al., 2018
7	Infrared Spectroscopy	Functional group identification	Zhao X. et al., 2018; Pugazhendhi A. et al., 2018
8	Atomic Absorption Spectroscopy (AAS)	Quantitative determination	Vyhnanovsky J. et al., 2018; Martinez Andrade J.M. et al., 2018
9	Raman Spectroscopy	Luminescent properties Biomarker detection	Guzman M. et al., 2018; Schechinger M. et al., 2018
10	Mass Spectroscopy (MS)	Mass	Sipe D.M. et al., 2018; Luo Z. et al., 2018
11	Circular Dichroism	Chiro-optical effect	Kong X.T. et al., 2018; Sett A. et al., 2018
12	Dynamic Light Scattering (DLS)	Stability	Sun Y. et al., 2017; Kim D.Y. et al., 2018
13	Zeta Potential	Surface charge	Purohit M.P. et al., 2017; Sharma A.K. et al., 2017
14	X-ray Diffraction (XRD)	Crystalline structure	Dhiman N. et al., 2017; Varadavenkatesan T. et al., 2017
15	Thermal Gravimetric Analysis (TGA)	Thermal property	Koutsopoulos S. et al., 2017; Ayumi N.S. et al., 2018
16	Differential Scanning Calorimetry (DSC)	Compatibility of the formulation	Daneshmand S. et al., 2018; Can H. K. et al., 2017
17	Vibrating-sample Magnetometer (VSM)	Magnetic properties	Lassoued A. et al., 2018; Nosrati H. et al., 2018
18	Analytical Ultracentrifugation (AUG)	Size wise particle distribution	Ye J.C. et al., 2017; Pandey M. et al., 2018
19	Brunauer–Emmett–Teller (BET) Surface Analyzer`	Total specific surface area	Sirdeshpande K.D. et al., 2018; Mehrizadeh H. et al., 2017
20	Nuclear Magnetic Resonance (NMR)	Interaction of NPs in synthesized formulation	Khartchenko A. et al., 2018; Ali S. et al., 2018

The information regarding the degree of aggregation, dispersion, and uniformity can be obtained using SEM. It is also possible to assess the location of NPs when they are part of secondary and tertiary nanostructures (Moroz, 2011). SEM is also advantageous and routinely used in selected point or spot analysis on the nanomaterials to deduce qualitative or semi-quantitative chemical compositions (energy dispersive X-ray spectroscopy or EDX), crystalline structure, and crystal orientations (EBSD). (Lim et al., 2006, synthesized and characterized the up-converting phosphor NPs non-aggregated nanocrystals of size range 50–200 nm using high-spatial-resolution SEM. A few examples highlighting the utility of the SEM technique are described here. In a study by (Sokolova et al., 2011), isolated exosomes from three different cell types (HEK 293T, ECFC, and MSC) were characterized. Another study by (Guehrs et al., 2017) uses a novel approach where NPs quantifications in individual cells were made possible with high-resolution imaging of their intracellular distribution based on focused ion beam/SEM(FIB/SEM). (Dillingham, Stufflebeam, and Porter, 2017) performed an elemental mapping and surface morphology of synthesized thin films of polyvinylidene fluoride containing ceramic titanium dioxide NPs.

1.4.1.3 Transmission Electron Microscopy (TEM)

Unlike SEM, TEM is used for imaging the materials at atomic resolution. The greatest advantage of TEM is its ability to provide structural information at high magnification, ranging from 50 to 106× for a single sample. In TEM, a coherent beam of electrons emitted by electron gun (W or LaB_6) passes through the condenser aperture and incident onto the sample, which leads to the scattering of signals. These signals depend upon the electron transparency and thickness of the samples. The scattered electrons are focused by the objective lens, and further amplified by the magnifying projector lens into a phosphor screen or charge coupled device (CCD) camera, generating the image. Here, inelastic interactions with energy loss between primary electrons and the sample electrons, arising due to heterogeneities, can cause complex absorption and scattering effects, leading to a spatial variation in the intensity of the transmitted electrons (Reimer, 2013). Akin to SEM, various detectors in the TEM instrument allow elemental and chemical analysis. In TEM, one can switch between imaging the sample and viewing its diffraction pattern by changing the strength of the intermediate lens.

Typically, TEM instruments have resolutions better than 0.2 nm spatial resolution. This has the additional advantage of greater electron transparency, as electrons having high kinetic energy interact less strongly with the specimen sample compared to low-energy electrons. Therefore, working with denser samples is feasible using high-voltage TEM; however, the chances of irradiation damages to the samples are worse at higher acceleration voltages. There are a plethora of reports in which TEM analysis has been performed to identify and quantify samples for their shape and size. One example highlighting the usability of TEM is the study done by (Wang et al., 2008) to identify the nanoparticle-induced surface reconstruction of phospholipid membranes. Similarly, the shape and size of platinum NPs was explored to gauge their catalytic reduction of oxygen (Chao et al., 2008). Another study (Mühlfeld et al., 2007) exploited the stereological approach of TEM to analyze the distributions of NPs in the respiratory tract, to reveal the localization of NPs within tissues and cells and to investigate the 3D nature of NPs–lung interactions.

1.4.1.4 Scanning Tunneling Microscopy (STM)

STM is an indispensable technique in the area of avant-garde nano electronics research. This technique was developed by Binnig and Rohrer, at the IBM Zurich Research Laboratory in 1982, to observe the (110) surfaces of $CaIrSn_4$ and gold NPs (Binnig et al., 1982). STM and scanning tunneling spectroscopy (STS) are relatively new techniques to simultaneously obtain both a single-dot image and its local conductivity. The atomic resolution of STM relies on two physical principles: the quantum tunneling of electrons across the potential barrier and the piezoelectric effects (Kano, Tada, and Majima, 2015).

The conductivity of carbon nanomaterials and organic molecules has been determined by STS. The influence of point defects of carbon nanomaterials was theoretically studied by applying a tight binding model and Green's function to simulate the local density of states (Wei, Zhu, and Chen, 2000). Essentially, STM counts the tunneling current, I, produced through bias voltage, V, between the atomically sharp STM tip and the sample specimen surface. This tunneling current bears an indirect correlation with the distance, i.e., I increases as distance decreases. The distance in the xyz-directions between sample stage and the STM probe is controlled by a piezoelectric scanner, which measures angstrom order changes in the distance; additionally, a feedback loop controls the z-direction and keeps the tunneling current constant. By observing the tunneling current as the tip scans a surface, a two-dimensional profile of the surface builds up. This technique is very flexible, as measurements can be done at room or low temperature after applying ultra-high vacuum conditions. This tunneling current reflects information about various electrical properties of the analyzed specimen, such as the local density of states from dI/dV and the local barrier height from dI/dz.

(Kano, Tada, and Majima, 2015), using STM and STS, have studied the electrical and photonic properties on NPs by investigating a single-electron transport on individual NPs. (Biscarini et al., 2013) have used STM to determine the architecture of ligand-protected gold NPs by depositing onto gold/mica. They analyzed the topographical power spectral density (PSD) to extract the characteristic length scales of the features exhibited by NPs in STM images. They found the characteristic length 1.2 $\pm$ 0.1 nm for mixed ligand protected gold NPs, whereas for the homo ligand gold NPs this was 0.75 $\pm$ 0.05 nm. In another study, after synthesizing thiol-protected gold NPs, (Ong et al., 2013) used STM imaging to identify various structural possibilities such as ligand shells, striped, patchy, or Janus domains structures. The STM technique was applied by (Dai et al., 2002) to find out the diameter of $CoFe_2O_4$ Ferro fluids dispersed in acidic water. The majority of NPs possessed a regular spherical shape with a diameter of approximately 2–6 nm diameter. They also observed aggregates within this colloid depending on forces through which it held together.

1.4.1.5 Atomic Force Microscopy (AFM)

A lot of attention is being paid to AFM to circumvent the drawbacks of the STM technique, which is only suitable for conducting or semiconducting surfaces. The AFM, however, can image almost any type of surface, including polymer, ceramics, metallic, composites, glass, and even biological samples. Furthermore, it can also be used to examine NPs morphology, volume, height, size, and chemical, mechanical (modulus, stiffness, viscoelastic, frictional, scratching, wear, and boundary lubricant), electrical, and magnetic properties (Starostina et al., 2008). It has many advantages over other microscopic techniques as it is cost-effective, simple to operate, and able to image with atomic resolution. Moreover, it is a nondestructive technique to analyze the sample at ambient conditions (Mansfield et al., 2014). A comparison study had divulged that the AFM technique is more reliable and accurate for NPs characterization than the dynamic laser scattering technique (DLS) (Hoo et al., 2008). *In situ* surface characterization of local deformation of materials and thin coatings can be carried out using a tensile stage inside an AFM (Bhushan, 2008). (Pompeo et al., 2005) have studied lipid multilayers through AFM, which showed an initial layer of ordered vesicles followed by fusion in the vesicles.

1.4.2 Spectroscopic Techniques

Spectroscopy was originally the study of the how electromagnetic waves interact and behave with matter as a function of wavelength (λ). Spectroscopy/spectrometry is often used in physical and analytical chemistry for analyzing the materials and liquids by the amount of light emitted from, absorbed, or scattered by them. When an appropriate wavelength of light hits a sample, the energy of the radiation is absorbed by the ground-state electrons that are excited and jump to upper energy orbitals. This phenomenon is called absorption, and electrons in the upper unstable energy state often tend to come down to the ground level. Upon returning to their ground state, electrons lose their extra energy through emission by adopting several paths, which explains phenomena such as fluorescence, luminescence, and phosphorescence. The various spectroscopic techniques are discussed below.

1.4.2.1 Optical Spectroscopy

In the case of noble-metal NPs, the confined electrons oscillate creating an electric field around the NPs. When these oscillations are in resonance with the frequencies of incident light, they induce strongly enhanced surface plasmon resonance (SPR). This unique feature makes noble metallic NPs excellent scatterers and absorbers of visible light. Although optical spectroscopy offers a high degree of chemical sensitivity, its resolution is restricted by diffraction to about half the wavelength, thus preventing chemical mapping at nanoscale. Among the noble metallic NPs, especially, gold NPs offer enhanced absorption and scattering with the additional advantage of conjugation to a variety of biological ligands and other targeting moieties. Therefore, it is widely used in the application of biochemical sensing and detection and diagnostics, as well as in therapy. Several bio-affinity sensors have been developed based on the plasmon absorption and scattering of NPs and their assemblies. (El-Sayed, Abu-Farha, and Kanan, 2014) have confirmed the differentiation of cancerous cells from non-cancerous cells by dark-field light-scattering imaging and absorption spectroscopy of ~40 nm solid gold nanospheres. When designing NPs for the plasmon application, it is to be kept in mind that the plasmon resonance of metal NPs is strongly correlated to the size, shape, and dielectric properties of the surrounding medium (Mahmoud et al., 2012).

1.4.2.2 Ultraviolet-Visible (UV-Vis) Spectroscopy

UV-Vis spectroscopy is useful for the quantification of the light absorbed and scattered by a sample. Typically, a sample is placed between a light source and a detector, and the intensity of incident and refracted light is measured, leading to either absorption or reflectance spectroscopy. The data is typically plotted as a function of wavelength (Skoog, Holler, and Crouch, 2017). Generally, the light source consists of a tungsten (W) filament (300–2500 nm for visible range), a deuterium arc lamp (UV range 190–400 nm), or a xenon arc lamp, which is continuous from 160 to 2,000 nm. Recently, light-emitting diodes (LED) have also been used for the light source. The detector may be composed of a photomultiplier tube, a photodiode, a photodiode array, or a charge-coupled device (CCD). UV-Vis absorption spectroscopy is helpful to determine contamination by observing additional peaks due to impurities in the sample, after it is compared with the standard. It is a valuable tool for identifying, characterizing, and studying the nanomaterials because NPs have optical properties that are sensitive to size, shape, concentration, agglomeration state, and refractive index near the NPs surface. (Amendola and Meneghetti, 2009) presented a method for the evaluation of the average size of gold NPs based on the fitting of their UV-Vis spectra by the Mie model for spheres. Another study was done by (Haiss et al., 2007) on UV-Vis to determine the size and concentration of gold NPs. (Pestryakov et al., 2004) did a study in which the different nature of the copper NPs formation is studied through the methods of UV-Vis spectroscopy of diffuse reflectance, XRD, and electron microscopy. Similarly, (Kale and Jagtap, 2018) biosynthesized NPs and characterized them using UV-Vis spectrophotometer.

As discussed in Section 1.3.2 (Surface Property), the protein corona generation leads to size alterations of NPs and consequently changes in their absorption profiles. These changes can be used to evaluate the binding efficiency of the proteins. It was shown that there is a blue shift in the absorption spectra of fetal bovine serum (FBS)-gold nanorods compared to the naked nanorods (Mahmoudi et al., 2013). However, for carbonaceous NPs, such as carbon nanotubes (CNT), the phenomenon of the shift and broadening of the plasmon band of the complex is not regular. For example, the absorption spectrum of the single-walled carbon nanotubes (SWCNT)-BSA complex showed spectra that were identical to the overlap of the spectra of SWCNT and BSA. Compared to other methods, UV-Vis is faster and more flexible with greater ease of operation. Interestingly, a gamut of information about different characters NPs may be deduced. (Pandey et al., 2015) synthesized cobalt oxalate NPs by Nd:YAG pulsed laser operating at wavelength 1064 nm for the ablation of cobalt. The aging and agglomeration of synthesized NPs have been studied by UV-Vis absorption spectroscopy with respect to different concentration and time. (Cao and Shen, 2011) synthesized quaternary kestrite type Cu_2ZnSnS_4 NPs where the UV-Vis absorption spectra measurement indicated that the band gap of the synthesized NPs was about 1.5eV, which was near the optimum value for photovoltaic solar conversion within a single band gap device.

1.4.2.3 Fluorescence Spectroscopy

This is a simple, inexpensive, and rapid method to quantitatively assess the concentration of analytes on their fluorescent properties. Fluorescence emission can also be detected from NPs because of its innate intrinsic property, or when labeled with fluorescence probes (Khalid et al., 2016; Wolfbeis, 2015). The sensitivity of the technique allows researchers to study the kinetics of many biological phenomena due to the fact that the excited fluorescent state persists for nanoseconds, which is well matched to the time scale of many significant biological activities. This in turn aids in understanding the rotational motion of protein side chains, molecular binding, protein conformational changes, etc. Moreover, NP-analyte binding can be examined by steady-state or time-resolved fluorescence spectroscopy, fluorescence resonance energy transfer (FRET), or stepwise single-molecule photobleaching.

1. *Fluorescence resonance energy transfer (FRET)*: FRET is also known as Förster resonance energy transfer, which arises through non-radiative dipole–dipole coupling when a donor fluorophore in its excited electronic state transfers energy to an acceptor fluorophore and in this way excites it. There are three basic requirements for FRET to occur: (i) the probes have to be fluorescent; (ii) one of the probes' excitation wavelength should be within the binding partner's emission region, and (iii) the donor-acceptor probes should be in close proximity to each other. It is also possible to collect multicolor emission under single wavelength excitation. (Wang and Tan, 2006) prepared Si NPs encapsulated with three tandem organic dyes with varied doping ratio. They observed that FRET-mediated emission signatures can be tuned to have the NPs exhibit multiple colors under one single wavelength excitation. (Sanchez-Gatayan et al., 2017) used FRET to successfully visualize the real-time NPs self-assembly.
2. *Steady-state and time-resolved fluorescence spectroscopy*: Steady-state fluorescence spectroscopy is applied to investigate the prolonged average fluorescence of molecules when they are excited by a constant electromagnetic radiation and the emitted photons are

detected as a function of wavelength. In this aspect, the NP-protein interactions alter the local chemical environment of fluorophores that quenches the fluorescence of proteins as a consequence. This quenching of the fluorescence intensity by the quencher concentration is governed by the Stern–Volmer equation (Buboltz et al., 2007).

Besides the binding affinity, FRET can also throw light on the number of binding sites and binding constants, and on the degree of cooperativity of NPs-protein binding (Hill coefficient). In a study done by (Joshi, Varma, and Pant, 2013) steady-state fluorescence was applied to monitor the interaction of quinine sulfate with sodium dodecylsulphate surfactant at different premicellar, micellar, and postmicellar concentrations in aqueous phase. In another report, bovine serum albumin (BSA) and silver NPs interaction were studied by steady-state fluorescence (Ni et al., 2017), wherein they showed that silver NPs strongly interacted with the Trp residue of BSA and induced a blue shift from 335 to 332.5 nm. Time-resolved fluorescence provides single fluorescence life span of BSA around 6 ns, which changes into two fluorescence life spans after adding silver NPs. Furthermore, a fluorescence study was applied to determine biomacromolecule interactions with fluorescent NPs through stepwise photobleaching, such as determining membrane protein stoichiometry. The pre-requisite of this technique is that the NPs should have intrinsic fluorescence such as quantum dots or rare-earth-doped oxide NPs. The protein–NP ratio can be quantified by counting the steps of NPs photobleaching. For instance, the stepwise photobleaching of Alexa488-labeled protein attached to lanthanide-ion-doped oxide NPs was used to quantify the NP-protein (α-bungarotoxin) ratio for each NP as well as its distribution (Casanova et al., 2007).

1.4.2.4 Fluorescence Correlation Spectroscopy (FCS)

Another technique that uses the fluorescence principle is the FCS technique, according to which, in any temporal perturbations of the fluorescence emission intensity caused by single fluorophores passing, the detection volume is analyzed. This fluorescent intensity fluctuation is correlated to determine the average number of fluorescent particles in detecting volume, their diffusion coefficients, hydrodynamic radius, shape, viscosity, and concentrations. As this technique possesses single-molecule sensitivity, it can be applicable to quantify the hydrodynamic radius of NPs in extreme dilutions where particle aggregation due to hydrophobicity is rarely seen. The essential components of FCS are laser source, dichroic mirror, and single photon sensitive detector. In a report, (Xie et al., 2015) prepared polystyrene core and poly(N,N-diethylaminoethyl methacrylate) (PDEA) shell fluorescent NPs with an average hydrodynamic radius of 20 nm. The pH-induced change in the particles size caused by the swelling or the collapse of the PDEA hairs was assessed by FCS.

(Shang and Nienhaus, 2017) pioneered the use of FCS as an *in situ* technique, suitable in analyzing protein–NP interactions when the latter is suspended in biological fluids. (Xu et al., 2014) studied the fluorescence properties and diffusion behavior of gold NPs in solution by using FCS. Further extending their work on the unique fluorescence properties and diffusion behaviors of gold NPs, they developed a sensitive thrombin assay.

Using FCS, (Wang et al., 2011) were able to deduce the single molecule diffusion behavior of gold NPs as a function of concentration and strongly influenced by the properties of gold NPs. Likewise, using FCS, (Weyermann et al., 2004) characterized the cationic acrylate NPs prepared from diethylaminoethyl dextran (DEAE-dextran) and poly(n-butyl-2-cyanoacrylate) (PBCA), whose size were in a range of 130–140 nm. An excellent work on fluorescence correlation spectroscopy in colloid and interface science was presented by (Koynov and Butt, 2012).

1.4.2.5 Confocal Correlation Spectroscopy (CCS)

The actual size determination at low concentration in small volumes is very critical but useful for a wide range of applications. In this concept, CCS is useful in determining the sizes of both fluorescent as well as non-fluorescent NPs. Fluorescent NPs, such as quantum dots, gold colloids, polymer beads, fluorescent beads, etc., and non-fluorescent NPs, such as biological particles (vesicles and viruses), have been successfully characterized using this technique. This method is an extension of FCS. (Kuyper et al., 2006) used CCS for the real-time monitoring of the size of quantum dots, gold colloids, and fluorescent and non-fluorescent beads.

1.4.2.6 Infrared (IR) Spectroscopy

IR spectroscopy is the study of the oscillations induced by certain vibrational frequencies of infrared electromagnetic radiation with material. This frequency range is responsible for the vibrational and rotational level excitation of a particular type of bond and a group of atoms. The whole IR spectrum is classified into three regions: near (0.8–2.5 μm wavelength), mid (2.5–25 μm), and far (25–1000 μm) IR regions. In the IR spectroscopy, a plot of relative intensity versus % T or % R versus frequency is acquired in the infrared spectrum. IR is a nondestructive characterization technique in which the atomic weight and rigidity of bonds decide the vibrational frequency of a functional group. Often "FTIR" (Fourier-transform infrared spectroscopy) terminology is used in lieu of IR, which is nothing but the Fourier transformation the intensity-time output of the interferometer into the intensity versus frequency. The FTIR spectrum is helpful in determining the chemical footprints in terms of identities, surrounding micro-environments, and atomic arrangements of chemical bonds present in analyte with sufficient spatial resolution.

In the case of NPs, the IR signal helps to identify functional groups on the surface of functionalized NPs. It can also come to aid for studying the physisorption of molecules on surface of nanomaterials (Baer et al., 2013). Furthermore, IR spectra of suspension were used to extract information about specific surface area, surface charge density and concentration of nanomaterials. (Kiefer, Grabow, Kurland, and Müller, 2015), has used solvent infrared spectroscopy for characterization of surface chemistry of various NPs. They recorded IR spectra in different solvents that enables the characterization of the nanomaterials surface chemistry *via* the molecular interactions. FT-IR spectroscopy is also used to illustrate structures of NP conjugated proteins based on the absorption of amide bonds. Amid amide I, II and III bands, the amide I vibrational band (1700-1600 cm^{-1}) is most sensitive and used frequently to obtain valuable information about the protein conformation. Recently, The "NanoFTIR" hyphenated technique, which combines scattering near-field optical microscopy (s-SNOM) and has been a well explored method for fast and reliable chemical identification of IR-active materials in the nanometer scale. This technique allows better spatial resolution more than a factor of 300 than conventional spectroscopy. (Huth et al., 2012). Another such hyphenated technique, TGA-FTIR, is used to simultaneously monitor the variations of the free carrier density obtained from the correlation between the surface reactions and the changes in the infrared absorbance under gas adsorption/desorption cycles. It gives information on the chemical phenomena responsible for electrical conductivity variations. A study by (Baraton and Merhari, 2007) applied FTIR to illustrate the interaction of CO and NOx with tin oxide NPs.

1.4.2.7 Raman Scattering (RS)

The Raman spectroscopy technique is based on the shift of the inelastic scattering of monochromatic radiation revealing structural and chemical information. This shift in the frequency of the reemitted photons can be higher or lower depending upon the various energy states such as rotational, vibrational, and electronic states of the sample specimen (Kim, 2015). The applicability of Raman spectroscopy across all sample types (solid, liquid, and gaseous) makes it a preferred technique for sample characterization. In principle, Raman effects depend on the molecular deformations happening in an electric field "E" determined by molecular polarizability "α". The incident laser beam induces an electric dipole moment $P = \alpha E$, due to its oscillating-electromagnetic-wave nature with electrical vector "E", resulting in the deformation of molecules. Consequently, these periodical deformations lead molecules to start vibrating with the characteristic frequency υm.

In a study, Raman spectroscopy was applied to characterize titania NPs produced by flame pyrolysis and to evaluate the influence of size and stoichiometry (Bassi et al., 2005). One of the limitations of Raman spectroscopy is its weak signals, which can be enhanced by several orders of magnitude for molecules when adsorbed on roughened surfaces, colloid particles, and noble metals such as silver and gold. This enhancement in signals is due to the generation of hot spots that have intense localized regions believed to be originated by local surface plasmon resonances (LSPR). The enhancement is called surface-enhanced Raman spectroscopy (SERS), and the amplification results from the interaction between the electromagnetic field of the laser excitation and the surface plasmon of the metal (Maher, 2012; Kleinman et al., 2013). It is often a better alternative to fluorescent probes for biological labeling because of their photostability and multiplexing capabilities (Navas-Moreno et al., 2017). Another application of Raman spectroscopy is state-of-the-art tip-enhanced Raman spectroscopy (TERS), which combines the chemical sensitivity of surface-enhanced Raman spectroscopy (SERS) with the high spatial resolution of scanning probe microscopy (SPM) and enables chemical imaging of surfaces at the nanometer length scale. Hot spots are recognized as being a pre-requisite for the observation of single-molecule SERS. In a study, a comprehensive review of SERS studies, including enhancement on the lattice vibrations, molecule detection, and photochemical reactions by coupling semiconductor nanostructures with LSPR, was apprised (Itoh et al., 2016; Li et al., 2016).

1.4.2.8 Nuclear Magnetic Resonance (NMR)

NMR spectroscopy is yet another handy tool used for elucidating and describing the molecular structure of molecules bound to the surfaces of nanomaterials, along with the size of NPs in solution phase. This technique is useful to investigate the surface chemistry and physical properties of nanomaterials. It was applied to characterize the nanostructural organization of a new drug nanocarrier composed of tripalmitin, lecithin, and poly(ethylene glycol) (PEG)-stearate in a study done by (Garcia-Fuentes et al., 2004). The NMR results showed that tripalmitin, present in the core of the NPs, is the main component of these systems, whereas PEG-stearate is firmly attached to the surface of the NPs, forming a hydrated polymeric layer. This characterization by NMR provided very useful information about the architectural organization. Moreover, various NMR techniques were used to characterize micellar systems, colloids of pharmaceutical interest NPs, and nano capsules. (Gomez et al., 2009) used ^1H NMR to find out the size of dendrimer-encapsulated palladium NPs. Palladium NPs were shown to contain 55, 147, 200, or 250 atoms, and there is a loss of proton signals residing in close proximity with metal NPs as the particle size increases.

Solid-state NMR spectroscopy is a powerful structural elucidation technique for powdered solid samples and provides very detailed structural and dynamics information. This technique has wide applications across both the physical and biological sciences (Nonappa and Kolehmainen, 2016). To overcome the limitations of the conventional ^1H NMR technique, an electron source consisting of either stable radicals or transition metals having high-spin ions has been used as source to amplify the solid-state NMR signal by multiple orders of magnitude, known as dynamic nuclear polarization (DNP) (Ni et al., 2017). (Lee et al., 2012) functionalized SiO_2 NPs to produce a surface covered with organosilanes and employed ^{29}Si solid-state NMR to find the surface information on whether the organosilanes polymerization proceeded parallelly or perpendicularly. Using solid-state NMR, (Piveteau et al., 2015) defined the structure of different organic- and inorganic-capped quantum dots (CdSe, CdTe, InP, PbSe, PbTe, $CsPbBr_3$) through DNP.

Solution NMR investigations can provide a wealth of information about functionalized NPs for small molecules and peptides covalently linked through flexible spacers. (Schuetze et al., 2016) functionalized gold NPs by aliphatic and aromatic mercapto-functionalized carboxylic acids as well as by two small peptides (CG and CGGRGD) and studied through ^1H-NMR diffusion ordered spectroscopy (DOSY) to obtain the hydrodynamic diameter of the particles between 1.8 and 4.4 nm. Therefore, DOSY is a valuable and non-destructive tool to determine the hydrodynamic diameter of dispersed NPs and to assess the purity of the final particles. In another work, (Guo, Holland, and Yarger, 2016) investigated the lysine adsorption behavior on fumed Si NPs by solid-state NMR spectroscopy. They used a sequence of ^1H, ^{13}C, and ^{15}N solid-state magic angle spinning (MAS) NMR techniques to elucidate the adsorption pattern *via* strong hydrogen-bonding interaction between the protonated side-chain amine group and the silanol group on Si NPs surfaces.

1.4.2.9 Mass Spectrometry (MS)

A well-documented piece of work pertaining to the analytical capabilities of different MS-based techniques, including elemental and molecular detection, and hyphenated techniques derived from their coupling with different separation approaches, to identify, characterize, and quantify NPs, is summarized by (Fernandez and Garcia-Reyes, 2017). MS is gaining exclusive attention in NPs characterization for its ease in quantitation at extremely low levels, compatibility with any type of sample, high sensitivity, and ease of hyphenation with separation techniques. The elemental MS techniques can determine the chemical composition, core/surface ligand stoichiometries, size, number, and concentration of NPs. The hyphenated separation technique is helpful to characterize the polydispersity of samples. Since the suitability of NPs in a wide range of applications in various research domains requires some kind of surface tailoring to reinforce its performance, the detection and quantification of NPs in complex matrices necessitate a sophisticated analysis approach. Nowadays, inductively coupled plasma-mass spectrometry (ICP-MS) has become the preferred technique to analyze and detect elemental compositions of NPs, due to its high sensitivity and robust nature. There are two basic components in an assembly of ICP-MS: (i) inductively coupled plasma, which introduces efficient vaporization, atomization, and ionization of most elements, and (ii) mass spectrometry, which detects the extracted ions on the basis of mass to charge ratio. An analysis of NPs through ICP-MS can give a complete picture of size, distribution, chemical composition, mass, and number NPs concentrations, and even about the presence of impurities in NPs. Generally, two different strategies are adopted for NPs ICP-MS analysis: the first is single-particle ICP-MS (spICP-MS), which is able to determine the particle number concentration (particle/mL), mass, and size distribution of NPs. However, information about the morphology/shape of particles cannot be deduced. The second approach is the hyphenated separation techniques with ICP-MS such as hydrodynamic chromatography (HDC) and field flow fractionation (FFF), which determine the total NPs concentration as a function of particle size fraction. (Mitrano et al., 2012) made a comparison between spICP-MS and asymmetric field flow fractionation ICP-MS to evaluate monodisperse silver NPs with respect to size/concentration detection limits, resolution, and multi-form elemental analysis. They found spICP-MS is superior to flow field flow fractionation ICP-MS. In another work, (Helfrich, Brüchert, and Bettmer, 2006) coupled ICP-MS with liquid chromatography (LC) and gel electrophoresis to examine the size and chemical structure of the surface of gold NPs and showed that LC-ICP-MS provides good reproducibility (RSD, 1%) in a size-dependent retention behavior. Another unique application of mass spectrometry is the label-free quantitative analysis of the uptake mechanisms and intracellular NPs transport. Yet another advantageous application that utilizes the electrothermal vaporization (ETV) technique provides fast volatilization of analytes for atomization, ionization, and excitation. ETV-ICP-MS offers a robust, particle-size independent method for quantifying metal NPs and their agglomerates in colloidal and real environmental samples, even in low sample volumes. (Yang et al., 2007) used ICP-MS for the quantitative estimation of quantum dot QD705 (Cd/Se/Te based QDs) in blood and tissues.

1.4.2.10 Circular Dichroism (CD)

Many biomolecules such as nucleic acids and proteins are chiral in nature, i.e., they can polarize the incident light beam. The chirality of a biomolecule can be studied optically by using light pulses of opposite circular polarizations and is absorptive in nature. The difference in the absorbance of right and left circularly polarized light is called circular dichroism (CD). Naturally, abundant biomolecules exist in a wide range of stereo-conformers and have their own characteristic intense CD signals in the UV region (200–300 nm). On the other hand, metallic (silver, gold) NPs exhibit strong absorption in the visible range but are achiral with no inherent chiroptical properties. When chiral biomolecules are conjugated to these NPs, biomolecules can impart chirality to the particles and exhibit a plasmon-induced CD signal in the visible spectral region.

CD spectroscopy has been utilized to describe the interactions between protein and NPs after quantifying the amount of free protein in solution (Treuel et al., 2010). (Slocik, Govorov, and Naik, 2011) demonstrated the chiroptical properties of peptide-functionalized gold NPs and observed the appearance of a moderately strong visible CD signal ~520 nm for peptide-functionalized gold NPs. The aggregation of peptide-functionalized NPs is also assessed through a bathochromic shift in CD spectra (Slocik, Govorov, and Naik, 2011). (Billsten et al., 1995) were successful in observing the changes in the secondary structure of T4 lysozyme upon adsorption to Si NPs by using CD spectra. In a study done by (Li et al., 2004), hydrogen bonding was shown to be essential in order to generate the CD spectra. They conjugated cysteine, glutathione, penicillamine, lysine, and glutamine with citrate-capped silver NPs and found that the lysine and glutamine did not show any CD signal changes, while others showed significant spectral changes.

1.4.3 Miscellaneous Techniques

1.4.3.1 Dynamic Light Scattering (DLS)

This is a technique for measuring dynamic properties and the size distribution of particles typically in the sub-micron region suspended in a solution. These particles can be composed of a wide variety of physical, chemical, and biological systems. Similar techniques, such as photon correlation spectroscopy (PCS) or quasi-elastic light scattering (QELS), are also based on the DLS principle. It is a very powerful tool for studying the diffusion behavior of NPs within a solution phase. Depending on the size and shape of nanomaterials, the diffusion coefficient and the hydrodynamic radii can be easily obtained. The basic working principle of DLS is: a laser beam, when incident on the colloidal suspension of NPs, causes a Doppler shift due to Brownian motion of NPs. When the light hits the moving particle, the shifts of the scattered light are detected at a known scattering angle θ by a fast photon detector (Germain, Leocmach, and Gibaud, 2016). Several types of illuminating sources are also used, such as HeNe (632.8 nm), laser diodes (635–780 nm), air-cooled Ar (488–514.5 nm), and water-cooled Ar (488–514.5 nm).

The complete particle size distribution can be determined by elaborated multiangle instruments. One of such techniques is DLS, which offers a slew of advantages over other methods and is widely preferred as a first line of size characterization of NPs. Typically, it is used to characterize various particles, including emulsions, micelles, polymers, proteins, NPs, or colloids in a diluted solution (Stetefeld, McKenna, and Patel, 2016). One of the main applications of DLS is to measure the hydrated diameter of the nanomaterials, which is dependent on various factors such as sample concentration, ionic strength of suspending media, and electrical double layer (Mansfield et al., 2014). In a study by (Borissevitch et al., 2013), the formation of complexes between porphyrins and quantum dots was characterized by DLS to explain observed fluorescence-suppression phenomena.

On metallic cores, polymer coating layers have been observed through microscopy techniques; however, the variation in diameter can be measured by DLS. For example, in a study by (Jana, Earhart, and Ying, 2007), the silica layer coatings of gold and silver metallic NPs could not be observed by high-resolution TEM (HR-TEM). Thus, they concluded that the sizes measured by HR-TEM correspond only to the metallic cores, while the results obtained by DLS correspond to the total size of the metallic core and the coating layer. Particle stability of nanomaterials can also be studied by DLS (Ramos, 2017). However, on the flip side, DLS is unable to provide morphological information of NPs, and multiple scattering also affects the data analysis. The pre-requisite of a good DLS measurement is that the NPs suspension should be free from artifacts or dusts to avoid erroneous data interpretation.

1.4.3.2 Zeta Potential

The zeta potential (ζ) tells us about the surface charge of nanomaterials on their interaction with surroundings. It is a key feature to describe the surface behavior of nanomaterials. Many studies have addressed the fate of NPs uptake *via* protein adsorption, such as (Patil et al., 2007), where they found that charged CeO_2 NPs adsorbed more BSA compared to negatively charged NPs. The positive or negative charged surface of the NPs play a vital role in the distribution and diffusion of ions in the surrounding interfacial region; consequently, the counter-ion effect predominates near the surface, inducing an electrical double layer around each particle due to this close proximity. Charged NPs in a liquid system experience two phenomena. One arises because of the inner layer (Stern plane) where the solvated ions are strongly bound, and the second is due to the outer layer (diffuse) where the above interactions are weak. Within the diffused layer there is a theoretical boundary inside which the ions and particles coexist in a stable entity and particle motion is responsible for the movement of ions within the boundary. The potential at this hydrodynamic shear is the zeta potential. Generally, solution pH and/or ionic strength have an influence on the variation of zeta potential values. Sometimes, at a particular pH, the zeta potential of NPs might attain zero value and this point is referred to as the "isoelectric point" (IEP) of the colloidal system. When an electric field is applied across an electrolyte containing dispersed NPs, the charged NPs tend to move towards the oppositely charged electrode due to electrostatic attraction. This movement is governed by various factors, namely the size of the NPs, the strength of the electric field, the dielectric constant of the medium, the zeta potential of the particle, and the viscosity of the dispersed medium, which the particles experience in opposite directions. When the electrostatic force and the viscosity force balance each other, the particle movement occurs with constant velocity. Theoretically, the velocity of a particle in an electric field unit is referred to as its electrophoretic mobility. Zeta potential is related to the electrophoretic mobility by the Henry equation (Makino and Ohshima, 2010). The zeta potential of a sample is measured by an instrument known as "Zetasizer", which combines laser Doppler velocimetry and phase analysis light scattering (PALS) to measure the NPs charge. By detecting the fluctuate intensity of incident light due to electrophoretic mobility, potential can be calculated. As a rule of thumb, if a NP has a zeta potential value of ±10 mV, it is considered to be neutral, while a value of ±30 mV strongly exhibits cationic or anionic NPs (Clogston and Patri, 2011). The higher the potential (beyond ±30 mV), the higher the colloidal stability for NPs.

1.4.3.3 X-Ray Diffraction (XRD)

XRD is an effective technique that has long been used to address numerous issues, in particular, the assessment of the crystal structures of solids, including lattice constants, geometry, identification of unknown materials, orientation of single crystals, preferred orientation of polycrystals, defects, stresses, etc. (Whetten et al., 1996). Predominantly, it is an analytical technique used for the phase identification of a crystalline material and can provide data about unit cell dimensions (Liss et al., 2003). The X-rays wavelength lies on the atomic scale, and hence, XRD is the tool of choice for probing the structure of nanomaterials (Moroz, 2011). The average inter-particle distance is assessed by small-angle scattering, while the refinement of the atomic structure can be measured by wide-angle diffraction. There is a close relation between size, size distribution, defects, and strain in nanocrystals with the widths of the diffraction lines and it gets broader when the size of the nanocrystals decreases. This occurs due to loss of long-range order relative to the bulk. An estimation of the size of the particle could be obtained through XRD analysis by using the Debye Scherrer formula (Speakman, 2014).

$$D = 0.9\,\lambda / B \cos\theta \tag{1.2}$$

where D is the nanocrystal diameter, λ is the wavelength of light, β is the full width half at the maximum (FWHM) of the peak in radians, and θ is the Bragg angle (Holzwarth and Gibson, 2011). The crystallinity of materials often dictates their most obvious applications; however, important information can also be obtained from the semi-crystalline and even amorphous nature of materials. As widely described in the different research articles, typical examples of the nanomaterials investigated by hard X-rays are: nanoclusters, polymers/fibers with partly crystalline and partly amorphous sections, nanomaterials, or nanostructures, and NPs coated with proteins. Atoms are organized in periodic arrays (nanocrystals) or can be located in random assemblies (amorphous) in all of these materials, so they can be studied with X-rays at different scales, ranging from the atomic structure to the nanoscale, up to the mesoscale (hundreds of nanometers) (Giannini et al., 2016).

The working principle of X-ray diffraction is based on the constructive interference of single-wavelength X-rays and a crystalline sample. They are collimated to concentrate, and directed towards the sample. When conditions satisfy Bragg's Law ($n\lambda = 2d \sin \theta$), constructive interferences are produced as a result of the interaction of the incident radiations with the sample. The diffracted X-rays are detected, processed, and counted by scanning the sample through a range of 2θ angles. The identification of unknown materials can be done by comparing the d-spacing with standard reference patterns (Sharma et al., 2012). An XRD instrument consists of three basic elements: an X-ray tube, a sample holder, and an X-ray detector. X-ray spectra are produced when electrons acquire enough energy to get dislodged from the inner shell electrons of the target material. The most commonly used X-ray diffraction technique is powder diffraction. It can be employed to determine the average crystallite size of a nanocrystalline material.

(Cao and Shen, 2011) demonstrated quaternary kesterite-type Cu_2ZnSnS_4 (CZTS) NPs for low-cost thin-film solar cells. The crystallinity of CZTS NPs was greatly improved by annealing in H_2S (5%)/Ar mixed gases analyzed by XRD. Further, high-resolution X-ray photo-emission spectroscopy (XPS) analysis of the four constituent elements confirmed the purity and composition of CZTS NPs. XPS is a surface-sensitive technique for quantitative elemental composition. The XPS spectra is obtained by irradiating a material with X-rays while simultaneously measuring the kinetic energy and number of electrons present on surface. (Oezaslan, Hasché, and Strasser, 2011) studied the formation kinetics, time scales of individual processes, and particle growth rates of a Pt:Cu = 1:3 bimetallic alloy using *in situ* high-temperature XRD. Similarly, an earlier study by (Rose et al., 2003) investigated the palladium hydride phases of carbon-supported palladium NPs as a function of applied potential with the help of *in situ* extended X-ray absorption, XRD, and electrochemical studies. In another study, Yi-Chun Lu et al. (2009) probed the origin of the enhanced stability of "$AlPO_4$" NPs-coated $LiCoO_2$ with the help of combined XRD and XPS studies (Lu et al., 2009).

1.4.3.4 Thermal Gravimetric Analysis (TGA)

TGA is yet another analytical technique to measure the amount of weight change of a material as a function of either temperature or time. Its unique advantage is that it does not require any special sample preparation beyond the drying of the sample to remove humidity. In past decades, many researchers focused on the usability of TGA to evaluate the purity, thermal stability, and characterization of nanomaterials. For TGA, materials are heated to elevated temperatures while monitoring the mass loss of the sample to yield the decomposition curve. The analysis of the decomposition curve provides meaningful information about the oxidation temperature and residual mass of the sample. The oxidation temperature is the temperature at which the bulk of the material decomposes, while the residual mass could be due to residual metal catalysts from synthesis, or impurities within the sample. (Pang, Saxby, and Chatfield, 1993) examined the TGA of carbon nanotubes and NPs. The result reflected that nanotubes and NPs are more resistant towards oxidation than other allotropic forms of carbon such as diamond and graphite. Further, (Bom et al., 2002) concluded from TGA that defect sites along the wall and at the ends of the raw multiwall nanotubes facilitate the thermal oxidative destruction of the nanotubes. However, the destructive nature of the technique and it's limitation to laboratory scale samples makes it less suitable for a wide range of analysis.

1.4.3.5 Quartz Crystal Microbalance (QCM)

The use of the quartz crystal microbalance (QCM) method is an extension of TGA and can be defined as a microscale thermogravimetric analysis, or μ-TGA. The purity of NPs, as well as the presence of any surface coatings of nanomaterials, can be measured using the same thermal decomposition principle as TGA. The QCM is more sensitive as it offers detection limits far better than those available in the conventional TGA, i.e., in the range of nanograms. The μ-TGA is used to detect the mass of nanomaterials *via* piezoelectricity. When nanomaterials are deposited on the active area of the QCM electrode, the resonant frequency is dampened. The shift in the resonant frequency can be related to a shift of mass by using the equation developed by (Sauerbrey, 1959).

(Mansfield et al., 2014) used μ-TGA to determine the presence and amount of surface-bound ligand coverage on gold NPs and to confirm the presence of a poly(ethylene glycol) coating on SiO_2 NPs. The results were compared to traditional analytical techniques to demonstrate the reproducibility and validity of μ-TGA for determining the presence of NP surface coatings. The findings showed that μ-TGA is a valid method for quantitative determination of the coatings on NPs. In some cases, it can provide the purity and compositional data of the NPs.

1.4.3.6 Differential Scanning Calorimetry (DSC)

Calorimetry is a thermo-analytical technique for quantifying enthalpy into or out of a sample that establishes a connection between the temperature and specific physical properties of substances. Calorimeters are used frequently to measure thermodynamic properties such as melting point, glass transitions, or exothermic decompositions of nano-sized materials (Gill, Moghadam, and Ranjbar, 2010).

Based on the mode of operation, DSCs can be classified into heat-flux DSCs and power-compensated DSCs. In a heat-flux DSC, the sample material, enclosed in a pan, and an empty reference pan are placed on a thermoelectric disk surrounded by a furnace. The furnace is heated at a linear heating rate, and the heat is transferred to the sample and reference pan through the thermoelectric disk. Power-compensated DSCs are designed to calculate the direct and exact enthalpy change of the sample holder as a function of time (Tanaka, 1992). From the application point of view, (Sarmento et al., 2006) assessed insulin-loaded alginate NPs by DSC to interpret the insulin-polyelectrolyte interaction. The same way, (Siekmann and Westesen, 1994) employed DSC to analyze the recrystallization behavior and polymorphic transitions of melt-homogenized glyceride NPs.

In a study by (Castelli et al., 2005), indomethacin (IND)-loaded solid lipid NPs (SLN) and nanostructured lipid carriers (NLC) were prepared and the organization and distribution of the different ingredients originating each type of NP system were studied by DSC. DSC static and dynamic measurements performed on SLN and NLC revealed that oil nano compartments incorporated into NLC solid matrix influenced the IND distribution in a positive way.

1.4.3.7 Vibrating Sample Magnetometer (VSM)

VSM is an instrument to measure the magnetic properties of NPs. Here, the sample is firstly magnetized in an invariable magnetic field and sinusoidally vibrated, through the use of piezoelectric material. The induced voltage is correlated to the material's magnetic moment. The VSM pattern provides knowledge about the magnetization (emu/g), coercivity, H_c (the resistance of the magnetic material to changes in magnetization), remanent magnetization (M_r), and saturation magnetization (M_s) (Andrade et al., 2012).

$MgFe_2O_4$ NPs were synthesized by (Pradeep, Priyadharsini, and Chandrasekaran, 2008) using a sol-gel auto-combustion method. The M-H loop of $MgFe_2O_4$ was traced using the VSM. In another study, magnetic chitosan NPs were prepared for application in magnetic carrier technology. They used a co-precipitation of $FeCl_2$ and $FeCl_3$ solution in base medium for the preparation of the magnetic chitosan. The magnetic properties of chitosan magnetic NPs were analyzed by VSM, and Ms around 15 emu/g (Dung et al., 2009).

In another work by (Pham et al., 2016), curcumin-loaded superparamagnetic Fe_3O_4 NPs modified by chitosan were assessed. The saturation magnetization (Ms) values of the magnetic Fe_3O_4, chitosan-Fe_3O_4, and curcumin-chitosan-Fe_3O_4 NPs were 90, 50, and 18 emu g^{-1}, respectively. The decrease in Ms values of chitosan-Fe_3O_4 and curcumin-chitosan-Fe_3O_4 NPs compared to that of magnetic uncoated Fe_3O_4 NPs is due to the polymer layer coated on the surface of the particles enhancing the size, and this also leads to a decrease in the value of magnetic saturation.

1.4.3.8 Analytical Ultracentrifugation (AUG)

AUG is a classical absolute technique for the characterization of NPs by the pioneering work of Svedberg and coworkers (Cölfen, 2004a). It is a fractionation method to determine molar mass, particle size, and particle density distribution with high precision and resolution. The density gradient method allows to fractionate heterogeneous samples based on their chemical nature (Mächtle and Börger, 2006). (Walter et al., 2014) used AUG with a multiwavelength detector to address multidimensional NP analysis. They employed a high-performance optical setup and data acquisition software to obtain information about size, shape anisotropy, and optical properties. They were able to distinguish between different species of single-wall carbon nanotubes in a single experiment using the wavelength-dependent sedimentation coefficient distribution in a time-saver manner.

(Mehn et al., 2017) used the sedimentation velocity data from the AUC to characterize nanocarrier drug delivery systems. They were successful in determining the nanocarrier size distribution and the ratio of free versus NPs-encapsulated drugs in a commercially available liposomal doxorubicin formulation using interference and absorbance-based AUC measurements. In surface-engineered or multifunctional NPs, the cores are often polydispersed and are coated by stabilizing the shell, targeting ligands that vary in size and composition, as no single technique is adequate to determine the size distribution and the nature of the shell. Advances in AUG permit to determine sedimentation and diffusion coefficients. A single experimental run is sufficient for a complete NPs characterization, without the need for standards or other auxiliary measurements. (Carney et al., 2011) were able to determine NPs size distribution together with density or molecular weight by 2D AUG. The combination of AUG with other techniques such as light scattering or FFF are often applied to gain a deep insight into complex hybrid particle mixtures (Cölfen, 2004b).

1.4.3.9 Brunauer–Emmett–Teller (BET)

Surface area and porosity are the two unique features that impact the quality and utility of nanomaterials. Gas sorption analysis (both adsorption and desorption) is commonly used to determine the surface area and porosity of porous NPs. This physical adsorption of gas molecules on solid surface can be explained by using the BET theory (Gelb and Gubbins, 1998; Naderi, 2015). In a gas sorption experiment, several cycles of controlled doses of an inert gas, such as nitrogen, krypton, or argon are introduced to adsorb, and later, withdrawn and desorbed forming a monolayer or multilayers of gas molecules adsorbed on the NPs surface. The sample material is placed in a vacuum chamber at a constant and very low temperature (liquid nitrogen (77.4 K)), and subjected to a wide range of pressures, to generate sorption isotherms. The amounts of gas molecules adsorbed or desorbed are determined by the pressure variations due to the adsorption or desorption of the gas molecules by the material. By knowing the area occupied by one adsorbate molecule, and using an adsorption model, the total surface area of the material can be determined. Chemisorption analyses can provide much of the information needed to evaluate catalyst materials in the design and production phases, as well as after a period of use. Unlike physisorption, chemisorption, however, is highly selective and occurs only between certain adsorptive and adsorbent species, and only if the chemically active surface is cleaned of previously adsorbed molecules. For example, a CuO NPs prepared for potent antibacterial applications has a mean surface area of 15.69 m^2/g determined by BET (Ren et al., 2009). (Jörg et al., 2012) synthesized spherical ordered mesoporous carbon NPs for applying as cathode material in Li–S batteries. These NPs have high inner porosity of 2.32 $cm^3\ g^{-1}$ and a surface area of 2445 $m^2\ g^{-1}$ for reversible capacity and better cycling efficiency.

Conclusion

Nanomaterials are versatile candidates with multifunctional characteristics and have a wide spectrum of applications. Thus, the research in the area of design and fabrication of analytical tools to characterize nanomaterials has become interesting to scientific and technologist groups. Microscopic techniques based on the absorption of light make use of simple, user-friendly colorimetric and size-dependent analysis of NPs. Among the imaging techniques, confocal Raman microscopy creates high resolution, label-free, and non-invasive mapping. Hyphenated or combinatory techniques offer new opportunities because they save time and are less laborious, more rapid and more cost effective due to the possible integration at single platform, such as inductively coupled plasma time-of-flight mass spectrometer (ICP-TOF-MS) in combination with a microdroplet generator for the mass quantification of varied NPs in complex. The heat transfer is a critical insight feature in nanofluids, which can be calculated by several techniques that follow thermal comparator principles.

Additionally, electrochemical techniques such as cyclic voltammetry, electrochemical impedance spectroscopy, polarography, and potentiometry provide a standard addition in analyzing appliance. The electrocatalytic properties of nanomaterials can be assessed through cyclic voltammetry as well as BET. This technique was also applied to evaluate the surface traits of NPs. Information about mechanical properties such as stress, strain, or elasticity of nanofibers can be achieved by nano-indentation. In brief, there is a huge possibility to control or manipulate the physiochemical characteristics of nano-based products, and this urges us to look at robust techniques with the latest trends.

REFERENCES

Ahmadi, Amirhossein, and Sanam Arami. 2014. "Potential applications of nanoshells in biomedical sciences." *Journal of Drug Targeting* 22 (3):175–190.

Albanese, Alexandre, Peter S. Tang, and Warren C. W. Chan. 2012. "The effect of nanoparticle size, shape, and surface chemistry on biological systems." *Annual Review of Biomedical Engineering* 14:1–16.

Ali, Shahid, Safyan A. Khan, Julian Eastoe, Syed R. Hussaini, Mohamed A. Morsy, and Zain H. Yamani. 2018. "Synthesis, characterization, and relaxometry studies of hydrophilic and hydrophobic superparamagnetic Fe3O4 nanoparticles for oil reservoir applications." *Colloids and Surfaces A: Physicochemical and Engineering Aspects* 543:133–143.

Amendola, Vincenzo, and Moreno Meneghetti. 2009. "Size evaluation of gold nanoparticles by UV–vis spectroscopy." *The Journal of Physical Chemistry C* 113 (11):4277–4285.

Andrade, Ângela L., Manuel A. Valente, José M. F. Ferreira, and José D. Fabris. 2012. "Preparation of size-controlled nanoparticles of magnetite." *Journal of Magnetism and Magnetic Materials* 324 (10):1753–1757.

Attota, Ravikiran, Richard Kasica, Lei Chen, Purushotham Kavuri, Richard Silver, and Andras Vladar. 2010. "Nanoparticle size and shape evaluation using the TSOM optical microscopy method." *Proc. NSTI-Nanotech.*

Ayumi, Nursyafiqah Sahrum, Shariza Sahudin, Zahid Hussain, Mumtaz Hussain, and Nor Hayati Abu Samah. 2018. "Polymeric nanoparticles for topical delivery of alpha and beta arbutin: Preparation and characterization." *Drug Delivery and Translational Research*:1–15.

Baer, Donald R., Mark H. Engelhard, Grant E. Johnson, Julia Laskin, Jinfeng Lai, Karl Mueller, Prabhakaran Munusamy, Suntharampillai Thevuthasan, Hongfei Wang, and Nancy Washton. 2013. "Surface characterization of nanomaterials and nanoparticles: Important needs and challenging opportunities." *Journal of Vacuum Science and Technology A: Vacuum, Surfaces, and Films* 31 (5):050820.

Baraton, Marie-Isabelle, and Lhadi Merhari. 2007. "Dual contribution of FTIR spectroscopy to nanoparticles characterization: Surface chemistry and electrical properties." *Nanomaterials Synthesis, Interfacing, and Integrating in Devices, Circuits, and Systems II.*

Bassi, A. L., D. Cattaneo, V. Russo, C. E. Bottani, E. Barborini, T. Mazza, P. Piseri, P. Milani, F. O. Ernst, K. Wegner, and S. E. Pratsinis. 2005. "Raman spectroscopy characterization of titania nanoparticles produced by flame pyrolysis: The influence of size and stoichiometry." *Journal of Applied Physics* 98 (7):074305. doi: 10.1063/1.2061894.

Behzadi, Shahed, Vahid Serpooshan, Wei Tao, Majd A. Hamaly, Mahmoud Y. Alkawareek, Erik C. Dreaden, Dennis Brown, Alaaldin M. Alkilany, Omid C. Farokhzad, and Morteza Mahmoudi. 2017. "Cellular uptake of nanoparticles: Journey inside the cell." *Chemical Society Reviews* 46 (14):4218–4244. doi: 10.1039/c6cs00636a.

Bhushan, Bharat. 2008. "Nanotribology, nanomechanics and nanomaterials characterization." *Philosophical Transactions of the Royal Society of London A: Mathematical, Physical and Engineering Sciences* 366 (1869):1351–1381.

Biener, Jürgen, Arne Wittstock, Theodore Baumann, Jörg Weissmüller, Marcus Bäumer, and Alex Hamza. 2009. "Surface chemistry in nanoscale materials." *Materials* 2 (4):2404–2428.

Billsten, Peter, Marie Wahlgren, Thomas Arnebrant, Joseph McGuire, and Hans Elwing. 1995. "Structural changes of T4 lysozyme upon adsorption to silica nanoparticles measured by circular dichroism." *Journal of Colloid and Interface Science* 175 (1):77–82. doi: 10.1006/jcis.1995.1431.

Binnig, Gerd, Heinrich Rohrer, Ch Gerber, and Edmund Weibel. 1982. "Surface studies by scanning tunneling microscopy." *Physical Review Letters* 49 (1):57.

Biscarini, Fabio, Quy Khac Ong, Cristiano Albonetti, Fabiola Liscio, Maria Longobardi, Kunal S. Mali, Artur Ciesielski, Javier Reguera, Christoph Renner, and Steven De Feyter. 2013. "Quantitative analysis of scanning tunneling microscopy images of mixed-ligand-functionalized nanoparticles." *Langmuir* 29 (45):13723–13734.

Boles, Michael A., Daishun Ling, Taeghwan Hyeon, and Dmitri V. Talapin. 2016. "The surface science of nanocrystals." *Nature Materials* 15:141. doi: 10.1038/nmat4526.

Bom, David, Rodney Andrews, David Jacques, John Anthony, Bailin Chen, Mark S. Meier, and John P. Selegue. 2002. "Thermogravimetric analysis of the oxidation of multiwalled carbon nanotubes: Evidence for the role of defect sites in carbon nanotube chemistry." *Nano Letters* 2 (6):615–619.

Borissevitch, I. E., G. G. Parra, V. E. Zagidullin, E. P. Lukashev, P. P. Knox, V. Z. Paschenko, and A. B. Rubin. 2013. "Cooperative effects in CdSe/ZnS-PEGOH quantum dot luminescence quenching by a water soluble porphyrin." *Journal of Luminescence* 134:83–87.

Buboltz, Jeffrey T., Charles Bwalya, Santiago Reyes, and Dobromir Kamburov. 2007. "Stern-Volmer modeling of steady-state Förster energy transfer between dilute, freely diffusing membrane-bound fluorophores." *The Journal of Chemical Physics* 127 (21):12B601.

Burda, Clemens, Yongbing Lou, Xiaobo Chen, Anna C. S. Samia, John Stout, and James L. Gole. 2003. "Enhanced nitrogen doping in TiO2 nanoparticles." *Nano Letters* 3 (8):1049–1051. doi: 10.1021/nl034332o.

Can, Hatice Kaplan, Serap Kavlak, Shahed ParviziKhosroshahi, and Ali Güner. 2018. "Preparation, characterization and dynamical mechanical properties of dextran-coated iron oxide nanoparticles (DIONPs)." *Artificial Cells, Nanomedicine, and Biotechnology* 46 (2):421–431.

Cao, M, and Y. Shen. 2011. "A mild solvothermal route to kesterite quaternary Cu2ZnSnS4 nanoparticles." *Journal of Crystal Growth* 318 (1):1117–1120.

Carney, Randy P., Jin Young Kim, Huifeng Qian, Rongchao Jin, Hakim Mehenni, Francesco Stellacci, and Osman M. Bakr. 2011. "Determination of nanoparticle size distribution together with density or molecular weight by 2D analytical ultracentrifugation." *Nature Communications* 2:335.

Casanova, Didier, Domitille Giaume, Mélanie Moreau, Jean-Louis Martin, Thierry Gacoin, Jean-Pierre Boilot, and Antigoni Alexandrou. 2007. "Counting the number of proteins coupled to single nanoparticles." *Journal of the American Chemical Society* 129 (42):12592–12593.

Castelli, Francesco, Carmelo Puglia, Maria Grazia Sarpietro, Luisa Rizza, and Francesco Bonina. 2005. "Characterization of indomethacin-loaded lipid nanoparticles by differential scanning calorimetry." *International Journal of Pharmaceutics* 304 (1–2):231–238.

Chao, Wang, Daimon Hideo, Onodera Taigo, Koda Tetsunori, and Sun Shouheng. 2008. "A general approach to the size- and shape-controlled synthesis of platinum nanoparticles and their catalytic reduction of oxygen." *Angewandte Chemie International Edition* 47 (19):3588–3591. doi: 10.1002/anie.200800073.

Chen, Liang-Chia, and Chie-Chan Ho. 2008. "Development of nanoparticle shape measurement and analysis for process characterization of TiO2 nanoparticle synthesis." *Review on Advanced Material Science.*

Cheng, Qilin, Chunzhong Li, Vladimir Pavlinek, Petr Saha, and Huanbing Wang. 2006. "Surface-modified antibacterial TiO2/Ag+ nanoparticles: Preparation and properties." *Applied Surface Science* 252 (12):4154–4160.

Choi, M. J., S. Briançon, J. Andrieu, S. G. Min, and H. Fessi. 2004. "Effect of freeze-drying process conditions on the stability of nanoparticles." *Drying Technology* 22 (1–2):335–346. doi: 10.1081/DRT-120028238.

Chrissafis, K., and D. Bikiaris. 2011. "Can nanoparticles really enhance thermal stability of polymers? Part I: An overview on thermal decomposition of addition polymers." *Thermochimica Acta* 523 (1):1–24. doi: 10.1016/j.tca.2011.06.010.

Clogston, Jeffrey D., and Anil K. Patri. 2011. "Zeta potential measurement." In *Characterization of Nanoparticles Intended for Drug Delivery*, pp. 63–70. Springer.

Cölfen, Helmut. 2004a. "Analytical ultracentrifugation of nanoparticles." *Polymer News* 29 (4):101–116.

Cölfen, Helmut. 2004b. "Feature article: Analytical ultracentrifugation of nanoparticles." *Polymer News* 29 (4):101–116. doi: 10.1080/00323910490980840.

Dai, Dalin, Jian Li, Linhai Xiang, Yi Wen, Wenshong Zhang, Guo Li, and Samuel H. Cohen. 2002. "Comparative scanning tunneling microscopy studies of $CoFe_2O_4$ nanoparticles of ferrofluids in acidic medium." In *Atomic Force Microscopy/Scanning Tunneling Microscopy 3*, pp. 175–179. Springer.

Daneshmand, Sara, Mahmoud Reza Jaafari, Jebrail Movaffagh, Bizhan Malaekeh-Nikouei, Mehrdad Iranshahi, Atieh Seyedian Moghaddam, Zahra Tayarani Najaran, and Shiva Golmohammadzadeh. 2018. "Preparation, characterization, and optimization of auraptene-loaded solid lipid nanoparticles as a natural anti-inflammatory agent: In vivo and in vitro evaluations." *Colloids and Surfaces B: Biointerfaces* 164:332–339.

De Falco, Gianluigi, Mario Commodo, Patrizia Minutolo, and Andrea D'Anna. 2015. "Flame-formed carbon nanoparticles: Morphology, interaction forces, and hamaker constant from AFM." *Aerosol Science and Technology* 49 (5):281–289. doi: 10.1080/02786826.2015.1022634.

Desai, Neil. 2012. "Challenges in development of nanoparticle-based therapeutics." *The AAPS Journal* 14 (2):282–295. doi: 10.1208/s12248-012-9339-4.

Dhiman, Nitesh, Faimy Fatima, Prem N. Saxsena, Somendu Roy, Prashant K. Rout, and Satyakam Patnaik. 2017. "Predictive modeling and validation of arsenite removal by a one pot synthesized bioceramic buttressed manganese doped iron oxide nanoplatform." *RSC Advances* 7 (52):32866–32876.

Dhiman, Nitesh, Amrita Singh, Neeraj K. Verma, Nidhi Ajaria, and Satyakam Patnaik. 2017. "Statistical optimization and artificial neural network modeling for acridine orange dye degradation using in-situ synthesized polymer capped ZnO nanoparticles." *Journal of Colloid and Interface Science* 493:295–306.

Dillingham, T. Randy, Terry Stufflebeam, and Tim Porter. 2017. "Investigation of PVDF-TiO2 nanoparticle composite thin films by XPS, SEM and EDS for use in the capacitive storage of energy." APS March Meeting *Abstracts.*

Dulog, L., and T. Schauer. 1996. "Field flow fractionation for particle size determination." *Progress in Organic Coatings* 28 (1):25–31. doi: 10.1016/0300-9440(95)00584-6.

Dung, Doan Thi Kim, Tran Hoang Hai, Bui Duc Long, and Phan Nha Truc. 2009. "Preparation and characterization of magnetic nanoparticles with chitosan coating." *Journal of Physics: Conference Series.*

El-Sayed, Yehya, Nedal Abu-Farha, and Sofian Kanan. 2014. "Synthesis and characterization of porous WO3–SnO2 nanomaterials: An infrared study of adsorbed pyridine and dimethyl methylphosphonate." *Vibrational Spectroscopy* 75:78–85.

Elzey, Sherrie Renee. 2010. "Applications and physicochemical characterization of nanomaterials in environmental, health, and safety studies."

Fernandez, Facundo M., and Juan F. Garcia-Reyes. 2017. "Ambient mass spectrometry." *Analytical Methods* 9 (34):4894–4895. doi: 10.1039/C7AY90107K.

Filipe, Vasco, Andrea Hawe, and Wim Jiskoot. 2010. "Critical evaluation of nanoparticle tracking analysis (NTA) by NanoSight for the measurement of nanoparticles and protein aggregates." *Pharmaceutical Research* 27 (5):796–810.

Garcia-Fuentes, Marcos, Dolores Torres, Manuel Martín-Pastor, and Maria Alonso. 2004. "Application of NMR spectroscopy to the characterization of PEG-stabilized lipid nanoparticles." *Langmuir.*

Gelb, Lev D., and K. E. Gubbins. 1998. "Characterization of porous glasses: Simulation models, adsorption isotherms, and the Brunauer–Emmett–Teller analysis method." *Langmuir* 14 (8):2097–2111.

Germain, David, Mathieu Leocmach, and Thomas Gibaud. 2016. "Differential dynamic microscopy to characterize Brownian motion and bacteria motility." *American Journal of Physics* 84 (3):202–210.

Ghule, Kalyani, Anil Vithal Ghule, Bo-Jung Chen, and Yong-Chien Ling. 2006. "Preparation and characterization of ZnO nanoparticles coated paper and its antibacterial activity study." *Green Chemistry* 8 (12):1034–1041.

Giannini, Cinzia, Massimo Ladisa, Davide Altamura, Dritan Siliqi, Teresa Sibillano, and Liberato De Caro. 2016. "X-ray diffraction: A powerful technique for the multiple-length-scale structural analysis of nanomaterials." *Crystals* 6 (8):87.

Gill, Pooria, Tahereh Tohidi Moghadam, and Bijan Ranjbar. 2010. "Differential scanning calorimetry techniques: Applications in biology and nanoscience." *Journal of Biomolecular Techniques* 21 (4):167.

Goldstein, Joseph. 2012. *Practical Scanning Electron Microscopy: Electron and Ion Microprobe Analysis*: Springer Science & Business Media.

Gomez, M. Victoria, Javier Guerra, V. Sue Myers, Richard M. Crooks, and Aldrik H. Velders. 2009. "Nanoparticle size determination by 1H NMR spectroscopy." *Journal of the American Chemical Society* 131 (41):14634–14635.

Guehrs, Erik, Michael Schneider, Christian M. Günther, Piet Hessing, Karen Heitz, Doreen Wittke, Ana López-Serrano Oliver, Norbert Jakubowski, Johanna Plendl, and Stefan Eisebitt. 2017. "Quantification of silver nanoparticle uptake and distribution within individual human macrophages by FIB/SEM slice and view." *Journal of Nanobiotechnology* 15 (1):21.

Gun'ko, Vladimir M., and Vladimir V. Turov. 2013. *Nuclear Magnetic Resonance Studies of Interfacial Phenomena*. Vol. 154. CRC Press.

Guo, Chengchen, Gregory P. Holland, and Jeffery L. Yarger. 2016. "Lysine-capped silica nanoparticles: A solid-state NMR spectroscopy study." *MRS Advances* 1 (31):2261–2266.

Guzman, Maribel, Betty Flores, Loic Malet, and Stephane Godet. 2018. "Synthesis and characterization of zinc oxide nanoparticles for application in the detection of fingerprints." *Materials Science Forum*.

Haase, Andrea, Heinrich F. Arlinghaus, Jutta Tentschert, Harald Jungnickel, Philipp Graf, Alexandre Mantion, Felix Draude, Sebastian Galla, Johanna Plendl, Mario E. Goetz, Admir Masic, Wolfgang Meier, Andreas F. Thünemann, Andreas Taubert, and Andreas Luch. 2011. "Application of laser postionization secondary neutral mass spectrometry/time-of-flight secondary ion mass spectrometry in nanotoxicology: Visualization of nanosilver in human macrophages and cellular responses." *ACS Nano* 5 (4):3059–3068. doi: 10.1021/nn200163w.

Haiss, Wolfgang, Nguyen T. K. Thanh, Jenny Aveyard, and David G. Fernig. 2007. "Determination of size and concentration of gold nanoparticles from UV–Vis spectra." *Analytical Chemistry* 79 (11):4215–4221.

Hara, Shuta, Jumpei Aisu, Masahiro Kato, Takashige Aono, Kosuke Sugawa, Kouichi Takase, Joe Otsuki, Shigeru Shimizu, and Hiroki Ikake. 2018. "One-pot synthesis of monodisperse CoFe2O4@Ag core-shell nanoparticles and their characterization." *Nanoscale Research Letters* 13 (1):176.

Helfrich, Andreas, Wolfram Brüchert, and Jörg Bettmer. 2006. "Size characterisation of Au nanoparticles by ICP-MS coupling techniques." *Journal of Analytical Atomic Spectrometry* 21 (4):431–434.

Holzwarth, Uwe, and Neil Gibson. 2011. "The Scherrer equation versus the 'Debye-Scherrer equation'." *Nature Nanotechnology* 6 (9):534.

Hoo, Christopher M., Natasha Starostin, Paul West, and Martha L. Mecartney. 2008. "A comparison of atomic force microscopy (AFM) and dynamic light scattering (DLS) methods to characterize nanoparticle size distributions." *Journal of Nanoparticle Research* 10 (1):89–96. doi: 10.1007/s11051-008-9435-7.

Hossain, Mohammad Kamal, Masahiro Kitajima, Kohei Imura, and Hiromi Okamoto. 2014. "Near-field scanning optical microscopy: Single channel imaging of selected gold nanoparticles through two photon induced photoluminescence." *Advanced Materials Research* 938.

Hu, Jianyu, Dongyan Deng, Rui Liu, and Yi Lv. 2018. "Single nanoparticle analysis by ICPMS: A potential tool for bioassay." *Journal of Analytical Atomic Spectrometry* 33 (1):57–67. doi: 10.1039/c7ja00235a.

Huth, Florian, Alexander Govyadinov, Sergiu Amarie, Wiwat Nuansing, Fritz Keilmann, and Rainer Hillenbrand. 2012. "Nano-FTIR absorption spectroscopy of molecular fingerprints at 20 nm spatial resolution." *Nano Letters* 12 (8):3973–3978.

Hwang, Yujin, Jae-Keun Lee, Jong-Ku Lee, Young-Man Jeong, Seong-ir Cheong, Young-Chull Ahn, and Soo H. Kim. 2008. "Production and dispersion stability of nanoparticles in nanofluids." *Powder Technology* 186 (2):145–153. doi: 10.1016/j.powtec.2007.11.020.

Itoh, Tamitake, George C. Schatz, Duncan Graham, and Yukihiro Ozaki (eds.). 2016. *Frontiers of Plasmon Enhanced Spectroscopy Volume 1*. Vol. 1245, *ACS Symposium Series*. American Chemical Society.

Jain, Tapan K., John Richey, Michelle Strand, Diandra L. Leslie-Pelecky, Chris A. Flask, and Vinod Labhasetwar. 2008. "Magnetic nanoparticles with dual functional properties: Drug delivery and magnetic resonance imaging." *Biomaterials* 29 (29):4012–4021. doi: 10.1016/j.biomaterials.2008.07.004.

Jana, Nikhil R., Christopher Earhart, and Jackie Y. Ying. 2007. "Synthesis of water-soluble and functionalized nanoparticles by silica coating." *Chemistry of Materials* 19 (21):5074–5082.

Jores, Katja, Wolfgang Mehnert, and Karsten Mäder. 2003. "Physicochemical investigations on solid lipid nanoparticles and on oil-loaded solid lipid nanoparticles: A nuclear magnetic resonance and electron spin resonance study." *Pharmaceutical Research* 20 (8):1274–1283.

Jörg, Schuster, He Guang, Mandlmeier Benjamin, Yim Taeeun, Lee Kyu Tae, Bein Thomas, and Nazar Linda F. 2012. "Spherical ordered mesoporous carbon nanoparticles with high porosity for lithium–sulfur batteries." *Angewandte Chemie International Edition* 51 (15):3591–3595. doi: 10.1002/anie.201107817.

Jortner, Joshua, and C. N. R. Rao. 2002. "Nanostructured advanced materials. Perspectives and directions." *Pure and Applied Chemistry* 74 (9):1491–1506.

Joshi, Sunita, Tej Varma, and Debi D. Pant. 2013. "Steady state and time-resolved fluorescence spectroscopy of quinine sulfate dication in ionic and neutral micelles: Effect of micellar charge on photophysics." *Colloids and Surfaces A: Physicochemical and Engineering Aspects* 425:59–67.

Kaasalainen, Martti, Vladimir Aseyev, Eva von Haartman, Didem Şen Karaman, Ermei Mäkilä, Heikki Tenhu, Jessica Rosenholm, and Jarno Salonen. 2017. "Size, stability, and porosity of mesoporous nanoparticles characterized with light scattering." *Nanoscale Research Letters* 12 (1):74. doi: 10.1186/s11671-017-1853-y.

Kale, Ravindra D., and Priyanka Jagtap. 2018. "Biogenic synthesis of silver nanoparticles using citrus limon leaves and its structural investigation." In *Advances in Health and Environment Safety*, pp. 11–20. Springer.

Kano, Shinya, Tsukasa Tada, and Yutaka Majima. 2015. "Nanoparticle characterization based on STM and STS." *Chemical Society Reviews* 44 (4):970–987.

Kelly, K. Lance, Eduardo Coronado, Lin Lin Zhao, and George C. Schatz. 2003. "The optical properties of metal nanoparticles: The influence of size, shape, and dielectric environment." *ACS Publications*.

Khalid, A., Phong A. Tran, Romina Norello, David A. Simpson, Andrea J. O'Connor, and Snjezana Tomljenovic-Hanic. 2016. "Intrinsic fluorescence of selenium nanoparticles for cellular imaging applications." *Nanoscale* 8 (6):3376–3385.

Khan, Ibrahim, Khalid Saeed, and Idrees Khan. 2017. "Nanoparticles: Properties, applications and toxicities." *Arabian Journal of Chemistry*. doi: 10.1016/j.arabjc.2017.05.011.

Kharchenko, A., V. Zholobenko, A. Vicente, C. Fernandez, H. Vezin, V. De Waele, and S. Mintova. 2018. "Formation of copper nanoparticles in LTL nanosized zeolite: Spectroscopic characterization." *Physical Chemistry Chemical Physics* 20 (4):2880–2889.

Kiefer, Johannes, Janet Grabow, Heinz-Dieter Kurland, and Frank A. Müller. 2015. "Characterization of nanoparticles by solvent infrared spectroscopy." *Analytical Chemistry* 87 (24):12313–12317. doi: 10.1021/acs.analchem.5b03625.

Kim, Dae-Young, Rijuta Ganesh Saratale, Surendra Shinde, Asad Syed, Fuad Ameen, and Gajanan Ghodake. 2018. "Green synthesis of silver nanoparticles using Laminaria japonica extract: Characterization and seedling growth assessment." *Journal of Cleaner Production* 172:2910–2918.

Kim, Hyung Hun. 2015. "Endoscopic Raman spectroscopy for molecular fingerprinting of gastric cancer: Principle to implementation." *BioMed Research International* 2015.

Kleinman, Samuel L., Renee R. Frontiera, Anne-Isabelle Henry, Jon A. Dieringer, and Richard P. Van Duyne. 2013. "Creating, characterizing, and controlling chemistry with SERS hot spots." *Physical Chemistry Chemical Physics* 15 (1): 21–36.

Kong, Xiang-Tian, Larousse Khosravi Khorashad, Zhiming Wang, and Alexander O. Govorov. 2018. "Photothermal circular dichroism induced by plasmon resonances in chiral metamaterial absorbers and bolometers." *Nano Letters* 18 (3):2001–2008.

Korin, Efrat, Natalya Froumin, and Smadar Cohen. 2017. "Surface analysis of nanocomplexes by X-ray photoelectron spectroscopy (XPS)." *ACS Biomaterials Science and Engineering* 3 (6):882–889. doi: 10.1021/acsbiomaterials.7b00040.

Kothamasu, Pavankumar, Hemanth Kanumur, Niranjan Ravur, Chiranjeevi Maddu, Radhika Parasuramrajam, and Sivakumar Thangavel. 2012. "Nanocapsules: The weapons for novel drug delivery systems." *BioImpacts* 2 (2):71–81. doi: 10.5681/bi.2012.011.

Koutsopoulos, Sotirios, Rasmus Barfod, K. Michael Eriksen, and Rasmus Fehrmann. 2017. "Synthesis and characterization of iron-cobalt (FeCo) alloy nanoparticles supported on carbon." *Journal of Alloys and Compounds* 725:1210–1216.

Koynov, Kaloian, and Hans-Jürgen Butt. 2012. "Fluorescence correlation spectroscopy in colloid and interface science." *Current Opinion in Colloid and Interface Science* 17 (6):377–387.

Kuang, Hua, Wei Chen, Wenjing Yan, Liguang Xu, Yingyue Zhu, Liqiang Liu, Huaqin Chu, Chifang Peng, Libing Wang, Nicholas A. Kotov, and Chuanlai Xu. 2011. "Crown ether assembly of gold nanoparticles: Melamine sensor." *Biosensors and Bioelectronics* 26 (5):2032–2037. doi: 10.1016/j.bios.2010.08.081.

Kusch, Patryk, Nieves Morquillas Azpiazu, Niclas Sven Mueller, Stefan Mastel, Jose Ignacio Pascual, and Rainer Hillenbrand. 2018. "Combined tip-enhanced Raman spectroscopy and scattering-type scanning near-field optical microscopy." *The Journal of Physical Chemistry C*.

Kuyper, Christopher L., Bryant S. Fujimoto, Yiqiong Zhao, Perry G. Schiro, and Daniel T. Chiu. 2006. "Accurate sizing of nanoparticles using confocal correlation spectroscopy." *The Journal of Physical Chemistry B* 110 (48):24433–24441.

Labille, Jérôme, and Jonathan Brant. 2010. "Stability of nanoparticles in water." *Nanomedicine* 5 (6):985–998. doi: 10.2217/nnm.10.62.

Lassoued, Abdelmajid, Mohamed Saber Lassoued, Brahim Dkhil, Salah Ammar, and Abdellatif Gadri. 2018. "Synthesis, structural, morphological, optical and magnetic characterization of iron oxide (α-Fe 2 O 3) nanoparticles by precipitation method: Effect of varying the nature of precursor." *Physica E: Low-Dimensional Systems and Nanostructures* 97: 328–334.

Lee, Daniel, Hiroki Takahashi, Aany S. L. Thankamony, Jean-Philippe Dacquin, Michel Bardet, Olivier Lafon, and Gaël De Paëpe. 2012. "Enhanced solid-state NMR correlation spectroscopy of quadrupolar nuclei using dynamic nuclear polarization." *Journal of the American Chemical Society* 134 (45):18491–18494. doi: 10.1021/ja307755t.

Li, Guangshe, Liping Li, Juliana Boerio-Goates, and Brian F. Woodfield. 2005. "High purity anatase TiO2 nanocrystals: Near room-temperature synthesis, grain growth kinetics, and surface hydration chemistry." *Journal of the American Chemical Society* 127 (24):8659–8666.

Li, Taihua, Hyun Gyu Park, Hee-Seung Lee, and Seong-Ho Choi. 2004. "Circular dichroism study of chiral biomolecules conjugated with silver nanoparticles." *Nanotechnology* 15 (10):S660.

Li, Xiaowei, Hiro Minamimoto, Satoshi Yasuda, and Kei Murakoshi. 2016. "Surface-enhanced Raman spectroscopy for the characterization of semiconductor nanostructure surfaces." In *Frontiers of Plasmon Enhanced Spectroscopy Volume 1*, pp. 163–180. American Chemical Society.

Liang, L. H., C. M. Shen, S. X. Du, W. M. Liu, X. C. Xie, and H. J. Gao. 2004. "Increase in thermal stability induced by organic coatings on nanoparticles." *Physical Review B* 70 (20):205419. doi: 10.1103/PhysRevB.70.205419.

Lim, Shuang Fang, Robert Riehn, William S. Ryu, Nora Khanarian, Chih-kuan Tung, David Tank, and Robert H. Austin. 2006. "In vivo and scanning electron microscopy imaging of upconverting nanophosphors in *Caenorhabditis elegans*." *Nano Letters* 6 (2):169–174.

Lin, J. Q., X. K. Li, W. L. Yang, Z. Z. Li, S. J. Lu, and Q. Q. Lei. 2015. "The effects of nanoparticles with different shapes on the interface bond energy in a polyimide/KTa0·5Nb0·5O3 composite." *Materials Research Innovations* 19 (sup5):S5-292–S5-295. doi: 10.1179/1432891714Z.0000000001095.

Lin, Ping-Chang, Stephen Lin, Paul C. Wang, and Rajagopalan Sridhar. 2014. "Techniques for physicochemical characterization of nanomaterials." *Biotechnology Advances* 32 (4):711–726.

Liss, Klaus-Dieter, Arno Bartels, Andreas Schreyer, and Helmut Clemens. 2003. "High-energy X-rays: A tool for advanced bulk investigations in materials science and physics." *Texture, Stress, and Microstructure* 35 (3–4):219–252.

Liu, Fu-Ken, Fu-Hsiang Ko, Pei-Wen Huang, Chien-Hou Wu, and Tieh-Chi Chu. 2005. "Studying the size/shape separation and optical properties of silver nanoparticles by capillary electrophoresis." *Journal of Chromatography A* 1062 (1):139–145.

Liu, Jing-fu, Su-juan Yu, Yong-guang Yin, and Jing-bo Chao. 2012. "Methods for separation, identification, characterization and quantification of silver nanoparticles." *TrAC Trends in Analytical Chemistry* 33:95–106. doi: 10.1016/j.trac.2011.10.010.

Longmire, Michelle, Peter L. Choyke, and Hisataka Kobayashi. 2008. "Clearance properties of nano-sized particles and molecules as imaging agents: Considerations and caveats." *Nanomedicine (London, England)* 3 (5):703–717. doi: 10.2217/17435889.3.5.703.

Lu, Yi-Chun, Azzam N. Mansour, Naoaki Yabuuchi, and Yang Shao-Horn. 2009. "Probing the origin of enhanced stability of "AlPO4" nanoparticle coated LiCoO2 during cycling to high voltages: Combined XRD and XPS studies." *Chemistry of Materials* 21 (19):4408–4424.

Luo, Zhi, Domenico Marson, Quy K. Ong, Anna Loiudice, Joachim Kohlbrecher, Aurel Radulescu, Anwen Krause-Heuer, Tamim Darwish, Sandor Balog, and Raffaella Buonsanti. 2018. "Quantitative 3D determination of self-assembled structures on nanoparticles using small angle neutron scattering." *Nature Communications* 9 (1):1343.

Ma, Lin, Kenville E. Hendrickson, Shuya Wei, and Lynden A. Archer. 2015. "Nanomaterials: Science and applications in the lithium–sulfur battery." *Nano Today* 10 (3):315–338. doi: 10.1016/j.nantod.2015.04.011.

Mächtle, Walter, and Lars Börger. 2006. *Analytical Ultracentrifugation of Polymers and Nanoparticles*. Springer Science & Business Media.

Maher, Robert C. 2012. "SERS hot spots." In *Raman Spectroscopy for Nanomaterials Characterization*, pp. 215–260. Springer.

Mahmoud, Mahmoud A., Maysamreza Chamanzar, Ali Adibi, and Mostafa A. El-Sayed. 2012. "Effect of the dielectric constant of the surrounding medium and the substrate on the surface plasmon resonance spectrum and sensitivity factors of highly symmetric systems: Silver nanocubes." *Journal of the American Chemical Society* 134 (14):6434–6442.

Mahmoudi, Morteza, Samuel E. Lohse, Catherine J. Murphy, Arman Fathizadeh, Abbas Montazeri, and Kenneth S. Suslick. 2013. "Variation of protein corona composition of gold nanoparticles following plasmonic heating." *Nano Letters* 14 (1):6–12.

Maity, Pralay, Madhubanti Bepari, Ananya Pradhan, Rathindranath Baral, Sumita Roy, and Sujata Maiti Choudhury. 2018. "Synthesis and characterization of biogenic metal nanoparticles and its cytotoxicity and anti-neoplasticity through the induction of oxidative stress, mitochondrial dysfunction and apoptosis." *Colloids and Surfaces B: Biointerfaces* 161:111–120.

Makino, Kimiko, and Hiroyuki Ohshima. 2010. "Electrophoretic mobility of a colloidal particle with constant surface charge density." *Langmuir* 26 (23):18016–18019. doi: 10.1021/la1035745.

Mansfield, Elisabeth, Katherine M. Tyner, Christopher M. Poling, and Jenifer L. Blacklock. 2014. "Determination of nanoparticle surface coatings and nanoparticle purity using microscale thermogravimetric analysis." *Analytical Chemistry* 86 (3):1478–1484.

Martinez-Andrade, Juan M., Miguel Avalos-Borja, Alfredo R. Vilchis-Nestor, Luis O. Sanchez-Vargas, and Ernestina Castro-Longoria. 2018. "Dual function of EDTA with silver nanoparticles for root canal treatment–A novel modification." *PloS one* 13 (1):e0190866.

Masoudipour, Elham, Soheila Kashanian, Abbas Hemati Azandaryani, Kobra Omidfar, and Elham Bazyar. 2017. "Surfactant effects on the particle size, zeta potential, and stability of starch nanoparticles and their use in a pH-responsive manner." *Cellulose* 24 (10):4217–4234. doi: 10.1007/s10570-017-1426-3.

McNeil Scott, E. 2005. "Nanotechnology for the biologist." *Journal of Leukocyte Biology* 78 (3):585–594. doi: 10.1189/jlb.0205074.

Mehn, Dora, Patrizia Iavicoli, Noelia Cabaleiro, Sven Even Borgos, Fanny Caputo, Otmar Geiss, Luigi Calzolai, François Rossi, and Douglas Gilliland. 2017. "Analytical ultracentrifugation for analysis of doxorubicin loaded liposomes." *International Journal of Pharmaceutics* 523 (1):320–326.

Mehrizadeh, Habib, Aligholi Niaei, Hui-Hsin Tseng, Dariush Salari, and Alireza Khataee. 2017. "Synthesis of ZnFe2O4 nanoparticles for photocatalytic removal of toluene from gas phase in the annular reactor." *Journal of Photochemistry and Photobiology A: Chemistry* 332:188–195.

Mitrano, Denise M., Angela Barber, Anthony Bednar, Paul Westerhoff, Christopher P. Higgins, and James F. Ranville. 2012. "Silver nanoparticle characterization using single particle ICP-MS (SP-ICP-MS) and asymmetrical flow field flow fractionation ICP-MS (AF4-ICP-MS)." *Journal of Analytical Atomic Spectrometry* 27 (7):1131–1142.

Mittal, Sandeep, Veeresh Kumar, Nitesh Dhiman, Lalit Kumar Singh Chauhan, Renu Pasricha, and Alok Kumar Pandey. 2016. "Physico-chemical properties based differential toxicity of graphene oxide/reduced graphene oxide in human lung cells mediated through oxidative stress." *Scientific Reports* 6:39548. doi: 10.1038/srep39548.

Mock, J. J., M. Barbic, D. R. Smith, D. A. Schultz, and S. Schultz. 2002. "Shape effects in plasmon resonance of individual colloidal silver nanoparticles." *The Journal of Chemical Physics* 116 (15):6755–6759.

Molleman, Bastiaan, and Tjisse Hiemstra. 2018. "Size and shape dependency of the surface energy of metallic nanoparticles: Unifying the atomic and thermodynamic approaches." *Physical Chemistry Chemical Physics* 20 (31):20575–20587. doi: 10.1039/c8cp02346h.

Moradi, Golshan, Sirus Zinadini, Laleh Rajabi, and Soheil Dadari. 2018. "Fabrication of high flux and antifouling mixed matrix fumarate-alumoxane/PAN membranes via electrospinning for application in membrane bioreactors." *Applied Surface Science* 427:830–842.

Morga, Maria, Zbigniew Adamczyk, Dominik Kosior, and Magdalena Oćwieja. 2018. "Hematite/silica nanoparticle bilayers on mica: AFM and electrokinetic characterization." *Physical Chemistry Chemical Physics* 20 (22):15368–15379.

Moroz, Ella M. 2011. "X-Ray diffraction structure diagnostics of nanomaterials." *Russian Chemical Reviews* 80 (4):293.

Motin, Abdul Md, Thomas Haunold, Andrey V. Bukhtiyarov, Abhijit Bera, Christoph Rameshan, and Günther Rupprechter. 2018. "Surface science approach to Pt/carbon model catalysts: XPS, STM and microreactor studies." *Applied Surface Science* 440:680–687.

Mourdikoudis, Stefanos, Roger Molto Pallares, and Nguyen Thi Kim Thanh. 2018. "Characterization techniques for nanoparticles: Comparison and complementarity upon studying nanoparticle properties." *Nanoscale*.

Mueller, Nicole C., and Bernd Nowack. 2008. "Exposure modeling of engineered nanoparticles in the environment." *Environmental Science and Technology* 42 (12):4447–4453. doi: 10.1021/es7029637.

Mühlfeld, Christian, Barbara Rothen-Rutishauser, Dimitri Vanhecke, Fabian Blank, Peter Gehr, and Matthias Ochs. 2007. "Visualization and quantitative analysis of nanoparticles in the respiratory tract by transmission electron microscopy." *Particle and Fibre Toxicology* 4:11–11. doi: 10.1186/1743-8977-4-11.

Mukhopadhyay, Piyasi, Subhajit Maity, Sudipto Mandal, Abhay Sankar Chakraborti, A. K. Prajapati, and P. P. Kundu. 2018. "Preparation, characterization and in vivo evaluation of pH sensitive, safe quercetin-succinylated chitosan-alginate core-shell-corona nanoparticle for diabetes treatment." *Carbohydrate Polymers* 182:42–51.

Naderi, Majid. 2015. "Surface area: Brunauer–Emmett–Teller (BET)." In *Progress in Filtration and Separation*, pp. 585–608. Elsevier.

Navas-Moreno, Maria, Majid Mehrpouyan, Tatyana Chernenko, Demet Candas, Ming Fan, Jian Jian Li, Ming Yan, and James W. Chan. 2017. "Nanoparticles for live cell microscopy: A surface-enhanced Raman scattering perspective." *Scientific Reports* 7 (1):4471. doi: 10.1038/s41598-017-04066-0.

Nayak, Manoj K., Jaswant Singh, Baljit Singh, Shilpa Soni, Vidhu S. Pandey, and Sachin Tyagi. 2017. "Chapter 1 – Introduction to semiconductor nanomaterial and its optical and electronics properties." In *Metal Semiconductor Core-Shell Nanostructures for Energy and Environmental Applications*, edited by Raju Kumar Gupta and Mrinmoy Misra, pp. 1–33. Elsevier.

Naydenov, Borislav, Samuel Torsney, Alejandro Santana Bonilla, Mohamed El Garah, Artur Ciesielski, Andrea Gualandi, Luca Mengozzi, Pier Giorgio Cozzi, Rafael Gutierrez, and Paolo Samorì. 2018. "Self-assembled 2D supramolecular networks characterized by STM/STS in air and under vacuum." *Langmuir*.

Ni, Qing Zhe, Fengyuan Yang, Thach V. Can, Ivan V. Sergeyev, Suzanne M. D'Addio, Sudheer K. Jawla, Yongjun Li, Maya P. Lipert, Wei Xu, and R. Thomas Williamson. 2017. "In situ characterization of pharmaceutical formulations by dynamic nuclear polarization enhanced MAS NMR." *The Journal of Physical Chemistry B* 121 (34):8132–8141.

Nonappa, and E. Kolehmainen. 2016. "Solid state NMR studies of gels derived from low molecular mass gelators." *Soft Matter* 12 (28):6015–6026. doi: 10.1039/c6sm00969g.

Nosrati, Hamed, Marziyeh Salehiabar, Manjili Hamidreza Kheiri, Soodabeh Davaran, and Hossein Danafar. 2018. "Theranostic nanoparticles based on magnetic nanoparticles: Design, preparation, characterization and evaluation as novel anticancer drug carrier and MRI contrast agent." *Drug Development and Industrial Pharmacy* (just accepted):1–29.

Oezaslan, Mehtap, Frédéric Hasché, and Peter Strasser. 2011. "In situ observation of bimetallic alloy nanoparticle formation and growth using high-temperature XRD." *Chemistry of Materials* 23 (8):2159–2165.

Ong, Quy Khac, Javier Reguera, Paulo Jacob Silva, Mauro Moglianetti, Kellen Harkness, Maria Longobardi, Kunal S. Mali, Christoph Renner, Steven De Feyter, and Francesco Stellacci. 2013. "High-resolution scanning tunneling microscopy characterization of mixed monolayer protected gold nanoparticles." *ACS Nano* 7 (10):8529–8539.

Oshikane, Yasushi, Toshihiko Kataoka, Mitsuru Okuda, Seiji Hara, Haruyuki Inoue, and Motohiro Nakano. 2007. "Observation of nanostructure by scanning near-field optical microscope with small sphere probe." *Science and Technology of Advanced Materials* 8 (3):181.

Pace, Heather E., Nicola J. Rogers, Chad Jarolimek, Victoria A. Coleman, Christopher P. Higgins, and James F. Ranville. 2011. "Determining transport efficiency for the purpose of counting and sizing nanoparticles via single particle inductively coupled plasma mass spectrometry." *Analytical Chemistry* 83 (24):9361–9369.

Pandey, B. K., A. Sukla, A. K. Sinha, and R. Gopal. 2015. "Synthesis and characterization of cobalt oxalate nanomaterial for Li-Ion battery." *Materials Focus* 4 (5):333–337.

Pandey, Manisha, Hira Choudhury, Tarakini A. P. Gunasegaran, Saranyah Shanmugah Nathan, Shadab Md, Bapi Gorain, Minaketan Tripathy, and Zahid Hussain. 2018. "Hyaluronic acid-modified betamethasone encapsulated polymeric nanoparticles: Fabrication, characterisation, in vitro release kinetics, and dermal targeting." *Drug Delivery and Translational Research*:1–14.

Pang, Louis S. K., John D. Saxby, and S. Peter Chatfield. 1993. "Thermogravimetric analysis of carbon nanotubes and nanoparticles." *The Journal of Physical Chemistry* 97 (27):6941–6942. doi: 10.1021/j100129a001.

Parang, Z., A. Keshavarz, S. Farahi, S. M. Elahi, M. Ghoranneviss, and S. Parhoodeh. 2012. "Fluorescence emission spectra of silver and silver/cobalt nanoparticles." *Scientia Iranica* 19 (3):943–947. doi: 10.1016/j.scient.2012.02.026.

Patil, Swanand, Amanda Sandberg, Eric Heckert, William Self, and Sudipta Seal. 2007. "Protein adsorption and cellular uptake of cerium oxide nanoparticles as a function of zeta potential." *Biomaterials* 28 (31):4600–4607.

Peng, Xiaohui, Shelagh Palma, Nicholas S. Fisher, and Stanislaus S. Wong. 2011. "Effect of morphology of ZnO nanostructures on their toxicity to marine algae." *Aquatic Toxicology* 102 (3):186–196. doi: 10.1016/j.aquatox.2011.01.014.

Pestryakov, Alexey Nikolaevich, Vitalii P. Petranovskii, Andrey Kryazhov, Oleg Ozhereliev, Norbert PfÃ¤nder, and Axel Knop-Gericke. 2004. "Study of copper nanoparticles formation on supports of different nature by UV-Vis diffuse reflectance spectroscopy." *Chemical Physics Letters* 385 (3–4):173–176.

Pham, Xuan Nui, Tan Phuoc Nguyen, Tuyet Nhung Pham, Thi Thuy Nga Tran, and Thi Van Thi Tran. 2016. "Synthesis and characterization of chitosan-coated magnetite nanoparticles and their application in curcumin drug delivery." *Advances in Natural Sciences: Nanoscience and Nanotechnology* 7 (4):045010.

Piveteau, Laura, Ta-Chung Ong, Aaron J. Rossini, Lyndon Emsley, Christophe Coperet, and Maksym V. Kovalenko. 2015. "Structure of colloidal quantum dots from dynamic nuclear polarization surface enhanced NMR spectroscopy." *Journal of the American Chemical Society* 137 (43): 13964–13971.

Pompeo, G., M. Girasole, A. Cricenti, F. Cattaruzza, A. Flamini, T. Prosperi, J. Generosi, and A. Congiu Castellano. 2005. "AFM characterization of solid-supported lipid multilayers prepared by spin-coating." *Biochimica et Biophysica Acta (BBA)-Biomembranes* 1712 (1):29–36.

Pradeep, A., P. Priyadharsini, and G. Chandrasekaran. 2008. "Sol–gel route of synthesis of nanoparticles of MgFe2O4 and XRD, FTIR and VSM study." *Journal of Magnetism and Magnetic Materials* 320 (21):2774–2779. doi: 10.1016/j.jmmm.2008.06.012.

Prieto, Gonzalo, Jovana Zečević, Heiner Friedrich, Krijn P. de Jong, and Petra E. de Jongh. 2012. "Towards stable catalysts by controlling collective properties of supported metal nanoparticles." *Nature Materials* 12:34. doi: 10.1038/nmat3471.

Pugazhendhi, Arivalagan, Desika Prabakar, Jaya Mary Jacob, Indira Karuppusamy, and Rijuta Ganesh Saratale. 2018. "Synthesis and characterization of silver nanoparticles using Gelidium amansii and its antimicrobial property against various pathogenic bacteria." *Microbial Pathogenesis* 114:41–45.

Purohit, Mahaveer P., Neeraj K. Verma, Aditya K. Kar, Amrita Singh, Debabrata Ghosh, and Satyakam Patnaik. 2017. "Inhibition of thioredoxin reductase by targeted selenopolymeric nanocarriers synergizes the therapeutic efficacy of doxorubicin in MCF7 human breast cancer cells." *ACS Applied Materials and Interfaces* 9 (42):36493–36512.

Ramos, Ana P. 2017. "Dynamic light scattering applied to nanoparticle characterization." In *Nanocharacterization Techniques*, pp. 99–110. Elsevier.

Rao, C. N. R., G. U. Kulkarni, P. John Thomas, and Peter P. Edwards. 2002. "Size-dependent chemistry: Properties of nanocrystals." *Chemistry – A European Journal* 8 (1):28–35. doi: 10.1002/1521-3765(20020104)8:1<28::aid-chem28>3.0.co;2-b.

Reimer, Ludwig. 2013. *Transmission Electron Microscopy: Physics of Image Formation and Microanalysis*. Vol. 36. Springer.

Ren, Guogang, Dawei Hu, Eileen W. C. Cheng, Miguel A. Vargas-Reus, Paul Reip, and Robert P. Allaker. 2009. "Characterisation of copper oxide nanoparticles for antimicrobial applications." *International Journal of Antimicrobial Agents* 33 (6):587–590. doi: 10.1016/j.ijantimicag.2008.12.004.

Ristau, Roger, Ramchandra Tiruvalam, Patrick L. Clasen, Edward P. Gorskowski, Martin P. Harmer, Christopher J. Kiely, Irshad Hussain, and Mathias Brust. 2009. "Electron microscopy studies of the thermal stability of gold nanoparticle arrays." *Gold Bulletin* 42 (2):133–143. doi: 10.1007/bf03214923.

Robel, István, Masaru Kuno, and Prashant V. Kamat. 2007. "Size-dependent electron injection from excited CdSe quantum dots into TiO2 nanoparticles." *Journal of the American Chemical Society* 129 (14):4136–4137. doi: 10.1021/ja070099a.

Robertson, James D., Loris Rizzello, Milagros Avila-Olias, Jens Gaitzsch, Claudia Contini, Monika S. Magoń, Stephen A. Renshaw, and Giuseppe Battaglia. 2016. "Purification of nanoparticles by size and shape." *Scientific Reports* 6:27494.

Rose, Abigail, Stephanie Maniguet, Rebecca J. Mathew, Claire Slater, Jun Yao, and Andrea E. Russell. 2003. "Hydride phase formation in carbon supported palladium nanoparticle electrodes investigated using in situ EXAFS and XRD." *Physical Chemistry Chemical Physics* 5 (15):3220–3225.

Sanchez-Gaytan, Brenda L., Fay François, Hak Sjoerd, Alaarg Amr, Fayad Zahi A., Pérez-Medina Carlos, Mulder Willem J. M., and Zhao Yiming. 2017. "Real-time monitoring of nanoparticle formation by FRET imaging." *Angewandte Chemie International Edition* 56 (11):2923–2926. doi: 10.1002/anie.201611288.

Sapsford, Kim E., Katherine M. Tyner, Benita J. Dair, Jeffrey R. Deschamps, and Igor L. Medintz. 2011. "Analyzing nanomaterial bioconjugates: A review of current and emerging purification and characterization techniques." *Analytical Chemistry* 83 (12):4453–4488.

Sarmento, Bruno, Domingos Ferreira, Francisco Veiga, and Antonio Ribeiro. 2006. "Characterization of insulin-loaded alginate nanoparticles produced by ionotropic pre-gelation through DSC and FTIR studies." *Carbohydrate Polymers* 66 (1):1–7.

Sauerbrey, G. Z. 1959. "The use of quarts oscillators for weighing thin layers and for microweighing." *Zeitschrift für Angewandte Physik* 155:206–222.

Schechinger, Monika, Haley Marks, Andrea Locke, Mahua Choudhury, and Gerard Cote. 2018. "Development of a miRNA surface-enhanced Raman scattering assay using benchtop and handheld Raman systems." *Journal of Biomedical Optics* 23 (1):017002.

Schuetze, Benjamin, Christian Mayer, Kateryna Loza, Martin Gocyla, M. Heggen, and Matthias Epple. 2016. "Conjugation of thiol-terminated molecules to ultrasmall 2 nm-gold nanoparticles leads to remarkably complex 1 H-NMR spectra." *Journal of Materials Chemistry B* 4 (12):2179–2189.

Sett, Ayantika, Swagata Dasgupta, and Sunando DasGupta. 2018. "Rapid estimation of the β-sheet content of human serum albumin from the drying patterns of HSA-nanoparticle droplets." *Colloids and Surfaces A: Physicochemical and Engineering Aspects* 540:177–185.

Shang, Li, and G. Ulrich Nienhaus. 2017. "In situ characterization of protein adsorption onto nanoparticles by fluorescence correlation spectroscopy." *Accounts of Chemical Research* 50 (2):387–395.

Sharma, A., S. V. Madhunapantula, and G. P. Robertson. 2012. "Toxicological considerations when creating nanoparticle-based drugs and drug delivery systems." *Expert Opinion on Drug Metabolism and Toxicology* 8 (1):47–69. doi: 10.1517/17425255.2012.637916.

Sharma, Anuj Kumar, Vikas Singh, Ruchi Gera, Mahaveer Prasad Purohit, and Debabrata Ghosh. 2017. "Zinc oxide nanoparticle induces microglial death by NADPH-oxidase-independent reactive oxygen species as well as energy depletion." *Molecular Neurobiology* 54 (8):6273–6286.

Sharma, Ravi, D. P. Bisen, Usha Shukla, and B. G. Sharma. 2012. "X-ray diffraction: A powerful method of characterizing nanomaterials." *Recent Research in Science and Technology* 4 (8).

Shuai, Zhang. 2016. "The relationship between the size-dependent melting point of nanoparticles and the bond number." *Nanomaterials and Energy* 5 (2):125–131. doi: 10.1680/jnaen.16.00008.

Siekmann, Britta, and Kirsten Westesen. 1994. "Thermoanalysis of the recrystallization process of melt-homogenized glyceride nanoparticles." *Colloids and Surfaces B: Biointerfaces* 3 (3):159–175.

Sipe, David M., Logan D. Plath, Alexander A. Aksenov, Jonathan S. Feldman, and Mark E. Bier. 2018. "Characterization of mega-dalton-sized nanoparticles by superconducting tunnel junction cryodetection mass spectrometry." *ACS Nano* 12 (3):2591–2602.

Sirdeshpande, Karthikey Devadatta, Anushka Sridhar, Kedar Mohan Cholkar, and Raja Selvaraj. 2018. "Structural characterization of mesoporous magnetite nanoparticles synthesized using the leaf extract of *Calliandra haematocephala* and their photocatalytic degradation of malachite green dye." *Applied Nanoscience*:1–9.

Sivasankar, Palaniappan, Palaniappan Seedevi, Subramaniam Poongodi, Murugesan Sivakumar, Tamilselvi Murugan, Loganathan Sivakumar, Kannan Sivakumar, and Thangavel Balasubramanian. 2018. "Characterization, antimicrobial and antioxidant property of exopolysaccharide mediated silver nanoparticles synthesized by Streptomyces violaceus MM72." *Carbohydrate Polymers* 181:752–759.

Skoog, Douglas A., F. James Holler, and Stanley R. Crouch. 2017. *Principles of Instrumental Analysis*. Cengage Learning.

Slocik, Joseph M., Alexander O. Govorov, and Rajesh R. Naik. 2011. "Plasmonic circular dichroism of peptide-functionalized gold nanoparticles." *Nano Letters* 11 (2):701–705.

Sokolova, Viktoriya, Anna-Kristin Ludwig, Sandra Hornung, Olga Rotan, Peter A. Horn, Matthias Epple, and Bernd Giebel. 2011. "Characterisation of exosomes derived from human cells by nanoparticle tracking analysis and scanning electron microscopy." *Colloids and Surfaces B: Biointerfaces* 87 (1):146–150.

Speakman, Scott A. 2014. "Estimating crystallite size using XRD." *MIT Center for Materials Science and Engineering*.

Srivastav, Anurag Kumar, Nitesh Dhiman, Hafizurrahman Khan, Ankur Kumar Srivastav, Sanjeev Kumar Yadav, Jyoti Prakash, Nidhi Arjaria, Dhirendra Singh, Sanjay Yadav, Satyakam Patnaik, and Mahadeo Kumar. 2019. "Impact of surface-engineered ZnO nanoparticles on protein corona configuration and their interactions with biological system." *Journal of Pharmaceutical Sciences* 108 (5):1872–1889. doi: 10.1016/j.xphs.2018.12.021.

Starostina, N., M. Brodsky, S. Prikhodko, C. M. Hoo, M. L. Mecartney, and P. West. 2008. "AFM capabilities in characterization of particles and surfaces: From angstroms to microns." *Journal of Cosmetic Science* 59 (3):225–232.

Stetefeld, Jörg, Sean A. McKenna, and Trushar R. Patel. 2016. "Dynamic light scattering: A practical guide and applications in biomedical sciences." *Biophysical Reviews* 8 (4):409–427.

Sun, Mengmeng, Fei Liu, Yukun Zhu, Wansheng Wang, Jin Hu, Jing Liu, Zhifei Dai, Kun Wang, Yen Wei, Jing Bai, and Weiping Gao. 2016. "Salt-induced aggregation of gold nanoparticles for photoacoustic imaging and photothermal therapy of cancer." *Nanoscale* 8 (8):4452–4457. doi: 10.1039/c6nr00056h.

Sun, Minjie, Bin Sun, Yun Liu, Qun-Dong Shen, and Shaojun Jiang. 2016. "Dual-color fluorescence imaging of magnetic nanoparticles in live cancer cells using conjugated polymer probes." *Scientific Reports* 6:22368.

Sun, Yiling, Yan Nie, Zhenjiong Wang, Chun Hua, Renlei Wang, and Juan Gao. 2017. "Biomacromolecule-directed synthesis and characterization of selenium nanoparticles and their compatibility with bacterial and eukaryotic cells." *Nanoscience and Nanotechnology Letters* 9 (12):1987–1991.

Tanaka, Satohiro. 1992. "Theory of power-compensated DSC." *Thermochimica Acta* 210:67–76.

Techane, Sirnegeda D., Lara J. Gamble, and David G. Castner. 2011. "X-ray photoelectron spectroscopy characterization of gold nanoparticles functionalized with amine-terminated alkanethiols." *Biointerphases* 6 (3):98–98. doi: 10.1116/1.3622481.

Teetsov, Julie, and David A. Vanden Bout. 2000. "Near-field scanning optical microscopy (NSOM) studies of nanoscale polymer ordering in pristine films of poly (9, 9-dialkylfluorene)." *The Journal of Physical Chemistry B* 104 (40):9378–9387.

Thomas, John Meurig, Richard D. Adams, Erin M. Boswell, Burjor Captain, Henrik Grönbeck, and Robert Raja. 2008. "Synthesis, characterization, electronic structure and catalytic performance of bimetallic and trimetallic nanoparticles containing tin." *Faraday Discussions* 138:301–315.

Treuel, Lennart, Marcelina Malissek, Julia Susanne Gebauer, and Reinhard Zellner. 2010. "The influence of surface composition of nanoparticles on their interactions with serum albumin." *ChemPhysChem* 11 (14):3093–3099.

Trigueros, Sònia, Elena B. Domènech, Vasileios Toulis, and Gemma Marfany. 2019. "In vitro gene delivery in retinal pigment epithelium cells by plasmid DNA-wrapped gold nanoparticles." *Genes* 10 (4):289. doi: 10.3390/genes10040289.

Tumkur, Thejaswi, Xiao Yang, Chao Zhang, Jian Yang, Yue Zhang, Gururaj V. Naik, Peter Nordlander, and Naomi J. Halas. 2018. "Wavelength-dependent optical force imaging of bimetallic Al–Au heterodimers." *Nano Letters* 18 (3):2040–2046.

Varadavenkatesan, Thivaharan, Ramesh Vinayagam, and Raja Selvaraj. 2017. "Structural characterization of silver nanoparticles phyto-mediated by a plant waste, seed hull of *Vigna mungo* and their biological applications." *Journal of Molecular Structure* 1147:629–635.

Verma, Neeraj K., Mahaveer P. Purohit, Danish Equbal, Nitesh Dhiman, Amrita Singh, Aditya K. Kar, Jai Shankar, Sarita Tehlan, and Satyakam Patnaik. 2016. "Targeted smart pH and thermoresponsive N,O-carboxymethyl chitosan conjugated nanogels for enhanced therapeutic efficacy of doxorubicin in MCF-7 breast cancer cells." *Bioconjugate Chemistry* 27 (11):2605–2619.

Vernon-Parry, K. D. 2000. "Scanning electron microscopy: An introduction." *III-Vs Review* 13 (4):40–44.

Vyhnanovský, Jaromír, Jan Kratzer, Oldřich Benada, Tomáš Matoušek, Zoltán Mester, Ralph E. Sturgeon, Jiří Dědina, and Stanislav Musil. 2018. "Diethyldithiocarbamate enhanced chemical generation of volatile palladium species, their characterization by AAS, ICP-MS, TEM and DART-MS and proposed mechanism of action." *Analytica Chimica Acta* 1005:16–26.

Walter, Dirk. 2013. "Primary particles–agglomerates–aggregates." *Nanomaterials*:9–24.

Walter, Johannes, Konrad Löhr, Engin Karabudak, Wieland Reis, Jules Mikhael, Wolfgang Peukert, Wendel Wohlleben, and Helmut Cölfen. 2014. "Multidimensional analysis of nanoparticles with highly disperse properties using multiwavelength analytical ultracentrifugation." *ACS Nano* 8 (9):8871–8886.

Wang, Bo, Liangfang Zhang, Sung Chul Bae, and Steve Granick. 2008. "Nanoparticle-induced surface reconstruction of phospholipid membranes." *Proceedings of the National Academy of Sciences of the United States of America* 105 (47):18171–18175. doi: 10.1073/pnas.0807296105.

Wang, John, Julien Polleux, James Lim, and Bruce Dunn. 2007. "Pseudocapacitive contributions to electrochemical energy storage in TiO2 (Anatase) nanoparticles." *The Journal of Physical Chemistry C* 111 (40):14925–14931. doi: 10.1021/jp074464w.

Wang, Lin, and Weihong Tan. 2006. "Multicolor FRET silica nanoparticles by single wavelength excitation." *Nano Letters* 6 (1):84–88. doi: 10.1021/nl052105b.

Wang, Qingyan, Guowei Lu, Lei Hou, Tianyue Zhang, Chunxiong Luo, Hong Yang, Grégory Barbillon, Franck H. Lei, Christophe A. Marquette, and Pascal Perriat. 2011. "Fluorescence correlation spectroscopy near individual gold nanoparticle." *Chemical Physics Letters* 503 (4–6):256–261.

Wei, Feng, Jia-Lin Zhu, and Hao-Ming Chen. 2000. "STS properties of defective metallic carbon nanotubes." *Journal of Physics: Condensed Matter* 12 (40):8617.

Weyermann, Jörg, Dirk Lochmann, Christiane Georgens, Isam Rais, Jörg Kreuter, Michael Karas, Markus Wolkenhauer, and Andreas Zimmer. 2004. "Physicochemical characterisation of cationic polybutylcyanoacrylat-nanoparticles by fluorescence correlation spectroscopy." *European Journal of Pharmaceutics and Biopharmaceutics* 58 (1):25–35. doi: 10.1016/j.ejpb.2004.02.011.

Whetten, Robert L., Joseph T. Khoury, Marcos M. Alvarez, Srihari Murthy, Igor Vezmar, Z. L. Wang, Peter W. Stephens, Charles L. Cleveland, W. D. Luedtke, and Uzi Landman. 1996. "Nanocrystal gold molecules." *Advanced Materials* 8 (5):428–433.

Wolfbeis, Otto S. 2015. "An overview of nanoparticles commonly used in fluorescent bioimaging." *Chemical Society Reviews* 44 (14):4743–4768.

Wu, Libo, Jian Zhang, and Wiwik Watanabe. 2011. "Physical and chemical stability of drug nanoparticles." *Advanced Drug Delivery Reviews* 63 (6):456–469. doi: 10.1016/j.addr.2011.02.001.

Xie, Jing, Keita Nakai, Sayaka Ohno, Hans-Juergen Butt, Kaloian Koynov, and Shin-ichi Yusa. 2015. "Fluorescence correlation spectroscopy monitors the hydrophobic collapse of pH-responsive hairy nanoparticles at the individual particle level." *Macromolecules* 48 (19):7237–7244.

Xu, Zhancheng, Xiangyi Huang, Chaoqing Dong, and Jicun Ren. 2014. "Fluorescence correlation spectroscopy of gold nanoparticles, and its application to an aptamer-based homogeneous thrombin assay." *Microchimica Acta* 181 (7–8):723–730.

Yang, Raymond S. H., Louis W. Chang, Jui-Pin Wu, Ming-Hsien Tsai, Hsiu-Jen Wang, Yu-Chun Kuo, Teng-Kuang Yeh, Chung Shi Yang, and Pinpin Lin. 2007. "Persistent tissue kinetics and redistribution of nanoparticles, quantum dot 705, in mice: ICP-MS quantitative assessment." *Environmental Health Perspectives* 115 (9):1339.

Ye, Manping, Yarong Shi, and Huacai Chen. 2016. "Steady-state and time-resolved fluorescence spectroscopic studies on the interaction between bovine serum albumin and Ag-nanoparticles." In *Optical Measurement Technology and Instrumentation.*

Yu, Hongtao, and Stephanie L. Brock. 2008. "Effects of nanoparticle shape on the morphology and properties of porous CdSe assemblies (aerogels)." *ACS Nano* 2 (8):1563–1570. doi: 10.1021/nn8002295.

Zhang, Feng, Peng Wang, J. Koberstein, S. Khalid, and Siu-Wai Chan. 2004. "Cerium oxidation state in ceria nanoparticles studied with X-ray photoelectron spectroscopy and absorption near edge spectroscopy." *Surface Science* 563 (1):74–82. doi: 10.1016/j.susc.2004.05.138.

Zhang, Haijun, and Naoki Toshima. 2013. "Glucose oxidation using Au-containing bimetallic and trimetallic nanoparticles." *Catalysis Science and Technology* 3 (2):268–278.

Zhang, Yang, Yongsheng Chen, Paul Westerhoff, and John Crittenden. 2009. "Impact of natural organic matter and divalent cations on the stability of aqueous nanoparticles." *Water Research* 43 (17):4249–4257. doi: 10.1016/j.watres.2009.06.005.

Zhao, Xiaoyu, Wenshuai Hu, Yanfei Wang, Liang Zhu, Libin Yang, Zuoliang Sha, and Juankun Zhang. 2018. "Decoration of graphene with 2-aminoethanethiol functionalized gold nanoparticles for molecular imprinted sensing of erythrosine." *Carbon* 127:618–626.

Zhu, Lili, Qiang Gu, Pingchuan Sun, Wei Chen, Xiaoliang Wang, and Gi Xue. 2013. "Characterization of the mobility and reactivity of water molecules on TiO2 nanoparticles by 1H solid-state nuclear magnetic resonance." *ACS Applied Materials and Interfaces* 5 (20):10352–10356.

Ziegler-Borowska, Marta, Dorota Chełminiak, and Halina Kaczmarek. 2015. "Thermal stability of magnetic nanoparticles coated by blends of modified chitosan and poly(quaternary ammonium) salt." *Journal of Thermal Analysis and Calorimetry* 119 (1):499–506. doi: 10.1007/s10973-014-4122-7.

Zordan, Christopher A., M. Ross Pennington, and Murray V. Johnston. 2010. "Elemental composition of nanoparticles with the nano aerosol mass spectrometer." *Analytical Chemistry* 82 (19):8034–8038. doi: 10.1021/ac101700q.

2

Facile Chemical Fabrication of Designer Biofunctionalized Nanomaterials

A.H. Sneharani and K. Byrappa

CONTENTS

2.1 Introduction

Nanomaterials have become indispensable for science and technology. Any object with dimensions between 1 and 100 nm is defined as 'nanomaterial'. Metals, metal oxides, semiconductors, and core-shell hybrids are a few varieties of core materials used in the synthesis of nanomaterials. These nanomaterials can be made of a variety of materials, including noble metals (gold Au, silver Ag, palladium Pd), semiconductors (group IIVI compounds such as cadmium telluride, CdTe; palladium telluride, PbTe; indium phosphide, InP; lead sulphide, PbS; titanium oxide, TiO2; silicon, Si; silver–indium–antimony–tellurium AgInSbTe), quartz (SiO2), zincite (ZnO), rutile (tetragonal system, TiO2), anatase (tetragonal, TiO2), magnetic compounds (hematite, Fe2O3; magnetite, Fe3O4; Cobalt ferrite, CoFe2O4), and their combinations (core-shell nanoparticles and other composite structures). Various properties of nanomaterials, such as photoluminescence (semiconductor quantum dots), magnetic moment, high electron density, and strong optical absorption (Au) are decided by their composition. These functional properties are achieved by surface functionalization. The fabrication and fine tuning of the physical properties of nanoscale materials allow to achieve novel photonic and magnetic properties, providing excellent platforms for the design of advanced nanomaterials, with biomedical applications integrating biology and synthetic materials (Bhushan, 2010).

Over the last three decades, the field of nanotechnology has observed an incredible development in advanced biofunctionalized nanomaterials involving new and efficient synthetic routes. The name 'advanced biofunctional nanomaterial' refers to any chemical, organic, or inorganic substance possessing tailor-made physical, chemical, and biological features with specific functional properties. The biofunctionalization of nanomaterials allows them to be biocompatible with the immobilization of biomolecules or cells with high specificity, providing significant applications in detection/sensing, molecular imaging, theranostics, and in more complex tasks such as in targeted drug or gene delivery (Baptista, 2009). Significant advancements in the field of nanomedicine are made in early diagnosis of chronic disease, bioimaging, and targeted therapy. However, there are many challenging limitations that cannot be overlooked; for instance, smaller size nanoparticles tend to become aggregated, losing their surface chemistry; also, the bioavailability of biofunctionalized nanoparticles is achieved when the nanoparticles do not elicit the immune response, thus becoming available to the system (Shvedova, Kagan, and Fadeel, 2010).

2.2 Synthesis of Nanoparticles

A variety of nanoparticles can be synthesized with different elemental composition, differing in their size, shape, and chemical or physical properties. The physical, chemical, and biological methods, amongst others, are often used in the synthesis of nanoparticles. However, all these methods can be broadly grouped into two types: (i) bottom-up approach and (ii) top-down approach. Figure 2.1 shows the graphical representation of these two methods (Reverchon and Adami, 2006).

In the top-down method, the bulk materials are brought down to nanoscale using milling or nanomilling. The bottom-up approach involves a molecular assembly of atoms and molecules into small particles consisting of very few to hundreds of atoms referred as clusters, which in a later stage are converted into nanoparticles. The bottom-up approach gives more flexibility to incorporate desired properties; hence, it is popularly used in the preparation of nanoparticles and thereof conjugation with various biomacromolecules to produce biofunctionalized advanced nanomaterials. The bottom-up method is the most popularly used as it facilitates the tuning of the desired shape and size in the nanoparticles. In addition, *in situ* surface modification allows alteration of the surface chemistry and selective doping to alter the chemical properties of the nanoparticles. The biomimetic and bacterial synthesis routes are considered as green processes as they imitate the formation of nanomaterials at ambient or near-ambient conditions without involving high-temperature, high-pressure, hot-pressing, or high-pressure processing. For example, a seashell or molluscan shell or gastropod shell is made up of calcium carbonate and is formed in aqueous media under ambient conditions. These shells are very hard to break and are formed at temperatures not higher than 40 °C. If similar hard materials made up of calcium carbonate were to be formed in the laboratory, high temperature and high pressure would be required. Therefore, biological synthesis or biomolecule-mediated synthesis are energetically feasible for the formation of such hard shells under normal ambient conditions. Researchers mimic similar processes in the laboratory, imitating the natural biological synthesis; therefore, this is termed biomimetic process or synthesis. Synthetic methods commonly used in the preparation of nanoparticles include colloidal chemistry approach, hydrothermal technique, sol-gel technique, mechanical grinding, mechanical milling, chemical vapor deposition, and biological/biomimetic techniques. Key parameters that influence the formation of nanoparticles are the melting point and the solubility of its materials. The other controlling parameters include temperature, pressure, *pH*, surfactants, and dopants (if impurities are present).

Among these techniques, the hydrothermal technique has its own advantages over conventional techniques. Its advantages include high product yield, purity, and homogeneity with narrow particle-size distribution, low input energy, simple equipment, fast reaction times, and growth of nanoparticles with differing solubility. Such developments have made the hydrothermal technique unique in terms of its ability to fabricate functional products with *in situ* control over their growth. A hydrothermal technique is any homogeneous or heterogeneous chemical reaction in the presence of a solvent (whether aqueous or non-aqueous) above room temperature and at a pressure greater than 1 atmosphere in a closed system (Byrappa and Yoshimura, 2013). The size and morphology of the metal oxides play a significant role in various applications. Properties tuning in these metal oxides can be achieved through an appropriate surface modification and doping. The synthesis of stable nanoparticles generally requires the addition of surfactant molecules that bind to the surface of nanoparticles, thus stabilizing the nuclei and larger particles against the aggregation by a repulsive force. Surfactants play an important role in the preparation of nanoparticles, and they not only control the nucleation of a given phase or compound, but also provide controlled size, shape, dispersibility, and properties (Byrappa and Yoshimura, 2001). The surface modifiers inhibit the particle growth, facilitating smaller particle size with narrow particle-size distribution. They can make the nanoparticles either hydrophobic or hydrophilic. The surface chemistry of the nanoparticles is altered by an appropriate surfactant. Surfactants are fatty organic acids that do not mix in aqueous solvent at room temperature, but at high temperature these surfactants form a homogenous mixture, facilitating size reduction, dispersibility, and shape and surface chemistry. Commonly used surfactants are caprylic acid, oleic acid, gluconic acid, and n-butylamine. Colloidal nanoparticles are dispersed in either aqueous or organic solvent owing to the nature of hydrophilic or hydrophobic particles respectively, while amphiphilic particles work well with either solvent.

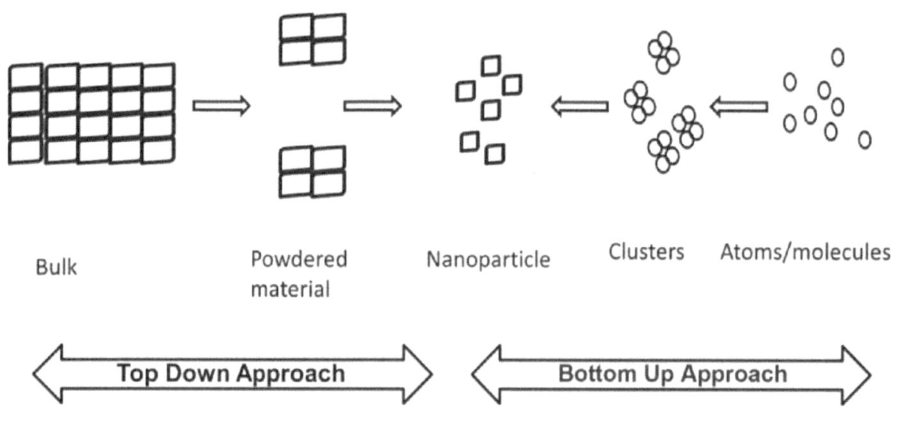

FIGURE 2.1 Schematic representation of top-down and bottom-up methods.

In addition to the types of nanoparticle material and surfactant molecules, the other major determinant for controlling the shape of nanoparticles is the use of ligand. Bound ligand molecules, in addition to controlling the shape and growth of nanoparticles, prevent the aggregation of nanoparticles due to repulsion between the particles. The repulsion may be due to electrostatic or steric hindrance, or to the hydration layer on the surface of the nanoparticle. Various functional groups of the ligands interact with the nanoparticle, for example thiol or poly(ethylene) glycol (PEG) (Liu and Thierry, 2012).

The pH of the reaction medium also influences the dispersity and stability of the nanoparticle formation. The selection of an appropriate surfactant is done based on the pH of the medium, the isoelectric point (pI), and the dissociation constant (pKa) of the modifiers. Either extreme in the pHs of the reaction medium leads to the synthesis of polydispersed particles consisting of a mixture of smaller and larger size particles. Therefore, in surface modification, the pH of the medium, the isoelectric point, and the dissociation constant of the modifiers play an important role, which has to be considered.

2.3 Methods of Surface Functionalization

Surface functionalization allows the nanoparticles to bear one or more functional groups to conjugate biologically active molecules for specific biological application. There are two methods for achieving surface functionalization: (i) *In situ* surface functionalization, (ii) Post-synthesis surface functionalization.

In situ surface functionalization

Surface functionalization of nanoparticles is commonly achieved during the synthesis of nanoparticles *per se*. This is a one-pot synthesis method and allows the simultaneous synthesis of morphologically similar functionalized small-size particles with narrow particle-size distribution in a single step. In this method, initially the starting material or the precursors are mixed along with the surface coating material or dopants to allow the nucleation process. As soon as the nucleation begins, dopants cover over the particle, forming a coat that further inhibits the growth of particles. The dopant coat and charge distribution over the particle determine the stability of nanoparticles against aggregation. Carboxylates, phosphates, and thiol groups are the most commonly used functional groups for surface functionalization.

Post-synthesis surface functionalization

In post-synthesis surface functionalization, Post synthesis surface functionalization involves the surface functionalization on nanoparticles in two consecutive steps; first step involves the synthesis of nanoparticles and in the second step functionalization of biomolecules on thus synthesised nanoparticles are made. Bifunctional ligands carrying two functional groups are used which allow the attachment of ligand to nanoparticles on one end, and the other functional group is amenable for conjugation with the biomolecule. A wide range of coupling agents is available for choice. (Figure 2.2).

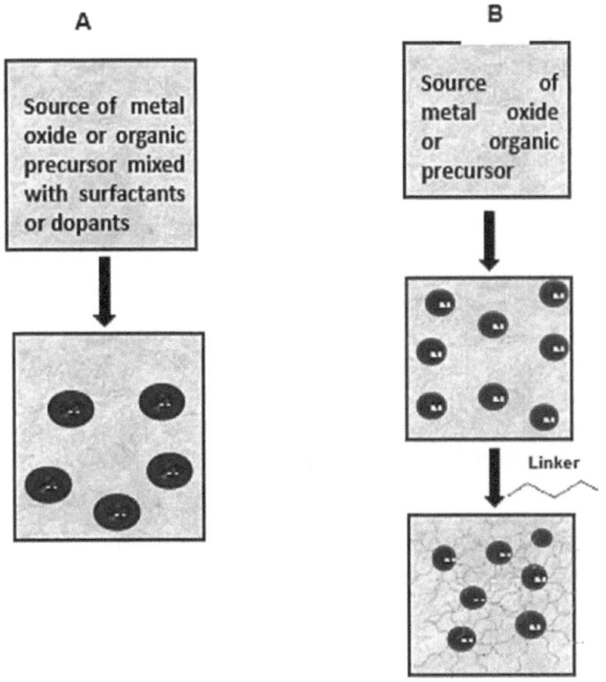

FIGURE 2.2 Methods of surface functionalization: (A) *In situ* surface functionalization, (B) Post-synthesis surface functionalization.

2.4 Coupling Strategies

Fundamentally, two different coupling strategies are used for the synthesis of designer biofunctionalized nanomaterials. Nanomaterials are functionalized by assembling various biomolecules, producing hybrid systems interacting through either of the following: (i) covalent coupling, and (ii) noncovalently through electrostatic, hydrophobic, or specific recognition. An important criterion for the biofunctionalization of nanoparticles with biomolecules during surface modification is to minimize the inadvertent physical, chemical, and biological properties which would otherwise modify the size, charge, hydrophobicity, and specificity. These properties would affect the stability of the nanoparticles.

2.4.1 Covalent Coupling

Stable binding of biological molecules on to nanoparticles during biofunctionalization if often achieved by covalent coupling involving the covalent bond linkage between biomolecules and nanoparticles. A stable and irreversible covalent binding of biomolecules to nanoparticles is advantageous in complex biological media which would be stable towards interfering species, ligands, or a change in pH, salt concentration, or temperature.

Nanoparticles surfaces functionalized with sulphydryl, amine aldehydes, and carboxylic, phosphate, or epoxide functional groups are conjugated with biomolecules. The process of conjugation of biomolecule is often mediated using linker molecules. The linker-mediated conjugation avoids the loss of structure of the biomolecule, which is rather crucial for its biological activity.

Nanoparticles with carboxyl groups can be readily bonded covalently to biomolecules bearing primary amine groups though carbodiimde chemistry. Carbodiimide coupling agents commonly used are a zero-length crosslinking agent, 1-ethyl-3-(3-dimethylaminopropyl)-carbodiimide (EDC) and N, N'-dicyclohexylcarbodiimide (DCC), used along with the solvent dimethylformamide (DMF) (Grabarek and Gergely, 1990). These carbodiimdes are popular crosslinking agents because of their high water solubility and the ease of removal of excess reagents and by-products. As the reactive ester rapidly formed hydrolyzes in aqueous solution, the coupling reaction is carried out in the presence of N-hydroxysuccinimide (NHS) or N-hydroxysulfosuccinimide (sulfo-NHS). NHS or sulfo-NHS reacts with carboxylic groups to form an active intermediate ester and further assist in amide-bond formation in the presence of EDC (Figure 2.3).

Successful coupling of proteins andamine-terminated DNA to water soluble nanoparticles using EDC or its derivatives as the main coupling reagent have been functionalized to nanoparticles. Many nanomaterials are fabricated using this strategy for various biomedical applications.EDC chemistry is used to attach methoxypolyethylene glycol amine to carboxylated magnetic nanoparticles. The number of PEG molecules on nanoparticle could be adjusted by varying the ratio of EDC/ nanoparticle. Similarly, Amine-PEG-Amine is coupled on to gold nanoparticle using EDC chemistry (Sanz et al., 2012). PEG molecules possessing –COOH and – NH2 group on their termini could be biofunctiolaized to produce water soluble gold nanoparticle. Maleimide is used to conjugate primary amines to sulfhydryl group. At pH 6.5–7.5, maleimide reacts with thiol group to yield a stable 3-thiosuccinimidyl ether linkage. Maleimide coupling is vastly used to couple various biomolecules on to gold nanoparticles, such as peptides and DNA (Brinkely, 1992).

2.4.1.1 Click-Chemistry Approach

Click chemistry is used to describe a class of reactions which include cycloadditions, nucleophilic substitutions, nonaldol carbonyl formation and additions of carbon–carbon bonds such as epoxidation. Click reaction have shown great potential in biofunctionalizing nanoparticles. Cuprous-mediated [3 + 2] azide–alkyne cycloaddition (CuAAC) referred as a click reaction, is a highly specific, facile reaction compatible with aqueous media facilitating the bioconjugation under mild reaction conditions (Figure 2.3). The popularity of this reaction derives from cuprous as an efficient catalyst accelerating the reaction ~10^8 fold, regioselectively forming 1,4-disubstituted 1,2,3-triazole (Qian et al., 2003). The copper catalyzed azide–alkyne cycloaddition occurs between an organic azide and acetylene forming triazole, a cyclic product. The CuAAC reaction has been extensively used for the functionalization of gold, silica, quantum dots, and carbon nanotubes. Shortcomings that limit its versatility include the instability and toxicity of the cuprous catalyst. Excessive intake

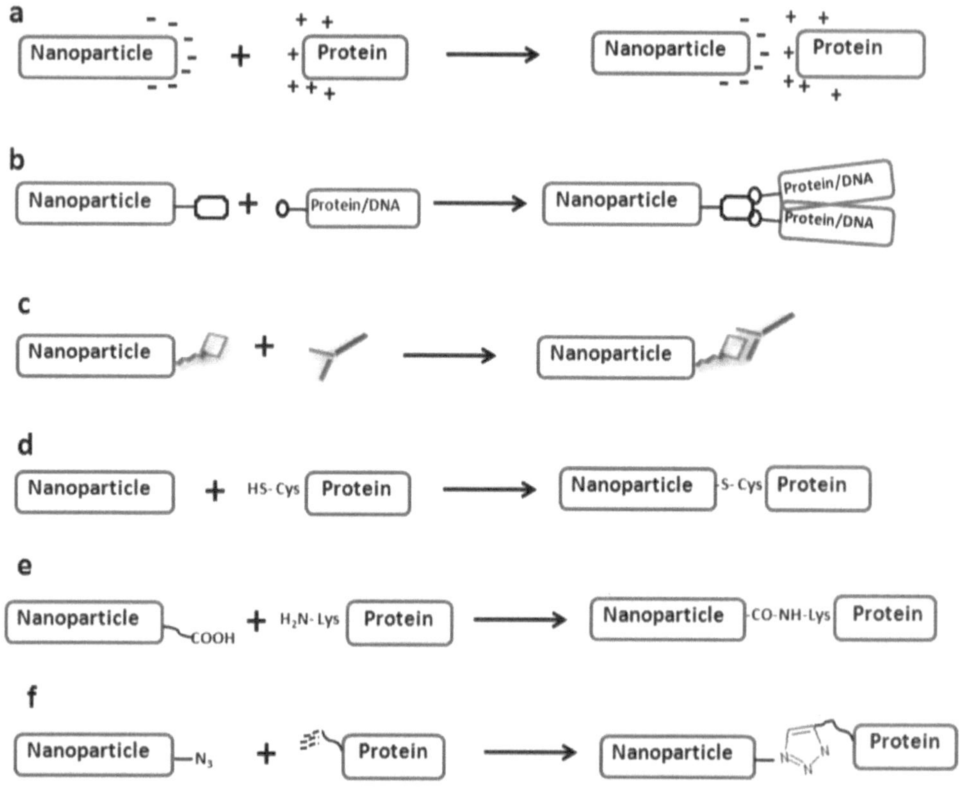

FIGURE 2.3 General schematic representation of biofunctionalization of NPs demonstrated by the use of a protein. (a) Assembly of NP-protein by electrostatic interaction, (b) Formation of nanoparticle-protein bioconjugate by affinity interaction using streptavidin–biotin binding, or (c) Antigen-antibody interaction, (d) Adsorption of nanoparticle-protein mediated via sulfhydryl group of cysteine amino acid, (e) Carboxylate–amine coupling of carboxylated nanoparticle and amine group presented by lysine amino acid in protein. NHS/EDC coupling agents are used during covalent coupling, (f) Click chemistry reaction with azide-tagged nanoparticle and protein with alkyne tail.

of copper can lead to drastic consequences on humans (kidney diseases, hepatitis, neurological disorders, and Alzheimer's disease), and an appropriate dosage should be formulated according to the body weight of the subject (Wang and Guo, 2006).

Click chemistry reaction has been used to couple gold nanoparticles to enzymes, proteins, fluorophores, polymers, and small molecules such as folate and biotin. Alkyne- functionalized gold nanoparticles are synthesized to detect cupric ions in aqueous solution. This method allows the visualization of Cu^{2+} ions by the aggregation of gold nanoparticles as a result of Cu(I)-catalyzed conjugation between the alkyne and azide functional groups. Azide-modified quantum dots are synthesized by strain-promoted azide–alkyne cycloaddition (SPACC) with a bifunctional linker conjugating the cyclooctyne group onto metalloprotein transferrin and azide-modified quantum dots. Fluorescent cadmium selenium/zinc sulphide core-shell quantum dots-transferrin conjugates were checked for their biological activity by monitoring their uptake in the transferrin-receptor (TfR) expressing tumor cells (Jańczewski et al., 2011).

2.4.2 Noncovalent Coupling

The biofunctionalization of nanoparticles by a noncovalent approach include electrostatic, hydrophobic, and specific biorecognition or affinity interactions. This approach of interaction features advantages such as the simplicity, speed, and the process does not involve the use of harsh chemicals, which compromise the native structure of the biomolecule affecting the biological activity. However, noncovalent coupling is less stable and less reproducible compared to the covalent approach.

Ionic coupling involves an electrostatic interaction, extensively used to functionalize nanoparticles to biomolecules, such as DNA, peptides, and antibodies (Brewer et al., 2005). This mode of interaction is between oppositely charged species on the nanoparticle surface and the biomolecule of interest (Figure 2.3). As the binding rate mainly depends on the density of the charge on the interacting species, the pH, ionic strength of the medium, and isoelectric point (pI) of the protein have to be considered.

Inorganic nanoparticles such as silver, gold, and quantum dots have been functionalized with oppositely charged biomolecules. Cationic nanoparticles are biofunctionalized with DNA possessing the negatively charged phosphate backbone with hyperthermia (Catherine et al., 2001). Citrate-stabilized gold nanomaterials could be coupled with antibodies to selectively target and photothermally destroy cancer cells. Self-assembled anionic hyaluronic acid (HA) are coupled on positively charged quantum dots through electrostatic interactions. HA-quantum dot conjugate has found its place in diagnostic and imaging applications (Bhang et al., 2009).

Hydrophobic interactions (Figure 2.3) have been widely exploited for the loading of drugs to nanoparticles. Gold nanoparticles are functionalized with zwitterionic ligands, forming stable complexes that allow the encapsulation of hydrophobic drugs in their hydrophobic pockets. Inorganic nanomaterials can be coated with hydrophobic polymeric structures and used for drug encapsulation and controlled drug release (Kim et al., 2009). The polymers used in this lipid-polymer hybrid nanoparticles include poly(lactic-co-glycolic acid) (PLGA), cyclodextrin or PEG shell. Drug molecules such as tamoxifen and β-lapachone are encapsulated in Au nanoparticles, mimicking the micelle structure. Doxorubicin, vincristine, and several other drugs have been encapsulated to liposomes and tested for their efficacy (Yallapu et al., 2011).

2.5 Affinity Interactions

Affinity interactions are based on the biomolecular interaction between two molecules having complementary recognition sites or very specific sequences. Unlike other linkages, the affinity interaction is least affected by variations in charge or ionic strength. Of all the noncovalent linkages, affinity interaction is the most stable and the strongest. Therefore, specific affinity interaction is very effective in establishing the biofunctionalization of nanoparticles using biomolecules such as antigen–antibody, avidin/streptavidin–biotin, lectin–carbohydrate, and nucleic acid–nucleic acid (Schmitt et al 2000). Complementary pairing of nucleotides in double-strand DNA is used for reversible assembly of DNA conjugated on nanoparticles.

In immunoassays for the detection of antigens, antibodies are conjugated on the nanoparticle surface. Immunoglobulin G is immobilized on the nanoparticle surface for the specific *in vitro* recognition of antigenic protein G present on the cell surface of staphylococcus bacteria. Similarly, protein tags are employed in purifying proteins, expressing an additional fusion protein or short amino acid sequence modified by a genetic engineering technique which acts like a tag. A well-known example is the polyhistidine tag where six or more histidine residues are tagged at any one terminus of protein. These his-tagged proteins bind specifically and strongly to nitrilotriacetic acid (NTA) or iminodiacetic acid (IDA) via a chelation complex with Ni^{2+} or other bivalent metal ions (Li et al., 2007). The resins (sepharose or agarose) functionalized with Ni-NTA or Ni-IDA are packed in the column and used for the purification of proteins containing the 'his' tag. Due to the strong affinity of he timidazole group of histidine with the NTA complex, this strategy is used for the purification of proteins.

Biotinylation of nanoparticles with avidin is very commonly found in biological application. This is due to the fairly high affinity of avidin–biotin interactions, which is around 10^{14}. The surfaces of nanoparticles can be biofunctionalized with avidin molecules which specifically bind to four molecules of biotin. For example, avidin-functionalized Ce/Tb-doped LaPO4 nanoparticles are used for the assembly of biotinylated proteins, which makes them an ideal crosslinker (Kim et al., 2008). Biotin–avidin interaction is explained elaborately in the following section. The conjugation of biotinylated polyethylene glycol or phospholipids to the surfaces of nanoparticles can interact with avidin or avidin-functionalized species, making them applicable as good biosensing agents.

2.5.1 Poly(ethylene glycol)

Polymers are used to coat the nanoparticles as they form a physical barrier to prevent the core nanoparticles from coming into contact, thereby preventing their aggregation. This property of polymers improves the stabilization and biodistribution when used for *in vivo* applications. There are many polymeric ligands used to prepare water-soluble nanoparticles, the most common being poly(ethylene glycol) (PEG) and carbohydrates such as dextran and chitosan.

PEG is a linear polymer made up of repeating units of (–CH2 –CH2 –O –)n. It is soluble in polar, apolar, and aqueous solvent, and forms a coiled structure with large molecular weight. Owing to its structure, PEG is an inert, biocompatible, and chemically stable polymer, which makes it an excellent ligand for use in biomedical application. Due to its non-toxic nature, it is a widely used macromolecule for the development of surface-modified and long-circulating nanoparticles as they escape from macrophage-mediated uptake/opsonization and removal from systemic circulation (Veronese, 2001; Zareie et al., 2008). As these coated nanoparticles have long plasma half-lives, they circulate in the body for a long period of time, thus targeting an organ to deliver therapeutic peptides or drugs or as the carriers of DNA in gene therapy. Due to steric hindrance, PEG prevents other molecules to bind on the surface of nanoparticles, producing an inert hydrophilic surface in the biological environment. PEG-modified proteins or drugs show increased aqueous solubility, resulting in the increased half-life (Daou et al., 2009; Liu et al., 2007). PEGmodifed nanoparticles are stable towards high salt concentration, and pH.PEGs with higher molecular weight (>3000 Da) are cleared from the immune system after consumption when compared with PEG-600-coated nanoparticles (Xie et al., 2007). Likewise, the colloidal stability and size of nanoparticles is also highly dependent on the chain length and density of the PEG molecule used in nanoparticles. The modification of nanoparticles with PEG is often referred to as 'PEGylation'. Conjugating and altering the head groups of PEG allow selective attachment to nanoparticle surface for biofunctionalization. For example in PEGylation of proteins, free amino group from lysine or sulfhydryl group from cysteine present in the proteins are often targeted for coupling with PEG.

Nanoparticles are functionalized through different strategies. Au nanoparticles are biofunctionalized by assembling PEGs and mixed biomolecule/PEG monolayers on the nanoparticles surface. The sequential addition of PEGs followed by the addition of peptide, oligonucleotide, or any biomolecule is reported to improve the stability of nanoparticles. The addition of biomolecules onto the Au nanoparticle is controlled by ionic and weak interactions positioning the entry of biomolecules through the PEG layer. PEGs are used as spacers to link the biomolecules on Au nanoparticles. The spacer PEGs are bifunctionalized; they contain thiol at one end and a suitable functional amino or carboxylated moiety on the other end. Using a combination of different bifunctional PEG spacers, gold nanoplatforms can be multifunctionalized with a variety of biologically relevant ligands (Feng et al., 2013).

2.5.2 Bioconjugation Using Biomolecules

The bioconjugation of nanoparticles is the functionalization of nanoparticles with a biomolecule using any of the above-described coupling strategies. A variety of natural organic molecules that play a vital role in supporting the structure or function of biological processes and cells can be selected and customized for bioconjugation to serve the purpose of biofunctionalization. Biomolecules include small molecules such as lipids, peptides, antibiotics, drugs, or vitamins, and large molecules include natural polymers such as proteins, enzymes, and nucleic acids (DNA/RNA). Biomolecule–nanoparticle conjugates form the hybrid structure amalgamating unique properties and functionality of both the inorganic particle and its counterpart, the functional biomolecule. Physico-chemical properties of the biomolecule, like fluorescent or magnetic property and functional properties like molecular recognition or catalytic property are exploited. (Figure 2.4).

2.5.3 Biotin–Avidin

The biotin–avidin system has the widest application in bionanotechnology due to the strong bond and specificity between biotin and avidin. This has allowed the researchers to develop many conjugates of biotin or avidin of biomolecules such as DNA oligomers, peptides, proteins, and antibodies. Nanoparticles modified with biofunctionalized biotin or avidin offer a good platform in many biomedical applications.

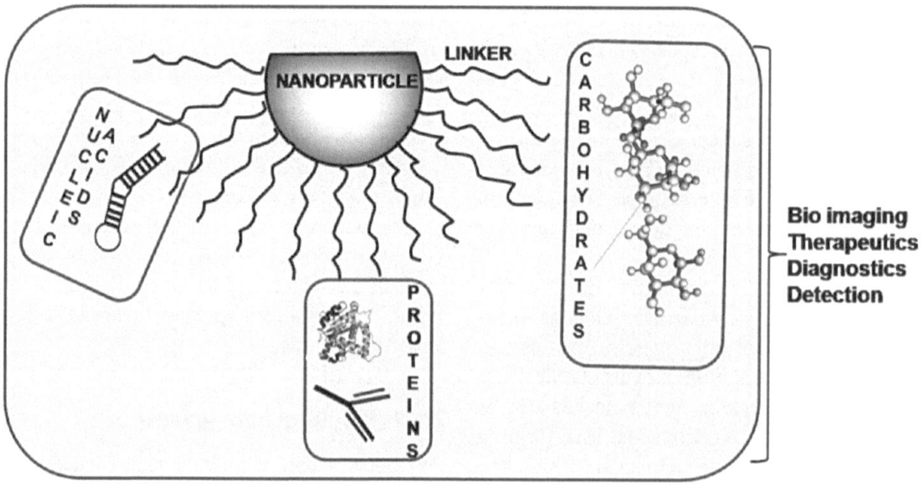

FIGURE 2.4 Biomolecules such as proteins (antibodies, enzymes, protein/peptide), nucleic acid (DNA/RNA), or carbohydrates are conjugated to nanoparticles via linker molecule. The biofunctionalized nanoparticle employed for therano-diagnostic applications in the field of bionanomedicine.

Biotin (vitamin H) is a small molecule with a free carboxylic group which is available for modification to $-NH_2$ or NHS. Biotin is bound to the nanoparticle surface by covalent conjugation chemistry or attached directly to the inorganic particle surface.

Avidin is a receptor globular glycoprotein with four binding pockets for biotin and is present abundantly in egg white. Though the bond between biotin and avidin is not covalent, the dissociation constant is of the order of 10^{15} M and the bond is extremely stable, resisting harsh chemical conditions and high temperature. Besides the egg protein, analogue proteins are expressed in bacteria called streptavidin or overexpressed as recombinant proteins without carbohydrate moiety (neutravidin). For the conjugation of avidin and related proteins onto the surface of nanoparticles, the characteristics of proteins, such as molecular weight, degree of hydrophobicity, free functional groups on the amino acid side chain, and isoelectric point have to be considered.

By using the EDC/NHS chemistry, streptavidin though carboxylic acid or amine group is covalently linked to quantum dots. Proteins are zwitterions at their isoelectric point and are charged positively or negatively below or above their isoelectric points respectively. This offers the possibility for negatively charged nanoparticles to electrostatically adsorb to the positively charged avidin molecules. Additional covalent crosslinking could be done to improve the stability of the nanoparticle–avidin complex (Wu et al., 2009).

2.5.4 DNA/Nucleic Acids

Among all the biomaterials, DNA offers a lot as a nanoplatform material by making significant contributions in the fields of biosensing, diagnostics, therapeutics, and drug delivery. This enormous range of application of DNA is due to the central role played by the DNA as the carrier of genetic information. Watson and Crick first described the structure of natural DNA as the double helix where two chains of linear deoxyribonucleic acid are wound around a common axis in a helical fashion to yield a double-strand DNA. Each chain of deoxyribonucleic acid is a polymer of repeating units called nucleotides with a backbone of sugar and phosphate groups joined by ester bonds that run in opposite direction. The nucleotide comprises of any one of the four nitrogenous bases: adenine (A), cytosine (C), guanine (G), and thymine (T). The order of the arrangement of nucleotides bases is called DNA sequence and it is this sequence that encodes the genetic information. T and C are pyrimidines, while A and G are purines, which are fused rings of pyrimidine and imidazole rings. A and T or G and C bases on the opposite strand of double helix are complementary and interact specifically with double or triple hydrogen bonds respectively. The duplex formation with a strand of complementary sequences represents a very specific mechanism of molecular recognition; DNA can be used as a 'programmable' chip by a combination of a large number of sequences. A great advancement in synthetic chemistry has allowed to synthesize oligonucleotides, which have taken an important place in the labs of molecular biology, nanomedicine, and nanobiotechnology. Synthetic oligomers are short sequences of nucleotides, made up of 12–30 nucleotides. The Mirkin Group at Northwestern University and the Alivisatos Group at U.C. Berkeley (Mirkin et al., 1996; Alivisatos et al., 1996) were the very first to demonstrate the functionalization of DNA on gold nanoparticles. The strong affinity of alkyl thiols or disulfide groups towards gold surfaces forming quasi-covalent bonds allows the functionalization of DNA to gold nanoparticles. By modifying the salt and DNA concentration, tailored DNA–Au nanoparticles of desirable shape and size are made possible with enhanced colloidal stability. The assembly of DNA on the Au nanoparticles can be fine-tuned by choosing appropriate DNA sequences (usually thymine rich), salt concentration with proper sonication during loading, and the use of spacers such as polyethylene glycol between the thiol and nucleotides of the DNA. The preparations of DNA–Au nanoparticles under different conditions are elaborated in the References (Deka et al., 2015; Zhao et al., 2009; Xu et al., 2011).

Besides for directing the binding of DNA to the surface of the nanoparticle, covalent coupling can be employed to conjugate DNA to the nanoparticle through the functional groups. Amino-functionalized DNA can be bound to nanoparticles with carboxylic acid or thiol-modified DNA to maleimide through EDC chemistry. Gold, magnetic nanoparticles, and quantum dots are covalently functionalized with DNA using covalent chemistry.

Because of the phosphate backbone, DNA and RNA are negatively charged and electrostatic adsorption occurs on positively charged surfaces, such as quaternary or tertiary amines respectively. In this method, however, non-specific adsorption of nucleic acids to nanoparticles is observed, especially when high stoichiometric excess is used.

The detection of nucleic acid in many DNA biosensors is based on the strategy that two DNA functionalized Au nanoparticles can be linked if the DNA presented on their surfaces is complementary to the other or to another third 'bridging' DNA. Since each nanoparticle carries numerous DNA strands, the hybridization with the complementary DNA from the second nanoparticle or the bridging DNA strand leads to aggregation, which is reflected in their surface plasmon absorption band, leading to a change in color and quantitatively detected by the colorimetric method. This relatively simple detection method allows detecting up to 10 nM, which can also be perceived by the naked eye, and is a better detection method over PCR or other fluorophore-based assays (Thaxton et al., 2006; Bai et al., 2010). Au nanoparticles functionalized with single-stranded DNA are capable of specifically hybridizing a complementary target for the detection of particular nucleic acid sequences in biological samples, allowing the specific detection of bioanalytes of clinical interest, which makes them the ideal candidates for developing biomarker platforms. Au nanoparticles carrying multiple copies of aptamer specifically recognizing proteins such as thrombin or lysozyme are developed as biosensors for detecting related diseases (Wang et al., 2011; Pavlov et al., 2004).

DNA-functionalized Au nanoparticles are employed as an intracellular probe for detecting and quantifying mRNA expression in the cells. These densely coated 13 nm Au nanoparticles called 'nanoflares' are functionalized with thiolated DNA which has 18 base oligonucleotide stretches and recognizes specific mRNA sequences. In addition, the DNA on Au nanoparticles is partially hybridized with the short sequence of DNA modified with Cy5 dye. In the absence of target mRNA, due to the proximity of Au surface, the dye is quenched; whereas, in the presence of target mRNA, mRNA displaces the dye-modified DNA with the emission of fluorescence. Based on same principle, multiplexed nanoflare is developed to detect two different mRNA sequences at a time (Dwilight et al., 2007).

Quantum dots are luminescent semiconductor nanocrystals with excellent tunable, photophysical properties. Quantum dots functionalized with nucleic acid are used as sensors to test target DNA based on FRET. For the detection of target DNA, two different-wavelength yet overlapping signals are emitted from the quantum dot and DNA target-probe complexes. Based on this concept of specificity of hybridization and signal emission, two approaches are designed. In the first approach, two quantum dots with distinct emission wavelengths are modified with two single-stranded DNA probes via biotin–streptavidin interaction. The DNA probes hybridize on different sites of target DNA strand, thus cross-linking the quantum dots. In the second approach, two single-stranded DNA probes complementary to the target DNA are used. One is a biotinylated single-stranded DNA probe, and the other is conjugated to an organic fluorophore. These are mixed with the target DNA to form a sandwich hybridization which is latter captured by streptavidin-modified quantum dot (Yeh et al., 2005).

2.5.5 Proteins and Peptides

Proteins and peptides are polymers of amino acids linked to one another via amide bond. Peptides are distinguished from proteins in their size, containing 50 or fewer amino acids. Peptides exist naturally or are custom-synthesized, while proteins are structurally complex molecules exhibiting hierarchy in their structure. Twenty naturally occurring different-standard amino acids with different side groups are linked through peptide bond with the involvement of carboxylic and amino groups from each amino acid. The sequence of arrangement of amino acids determines the unique properties associated with the protein. The functional properties of proteins are further defined, as the sequence of amino acids adopts a secondary and tertiary structure. The interior of the 3D structure of a soluble protein is hydrophobic, while the exterior is hydrophilic. The exterior of the protein harbors unique motifs of folded amino acid sequence, allowing recognition and binding of substrates or ligands involved in catalysis or signalling. The side chains of exposed amino acids such as lysine with free amino group or cysteine with thiol group can be exploited as the anchor groups for the attachment of proteins or peptide onto the surface of nanoparticles. These compositional and structural features make the protein a potential candidate for the stabilization and biofunctionalization of nanoparticles. Proteins include invariably different molecules such as structural proteins, signaling or carrier proteins, enzymes, and antibodies. Enzymes are biological catalyst, while antibodies also known as immunoglobulins are proteins functioning in the defense of the body against infectious diseases. The specific functional and targeted functions of proteins make them ideal for application for imaging, diagnostic, or delivery purposes, when conjugated with nanoparticles.

Peptides, which are shorter than proteins, can be custom-synthesized by the addition of amino acids possessing the functional groups or by introducing the functional groups ideal for the anchoring to the nanoparticles. The stability of the peptides depends on the length, hydrophobicity, and charge of the peptide molecules. The conjugation of proteins or peptides to nanoparticles is achieved by electrostatic interaction when the nanoparticles and the proteins are oppositely charged, or by non-specific interaction through van der Waals forces, hydrogen bonding, hydrophobic interactions, or covalent coupling. These flexible properties of proteins and peptides make the combination of these molecules with inorganic nanoparticles very interesting.

Multiple functional peptides-stabilized Au nanoparticles are obtained by one-step surface coating and used for diagnostic purposes on microarray. These particles with heterogenous functional peptides exhibit the advantageous property of being very specific in binding without particle aggregation. Certain specific peptides functioning as therapeutic peptides are conjugated over the nanoparticles for their uptake by cells. Peptide sequences biofunctionalized on Au nanoparticles are used as 'membrane translocation signals', such as the HIV–TAT peptide sequence capable of transporting nanoscale materials across the cell membrane (Fuente and Berry, 2005). Quantum dots coated with specific peptides having recognition site on tumors are used in imaging. Quantum dots modified with streptavidin can be conjugated to different biotinylated peptides.

Enzymes are proteins functioning as biological catalysts which are commonly used in biofunctionalization due to their potential application in biotechnological industries and biomedicine. Urease is an important enzyme used in the field of biomedicine as a biosensor for blood urea estimation. It is also used for bioremediation. Urease is immobilized on magnetic nanoparticles through carbodiimide reaction using a linker molecule. N-phosphonomethyl iminodiacetic acid acts as a linker molecule with the free carboxylic acid which is modified on magnetic nanoparticles, and these surface-coated particles are used for the immobilization of urease enzyme. The immobilization of cellulase enzyme on magnetic nanoparticles is carried by using EDS/NHS coupling reaction. The immobilization of enzymes on magnetic nanoparticles offers the advantage of an easy separation of immobilized enzymes after their use by employing a magnetic field. The immobilized magnetic nanoparticles can be recycled (Bohara et al., 2016).

Peptide-capped quantum dots are used for monitoring enzymatic activities which employed used as diagnostic tools. For example, quantum-dot bioconjugates are developed to detect proteolytic activity and the strategy is that of attaching dye- or Au nanoparticle-labelled peptide structures containing appropriate cleavage sequences of proteases to quantum dots. Proteases activities such as elastase, collagenase, and thrombin could be detected upon specific cleavage of the peptide and alteration of FRET (fluorescence resonance energy transfer) signal (Zhou et al., 2014).

2.5.6 Carbohydrates

Carbohydrates are one among the other three major classes (lipids, proteins, and nucleic acids) of biomolecules. They have unique application in nanobiomedicine due to their biocompatibility, biodegradability, and water solubility, and are a natural ligand molecule for various cellular receptors mediating cellular events such as cell adhesion, normal tissue growth, signal transducer, and antigenic nature in viral/bacterial infection. The chemically defined structures of carbohydrates with unique physical, chemical, biological, optical, and stereochemical properties offer an advantage to biofunctionalize and design nanomaterials for biological applications. Surface-functionalized silica with mannose attached via thiol group shows specific binding to MCF-7 human breast-cancer cells, and this could be used to design a nanocarrier for drug delivery. Carboxylated forms of carbohydrates such as mannose, galactose, fucose, and sialic acid are coupled to amino-functionalized magnetic nanoparticles via an amide linkage and used to image malignant cells by magnetic resonance imaging. Oligomannosides are functionalized with thiols coupled to gold nanoparticles to produce glycosylated Au nanoparticles. These glycosylated Au nanoparticles bind strongly to a C-type lectin on the surface of dendritic cells compared to gp120, which is a protein essential for the entry of HIV virus into cells and could serve as a potential carbohydrate-based drug against HIV. Rhamnose is a mannose-related 6-deoxy hexose and occurs naturally in the cell walls of bacteria and plants. Phosphonate strongly binds to metal in magnetic nanoparticles through which rhamnose is anchored. These rhamnose biofunctionalized magnetic nanoparticles bind to specific cell types, and due to the paramagnetic property of iron oxide nanoparticles they can be used as MRI contrast agents with specific cell targeting (Kang et al., 2015).

2.5.7 Phospholipids

Phospholipids are the major component of cell membranes. They are amphiphilic and contain a hydrophobic tail made up of two chains of fatty acid and a hydrophilic head constituting the phosphate group. As they can form lipid bilayer structures, phospholipids are inherently suited to encapsulate hydrophobic or hydrophilic molecules, allowing them to form an envelope on the nanocomposite structure. These drug-loaded nanoparticles are used in applications of drug delivery and hypothermia. Further, the head group can be biofunctionalized, for instance, for biotinylation, paving the ways for further conjugation steps exploiting biotin–avidin interactions.

Conclusion

The fabrication of nanoscale materials with surface functionalization allows the fine tuning of physical, photonic, optical, and magnetic properties, providing excellent platforms for the design of advanced nanomaterials with biomedical applications integrating biology and synthetic materials. The incredible development in advanced biofunctionalized nanomaterials involving new and efficient synthetic routes has been employed in recent days in the field of nanotechnology for the design of tailor-made biofunctionalized materials with specific functional properties. The biofunctionalization of nanomaterials allows them to be biocompatible with the immobilization of biomolecules, providing significant applications in detection/sensing and molecular imaging. The coupling of functional groups onto the nanomaterials depends on the chosen starting materials for various applications. The coupling strategy is flexible and achieved by either the noncovalent or the covalent approach, which could be mediated by using the linker molecules. Biomolecules used for biofunctionalization include deoxyribonucleic acid, ribonucleic acid, oligonucleotide, protein, peptide, antibody, hormone, enzyme, and carbohydrates such as monosaccharide or complex sugars such as hyaluronic acid. The merging of synthetic chemistry, nanotechnology, and biology has led to astounding developments in the field, creating various biofunctionalized molecules for a wide range of biomedical applications, including theranostics, imaging, and drug delivery.

REFERENCES

Alivisatos, A. P., Johnsson, K. P., Peng, X., Wilson, T. E., Loweth, C. J., and M. P. Bruchez. 1996. Organization of "nanocrystal molecules" using DNA. *Nature.* 382:609–11.

Bai, X., Shao, C., Han, X., Li, Y., Guan, Y., and Z. Deng. 2010. Visual detection of sub-femtomole DNA by a gold nanoparticle seeded homogeneous reduction assay: Toward a generalized sensitivity-enhancing strategy. *Biosens Bioelectron.* 25(8):1984–8.

Baptista, P. 2009. Cancer nanotechnology – Prospects for cancer diagnostics and therapy. *Curr Cancer Ther Rev.* 5(2):80–8.

Bhang, S. H., Won, N., Lee, T. J., Jin, H., Nam, J., Park, J., et al. 2009. Hyaluronic acid–quantum dot conjugates for *in vivo* lymphatic vessel imaging. *ACS Nano.* 23:3(6):1389–98.

Bhushan, B. 2010. *Springer Handbook of Nanotechnology* [Internet]. Springer, Berlin, Heidelberg.

Bohara, R. A., Thorat, N. D., and S. H. Pawar. 2016. Immobilization of cellulase on functionalized cobalt ferrite nanoparticles. *Korean J Chem Eng.* 33(1):216–22.

Brewer, S. H., Glomm, W. R., Johnson, M. C., Knagand, M. K., and S. Franzen. 2005. Probing BSA binding to citrate-coated gold nanoparticles and surfaces. *Langmuir.* 21:9303–07.

Brinkley, M. 1992. A brief survey of methods for preparing protein conjugates with dyes, haptens and crosslinking reagents. *Bioconjug Chem.* 3(1):2–13.

Byrappa, K. and M. Yoshimura. 2001. *Handbook of Hydrothermal Technology: A Technology for Crystal Growth and Materials Processing.* Noyes Publications.

Byrappa, K. and M. Yoshimura. 2013. *Handbook of Hydrothermal Technology.* William Andrew, Elsevier Science Publishers.

Catherine, M., Edward, A., Esposito, I., Boal, A. K., Simard, J. M., Martin, C. Y., and V. M. Rotello. 2001. Inhibition of DNA transcription using cationic mixed monolayer protected gold clusters. *J Am Chem Soc.* 123:7626–29.

Daou, T. J., Li, L., Reiss, P., Josserand, V., and I. Texier. 2009. Effect of poly(ethylene glycol) length on the in vivo behavior of coated quantum dots. *Langmuir.* 25(5):3040–44.

Deka, J., Měch, R., Ianeselli, L., Amenitsch, H., Cacho-Nerin, F., and P. Parisse. 2015. Surface passivation improves the synthesis of highly stable and specific DNA-functionalized gold nanoparticles with variable DNA density. *ACS Appl Mater Interfaces.* 7(12):7033–40.

Dwight, S., Seferos, D. A., Hill, D. H., Prigodich, A. E., and C. A. Mirkin. 2007. Nano-flares: Probes for transfection and mRNA detection in living cells. *J Am Chem Soc.* 129 :15477–15479.

Feng, C. L., Dou, X. Q., Liu, Q. L., Zhang, W., Gu, J. J., Zhu, S. M., et al. 2013. Dual-specific interaction to detect DNA on gold nanoparticles. *Sensors.* 13(5):5749–56.

Fuente, J. M. and C. C. Berry. 2005. Tat peptide as an efficient molecule to translocate gold nanoparticles into the cell nucleus. *Bioconjug Chem.* 16:1176–80.

Grabarek, Z. and J. Gergely. 1990. Zero-length crosslinking procedure with the use of active esters. *Anal Biochem.* 185(1):131–5.

Jańczewski, D., Tomczak, N., Han, M. Y., and G. J. Vancso. 2011. Synthesis of functionalized amphiphilic polymers for coating quantum dots. *Nat Protoc.* 6(10):1546–53.

Kang, B., Opatz, T., Landfester, K., and F. R. Wurm. 2015. Carbohydrate nanocarriers in biomedical applications: Functionalization and construction. *Chem Soc Rev.* 44(22):8301–25.

Kim, C. K., Ghosh, P., Pagliuca, C., Zhu, Z. J., Menichetti, S., and V. M. Rotello. 2009. Entrapment of hydrophobic drugs in nanoparticle monolayers with efficient release into cancer cells. *J Am Chem Soc.* 131(4):1360–11.

Kim, J., Park, H. Y., Kim, J., Ryu, J., Kwon, D. Y., Grailhe, R., et al. 2008. Ni– nitrilotriacetic acid-modified quantum dots as a site-specific labeling agent of histidine-tagged proteins in live cells. *Chem Commun.* 28(16):1910–16.

Li, Y. C., Lin, Y. S., Tsai, P. J., Chen, C, T., Chen, W. Y., and Y. C. Chen. 2007. Nitrilotriacetic acid coated magnetic nanoparticles as affinity probes for enrichment of histidine tagged proteins and phosphorylated peptides. *Anal Chem.* 79:7519–25.

Liu, T. and B. Thierry. 2012. A solution to the PEG Dilemma: Efficient bioconjugation of large gold nanoparticles for biodiagnostic applications using mixed layers. *Langmuir.* 28(44):15634–42.

Liu, Y., Mathew, K. S., Ryan, J., Kaufman, E. D., Franzenand, S., and D. L. Feldheim. 2007. Synthesis, stability, and cellular internalization of gold nanoparticles containing mixed peptide–poly(ethylene glycol) monolayers. *Anal Chem.* 79:2221–29.

Mirkin, C. A., Letsinger, R. L., Mucic, R. C., and J. J. Storhoff. 1996. A DNA-based method for rationally assembling nanoparticles into macroscopic materials. *Nature.* 382:607–9.

Pavlov, V., Xiao, Y., Shlyahovsky, B., and I. Willner. 2004. Aptamer-functionalized Au nanoparticles for the amplified optical detection of thrombin. *J Am Chem Soc.* 126:11768–69.

Qian, W., Timothy, R. C., Hilgraf, R., Valery, V., Fokin, K., Sharplessand, B., and M. G. Finn. 2003. Bioconjugation by copper(I)-catalyzed azide-alkyne [3 + 2] cycloaddition. *J Am Chem Soc.* 125:3192–3193.

Reverchon, E. and R. Adami. 2006. Nanomaterials and supercritical fluids. *J Supercrit Fluids.* 37(1):1–22.

Sanz, V., Conde, J., Hernández, Y., Baptista, P. V., Ibarra, M. R., and J. M. de la Fuente. 2012. Effect of PEG biofunctional spacers and TAT peptide on dsRNA loading on gold nanoparticles. *J Nanoparticle Res.* 14(6):917.

Schmitt, L., Ludwig, M., Gaub, H. E., and R. Tampé. 2000. A metal chelating microscopy tip as a new toolbox for single-molecule experiments by atomic force microscopy. *Biophys J.* 78(6):3275–85.

Shvedova, A. A., Kagan, V. E., and B. Fadeel. 2010. Close encounters of the small kind: Adverse effects of man-made materials interfacing with the nano-cosmos of biological systems. *Annu Rev Pharmacol Toxicol.* 50(1):63–88.

Thaxton, C. S., Georganopoulou, D. G., and C. A. Mirkin. 2006. Gold nanoparticle probes for the detection of nucleic acid targets. *Clin Chim Acta.* 363:120–26.

Veronese, F. M. 2001. Peptide and protein PEGylation: A review of problems and solutions. *Biomaterials.* 22(5):405–17.

Wang, T. and Z. Guo. 2006. Copper in medicine: Homeostasis, chelation therapy and antitumor drug design. *Curr Med Chem.* 13(5):525–37.

Wang, X., Xu, Y., Chen, Y., Li, L., Liu, F., and N. Li. 2011. The gold-nanoparticle- based surface plasmon resonance light scattering and visual DNA aptasensor for lysozyme. *Anal Bioanal Chem.* 400(7):2085–91.

Wu, S. C., Ng, K. K. S., and S. L. Wong. 2009. Engineering monomeric streptavidin and its ligands with infinite affinity in binding but reversibility in interaction. *Proteins Struct Funct Bioinforma.* 77(2):404–12.

Xie, J., Xu, C., Kohler, N., Hou, Y., and S. Sun. 2007. Controlled PEGylation of monodisperse Fe3O4 nanoparticles for reduced non-specific uptake by macrophage cells. *Adv Mater.* 19(20):3163–66.

Xu, S., Yuan, H., Xu, A., Wang, J., and L. Wu. 2011. Rapid synthesis of stable and functional conjugates of DNA/gold nanoparticles mediated by Tween 80. *Langmuir.* 27(22):13629–34.

Yallapu, M. M., Othman, S. F., Curtis, E. T., Gupta, B. K., Jaggi, M., and S. C. Chauhan. 2011. Multi-functional magnetic nanoparticles for magnetic resonance imaging and cancer therapy. *Biomaterials.* 32(7):1890–905.

Yeh, H. C., Ho, Y. P., and T. H. Wang. 2005. Quantum dot-mediated biosensing assays for specific nucleic acid detection. *Nanomed: Nanotechnol Biol Med.* 1(2):115–21.

Zareie, H. M., Boyer, C., Bulmus, V., Nateghi, E., and T. P. Davis. 2008. Temperature-responsive self-assembled monolayers of oligo(ethylene glycol): Control of biomolecular recognition. *ACS Nano.* 2(4):757–65.

Zhao, W., Lin, L., and I. M. Hsing. 2009. Rapid synthesis of DNA functionalized gold nanoparticles in salt solution using mononucleotide-mediated conjugation. *Bioconjug Chem.* 20(6):1218–22.

Zhou, G., Chen, C., Zhang, L., Guo, X., Wang, H., and X. Ji. 2014. Robust aqueous quantum dots capped with peptide ligands as biomaterials: Facile preparation, good stability, and multipurpose application. *Part Syst Charact.* 31(3):382–9.

3

Functionalized Nanogold: Its Fabrication and Needs

Biswajit Choudhury

CONTENTS

3.1 Introduction

Metal nanoparticles show fascinating electronic and optical properties entirely different from those of their bulk counterparts (Zhao et al., 2013; Shan and Tenhu, 2007). The various properties exhibited by the metal nanoparticles depend on their size and shape; the fabrication methods largely govern the structural and morphological changes. Over time, metal nanoparticles of gold (Gold), silver (Ag), copper (Cu), aluminum (Al), palladium (Pd), and platinum (Pt) are synthesized, and their various physical properties are investigated. Out of all these metal nanostructures, gold nanostructures have found a promising place in multiple fields, including biomedical, chemical, and biological sensing; surface-enhanced Raman scattering, catalysis, and photonics (Saha et al., 2012). The widespread applications of gold nanostructures are mainly due to their chemically inert nature, unique surface plasmon resonance absorption, and superior electronic conductivity. Gold nanoparticles functionalized with citrate ions, thiols, amines, polymers, and oligonucleotides are used for colorimetric and fluorescence sensing of toxic metal ions such as mercury (Hg^{2+}), lead (Pb^{2+}), arsenic (As^{3+}, As^{5+}), and cadmium (Cd^{2+}) present in polluted water. These colorimetric and fluorescence sensing methods of functionalized gold are exploited for detecting biomolecules such as oligonucleotides (Saha et al., 2012). Another diversified and widely explored area of functionalized gold nanostructures is catalysis (Mikami et al., 2013). Gas-phase reduction of carbon monoxide, nitrogen oxide, methanol oxidation, hydrogenation of alkyne, alkene, and various C–C coupling reactions are successfully performed over a functionalized gold catalyst (Villa et al., 2016). Very recently, the surface plasmon resonance properties of gold nanostructures have been exploited to enhance the photocatalytic reduction of carbon dioxide, water splitting, and organic pollutant degradation (Wang et al., 2012; Biroju et al., 2016; Rajender et al., 2017).

The first extraction and use of gold were reported around 1200–1300 B.C. in Egypt. Later on, its use was reported in Turkey and China (Daniel and Astruc, 2004; Yang et al., 2015). In Lydian, modern Turkey, around 560–547 B.C., King Croesus made a gold coin called Croesid (Yang et al., 2015). Another famous example of the use of gold dates back to the 4th century during the Roman empire (Freestone et al., 2007). The Romans used a particular type of colored vessel known as the Lycurgus cup, which shows ruby red in transmitted light and green under reflected light. Later on, scientists realized that the origin of the different coloring under light is due to the presence of colloidal gold nanoparticles.

The modern method of fabrication of gold nanostructures first appeared in the seminal work of Michael Faraday in 1857. In his work, tetrachloroaurate was reduced by phosphorous in carbon disulfide solvent to produce colloidal gold (Faraday, 1857). In 1939, Pauli et al. improvised this method, and a few years later Turkevich proposed the modern and most widely adopted synthesis method of colloidal gold (Pauli et al., 1939; Turkevich et al., 1951). French was successful in controlling the size of colloidal gold (Frens, 1973). The various physical methods involved in the fabrication of gold nanostructures are lithography (Coribierre et al., 2005), laser ablation (Mafuné et al., 2002), chemical vapor deposition (Manna et al., 2016; Palgrave and Parkin, 2007), physical vapor deposition (Cross et al., 2007), and solution plasma sputtering and processing (Hu et al., 2013; Bratescu et al., 2011).

In this chapter, both physical and chemical methods employed for the fabrication of gold nanostructures will be discussed. A thorough discussion will be made on the application of the fabricated functionalized nanostructures for chemical and biological sensing, and catalysis.

3.2 Fabrication of Functionalized Gold Nanostructures

3.2.1 Physical Techniques of Fabrication

Lithographic techniques can fabricate arrays of nanoparticles that have excellent control over the size and shape of the nanoparticles. One such experiment shows the preparation of sub-50 nm patterns composed of arrays of gold nanoparticles (Corbierre et al., 2005). Gold film functionalized with thiol-group is kept under a field-emission gun scanning electron microscope with beam current and accelerating voltage set at 425 pA and 20 kV. An electron beam is supplied over the film area to create patterns. The unexposed part of the film is removed by immersing it in chloroform and drying it in nitrogen. The organic residue from gold-thiolate is removed by thermolysis. Depending on the electron dose, exposure time, and film thickness, the nanoparticle diameter varies between 2 and 6 nm (Corbierre et al., 2005). In the nanosphere lithography technique, polystyrene spheres of diameter 356 and 1053 nm are spin-coated over a Si (100) wafer which self-assembled into a hexagonal close-packed array forming a nanosphere mask (Tan et al., 2005). The exposure of this mask to reactive ion etching reduces the dimension of the masking nanosphere. Gold nanoparticles were then deposited in the etched pores. This technique results in size-tunable and ordered arrays of gold nanoparticles. The approximate size of the gold nanoparticles is 100 nm. It is predicted that the gold nanoparticles confined in the nanopores of the substrate could act as a catalyst for the growth of ordered nanomaterials (Tan et al., 2005). Physical vapor deposition is another suitable method for the fabrication of the ordered array of gold nanoparticles. One such result shows successful fabrication of one-dimensional arrays of gold nanoparticles of 2–15 nm by physical vapor deposition on a highly oriented pyrolytic graphite (HOPG) (Cross et al., 2007). This one-dimensional ensemble of gold nanoparticles acts as a nucleation site to catalyze the electrodeposition of gold nanowires having 70–90 nm diameter. The catalyst-free chemical vapor deposition (CVD) method can result in the growth of anisotropic gold nanocrystals over Si substrates (Manna et al., 2016). At high-temperature (470 °C) gold chloride vaporizes and reduced to pure gold crystals over Si (100) substrates under flowing argon gas. The formation of gold nanocrystals follows the nucleation and growth process with a progressive change in morphology from icosahedrons to decahedrons to tripyramids as reaction time and temperature increase. Above 420 °C, complete reduction of gold chloride is achieved with a single-crystal gold nanostructure formation following Volmer–Weber growth (Manna et al., 2016). Atmospheric chemical vapor deposition uses volatile precursors for the growth of nanocrystals, which sometimes become difficult to handle because of their high moisture sensitivity and reactivity. Aerosol-assisted chemical vapor deposition (AACVD) is an alternate method to synthesize gold nanoparticle film by using soluble precursors, such as gold chloride dissolved in methanol (Palgrave et al., 2007). The method uses a liquid-gas aerosol to carry precursors to a heated substrate under atmospheric conditions. During fabrication, the substrate temperature and precursor concentration have substantial influence over the growth of particulate or island morphology. In an arc-discharge method of preparation of gold nanoparticles, a pulse voltage of 70–100 V is produced for 2–3 μs between two gold electrodes submerged in water (Lung et al., 2007). Then a pulse of 20–40 V is maintained for nearly 10 μs. The surface of the gold wire is evaporated and condensed in water forming a gold suspension. The prepared gold nanoparticles have an average diameter of 20 nm. In solution plasma processing, gold nanoparticles of size 1–10 nm can be synthesized by generating pulsed plasma in an aqueous gold solution with cetrimonium chloride (CTAC) as a surfactant (Bratescu et al., 2011). The plasma is generated at a frequency of 10 kHz; electric field of plasma at the inter-electrode space was 3×10^7 Vm^{-1}. The hydrogen radicals generated in the gas phase reduce gold (3+) ions to gold (0) in solution. Plasma discharge time and the applied voltage have a substantial impact in determining the particle shape and size distribution (Saito et al., 2009). Plasma discharge time for one min can result in a dendrite-shaped nanoparticle of a size around 150 nm. As the discharge time continues, progressive size reduction and shape transformation to triangular, pentagonal, and hexagonal forms take place. An increase in the discharge time to 45 min results in 20 nm-sized gold nanoparticles. The size reduction is more prominent at an applied voltage of 3200 V than at 1600 V (Saito et al., 2009). Gold nanoparticles prepared by radio-frequency magnetron sputtering acquire size and morphological changes (island, nanoparticle, film) as the sputtering gas, pressure, radio-frequency input power, and sputtering time change (Terauchi et al., 1995). A combinatory approach of argon ion sputtering on gold foil to synthesize a small cluster of gold atoms and subsequent deposition in an ionic liquid (capture medium) of low vapor pressure is an elegant way to generate nanoparticles of uniform size (Hatakeyama et al., 2011a). The ionic liquid is 1-butyl-3-methylimidazolium tetrafluoroborate. The work considers the influence of several factors on the size distribution of gold nanoparticles. These include the distance between the target and the capture medium, the temperature of the capture medium, the target temperature, and the applied voltage. Polyethylene glycol (PEG) can function both as capture medium and stabilizer during the growth of gold nanoparticles by sputter deposition (Hatakeyama et al., 2011b). The method involves the sputter deposition of gold nanoparticles from the gold target over liquid PEG cast on a stainless-steel plate. The sputtering time is 50 min, and the nanoparticle size variation is 2–8 nm. The difference between ionic liquid and PEG as capture medium is that the size of the nanoparticles is smaller in the former than in the latter. During solution plasma sputtering, the concentration of stabilizer strongly affects the size distribution of the prepared gold nanoparticles. One such study has shown plasma discharging between two gold electrodes submerged in aqueous alginate (Alg) gel matrix (Watthanaphanit et al., 2014). During plasma generation, the pulse width and frequency are maintained at 2 μs and 15 kHz, respectively. There is an increase in the concentration of sputtered gold ions in solution with an increase in the discharge time from 0 to 10 min. In the absence of alginate, the synthesized gold nanoparticles easily agglomerate, whereas size reduction is easily achieved if the concentration of alginate increases during sputtering. The size of gold nanoparticles reduces from ~4.32 nm (water) to ~3.54 nm (0.5 % Alg), to 2.87 nm (0.9 % Alg) (Watthanaphanit et al., 2014). Thiol-containing ligand (1-octadecanethiol) is an excellent stabilizer for maintaining proper interparticle separation and size distribution of gold nanoparticles. Solution sputtering of gold foil dipped in 1-octadecanethiol-silicon oil matrix can suitably control the size of gold nanoparticles between 1.3 and 4.9 nm (Ishida et al., 2015). Solution plasma sputtering of gold wire at atmospheric pressure in liquid N_2 medium can generate gold clusters of ~2 nm diameter in water (Hu et al., 2012). The water is injected into liquid nitrogen containing discharged gold ions. Water freezes quickly with gold clusters trapped inside. The melting of ice releases gold clusters. The formation of gold clusters involves the following steps: discharging, plasma particle bombardment, atom vapor deposition, and condensation (Figure 3.1).

The laser ablation method of preparation of gold nanoparticles shows good control over size and geometry. Surfactant-free gold nanoparticles are synthesized by pulsed laser ablation of an acidic solution containing sulfuric acid, nitric acid, and hydrogen chloride. The nanoparticles are stabilized by a negative surface potential with minimal agglomeration of nanoparticles (Palazzo et al., 2017). The exciting laser was a 1064 nm Nd:YAG laser with a frequency of 10 Hz and a pulse duration of 8 ns. The laser ablation time was 6 min. The gold nanoparticles are stabilized by adsorption of anions such as chloride ions, sulphate ions, and nitrate ions. The prepared nanoparticles have a diameter of 13 ± 3 nm.

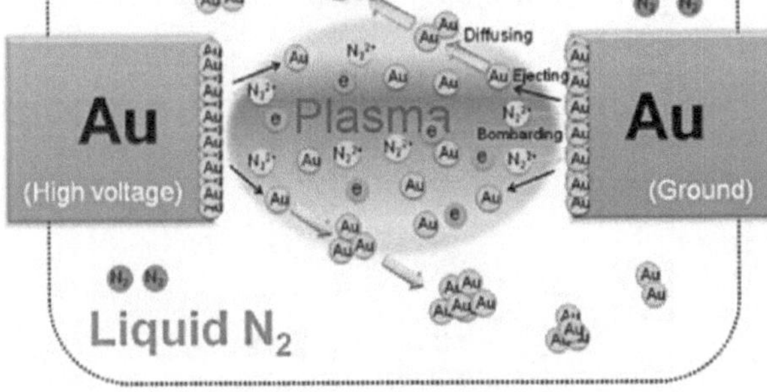

FIGURE 3.1 Schematic showing the formation of gold clusters by plasma sputtering of gold wire electrode under liquid nitrogen. Reproduced with permission from Ref. (Hu et al., 2012). Copyright American Chemical Society.

Laser ablation with 1064 nm followed by irradiation with 532 nm is an improved strategy to obtain size reduction of gold nanoparticles. Irradiation with 532 nm results in the fragmentation of the gold nanoparticles to a dimension of 1–5 nm (Mafuné et al., 2001).

3.2.2 Chemical Synthesis Methods for Functionalized Gold

The wet chemical synthesis method is the most widely adopted method for the fabrication of functionalized gold nanostructures. In this section we will discuss the fabrication of gold nanostructures functionalized with citrate, thiols, and polymers. We will also discuss the fabrication of anisotropic gold nanostructures.

3.2.2.1 Citrate Stabilized Gold Nanoparticles

Stability and reactivity are the two issues that cause the greatest concern when gold sols are prepared in any media. Surface protection of gold sols stabilizes the system against aggregation in any solution media. The chemical reduction method is more than 150 years old, and it started with the pioneering work of Michael Faraday (1857). In his work, gold colloid was synthesized by the reduction of an aqueous solution of tetrachloroaurate by phosphorous in carbon disulfide. Pauli et al. (1939) performed an alcohol reduction of a mixture of gold salt and potassium carbonate to prepare large quantities of gold sol. The prepared sol was further purified by electro-decantation. Turkevich proposed an improved synthesis method of gold nanoparticles by reacting an aqueous solution of hydrogen tetrachloroaurate ($HAuCl_4$) with sodium citrate, where the citrate ions function both as a reducing and as a stabilizing agent (Turkevich et al., 1951; Turkevich et al., 1953). Turkevich's is known to be the most widely used method for the synthesis of gold nanostructures. In 1973, French went a step farther and controlled the size of the synthesized gold nanoparticles by varying the gold salt-to-citrate ratio (Frens, 1973). The work showcases the reduction in particle size at a high concentration of citrate ions and vice-versa. The reaction is rapid at a higher citrate concentration, and the conversion of gold (+3) to gold (0) sol completes within 5 min. The reaction rate slows down and the reaction takes nearly 30 min for completion as the citrate concentration decreases. A high citrate concentration favors the formation of stabilized gold nanocrystals, whereas a low citrate concentration favors the aggregation of the nanocrystals, and destabilizes the system (Frens, 1973). Similarly, a gold chloride concentration above 0.8 mM could favor particle destabilization and aggregation (Kimling et al., 2006). Zabetakis et al. varied the gold concentration up to 2 mM and the citrate/gold(III) molar ratio was varied up to 18:1. The results showed that monodisperse gold nanoparticles were formed at an optimum 0.6 mM gold(III) concentration (Zabetakis et al., 2012). One group adopted Turkevich's method and modified a few parameters to synthesize gold nanoparticles (Li et al., 2011). The reaction was performed in an alkaline medium at a gold(III) concentration of 2.5 mM. This condition accelerated the reaction rate and resulted in gold nanoparticles having sizes in the range of 12–13 nm. To overcome the possibility of aggregation and destabilization of the system, the reaction was performed at low temperature

(70 °C) with pH controlled by varying the NaOH concentration (Li et al., 2011). Turkevich monitored the growth of nanocrystals at various stages with an electron microscope and observed that the growth of nanocrystals follows a nucleation and growth model in which, during the start of the reaction, smaller nuclei are formed. These, as time progresses, aggregate to form bigger gold sol (Turkevich et al., 1953). A reverse process has also been reported in which the temporal evolution of gold nanoparticles is monitored as a function of reaction time (Biggs et al., 1993). Initially, large aggregates of gold sols of size 200–300 nm are formed due to preferential adsorption of $AuCl_4^-$ at the gold sol/water interface. As the reaction progresses, the accumulation of a large number of citrate ions at the interface increases the electrostatic repulsion, resulting in the fragmentation of the large particles to 10 nm gold nanoparticles. In addition to the reaction time, the pH of the medium takes an active part in the size evolution of nanoparticles. By varying the pH of the medium, gold nanoparticles of sizes between 20 and 40 nm are synthesized (Ji et al., 2007). It is observed that, besides playing the role of stabilizer and reducing agents, citrate ions also act as pH mediators. The pH value of 6.5 is the optimum value above or below which the growth mechanism follows a different pathway (Figure 3.2). At pH above 6.5, the growth of nanocrystals follows a nucleation and diffusion-controlled mechanism. At pH lower than 6.5, the growth of nanocrystals follows nucleation and aggregation of gold clusters to form an extended network of nanowires. As the reaction progresses, the nanowires' diameter increases and finally intra-particle ripening occurs to produce gold nanocrystals (Pong et al., 2007).

The kinetics of gold nanoparticles formation appears to be faster with small nanoparticles and under an acidic medium at pH ~ 4.7 (~3 min, ~16.1 ± 3.3 nm) than under mildly acidic conditions at pH ~ 5.6 (~5 min, ~20.4 ± 4.1 nm). At neutral pH ~ 6.5, the reaction takes more than 7 min for completion with an increase in nanoparticle size to 36.6 ± 6.8 nm (Ojea-Jiménez and Campanera, 2012). In order to establish the relation between color transformation and the intermediate complex formed during the reaction course, Mikhlin and his team performed a series of in-situ and ex-situ measurements, including atomic force microscopy (AFM), dynamic light scattering (DLS), small-angle X-ray scattering (SAXS), and transmission electron microscopy (TEM) (Mikhlin et al., 2011). It is reported that the intermediate complex is a sub-micrometer particle consisting of many liquid droplets having sizes between 30 and 50 nm. At the initial stage of the reaction, the as-formed gold nuclei with reduced gold(III) species are trapped inside the liquid droplet. As the reaction accelerates, the reactants in the droplet take part in the growth of nanoparticles followed by disintegration and reduction in particle size. The color transformation is associated with shape evolution such as wires, wormlike nanoparticles, and islands (Mikhlin et al., 2011). The conclusive results of the growth of nanoparticles as studied with SAXS show that the growth of nanoparticles involves four steps: nucleation, coalescence, slow growth, and finally fast growth (Polte et al., 2010). Turkevich in his organizer model of the growth of gold nanoparticles remarked that $AuCl_4^-$ formed an acetone dicarboxylate (ADC) molecular complex which decomposed to give gold sols (Turkevich et al., 1953). Following this, Kumar and his colleague proposed a theoretical model which stated that gold nanoparticles formation followed a balance between nucleation rate and the decomposition of the molecular complex (Kumar et al., 2007). A fast nucleation can be achieved by increasing the concentration of the ADC complex. This can be achieved if an inverse Turkevich method is approached. In an inverse Turkevich method, gold(III) precursors are added to a boiling aqueous solution of citrate ions (Ojea-Jiménez et al., 2011). The inverse way promotes thermal oxidation of sodium citrate and formation of ADC, which acts as the nucleation site for the formation of monodispersed gold nanoparticles. Schulz proposed an ADC-mediated formation of gold nanoparticles by optimizing the pH of the medium and the citrate/gold ratio. The gold nanoparticles are synthesized in large volume, having narrow size distribution and uniform dispersion by the addition of EDTA (Schulz et al., 2014). The role of chloride ions on the size tunability of gold nanoparticles has been studied in detail. The concentration of chloride ions is controlled by the addition of different molar concentrations of sodium chloride (Zhao et al., 2012). Size and shape evolution at different reaction times (0.5–20 min) show that the gold particles aggregate and form wire-like structures, which finally transform to particles. The added chloride ions help in the aggregation of the particles. The numbers and the position of the chloride ions in the intermediate complex can determine the reaction pathway to gold nanoparticle formation (Ojea-Jiménez et al., 2012). Based on the activation energy consideration, which determines the kinetics of the reaction, the reaction goes to completion following four distinct steps: (i) Chloride ion substitution in gold chloride ($AuCl_4^-$) by the citrate ions and formation of an Au(III)-citrate complex. (ii) Deprotonation of the acidic group in the citrate group. (iii) Internal conversion and hydroxyl group addition to gold

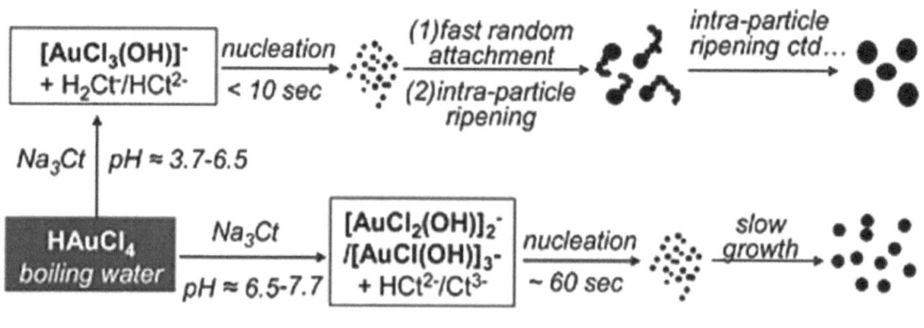

FIGURE 3.2 The mechanistic pathway of gold nanostructure formations under different reaction conditions following the Turkevich method. Reprinted with permission from Ref. (Ji et al., 2007). Copyright American Chemical Society.

species. (iv) Transition state (TS) formation, decarboxylation, and electron transfer.

3.2.2.2 Thiol-Protected Gold Nanostructures

More than a decade ago, Brust and his co-workers developed a two-phase method for the synthesis of thiol-protected stable gold nanoparticles with a size of 1–3 nm. This synthesis method, more popularly known as the Brust–Schiffrin method (BSM), is widely used for the fabrication of small nanoparticles. Unlike the Turkevich method, the Brust–Schiffrin method employs a phase transfer catalyst to transfer gold precursors from the aqueous phase into the organic phase (Brust et al., 1995). The Brust–Schiffrin method involves the following steps. (1) Au(III) Cl_4^- species in the aqueous phase is transferred to the organic phase in solvents such as toluene, chloroform, and benzene by the use of a phase transfer agent, tetraoctylammonium halide. (2) Addition of thiols to the organic phase transfers the gold(III) species to gold(I). (3) Finally, $NaBH_4$ is added to reduce gold(I) ions to metallic gold nanoparticles with uniform size distribution. Brust–Schiffrin also considered single-phase fabrication of gold nanoparticles stabilized with p-mercaptophenol (Brust et al., 1995).

Several factors contribute to size variation and dispersity of gold nanoparticles, including thiol/gold ratio, length of the ligand chain, ligand other than thiol, and halide group attached to the phase transfer agent (Alvarez et al., 1997; Soliwoda et al., 2014; Zaluzhna et al., 2012; Booth et al., 2017). Long-chain alkylthiols and sulfur to gold species ratio of 3 : 1 are the optimum conditions for the recovery of fine gold nanoparticles (Alvarez et al., 1997). Similarly, alkyl amine stabilized gold nanoparticles are also known to form stable gold nanoparticles dispersion in nonpolar solvents (Figure 3.3). However, the length of alkyl chain governs the feasibility of the phase transfer process of gold species from water to organic phase (Soliwoda et al., 2014).

Hydrocarbon chains with twelve carbon atoms stabilize small gold nanoparticles and provide easy transfer from one phase to the other. Hydrocarbon chains with eight carbon atoms result in an unstable gold sol. The instability is caused by the interaction and penetration of small alkyl chains present on the nearest gold nanoparticles. Therefore, the distances between metal nanoparticles are reduced; strong metal–metal interaction results in the loss of stability (Soliwoda et al., 2014). The use of selenide containing groups in the ligand, e.g., didodecyl diselenide, and

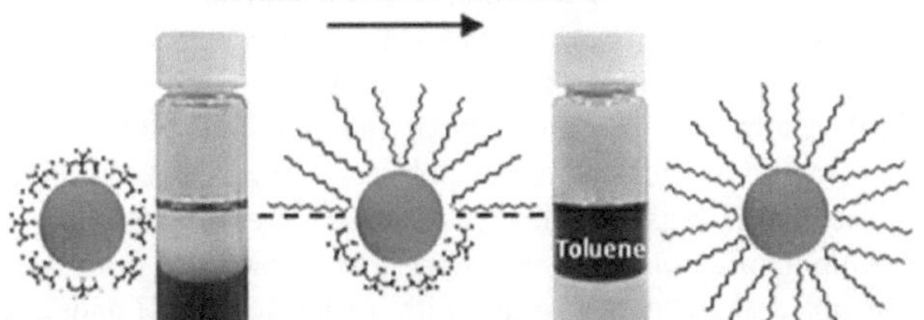

FIGURE 3.3 Gold nanoparticles formation at the water-toluene interface following the Brust-Schiffrin protocol. Reprinted with permission from Ref. (Soliwoda et al., 2014). Copyright American Chemical Society.

dioctyl-diselenides instead of thiol can be useful for the fabrication of ultrafine gold nanoparticles in a single phase instead of two phases, where size dispersity increases (Zaluzhna et al., 2012; Yee et al., 2003). Dioctyl-diselenides ($(C8Se)_2$) as the ligand in the single-phase reaction medium can change the gold coloration from wine-red to yellow; the color remains unchanged if the synthesis is performed in a single phase. Only a single reaction intermediate is formed in a single-phase reaction, whereas more than one reaction intermediates are identified in a two-phase reaction (Zaluzhna et al., 2012). Selenolate is considered to form strong bonding with gold rather than thiolate and forms better self-assembled monolayers over gold surface (Ossowski et al., 2015). The use of tetraoctylammonium bromide is preferred over the corresponding chloride in converting gold(III)bromide to gold(I)bromide, with the latter being stable without undergoing instant reaction with the thiol. However, gold(I)chloride is highly unstable and readily precipitates by reacting with thiol (Booth et al., 2017). Although the method is widely adopted to synthesize thiol-functionalized gold nanoparticles, the mechanism involving the identification of the intermediates is still under debate. Many earlier reports assumed that gold(I)-thiolate polymer is the intermediate precursor that governs the formation of gold nanoparticles. However, another group of scientists remarked that gold-thiolate is not the intermediate species; the tetraoctylammonium (TOA) metal (I) halide (X) complex ([TOA] + [M(I) X2]) is the precursor for Brust–Schiffrin (Goulet and Lennox, 2010; Li et al., 2011a; Li et al., 2011b). This precursor is formed following the reaction

$$\left[\text{TOA}\right]\left[\text{M(III)X4}\right] + 2\text{RSH} \rightarrow \left[\text{TOA}\right]\left[\text{M(I)X}_2\right] + \text{RSSR} + 2\text{HX}.$$

A recent study shows that the sulfur/gold ratio determines the formation and types of intermediate in the Brust–Schiffrin method. The intermediate complex is $[TOA]^+[AuX_2]^-$ when sulfur/gold <2, or a mixture of $[TOA]^+[AuX_2]^-$ and polymeric $[Au(I)–SR]_n$ when sulfur/gold >2 (Zhu et al., 2013). This mixture of complexes is one reason for the polydispersity of synthesized gold nanoparticles. It is understood that $[TOA]^+[AuX_2]^-$ is the starting complex that governs the fate of the chemical reaction. However, the mechanism as to how the phase transfer catalyst transfers AuX_2 from aqueous to organic phase is not understood. There are two contradicting mechanisms for phase transfer: (1) Transfer through inverse micelles. (2) Transfer through complexation (Goulet and Lennox, 2010; Li et al., 2011b; Li et al., 2011c; Perala et al., 2013). The inverse-micelle synthesis was proposed by Li and co-workers with ^{1}H NMR spectroscopy (Li et al., 2011b; Li et al., 2011c). The organic phase $[TOA]^+$ in contact with aqueous $AuCl_4^-$ forms an inverse micelle with encapsulated water. The $[TOA]^+AuCl_4^-$ complex along with gold(III) species transfer from aqueous to the water core in the inverse micelle. This complex in the micelle acts as a metal nucleation center. It is further stated that a high amount of water leads to the formation of small gold nanoparticles with lower polydispersity. This inverse-micelle pathway for gold nanoparticles synthesis was nullified by another group of scientists, who by employing different instrumental tools verified that gold nanoparticles formation is preceded by the initial complex aggregation between TOA and $AuCl_4^-$ (Perala et al., 2013). However, the transfer of $AuCl_4^-$ from the aqueous to the organic phase through the two immiscible liquid interfaces is not a simple ion transfer reaction. The process of ion transfer is accelerated only when a small amount of TOA^+ is present in the organic phase (Duong et al., 2015). Electrochemical measurements show that under the influence of an electric field, $AuCl_4^-$ and TOA^+ start migrating to the liquid–liquid interface boundary. The two different ions form a short-lived ion pair which, under the influence of the electric field, migrates toward the bulk of the organic phase, and the pair dissociates, leaving $AuCl_4^-$ in the organic phase. TOA^+ then continues this process and migrates to the interface for the next $AuCl_4^-$ species, and the process continues depending on the concentration of TOA^+ and $AuCl_4^-$ in the medium (Duong et al., 2015). The final stage of Brust–Schiffrin synthesis is the formation of ligand-protected gold nanoparticles. This is enabled by the addition of $NaBH_4$. As we understand, the organic phase contains $TOA^+AuCl_4^-$, $AuCl_2^-$, or polymeric $[Au(I)SR]_n$ species. The addition of aqueous sodium borohydride to this complex initiates a heterogeneous redox reaction between ionic gold species and BH_4^-. During the reaction, BH_4^- transfers from the aqueous to the organic phase and chloride ion transfers from the organic to the aqueous phase to balance charge electroneutrality (Uehara et al., 2015).

3.2.2.3 Polymer-Stabilized Gold Nanostructures

Polymers act as a scaffold for the adhesion of nanoparticles and provide steric stabilization, preventing them from agglomeration. The first report of the use of polymer stabilizer in gold sol preparation dates back to 1718, where starch was used as the stabilizer for gold sol (Shan et al., 2007). The protection of gold nanoparticles with polymers is achieved either by physisorption or chemisorption of polymers over the surface of gold nanoparticles. Stabilization of gold sol via chemisorption involves the use of polymers with specific atoms/groups (e.g., thiol) containing end groups that stabilize the gold nanoparticles through covalent linkages. On the other hand, stabilization through physisorption involves polymers without specific end groups, such as block polymers. There are two ways to achieve functionalization of the gold surface via covalent binding of polymers: "grafting from" and "grafting to" techniques. In the grafting-from technique, a small polymer initiator is immobilized on the surface of a synthesized nanoparticle. The addition of monomer results in chain growth polymerization and forms high-density polymer brushes over the gold surface (Gann and Yan, 2008). Grafting-from is the in-situ formation of polymer chains over preformed gold surface. Examples of polymers coated over gold nanoparticles by grafted-from techniques include poly(n-isopropylacrylamide), poly(n-butyl acrylate), and poly(methyl methacrylate) (Shan et al., 2005). Grafting-to is a process in which prepared polymers are added to a gold chloride solution; a method by which polymers protect the prepared gold nanoparticles. The grafting-to technique involves the chemisorption of end functionalized polymers during in-situ growth of gold nanoparticles. The end group of the polymer is terminated with a sulfur-containing group (thiol, thioester, disulfide). The polymer without sulfur-terminated groups can cover up gold surface by physisorption. The polymers in this category are poly(n-vinyl pyrrolidone), poly(vinly alcohol), poly(ethylene glycol), poly(vinyl pyridine), poly(vinyl methyl ether). Gold nanoparticles coated with sulfur-terminated polymers are stabilized through gold-sulfur bonds. On the other hand, sulfur-free polymers cannot stabilize the gold nanoparticles, and, as a result, particle aggregation readily occurs (Uehara, 2010). Grafting-to techniques provide a better possibility of nanoparticles size control and stabilization than grafting-from techniques. One-step synthesis of gold nanoparticles is achieved in the presence of poly(dimethylaminoethyl methacrylate) homopolymer (PDMAEMA) with tertiary amino acids, which act both as reducing as well as stabilizing agents (Mountrichas et al., 2014). Unlike the citrate-stabilized synthesis of gold nanoparticles, where the reaction is performed at a boiling temperature, PDMAEMA-stabilized gold nanoparticles synthesis is performed at a temperature range of 30–60 °C. Polymer-coated gold nanoparticles can show thermally reversible aggregation and re-dispersion in a polymer matrix if the interaction of the ligand–matrix–particle surface is stabilized via hydrogen (H) bonding (Heo et al., 2013). In this way, high-molecular-weight polymer-coated gold surface can be well dispersed. At low reaction temperature, the system shows good dispersion through H-bonding. However, at high temperature, the rupture of the H-bonding provides aggregation of the nanoparticles. Poly(styrene-r-2-vinylpyridine)- or P(S-r-2VP)-coated gold nanoparticles are stabilized over the matrix Poly(styrene-r-4-vinylphenol) or P (s-r-4VPh) through hydrogen bonding between pyridine and phenol moiety present in stabilizer and matrix. The sample, when heated at 200 °C, shows aggregation due to the breaking of the hydrogen bonding. Re-dispersion again occurs due to the re-formation of hydrogen bonding at a reduced temperature (Heo et al., 2013). A grafting-to technique involving ligand exchange reactions forms small and stable gold nanoparticles. For example, 4-(N,N-dimethyl amino) pyridine (DMAP)-coated gold nanoparticles can be stable when kept in a solution containing thiol-terminated ligands, such as poly(ethylene oxide) (PEOnSH) and polystyrene (PSnSH). The ligand exchange reactions result in the displacement of DMAP with thiol (Rucareanu et al., 2008). Copolymers-coated gold nanoparticles provide better control over particle formation and stabilization over homopolymer-coated gold surface (Lowe et al., 2002). One of the synthesis methods of copolymer-stabilized gold is a reversible addition-fragmentation chain transfer (RAFT). Amphiphilic gold nanoparticles can be synthesized by an in-situ method in a polymer matrix comprising of a hydrophilic N-isopropylacrylamide (NIPAM) and hydrophobic polystyrene in tetrahydrofuran solvent with different chain lengths (Shan et al., 2005). The ratio of PNIPAM to polystyrene influences the size of the gold nanoparticles as well as the conformation changes in the polymer during the making of a film at the air–water interface (Figure 3.4). Gold nanoparticles prepared with a 5 : 1 ratio of PNIPAM : PS have a size of 2.5 ± 0.7 nm and those prepared at 2 : 1 ratio have a size of 3 ± 1 nm. Furthermore, the 2 : 1 ratio is the optimum ratio that initiates the conformational changes in the polymer from pancake to high-density brush transitions at various surface pressure.

Amphiphilic block copolymers are known to form self-assembled aggregates in water with a variation in pH and temperature. Shell cross-linking of polymers is a suitable strategy to build up a core-shell structure of polymer–gold nanoparticles with much higher colloidal stability (Luo et al., 2005). By adopting this strategy, the hydrophobic surface of gold cores can be well encapsulated inside the micellar structure with complete

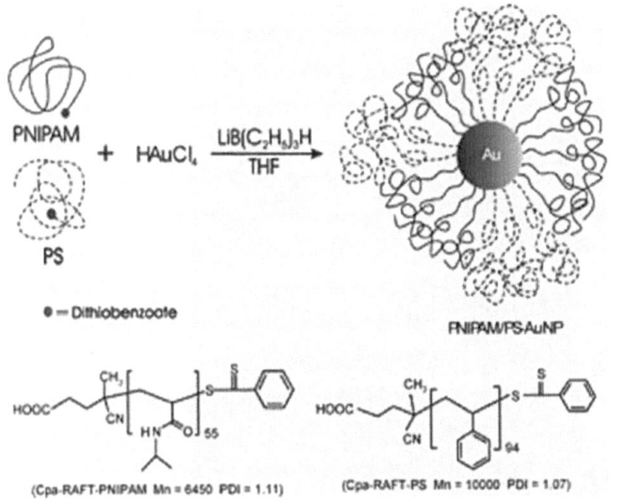

FIGURE 3.4 Scheme showing the preparation of gold nanoparticles grafted with poly(N-isopropylacrylamide) (PNIPAM), PS chains, and structures of the corresponding homopolymers. Reprinted with permission from Ref. (Shan et al., 2005). Copyright American Chemical Society.

protection of the nanoparticles from possible aggregation in different solvents. One such approach shows the preparation of thiol-terminated diblock copolymers of poly(2-dimethylamino) ethyl methacrylate (PDMA) and polyethylene oxide (PEO). The diblock copolymer is then coated over gold core containing PDMA in the inner shell and PEO in the outer tail. The polymer is finally cross-linked by adding a suitable reagent (Luo et al., 2005). By the cross-linking of polymers, a permanent core-shell structure with high stability can be formed.

3.2.2.4 Anisotropic Gold Nanostructures

Based on the synthesis protocol adopted, gold colloids can attain several anisotropic forms including nanorods, nanowires, nanotubes, prisms, hexagons, pentagons, star, and many more. Sometimes the growth conditions result in the formation of a mixture of the different forms. Among these different anisotropic forms of gold, synthesis strategy leading to nanorods formation is highly explored. We will mostly discuss the fabrication methods for gold nanorods and will touch upon the synthesis of other anisotropic forms of gold nanostructures. In the seed-mediated growth of gold nanorods, gold seeds are synthesized first, and act as nucleation sites. The seeds are added to a growth solution of gold chloride to form anisotropic nanostructures. Jana et al. (2001a) synthesized 12 nm gold seeds by reducing gold chloride with sodium citrate. If the sodium citrate solution is added at once during seed formation, it helps to obtain narrow size and a large number of gold seeds, giving more nucleation sites (Scarabelli et al., 2015). In place of citrate, sodium borohydride can be used as a reducing agent. Although the reaction occurs at a low temperature, the heterogeneity of seeds increases. After the preparation of gold seeds, gold chloride solution is added, followed by a fast addition of ascorbic acid. Depending on the rate of addition of reducing agent and metal salt to seed concentration ratio, the additional nucleation sites beyond seed formation can be controlled. For example, a fast addition of reducing agent initiates further nucleation, whereas no nucleation happens if the addition occurs slowly. Sodium citrate is a weak reducing agent, and the reaction requires boiling temperature. But the seeds thus fabricated are nearly monodispersed, which is a requisite for the gold nanorods formation. The seminal work by Nikoobakht and El-Sayed (2003) showed that the problem associated with citrate-capped gold seeds, e.g., creation of non-cylindrical gold nanorods, could be overcome if CTAB had been used as the capping agent instead of citrate. Seed solution is prepared by adding CTAB as the capping agent and ice-cold $NaBH_4$ as a reducing agent to the gold chloride solution. The seed solution was added to the growth solution, which comprises of CTAB, ascorbic acid, silver nitrate, and gold chloride. The length (L), diameter (D), and overall aspect ratio (L/D) of the gold nanorods can be tuned by controlling silver nitrate concentration, CTAB concentration, and use of co-surfactant with CTAB, and by increasing gold content, pH of the medium, and ascorbate ion concentration. The seed-mediated growth process gives nanostructures of various forms other than nanorods. The aspect ratio can be suitably tailored by changing the seed/metal salt ratio and performing the reaction in a two-step or three-step seeding method (Jana et al., 2001b; Busbee et al., 2003). When the reaction is performed in three steps, with similar reactants to those required for the two-steps method, more and more nucleation centers generate which are arranged to give a large proportion of gold nanorods. The pH increase in the medium from 2.8 to 5.6 can dramatically change the aspect ratio from 18.3 ± 1.3 nm to 25.1 ± 5.1 nm. An increase in the pH can simultaneously increase the concentration of ascorbate ions over the CTAB-gold in the solution, which reduces more numbers of Au^{3+} to Au^{+} (Busbee et al., 2003). The pH of the medium can also be changed by adding alkali or acids. An increase in the pH by adding sodium hydroxide can accelerate the reaction. However, a high pH also increases the aspect ratio with a concomitant increase in the polydispersity (Wang et al., 2013; Wu et al., 2005). The nanorods are short under basic conditions. However, under acidic conditions, with varying concentrations of nitric acid during the seed-mediated growth, nanorods with length up to 4.5 μm can be obtained. The use of nitric acid can increase the production of gold nanorods, along with the formation of nanostructures of other morphology as side products. The percentage of side products increases with hydrogen chloride, sulfuric acid, and phosphoric acid. It is speculated that nitrate ion plays an important role in controlling the growth of gold nanorods. The seed size also governs whether single crystalline or twinned gold nanostructures result. The aspect ratio of crystalline nanorods increases with an increase in the seed size (Gole and Murphy, 2004). In the seed-mediated growth process, initially a pentatwinned structure is formed with different facets (Johnson et al., 2002). As the reaction progresses, Au–CTAB–Ag^{+} are adsorbed over different facets, and through a twinning mechanism the pentatwinned crystals undergo an anisotropic growth along [110] crystal direction. Ag^{+} ions play an important role in controlling the growth of gold nanorods. Ag^{+} addition to the seed can slow down the reaction, but it helps gold atoms to grow over the seed without the creation of any defects. This, as a result, produces large numbers of gold nanorods (Figure 3.5a). The reason behind this is the underpotential monolayer deposition of Ag(I) over {110} and {111} facets of gold surface, which provides nanorod growth along [100] direction (Liu and

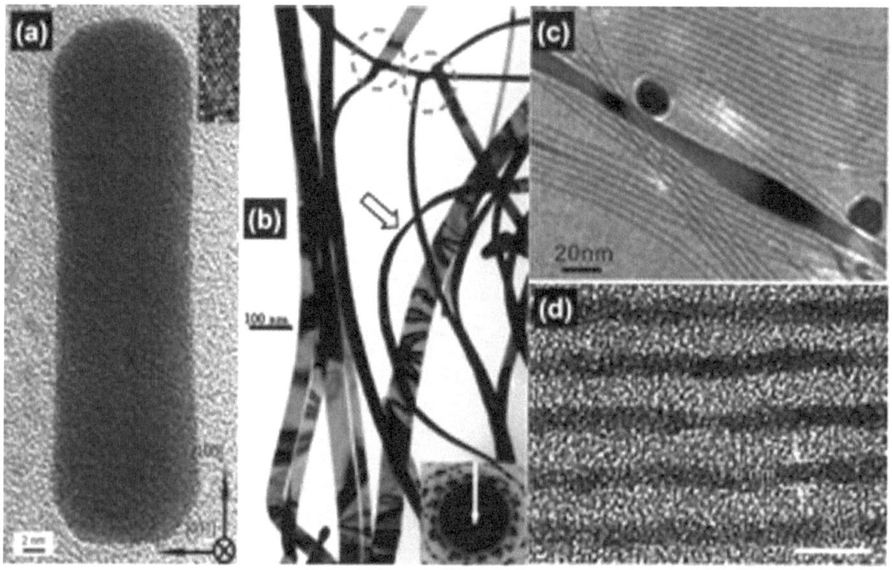

FIGURE. 3.5 Transmission electron microscope images of (a) nanorods, (b) nanoribbons, (c and d) and nanowires. Reprinted with permission from Ref. (Liu and Guyot-Sionnest, 2005); (Bakshi et al., 2008); and (Huo et al., 2008). Copyright American Chemical Society.

Guyot-Sionnest, 2005). The presence of bromide ions, iodide ions impurity in CTAB, and water quality can significantly affect the uniformity and shapes of gold nanorods (Garg et al., 2010; Scarabelli et al., 2015; Smith et al., 2009; Millstone et al., 2008). It is reported that the use of binary surfactant of sodium oleate along with CTAB can result in excellent monodispersity with <0.5 % shape impurities (Ye et al., 2013). The method uses a very low concentration of CTAB (<0.037 M), which is nearly three times lower than the concentration of CTAB used in conventional techniques without co-surfactant. The diameter and the length of the nanorods can vary depending on variations in the concentration of silver nitrate, hydrogen chloride, and seed. Liu et al. (2017) studied the effect of the reaction temperature on the bi-surfactant-mediated growth of gold nanorods. The temperature is varied between 21 and 35 °C; the resultant product contains 90–92 % gold nanorods, and the rest includes particles of other morphologies. A nearly 99 % yield of gold nanorods is found at a reaction temperature of 25 °C. A rise in temperature increases both length and diameter, with an overall decrease in the aspect ratio.

Gold nanobelt/nanoribbon can also be synthesized by a seed-mediated growth process using a gemini surfactant, such as bis (tetradecyldimethyl-ammonium bromide), as the template for the growth of gold nanoribbon (Bakshi et al., 2008). The synthesis procedure involves the formation of gold seed. The growth solution is prepared with gemini surfactant, gold chloride, and ascorbic acid. During the reaction, the gemini surfactants undergo several phase transformations: micelle to rod/wormlike to vesicles and multilayer. Figure 3.5b shows the Au nanobelt following the report of (Bakshi et al., 2008).

The synthesis procedure of gold nanobelt accompanies the formation of nanobelt-dimers and branched nanostructures (Payne et al., 2013; Payne et al., 2014; Zhao et al., 2008). A mixture of cationic and anionic surfactants can increase the yield of nanobelts over other shapes. The use of anionic SDS along with cationic CTAB in water forms micelles. The synthesis protocol involves the addition of CTAB, SDS, and gold chloride in water followed by the addition of ascorbic acid. The reaction period and temperature vary between 4 and 38 °C, and the reaction period is from 12 to 24 h. If the reaction is performed at 27 °C, gold nanobelts are formed with length in micrometers, diameter from 40 to 200 nm, and thickness from 20 to 30 nm, respectively. However, if the reaction is initially performed at 4 °C for 0.5 h and then continued for another 12 h at 27 °C, nanocombs made up of nanobelts are formed (Zhao et al., 2008). Payne et al. (2014) studied the effect of reaction temperature on the micellar growth and the associated shape transformation of gold nanobelts. Micellar forms wormlike structure in CTAB–SDS surfactant mixtures, which act as the template for gold nanobelts. This wormlike micelle is formed in the temperature range of 10–30 °C, and the spherical micelle structure is produced above 30 °C. In the temperature range of 22–30 °C, spherical, tripod-type, and branch-like gold nanostructures are formed. The temperature of 27 °C is optimum for the growth of gold nanobelts. The nanobelt has <110> as the preferred growth direction.

Seed-mediated synthesis procedures with longer aging time can result in gold nanowires (Kim et al., 2008). Transmission electron micrograph of gold nanowires is shown in Figures 3.5c–d (Huo et al., 2008). Gold seed of 3–4 nm size is synthesized by the rapid addition of sodium borohydride to aqueous gold chloride. The seed particles are added to a growth solution containing gold chloride, ascorbic acid, and CTAB. The three-step reduction reaction performed at 20–60 °C is allowed to continue for an extended period for the growth of nanowires. Control of pH with HNO_3 in the reaction medium is essential for maintaining the uniformity of gold nanowires. A high pH gives nanorod morphology, whereas a low pH gives thinner nanowire morphology. Hairy gold nanowires are fabricated at 100 °C through the reduction of aqueous gold(III) chloride by 2-mercaptosuccinic acid without the use of any surfactant (Vasilev et al., 2005). The hairy-like morphology is expected to be formed due to the fusion of the large gold nanoparticles. Use of oleylamine (OA) during the seed-mediated growth process can transform the gold nuclei to gold nanowires. Oleylamine can serve as a reducing agent only or can be used both as a reducing agent and as a solvent (Halder and Ravishankar, 2007; Huo et al., 2008). Halder et al. (2007) fabricated gold nanowires from gold seeds of 2 nm. The reaction medium contains toluene as the solvent, and gold chloride, oleylamine, and oleic acid as the reactants. The reaction is performed at 120 °C. Finally, ascorbic acid addition results in the formation of gold nanowires. During the seeded growth reaction and after ascorbic acid addition, the reaction is aged for a sufficiently long period (in days). Aging for a long time can help the recovery of a significant fraction of nanowires with <10 % possibility for the formation of nanoparticles of other shapes (Halder et al., 2007). The growth of gold nanowires occurs by the oriented attachment of gold nanoparticles along the growing <111> direction. A high temperature favors the growth of nanowires, whereas a low temperature results in branched nanostructures. The oleylamine–gold chloride complex results in the growth of gold nanowires if the reaction is performed at ~85 °C (Kura and Ogawa, 2010). A temperature lower than this (70–75 °C) does not result in any nanostructure formation, and a temperature higher than 100 °C gives only nanoparticles. He proposed that the nanoparticles aggregate should undergo reconstruction by atomic diffusion to give nanowire structures. Huo et al. (2008), however, used oleylamine both as reducing agent and solvent. The reaction is performed at a temperature of 20–25 °C and aged for 4 days, followed by washing and final dispersion in hexane or chloroform. They proposed that the growth of nanowires at a low temperature is due to the formation of the Au^+–oleylamine complex. The interaction leads to a mesostructure formation which acts as the template for the growth of nanowires. Lu et al. (2008) reported the presence of an amphiphilic interaction in the oleylamine–Au^+Cl^- complex. The use of silver nanoparticles in the growth solution can speed up the reaction and reduce Au^+ to Au^0. The addition of silver nanoparticles can achieve a high yield of nanowires. In the absence of silver nanoparticles, the yield of nanowires is only ~20 %. The aspect ratio of gold nanowires can be changed by changing the volume ratio of oleylamine/Au^{3+}. This practice can result in nanowires of diameter ~1.6 nm and length between 10 nm to 3.5 μm. The reaction is performed at room temperature.

In the process of seed-mediated growth of gold nanorods, the final product contains a fraction of gold nanoprisms along with nanorods. The percentage of nanoprisms can vary during the reaction of seed and growth solution. The presence of iodide and bromide ions can significantly alter the anisotropic structure of gold. Iodide ions prevent single preferential growth of gold(III) nanorods and result in faceted gold nanoprisms (Ha et al., 2007). In the seeded growth approach for the synthesis of gold nanoprisms, citrate-capped gold nanoparticles (<5 nm) are synthesized by an aqueous phase reaction of gold chloride, trisodium citrate, and ice-cold sodium borohydride followed by heating at 40–45 °C for 15 min to ensure hydrolysis of borohydride (Matthew et al., 2013). The growth solution is prepared by mixing CTAB/sodium iodide, gold chloride, sodium hydroxide, and ascorbic acid in appropriate quantities. Finally, the seed nanoparticles and the growth solution are mixed to obtain the nanoprism. The dimension of the nanoprism and the fraction of nanosphere as side products are primarily dependent on the seed concentration considered during synthesis. The sphere particles can be separated from the prism by a depletion-force-mediated procedure.

A seedless approach can be considered to synthesize gold nanotriangles (Kuttner et al., 2018). The process involves reacting an appropriate amount of gold chloride and benzyl dimethyl hexadecyl ammonium chloride (BDAC), followed by heating the mixture to a variable temperature of 70–95 °C. 3-barbituric acid (3BA) was added while maintaining the ratio [3BA] / [gold chloride] = 45.6. The final product comprises of nanotriangle and nanooctahedra, which are collected by centrifugation removing excess 3BA and BDAC. The gold nanotriangle and nanooctahedra are separated by a depletion-induced separation approach using CTAC. The mixture of particles is centrifuged, and the supernatant are discarded. The precipitate was re-dispersed in CTAC, and after 4 h the precipitate was discarded. The supernatant was centrifuged for 30 min. The precipitate was re-dispersed again in CTAC, and after 4 h the supernatant was discarded and the precipitate with gold nanotriangle was re-dispersed in BDAC. Overgrowth of gold is achieved taking gold nanotriangle as the seeds. For the overgrowth, a various ratio of [gold chloride] / [gold nanotriangle] and [3BA] / [gold chloride] was taken. Gold chloride and BDAC were mixed at 70 °C followed by the addition of 3-BA unless the solution became colorless. Gold nanotriangle seeds were added to the colorless solution while continuing stirring.

Gold nanoprisms function as the precursor for the synthesis of circular disks and hexagonal nanoplates of gold (O'Brien et al., 2015; Hong et al., 2011). Figure 3.6 shows the transformation of gold nanoprisms to circular disks (O'Brien et al., 2015). CTAB coated over gold nanoprisms is dissolved by heating. The solution is cooled, and 2M NaCl is added to to a 10 mL of the cold solution in order allow separation of nanoparticles and assembly of nanoprisms. The solution was vortexed and settled for 2 h. After this time, the nanoparticle solutions were centrifuged, and the supernatant was removed. Then 10 mL and 50 mM concentration of CTAB was added to the solution.

For the fabrication of hexagonal nanoplates, a certain volume of prism solution and gold chloride is added to a solution containing CTAB and sodium iodide (Hong et al., 2011). Ascorbic acid is added dropwise to the solution for 45 min under vigorous stirring. This results in a color change from yellow to greenish brown. The solution is then allowed to settle. A variation in the amount and concentration can change the thickness of the nanoplates.

The preparation of gold octahedron involves mixing up polyvinyl pyrrolidone, polyethylene glycol, sodium borohydride, and gold chloride at room temperature (Li et al., 2007). The mixed solution is then heated in an oil bath at 75 °C for more than 24 h, followed by further heating at 125 °C to perform the reduction of metal salts by PEG. The products were collected by centrifugation. A hydrothermal method can also be used to fabricate gold octahedral (Chang et al., 2008). An appropriate quantity of CTAB, gold chloride, and trisodium citrate is added to a hydrothermal reactor and then kept at 110 °C. The reaction time was adjusted at 6, 12, 24, 48, and 72 h for attaining gold octahedral sizes of 30, 60, 90, 120, and 150 nm, respectively. This process yields 90 % gold octahedron with {111} as the reactive facets. The [CTAB] / [$AuCl_4^-$] ratio governs the amount of yield and

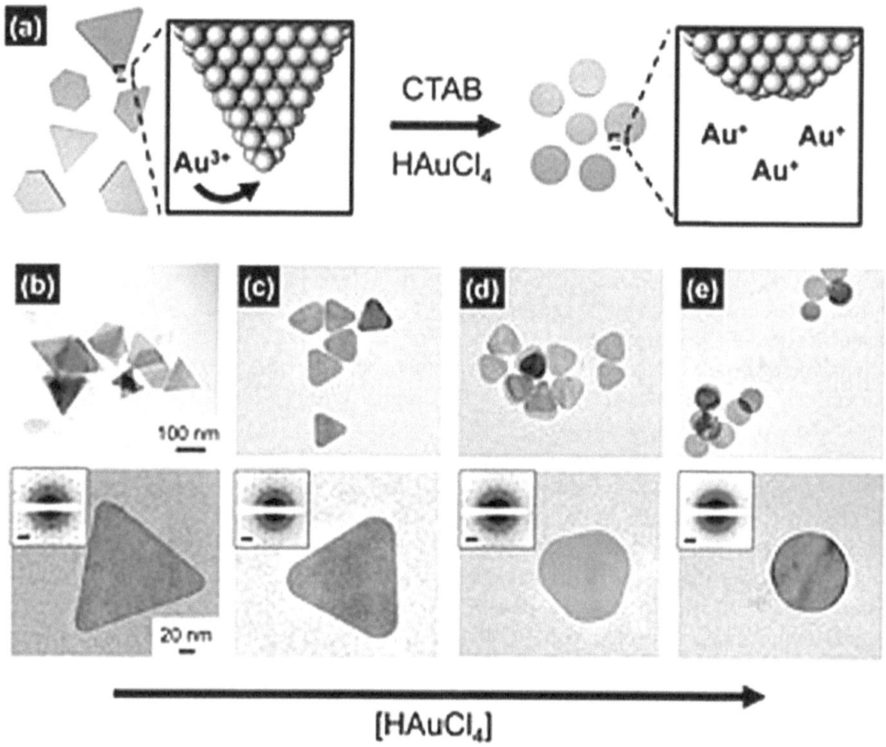

FIGURE 3.6 (a) Formation of gold nanodisk from gold nanoprism via a conproportionation reaction. Oxidation of triangular nanoprisms by $HAuCl_4$ in presence of CTAB. (b–e) TEM images taken during the transformation of nanoprisms to disk at various oxidizing concentrations of $HAuCl_4$. Reproduced with permission from Ref. (O'Brien et al., 2015). Copyright American Chemical Society.

the shape of the nanocrystals. The molar ratio of 7.5 gives nanoplates, decahedron, and icosahedra. An increase in the molar ratio to 15 gives octahedral gold with diameters of 150–200 nm along with triangular nanoplates. The molar ratio of 100 gives octahedral gold with size ~35 nm.

3.2.3 Electrochemical and Photochemical Synthesis

The electrochemical method for the fabrication of gold nanostructures was reported by Yu et al. (1997). In that method, a gold metal plate acts as an anode and a platinum plate as the cathode. These two electrodes are inserted into an electrolytic solution of hexadecyl trimethyl ammonium bromide (C16TAB), a shape-directing co-surfactant tetraoctyl ammonium bromide (TC8AB), and acetone. During the electrolysis, gold nanoparticles are deposited at the cathode and in the electrolyte solution. The electrolysis is conducted for 30 min at a temperature of 38 °C while continuing ultrasonication during the synthesis. This method results in gold nanorods with aspect ratio varied from 1.8 to 5.2. In the reaction medium, the use of acetone loosens the micellar structure, while cyclohexane facilitates the growth of the gold nanorods. A silver plate is also inserted into the reaction medium. The silver ions from the plate favor the growth of gold nanorods along its major axis (Chang et al., 1999). The use of PVP during the electrochemical reduction of gold metal plate stabilizes and protects the gold nanoparticles. The nitrogen and oxygen coordinating sites of PVP attach with gold ions and prevent the metal ion aggregation (Huang et al., 2005). The formation of gold nanocrystals follows a stepwise reduction process. PVP can coordinate with gold ions in solution during electrochemical oxidation and forms the Au^{3+}–PVP complex. The electrochemical reduction of the complex at the cathode/electrolyte interface generates Au^0 nanoparticles through the nucleation and growth of atoms and clusters. During the reduction process, numerous gold nuclei are formed along the PVP chain, which finally grows into nanocrystals. If the chain length of polymer increases, the gold nanocrystals self-assemble to give one-dimensional nanostructures or nanorings (Ma et al., 2006). Gold nanoflowers can be suitably deposited over an indium–tin–oxide (ITO) substrate following the electrochemical method (Duan et al., 2006; Guo et al., 2007). The process involves gold chloride as the electrolyte, PVP as the stabilizer, ITO as the cathode, and graphite sheet as the counter electrode. The formation of nanoflowers proceeds via the process of electroreduction and crystallization. The chloride ions in solution are adsorbed on the grown {111} facets and do not allow the preferential growth along <111> direction. The formation of the metal–ligand complex as the intermediate and the stepwise reduction of metal ions is reported during the electrochemical reduction of the Au^{3+}–pyridine complex, wherein 4-methoxy pyridine is used as the source of pyridine (Biggs et al., 2016). The two-step reduction, recorded in voltammogram, involves the conversion of Au(III) to Au^+ intermediate; then its subsequent conversion to Au^0 nanoparticles over ITO. Electrochemical deposition of Au^+ gives anisotropic nanorod formation.

The photochemical synthesis of gold nanostructures involves exposure of a gold salt–ligand complex to UV irradiation. For example, rod-like gold nanoparticles are formed by the UV irradiation of a micellar solution containing CTAC and gold chloride

(Esumi et al., 1995). Irradiation is performed at 253 nm with a radiation power of 250 W. The irradiation performed with an increase in time from 3 to 24 h shows progressive shape transformation from sphere–rod–prism to finally elongated rod-like structures. Many other syntheses consider CTAB in place of CTAC (Kim et al., 2002). In the reaction medium, gold chloride binds with CTAB and forms a micelle. The ligand-exchange reaction of chloride with bromide results in the formation of $AuBr_4^-$ and final conversion to $AuBr_2^-$. Under UV irradiation, acetone forms ketyl radical, which transfers electrons to Au^+ ions and reduces to Au^0 attaining a nanorod morphology (Nishioka et al., 2007; Marin et al., 2008). The use of Ag^+ ions during photoirradiation provides an end-to-end assembly of gold nanoparticles. Ag^+ is easily reduced to Ag^0 under light and deposit over {111} and {100} distant end-planes of gold. The deposited silver acts as a linker between the nearest nanoparticles and forms chains and networks of gold nanostructures (Shu Jun et al., 2013).

3.3 Surface Plasmon Resonance Properties of Gold Nanostructures

Gold nanostructures display unique optical absorption in the UV-visible and the near infra-red region when they interact with light. The phenomena, known as surface plasmon resonance (SPR), refers to the collective excitation of conduction electrons in resonance with the frequency of the incident light (Ghobadi Turkan Gamze et al., 2017; Kim et al., 2017; Zhang et al., 2018; Wang et al., 2012). SPR induces strong localization of electric field in the near-surface region and intensifies the absorption intensity. The position of the surface plasmon resonance peak is dependent on the size and morphology of gold nanostructures. The plasmon energy can be utilized within the stipulated time frame for driving various chemical reactions on the gold surface. Because of the SPR properties, gold nanostructures have found widespread application in photochemical reactions, optical sensing, and biomedical applications.

The localized surface plasmon resonance has a very short lifetime, within 1–100 fs (Zhang et al., 2018). The plasmon energy can be transferred to the nearest array of plasmonic nanostructures, to an attached molecule, or to a semiconductor through various ways. These methods include plasmon resonant energy transfer (PRET), optical near-field coupling, scattering, and hot electron injection (Ghobadi Turkan Gamze et al., 2017). In PRET, the plasmon energy directly excites electron-hole pairs in the nearest semiconductor via plasmon-dipole and semiconductor excitons dipole coupling. In optical near-field coupling, the electric field is amplified at the narrow gap between metal nanoparticles or the metal-semiconductor interface. The strong accumulation of plasmon energy is transferred to an attached molecule or semiconductor. Plasmonic scattering is prominent for metal nanoparticles with sizes between 50 and 100 nm. Hot carrier injection and extraction is another method by which plasmon energy can be exploited to drive chemical reactions. In the case of a metal-semiconductor junction, the hot carriers are either transferred indirectly by crossing the Schottky junction at the interface or through a direct injection to the semiconductor.

3.4 Application of Gold Nanostructures

3.4.1 Chemical Sensing

Absorption-based colorimetric detection of heavy metal ions in solution can be performed with gold nanostructures, since the latter shows size- and shape-selective optical responses (Qian et al., 2013). Selective detection of metal ions can provide an accurate measurement of the metal ion concentration in solution. Some ligand-protected gold nanoparticles show an affinity for a wide variety of metal ions present in water. For example, 3 mercaptopropionic acid (MPA)-functionalized gold nanoparticles can effectively bind with Hg^{2+} and provide aggregation-mediated color change (Huang and Chang, 2007). The drawback is that aggregation also occurs because the acid part on the ligand has the affinity to bind with other toxic metal ions, such as Ca^{2+}, Sr^{2+}, Mn^{2+}, Cd^{2+}, and Pb^{2+}. Thus, selective colorimetric determination of Hg^{2+} becomes difficult. The use of 2,6-pyridinedicarboxylic acid (PDCA) with MPA-gold can form a complex with Hg^{2+} ions and improve the selectivity for Hg^{2+} ions, masking the interference from the other metal ions. When Tween-20-stabilized gold nanoparticles in water containing Hg^{2+} and Ag^+ are reduced with ascorbic acid, the reduced Hg^{2+} forms an Hg–Au alloy and silver is deposited over the Tween-stabilized gold nanoparticles, forming a layer on the surface (Lou et al., 2011). The use of mercapto ligand at this stage cannot bind with the gold surface. Thus, the gold nanoparticles are not aggregated in solution and there is no change in the red color. However, in the absence of Hg^{2+} and Ag^+, mercapto ligand can bind with the gold surface via gold–sulfur bonds, resulting in self-assembled aggregation of gold nanoparticles giving a blue color. By this method, 5 nM concentration of Hg^{2+} and 10 nM concentration of Ag^+ can be detected in pure water. Pyrazole-functionalized gold nanorods show a distinct color change and a red-shift of the longitudinal plasmonic band in a solution containing 0.03 ppm of Hg^{2+} (Placido et al., 2013). Though effective, this method cannot selectively determine Hg^{2+} ions in water. 11-mercapto-undecyl-trimethyl ammonium (MTA) with quaternary ammonium (QA)-terminated thiols can bind strongly with gold nanoparticles in solution via gold-sulfur bond via ligand exchange (Liu et al., 2010a). In acidic conditions, QA-gold nanoparticles are stabilized and well dispersed in water because of the repulsion of cationic QA and H^+ giving a red color. However, when Hg^{2+} is present in the solution, thiols on the gold surface can bind with Hg^{2+}. This leads to the ligand abstraction from the gold surface and aggregation of nanoparticles with a simultaneous color change from red to blue (Figure 3.7) (Liu et al., 2010a).

This method, tested with other metal ions such as Al^{3+}, Ba^{2+}, Ca^{2+}, Cd^{2+}, Co^{2+}, Cr^{2+}, Cu^{2+}, Fe^{2+}, Fe^{3+}, Hg^{2+}, K^+, Mg^{2+}, Mn^{2+}, Na^+, Ni^{2+}, Pb^{2+}, and Zn^{2+}, shows that Hg^{2+} can selectively interact with QA–gold nanoparticles and destabilize the system by removing QA (Liu et al., 2010a). This method shows efficiency to detect 1.0 μM Hg^{2+} concentration from the absorption spectra. The detection limit is further decreased to 30 nM when the system is irradiated with solar light. The detection of Hg^{2+} with deoxyribonucleic acid (DNA)-functionalized gold nanoparticles relies on the thymidine–Hg^{2+}–thymidine (T–Hg–T) coordination between two complementary DNA strands with T-mismatch in

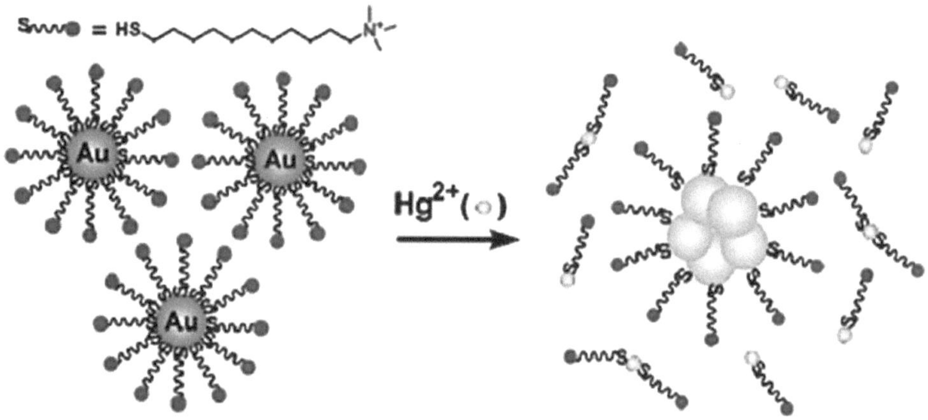

FIGURE 3.7 Detection of Hg^{2+} by functionalized gold nanoparticles. Quaternary ammonium group terminated with thiols stabilizes red-colored gold nanoparticles in water (left). Thiols abstraction by Hg^{2+} can result in particle aggregation and the color of the solution changes to blue (right). Reproduced with permission from Ref. (Liu et al., 2010a). Copyright American Chemical Society.

complementary strands (Lee et al., 2007). Hg^{2+} binds with two T residues in the duplex to form a T–Hg–T complex and results in the aggregation of functionalized gold nanoparticles. The aggregate has a higher melting temperature (T_m > 46 °C) with a color change from red to purple. T_m can reach a temperature of ~56 °C at 2 μM Hg^{2+}. In the absence of Hg^{2+}, melting of aggregates occurs at 46 °C. This method can selectively detect Hg^{2+} ions. Optimizing the amount of T residues in the T–T mismatch, it becomes possible to measure the colorimetric detection at room temperature (Xue et al., 2008). Figure 3.8 shows colorimetric Hg^{2+} detection by DNA/Au nanoparticles conjugate.

DNA-gold complex with mismatched T–T strand can effectively protect gold nanoparticles from salt-induced aggregation. The presence of Hg^{2+} ions can weaken this protection by forming a T–Hg–T complex leading to the aggregation of gold nanoparticles in the salt with an associated color change from red to blue (Xu et al., 2009). The use of small thymine (T) derivative in place of the oligonucleotide can make the colorimetric detection operate at ambient temperature (Lou et al., 2012). Citrate-stabilized gold nanoparticles attach with small thymine (T) molecules. In a Hg^{2+}-free condition, citrate-stabilized gold undergoes ligand exchange with thymine nitrogen, forming gold–nitrogen bonds, and resulting in the aggregation of nanoparticles with a color change from red to blue. The red color of gold nanoparticles remains the same in a solution containing Hg^{2+}. The thymine residues in the Hg–T complex cannot easily undergo ligand exchange with citrate-stabilized gold nanoparticles, resulting in no aggregation of nanoparticles (Lou et al., 2012). A peptide-modified gold nanoparticles probe can simultaneously detect Cd^{2+}, Co^{2+}, and Ni^{2+} ions in water (Zhang et al., 2012). These ions can bind with the peptide ligand of gold nanoparticles in solution and induce a color change from red–blue to purple due to the aggregation of gold nanoparticles. The detection limits for Cd^{2+}, Ni^{2+}, and Co^{2+} are 0.05 μM, 0.3 μM and 2 μM, respectively. The selective detection of a single ion can be possible if interference from the other ions is masked. Use of EDTA–imidazole can mask Co^{2+} and Ni^{2+} and detect only Cd^{2+}. Glutathione and EDTA can mask Cd^{2+} and Co^{2+} and detect only Ni^{2+}. Glutathione and imidazole can mask Cd^{2+} and Ni^{2+} and detect only Co^{2+} (Zhang et al., 2012). Gold nanoparticles, co-functionalized with 6-mercaptonicotinic acid and L-cysteine, show colorimetric response to Cd^{2+} ions to a level of 10^{-7} M (Xue et al., 2011). In an aqueous solution containing Pb^{2+}, glutathione-protected gold nanoparticles undergo aggregation and a color change from red to violet. A red-shift in the plasmon peak accompanies this color change (Chai et al., 2010). This method can selectively detect Pb^{2+} ions with a lower detection limit of 100 nM. A similar red-shift in the plasmonic band is reported while sensing Cu^{2+} ions and Pb^{2+} ions in a solution (Liu et al., 2011; Cai et al., 2014). Meso-2,3-dimercaptosuccinic acid (DMSA) with dithiol groups has a strong affinity to bind with As (+3, +5), and thus can detect these metal ions in solution at a concentration of ~1 ppb. As^{3+} binds strongly with the – SH functionalized gold nanoparticles along the length and side. As a result, a strong red-shift in the longitudinal band is monitored. As^{3+} binds strongly on the – SH functionalized edge-site along the rod resulting in a strong plasmonic shift (Priyadarshni et al., 2018).

The fluorescence-based sensing of metal ions using functionalized gold nanoparticles is due to the high molar extinction coefficient of the nanoparticles. Gold nanoparticles, when bound

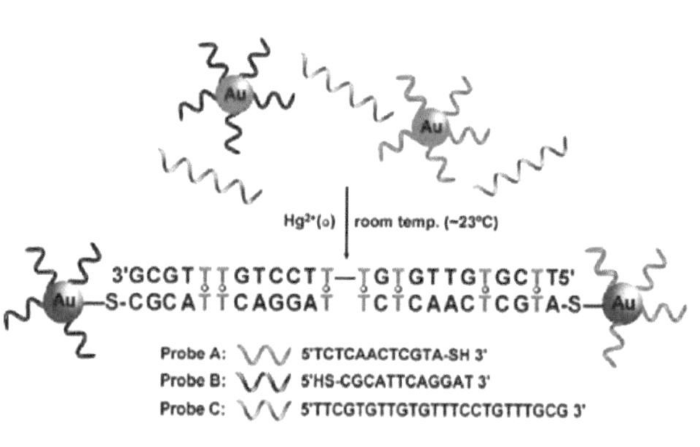

FIGURE 3.8 Oligonucleotide-functionalized gold nanoparticles and a linker oligonucleotide with thymine–thymine (T–T) mismatch are introduced with Hg^{2+} ions in aqueous solution. The high selectivity of thymine to Hg^{2+} forms T–Hg^{2+} –T base pairs and an aggregation of the nanoparticles showing colorimetric changes. Reproduced with permission from Ref. (Xue et al., 2008). Copyright American Chemical Society.

to fluorescent chromophore perylene bisimide, can reduce the fluorescent intensity of the latter by acting as emission quencher through a process called fluorescence resonance energy transfer (FRET), in which an energy transfer occurs from chromophore to gold nanoparticles (He et al., 2005). The chromophore, having pyridyl moieties, can bind weakly to gold nanoparticles via the gold–nitrogen bond. When Cu^{2+} is present, a strong coordination is established between Cu^{2+} and pyridyl, displacing gold nanoparticles from the coordination. This way, the fluorescence is again turned on. Rhodamine B (RhB) is a colored organic dye showing strong visible fluorescence. The fluorescent intensity quenches in contact with gold nanoparticles by the FRET mechanism. The fluorescence is again turned on by the coordination of Hg^{2+} on the surface of RhB–gold (Huang and Chang, 2006). The fluorescence of $[Ru(bpy)_3]^{2+}$ quenches in a solution containing N(-2-mercaptopropionyl) glycine (tiopronin)-coated gold nanoparticles (Huang and Murray, 2002). A complex is formed between the chromophore and the ligand by electrostatic interaction. The emission intensity is quenched when this complex is formed. This complex dissociates, and the emission intensity gradually restores when an electrolyte containing K^+, Bu_4N^+, and Ca^{2+} is added to the solution containing the complex. Bovine serum albumin (BSA)-stabilized gold nanocrystals have a deep-brown color and display a bright-red fluorescence on UV excitation (Liu et al., 2010b). In the presence of cyanide ions, gold–cyanide complex is formed, and the brown color becomes colorless. Under UV, blue fluorescence is observed. Selective detection of I^- and CN^- ions in highly saline water and edible salt is possible with BSA-labeled fluorescein isothiocyanate (FITC)-capped gold nanoparticles (Wei et al., 2012). FITC fluorescence is quenched in association with BSA–gold. When BSA–FITC-capped gold nanoparticles are present with I^- and CN^- ions, the anions attach on the gold surface and replace the FITC molecules. This way, the fluorescence of FITC is recovered and the presence of anions is detected. The use of $S_2O_8^{2-}/Pb^{2+}$ or H_2O_2 can enhance the selective detection of cyanide ions over other anions. Iodide ions detection can go down to a limit of 50 nM. Meso-(4-pyridinyl)-substituted BODIPY (4,4-difluro-4-bora-3a,4a-diaz a-s-indacene) coordinate on the surface of gold nanoparticles via a weak gold–nitrogen bond (Xu et al., 2016). Due to FRET, this ligand–metal complex shows a reduced fluorescence intensity. In the presence of thiols, the gold–nitrogen bond becomes weak, and the thiols attach over the surface, resulting in the detachment of the chromophore into the solution. The detached chromophore shows strong fluorescence intensity, and the intensity shows a direct correlation with the presence of thiols. By adopting this sensing method, the presence of thiols within living HeLa cells is detected.

Functionalized gold nanostructures displaying surface-enhanced Raman scattering (SERS) can be used for the sensing of metal ions in solution. SERS gives spectral fingerprint signature of analytes and experiences no signature of the presence of water. The SERS intensity enhancement depends on the size, shape, and charge transfer in NPs aggregates, and on the interparticle separation of nanoparticles (Krajczewski et al., 2017; Li et al., 2015; Li et al., 2013). 4-mercaptobenzoic acid (MBA)-capped gold nanoparticles are used as the SERS sensor for the detection of Hg^{2+} (Liu et al., 2014). The carboxylic group of MBA coordinates with Hg^{2+}, leading to the aggregation forming dimer and trimer. The interparticle plasmonic hot spot increases the electromagnetic field intensity and a fingerprint SERS peak at 374 cm^{-1} appears due to Hg^{2+} –Au–MBA complex formation. The presence of other metal ions, such as Pb^{2+}, As^{2+}, and Cd^{2+} also contributes to 374 cm^{-1} peaks. The addition of 2,6 pyridine dicarboxylic acid (PDCA) can mask the SERS contribution of the other metal ions and can selectively detect the presence of Hg^{2+} to a detection limit of 5 ppt. A molecular probe 2,4,6-trimercapto 1,3,5-triazine species, when bound to gold nanoparticles, forms tridentate coordination with thiol, and nitrogen binds on the surface of gold nanoparticles (Zamarion et al., 2008). The addition of NaCl helps in the aggregation of nanoparticles, and as a result, strong SERS signals are recorded. Upon Hg^{2+} addition to the thiol-protected gold nanoparticles, the SERS signal of C–S diminishes due to the introduction of a bidentate N–Hg–S bonding, and the signal from the triazine ring of the probe increases. The addition of Cd^{2+} intensifies the triazine ring vibrational signal due to the strong Cd–N bonding with ring nitrogen.

3.4.2 Biosensing

The colorimetric detection of oligonucleotides has been possible because of the aggregation of the gold nanoparticles in the presence of an oligonucleotide with an associated color transformation. Using DNA-functionalized gold nanoparticles, a colorimetric screening of triplex DNA binders has been possible (Han et al., 2006). Two non-complimentary pyrimidine-rich thiol-modified DNA strands attach on the gold nanoparticles and a free DNA strand links up with them to form a triplex structure. Because of the low stability of the triplex structure, the oligonucleotide-functionalized gold nanoparticles do not aggregate at room temperature. The use of a triplex specific binder (benzo [e] pyridoinde) can induce reversible aggregation of the functionalized gold nanoparticles with a color change from red to blue and a red-shift in the plasmonic peak (Han et al., 2006). Single-stranded DNA (ss-DNA) and double-stranded DNA (ds-DNA) undergo different electrostatic interactions with citrate-stabilized gold nanoparticles (Li and Rothberg, 2004). In this assembly, ss-DNA is observed to have better interaction with gold nanoparticles due to the partial uncoiling of the bases exposed to the gold surface. Duplex DNA does not uncoil to expose bases to gold nanoparticles. When NaCl salt is added to the ss-DNA-stabilized gold nanoparticles, no aggregation of gold nanoparticles occurs, and the color remains pink. On the other hand, ds-DNA does not secure the gold nanoparticles against aggregation with a resulting color change from pink to blue (Li and Rothberg, 2004). Non-cross linking of a probe and a complementary target DNA strand over the surface of gold nanoparticles, in a 0.5 M salt, results in rapid aggregation of nanoparticles and color transformation to purple (Sato et al., 2003). In the absence of target DNA, probe ss-DNA-coated gold nanoparticles do not undergo any aggregation, and the color remains stable.

An interparticle distance-dependent colorimetric change can provide sensing of an oligonucleotide in an ss-DNA–gold aggregate. Mercaptoalkyloligo nucleotide-modified gold nanoparticles (~13 nm) are used as a probe molecule to target ss-DNA (Elghanian et al., 1997). This interaction results in an extended polymeric network formed by multiple duplex DNA linkages. If the interparticle distance is substantially larger than the average particle diameter, the color of the aggregate appears red. Interparticle distance less than the average particle diameter results in a color change from red to blue. Peptide nucleic acid (PNA) attached to citrate-stabilized gold nanoparticles undergoes self-assembly by PNA–PNA interaction and acts as a probe for sensing target DNA within a time shorter than 10 min (Chakrabarti and Klibanov, 2003). At a salt concentration less than 40 mM, PNA–PNA hybridization is effective. The PNA–gold conjugate, when attached with the target DNA with matching hybridization, remains stable at a Na^+ concentration above 200 mM. The plasmonic shift is higher in PNA–DNA hybridized gold nanoparticles than in PNA–PNA conjugate systems.

Aptamer-based colorimetric sensing of protein using gold nanoparticles involves conformational modification of aptamers binding over gold surfaces and removal of the excess aptamer from the system. Unmodified aptamer–gold nanoparticles can probe colorimetric sensing of complex protein molecules. Interaction of thrombin with its ss-DNA aptamer results in a folded structure formation; a G-quadruplex/duplex (Wei et al., 2007). The coating of unfolded aptamer ss-DNA onto the gold surface stabilizes its aggregation and the complex attains a red color. However, when ss-DNA–gold interacts with thrombin, a G-quadruplex DNA strand forms. Because of the ligand structure, the quadruplex cannot properly bind gold nanoparticles. Thus, at a high salt concentration, aggregation occurs and the color changes to blue (Wei et al., 2007). Gold NPs-padlock ss-DNA conjugates can probe alpha-fetoprotein with a deduction limit of 33.45 $pmml^{-1}$. The conjugate in the presence of protein results in a color change from red to purple (Chen et al., 2015). Click chemistry technique can provide a quantitative estimation of bovine serum albumin in milk products using azide and alkyne-functionalized gold nanoparticles (Zhu et al., 2012). When Cyt-c protein binds to gold nanoparticles and gold film, the protein–gold aggregate shows a color change. The color change is associated with a conformational change of the protein at different pH. The protein unfolds at low pH and refolds at a higher pH (Chah et al., 2005). Colorimetric sensing of antibodies in serum samples is based on the use of protein-coated gold nanoparticles with an anti-protein molecule. The aggregation leads to a color change from red to purple (Thanh and Rosenzweig, 2002). The nanoparticle aggregation gets stronger at pH ~ 7 and favors 20 °C to 37 °C.

SERS-based biosensors can be developed utilizing the plasmonic properties of gold nanoparticles. For biosensing using SERS, the gold nanoparticles are functionalized with a Raman-sensitive dye. Oligonucleotides are labeled with Raman dye to distinguish oligonucleotide sequences. This Raman-sensitive dye labeled oligonucleotide is functionalized with gold nanoparticles of average size 13 nm. A three-component sandwich assay is used in a microarray format. Gold nanoparticles are capped with a Cys-3 oligonucleotide strand containing alky thiol group to probe target DNA (Cao et al., 2002). For the detection of the SERS signal, the Raman dye-tagged oligonucleotide–gold nanoparticles complex is inserted in a solution containing Ag^+. The silver nanoparticles grow around the nanoparticles Cys 3-labelled probe leading to SERS enhancement. A gold nanoparticle-on-wire structure constructed by the self-assembly of gold nanoparticles onto gold nanowires through hybridization with DNA can act as a SERS active probe molecule for sensing pathogenic DNA (Kang et al., 2010). Gold nanowires on Si substrate undergo modification with thiolated-probe DNAs. The gold nanowires–DNA substrate are modified with reporter DNA and incubated with a target DNA. The gold nanowires with a probe and target DNA are immersed into gold nanoparticles modified with Raman dye-labeled reporter DNA. The SERS intensity from the dye-labeled probe increases with the concentration of target DNA. No SERS signal is observed in the absence of target DNA. This method can detect a DNA concentration of 10 pM (Kang et al., 2010).

A SERS-based probe with non-fluorescent Raman tags can efficiently determine DNA markers (Sun et al., 2007). The probe is designed by tagging non-fluorescent Raman dyes over DNA-functionalized gold nanoparticles. The SERS intensity of the probe molecule can be controlled by the surface coverage of the Raman tags on gold nanoparticles. The sensitivity of the detection is optimized by controlling the concentration of DNA molecules and Raman tags (Sun et al., 2007). SERS can be enhanced by long-range plasmonic coupling for the detection of the oligonucleotide clustered reception on cell membranes (Qian et al., 2008). A surface-enhanced Raman molecular beacon is constructed by tagging gold nanoparticles with a reporter molecule (malachite green) and then functionalized with a thiolated-DNA probe. The signals from SERS beacons can be turned on and off by bimolecular binding and dissociation events (Qian et al., 2008).

3.4.3 Catalysis

Gold nanoparticles act as an active catalyst for carbon monoxide oxidation. The catalytic activity is evaluated as an unsupported catalyst or with metal oxides as support. Bulk gold is not as effective as a catalyst. The importance of gold nanoparticles supported on a metal oxide in the oxidation of carbon monoxide is reported by Haruta et al. (1987). The various metal oxide supports are MgO, Al_2O_3, SiO_2, TiO_2, ZnO, and Fe_2O_3. Adsorption of carbon monoxide on gold catalyst forms carbonate, bicarbonate, and carboxylate in the presence of adsorbed water molecules (Tseng et al., 2009). The carbon monoxide adsorption over a monolayer of gold islands on FeO (111) (0.01 nm thickness) is different than that over bulk gold. If the thickness increases to 0.2 nm, the deposited gold nanoparticles (7 nm) form 4–5 numbers of gold layers (Lemire et al., 2003). Temperature-programmed desorption (TPD) data shows that the carbon monoxide desorption on gold film occurs from different coordination sites such as a terrace and edge. The layer thickness does not accelerate carbon monoxide desorption, but provides different coordination sites for the gas adsorption (Lemire et al., 2003). The extent of carbon monoxide adsorption varies on various facets of the gold surface. Preferential carbon monoxide adsorption occurs on the gold (111) facet as well as on the undercoordinated sites. Low-temperature infrared reflection spectra at 30 K show that carbon monoxide frequency between 2128 and 2132 cm^{-1} is due to carbon monoxide adsorbed on the gold (111) terrace, whereas the frequency 2110–2125 cm^{-1} is due to the carbon monoxide adsorbed on the undercoordinated sites (Pischel and Pucci, 2015). Studies have demonstrated that the size of gold nanopowder has a large impact on the carbon monoxide oxidation reaction. Carbon monoxide to carbon dioxide oxidation occurs efficiently over 70 nm gold nanoparticles (Mikami et al., 2013; Ketchie et al., 2007).

However, there is a considerable decrease in carbon monoxide oxidation activity if the size of the gold particles is in the range of 500–800 nm. As compared to perfect gold (110) crystal, powdered gold has large numbers of catalytically active sites on the terrace and edges (Mikami et al., 2013). Therefore, enhanced carbon monoxide oxidation occurs over powdered gold.

Gold nanotube is another efficient catalyst for carbon monoxide oxidation. Template-mediated electroless deposition of gold nanotube occurs on a 10 μm-thick track-etched polycarbonate membrane containing 200 nm-diameter pores (Sanchez-Castillo Marco et al., 2004). Partial etching of 2.3 μm size of the polycarbonate membrane exposes gold nanotubes on one side of the membrane for interaction with a gas stream containing carbon monoxide and oxygen. The other side of the membrane comprising gold nanotube is allowed to interact in a basic aqueous solution maintained at different pH. Treatment with basic water generates a negatively charged hydroxyl group formed by the heterolytic dissociation of water. The hydroxyl group reacts with carbon monoxide, converting it to carbon dioxide (Sanchez-Castillo Marco et al., 2004). The presence of water in the reaction medium initially converts carbon monoxide to a carboxylic acid; the latter then activates molecular oxygen for promoting carbon monoxide oxidation. The detailed reaction pathways in CO oxidation are shown in Figure 3.9 (Sun et al., 2018).

The kinetic study reveals that carbon monoxide oxidation is many times faster with water than without (Sun et al., 2018). During carbon monoxide conversion on the unsupported gold surface, there is a competing interaction between carbon monoxide and oxygen adsorption, leading to a weaker interaction of the adsorbates with gold (Haruta, 2007). Carbon monoxide adsorption can take place over low coordination sites, but oxygen does not adsorb on a pure gold crystal. It is noticed that the presence of a minute amount of silver facilitates binary gold–silver bimetallic formation and enhances oxygen adsorption without any dependence on the particle size (Sanchez-Castillo Marco et al., 2004; Wittstock et al., 2009).

Gold catalyst supports C–C coupling reaction between acids and unsaturated ketones. 1.4 addition of arylboronic acids to 2 cyclohexenone in toluene is achieved over TiO_2-supported gold catalyst (Moragues et al., 2015). The hydrogenation of alkynes to alkenes on gold catalyst could be facilitated if a nitrogen-containing base (e.g., piperazine) preadsorbed on the gold surface (Fiorio et al., 2017). This process results in the formation of a frustrated Lewis pair interface, which provides heterolytic hydrogen activation on the gold surface. Adsorption of alkynes on the gold surface favors hydrogenation of alkynes, abstraction of H^+ from piperazine, and desorption of alkene from the metal surface (Fiorio et al., 2017). An unsupported gold catalyst can also serve as the site for hydrogen dissociation and selective hydrogenation of C ≡ C, C = C, C = N, C = O (Takale et al., 2016). Gold–silver alloy favors stronger hydrogen dissociation. Gold catalyst shows high chemoselectivity in the reduction of terminal alkynes and reduction of aldehydes to alcohols with excellent product recovery.

Gold nanoparticles can also catalyze alcohol oxidation. Aerobic oxidation of unsupported 1-phenyl butanol to ketone takes place on nanoporous gold in methanol. The nanoporous gold is a stable catalyst and can be reused with high catalytic efficiency. The ketone yield is 96 % for both the fresh and the reused one (Asao et al., 2012). Methanol oxidation is studied in neutral water, in basic water, and in the gas phase. Methanol interacts with the preadsorbed oxygen on the catalyst surface via the –OH group present on methanol (Muñoz-Santiburcio et al., 2018). There is a sequence of steps involving methanol oxidation to a ketone. The steps include: oxygen dissociation on catalyst promoting methanol oxidation; hydroxyl hydrogen addition to one of the dissociated oxygens on the catalyst surface; and interaction of aliphatic hydrogen of methanol with gold nanoparticles. The reaction kinetics depends on O_2 dissociation on the catalyst surface; it is different in neutral, basic, and gas phases. O_2 dissociation is faster in basic water than in neutral water. In the liquid phase, the total charge on the gold is very high. This is because of the higher solvation of gold surface resulting in O_2 activation and dissociation. The methoxy intermediate accelerates the catalytic process of methanol oxidation to formaldehyde (Muñoz-Santiburcio et al., 2018).

3.4.3.1 Plasmonic Photocatalysis

By their strong visible-light absorbing capacity, plasmonic nanostructures can concentrate the incident light energy on nanoscale confinement to drive photochemical reactions. There are two

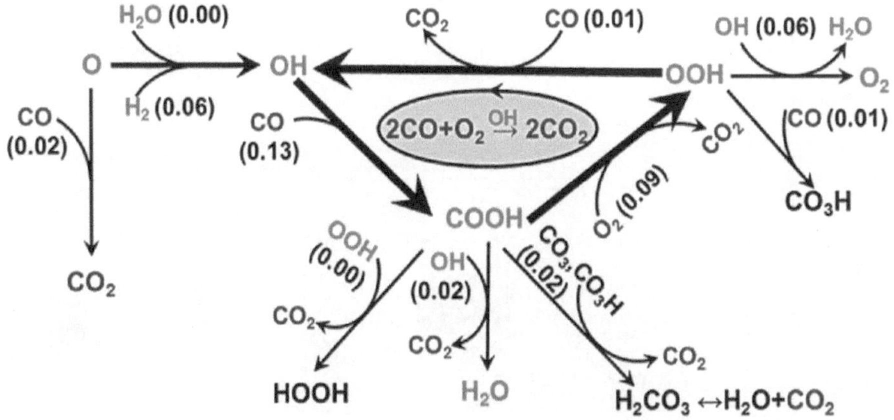

FIGURE 3.9 Reaction mechanism of CO oxidation on an Au (100) surface with the help of water or H_2. The numbers in parentheses show the corresponding reaction barriers in units of eV. Reproduced with permission from Ref. (Sun et al., 2018). Copyright American Chemical Society.

ways by which the solar-to-chemical energy conversion takes place on gold nanostructures (Kale et al., 2014):

(1) Direct photocatalysis, in which the plasmonic nanostructure itself act as the catalytic site.
(2) Indirect photocatalysis, in which the plasmonic energy concentrated on the metal nanostructures is utilized by the nearest surface-adsorbed molecules or attached semiconductors to perform the photocatalysis.

In this section, the emphasis will be on the plasmonic photocatalysis exhibited by unsupported gold nanostructures. Plasmonic gold nanostructures can be used for water splitting, carbon dioxide reduction, dye degradation, and several other applications. By harvesting visible light, gold nanostructures can induce efficient carbon dioxide conversion to ethane and methane via a multi-electron multi-proton involvement during the reduction (Yu et al., 2018). Methane production is recorded at 488 nm and 532 nm excitation. Methane production is higher at 488 nm than at 532 nm. The production of ethane is recorded only at 488 nm. It is speculated that the higher carbon dioxide reduction at 488 nm is due to the interband transition in gold with plasmon-excited electrons in the sp band and the holes in the d-band with an electron-hole separation lifetime of 1 ps. Photoexcitation at 532 nm favors intraband transition, with excited electrons and holes lying within the same sp band. Because of their small carrier lifetime (400 fs), the carbon dioxide reduction is relatively inefficient under 532 nm plasmonic excitation (Yu et al., 2018).

Plasmonic excitation generates low-density hot electrons. However, interband *d-sp* excitation forms a high density of hot-hole carriers. Gold nanorods are easily etched away by a solution of ferric chloride ($FeCl_3$) with the formation of cationic gold species (Zhao et al., 2017). Gold nanorods immersed in ferric chloride solution are subjected to interband as well as surface plasmon excitation. The etching of gold tips and the formation of cationic gold is 3–4 orders of magnitude higher with interband excitation than at plasmonic excitation. Under interband excitation, the adsorbed photons generate hot electrons which reduce Fe^{3+} on the surface of gold nanorods and generate Fe^{2+}. This leaves gold (3+) charge on the gold nanorods. Gold (3+) ions further react with gold nanorods and produce gold (+). This process is much slower under plasmonic excitation (Zhao et al., 2017).

Similarly, the cyclisation reaction of 2-(phenyl ethynyl) phenol under interband excitation at 390–420 nm gives 99 % yield of 2-phenyl benzofuran (Zhao et al., 2017). This yield is much lesser under plasmonic excitation. Octahedral gold nanoparticles show superior photocatalytic oxygen evolution reaction under UV excitation, which corresponds to the interband transition. Plasmon excitation triggers sp intraband transition with gold (+) near the *sp*-occupied states. These *sp* holes cannot perform the oxygen evolution reaction because of the potential barrier. Interband transition promotes *d-sp* transition and forms holes in the d-states. Hole transfer from 5d to oxygen evolution reaction potential is energetically favored since there is no reaction potential barrier involves (Zhao et al., 2017).

The gold nanocrystals show a reduction peak at −0.63 V. Two oxidation peaks appear at +0.97 V and +1.51 V vs. reversible hydrogen electrodes. Thus, gold exhibits the ability to induce hydrogen generation by water splitting under visible light (Abbas et al., 2018). A photoelectrolytic cell is constructed for studying water splitting with gold photoelectrode. The electrode is connected with a hole transporting layer for efficient carrier separation (Robatjazi et al., 2015). Strong light absorption and hot electron injection result in the photolysis of water. The study remarks that hydrogen generation is efficient when excited at the plasmonic band and weak at interband transition. Plasmonic hot electrons can be efficiently utilized for the photocatalytic dissociation of hydrogen over a gold surface (Mukherjee et al., 2013). Hot electron energy is transferred to hydrogen (H_2) and forms a transition state. The excited H_2^- state releases the energy to gold and finally undergoes dissociation to hydrogen atoms. This process is shown in Figure 3.10 (Mukherjee et al., 2013).

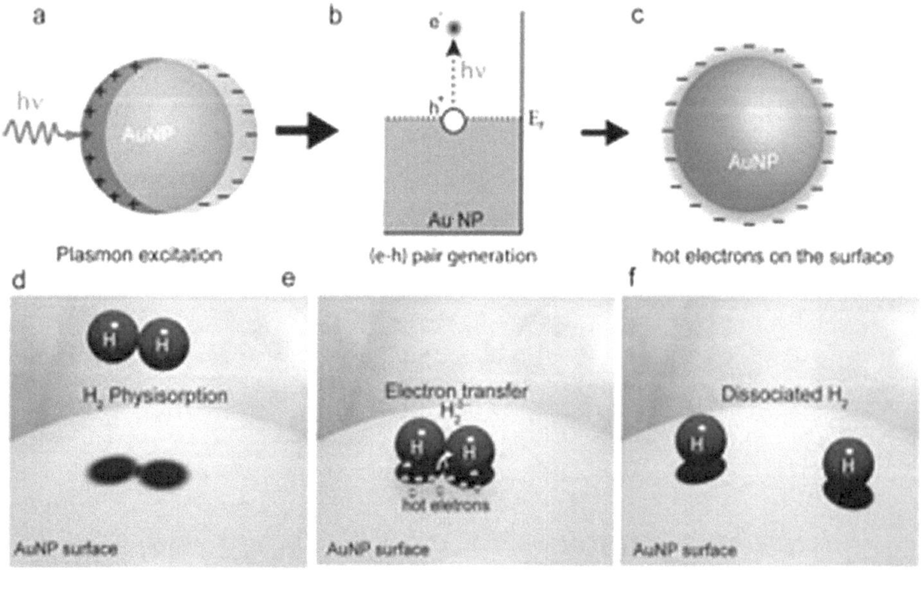

FIGURE 3.10 Schematics of plasmon-induced hot electron generation on Au nanoparticles and mechanistic representation of H_2 dissociation on the Au nanoparticle surface. Reproduced with permission from Ref. (Mukherjee et al., 2013). Copyright American Chemical Society.

It is reported that the presence of protecting ligand can help improve the photocatalytic activity of gold. For example, thiol-stabilized gold is reported to show superior activity in hydrogen generation (Abbas et al., 2018). Similarly, CTAB coated over gold nanorods triggers photoreduction of p-aminothiophenol to p-nitrothiophenol (Kafle et al., 2017). In order to monitor the plasmonic hot spots and hot electron-mediated photochemical reactions, combined instrumental techniques of single-molecule super-resolution fluorescence microscopy with scanning electron microscopy have been employed (Zou et al., 2018). Photocatalytic reduction of resazurin to fluorescent resorufin by NH_2OH is observed over a gold nanorod dimer. The interparticle spacing of the dimmer is ~ 8.8 nm. Catalytic reduction is stronger at the interparticle gap or at the plasmonic hot spot. At the hot spot, strong electromagnetic field enhancement occurs with the generation of sufficient hot electrons. These hot electrons perform the reduction reaction to resorufin.

Conclusion

Functionalized gold nanostructures of various forms are fabricated using a wide variety of techniques. Shape transformation from sphere to other anisotropic structures can be achieved by changing the precursor concentration, temperature, solvents, pH of the medium, capping agents, and different reducing agents. It is the type of functional groups over gold that sometimes determines the usefulness of the functionalized gold for sensing, biomedical, and photocatalysis applications. Thiol-stabilized gold nanostructures are highly effective in chemical sensing. Similarly, a biopolymer such as DNA or RNA-functionalized gold nanostructures can effectively sense inorganic ions as well as biomolecules. These functionalized gold nanostructures provide reactive sites for performing gas-phase carbon monoxide oxidation, hydrogen dissociation, alcohol oxidation, and hydrogenation reactions. It is mentioned that thiol-stabilized gold nanostructures show superior hydrogen evolution activity under visible light. This plasmonic photocatalytic activity is sufficient for carbon dioxide reduction, hydrogen dissociation, water treatment, and water splitting reaction.

REFERENCES

Abbas, M. A., P. V. Kamat, and J. H. Bang. 2018. Thiolated gold nanoclusters for light energy conversion. *ACS Energy Lett* 3:840–54.

Alvarez, M. M., J. T. Khoury, T. G. Schaaff, et al. 1997. Critical sizes in the growth of Au clusters. *Chem Phys Lett* 266:91–98.

Asao, N., N. Hatakeyama, Menggenbateer, et al. 2012. Aerobic oxidation of alcohols in the liquid phase with nanoporous gold catalysts. *Chem Commun* 48:4540–42.

Bakshi, M. S., F. Possmayer, and N. O. Petersen. 2008. Aqueous-phase room-temperature synthesis of gold nanoribbons: Soft template effect of a gemini surfactant. *J Phys Chem C* 112:8259–65.

Biggs, S., M. K. Chow, C. F. Zukoski, et al. 1993. The role of colloidal stability in the formation of gold sols. *J Colloid Interface Sci* 160:511–13.

Biroju, R. K., B. Choudhury, and P. K. Giri. 2016. Plasmon-enhanced strong visible light photocatalysis by defect engineered CVD graphene and graphene oxide physically functionalized with Au nanoparticles. *Catal Sci Technol* 6:7101–12.

Booth, S. G., A. Uehara, S. Y. Chang, et al. 2017. The significance of bromide in the Brust–Schiffrin synthesis of thiol protected gold nanoparticles. *Chem Sci* 8:7954–62.

Bratescu, M. A., S.-P. Cho, O. Takai, et al. 2011. Size-controlled gold nanoparticles synthesized in solution plasma. *J Phys Chem C* 115:24569–76.

Brust, M., J. Fink, D. Bethell, et al. 1995. Synthesis and reactions of functionalised gold nanoparticles. *J Chem Soc, Chem Commun* 0:1655–56.

Busbee, B. D., S. O. Obare, and C. J. Murphy. 2003. An improved synthesis of high-aspect-ratio gold nanorods. *Adv Mater* 15:414–16.

Cai, H.-H., D. Lin, J. Wang, et al. 2014. Controlled side-by-side assembly of gold nanorods: A strategy for lead detection. *Sens Actuators B Chem* 196:252–59.

Cao, Y. C., R. Jin, and C. A. Mirkin. 2002. Nanoparticles with Raman spectroscopic fingerprints for DNA and RNA detection. *Science* 297:1536.

Chah, S., M. R. Hammond, and R. N. Zare. 2005. Gold nanoparticles as a colorimetric sensor for protein conformational changes. *Chem Biol* 12:323–28.

Chai, F., C. Wang, T. Wang, et al. 2010. Colorimetric detection of Pb^{2+} using glutathione functionalized gold nanoparticles. *ACS Appl Mater Interfaces* 2:1466–70.

Chakrabarti, R., and A. M. Klibanov. 2003. Nanocrystals modified with peptide nucleic acids (PNAs) for selective self-assembly and DNA detection. *J Am Chem Soc* 125:12531–40.

Chang, C.-C., H.-L. Wu, C.-H. Kuo, et al. 2008. Hydrothermal synthesis of monodispersed octahedral gold nanocrystals with five different size ranges and their self-assembled structures. *Chem Mater* 20:7570–74.

Chang, S.-S., C.-W. Shih, C.-D. Chen, et al. 1999. The shape transition of gold nanorods. *Langmuir* 15:701–09.

Chen, C., M. Luo, T. Ye, et al. 2015. Sensitive colorimetric detection of protein by gold nanoparticles and rolling circle amplification. *Analyst* 140:4515–20.

Corbierre, M. K., J. Beerens, and R. B. Lennox. 2005. Gold nanoparticles generated by electron beam lithography of gold(I)–thiolate thin films. *Chem Mater* 17:5774–79.

Cross, C. E., J. C. Hemminger, and R. M. Penner. 2007. Physical vapor deposition of one-dimensional nanoparticle arrays on graphite: Seeding the electrodeposition of gold nanowires. *Langmuir* 23:10372–79.

Daniel, M.-C., and D. Astruc. 2004. Gold nanoparticles: Assembly, supramolecular chemistry, quantum-size-related properties, and applications toward biology, catalysis, and nanotechnology. *Chem Rev* 104:293–346.

Duan, G., W. Cai, Y. Luo, et al. 2006. Electrochemically induced flowerlike gold nanoarchitectures and their strong surface-enhanced Raman scattering effect. *Appl Phys Lett* 89: 211905.

Duong, Q., Y. Tan, J. Corey, et al. 2015. Mechanism of the transfer of $AuCl_4$– and TOA^+ ions across the liquid/liquid interface. *J Phys Chem C* 119:10365–69.

Elghanian, R., J. J. Storhoff, R. C. Mucic, et al. 1997. Selective colorimetric detection of polynucleotides based on the distance-dependent optical properties of gold nanoparticles. *Science* 277:1078.

Esumi, K., K. Matsuhisa, and K. Torigoe. 1995. Preparation of rod-like gold particles by UV irradiation using cationic micelles as a template. *Langmuir* 11:3285–87.

Faraday, M. 1857. The Bakerian lecture: Experimental relations of gold (and other metals) to light. *Phil Trans R Soc* 147:145–81.

Fiorio, J. L., N. López, and L. M. Rossi. 2017. Gold–ligand-catalyzed selective hydrogenation of alkynes into cis-alkenes via H_2 heterolytic activation by frustrated Lewis pairs. *ACS Catal* 7:2973–80.

Freestone, I., N. Meeks, M. Sax, et al. 2007. The Lycurgus cup – A Roman nanotechnology. *Gold Bull* 40:270–77.

Frens, G. 1973. Controlled nucleation for the regulation of the particle size in monodisperse gold suspensions. *Nat Phys Sci* 241:20.

Gann, J. P., and M. Yan. 2008. A versatile method for grafting polymers on nanoparticles. *Langmuir* 24:5319–23.

Garg, N., C. Scholl, A. Mohanty, et al. 2010. The role of bromide ions in seeding growth of Au nanorods. *Langmuir* 26:10271–76.

Ghobadi Turkan Gamze, U., A. Ghobadi, E. Ozbay, et al. 2017. Strategies for plasmonic hot-electron-driven photoelectrochemical water splitting. *ChemPhotoChem* 2:161–82.

Gole, A., and C. J. Murphy. 2004. Seed-mediated synthesis of gold nanorods: Role of the size and nature of the seed. *Chem Mater* 16:3633–40.

Goulet, P. J. G., and R. B. Lennox. 2010. New insights into Brust–Schiffrin metal nanoparticle synthesis. *J Am Chem Soc* 132:9582–84.

Guo, S., L. Wang, and E. Wang. 2007. Templateless, surfactantless, simple electrochemical route to rapid synthesis of diameter-controlled 3D flowerlike gold microstructure with "clean" surface. *Chem Commun* 0:3163–65.

Ha, T. H., H.-J. Koo, and B. H. Chung. 2007. Shape-controlled syntheses of gold nanoprisms and nanorods influenced by specific adsorption of halide ions. *J Phys Chem C* 111:1123–30.

Halder, A., and N. Ravishankar. 2007. Ultrafine single-crystalline gold nanowire arrays by oriented attachment. *Adv Mater* 19:1854–58.

Han, M. S., A. K. R. Lytton-Jean, and C. A. Mirkin. 2006. A gold nanoparticle-based approach for screening triplex DNA binders. *J Am Chem Soc* 128:4954–55.

Haruta, M. 2007. New generation of gold catalysts: Nanoporous foams and tubes – is unsupported gold catalytically active? *ChemPhysChem* 8:1911–13.

Haruta, M., T. Kobayashi, H. Sano, et al. 1987. Novel gold catalysts for the oxidation of carbon monoxide at a temperature far below 0° C. *Chem Lett* 16:405–08.

Hatakeyama, Y., T. Morita, S. Takahashi, et al. 2011b. Synthesis of gold nanoparticles in liquid polyethylene glycol by sputter deposition and temperature effects on their size and shape. *J Phys Chem C* 115:3279–85.

Hatakeyama, Y., K. Onishi, and K. Nishikawa. 2011a. Effects of sputtering conditions on formation of gold nanoparticles in sputter deposition technique. *RSC Adv* 1:1815–21.

He, X., H. Liu, Y. Li, et al. 2005. Gold nanoparticle-based fluorometric and colorimetric sensing of copper(II) ions. *Adv Mater* 17:2811–15.

Heo, K., C. Miesch, T. Emrick, et al. 2013. Thermally reversible aggregation of gold nanoparticles in polymer nanocomposites through hydrogen bonding. *Nano Lett* 13:5297–302.

Hong, S., K. L. Shuford, and S. Park. 2011. Shape transformation of gold nanoplates and their surface plasmon characterization: Triangular to hexagonal nanoplates. *Chem Mater* 23:2011–13.

Hu, X., S.-P. Cho, O. Takai, et al. 2012. Rapid synthesis and structural characterization of well-defined gold clusters by solution plasma sputtering. *Cryst Growth Des* 12:119–23.

Hu, X. L., O. Takai, and N. Saito. 2013. Synthesis of gold nanoparticles by solution plasma sputtering in various solvents. *J Phys Conf Ser* 417:012030.

Huang, C.-C., and H.-T. Chang. 2006. Selective gold-nanoparticle-based "turn-on" fluorescent sensors for detection of mercury(II) in aqueous solution. *Anal Chem* 78:8332–38.

Huang, C.-C., and H.-T. Chang. 2007. Parameters for selective colorimetric sensing of mercury(ii) in aqueous solutions using mercaptopropionic acid-modified gold nanoparticles. *Chem Commun* 1215–17.

Huang, S., H. Ma, X. Zhang, et al. 2005. Electrochemical synthesis of gold nanocrystals and their 1D and 2D organization. *J Phys Chem B* 109:19823–30.

Huang, T., and R. W. Murray. 2002. Quenching of $[Ru(bpy)3]^{2+}$ fluorescence by binding to Au nanoparticles. *Langmuir* 18:7077–81.

Huo, Z., C.-k. Tsung, W. Huang, et al. 2008. Sub-two nanometer single crystal Au nanowires. *Nano Lett* 8:2041–44.

Ishida, Y., T. Sumi, and T. Yonezawa. 2015. Sputtering synthesis and optical investigation of octadecanethiol-protected fluorescent Au nanoparticles. *New J Chem* 39:5895–97.

Jana, N. R., L. Gearheart, and C. J. Murphy. 2001a. Evidence for seed-mediated nucleation in the chemical reduction of gold salts to gold nanoparticles. *Chem Mater* 13:2313–22.

Jana, N. R., L. Gearheart, and C. J. Murphy. 2001b. Wet chemical synthesis of high aspect ratio cylindrical gold nanorods. *J Phys Chem B* 105:4065–67.

Ji, X., X. Song, J. Li, et al. 2007. Size control of gold nanocrystals in citrate reduction: The third role of citrate. *J Am Chem Soc* 129:13939–48.

Johnson, C. J., E. Dujardin, S. A. Davis, et al. 2002. Growth and form of gold nanorods prepared by seed-mediated, surfactant-directed synthesis. *J Mater Chem* 12:1765–70.

Jones Matthew, R., and A. Mirkin Chad. 2013. Bypassing the limitations of classical chemical purification with DNA-programmable nanoparticle recrystallization. *Angew Chem Int Ed* 52:2886–91.

Kafle, B., M. Poveda, and T. G. Habteyes. 2017. Surface ligand-mediated plasmon-driven photochemical reactions. *J Phys Chem Lett* 8:890–94.

Kale, M. J., T. Avanesian, and P. Christopher. 2014. Direct photocatalysis by plasmonic nanostructures. *ACS Catal* 4:116–28.

Kang, T., S. M. Yoo, I. Yoon, et al. 2010. Patterned multiplex pathogen DNA detection by Au particle-on-wire SERS sensor. *Nano Lett* 10:1189–93.

Ketchie, W. C., Y.-L. Fang, M. S. Wong, et al. 2007. Influence of gold particle size on the aqueous-phase oxidation of carbon monoxide and glycerol. *J Catal* 250:94–101.

Kim, F., K. Sohn, J. Wu, et al. 2008. Chemical synthesis of gold nanowires in acidic solutions. *J Am Chem Soc* 130:14442–43.

Kim, F., J. H. Song, and P. Yang. 2002. Photochemical synthesis of gold nanorods. *J Am Chem Soc* 124:14316–17.

Kim, M., M. Lin, J. Son, et al. 2017. Hot-electron-mediated photochemical reactions: Principles, recent advances, and challenges. *Adv Opt Mater* 5:1700004.

Kimling, J., M. Maier, B. Okenve, et al. 2006. Turkevich method for gold nanoparticle synthesis revisited. *J Phys Chem B* 110:15700–07.

Krajczewski, J., K. Kołątaj, and A. Kudelski. 2017. Plasmonic nanoparticles in chemical analysis. *RSC Adv* 7:17559–76.
Kumar, S., K. S. Gandhi, and R. Kumar. 2007. Modeling of formation of gold nanoparticles by citrate method. *Ind Eng Chem Res* 46:3128–36.
Kura, H., and T. Ogawa. 2010. Synthesis and growth mechanism of long ultrafine gold nanowires with uniform diameter. *J Appl Phys* 107:074310.
Kuttner, C., M. Mayer, M. Dulle, et al. 2018. Seeded growth synthesis of gold nanotriangles: Size control, SAXS analysis, and SERS performance. *ACS Appl Mater Interfaces* 10:11152–63.
Lee, J.-S., M. S. Han, and A. Mirkin Chad. 2007. Colorimetric detection of mercuric ion (Hg^{2+}) in aqueous media using DNA-functionalized gold nanoparticles. *Angew Chem Int Ed* 46:4093–96.
Lemire, C., R. Meyer, S. Shaikhutdinov, et al. 2003. Do quantum size effects control co adsorption on gold nanoparticles? *Angew Chem Int Ed* 43:118–21.
Li, C., D. Li, G. Wan, et al. 2011. Facile synthesis of concentrated gold nanoparticles with low size-distribution in water: Temperature and pH controls. *Nanoscale Res Lett* 6:440.
Li, C., K. L. Shuford, Q. H. Park, et al. 2007. High-yield synthesis of single-crystalline gold nano-octahedra. *Angew Chem Int Ed* 46:3264–68.
Li, H., and L. Rothberg. 2004. Colorimetric detection of DNA sequences based on electrostatic interactions with unmodified gold nanoparticles. *Proc Natl Acad Sci USA* 101:14036.
Li, M., S. K. Cushing, and N. Wu. 2015. Plasmon-enhanced optical sensors: A review. *Analyst* 140:386–406.
Li, M., H. Gou, I. Al-Ogaidi, et al. 2013. Nanostructured sensors for detection of heavy metals: A review. *ACS Sustain Chem Eng* 1:713–23.
Li, Y., O. Zaluzhna, and Y. J. Tong. 2011a. Identification of a source of size polydispersity and its solution in Brust–Schiffrin metal nanoparticle synthesis. *Chem Commun* 47:6033–35.
Li, Y., O. Zaluzhna, and Y. J. Tong. 2011b. Critical role of water and the structure of inverse micelles in the Brust–Schiffrin synthesis of metal nanoparticles. *Langmuir* 27:7366–70.
Li, Y., O. Zaluzhna, B. Xu, et al. 2011c. Mechanistic insights into the Brust–Schiffrin two-phase synthesis of organo-chalcogenate-protected metal nanoparticles. *J Am Chem Soc* 133:2092–95.
Liu, D., W. Qu, W. Chen, et al. 2010a. Highly sensitive, colorimetric detection of Mercury(II) in aqueous media by quaternary ammonium group-capped gold nanoparticles at room temperature. *Anal Chem* 82:9606–10.
Liu, J.-M., H.-F. Wang, and X.-P. Yan. 2011. A gold nanorod based colorimetric probe for the rapid and selective detection of Cu^{2+} ions. *Analyst* 136:3904–10.
Liu, M., and P. Guyot-Sionnest. 2005. Mechanism of silver(I)-assisted growth of gold nanorods and bipyramids. *J Phys Chem B* 109:22192–200.
Liu, M., Z. Wang, S. Zong, et al. 2014. SERS detection and removal of mercury(II)/silver(I) using oligonucleotide-functionalized core/shell magnetic silica sphere@Au nanoparticles. *ACS Appl Mater Interfaces* 6:7371–79.
Liu, X., J. Yao, J. Luo, et al. 2017. Effect of growth temperature on tailoring the size and aspect ratio of gold nanorods. *Langmuir* 33:7479–85.
Liu, Y., K. Ai, X. Cheng, et al. 2010b. Gold-nanocluster-based fluorescent sensors for highly sensitive and selective detection of cyanide in water. *Adv Funct Mater* 20:951–56.
Lou, T., L. Chen, C. Zhang, et al. 2012. A simple and sensitive colorimetric method for detection of mercury ions based on anti-aggregation of gold nanoparticles. *Anal Methods* 4:488–91.
Lou, T., Z. Chen, Y. Wang, et al. 2011. Blue-to-red colorimetric sensing strategy for Hg^{2+} and Ag^{+} via redox-regulated surface chemistry of gold nanoparticles. *ACS Appl Mater Interfaces* 3:1568–73.
Lowe, A. B., B. S. Sumerlin, M. S. Donovan, et al. 2002. Facile preparation of transition metal nanoparticles stabilized by well-defined (co)polymers synthesized via aqueous reversible addition-fragmentation chain transfer polymerization. *J Am Chem Soc* 124:11562–63.
Lu, X., M. S. Yavuz, H.-Y. Tuan, et al. 2008. Ultrathin gold nanowires can be obtained by reducing polymeric strands of oleylamine–AuCl complexes formed via aurophilic interaction. *J Am Chem Soc* 130:8900–01.
Lung, J.-K., J.-C. Huang, D.-C. Tien, et al. 2007. Preparation of gold nanoparticles by arc discharge in water. *J Alloys Compd* 434–435:655–58.
Luo, S., J. Xu, Y. Zhang, et al. 2005. Double hydrophilic block copolymer monolayer protected hybrid gold nanoparticles and their shell cross-linking. *J Phys Chem B* 109:22159–66.
Ma, H., S. Huang, X. Feng, et al. 2006. Electrochemical synthesis and fabrication of gold nanostructures based on poly(n-vinylpyrrolidone). *ChemPhysChem* 7:333–35.
Mafuné, F., J.-y. Kohno, Y. Takeda, et al. 2001. Formation of gold nanoparticles by laser ablation in aqueous solution of surfactant. *J Phys Chem B* 105:5114–20.
Mafuné, F., J.-y. Kohno, Y. Takeda, et al. 2002. Full physical preparation of size-selected gold nanoparticles in solution: Laser ablation and laser-induced size control. *J Phys Chem B* 106:7575–77.
Manna, S., J. W. Kim, Y. Takahashi, et al. 2016. Synthesis of single-crystalline anisotropic gold nano-crystals via chemical vapor deposition. *J Appl Phys* 119:174301.
Marin, M. L., K. L. McGilvray, and J. C. Scaiano. 2008. Photochemical strategies for the synthesis of gold nanoparticles from Au(III) and Au(I) using photoinduced free radical generation. *J Am Chem Soc* 130:16572–84.
Mikami, Y., A. Dhakshinamoorthy, M. Alvaro, et al. 2013. Catalytic activity of unsupported gold nanoparticles. *Catal Sci Technol* 3:58–69.
Mikhlin, Y., A. Karacharov, M. Likhatski, et al. 2011. Submicrometer intermediates in the citrate synthesis of gold nanoparticles: New insights into the nucleation and crystal growth mechanisms. *J Colloid Interface Sci* 362:330–36.
Millstone, J. E., W. Wei, M. R. Jones, et al. 2008. Iodide ions control seed-mediated growth of anisotropic gold nanoparticles. *Nano Lett* 8:2526–29.
Moragues, A., F. Neațu, V. I. Pârvulescu, et al. 2015. Heterogeneous gold catalyst: Synthesis, characterization, and application in 1,4-addition of boronic acids to enones. *ACS Catal* 5:5060–67.
Mountrichas, G., S. Pispas, and E. I. Kamitsos. 2014. Effect of temperature on the direct synthesis of gold nanoparticles mediated by poly(dimethylaminoethyl methacrylate) homopolymer. *J Phys Chem C* 118:22754–59.
Mukherjee, S., F. Libisch, N. Large, et al. 2013. Hot electrons do the impossible: Plasmon-induced dissociation of H_2 on Au. *Nano Lett* 13:240–47.

Muñoz-Santiburcio, D., M. Farnesi Camellone, and D. Marx. 2018. Solvation-induced changes in the mechanism of alcohol oxidation at gold/titania nanocatalysts in the aqueous phase versus gas phase. *Angew Chem Int Ed* 57:3327–31.

Nikoobakht, B., and M. A. El-Sayed. 2003. Preparation and growth mechanism of gold nanorods (NRs) using seed-mediated growth method. *Chem Mater* 15:1957–62.

Nishioka, K., Y. Niidome, and S. Yamada. 2007. Photochemical reactions of ketones to synthesize gold nanorods. *Langmuir* 23:10353–56.

O'Brien, M. N., M. R. Jones, K. L. Kohlstedt, et al. 2015. Uniform circular disks with synthetically tailorable diameters: Two-dimensional nanoparticles for plasmonics. *Nano Lett* 15:1012–17.

Ojea-Jiménez, I., N. G. Bastús, and V. Puntes. 2011. Influence of the sequence of the reagents addition in the citrate-mediated synthesis of gold nanoparticles. *J Phys Chem C* 115:15752–57.

Ojea-Jiménez, I., and J. M. Campanera. 2012. Molecular modeling of the reduction mechanism in the citrate-mediated synthesis of gold nanoparticles. *J Phys Chem C* 116:23682–91.

Ossowski, J., T. Wächter, L. Silies, et al. 2015. Thiolate versus selenolate: Structure, stability, and charge transfer properties. *ACS Nano* 9:4508–26.

Palazzo, G., G. Valenza, M. Dell'Aglio, et al. 2017. On the stability of gold nanoparticles synthesized by laser ablation in liquids. *J Colloid Interface Sci* 489:47–56.

Palgrave, R. G., and I. P. Parkin. 2007. Aerosol assisted chemical vapor deposition of gold and nanocomposite thin films from hydrogen tetrachloroaurate(III). *Chem Mater* 19:4639–47.

Pauli, W., J. Szper, and S. Szper. 1939. The structure and proprieties of highly purified reduction gold sols. *Trans Faraday Soc* 35:1178–83.

Payne, C. M., L. J. E. Anderson, and J. H. Hafner. 2013. Novel plasmonic structures based on gold nanobelts. *J Phys Chem C* 117:4734–39.

Payne, C. M., D. E. Tsentalovich, D. N. Benoit, et al. 2014. Synthesis and crystal structure of gold nanobelts. *Chem Mater* 26:1999–2004.

Perala, S. R. K., and S. Kumar. 2013. On the mechanism of phase transfer catalysis in Brust–Schiffrin synthesis of metal nanoparticles. *Langmuir* 29:14756–62.

Pischel, J., and A. Pucci. 2015. Low-temperature adsorption of carbon monoxide on gold surfaces: IR spectroscopy uncovers different adsorption states on pristine and rough Au(111). *J Phys Chem C* 119:18340–51.

Placido, T., G. Aragay, J. Pons, et al. 2013. Ion-directed assembly of gold nanorods: A strategy for mercury detection. *ACS Appl Mater Interfaces* 5:1084–92.

Polte, J., R. Erler, A. F. Thünemann, et al. 2010. SAXS in combination with a free liquid jet for improved time-resolved in situ studies of the nucleation and growth of nanoparticles. *Chem Commun* 46:9209–11.

Pong, B.-K., H. I. Elim, J.-X. Chong, et al. 2007. New insights on the nanoparticle growth mechanism in the citrate reduction of gold(III) salt: Formation of the Au nanowire intermediate and its nonlinear optical properties. *J Phys Chem C* 111:6281–87.

Priyadarshni, N., P. Nath, Nagahanumaiah, et al. 2018. DMSA-functionalized gold nanorod on paper for colorimetric detection and estimation of Arsenic (III and V) contamination in groundwater. *ACS Sustain Chem Eng* 6:6264–72.

Qian, H., A. Pretzer Lori, C. Velazquez Juan, et al. 2013. Gold nanoparticles for cleaning contaminated water. *J Chem Technol Biotechnol* 88:735–41.

Qian, X., X. Zhou, and S. Nie. 2008. Surface-enhanced Raman nanoparticle beacons based on bioconjugated gold nanocrystals and long range plasmonic coupling. *J Am Chem Soc* 130:14934–35.

Rajender, G., B. Choudhury, and P. K. Giri. 2017. In situ decoration of plasmonic Au nanoparticles on graphene quantum dots-graphitic carbon nitride hybrid and evaluation of its visible light photocatalytic performance. *Nanotechnology* 28:395703.

Robatjazi, H., S. M. Bahauddin, C. Doiron, et al. 2015. Direct plasmon-driven photoelectrocatalysis. *Nano Lett* 15:6155–61.

Rucareanu, S., M. Maccarini, J. L. Shepherd, et al. 2008. Polymer-capped gold nanoparticles by ligand-exchange reactions. *J Mater Chem* 18:5830–34.

Saha, K., S. S. Agasti, C. Kim, et al. 2012. Gold nanoparticles in chemical and biological sensing. *Chem Rev* 112:2739–79.

Saito, N., J. Hieda, and O. Takai. 2009. Synthesis process of gold nanoparticles in solution plasma. *Thin Solid Films* 518:912–17.

Sanchez-Castillo Marco, A., C. Couto, B. Kim Won, et al. 2004. Gold-nanotube membranes for the oxidation of CO at gas–water interfaces. *Angew Chem Int Ed* 43:1140–42.

Sato, K., K. Hosokawa, and M. Maeda. 2003. Rapid aggregation of gold nanoparticles induced by non-cross-linking DNA hybridization. *J Am Chem Soc* 125:8102–03.

Scarabelli, L., A. Sánchez-Iglesias, J. Pérez-Juste, et al. 2015. A "Tips and Tricks" practical guide to the synthesis of gold nanorods. *J Phys Chem Lett* 6:4270–79.

Schulz, F., T. Homolka, N. G. Bastús, et al. 2014. Little adjustments significantly improve the Turkevich synthesis of gold nanoparticles. *Langmuir* 30:10779–84.

Shan, J., M. Nuopponen, H. Jiang, et al. 2005. Amphiphilic gold nanoparticles grafted with poly(N-isopropylacrylamide) and polystyrene. *Macromolecules* 38:2918–26.

Shan, J., and H. Tenhu. 2007. Recent advances in polymer protected gold nanoparticles: Synthesis, properties and applications. *Chem Commun* 0:4580–98.

Shu Jun, Z., Z. Zhong Yue, L. Na, et al. 2013. UV light-induced self-assembly of gold nanocrystals into chains and networks in a solution of silver nitrate. *Nanotechnology* 24:055601.

Smith, D. K., N. R. Miller, and B. A. Korgel. 2009. Iodide in CTAB prevents gold nanorod formation. *Langmuir* 25:9518–24.

Soliwoda, K., E. Tomaszewska, B. Tkacz-Szczesna, et al. 2014. Effect of the alkyl chain length of secondary amines on the phase transfer of gold nanoparticles from water to toluene. *Langmuir* 30:6684–93.

Sun, K., M. Kohyama, S. Tanaka, et al. 2018. Roles of water and H_2 in CO oxidation reaction on gold catalysts. *J Phys Chem C* 122:9523–30.

Sun, L., C. Yu, and J. Irudayaraj. 2007. Surface-enhanced Raman scattering based nonfluorescent probe for multiplex DNA detection. *Anal Chem* 79:3981–88.

Takale, B. S., X. Feng, Y. Lu, et al. 2016. Unsupported nanoporous gold catalyst for chemoselective hydrogenation reactions under low pressure: Effect of residual silver on the reaction. *J Am Chem Soc* 138:10356–64.

Tan, B. J. Y., C. H. Sow, T. S. Koh, et al. 2005. Fabrication of size-tunable gold nanoparticles array with nanosphere lithography, reactive ion etching, and thermal annealing. *J Phys Chem B* 109:11100–09.

Terauchi, S., N. Koshizaki, and H. Umehara. 1995. Fabrication of Au nanoparticles by radio-frequency magnetron sputtering. *Nanostruct Mater* 5:71–78.

Thanh, N. T. K., and Z. Rosenzweig. 2002. Development of an aggregation-based immunoassay for anti-protein a using gold nanoparticles. *Anal Chem* 74:1624–28.

Tseng, C.-H., T. C. K. Yang, H.-E. Wu, et al. 2009. Catalysis of oxidation of carbon monoxide on supported gold nanoparticle. *J Hazard Mater* 166:686–94.

Turkevich, J., P. C. Stevenson, and J. Hillier. 1951. A study of the nucleation and growth processes in the synthesis of colloidal gold. *Discussions of the Faraday Society* 11:55–75.

Turkevich, J., P. C. Stevenson, and J. Hillier. 1953. The formation of colloidal gold. *J Phys Chem* 57:670–73.

Uehara, A., S. G. Booth, S. Y. Chang, et al. 2015. Electrochemical insight into the Brust–Schiffrin synthesis of Au nanoparticles. *J Am Chem Soc* 137:15135–44.

Uehara, N. 2010. Polymer-functionalized gold nanoparticles as versatile sensing materials. *Anal Sci* 26:1219–28.

Vasilev, K., T. Zhu, M. Wilms, et al. 2005. Simple, one-step synthesis of gold nanowires in aqueous solution. *Langmuir* 21:12399–403.

Villa, A., N. Dimitratos, C. E. Chan-Thaw, et al. 2016. Characterisation of gold catalysts. *Chem Soc Rev* 45:4953–94.

Wang, P., B. Huang, Y. Dai, et al. 2012. Plasmonic photocatalysts: Harvesting visible light with noble metal nanoparticles. *Phys Chem Chem Phys* 14:9813–25.

Wang, Y.-N., W.-T. Wei, C.-W. Yang, et al. 2013. Seed-mediated growth of ultralong gold nanorods and nanowires with a wide range of length tunability. *Langmuir* 29:10491–97.

Watthanaphanit, A., G. Panomsuwan, and N. Saito. 2014. A novel one-step synthesis of gold nanoparticles in an alginate gel matrix by solution plasma sputtering. *RSC Adv* 4:1622–29.

Wei, H., B. Li, J. Li, et al. 2007. Simple and sensitive aptamer-based colorimetric sensing of protein using unmodified gold nanoparticle probes. *Chem Commun* 0:3735–37.

Wei, S.-C., P.-H. Hsu, Y.-F. Lee, et al. 2012. Selective detection of iodide and cyanide anions using gold-nanoparticle-based fluorescent probes. *ACS Appl Mater Interfaces* 4:2652–58.

Wittstock, A., B. Neumann, A. Schaefer, et al. 2009. Nanoporous Au: An unsupported pure gold catalyst? *J Phys Chem C* 113:5593–600.

Wu, H.-Y., H.-C. Chu, T.-J. Kuo, et al. 2005. Seed-mediated synthesis of high aspect ratio gold nanorods with nitric acid. *Chem Mater* 17:6447–51.

Xu, J., H. Yu, Y. Hu, et al. 2016. A gold nanoparticle-based fluorescence sensor for high sensitive and selective detection of thiols in living cells. *Biosens Bioelectron* 75:1–7.

Xu, X., J. Wang, K. Jiao, et al. 2009. Colorimetric detection of mercury ion (Hg^{2+}) based on DNA oligonucleotides and unmodified gold nanoparticles sensing system with a tunable detection range. *Biosens Bioelectron* 24:3153–58.

Xue, X., F. Wang, and X. Liu. 2008. One-step, room temperature, colorimetric detection of mercury (Hg^{2+}) using DNA/nanoparticle conjugates. *J Am Chem Soc* 130:3244–45.

Xue, Y., H. Zhao, Z. Wu, et al. 2011. Colorimetric detection of Cd^{2+} using gold nanoparticles cofunctionalized with 6-mercaptonicotinic acid and l-cysteine. *Analyst* 136:3725–30.

Yang, X., M. Yang, B. Pang, et al. 2015. Gold nanomaterials at work in biomedicine. *Chem Rev* 115:10410–88.

Ye, X., C. Zheng, J. Chen, et al. 2013. Using binary surfactant mixtures to simultaneously improve the dimensional tunability and monodispersity in the seeded growth of gold nanorods. *Nano Lett* 13:765–71.

Yee, C. K., A. Ulman, J. D. Ruiz, et al. 2003. Alkyl selenide-and alkyl thiolate-functionalized gold nanoparticles: Chain packing and bond nature. *Langmuir* 19:9450–58.

Yu, S., A. J. Wilson, J. Heo, and C. R. C. Wang. 2018. Plasmonic control of multi-electron transfer and C–C coupling in visible-light-driven CO_2 reduction on Au nanoparticles. *Nano Lett* 18:2189–94.

Yu, S.-S. Chang, C.-L. Lee, et al. 1997. Gold nanorods: Electrochemical synthesis and optical properties. *J Phys Chem B* 101:6661–64.

Zabetakis, K., W. E. Ghann, S. Kumar, et al. 2012. Effect of high gold salt concentrations on the size and polydispersity of gold nanoparticles prepared by an extended Turkevich–Frens method. *Gold Bull* 45:203–11.

Zaluzhna, O., Y. Li, C. Zangmeister, et al. 2012. Mechanistic insights on one-phase vs. two-phase Brust–Schiffrin method synthesis of Au nanoparticles with dioctyl-diselenides. *Chem Commun* 48:362–64.

Zamarion, V. M., R. A. Timm, K. Araki, et al. 2008. Ultrasensitive SERS nanoprobes for hazardous metal ions based on trimercaptotriazine-modified gold nanoparticles. *Inorg Chem* 47:2934–36.

Zhang, M., Y.-Q. Liu, and B.-C. Ye. 2012. Colorimetric assay for parallel detection of Cd^{2+}, Ni^{2+} and Co^{2+} using peptide-modified gold nanoparticles. *Analyst* 137:601–07.

Zhang, Y., S. He, W. Guo, et al. 2018. Surface-plasmon-driven hot electron photochemistry. *Chem Rev* 118:2927–54.

Zhao, J., S. C. Nguyen, R. Ye, et al. 2017. A comparison of photocatalytic activities of gold nanoparticles following plasmonic and interband excitation and a strategy for harnessing interband hot carriers for solution phase photocatalysis. *ACS Cent Sci* 3:482–88.

Zhao, L., D. Jiang, Y. Cai, et al. 2012. Tuning the size of gold nanoparticles in the citrate reduction by chloride ions. *Nanoscale* 4:5071–76.

Zhao, N., Y. Wei, N. Sun, et al. 2008. Controlled synthesis of gold nanobelts and nanocombs in aqueous mixed surfactant solutions. *Langmuir* 24:991–98.

Zhao, P., N. Li, and D. Astruc. 2013. State of the art in gold nanoparticle synthesis. *Coord Chem Rev* 257:638–65.

Zhu, K., Y. Zhang, S. He, et al. 2012. Quantification of proteins by functionalized gold nanoparticles using click chemistry. *Anal Chem* 84:4267–70.

Zhu, L., C. Zhang, C. Guo, et al. 2013. New insight into intermediate precursors of Brust–Schiffrin gold nanoparticles synthesis. *J Phys Chem C* 117:11399–404.

Zou, N., G. Chen, X. Mao, et al. 2018. Imaging catalytic hotspots on single plasmonic nanostructures via correlated super-resolution and electron microscopy. *ACS Nano* 12:5570–79.

4

Biogenic Synthesis of Silver Nanoparticles and Their Applications

G. Krishna, V. Pranitha, Reeja Sundaram, and M.A. Singara Charya

CONTENTS

4.1 Nanotechnology

The field of nanotechnology is a tremendously developing field as a result of its wide-ranging applications in different areas of science and technology. The term nanotechnology is defined as the creation, exploitation, and synthesis of materials at a scale smaller than 1 mm. The word "nano" is derived from a Greek word meaning dwarf or extremely small (Rai *et al.*, 2008). The concept of nanotechnology was given by physicist Professor Richard Feynman in his historic talk "there's plenty of room at the bottom" (Feynman, 1959), though the term nanotechnology was introduced by Tokyo Science University Professor Norio Taniguchi (Taniguchi, 1974). Nanobiotechnology is a multidisciplinary field and involves research and development of technology in different fields of science such as biotechnology, nanotechnology, physics, chemistry, and material science (Huang *et al.*, 2007; Rai *et al.*, 2008). Nanoparticles are metal particles with size 1–100 nm and exhibit different shapes – they can be spherical, triangular, rod-shaped, etc. (Rai *et al.*, 2009a). Research on the synthesis of nanoparticles is the current area of interest due to the unique visible properties (chemical, physical, optical, etc.) of nanoparticles compared with bulk material (Rai *et al.*, 2009a; Sau and Rogach, 2010).

4.2 Nanomaterials and Nanoparticles

Nanomaterials are of interest because of their novel properties and of the functions attributable to their small size. First, they have greater surface area when compared to the same mass material in larger particles (Royal society and Royal Academy of Engineering, 2004). Nanomaterials might offer solutions to technological and environmental challenges within the areas of solar power conversion, chemical process drugs, and water treatment (Mandal *et al.*, 2006;

Sharma *et al.*, 2009). Nanoparticles are attracting immense attention from researchers in a wide range of fields: physics, physical chemistry, mineralogy, material science, biology, and medicine, amongst others. There are many terms for nanoparticles such as small particles, Q particles, quantum dots, aerosols, and hydrosols. Their unique size-dependent properties make nanomaterials superior and indispensable in many walks of human life. The environmental impacts on human and non-human biota and on the ecosystem are still being studied (Ahmad *et al.*, 2003). The use of nanoparticles is gaining impetus in the present century as they possess defined chemical, optical, and mechanical properties (Maharani, 2015). The metallic nanoparticles are the most promising as they show good antibacterial properties due to their large surface area to volume ratios, which draw growing interest from researchers due to increasing microbial resistance against metal ions, antibiotics and the development of resistant strains (Gong *et al.*, 2007). Among the noble metal nanoparticles, silver nanoparticles are an arch-product in the field of nanotechnology, having gained boundless interest because of their unique properties such as chemical stability, good conductivity, as well as catalytic – and, most importantly, antibacterial, anti-viral, antifungal, and anti-inflammatory – activities. These can be incorporated into composite fibres, cryogenic superconducting materials, cosmetic products, foods, and electronic components (Shakeel *et al.*, 2016).

4.3 Silver Nanoparticles

Silver has been known since ancient times for its many properties, which are useful to humans. Silver is a transition metal element having atomic number 47 and atomic mass 107.87. It is, however, an element of many faces. It is used as a precious commodity in currencies, ornaments, and jewelry. It has the highest electrical conductivity of any element, which is useful in electrical contacts and conductors.

Silver is a nontoxic, safe inorganic antibacterial agent. It has been used for centuries and is capable of killing about 650 types of disease-causing microorganisms. Silver has been described as being oligodynamic because of its ability to exert a bactericidal effect at minute concentrations. Silver ions ($Ag+$) and its compounds are highly toxic to microorganisms, exhibiting strong biocidal effects on many species of bacteria.

Silver nanoparticles have different catalytic properties compared with those attributed to the bulk form of the noble metal, such as surface plasmon resonance, large effective scattering cross section of individual silver nanoparticles, and strong toxicity to a wide range of microorganisms (Elechiguerra *et al.*, 2005). Silver nanoparticles show remarkable optical properties which depend on their size and shape (Zheng and Dickson, 2002; Peyser *et al.*, 2001).

In general, there are mainly two ways to synthesize silver nanoparticles, either "top down" or "bottom up". In the bottom-up approach, nanoparticles can be synthesized using chemical and biological methods by self-assembling atoms to new nuclei which grow into nanoscale particles. In the top-to-bottom approach, suitable bulk material is broken down into fine particles through size reduction with various lithographic techniques, e.g., grinding, milling, sputtering, and thermal/laser ablation.

Nanomaterials such as copper, zinc, titanium (Retchkiman *et al.*, 2006), magnesium, gold (Gu *et al.*, 2003), alginate (Ahmad *et al.*, 2005), and silver have been used, but among them silver nanoparticles have proved to be the most effective thanks to their many properties (Gong *et al.*, 2007). Silver is widely known as a catalyst for the oxidation of methanol to formaldehyde and ethylene to ethylene oxide (Nagy and Mestl, 1999). Colloidal silver is of particular interest because of its distinctive properties, such as good conductivity, chemical stability, and catalytic and antibacterial activity (Frattini *et al.*, 2005). Silver is a naturally occurring precious metal, most often as a mineral ore in association with other elements. It has been positioned as the 47th element in the periodic table, having an atomic weight of 107.8 and two natural isotopes 106.90 and 108.90 Ag with abundance 52 and 48%. It has been used in a wide variety of applications as it has some special properties like high electrical and thermal conductivity (Nordberg and Gerhardsson, 1988). Of all the different types of nanoparticles, silver nanoparticles are playing a major role in the fields of nanotechnology and nanomedicine (Sivagnanam and Jayanthi, 2013).

4.4 Publication Scenario on Silver Nanoparticles Synthesis

A literature search was primarily conducted by coining silver nanoparticles as a keyword in the 'ISI Web of Science' database, which yielded 37,634 articles from 2000 to 2016 (Figure 4.1). In order to concentrate on the biological synthesis of silver nanoparticles, the search was refined by the following keywords: 'biological synthesis of silver nanoparticles' under the search results of 'silver nanoparticles', which generated 1,079 articles. The search was again refined by the following keywords: 'synthesis of silver nanoparticles from fungi', which generated 220 articles. The search was refined further with the terms 'synthesis of silver nanoparticles from white-rot fungi', which generated only 7 articles (Figure 4.2).

China has the greatest number of publications on the AgNPs topic, with a total of 4,434 (23.6 %), followed by the USA at 3,809 (20.7 %), India at 1,842 (9.8 %), South Korea at 1,331 (7.1 %), Japan at 1,283 (6.8 %), Germany at 1,079 (5.7 %), France at 770

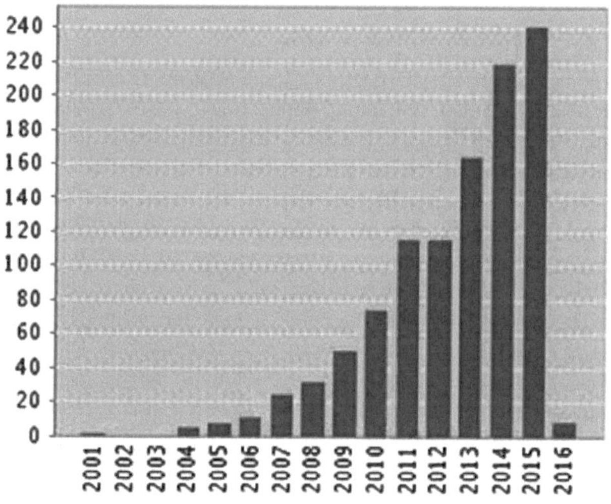

FIGURE 4.1 Literature on silver nanoparticles (from 2000 to 2016).

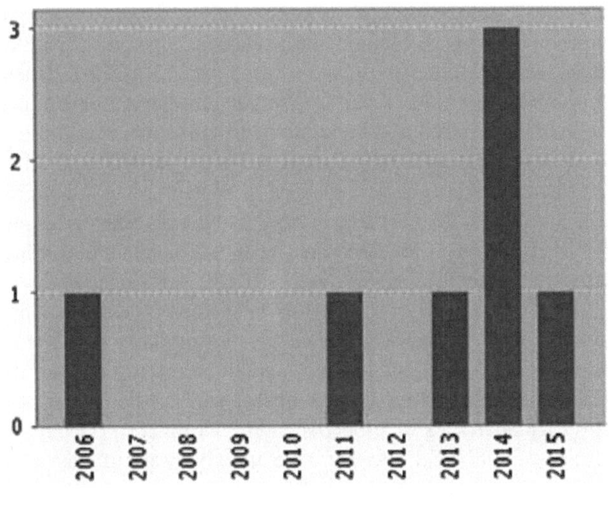

FIGURE 4.2 Synthesis of silver nanoparticles white-rot fungi (from 2006 to 2015).

(4.1 %), Taiwan at (3.6 %), Spain at 540 (2.8 %), Russia at 539 (2.8 %), and so on. A large number of practical applications utilizing AgNPs in consumer products is being developed in parallel with studies on these syntheses and properties.

Research in this field has been growing dramatically throughout the world over the last decade. The development of new materials of nanometer size, including nanoparticles, nanotubes, nanowires, etc., is the major activity. Among all, nanoparticles with special properties in the fields of chemistry, optics, electronics, and magnetics have led to an increasing interest in their synthesis. Various physical and chemical processes are implemented for nanoparticles synthesis; however, some chemical methods cannot avoid the use of toxic chemicals in the synthesis process (Azim *et al.*, 2009; Tapan and Andrey, 2009). Therefore, there is an urgent need to develop a green process of nanoparticle synthesis. Biological methods of nanoparticle synthesis using either microbes or plant extracts have offered a reliable, eco-friendly alternative to chemical and physical methods. It is well known that microorganisms have been used in the remediation of toxic metals through reduction of metal ions, but the interest in nanoparticle synthesis using microbes has emerged quite recently. Nanoparticles can be synthesized biologically by using microorganisms which grab target ions from their solutions, and then accumulate the reduced metal in its element form through enzymes generated by microbial cell activities. Synthesis can be categorized as intracellular or extracellular according to the place where the nanoparticles are formed (Simkiss and Wilbur, 1989; Mann, 1996).

The intracellular method consists in transporting ions into the microbial cell to form the nanoparticles in the presence of enzymes. The extracellular synthesis of nanoparticles involves trapping the metal ions on the surface of the cells and reducing ions in the presence of enzymes. So far, many microbes, such as magnetotactic bacteria (Blackmore, 1982), diatoms (Mann, 2001), S-layer bacteria (Pum and Sleytr, 1999), fungi (Bruins *et al.*, 2000), actinomycete (Ahmad *et al.*, 2003), and yeast (Mithila *et al.*, 2009) have been employed for generating nanostructured mineral crystals and metallic nanoparticles. The control of size, shape, composition, and monodispersity of particles was also studied. On the other hand, the effect of nanoparticles on microbes has also caught great interest. Nanoparticles are capable of assisting microbe activities. Several studies have been reported on the influence of nanoparticles on the microbiological reaction rates (Windt *et al.*, 2005; Shin and Cha, 2008).

4.4.1 Physical Approaches

Physical methods used for the synthesis of nanoparticles include evaporation-condensation and laser ablation. Various metal nanoparticles, such as silver, gold, lead sulfide, cadmium sulfide, and fullerene have previously been synthesized using the evaporation-condensation method. The absence of solvent contamination in the prepared thin films and the uniformity of nanoparticles distribution are the advantages of physical approaches in comparison with chemical processes (Kruis *et al.*, 2000; Magnusson *et al.*, 1999). It was demonstrated that silver nanoparticles could be synthesized via a small ceramic heater with a local heating source (Jung *et al.*, 2006). The evaporated vapor can cool at a suitably rapid rate, because the temperature gradient in the vicinity of the heater surface is very steep in comparison with that of a tube furnace. This makes possible the formation of small nanoparticles in high concentration. This physical method can be useful as a nanoparticle generator for long-term experiments for inhalation toxicity studies, and as a calibration device for nanoparticle measurement equipment. Silver nanoparticles can be synthesized by laser ablation of metallic bulk materials in solution (Mafune *et al.*, 2000; Mafune *et al.*, 2001; Kabashin, 2003; Dolgaev *et al.*, 2002; Sylvestre *et al.*, 2004). The ablation efficiency and the characteristics of produced nanosilver particles depend upon many factors, such as the wavelength of the laser impinging the metallic target, the duration of the laser pulses (in the femto-, pico- and nanosecond regime), the laser fluence, the ablation time duration, and the effective liquid medium, with or without the presence of surfactants (Link *et al.*, 2000; Kim *et al.*, 2005; Kawasaki and Nishimura, 2006; Tarasenko *et al.*, 2006). One important advantage of the laser ablation technique compared to other methods for the production of metal colloids is the absence of chemical reagents in solutions. Therefore, pure and uncontaminated metal colloids for further applications can be prepared through this technique (Tsuji *et al.*, 2002; Panigrahi, 2013).

4.4.2 Chemical Approaches

The most common approach for the synthesis of silver nanoparticles is chemical reduction by organic and inorganic reducing agents. In general, different reducing agents such as sodium citrate, ascorbate, sodium borohydride ($NaBH_4$), elemental hydrogen, polyol process, Tollens' reagent, N, N-dimethylformamide (DMF), and poly(ethylene glycol)-block copolymers are used for the reduction of silver ions (Ag+) in aqueous or non-aqueous solutions. The aforementioned reducing agents reduce silver ions (Ag+) and lead to the formation of metallic silver (Ag0), which is followed by agglomeration into oligomeric clusters. These clusters eventually lead to the formation of metallic colloidal silver particles (Wiley *et al.*, 2005; Merga *et al.*, 2007; Evanoff, 2004). It is important to use protective agents to stabilize dispersive nanoparticles during the course of metal nanoparticle preparation, and protect the nanoparticles that can be absorbed on or bind on to nanoparticle surfaces, avoiding their agglomeration

(Oliveira *et al.*, 2005). The presence of surfactants comprising functionalities (e.g. thiols, amines, acids, and alcohols) for interactions with particle surfaces can stabilize particle growth, and protect particles from sedimentation, agglomeration, or losing their surface properties. Recently, a simple one-step process, Tollens' method, has been used for the synthesis of silver nanoparticles with a controlled size. In the modified Tollens' procedure, silver ions are reduced by saccharides in the presence of ammonia, yielding silver nanoparticle films (50–200 nm), silver hydrosols (20–50 nm), and silver nanoparticles of different shapes (Ahmad *et al.*, 2003a).

4.4.3 Biological Synthesis of Silver Nanoparticles

The main problem with most of the chemical and physical methods of nanosilver production is that they are mostly expensive and also involve the use of toxic, hazardous chemicals, which may pose potential environmental and biological risks. It is an unavoidable fact that the silver nanoparticles synthesized have to be handled by humans and must be available at lesser rates for their effective utilization; thus, there is a need for an environmentally and economically feasible way to synthesize these nanoparticles. The quest for such a method has led to the need for biomimetic production of silver nanoparticles, whereby biological methods are used to synthesize them (Kalishwaralal *et al.*,2008). In most cases, the chemical synthesis methods lead to some chemically toxic substances being absorbed on the surface. This can hinder their usage in medical applications (Parashar *et al.*, 2009). There are three major sources to synthesize silver nanoparticles: bacteria, fungi, and plant extracts. The biosynthesis of silver nanoparticles is a bottom-up approach that mostly involves reduction/oxidation reactions. It is majorly the microbial enzymes or the plant phytochemicals with antioxidant or reducing properties that act on the respective compounds and give the desired nanoparticles. The three major components involved in the preparation of nanoparticles using biological methods are the solvent medium for synthesis, the environmentally friendly reducing agent, and a nontoxic stabilizing agent (Prabhu and Poulose, 2012).

4.5 Microbe-Assisted Synthesis of Silver Nanoparticles

The first evidence of bacteria synthesizing silver nanoparticles was established using the *Pseudomonas stutzeri* AG259 strain that was isolated from a silver mine (Haefeli *et al.*, 1984). There are few microorganisms that can survive metal ion concentrations and can also grow under those conditions, and this phenomenon is due to their resistance to that metal. The mechanisms involved in the resistance are efflux systems, alteration of solubility and toxicity via reduction or oxidation, biosorption, bioaccumulation, extracellular complex formation or precipitation of metals, and lack of specific metal transport systems (Husseiny *et al.*, 2006). There is also another aspect: though these organisms can grow at very lower concentrations, their exposure to higher concentrations of metal ions can induce toxicity. Popularly, the accepted mechanism of silver biosynthesis is the presence of the nitrate reductase enzyme. The enzyme converts nitrate into nitrite. In *in vitro* synthesis of silver using bacteria, the presence of alpha-nicotinamide adenine dinucleotide phosphate reduced form (NADPH)- dependent nitrate reductase would remove the downstream processing step that is required in other cases. During the reduction, nitrate is converted into nitrite and the electron is transferred to the silver ion; hence, the silver ion is reduced to silver ($Ag+$ to Ag^0). This has observed in *Bacillus licheniformis*, which is known to secrete NADPH and NADPH-dependent enzymes such nitrate reductase which effectively converts Ag+ to Ag0 (Vaidyanathan *et al.*, 2010). The mechanism was further confirmed by using purified nitrate reductase from *Fusarium oxysporum* and silver nitrate along with NADPH in a test tube, and the change in color of the reaction mixture to brown and further analysis confirmed that silver nanoparticles were obtained (Anil *et al.*, 2007). There are also cases which indicate that there are other ways to biosynthesize silver nanoparticles without the presence of enzymes. It was found that dried cells of *Lactobacillus sp.* A09 can reduce silver ions through the interaction of the silver ions with the groups on the microbial cell wall (Fu *et al.*, 2000).

4.6 Plant-Mediated Synthesis of Silver Nanoparticles

The advantage of using plant extracts for silver nanoparticle synthesis is that they are easily available, safe, and nontoxic in most cases. They also have a broad variety of metabolites that can aid in the reduction of silver ions. The most important mechanism involved in the process is plant-assisted reduction due to phytochemicals. The main phytochemicals involved are terpenoids, flavones, ketones, aldehydes, amides, and carboxylic acids. Flavones, organic acids, and quinones are water-soluble phytochemicals which are responsible for the immediate reduction of the ions. Studies have revealed that xerophytes contain emodin, an anthraquinone that undergoes tautomerization, leading to the formation of the silver nanoparticles. In the case of mesophytes, it was found that they contain three types of benzoquinones: dietchequinone, cyperoquinone, and remirin. It was mainly suggested that the phytochemicals are involved directly in the reduction of the ions and formation of silver nanoparticles (Jha *et al.*, 2009). Though the exact mechanism involved in each plant varies as the phytochemical involved varies, the major mechanism involved is the reduction of the ions.

4.7 Fungal-Derived Silver Nanoparticles

When in comparison with bacteria, fungi can produce large amounts of nanoparticles because they can secrete vast amounts of proteins which directly translate to higher productivity of the nanoparticles (Mohanpuria *et al.*, 2008). The mechanism of silver nanoparticle production by fungi is said to follow the following steps: trapping of Ag+ ions at the surface of the fungal cells, and subsequent reduction of the silver ions by the enzymes present in the fungal system (Mukherjee *et al.*, 2001). Extracellular enzymes such as naphthoquinones and anthraquinones are said to facilitate the reduction. Considering the example of *Fusarium oxysporum*, it is believed that the NADPH-dependent nitrate reductase and a shuttle quinine extracellular process are responsible for nanoparticle formation (Ahmad *et al.*, 2003b). Though the exact mechanism involved in silver nanoparticle production by fungi is not fully deciphered, it is believed that the abovementioned phenomenon is responsible for the process.

4.8 Superiority of Biological Methods

The biological method for the synthesis of nanoparticles employs biological agents such as bacteria, fungi, actinomycetes, yeast, and plants (Rai *et al.*, 2008; Thakkar *et al.*, 2010). Thus, the biological method creates a wide range of resources for the synthesis of nanoparticles. The rate of reduction of metal ions using biological agents is found to be much faster, and to take place at ambient temperature and pressure conditions. For example, when using *Aspergillus niger*, the synthesis of silver nanoparticles was observed within 2 h of treatment of fungal filtrate with silver salt solution (Gade *et al.*, 2008). Thus, the biological method requires minimum time for nanoparticles synthesis.

Shape- and size-controlled nanoparticles could be synthesized by modulating the temperature or pH of the reaction mixture. (Gericke and Pinches, 2006) obtained different-shape morphologies (triangles, hexagons, spheres, and rods) by modulating the pH of the reaction mixture to 3, 5, 7, and 9. (Riddin *et al.*, 2006) also explained that at a temperature of 65 °C, a smaller amount of nanoparticles was synthesized, whereas at 35 °C, a larger amount of nanoparticles was synthesized. The biological agents secrete a large amount of enzymes, which are capable of hydrolyzing metals and thus bring about enzymatic reduction of metals ions (Rai *et al.*, 2009b).

In the case of fungi, the enzyme nitrate reductase is found to be important and responsible for the synthesis of nanoparticles (Kumar *et al.*, 2007a). The biomass used for the synthesis of nanoparticles is simpler to handle and is easily disposed of in the environment; moreover, the downstream processing of the biomass is much easier. Synthesis can be carried out at ambient temperature and pressure conditions that require smaller amounts of chemicals (Ingle *et al.*, 2008). The synthesizing process is a less labor-intensive, low-cost technique; it is also nontoxic and has a greener approach. Thus, considering the above points, the biological method employed for the synthesis of nanoparticles proves to be superior compared with the physical and chemical methods of synthesis, due to its environment-friendly approach and lower cost (Rai and Duran, 2011).

4.9 Silver Nanoparticles from White-Rot Fungi

Currently, the use of white-rot fungi for the production of metal nanoparticles is a new approach of research with less than a decade of study (Rubilar *et al.*, 2008; Tortella *et al.*, 2008; Acevedo *et al.*, 2012). The small number of reports on the synthesis and use of different species of fungi has been limited to a few species, such as *Phanerochaete chrysosporium* (Vigneshwaran *et al.*, 2009), *Pleurotus sajorcaju* (Nithya and Ragunathan, 2009), *Coriolus versicolor* (Sanghi and Verma, 2009), *Pycnoporus sanguineus* (Chan and Don, 2012a), and *Schizophyllum commune* (Chan and Don, 2012b). In many ways, fungi are more advantageous compared to other microbes. Fungal mycelial mesh can withstand harsher flow pressure, agitation, and other conditions in bioreactors or other chambers compared to plant materials and bacteria. These are fastidious to grow, easy to handle, and of easy fabrication. The extracellular secretions of reductive proteins are more and can be easily handled in downstream processing. And also, since the nanoparticles precipitated outside the cell is devoid of unnecessary cellular components, it can be directly used in various applications. Due to the nonpathogenic nature of white-rot fungus, the application of these microorganisms for the production of metal nanoparticles could be seen an interesting alternative due to high biomass production and easy handling (Vigneshwaran *et al.*, 2007).

4.10 Silver Nanoparticles Synthesis

Silver nanoparticles, like their bulk counterpart, are an effective antimicrobial agent against various pathogenic microorganisms. Although various chemical and biochemical methods are being explored for the production of silver nanoparticles(Naik *et al.*, 2002), microbes are very much effective in this process. Various microbes are known to reduce the metals; most of them are found to be spherical particles, as reported earlier (Chen *et al.*, 2003; Ahmad *et al.*, 2003; Sastry *et al.*, 2003). The resistance conferred by bacteria to silver is determined by the 'sil' gene in plasmids, while a nitrate-dependent reductase and a shuttle quinone extracellular process were reported for the reduction of silver ions by several *Fusarium oxysporum* strains (Duran *et al.*, 2005).

4.11 Biosynthesis of Nanoparticles by Fungi

Fungi are recognized as eukaryotic organisms that reside in various ordinary lodgings and they typically form decomposer organisms. From an estimated sum of 1.5 million species of fungi on Earth, only about 70,000 species have been recognized. The estimation of more recent data shows that, according to high-throughput sequencing methods, approximately 5.1 million fungal species can be found (Blackwell, 2011). It is worth mentioning that these organisms have the ability to digest extracellular food, discharging particular enzymes to hydrolyze complicated compositions into easier molecules, which are soaked up and utilized as an energy resource. The exploration of the implications of using fungi in nanobiotechnology is considered important. In this regard, fungi have attracted attention regarding the research on biological production of metallic nanoparticles due to their toleration and metal bioaccumulation capabilities (Sastry *et al.*, 2003). The easiness of fungi scale-up is a further privilege when utilizing them in nanoparticle synthesis (e.g., utilizing a thin solid substrate fermentation technique). Fungi are very effective secretors of extracellular enzymes; therefore, achieving vast production of enzymes is feasible (Castro *et al.*, 2012). Economic livability and facility of employing biomass are further merits of the utilization of a green approach mediated by fungal species to synthesize metallic nanoparticles.

Moreover, a number of species grow fast; therefore, culturing and keeping them in the laboratory is very simple

(Castro *et al.*, 2011). High wall-binding and intracellular metal uptake are among the capacities of most fungi (Volesky and Holan, 1995). Fungi are able to produce metal nanoparticles and nanostructures by reducing enzymes intracellularly or extracellularly and though the biomimetic mineralization procedure (Ahmad, *et al.*, 2003; Duran *et al.*, 2005). As NP processes in nanotechnology, the study of fungal species is somewhat new. One of the primary investigations of the biosynthesis of metallic NPs by means of fungi illustrates the synthesis of silver NPs extracellularly by the filamentous fungus *Verticillium* sp. (Mukherjee *et al.*, 2001). With this aim, the filamentous fungus *Fusarium oxysporum* has been the most widely utilized species among the fungal ones identified for NP synthesis (Amin *et al.*, 2015).

4.12 Intracellular Synthesis of Nanoparticles by Fungi

The nanoparticles formed inside the organism could be smaller compared with the size of extracellularly reduced nanoparticles. The size limit might be related to the particles nucleating inside the organisms. (Mukherjee *et al.*, 2001) demonstrated the use of eukaryotic microorganisms in the biological synthesis of gold nanoparticles using *Verticillium* sp. (AAT-TS-4). Synthesis of gold nanoparticles was reported on the surface and on the cytoplasmic membrane of fungal mycelium, which were around 20 nm in diameter. These formed nanoparticles have well-defined dimensions and good dispersity. Upon TEM analysis, ultrathin sections of fungal mycelia showed spherical and a few triangular and hexagonal nanoparticles on the cell wall, and quasi-hexagonal morphology on the cytoplasmic membrane. *Trichothecium* sp. was found to accumulate gold nanoparticles intracellularly (Ahmad *et al.*, 2003). In addition, *Verticillium luteoalbum* and other isolates produced gold nanoparticles in 24 h. When *V. luteoalbum* was incubated at pH 3.0, spherical particles of <10 nm diameter but with pH 5.0 spheres and rods were also observed along with triangular and hexagonal morphologies (Gericke, 2006a). *Phoma* PT35 was able to selectively accumulate silver (Pighi *et al.*, 1989) and *Phoma* sp. 3.2883 was, in fact, a biosorbent suited for preparing silver nanoparticles (Chen *et al.*, 2003). The fungal biomass of *Verticillium sp.* on exposure to aqueous silver nitrate solution resulted in the accumulation of silver nanoparticles below the fungal cell surface with a negligible amount on the solution (Mukherjee *et al.*, 2001). Vigneshwaran *et al.* (2006) also showed that the use of *Aspergillus flavus* resulted in the accumulation of silver nanoparticles on the surface of its cell wall when incubated with silver nitrate solution for 72 h. The average particle size was found to be 8.92 nm (Table 4.1). Intracellular synthesis of gold nanoparticles produced by *Verticillium* sp. (Mukherjee *et al.*, 2001), *Trichothecium* sp. (Ahmad *et al.*, 2005), and *V. luteoalbum* (Gericke, 2006b), and of silver nanoparticles by *Verticillium sp.* (Senapati *et al.*, 2004; Mukherjee *et al.*, 2001) and *Aspergillus flavus* (Vigneshwaran *et al.*, 2007) showed morphologies of spherical, quasi-hexagonal, and rod shapes in the size range of 8.92–25 nm.

4.13 Extracellular Synthesis of Nanoparticles by Fungi

The synthesis of nanoparticles outside the cell, extracellularly, has many applications as it is void of unnecessary adjoining cellular components from the cell. Mostly, fungi are regarded as organisms that produce nanoparticles extracellularly because of their enormous secretory components, which are involved in the reduction and capping of nanoparticles. Bhainsa and D'Souza (2008) reported the use of *Aspergillus fumigatus* in the production of monodispersed silver nanoparticles of size 5–25 nm within 10 min when silver nitrate was added. TEM micrograph showed nanoparticles with variable shapes, predominantly spherical and triangular. This is the fastest reduction by a biological process, even compared with physical and chemical processes (Figure 4.3). Extracellular synthesis of the pyramidal morphology of silver nanoparticles was reported in white-rot fungus *P. chrysosporium* when challenged with silver nitrate. Environmental SEM analysis revealed that silver nanoparticles were in the size range of 50–200 nm on the surface of the mycelium. This demonstrated the presence of reductase enzymes on the surface of the mycelium, which reduced silver ions to silver nanoparticles (Vigneshwaran *et al.*, 2006). The presence of thick capping agent provided more stabilization to the polydispersed silver nanoparticles. Also, Basavaraja *et al.*, 2008) demonstrated that when culture filtrate of *F. semitectum* is treated with silver ions, it reduces to silver nanoparticles with size range of 10–60 nm with spherical morphology, indicating polydispersity. This was found to be stable for many weeks. Similarly, *Aspergillus niger* was isolated from soil-produced spherical silver nanoparticles of 20 nm in diameter. Elemental spectroscopy demonstrated the presence of fungal protein for the stabilization of the nanoparticles. The reduction of $Ag+$ ions occurred through the action of nitrate reductase enzyme and quinine in extracellular electron transfer (Gade *et al.*, 2008). In addition, *F. solani*, a phytopathogenic fungus of the onion, produced polydispersed spherical silver nanoparticles in the range of 16.23 nm (Ingle *et al.*, 2009).

4.14 Silver Nanoparticles from White-Rot Fungi

Since the last decade, white-rot fungi have been used intensively for bioremediation, as they have the capability to transform or mineralize a wide range of environmentally hazardous compounds (Paszczynski and Crawford, 1995; Cameron *et al.*, 2000) through oxidative enzymatic mechanisms. The discovery of ligninolytic enzymes such as lignin peroxidase (LiP), manganese-peroxidase (MnP), and laccase from white-rot fungi has triggered biochemical research on lignin biodegradation (Glenn *et al.*, 1983; Kuwahara *et al.*, 1984; Kertsen, 1990; Wesenberg *et al.*, 2003; Unyayar, 2005).

However, the mechanistic action of these fungal-secreted enzymes is nonspecific, non-stereospecific, and extracellular in nature (David and Steven, 1994). Although the application of these fungi involves mainly lignin degradation, they are known to degrade a wide range of hydrocarbons such as polyaromatic hydrocarbons (PAHs), chlorinated aromatic

TABLE 4.1
List of Fungi That Synthesize Nanoparticles

Microorganism	Nanoparticle	Localization/ Morphology	Size	Reference
Fungus-intracellular				
Verticillium	Au	Cell wall/spherical, cytoplasmic membrane/quasi-hexagonal	20±8 nm	(Mukherjee *et al.*, 2001)
Trichothecium sp	Au	ND	ND	(Ahmad *et al.*, 2005)
V. luteoalbum and isolate 6-3	Au	Spherical, spheres and rods	>10 nm	(Gericke, 2006; Gericke, 2006a)
Verticillium sp.	Ag	Cell wall, cytoplasmic membrane/ spherical	25 ± 12 nm	(Senapati *et al.*, 2001; Mukherje *et al.*, 2001)
Aspergillus flavus	Ag	Cell wall	8.92 nm	(Vigneshwaran *et al.*, 2006)
Fungus-extracellular				
Fusarium oxysporum	Au	Spherical, triangular	20–40 nm	(Mukherjee *et al.*, 2003)
Colletotrichum sp	Au	Spherical	20–40 nm	(Shankar *et al.*, 2003)
Trichothecium sp.	Au	Triangular, hexagonal	5–200 nm	(Ahmad *et al.*, 2005)
Trichoderma asperellum	Ag	ND	13–18 nm	(Mukherjee *et al.*, 2008)
T. viride	Ag	Spherical, rod-like	5–40 nm	(Fayaz *et al.*, 2010)
F. oxysporum	Ag	ND	5–50nm	(Senapati *et al.*, 2004
Phanerochaete chrysosporium	Ag	Pyramidal	5–200 nm	(Vigneshwaran *et al.*, 2006
F. solani USM 3799	Ag	Spherical	16.23 nm	(Ingle *et al.*, 2009)
F. semitectum	Ag	Spherical	10–60 nm	(Basavaraja *et al.*, 2008)
F. acuminatum Ell. & Ev.	Ag	Spherical	5–40 nm	(Ingle *et al.*, 2008)
A. fumigates	Ag	Spherical, triangular	5–25 nm	(Bhainsa and D'Souza, 2006)
Coriolus versicolor	Ag	Spherical	25–75 nm 444–491 nm	(Snghi, 2009)
Aspergillus niger	Ag	Spherical	20 nm	(Gade *et al.*, 2008)
Phoma glomerata	Ag	Spherical	60–80 nm	(Birla *et al.*, 2009)
Penicillium brevicompactum	Ag	ND	58.35 ± 17.88 nm	(Shaligram *et al.*, 2009)
Cladosporium cladosporioides	Ag	Spherical	10–100 nm	(Balaj *et al.*, 2009)
Penicillium fellutanum	Ag	Spherical	5–25 nm	(Kathiresan *et al.*, 2009)
F. oxysporum	Au–Ag	ND	8–14 nm	(Senapati *et al.*, 2004)
Volvariella volvacea	Au, Ag, Au–Ag	Spherical, hexagonal	20–150 nm	(Philip, 2009)
F. oxysporum	Si	Quasi-spherical	5–15 nm	(Bansal *et al.*, 2005)
F. oxysporum	Ti	Spherical	6–13 nm	(Bansal *et al.*, 2004)
F. oxysporum	Zr	Quasi-spherical	3–11 nm	(Bansal *et al.*, 2004)
F. oxysporum f. sp. lycopersici	Pt	Triangular, hexagonal, square, rectangular	10–50 nm	(Riddin *et al.*, 2006)
F. oxysporum	Magnetite	Quasi-spherical	20–50 nm	(Bharde *et al.*, 2006)
Verticillum sp.	Magnetite	Cubo-octahedral	100–400 nm	(Bharde *et al.*, 2006)
F. oxysporum	CdSe	Spherical	9–15 nm	(Kumar *et al.*, 2007)
F. oxysporum	SrCO3	Needle morphology	ND	(Rautaray *et al.*, 2004)
F. oxysporum	BaTiO3	Quasi-spherical	4 ± 1 nm	(Bansal *et al.*, 2006)
F. oxysporum	Bi2O3	Quasi-spherical	5–8 nm	(Uddin *et al.*, 2008)

hydrocarbons (CAHs) (Wang *et al.*, 2008), polycyclic aromatics, polychlorinated biphenols (Siripong *et al.*, 2009), polychlorinated dibenzo-p-dioxins, pesticides DDT, lindane, and azo dyes (Wesenberg *et al.*, 2003; Unyayar, 2005). As stated by a few researchers, the most common white-rot fungi used in bioremediation are *Phanerochaete chrysosporium* (Wang, *et al.*, 2008), *Schizophyllum commune* (El-Rahim *et al.*, 2009), and *Pycnoporus sanguineus* (Trovaslet and Enaud, 2007).

Although the white-rot fungi are commonly used in bioremediation, it was reported that these fungi can also serve as a platform for the bioreduction of Ag nanoparticles (AgNPs) (Vigneshwaran *et al.*, 2006). Bioreduction is reported to be a biomimetic synthesis which utilizes natural or biological principles, and can be implemented in engineering (Vigneshwaran *et al.*, 2006). The process involved absorption of metal ions onto the microbial surface by functional groups on the cellwall, and indirectly reduced by reducing sugars from hydrolysate of polysaccharides of the biomass in to metal atoms (Chan and Mashitah, 2013).

4.15 Applications of Silver Nanoparticles

Silver is well known to possess strong antimicrobial properties both in its metallic and nanoparticle forms; hence, it has found major application in the field of medicine. Ancient civilizations also used this precious metal in medicine, as well as to make

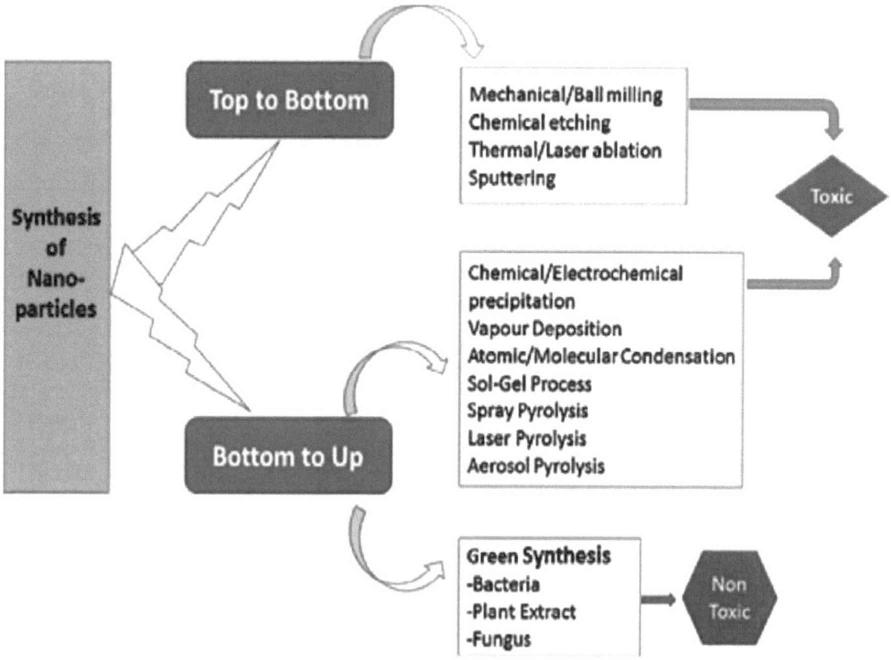

FIGURE 4.3 Methods for nanoparticle sythesis.

eating utensils, plates, cups, foods, food containers, jewelry, coins, clothes, building materials, disinfectants for water and human infections (Nagendra *et al.*, 2008). Silver nanoparticles were also used for the impregnation of polymeric medical devices to increase their antibacterial activity. Silver-impregnated medical devices such as surgical masks and implantable devices show significant antimicrobial efficiency (Furno *et al.*, 2004).

Nanotechnology has a wide range of applications in the fields of medicine, biology, optics, electricity, mechanics, optoelectronics, etc. Silver nanoparticles have also been used for various applications such as non-linear optics, spectrally selective coating for solar energy biolabeling, absorption, and antibacterial activities. Silver nanoparticles have shown promise against gram-positive bacterium *Staphylococcus aureus*. Nanoparticles have also been incorporated in cloth, which has shown promise to be sterile and thus help minimizing infections. Kumar *et al.*, 2008) reported that metal nanoparticle-embedded paints have been synthesized using vegetable oils and have been found to have good antibacterial activity.

Presently, researchers are studying the usage of magnetic nanoparticles in the detoxification of military personnel in case of biochemical warfare. It is hypothesized that by utilizing the magnetic field, gradient toxins can be removed from the body. The improved catalytic properties of the surfaces of nanoceramics or those of noble metals such as platinum and gold are used in the destruction of toxins and other hazardous chemicals (Salata, 2004).

Pankhurst *et al.*, 2003) reported that magnetic nanoparticles are also used in targeted therapy, where a cytotoxic drug is attached to a biocompatible magnetic nanoparticle. When these particles circulate in the bloodstream, external magnetic fields are used to concentrate the complex at a specific target site within the body. Once the complex is concentrated in the target, the drug can be released by enzymatic activity or by changes in pH or temperature, and is taken up by the tumor cells. Nowadays, porous nanoparticles are used in cancer therapy where the hydrophobic version of a dye molecule is trapped inside the Ormosil nanoparticle. The dye is used to generate atomic oxygen which is taken up more by the cancer cells than by healthy tissue. When the dye is not entrapped, it travels to the eyes and skin, making the patient sensitive to light. Entrapment of the dye inside the nanoparticle ensures that the dye does not migrate to other parts, and also that the oxygen-generating ability is not affected (Prathna, 2010)

Dressings have an important part to play in the management of wounds, whether they be open, sutured, or chronic wounds of many aetiologies which are healing by secondary intention. Nanocrystalline technology appears to give the highest sustained release of silver to a wound without clear risk of toxicity (Leaper, 2006).

4.16 Antimicrobial Activity

Nanotechnology refers broadly to a field of applied science and technology whose unifying theme is the control of matter on the atomic and molecular scale. Metal-microbe interactions have an important role in several biotechnological applications, including the fields of bioremediation, biomineralization, bioleaching, and microbial corrosion (Oza *et al.*, 2012). Studies have demonstrated that specially formulated metal oxide nanoparticles have good antimicrobial activity (Nithya *et al.*, 2009).

Since ancient times, it has been known that silver and its compounds have strong inhibitory and bactericidal effects as well as a broad spectrum of antimicrobial activities for bacteria, fungi, and viruses (Nasrollahi *et al.*, 2011). In recent years, due to the development of resistant strains, the resistance of bacteria to bactericides and antibiotics has increased. Some antimicrobial agents are extremely irritant and toxic and there is a vital need to find ways to formulate new types of safe and cost-effective biocidal materials (Sondi *et al.*, 2004). The antimicrobial activity of

Ag nanoparticles against microorganisms is evaluated through the agar disk diffusion method (Jain *et al.*, 2012).

The antimicrobial nature of silver nanoparticles is the most exploited feature of silver nanoparticles in the medical field (Swathy, 2014). Many reports clearly revealed the broad-spectrum antibacterial activity of AgNPs against both Gram-positive and Gram-negative bacteria, including multi resistant strains (Panacek *et al.*, 2009). It is worth mentioning that AgNPs are performing multiple modes of inhibitory action against microorganisms, rather than a single specific action like that of antibiotics (Gogoi *et al.*, 2006; Kim *et al.*, 2008; Jo *et al.*, 2009;). Interestingly, AgNPs are effective against *Candida sp.*,dermatophytes, and a few phytopathogenic fungi including *Bipolaris sorokiniana* and *Magnapothe grisea* (Morones *et al.*, 2005; Panacek *et al.*, 2006; Shrivastava *et al.*, 2007). On the other hand, many phytopathogenic fungi are not explored, although they are causing dreadful diseases on important crop plants thereby reducing the yield of agricultural products (Sachindra, 2014).

4.17 Anticandidal Activity

The *Candida* species belongs to the normal microbiota of an individual's gastrointestinal tract, mucosal oral cavity, and vagina (Shao *et al.*, 2007), and is responsible for many clinical manifestations, from mucocutaneous overgrowth to bloodstream infections (Eggimann *et al.*, 2003). These yeasts form symbiotic relationship in healthy humans and may cause systemic infections in immune-compromised states due to their great adaptability to different host niches (Sardi *et al.*, 2013). Presently, there is a sudden increase in the number of yeasts that are impervious to antifungal drugs worldwide; therefore, the use of *in vitro* laboratory tests may help the doctor in selecting an appropriate therapy. The effects of silver nanoparticles against fungal pathogens have received only marginal attention from the researchers. AgNps have exhibited significant antifungal activity against clinical isolate strains of *Candida* spp. that cause life-threatening fungal infections (Panacek *et al.*, 2009). According to Kim *et al.*, (2008) the AgNPs show potent fungistatic activity against clinical isolates and ATCC strains of *Trichophyton mentagrophytes* and *Candida spp.* Consequently, AgNPs are proving to be an alternative for the development of new antifungal drugs; nevertheless, much effort is needed to understand their effects against other fungal pathogens.

4.18 Application of Biogenic Silver Nanoparticles in Fabrics

With respect to the application of silver nanoparticles in the production of sterile materials, most studies in the literature about producing sterile fabrics with silver use silver ions or silver nanoparticles produced by chemical methods. However, some studies have applied biogenic silver nanoparticles in fabrics. (Duran *et al.*, 2007) studied the impregnation of biogenic silver nanoparticles in cotton and polyester fabrics. The particles were produced by *Fusarium oxysporum*, and 2 % silver nanoparticles impregnated in the fabrics were obtained. These fabrics exhibited high antibacterial effects against *S. aureus* (99.9 % bacterial reduction). The fabrics, after the antibacterial assay, were analyzed by SEM-EDS and showed the presence of a silver peak and the absence of contamination with bacteria. However, bacterial contamination on top of the fabric was observed in the samples without silver nanoparticles.

4.19 Anticancer Activity

Silver is an effective antibacterial agent with low toxicity, which is important especially in the treatment of burn wounds. Given its broad-spectrum activity, AgNPs have been the focus of increasing interest and are being used as an excellent candidate for therapeutic purposes (Bellantone *et al.*, 2000).

Cancer is an abnormal type of tissue growth in which the cells exhibit an uncontrolled division, in a relatively autonomous fashion, leading to a progressive increase in the number of dividing cells (Kanchana and Balakrishna, 2011). There is increasing demand for anticancer therapies (Unno *et al.*, 2005). *In vitro* cytotoxicity testing procedures reduce the use of laboratory animals (Abraham *et al.*, 2004); hence, the use of cultured tissues and cells has increased (Byrd *et al.*, 2000).

The discovery and identification of new antitumor drugs with low side effects on the immune system has become an essential goal in many studies of immuno-pharmacology (Xu *et al.*, 2009). With this aim, a lot of attentions has been paid to microorganisms. Many medically relevant nanoparticles such as AgNPs were investigated for their cytotoxicity aspect. AgNPs showed different degrees of *in vitro* cytotoxicity (Hsin *et al.*, 2008). The apoptotic effect of nanosilver is mediated by a ROS- and JNK-dependent mechanism involving the mitochondrial pathway in NIH3T3 cells (Park *et al.*, 2007 Saraniya *et al.*, 2012).

AgNPs have been proven to have great potential in anticancer activity because they are selectively involved in the disruption of the mitochondrial respiratory chain, which leads to the production of reactive oxygen species (ROS), and in the interruption of adenosine triphosphate (ATP) synthesis, thereby causing nucleic acid damage. Excess free radicals generated in the body play a key role in many degenerative diseases connected to aging, having antitumor and antioxidant properties (Vasanth *et al.*, 2014; Sherif *et al.*, 2015).

Many attempts have been made to use silver nanoparticles as an anticancer agent and they have all given positive results (Vaidyanathan *et al.*, 2010). The role of silver nanoparticles as an anticancer agent should open new doors in the field of medicine. AgNPs were also found to play an effective role in tumor control via their cytotoxic effects (Lara *et al.*, 2010; Sukirtha *et al.*, 2012). Therefore, the present study aims at evaluating the anticancer effects of biologically synthesized AgNPs through their cytotoxic activity.

4.20 Nanotechnology in Wood Protection

Wood protection nanotechnology can also be used to produce better wood products. Using nano-metals and nano-delivery systems to modify coatings at the molecular level results in products that make wood more durable and resistant to moisture changes, decay, fire, UV light, pests, and fungi. Some of these enhanced products are already on the market. Specifically, a company in Australia has created a water resistant coating called Nanoseal Wood, which is marketed as water repellant but a unique ability to bond to the wood surface, also protects it from UV light and decay. Protection in the form of biocides can be improved by metal nanomaterials, which promote even distribution while being non-degradable, odorless, and long lasting. Limitations in this area are mostly cost related, because nano-metals and nanodelivery mechanisms is not as cheap as employing nanocellulose. Moreover, in the category of wood protection, as noted earlier, biosensors can signal potential trouble related to zones of high-moisture content.

Conclusion

Nanomedicine is a burgeoning field of research with tremendous prospects for the improvement of the diagnosis and treatment of human diseases. Research is currently carried out by manipulating microorganisms at the genomic and proteomic levels. With the recent progresses and the ongoing efforts in improving particle synthesis efficiency and exploring their biomedical applications, it is hoped that the implementation of these approaches on a large scale and their commercial application in medicine and health care will take place in the coming years.

REFERENCES

Abraham, S.A., McKenzie, C., Masin, D., Harasym, T.O., Mayer, L.D., Bally, M.B. (2004). *In-vitro* and *in-vivo* characterization of doxorubicin and vincristine coencapsulated within liposomes through use of transition metal ion complexation and pH gradient loading. *J. Clin. Cancer Res.*, **10**: 728–738.

Acevedo, F., Pizzul, L., Castillo, M.P., González, M.E., Cea, M., Gianfreda, L., Diez, M.C. (2012). Degradation of polycyclic aromatic hydrocarbons by free and nanoclay-immobilized manganese peroxidase from *Anthracophyllum discolor*. *Chemosphere*, **80**: 271–278.

Ahmad, A., Mukherjee, P., Senapati, S., Mandal, D., Khan, M.I., Kumar, R., Sastry, M. (2003a). Extracellular biosynthesis of silver nanoparticles using the fungus *Fusarium oxysporum*. *Coll. Surf. B. Biointerf.*, **28**(4): 313–318.

Ahmad, A., Senapati, S., Khan, M.I., Kumar, R., Sastry, M. (2003b). Extracellular biosynthesis of monodisperse gold nanoparticles by a novel extremophilic actinomycete, *Thermomonospora* sp. *Langmuir*, **19**(8): 3550–3553.

Ahmad, A., Senapati, S., Khan, M.I., Kumar, R., Sastry, M. (2005). Extra-/intracellular biosynthesis of gold nanoparticles by an alkalotolerant fungus, *Trichothecium* sp. *J. Biomed. Nanotechnol.*, **1**(1): 47–53.

Ahmad, A., Senapati, S., Khan, M.I., Kumar, R., Ramani, R., Srinivas, V., Sastry, M. (2003c). Intracellular synthesis of gold nanoparticles by a novel alkalotolerant actinomycete, *Rhodococcus* sp. *Nanotechnology*, **14**(7): 824–828.

Ahmad, Z., Pandey, R., Sharma, S., Khuller, G.K. (2006). Alginate nanoparticles as antituberculosis drug carriers: Formulation development, pharmacokinetics and therapeutic potential. *Indian J. Chest Dis. Allied Sci.*, **48**(3): 171–176.

Amin, B.M., Farideh, N., Mona, M., Paridah, M.d.T., Susan, A., Rosfarizan, M. (2015). *Review* nanoparticles biosynthesized by fungi and yeast: A review of their preparation, properties, and medical applications. *Molecules*, **20**: 540–565.

Anil, K.S., Majid, K.A., Gosavi, S.W., Kulkarni, S.K., Pasricha, R., Ahmad, A., and Khan, M.I. (2007). Nitrate reductase mediated synthesis of silver nanoparticles from $AgNO_3$. *Biotechnol. Lett*, **29**: 439–445.

Azim, A., Davood, Z., Ali, F., Mohammad, R.M., Dariush, N., Shahram, T., Majid, M., Nasim, B. (2009). Synthesis and characterization of gold nanoparticles by tryptophane. *Am. J. Appl. Sci.*, **6**: 691–695.

Balaji, D.S., Basavaraja, S., Deshpande, R., Mahesh, B.D., Prabhakar, B.K., Venkataraman, A. (2009). Extracellular biosynthesis of functionalized silver nanoparticles by strains of *Cladosporium cladosporioides* fungus. *Coll. Surf. B. Biointerf.*, **68**: 88–92.

Balch, P.A. (2006). *Prescription for Nutritional Healing*. Penguin Group(USA) Inc., New York.

Bansal, V., Poddar, P., Ahmad, A., Sastry, M. (2006). Room-temperature biosynthesis of ferroelectric barium titanate nanoparticles. *J. Am. Chem. Soc.*, **128**(36): 58–63.

Bansal, V., Rautaray, D., Ahmad, A., Sastry, M.J. (2004). Biosynthesis of zirconia nanoparticles using the fungus *Fusarium oxysporum. Mater. Chem.*, **14**: 3303.

Bansal, V., Rautaray, D., Bharde, A., Ahire, K., Sanyal, A., Ahmad, A., Sastry, M. (2005). Fungus-mediated biosynthesis of silica and titania particles. *J. Mater. Chem.*, **15**(26): 83–89.

Barathmanikanth, S., Kalishwaralal, K., Sriram, M., Pandiyan, S.R.K., Youn, H.S., Eom, S., Gurunathan, S. (2010). Antioxidant effect of gold nanoparticles restrains hyperglycemic conditions in diabetic mice. *J. Nanobiotechnol.*, **8**: 16–20.

Basavaraja, S., Balaji, S.D., Lagashetty, A., Rajasab, A.H., Venkataraman, A. (2008). Extracellular biosynthesis of silver nanoparticles using the fungus, *Fusarium semitectum. Mat. Res. Bull.*, **43**(5): 1164–1170.

Bellantone, M., Coleman, N.J., Hench, L.L. (2000). Bacteriostatic action of a novel four-component bioactive glass. *J. Biomed. Mater. Res.*, **51**(3): 484–490.

Bhainsa, K.C., Dsouza, S.F. (2006). Extracellular biosynthesis of silver nanoparticles using the fungus *Aspergillus fumigatus*. *Coll. Surf. B. Interf.*, **47**: 160–164.

Bharde, A., Rautaray, D., Bansal, V., Ahmad, A., Sarkar, I., Yusuf, S. M., Sastry, M. (2006). Extracellular biosynthesis of magnetite using fungi. *Small*, **2**: 135–141.

Birla, S.S., Tiwari, V.V., Gade, A.K., Ingle, A.P., Yadav, A.P., Rai, M.K. (2009). Fabrication of silver nanoparticles by *Phoma glomerata* and its combined effect against *Escherichia coli*, *Pseudomonas aeruginosa* and *Staphylococcus aureus*. *Lett. Appl. Microbiol.*, **48**(2): 173–179.

Blackmore, R.P. (1982). Magneto tactic bacteria. *Annu. Rev. Microbiol.*, **36**: 217–238.

Blackwell, M. (2011). The fungi: 1, 2, 3 ... 5.1 million species? *Am. J. Bot.*, **98**: 426–438.

Bruins, R.M., Kapil, S., Oehme, S.W. (2000). Microbial resistance to metals in the environment. *Ecotoxicol. Environ. Saf.*, **45**(3): 198–207.

Byrd, J.C., Lucas, D.M., Mone, A.P., Kitner, J.B., Drabick, J.J., Grever, M.R. (2000). Anovel therapeutic agent with *in-vitro* activity against human B-cell chronic lymphocytic leukemia cells mediates cytotoxicity via the intrinsic pathway of apoptosis. *J. Hematol.*, **101**: 4547–4550.

Cameron, M., Timofeevski, S., Aust, S. (2000). Enzymology of *Phanerochaete chrysosporium* with respect to the degradation of recalcitrant compounds and xenobiotics. *Appl. Microbiol. Biotechnol.*, **54**(6): 751–788.

Castro, L.E., Moreno, V.S.D., Vilchis, N.A.R., Arenas, B.E., Avalos, B.M. (2012). Production of platinum nanoparticles and nanoaggregates using *Neurosporacrassa. J. Microbiol. Biotechnol.*, **22**(7): 1000–1004.

Castro, L.E., Vilchis, N.A.R., Avalos, B.M. (2011). Biosynthesis of silver, gold and bimetallic nanoparticles using the filamentous fungus *Neurospora crassa. Coll. Surf. B. Biointerf.*, **83**(1): 42–48.

Chan, Y.S., Don, M.M. (2012). Characterization of Ag nanoparticles produced by white-rot fungi and its in vitro antimicrobial activities. *Intern. Arab. J. Antimicrob. Agents*, **2**: 3–5.

Chan, Y.S., Don, M.M. (2013). Biosynthesis and structural characterization of Ag nanoparticles from white rot fungi. *Mat. Sci. Eng.*, **33**: 282–288.

Chan, Y.S., Don, M.M. (2013). Biosynthesis of silver nanoparticles from Schizophyllum commune and in-vitro antibacterial and antifungal activity studies. *J. Phys. Sci.*, **24**: 83–96.

Chen, J.C., Lin, Z.H., Ma, X.X. (2003). Evidence of the production of silver nanoparticles via pretreatment of Phoma sp.3.2883 with silver nitrate. *Lett. Appl. Microbiol.*, **37**(2): 105–108.

David, P., Steven, D. (1994). Mechanisms white-rot fungi use to degrade pollutants. *Environ. Sci. Technol.*, **28**: 79–87.

Dolgaev, S.I.S., Voronov, A.V., Shafeev, V.V., Bozon, V.F. (2002). Nanoparticles produced by laser ablation of solids in liquid environment. *Appl. Surf. Sci.*, **186**(1–4): 546–551.

Duran, N., Marcato, P.D., Alves, O.L., DeSouza, G.I.H., Esposito, E. (2005). Mechanistic aspects of biosynthesis of silver nanoparticles by several *Fusarium oxysporum* strains. *J. Nanobiotechnol.*, **3**: 3–8.

Eggen, T., Majcherczyk, A. (1998). Removal of polycyclic aromatic hydrocarbons (PAH) in contaminated soil by white rot fungus *Pleurotus ostreatus. Int. Biodeterior. Biodegrad.*, **41**(2): 111–117.

Eggimann, P., Garbino, J., Pittet, D. (2003). Epidemiology of *Candida* species infections in critically ill non-immunosuppressed patients. *Lancet Infect. Dis.*, **3**(11): 685–702.

Elechiguerra, J.L., Burt, J.L., Morones, J.R., Bragado, A.C., Gao, X., Lara, H.H., Yacaman, M.J. (2005). Interaction of silver nanoparticles with HIV-1. *J. Nanobiotechnol.*, **3**: 6.

El-Rahim, W. M. A., El-Ardy, O. A. M., Mohammad, F. H. (2009). The effect of pH on bioremediation potential for the removal of direct violet textile dye by Aspergillus niger. *Desalination*, **249**: 1206–1211.

Evanoff, J.C. (2004). Size-controlled synthesis of nanoparticles. 2. Measurement of extinction, scattering, and absorption cross sections. *J. Phys. Chem. B*, **108**: 57–62.

Fayaz, M., Balaji, K., Girilal, M., Yadav, R., Kalaichelvan, P.T., Venketesan, R. (2010). Biogenic synthesis of silver nanoparticles and their synergistic effect with antibiotics: A study against Gram-positive and Gram-negative bacteria. *Nanomedicine*, **6**(1): 103–109.

Feynman, R. (1959). *Lecture at the California Institute of Technology.* December 29.

Fogliano, V., Veronica, V., Giacomino, R., Alberto, R. (1999). Method for measuring antioxidant activity and its application to monitoring the antioxidant capacity of wines. *J. Agric. Food Chem.*, **47**(3): 1035–1040.

Frattini, A., Pellegri, N., Nicastro, D., Sanctis, O. (2005). Effect of amine groups in the synthesis of Ag nanoparticles using aminosilanes. *Mater. Chem. Phys.*, **94**(1): 148–152.

Fu, J.K., Liu, Y., Gu, P., Tang, D.L., Lin, Z.Y., Yao, B.X., Weng, S. (2000). Spectroscopic characterization on the biosorption and bioreduction of Ag(I) by *Lactobacillus* sp. A09. *Acta Phys. Chim. Sin.*, **16**: 770–782.

Furno, F., Morley, K.S., Wong, B., Sharp, B.L., Arnold, P.L., Howdle, S.M., Bayston, R., Brown, P.D., Winship, P.D., Reid, H.J. (2004). Silver nanoparticles and polymeric medical devices: A new approach to prevention of infection. *J. Antimicrob. Chemother.*, **54**(6): 1019–1024.

Gade, A.K., Bonde, P., Ingle, A.P., Marcato, P.D., Dura, N., Rai, M.K. (2008). Exploitation of *Aspergillus niger* for synthesis of silver nanoparticles. *J. Biobased Mater. Bioenergy*, **3**: 123–129.

Gericke, M., Pinches, A. (2006a). Biological synthesis of metal nanoparticles. *Hydrometallurgy*, **83**(1–4): 132.

Gericke, M., Pinches, A. (2006b). Microbial production of gold nanoparticles. *Gold Bull.*, **39**(1): 22–28.

Glenn, J., Morgan, M., Mayfield, M., Kuwahara, M., Gold, M. (1983). An extracellular H_2O_2-requiring enzyme preparation involved in lignin biodegradation by the white rot basidiomycete *Phanerochaete chrysosporium. Biochem. Biophys. Res. Commun.*, **114**(3): 1077–1083.

Gogoi, S.K., Gobinath, P., Paul, A., Ramesh, A., Ghosh, S.S., Chattopadhyay, A. (2006). Green fluorescent protein-expressing *Escherichia coli* as a model system for investigating the antimicrobial activities of silver nanoparticles. *Langmuir*, **22**(22): 9322–9328.

Gong, P., Li, H., He, X., Wang, K., Hu, J., Tan, W., Zhang, X., Yang, X. (2007). Preparation and antibacterial activity of Fe3O4@ Ag nanoparticles. *Nanotechnology*, **18**(28):285604.

Gu, H., Ho, P.L., Tong, E., Wang, L., Xu, B. (2003). Presenting vancomycin on nanoparticles to enhance antimicrobial activities. *Nano Lett.*, **3**(9): 1261–1263.

Gutteridge, J.M.C. (1993). Free radicals in disease processes: A compilation of cause and consequence. *Free Radic. Res. Commun.*, **19**(3): 141–158.

Haefeli, C., Franklin, C., Hardy, K. (1984). Plasmid-determined silver resistance in *Pseudomonas stutzeri* isolated from a silver mine. *J. Bacteriol.*, **158**(1): 389–392.

Hsin, Y.H., Chen, C.F., Huang, S., Shih, T.S., Lai, P.S., Chueh, P.J. (2008). The apoptotic effect of nanosilver is mediated by a ROS- and JNK-dependent mechanism involving the mitochondrial pathway in NIH3T3 cells. *Toxicol. Lett.*, **179**(3): 130–139.

Huang, J., Li, Q., Sun, D., Lu, Y., Su, Y., Yang, X., Hong, J. (2007). Biosynthesis of silver and gold nanoparticles by novel sundried *Cinnamomum camphora* leaf. *Nanotechnology*, **18**: 105–106.

Husseiny, O.M., Abdul-Aziz, G.M., El-Haroun, E.R., Goda, A.M.A.S. (2006). Fish meal replacer studies with Nile tilapia and mullet under polyculture conditions in Egypt. *Int. Aquafeed*, **9**: 20–29.

Inbathamizh, L., Mekalai, P.T., Jancy, M.E. (2013). In vitro evolution of antioxidant and anticancer potential of *Morinda pubescens* synthesized silver nanoparticles. *J. Pharm. Res.*, **6**: 32–38.

Ingle, A., Rai, M., Gade, A., Bawaskar, M. (2009). *Fusarium solani*: A novel biological agent for the extracellular synthesis of silver nanoparticles. *J. Nanopart. Res.*, **11**(8): 2079.

Ingle, A.P., Gade, A.K., Pierrat, S., Sonnichsen, C., Rai, M.K. (2008). Mycosynthesis of silver nanoparticles using the fungus *Fusarium acuminatum* and its activity against some human pathogenic bacteria. *Curr. Nanosci.*, **4**(2): 141–144.

Jain, P., Aggarwal, V. (2012). Synthesis, characterization and antimicrobial effects of silver nanoparticles from microorganisms. *Inter. J. Nano. Mat. Sci.*, **1**: 108–120.

Jha, A.K., Prasad, K., Prasad, K., Kulkarni, A.R. (2009). Plant system: Nature's nanofactory. *Coll. Surf. B. Biointerf.*, **73**(2): 219–223.

Jo, Y.K., Kim, B.H., Jung, G. (2009). Antifungal activity of silver ions and nanoparticles on phytopathogenic fungi. *Plant Dis.*, **93**(10): 1037–1043.

Jung, J.O., Noh, H., Ji, J., Kim, S. (2006). Metal nanoparticle generation using a small ceramic heater with a local heating area. *J. Aerosol Sci.*, **37**(12): 1662–1670.

Kabashin, A.V.M., Meunier, M. (2003). Synthesis of colloidal nanoparticles during femtosecond laser ablation of gold in water. *J. Appl. Phys.*, **94**(12): 41–43.

Kalishwaralal, K., Deepak, V., Ramkumarpandian, S., Nellaiah, H., Sangiliyandi, G. (2008). Extracellular biosynthesis of silver nanoparticles by the culture supernatant of *Bacillus licheniformis*. *Mater. Lett.*, **62**(29): 4411–4413.

Kanchana, A., Balakrishna, M. (2011). Anti-cancer effect of saponins isolated from *Solanum trilobatum* leaf extract and induction of apoptosis in human larynx cancer cell lines. *Int. J. Phar. Pharm. Sci.*, **3**: 356–364.

Kathiresan, K., Manivanan, S., Nabeel, M.A., Dhivya, B. (2009). Studies on silver nanoparticles synthesized by a marine fungus *Penicillium fellutanum* isolated from coastal mangrove sediment. *Coll. Surf. B. Biointerf.*, **71**: 133–137.

Kawasaki, M., Nishimura., N. (2006). 1064-nm laser fragmentation of thin Au and Ag flakes in acetone for highly productive pathway to stable metal nanoparticles. *Appl. Surf. Sci.*, **253**(4): 08–16.

Kertsen, P., Kalyanaraman, B., Hamel, K., Reinhammar, B., Kirk, T. (1990). Comparison of lignin peroxidase, horseradish peroxidase and laccase in the oxidation of methoxybenzenes. *Biochem. J.*, **268**(2): 475–480.

Kim, K.J., Sung, W.S., Moon, S.K., Choi, J.S., Kim, J.G., Lee, D.G. (2008). Antifungal effect of silver nanoparticles on dermatophytes. *J. Microbiol. Biotechnol.*, **18**(8): 1482–1484.

Kim, S.Y.B., Chun, K., Kang, W., Choo, J., Gong, M., Joo, S. (2005). Catalytic effect of laser ablated Ni nanoparticles in the oxidative addition reaction for a coupling reagent of benzylchloride and bromoacetonitrile. *J. Mol. Catal. A Chem.*, **226**: 231–234.

Kruis, F.F., Rellinghaus, B. (2000). Sintering and evaporation characteristics of gas-phase synthesis of size-selected PbS nanoparticles. *Mater Sci. Eng. B*, **69**: 329–324.

Kumar, A., Kumar, P., Ajayan, M.P., John, G. (2008). Silver-nanoparticle-embedded antimicrobial paints based on vegetable oil. *Nat. Mater.*, **7**(3): 236–241.

Kumar, A.S., Ansary, A.A., Ahmad, A., Khan, M.I. (2007a). Extracellular biosynthesis of CdSe quantum dots by the fungus, *Fusarium oxysporum*. *J. Biomed. Nanotechnol.*, **3**(2): 190–194.

Kumar, S.A., Abyaneh, M.K., Gosavi, S.W., Kulkarni, S.K., Pasricha, R., Ahmad, A., Khan, M.I. (2007b). Nitrate reductase-mediated synthesis of silver nanoparticles from AgNO3. *Biotechnol. Lett.*, **29**(3): 439–445.

Kuwahara, M., Glenn, J., Morgan, M., Gold, M. (1984). Separation and characterization of two extracellular H_2O_2-dependent oxidases from ligninolytic cultures of *Phanerochaete chrysosporium*. *FEBS Lett.*, **169**(2): 247–250.

Leaper, D.L. (2006). Silver dressings: Their role in wound management. *Int. Woundj.*, **3**(4): 282–294.

Link, S., Burda, C., Nikoobakht, B., El-Sayed, M.A. (2000). Laser-induced shape changes of colloidal gold nanorods using femtosecond and nanosecond laser pulses. *J. Phys. Chem. B*, **104**(26): 6152–6163.

Mafune, F.K., Takeda, J., Kondow, T., Sawabe, H. (2000). Structure and stability of silver nanoparticles in aqueous solution produced by laser ablation. *J. Phys. Chem.*, **104**: 33–37.

Mafune, F.K., Takeda, J., Kondow, Y., Sawabe, H. (2001). Formation of gold nanoparticles by laser ablation in aqueous solution of surfactant. *J. Phys. Chem. B*, **105**: 114–120.

Magnusson, M.D., Malm, K., Bovin, J., Samuelson, L. (1999). Gold nanoparticles: Production, reshaping, and thermal charging. *J. Nanopart. Res.*, **1**(2): 243–251.

Maharani, V. (2015). Extracellular Synthesis of Silver Nanoparticles by a Marine Escherichia coli vm1 and their Potential Bioactive Applications. *Nanobiotech.* **17**: 65–73.

Mandal, D., Bolander, M.E., Mukhopadhyay, D., Sarkar, G., Mukherjee, P. (2006). The use of microorganisms for the formation of metal nanoparticles and their application. *Appl. Microbiol. Biotechnol.*, **69**(5): 485–492.

Mann, S. (1996). *Biomimetic Materials Chemistry*. Wiley-VCH, New York.

Mann, S. (2001). *Biomineralization: Principles and Concepts in Bioinorganic Materials Chemistry*. Oxford University Press, Oxford.

Merga, O., Lynn, G.W.R., Milosavljevic, G., Meisel, D. (2007). Redox catalysis on "naked" silver nanoparticles. *J. Phys. Chem. C*, **111**: 20–26.

Mithila, A., Swanand, J., Ameeta, R.K., Smita, Z., Sulabha, K. (2009). Biosynthesis of gold nanoparticles by the tropical marine yeast *Yarrowia lipolytica* NCIM 3589. *Mater. Lett.*, **63**(15): 1231–1234.

Mohanpuria, P., Rana, K.N., Yadav, S.K. (2008). Biosynthesis of nanoparticles: Technological concepts and future applications. *J. Nanopart. Res.*, **10**(3): 507–517.

Morones, J.R., Elechiguerra, J.L., Camacho, A., Holt, K., Kouri, J.B., Ramirez, J.T., Yacaman, M.J. (2005). The bactericidal effect of silver nanoparticles. *Nanotechnology*, **16**(10): 46–53.

Mukherjee, P., Ahmad, A., Mandal, D., Senapati, S., Sainkar, S.R., Khan, M.I., Parishcha, R., Ajaykumar, P.V., Alam, M., Kumar, R., Sastry, M. (2001). Fungus-mediated synthesis of silver nanoparticles and their immobilization in the mycelial matrix: A novel biological approach to nanoparticle synthesis. *Nano Lett.*, **1**(10): 515–519.

Mukherjee, P., Roy, M., Mandal, B.P., Dey, G.K., Mukherjee, P.K., Ghatak, J. (2008). Green synthesis of highly stabilized nano crystalline silver particles by a non-pathogenic and agriculturally important fungus *T. asperellum*. *Nanotechnology*, **19**(34): 1–7.

Mukherjee, P., Senapati, S., Mandal, D., Absar, A., Islam, M., Rajiv, K., Murali, S. (2003). Extracellular synthesis of gold nanoparticles by the fungus *Fusarium oxysporum*. *Chem. Bio. Chem.*, **3**: 461–466.

Mukherjee, P., Senapati, S., Mandal, D., Absar, A., Islam, M., Rajiv, K., Murali, S. (2002). Extracellular synthesis of gold nanoparticles by the fungus *Fusarium oxysporum*. *Chem. Bio. Chem.*, **3**: 461–466.

Nagendra, R.P., Eladina, m., Peen, M., Josf, H. (2008). Silver or silver nanoparticles: A hazardous threat to the environment and human health – A review. *J. App. Biomed.*, **6**: 117–129.

Nagy, A., Mestl, G. (1999). High temperature partial oxidation reactions over silver catalysts. *Appl. Catal. A*, **188**(1–2): 337–353.

Naik, R.R., Stringer, S.J., Agarwal, G., Jones, S.E., Stone, M.O. (2002). Ferro electricity in a one-dimensional organic quantum magnet. *Nat. Matters*. 169–172.

Nanoscience and nanotechnologies: opportunities and uncertainties, The Royal Society & The Royal Academy of Engineering (2004).

Narayanan, Kannan Badri, Sakthivel, N. (2010). Biological synthesis of metal nanoparticles by microbes. *Adv. Coll. Inter. Sci.*, 156: 1–13.

Nasrollahi, A., Pourshamsian, K., Mansourkiaee, P. (2011). Antifungal activity of silver nanoparticles on some of fungi. *Inter. J. NANO Dimen.*, **1**: 233–239.

Nithya, R., Ragunathan, R. (2009). Synthesis of silver nanoparticle using *Pleurotus Sajor Caju* and its antimicrobial activity. *Dig. J. Nanomat. Biostr.*, **4**: 623–629.

Nordberg, G., Gerhardsson, L. (1988). Silver. In Seiler, H.G., Sigel, H., Sigel. A. (eds.), *Handbook on Toxicity of Inorganic Compounds*. Marcel Dekker, New York, pp. 619–624.

Oliveira, A.P., Nielsen, J., Jochen Forster, J. (2005). Modeling *Lactococcus lactis* using a genome-scale flux model. *BMC Microbiol.*, **5**: 39.

Oza, V., Pandey, S., Shah, V., Sharon, M. (2012). Extracellular fabrication of silver nanoparticles using *Pseudomonas aeruginosa* and its antimicrobial assay. *Pel. Res. Lib. Adv. Appl. Sci. Res.*, **3**: 1776–1783.

Panáček, A., Kolář, M., Večeřová, R., Prucek, R., Soukupová, J., Kryštof, V., Kvítek, L. (2009). Antifungal activity of silver nanoparticles against Candida spp. *Biomaterials*, **30**(31): 6333–6340.

Panacek, A., Kvitek, L., Prucek, R., Kolar, M., Vecerova, R., Pizurova, N., Sharma, V.K., Nevecna, T., Zboril, R. (2006). Silver colloid nanoparticles: Synthesis, characterization, and their antibacterial activity. *J. Phys. Chem. B*, **110**(33): 48–53.

Panigrahi, T. (2013). Synthesis and characterization of silver nanoparticles using leaf extract of *Azadirachta indica*. Project thesis, National Institute of Technology, Rourkela.

Pankhurst, Q.A., Connolly, J., Jones, S.K., Dobson, J. (2003). Applications of magnetic nanoparticles in biomedicine. *J. Phys. D: Appl. Phys.*, **36**(13): 167–181.

Parashar, U.K., Saxena, S.P., Srivastava, A. (2009). Bioinspired synthesis of silver nanoparticles. *Dig. Nanomat. J. Biostruct.*, **4**: 159–166.

Park, S., Lee, Y.K., Jung, M., Kim, K.H., Chung, N., Ahn, E.K., Lim, Y., Lee, K.H. (2007). Cellular toxicity of various inhalable metal nanoparticles on human alveolar epithelial cells. *Inhal. Toxicol.*, **19** Supplement 1: 59–65.

Paszczynski, A., Crawford, R. (1995). Potential for bioremediation of xenobiotic compounds by the white-rot fungus Phanerochaete chrysosporium. *Biotechnol. Prog.*, **11**(4): 368–379.

Peyser, L.A., Vinson, A.E., Bartko, A.P., Dickson, R.M. (2001). Photoactivated fluorescence from individual silver nanoclusters. *Science*, **291**(5501): 103–106.

Philip, D. (2009). Biosynthesis of Au, Ag and Au–Ag nanoparticles using edible mushroom extract. *Spectrochim. Acta A*, **73**(2): 374–381.

Pighi, L., Pumpel, T., Schinner, F. (1989). Selective accumulation of silver by fungi. *Biotechnol. Lett.*, **11**(4): 275–279.

Prabhu, S., Poulose, E. K. (2011). Silver nanoparticles: Mechanism of antimicrobial action, synthesis, medical applications, and toxicity effects. *Int. Nano Lett.*, **2**: 32–36.

Prathna, T.C., Lazar, M., Chandrasekaran, N., Ashok, M., Raichur, Amitava, M. (2010). Biomimetic synthesis of nanoparticles: Science, technology & applicability, biomimetics learning from nature. In Amitava Mukherjee (ed.), ISBN: 978-953-307-025-4, InTech. Available from: http://www.intechopen.com /books/biomimetics-learning-from-nature/biomimetic-synthesis-of-nanoparticlesscience-technology-amp-applicability.

Prior, R.L., Cao, G. (1999). In vivo total antioxidant capacity: Comparison of different analytical methods. *Free Rad. Biol. Med.*, **27**(11–12): 1173–1181.

Pum, D., Sleytr, U.B. (1999). The application of bacterial S-layers in molecular nanotechnology. *Trends Biotechnol.*, **17**(1): 8–12.

Quang, H. Tran, Nguyen, V. Q., Le, A. (2013). Review silver nanoparticles: Synthesis, properties, toxicology, applications and perspectives. *Adv. Nat. Sci. Nanosci. Nanotechnol.*, **4**: 1–20.

Rai, M., Yadav, A., Bridge, P., Gade, A. (2009a). Myconanotechnology: A new and emerging science. In M.K. Rai, P.D. Bridge (eds.), *Applied Mycology*, vol 14. CAB International, New York, pp. 258–267.

Rai, M., Yadav, A., Gade, A. (2008). CRC 675—Current trends in phytosynthesis of metal nanoparticles. *Crit. Rev. Biotechnol.*, **28**(4): 277–284.

Rai, M., Yadav, A., Gade, A. (2009b). Silver nanoparticles: As a new generation of antimicrobials. *Biotechnol. Adv.*, **27**(1): 76–83.

Rai, M., Nelson Duran. (2011) eds. Metal nanoparticles in microbiology. Springer Science & Business Media.

Rautaray, D., Sanyal, A., Adyanthaya, S.D., Ahmad, A., Sastry, M. (2004). Biological synthesis of strontium carbonate crystals using the fungus *Fusarium oxysporum*. *Langmuir*, **20**: 27–33.

Retchkiman, P.S.S., Canizal, G., Herrera-Becerra, R., Zorrilla, C., Liu, H.B., Ascencio, J.A. (2006). Biosynthesis and characterization of Ti/Ni bimetallic nanoparticles. *Opt. Mater.*, **29**(1): 95–99.

Riddin, T.L., Gericke, M., Whiteley, C.G. (2006). Analysis of the inter- and extracellular formation of platinum nanoparticles by *Fusarium oxysporum* sp. lycopersici using response surface methodology. *Nanotechnology*, **17**(14): 3482–3489.

Rubilar, O., Diezand, M.C., Gianfreda, L. (2008). Transformation of chlorinated phenolic compounds by white rot fungi. *Crit. Rev. Environ. Sci. Tech.*, **38**(4): 227–268.

Sachindra, R. (2014). Thesis on synthesis and characterization of biogenic silver nanoparticles and its various applications. Vels University.

Salata, O.V. (2004). Applications of nanoparticles in biology and medicine. *J. Nano. Biotech.*, **2**(1): 3–8.

Sanghi, R., Verma, P. (2009). Biomimetic synthesis and characterisation of protein capped silver nanoparticles. *Bioresour. Technol.*, **100**(1): 501–504.

Saraniya, D.J., Valentine, B., Krupa, R. (2012). *In vitro* anticancer activity of silver nanoparticles synthesized using the extract of *Gelidiella sp. Int. J. Pharm. Pharm. Sci.*, **4**: 710–715.

Sardi, J.C.O., Scorzoni, L., Bernardi, T., Fusco-Almeida, A.M., Giannini, M.M. (2013). Candida species: Current epidemiology, pathogenicity, biofilm formation, natural antifungal products and new therapeutic options. *J. Med. Microbiol.*, **62**(1): 10–24.

Sastry, M., Ahmad, A., Islam, K.M., Kumar, R. (2003). Biosynthesis of metal nanoparticles using fungi and actinomycete. *Curr. Sci.*, **85**: 162–170.

Sau, T.K., Rogach, A.L. (2010). Nonspherical noble metal nanoparticles: Colloid-chemical synthesis and morphology control. *Adv. Mater.*, **22**(16): 1781–1804.

Senapati, S., Ahmad, A., Khan, M.I., Sastry, M., Kumar, R. (2005). Extracellular biosynthesis of bimetallic Au–Ag alloy nanoparticles. *Small*, **1**(5): 517–520.

Senapati, S., Mandal, D., Ahmad, A., Khan, M.I., Sastry, M., Kumar, R. (2001). Fungus mediated synthesis of silver nanoparticles: A novel biological approach. *Ind. J. Phys.*, **78**: 101–105.

Senapati, S., Mandal, D., Ahmad, A., Khan, M.I., Sastry, M., Kumar, R. (2004). Fungus mediated synthesis of silver nanoparticles: A novel biological approach. *Ind. J. Phys.*, **78**: 101–105.

Shakeel, A., Ahmad, M., Swami, B.L., Ikram, S. (2016). A review on plants extract mediated synthesis of silver nanoparticles for antimicrobial applications: A green expertise. *J. Adv. Res.*, **7**(1): 17–28.

Shaligram, N.S., Bule, M., Bhambure, R., Singhal, R.S., Singh, S.K., Szakacs, G. (2009). Biosynthesis of silver nanoparticles using aqueous extract from the compacting producing fungi. *Proc. Biochem.*, **44**: 939–943.

Shankar, S.S., Ahmad, A., Pasricha, R., Sastry, M. (2003). Bioreduction of chloroaurate ions by geranium leaves and its endophytic fungus yields gold nanoparticles of different shapes. *J. Mat. Chem.*, **13**(7): 22–26.

Shao, L.C., Sheng, C.Q., Zhang, W.N. (2007). Recent advances in the study of antifungal lead compounds with new chemical scaffolds. *Yao Xue Xue Bao*, **42**: 1129–1136.

Sharma, V.K., Yngard, R.A., Lin, Y. (2009). Silver nanoparticles: Green synthesis and their antimicrobial activities. *Adv. Colloid Interface Sci.*, **145**(1–2): 83–96.

Sherif, M.H., Taher, A.S., Hend, A. (2015). Biosynthesis of size controlled silver nanoparticles by *Fusarium oxysporum*, their antibacterial and antitumor activities. *J. Basic Appl. Sci.*, **4**: 225–231.

Shin, K.H., Cha, D.K. (2008). Microbial reduction of nitrate in the presence of nanoscale zero-valent iron. *Chemosphere*, **72**(2): 257–262.

Shrivastava, S., Bera, T., Roy, A., Singh, G., Ramachandrarao, P., Dash, D. (2007). Characterization of enhanced antibacterial effects of novel silver nanoparticles. *Nanotechnology*, **18**(22): 103–108.

Simkiss, K., Wilbur, K.M. (1989). *Biomineralization*. Academic, New York.

Siripong, P., Oraphin, B., Sanro, T., Duanporn, P. (2009). Screening of fungi from natural sources in Thailand for degradation of polychlorinated hydrocarbons. *Am Eurasian J Agric Environ Sci*, **5**: 466–472.

Sivagnanam, S., Jayanthi, A. (2013). Biosynthesis of silver nanoparticles review. *Afr. J. Biotechnol.*, **12**: 3088–3098.

Sivanandham, V. (2011). Free radicals in health and diseases - A mini review. *Pharmacology online*, **1**: 1062–1077.

Sondi, I., Sondi, B. (2004). Silver nanoparticles as antimicrobial agent: A case study on *E. coli* as a model for Gram-negative bacteria. *J. Coll. Interface Sci.*, **275**(1): 177–182.

Swathy, B. (2014). A review on metallic silver nanoparticles. *Iosrphr .org*, **4**: 38–44.

Sylvestre, J.P.K., Sacher, A.V., Meunier, M., Luong, J.H.T. (2004). Stabilization and size control of gold nanoparticles during laser ablation in aqueous cyclodextrins. *J. Am. Chem. Soc.*, **126**: 76–77.

Taniguchi, N. (1974). On the basic concept of nano-technology. *Proceedings of the International Conference on Production Engineering Tokyo Part II Japan Society of Precision Engineering.*

Tapan, K.S., Andrey, L.R. (2009). Nonspherical noble metal nanoparticles: Colloidchemical synthesis and morphology control. *Adv. Mater.*, **21**: 1–24.

Tarasenko, N., Butsen, A., Nevar, E., Savastenko, N. (2006). Synthesis of nanosized particles during laser ablation of gold in water. *Appl. Surf. Sci.*, **252**(13): 4439.

Thakkar, K.N., Mhatre, S.S., Parikh, R.Y. (2010). Biological synthesis of metallic nanoparticles. *Nanomedicine*, **6**(2): 257–262.

Tiwari, A.K. (2004). Antioxidants: New generation therapeutic base for treatment of polygenic disorders. *Curr. Sci.*, **86**: 1092–1102.

Tortella, G.R., Rubilar, O., Gianfreda, L., Valenzuela, E., Diez, M.C. (2008). Enzymatic characterization of Chilean native woodrotting fungi for potential use in the bioremediation of polluted environments with chlorophenols. *World J. Microbiol. Biotechnol.*, **12**(12): 2805–2818.

Trovaslet, M., Enaud, E., Guiavarc'h, Y., Corbisier, A. M., Vanhulle, S. (2007). Potential of a Pycnoporus sanguineus laccase in bioremediation of wastewater and kinetic activation in the presence of an anthraquinonic acid dye. *Enzyme Microb. Technol.*, **41**: 368–376.

Tsuji, T., Iryo, K., Watanabe, N., Tsuji, M. (2002). Preparation of silver nanoparticles by laser ablation in solution: Influence of laser wavelength on particle size. *Appl. Surf. Sci.*, **202**(1–2): 80–86.

Uddin, I., Adyanthaya, S., Syed, A., Selvaraj, K., Ahmad, A., Poddar, P. (2008). Structure and microbial synthesis of sub-10 nm Bi_2O_3 nanocrystals. *J. Nanosci. Nanotechnol.*, **8**(8): 09–13.

Unno, Y., Shino, Y., Kondo, F., Igarashi, N., Wang, G., Shimura, R., Yamaguchi, T., Asano, T., Saisho, H., Sekiya, S., Shirasawa, H. (2005). Oncolytic viral therapy for cervical and ovarian cancer cells by Sindbis virus AR339 strain. *Clin. Cancer Res.*, **11**(12): 4553–4560.

Unyayar, A., Mazmanci, M., Erkurt, E., Atacag, H., Gizir, A. (2005). Decolorization kinetics of the azo dye Drimaren blue X3LR by laccase. *React. Kinet. Catal. Lett.*, **86**(1): 99–107.

Vaidyanathan, R., Gopalram, S., Kalishwaralal, K., Deepak, V., Pandian, S.R., Gurunathan, S. (2010). Enhanced silver nanoparticle synthesis by optimization of nitrate reductase activity. *Coll. Surf. B. Biointerf.*, **75**(1): 335–341.

Vasanth, K., Ilango, K., MohanKumar, R., Agrawal, A., Dubey, G. P. (2014). Anticancer activity of Moringa oleifera mediated silver nanoparticles on human cervical carcinoma cells by apoptosis induction. *Coll. Surf. B. Biointerf.*, **117**: 354–359.

Vigneshwaran, N., Ashtaputre, N.M., Varadarajan, P.V., Nachane, R.P., Paralikar, K.M., Balasubramanya, R.H. (2007). Biological synthesis of silver nanoparticles using the fungus *Aspergillus flavus*. *Mater. Lett.*, **61**(6): 1413–1418.

Vigneshwaran, N., Kathe, A.A., Varadarajan, P.V., Nachane, R.P., Balasubramanya, R.H. (2006). Biomimetics of silver nanoparticles by white rot fungus, *Phaenerochaete chrysosporium. Coll. Surf. B.*, **53**: 55–59.

Vigneshwaran, N., Kathe, A., Varadarajan, P.V., Nachane, R., Balasubramanya, R.H. (2009). Biomimetics of silver nanoparticles by white rot fungus, *Phaenerochaete chrysosporium. Coll. Surf. B. Biointerf.*, **53**(1): 55–59.

Volesky, B., Holan, Z.R. (1995). Biosorption of heavy metals. *Biotechnol. Prog.*, **11**(3): 235–250.

Wang, C., Xi, C., Hu, J., H., Wen, X. (2008). Biodegradation of gaseous chlorobenzene by white-rot fungus *Phanerochaete chrysosporium. Biomed. Environ. Sci.*, **21**(6): 474–478.

Wesenberg, D., Kyriakides, I., Agathos, S. (2003). White-rot fungi and their enzymes for the treatment of industrial dye effluents. *Biotechnol. Adv.*, **22**(1–2): 161–187.

Wiley, B.S.Y., Mayers, B., Xi, Y. (2005). Shape-controlled synthesis of metal nanostructures: The case of silver. *Chem. Eur.*, **11**: 454–463.

Windt, W., Peter, A., Willy, V. (2005). Bioreductive deposition of palladium (0) nanoparticles on *Shewanella oneidensis* with catalytic activity towards reductive dechlorination of polychlorinated biphenyls. *Environ. Microb.*, **7**(3): 314–325.

Xu, H., Yao, L., Sung, H., Wu, L. (2009). Chemical composition and antitumor activity of different polysaccharides from the roots *Actinidia eriantha. Carbohydr. Pol.*, **78**: 316–322.

Zheng, J., Dickson, R.M. (2002). Individual water-soluble dendrimer-encapsulated silver nanodot fluorescence. *J. Am. Chem. Soc.*, **124**(47): 82–83.

5

Nanostructure Thin Films: Synthesis and Different Applications

Ho Soon Min, Debabrata Saha, J.M. Kalita, M.P. Sarma, Ayan Mukherjee, Benjamin Ezekoye, Veronica A. Ezekoye, Ashok Kumar Sharma, Manesh A. Yewale, Ayaz Baayramov, and Trilok Kumar Pathak

CONTENTS

5.1 Introduction

Thin films of metal chalcogenides have good chemical, physical, optical, and electrical properties [Atan et al., 2010]. As a result, these materials could be used for a variety of applications including photo degradation, gas sensing, energy conversion, energy storage, and field-effect transistors. Several methods exist, such as the chemical method [Teo et al., 2010], and physical techniques have been developed to produce these films. Researchers found that each deposition technique has both advantages and disadvantages.

In this chapter, authors discuss the synthesis of various thin films using different deposition techniques. Research findings from various scientists are also reported.

5.2 Atomic Layer Deposition of Thin Film

Over the past two decades, atomic layer deposition (ALD), which is a unique modification of the conventional chemical vapor deposition (CVD) technique, has emerged as a promising method to grow metal oxides, nitrides, sulfides, selenides, tellurides, and pure metals [Riikka and Puurunena, 2005; George, 2010]. Thin-film deposition in ALD relies on the alternate exposure of precursors separated by an inert gas purging step [Riikka, 2005; George, 2010]. The complementary and self-limiting surface chemical reactions during film growth provide atomic level control over film thickness and uniformity over a large deposition area [Riikka, 2005; George, 2010]. In addition, ALD offers a number of industrially and technologically important characteristics, which include excellent conformal depositions on high aspect ratio structures, easy thickness control, pinhole-free dense coatings, and relatively low deposition temperature [Riikka, 2005; George, 2010].

Among different materials, oxide-based semiconductors and dielectrics are the most extensively studied using ALD [Riikka, 2005; George, 2010]. A variety of dielectrics such as hafnium (IV) oxide (HfO_2), titanium dioxide (TiO_2), aluminium oxide (Al_2O_3), zirconium dioxide (ZrO_2), tantalum pentoxide (Ta_2O_5), their amorphous nanolaminates (Al_2O_3/TiO_2, ZrO_2/HfO_2, Ta_2O_5/HfO_2, Ta_2O_5/ZrO_2), and mixed oxides have been investigated [Kaupo et al., 2002; Sandy et al., 2005; Kaupo et al., 2017; Lee et al., 2013]. The interfacial self-cleaning process during ALD of high-k HfO_2, Al_2O_3 on III-V compound is found to be particularly useful to achieve an atomically sharp interface between gate dielectric and channel layer [Martin et al., 2005; Chang et al., 2006; Hinkle et al., 2008]. Another leading area of research in which ALD technology has been widely explored is the growth of transparent conducting oxide (TCO) such as binary (zinc oxide, tin (IV) oxide, indium (III) oxide), and ternary ((aluminium, gallium, indium, titanium, boron) doped zinc oxide, Sn doped (zinc oxide, indium (III) oxide)) metal oxide semiconductors [Saha et al., 2013; Diana et al., 2015; Jeffrey et al., 2006]. However, the optimum electrical conductivity for ALD-grown *n*-type TCO is found to be much lower compared to the sputtered and other physical vapor-deposited films [Saha et al., 2014; Saha et al., 2016; Wu et al., 2013]. This is attributed to the delta-doping scheme in ALD in which dopants are incorporated in dense monolayer on the doping plane [Saha et al., 2014; Saha et al., 2016]. In order to alleviate this issue and to achieve higher doping efficiency, subsaturating exposure of dopant precursor and the use of larger ligand size precursor molecules (enhancing steric hindrance effect) have been proposed by several groups [Wu et al., 2013; Saha et al., 2016]. ALD deposition of oxides is usually carried out at 150–200 °C [Riikka, 2005; George, 2010]. However, using the plasma option, the deposition temperature can be significantly decreased even down to room temperature [Rowlette et al., 2009]. The low-temperature growth capability of ALD is exploited to deposit multicomponent amorphous oxide semiconductors such as zinc tin oxide (ZTO), indium zinc oxide (IZO), and indium gallium zinc oxide (IGZO) for flexible and stretchable electronic applications [Sheng et al., 2016; Illiberi et al., 2015]. The high aspect ratio coating capability of ALD has been efficiently used for template-based nanostructuring (using anodic alumina and aerogel templates) of high surface area ZnO and TiO_2 photoanodes for dye-sensitized solar cell applications [Thomas et al., 2008; Hamann et al., 2008].

Other than oxides, metal chalcogenides, especially sulfides, form a significant part of the materials which are investigated using the ALD [Riikka, 2005; George, 2010] method. ZnS is the first material that was grown using ALD by employing elemental source of Zn and S [Riikka, 2005]. However, typically H_2S is considered as the most efficient reactant because of its higher volatility and chemical reactivity towards most of the metal organic precursors [Riikka, 2005; George, 2010]. To date, a variety of binary metal sulfides including copper sulfide (CuS), bismuth (III) sulfide (Bi_2S_3), nickel sulfide (NiS), aluminium sulfide (Al_2S_3), indium sulfide (In_2S_3), iron sulfide (FeS), SnS, and antimony trisulfide (Sb_2S_3) are successfully grown using ALD [Riikka, 2005; George, 2010; Neil et al., 2015; Shannon et al., 2017; Elijah et al., 2012]. Multicomponent sulfides such as ternary (copper indium disulfide ($CuInS_2$), chalcostibite ($CuSbS_2$)) and quaternary (copper zinc tin sulfide (Cu_2ZnSnS_4)) are also deposited for photo absorber materials in solar cells [Neil et al., 2015; Shannon et al., 2017; Elijah et al., 2012]. The self-limiting and complementary deposition chemistry of these materials was confirmed by in-situ quartz crystal microbalance (QCM) and Fourier transform infrared (FTIR) spectroscopy measurements [Neil et al., 2015].

Recently, the potential of transition metal dichalcogenides (TMDs) has been explored as post-Si channel materials in field effect transistor applications [Schwierz et al., 2015]. However, in most of the reports, mechanically exfoliated flakes of micron sizes are used which limit their practical implementation [Schwierz et al., 2015]. In this perspective, ALD technology is found to be highly useful to realize a wafer-scale growth of ultra-thin 2D molybdenum disulfide (MoS_2) layers using different Mo precursors such as molybdenum (V) chloride ($MoCl_5$), molybdenum hexacarbonyl ($Mo(CO)_6$), molybdenum hexafluoride (MoF_6) and H_2S as sulfur source [Anil et al., 2018; Arturo et al., 2016]. However, in most cases, grown MoS_2 layers are found to be amorphous and turn into stoichiometric and crystalline phase after post-growth annealing at high temperature [Anil et al., 2018; Arturo et al., 2016].

ALD processes of various binary metal tellurides and selenides including zinc telluride (ZnTe), lead telluride (PbTe), bismuth telluride (Bi_2Te_3), zinc selenide (ZnSe), bismuth selenide (Bi_2Se_3), indium selenide (In_2Se_3), copper selenide (CuSe), and lead selenide (PbSe) are also developed using metal halides and alkylsilyl compounds of tellurium and selenium precursors [Viljami et al., 2009]. The inherent self-limiting growth mechanism of ALD is used to grow the nanolaminate-type structure of PbTe/PbSe in which individual nanolayers of alternately deposited materials are separated by atomically sharp interfaces [Zhang et al., 2014]. Such super lattice engineering is found to be particularly useful to enhance the thermoelectric performance of chalcogenide materials by tailoring thermal conductivity and electrical transport properties [Zhang et al., 2014].

Although materials growth in ALD is expected to proceed in monolayer-by-monolayer fashion, a deviation from this ideal characteristic is usually observed during the nucleation period of thin-film deposition [Zhang et al., 2014; Shrestha et al., 2010]. At the initial stage, adatom-to-adatom interactions dominate over the adatom-to-surface interactions, which results in discrete island formation (Volmer-Weber growth mode) instead of surface saturating growth. However, beyond the nucleation period, discrete islands conglomerate and deposit continuous film which continues to grow in typical layer-by-layer Frank-Van der Merwe mode with constant deposition rate [Zhang et al., 2014; Shrestha et al., 2010]. Nucleation in ALD is usually attributed to the presence of non-uniformly distributed nucleation sites on the substrate surface [Zhang et al., 2014; Shrestha et al., 2010]. Moreover, in-situ QCM and FTIR measurements during the depositions of a few metal sulfides using β-diketonate precursors (such as 2,2,6,6-Tetramethyl-3,5-heptanedione (thd) and 1,1,1,5,5,5-Hexafluoroacetylacetone (hfac)) reveal that steric hindrance (bulky ligands) and surface poisoning effect due to the readsorption of the H(thd) or H(hfac) species on the growing films surface plausibly result in an extended nucleation period [Kim and George, 2010].

5.3 Chemical Bath Deposition of Thin Film

The chemical bath deposition (CBD) method is a simple technique to prepare stable and uniform films [Ngai et al., 2011]. It is classified as a cost-effective technique to produce adherent films under large-scale deposition process [Lim et al., 2011]. Despite having several advantages, the CBD method also has a few disadvantages. The big challenge is the wastage of residual chemical solution after every deposition. Moreover, it is hard to achieve single crystalline films using this method and mostly polycrystalline films form onto the substrate.

In CBD, the growth of the films strongly depends on the growth conditions [Gary, 2002]. The concentration, composition, and temperature of the precursor solution, pH of the solution, duration of deposition, and chemical nature of the substrate influence the deposition process [Ho, 2014]. Apart from these, a chemical complexing agent used in the precursor solution also affects the quality of the films. In practice, the thickness of the films is adjusted by changing the deposition time. Figure 5.1 shows a schematic experimental setup for the CBD method. A pH meter and a thermometer are required to monitor the pH [Halim et al., 2010] and temperature of the solution. Some non-reacting solids are used as the substrate. In most cases, glass slide is used;

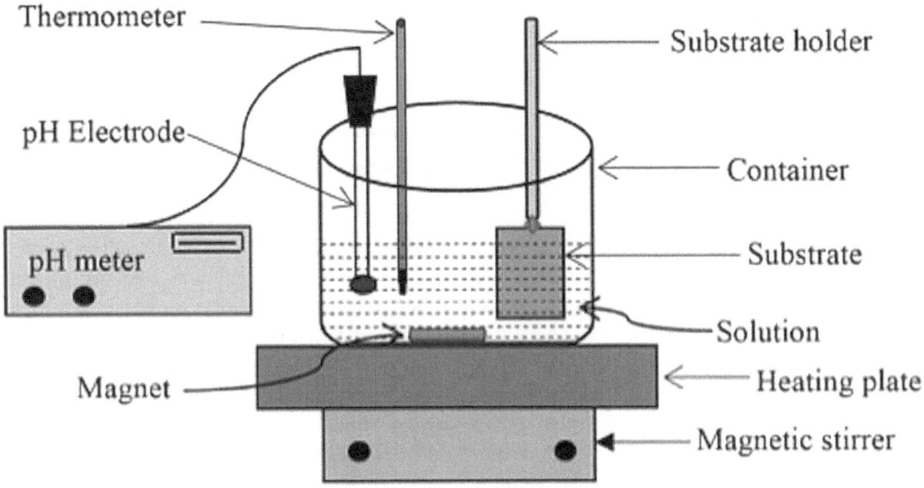

FIGURE 5.1 Schematic experimental setup used in CBD method.

however, depending on the precursor solution, ebonite, iron, steel, porcelain, and brass can also be used as a substrate. The substrate is vertically immersed in the solution using a substrate holder. The solution is stirred using a magnetic stirrer and, whenever necessary, the heating element embedded with the stirrer is used to increase the temperature of the solution.

Under specific growth conditions, when the value of the ionic product of anions and cations becomes equal or exceeds the solubility product of the solution, precipitation starts and the film begins to deposit on the substrate. To achieve such a critical state and to prevent bulk precipitation due to unstable chemical reactions, a suitable complexing agent is used. Two major steps are involved in CBD, namely, nucleation and particle growth. There are various processes, such as *ion-by-ion* and *cluster*, which take place during the deposition. In the *ion-by-ion* process, the film is deposited by sequential ionic reactions, whereas in the *cluster* process, colloidal particles are formed which absorb at the surface of the substrate to form the thin layer. In most cases, both processes run together and compete with each other to form the film. This competition is governed by the degree of homogeneous and heterogeneous nucleation.

Homogeneous nucleation occurs due to changes in local concentration, temperature, or other variables of the solution. Due to collision of ions, some thermodynamically unstable entities are formed. These are called 'embryos'. These embryos may re-dissolve in the solution or collapse themselves and grow up to some thermodynamically stable state. In addition, some subcritical embryos can also grow in the solution due to surface diffusion. These subcritical embryos or the individual ions can directly adsorb onto the surface of the substrate, which results in heterogeneous nucleation. The stable entity further forms bigger crystals via several processes, such as self-assembling and Ostwald ripening [Gary, 2002]. Several crystals combine to form particles of different sizes. When approaching each other, these particles form aggregates due to van der Waals attraction force. In addition, at adequately high temperature, coalescence may also occur, where two or more particles can combine due to surface diffusion to form a larger particle.

Three different patterns of film growth are observed, namely, *island* or Volmer-Weber growth, *layer-by-layer* or Frank-van der Merwe growth, and *mixed* or Stranski-Krastanov growth [Venables et al., 1984]. The growth patterns of each of them are described in Figure 5.2.

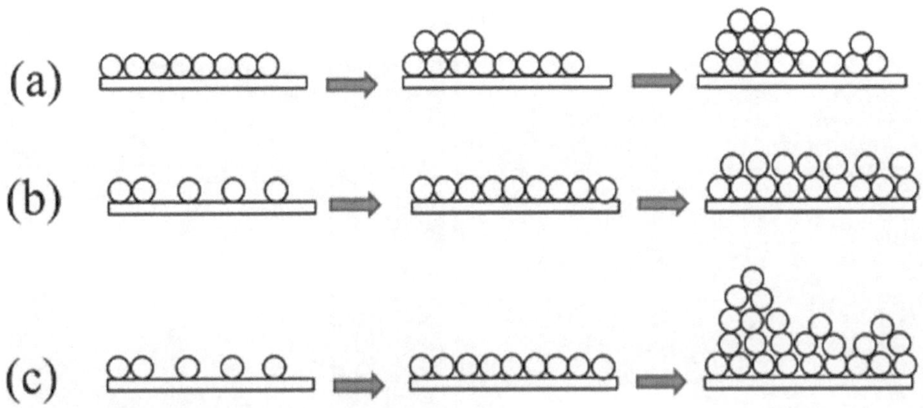

FIGURE 5.2 Growth patterns of thin film: (a) Shows the *island* growth in which the atoms or molecules are arranged in an island-like pattern, (b) shows the *layer-by-layer* growth where the atoms or molecules in a layer are arranged one above the other, and (c) shows the *mixed* growth representing the combined *layer-by-layer* and *island* growth patterns.

Zinc sulfide thin films were synthesized using zinc acetate (0.2 M) and thiourea (0.2 M) solutions. The cleaned glass slide was immersed into a chemical bath at 343 K, pH about 8–9, 40 minutes. The substrate was removed from the chemical bath after 24 hours. The film looks transparent on the surface of the glass subtrate. X-ray diffraction (XRD), scanning electron microscopy (SEM), atomic force microscopy (AFM), and X-ray fluorescence spectroscopy were selected to study the physical properties of films.

Sarma et al., (2017a) reported the physical characteristics of ZnS films synthesized by the CBD method described above. Figure 5.3 shows the XRD, SEM, and AFM results of ZnS films deposited at various solution concentrations (0.1, 0.15, and 0.2 M). The films were polycrystalline; the grains had a hexagonal rod-like shape and were uniformly deposited on the substrate.

TiO_2 films were produced in the presence of 3 M of titanium (IV) isopropoxide ($C_{12}H_{28}O_4Ti$) and 2-propanol ($CH_3CHOHCH_3$) solutions. These two solutions were then mixed in a beaker and stirred for 1 hour at 353 K. After stirring, a glass substrate was immersed in the solution for 2 hours at pH 2–2.3. The glass substrate was taken out after 24 hours and dried in air. A thin layer of the film can be seen on the substrate. Using this process, Sarma et al., (2017b) synthesized TiO_2 films at various concentrations such as 1.5, 2, 2.5, and 3 M. The films were polycrystalline, uniform, and adhered well to the surface.

Before the deposition process, the glass substrate was cleaned using a mixture of concentrated nitric acid and isopropyl alcohol in equal proportion. Thereafter, distilled water was used to wash the glass slide, and, lastly, it was dried in a muffle furnace at 333 K [Sarma et al., 2017a; Sarma et al., 2017b].

5.4 Electrodeposition of Thin Films

Currently, metal chalcogenide thin films are prepared through many physical and chemical methods. Researchers observed that there is wastage of chemicals and energy during the deposition process. Electrodeposition is one of the simplest chemical methods [Anuar et al., 2008] for the synthesis of metal chalcogenide thin films [Noraini et al., 2010].

Cadmium sulfide (CdS) thin films have been synthesized both in aqueous and non-aqueous solutions containing precursors of Cd and S elements. The cathodic electrodeposition of CdS from acidic solutions at 90 °C has been studied [Dennison, 1993]. Sulfur is produced by the decomposition of thiosulfate during the deposition process. The favorable conditions for the deposition of CdS were maintained by the pH of the reaction bath. Yamaguchi et al., (1998) deposited the CdS thin films on an indium tin oxide (ITO) glass substrate using the electrodeposition method in a non-aqueous medium such as dimethyl sulfoxide, ethylene glycol. The fine particle of CdS was agglomerated on the surface of the substrate. The film was developed through atom-by-atom deposition and formed hexagonal CdS crystals (film thickness of 500 nm). On the other hand, CdSe films were deposited by using the galvanostatic

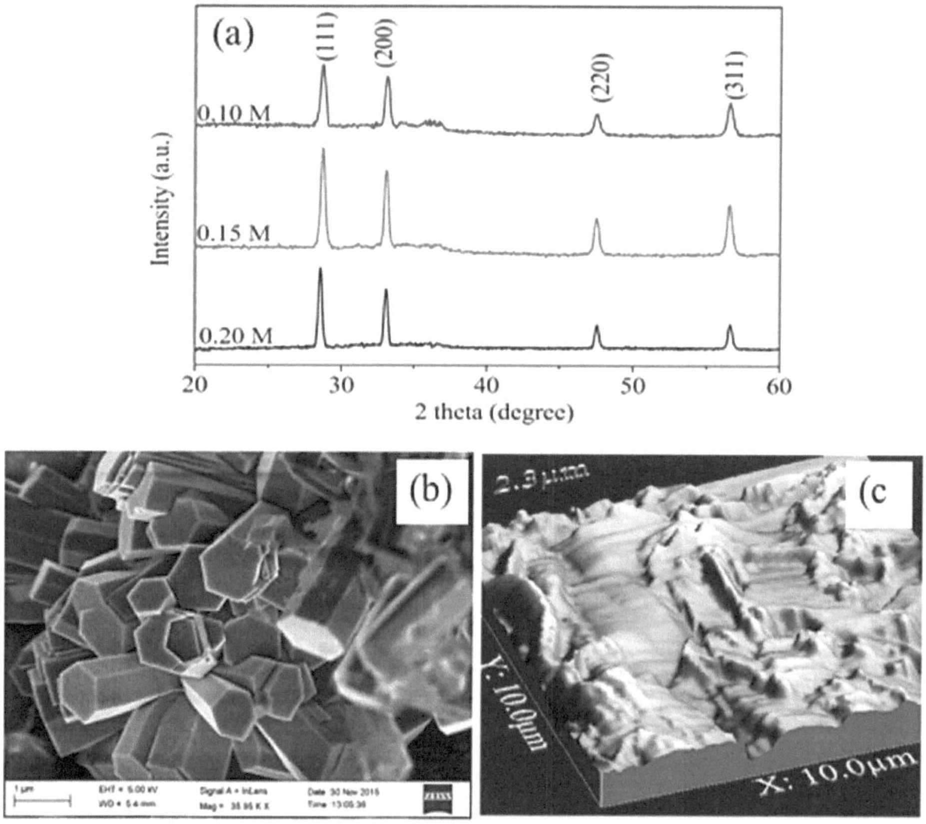

FIGURE 5.3 (a) XRD patters of 0.10, 0.15, and 0.20 M films, (b) SEM image of 0.20 M film, and (c) AFM image of 0.20 M film (reproduced from Sarma et al., (2017a) with permission from Elsevier).

method onto tin substrate as suggested by Tomkiewicz et al., (1982). The "cauliflower" morphology of cadmium selenide (CdSe) was obtained as prepared and annealed at 750 °C in an argon atmosphere by the authors. The cauliflower morphology, because of its high surface area, was used to study the liquid junction solar cell, and the power conversion efficiency was 6.4 %.

From experimental observation, it is evident that the nature of electrodeposited products such as Cu_2Se, Cu_2Se_3, and CuSe depended upon applied potential. If we made more negative potential, the phase of CuSe would change to Cu_2Se. Massaccesi et al., (1993) reported the acidic deposition of CuS and CuSe. The phase of the material changed with the precursor concentrations and the applied potential on conducting substrate.

Copper indium diselenide films were deposited through the electrodeposition method for the photovoltaic applications. The first copper indium diselenide thin films were prepared by Bhattacharya, (1983) in an acidic medium, by using $CuCl_2$, $InCl_3$, and SeO_2 as a source of Cu, In, and Se (ammonia and triethanolamine as a complexing agent). Many researchers synthesised copper indium diselenide thin films. They found that the microstructures of copper indium diselenide change with deposition potential and other experimental conditions. Ueno et al., prepared stoichiometric copper indium diselenide thin films at −0.8 V/SCE (saturated calomel electrode) from an acidic bath and without a complexing agent. Gomez et al., (1995) reported that the tetragonal chalcopyrite CIS was deposited on a glassy carbon disk substrate in the deposition potential of −0.7 to −0.9 V/SCE.

Brownson et al., (2006) described the deposition of tin sulfide (SnS) films onto FTO glass by the electrodeposition method in an aqueous medium in the presence of $SnCl_2$, $Na_2S_2O_3$, and tartaric acid (as a complexing agent). The prepared films showed orthorhombic polymorph of tin monosulfide (δ-SnS phase), which was changed to α-SnS phase (herzenbergite) after being annealed at 350 ° C in argon gas. The uniform electrodeposited SnS thin films were deposited at pH 1 and 2.5 [Brownson et al., 2008]. They reported that using tartaric acid as complexing agent can improve the adhesion of the film.

Electrodeposition of tin selenide (SnSe) thin films was described by Engelken and co-workers (1986). The $SnCl_2$ and SeO_2 solutions were used to produce smooth and uniform SnSe films from acidic conditions. Meanwhile, Baranski and Fawcett (1980) synthesised SnSe films in the non-aqueous medium. Researchers concluded that the properties of these films in both media were nearly the same.

Lead sulfide (PbS) film was prepared by using lead nitrate and sodium thiosulfate in the acidic aqueous medium on the different conducting substrates at room temperature. For example, the crystallite size of PbS was 13 nm on the tin substrate [Takahashi et al., 1993]. The obtained experimental results indicated that polycrystalline PbS thin film was successfully deposited onto various substrates, including titanium (Ti), aluminium (Al), and stainless steel (SS) by using the potentiostatic method [Sharon et al., 1997]. XRD pattern supported the prominent peaks corresponding to (200) and (100) planes for the films prepared onto Ti and Al substrates. However, a (200) plane was observed for the films deposited onto a stainless-steel substrate.

Lead telluride (PbTe) thin films were directly electrodeposited onto *n*-type silicon (Si) substrates, in the absence of any buffer solution in acidic medium (pH 1) at 20 °C. Compact morphology was observed in SEM analysis. These films with 50–100 nm indicated polycrystalline cubic structure as observed in XRD data. The energy-dispersive X-ray spectroscopy (EDX) displayed 51.2 % Pb and 48.8 % Te in sample. The band gap was 0.31 eV as calculated using an UV-visible spectrophotometer [Li and Nandhakumar, 2008].

5.5 Spray Pyrolysis Deposition of Thin Film

Many types of metal oxides, metal chalcogenides, metallic spinel oxides, and high-temperature superconducting oxide thin films have been successfully synthesized using the chemical spray pyrolysis technique. Some of the terms, such as spray pyrolysis, aerosol, atomization [Patil, 1999], suspension, and substrate will be explained for better understanding.

The spray pyrolysis method (SPM) is also called *pyrosol process* or *pyrosol deposition*, or *pyrolytic spray* [Korotcenkov and Cho, 2017]. It is a very versatile technique that can be employed to prepare thin films, thick films [Sahu, 2015], ceramic coatings, nanopowders, and multilayer materials [Perendnis and Gauckler, 2005] with numerous applications in solar cells and optoelectronics [Bang and Kenneth, 2010; Otto, 2012]. The properties of thin film can be easily manipulated, modified, or engineered by adjusting various experimental conditions to get thin films that match any desired applications [Patil, 1999]. SPM is a solution-based method [Korotcenkov and Cho, 2017] and is categorized under the chemical deposition methods. It is one of the two broad classifications of thin-film deposition methods. In the "family of solution methods" are spin coating, dip-coating, sol-gel drop casting, and spray pyrolysis. The spray pyrolysis process involves the formation of droplets from the precursor. The thin films are deposited onto a hot substrate surface at a given temperature [Patil, 1999]. The recombination of the constituent species causes the production of films [Korotcenkov and Cho, 2017] and nanopowder. The spray pyrolysis machine comprises of the following main features: (i) spray nozzle, (ii) pressure regulator, (iii) precursor container, (iv) air compressor (or gas propellant), (vi) electric heater for the substrate, and (vii) temperature controller [Patil, 1999; Korotcenkov and Cho, 2017]. The various types of atomizers used in the spray pyrolysis method are pressure, nebulizer, ultrasonic, and electrostatic [Perendis, 2003].

The droplet size (*D*) is given by Lang equation as [Bang and Kenneth, 2010]:

$$D=0.34\left(\frac{8\pi\gamma}{\rho f^2}\right)^{\frac{1}{3}}, \tag{5.1}$$

where D, γ, ρ, and f represent mean droplet diameter, surface tension (N/m), solution density (kg/m^3), and ultrasonic frequency (MHz), respectively. However, higher frequencies (>1 MHz) are required for nanopowder, but commercial nebulizers for spraying, drying, and painting use have frequencies in the region of 25 kHz [Bang and Kenneth, 2010].

The spray pyrolysis method (SPM) has many advantages [Tewari and Bhattacharjee, 2009; Gryteselv, 2016; Kozhukharou and Tchaoushev, 2013]. It is very simple and relatively cost-effective, it requires moderately low temperatures, it is a microcapsule reactor and an eco-friendly deposition method, it controls the properties of films, it is up-scalable for large-area applications, it is quite an empirical technique, and it gives homogenous results with good adhering layers. However, there are some disadvantages, as highlighted by researchers. For example, aerosol droplets produced in SPM behave differently depending on their sizes: smaller droplets will escape to the surroundings, thus causing environmental pollution, especially if the gas is toxic, resulting in wastage if too hot [Patil, 1999]. Undesirable oxidation of the films can take place during the film growth due to the open system; there are difficulties in precise surface temperature measurement during film growth and the number of precursors is limited since they must be soluble in the solvent used [Otto, 2012].

Thin films deposited by the spray pyrolysis method have matching properties suitable for a wide range of potential applications [Duta et al., 2014] in areas such as solar cells, anti-reflection coatings, solid oxide fuel cells, and gas sensors. For example, chalcogenide semiconductor compounds of groups II-VI and V-VI have applications in the precise temperature control of laser diodes, optical recording systems, electrochemical devices, strain gauges, and thermoelectric devices [Patil, 1999]. High-temperature superconductivity in oxides such as Y-Ba-Cu-O films could be used in electronic devices, superconducting quantum interference devices, flexible polymer light-emitting displays, optical coatings, supra-conductive films, magnetic films, low-energy window glass coating, heat prevention, and corrosion resistance.

Nanostructured CdS thin films were synthesized using the spray pyrolysis method [Kerimova et al., 2017]. During the deposition process, cadmium sulfide films were deposited onto soda lime glass using a mixture of cadmium acetate (0.025 M), thiourea (0.025 M), and ammonium acetate, at 400 °C soda lime glass substrates. The pH value was adjusted by using a NH_4OH solution. The influence of pH value on the properties of films was studied. Hexagonal wurtzite structures (pH 6.7, 9.5) and amorphous structures (pH 10.2) were obtained for the films prepared under various pH values. AFM results supported that the grain size (20 to 10 nm) and surface roughness values (8.62 to 4.63 nm) were reduced from 20 nm to 10 nm as pH was increased. Two peaks (520 and 600 nm) were observed for the films prepared at pH 6.7 and 9.5 as indicated in photoluminescence analysis. An additional peak (460 nm) appeared at pH 10.2 because of quantum size effect, indicating that blue shift of band edge of CdS films takes place at grain sizes below 10 nm [Banyai and Koch, 1993; Yoffe, 1993]. The analysis of spectroscopic ellipsometry revealed two transitions, namely real ε_1 (2.4 eV) and imaginary ε_2 (4.5 eV) parts of dielectric function for the films prepared at pH 10.2 [Susumu and Adachi, 1995].

5.6 Successive Ionic Layer Absorption and Reaction Deposition of Thin Film

Successive ionic layer absorption and reaction (SILAR) is a very useful method for the fabrication of thin films onto different substrates. The obtained thin films can be used in semiconductor industries, optoelectronics, gas sensors, and solar cells. It involves the adsorption of a layer of complex ions on the substrate followed by the reaction of the adsorbed ion layer.

In step [*a*], adsorption of cations is observed. Step [*b*] involves rinsing loosely bound cations in deionized. In step [*c*], adsorption of anions and reaction between pre-absorbed cation and newly absorbed anion is observed. In step [*d*], the excess species and the reaction by-product are removed by rinsing in deionized water. The formation of films onto the substrate can be observed by repeating these cycles [Pathan and Lokhande, 2004]. The SILAR method contains several advantages, such as the formation of films onto various substrates at low temperature. This is known as the wet chemical deposition technique, and the properties of films strongly depend on the experimental conditions. The disadvantages of the SILAR method including the different parameters have a significant impact on the film growth process. Therefore, all the parameters need to be chosen with extreme care and maintained well during the deposition process. The immersion-reaction cycle strongly depends on the substrate–thin film interface.

Nowadays, the synthesis and characterization of metal sulfide thin films via SILAR technique have attracted considerable attention. In general, SnS thin films are prepared through the SILAR method, in which tin chloride has been used as a source of tin ion and sodium sulfide has been used as a source of sulfur ion [Safonova et al., 2014; Ghosh et al., 2008; Mondal and Mitra, 2008]. Safanov et al., (2014) used sodium thiosulfate as anionic precursor and glass slide (substrate) during the SILAR deposition process. Meanwhile, Ghosh et al., (2008) used ITO as substrate for film deposition. Mondal and Mitra (2008) reported the preparation of SnS films by using a cationic bath (at 80 °C) and an anionic bath at room temperature. XRD data showed orthorhombic structures with particles ranging between 8 and 11 nm. An activation energy value of ~0.28 eV was obtained also. Ghosh et al., (2008) obtained SnS films (0.20 μm thickness) with a direct band gap of 1.43 eV. Morphology studies indicated that these films were uniform and strongly adherent to the substrate. Sankapal et al., (2000) used tin chloride and sodium sulfide solutions for SILAR deposition of tin sulfide. Film thickness was about 1 μm with a pure hexagonal structure. The band gap and electrical resistivity were 2.6 eV and 1000 Ω-cm, respectively. Mukherjee and Mitra (2015) deposited SnS thin films onto a glass slide substrate by using the SILAR method in the presence of tin chloride and ammonium sulfide. The obtained films have orthorhombic phase with particle size of about 45 nm. The obtained band gap was 1.63 eV, which indicates that SnS films can be used as an important material for solar cell applications.

In general, chemical bath deposition (CBD) has been broadly used to deposit cadmium sulfide (CdS) thin films; the SILAR method has been less exploited for this purpose. In 1985, Nicolau (1985) used the SILAR method to deposit CdS thin films over both glass and FTO substrate. Cadmium sulfate (pH ~ 8) and sodium sulfide (pH ~ 11.6) were used to deposit CdS thin films at room temperature by 165 cycles of dipping. The obtained films contain hexagonal phase with polycrystalline nature. Later, Mane et al., (2000) deposited CdS thin films on glass substrate from cadmium acetate and sodium sulfide with varying molar concentration at 80 °C by only 20 cycles of dipping. The obtained films have pure hexagonal phase with band gap in the range of 2.24–2.17 eV. Lokhande et al., (2001) synthesized CdS thin films over glass slide substrate by alternate dipping (160 times repetition) in 0.125 M cadmium acetate (pH ~ 5) and 0.05 M sodium sulfide (pH ~ 12). Both solutions were kept at a temperature of 30 °C. The as-deposited films show pure hexagonal phase with particle size ranging between 7 and 8 nm. The deposited CdS thin films were well adherent, uniform, dense, well stoichiometric, and homogenous without any well-visible porosity. Some research team members have also reported a deposition of Mn-doped and undoped CdS thin films onto a glass/FTO substrate using the SILAR method at room temperature. Researchers concluded that adherent thin films were obtained after 165 dipping cycles. The crystallinity increases with Mn doping along with its band gap value [Pathan and Lokhande, 2004]. Ates et al., (2007) deposited polycrystalline, well-adherent hexagonal CdS thin films over an ITO substrate using the SILAR technique. They observed that the band gap decreases (1.48 to 1.157 eV) with increasing temperature. Garadkar et al., (2010) deposited CdS thin films from 0.1 M cadmium acetate and 0.1 M thiourea by 125 dipping cycles. The obtained films had both hexagonal and cubic phase with hexagonal as a dominant phase. The band gap was about 2.39 eV based on UV-visible spectrophotometer spectra. Mukherjee et al., (2015) synthesized CdS thin films using the SILAR technique with ammonium sulfide. The obtained films had pure cubic structure with band gap about 2.56 eV. The particle size was about 5–15nm. It is evident that almost all the researchers used sodium sulfide as a source of sulfur ion, while Garadkar and other researchers (2010) have used thiourea as a source of sulfur ion during the deposition process.

5.7 RF Sputtering Deposition of Thin Films

In recent times, many deposition methods were used to produce micro and nanostructures as reported by many researchers [Kim et al., 2015; Guangming et al., 2008; Bandaranayake et al., 1995; Yang and Lieber, 1996; Gadenne et al., 1989]. For example, ultrasonic spray coating, sol-gel, DC/RF magnetron sputtering, chemical bath deposition, pulsed laser deposition, and thermal evaporation methods were selected to prepare nanostructured thin films, while the precipitation, hydrothermal synthesis, hydrolysis, and plant extract methods were used to produce nanoparticle samples [Gencer, 2016; Ghosh et al., 2005; Subramanyam et al., 2001; Ghimpu et al., 2013; Bulakhe and Lokhande, 2014; Zaien et al., 2013; Li et al., 2016]. The quality of the nanostructure was examined by structural, morphological, thermal, optical, and electrical properties using different characterization techniques. Among these deposition techniques, sputtering is considered as one of the most popular techniques to synthesize thin films.

The formation of thin films was carried out by using the sputtering method. It is grouped under the physical vapor deposition techniques. Generally, thin films are produced by ejecting atoms from target materials in a high-vacuum environment (Figure 5.4). These ejected atoms condense, and start to bind each other for production of precise layered thin-film structures [Minami et al., 2010; Kang and Joung, 2007; Al-Hardan et al., 2010].

In this chapter, we discuss the 12″ MSPT RF sputtering machine (Hind High Vacuum) for deposition purposes. The sputtering system contains two sections, namely vacuum and power supply. Vacuum is created in the deposition chamber to produce a low-pressure environment for gas ionization. In this machine, two vacuum pumps are used to obtain vacuum: rotary pump (up to 10^{-3} Torr) and diffusion pump (10^{-6} Torr). The RF power supply was used (SEREN Industrial Power Systems Inc., U.S.A. model R301) to provide 13.56 MHz, level-controlled radio frequency power output up to 300 Watts. It gives operator-accessible controls, a visual display of power supply status, and a control interface to the user's processing system. Other components of the sputtering system are water supply, pressure gauge, valves, gas cylinders, and substrate holder with heater.

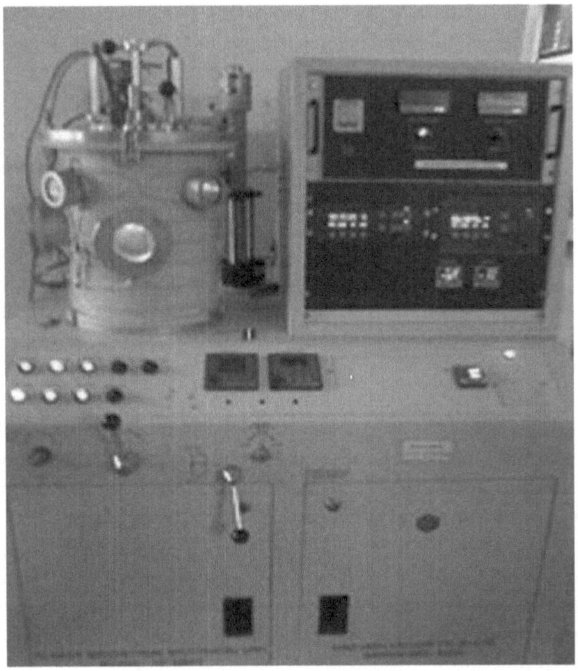

FIGURE 5.4 The mechanism of thin films formation by sputtering process.

Oxide materials such as zinc oxide, cadmium oxide, and copper oxide were successfully prepared using the RF sputtering method because it has larger band-gap value and high melting point. Generally, the powder of oxide materials was used as target substance. The ultrasonically cleaned glass substrate was chosen for coating in the sputtering chamber. It was pumped down from atmospheric pressure to a base pressure of 5×10^{-6} Torr. Following that, the argon gas (as sputtering gas) was added into the chamber. The power supply was turned on as the pressure of the chamber reached about 5×10^{-2} Torr. Lastly, the power was held at 160 W for 30 minutes before the formation of thin films.

Pathak et al., (2015) prepared zinc oxide thin films onto a glass substrate using the RF sputtering method. Deposition was carried out under high-purity ZnO target in the presence of N_2/Ar + N_2 ratios ranging from 0 to 1 at a pressure of 10^{-3}–10^{-2} Torr. They explain that nitrogen and argon were acted as reactive gas and sputtering enhancing gas, respectively. Before starting the experiment, the target was pre-sputtered in order to eliminate contamination on the surface.

Doped oxide thin films are also easily deposited by using the sputtering technique. The target is easily formed with different doping concentrations (weight in %) and thin films are deposited onto different substrates. Trilok et al., (2015) deposited Al-ZnO thin films on glass substrate. Aluminum (Al)-doped zinc oxide (ZnO) powders were formed using the solid-state reaction method. ZnO powder and Al_2O_3 were mixed using methanol for 4 hours by ball-milling, then dried. After drying at 100 ° C for 6 hours, powders were ground using a mortar and pestle. The AZO powder is ready to act as sputtering target. The sintered target was used for the synthesis of thin films by the RF sputtering coating unit.

Cadmium sulfide thin films have band gap of 2.42 eV and could be used in light-emitting diodes, solar cells, and optoelectronic devices. Cadmium-based thin films such as cadmium telluride [Wu, 2004; Britt and Ferekides, 1993], copper indium diselenide (CIS), and copper indium gallium diselenide (CIGS) [Ramanathan et al, 2005] have been studied by many researchers. Wu (2004) has described how better power conversion efficiency can be observed for the CdTe films prepared using the RF sputtering process if compared to CdS films. Another example was CdS:O thin films, which show higher band-gap value if compared to bulk CdS films. The influence of oxygen partial pressure on the CdS:O thin films was investigated. The XRD data confirmed the existence of amorphous and crystal structures for the films prepared using more that 3 % and less than 3 % oxygen partial pressure, respectively [Asaba et al., 2013].

Conclusion

Thin films have received great attention due to their remarkable and unique physical, morphological, and optical properties. These films can be categorized into binary, ternary, and quaternary compounds. Many physical and chemical deposition techniques have been used to produce thin films. The thin films obtained were characterized by using XRD, AFM, EDAX, and UV-visible spectrophotometers. Researchers concluded that these films could be used in solar-cell and optoelectronic applications.

Acknowledgments

This research work was supported by INTI International University (Ho SM).

REFERENCES

Al-Hardan, N.H., Abdullah, M.J., and Aziz, A.A. 2010. Sensing mechanism of hydrogen gas sensor based on RF-sputtered ZnO thin films. *International Journal of Hydrogen Energy* 35: 4428–4434.

Anil, U.M., Letourneau, S., Mandia, D., Liu, J., Libera, J., and Lei, Y. 2018. Atomic layer deposition of molybdenum disulfide films using MoF_6 and H_2S. *Journal of Vacuum Science and Technology A: Vacuum, Surfaces, and Films* 36. doi: 10.1116/1.5003423.

Anuar, K., Ho, S.M., Saravanan, N., Tan, W.T., Atan, S., and Kuang, M. 2008. Effects of bath temperature on the electrodeposition of Cu_4SnS_4 thin films. *Journal of Applied Sciences Research* 4: 1701–1707.

Arturo, V., Douglas, J.T., and John, F.C. 2016. Atomic layer deposition of two dimensional MoS_2 on 150 mm substrates. *Journal of Vacuum Science and Technology A: Vacuum, Surfaces, and Films*. doi: 10.1116/1.4941245.

Asaba, R., Wakita, K., Kitano, A., Shim, Y., Mamedov, N., Bayramov, A., Huseynov, E., and Hasanov, I. 2013. Structure and optical properties of CdS:O thin films. *Physica Status Solidi C* 10: 1098–1101.

Atan, S., Ho, S.M., Anuar, K., and Saravanan, N. 2010. X-ray diffraction and atomic force microscopy studies of chemical bath deposited FeS thin films. *Studia Universitatis Babes-Bolyai Chemia* 55: 5–11.

Ates, A., Yildirim, M.A., Kundakci, M., and Yildirim, M. 2007. Investigation of optical and structural properties of CdS thin films. *Chinese Journal of Physics* 45: 135–141.

Bandaranayake, R.J., Wen, G.W., Lin, J.Y., Jiang, H.X., and Sorensen, C.M. 1995. Structural phase behavior in II–VI semiconductor nanoparticles. *Applied Physics Letters* 67. doi: 10.1063/1.115458.

Bang, J.H., and Kenneth, S.S. 2010. Applications of ultrasound to the synthesis of nanostructured materials. *Advanced Materials* 22: 1039–1059.

Banyai, L., and Koch, S.W. 1993. Optical properties of small dots. *Semiconductor Quantum Dots*: 116–149. doi: 10.1142/9789814354417_0005.

Baranski, A.S., and Fawcett, W.R. 1980. The electrodeposition of metal chalcogenides. *Journal of the Electrochemical Society* 127: 766–767.

Bhattacharya, R.N. 1983. Solution growth and electrodeposited $CuInSe_2$ thin films. *Journal of the Electrochemical Society* 130: 2040–2042.

Britt, J., and Ferekides, C. 1993. Thin film CdS/CdTe solar cell with 15.8 % efficiency. *Applied Physics Letters* 62. doi: 10.1063/1.109629.

Brownson, J.R.S., Georges, C., Larramona, G., Jacob, A., Delatouche, B., and Lévy-Clément, C. 2008. Chemistry of tin monosulfide (δ-SnS) electrodeposition effects of pH and temperature with tartaric acid. *Journal of the Electrochemical Society* 155: D40–D46.

Brownson, J.R.S., Georges, C., and Lévy-Clément, C. 2006. Synthesis of a δ-SnS polymorph by electrodeposition. *Chemistry of Materials* 18: 6397–6402.

Bulakhe, R.N., and Lokhande, C.D. 2014. Chemically deposited cubic structured CdO thin films: Use in liquefied petroleum gas sensor. *Sensors and Actuators B* 200: 245–250.

Chang, C., Chiou, Y., Chang, Y., Lee, K., Lin, T., and Wu, T. 2006. Interfacial self-cleaning in atomic layer deposition of HfO_2 gate dielectric on $In_{0.15}Ga_{0.85}As$. *Applied Physics Letters*. doi: 10.1063/1.2405387.

Dennison, S. 1993. Studies of the cathodic electrodeposition of CdS from aqueous solution. *Electrochimica Acta* 38: 2395–2403.

Diana, G., Potts, S.E., Helvoirt, C.A.A., Verheijen, M., and Kessels, W.M.M. 2015. Atomic layer deposition of B-doped ZnO using triisopropyl borate as the boron precursor and comparison with Al-doped ZnO. *Journal of Materials Chemistry C* 3: 3095–3107.

Duta, D., Perniu, A., and Duta, A. 2014. Photocatalytic zinc oxide thin films obtained by surfactant assisted spray pyrolysis deposition. *Applied Surface Science* 306: 80–88.

Elijah, T., Riha, S., Baryshev, S., Martinson, A., Elam, J., and Pellin, M. 2012. Atomic layer deposition of the quaternary chalcogenide Cu_2ZnSnS_4. *Chemistry of Materials* 24: 3188–3196.

Engelken, R.D., Berry, A.K., Van, T.P., Boone, J.L., and Shahnazary, A. 1986. Electrodeposition and analysis of tin selenide films. *Journal of the Electrochemical Society* 133: 581–585.

Gadenne, P., Yagil, Y., and Deutscher, G. 1989. Transmittance and reflectance in situ measurements of semicontinuous gold films during deposition. *Journal of Applied Physics* 66. doi: 10.1063/1.344187.

Garadkar, K.M., Patil, A.A., Korake, P.V., and Hankare, P.P. 2010. Characterization of CdS thin films synthesized by SILAR method at room temperature. *Archives of Applied Science Research* 2: 429–437.

Gary, H. 2002. *Chemical Solution Deposition of Semiconductor Films*, Marcel Dekker, Inc., New York, Basel.

Gencer, A. 2016. Investigation of Al doping concentration effect on the structural and optical properties of the nanostructured CdO thin film. *Superlattices and Microstructures* 92: 278–284.

George, S. 2010. Atomic layer deposition: An overview. *Chemical Reviews* 110: 111–131.

Ghimpu, L., Tiginyanu, I., Lupan, O., Mishra, Y.K., Paulowicz, I., Gedamu, D., Cojocaru, A., and Adelung, R. 2013. Effect of Al Sn-doping on properties of zinc oxide nanostructured films grown by magnetron sputtering. *Proceedings of the International Semiconductor Conference*, Sinaia, Romania, IEEE, 1, CAS, 133–136.

Ghosh, B., Das, M., Banerjee, P., and Das, S. 2008. Fabrication of SnS thin films by the successive ionic layer adsorption and reaction (SILAR) method. *Semiconductor Science and Technology* 23. doi: 10.1088/0268-1242/23/12/125013.

Ghosh, P.K., Das, S., Kundoo, S., and Chattopadhyay, K.K. 2005. Effect of fluorine doping on semiconductor to metal-like transition and optical properties of cadmium oxide thin films deposited by sol-gel process. *Journal of Sol-Gel Science and Technology* 34: 173–179.

Gomez, H., Schrebler, R., Cordova, R., Ugarte, R., and Dalchielle, E.A. 1995. Nucleation and growth of $CuInSe_2$ on a glassy carbon electrode. *Electrochimica Acta* 40: 267–269.

Gryteselv, M. 2016. *Synthesis of $BaTiO_3$-nanoparticles by spray-pyrolysis of TiO_2-nanoparticles dispersed in $Ba(NO_3)_2$-solution.* M. Sc Thesis. Department of Materials Science and Engineering, Norwegian University of Science and Technology.

Guangming, L., Xinchang, W., Yinghua, W., Xinwei, S., Ning, Y., and Binglin, Z. 2008. Synthesis and field emission properties of ZnCdO hollow micro nano spheres. *Physica E: Low Dimensional Systems and Nanostructures* 40: 2649–2653.

Halim, A., Ho, S.M., Kassim, A., Tan, W.T., Jusoh, A.H., and Nagalingam, S. 2010. Effect of concentration on MnS_2 thin films deposited in a chemical bath. *Kasetsart Journal-Natural Science* 44: 446–453.

Hamann, T., Martinson, A., Elam, J., Pellin, M., and Hupp, J. 2008. Atomic layer deposition of TiO_2 on aerogel templates: New photoanodes for dye-sensitized solar cells. *The Journal of Physical Chemistry C* 112: 10303–10307.

Hinkle, C.L., Sonnet, A.M., Vogel, E.M., McDonnell, S., Hughes, G.J., Milojevic, M., Lee, B., Aguirre-Tostado, F.S., Choi, K.J., Kim, H.C., Kim, J., and Wallace, R.M. 2008. GaAs interfacial self-cleaning by atomic layer deposition. *Applied Physics Letters* 92. doi: 10.1063/1.2883956.

Ho, S.M. 2014. Influence of complexing agent ton the growth of chemically deposited $Ni_3Pb_2S_2$ thin films. *Oriental Journal of Chemistry* 30: 1009–1012.

Illiberi, A., Cobb, B., Sharma, A., Grehl, T., Brongersma, H., and Roozeboom, F. 2015. Spatial atmospheric atomic layer deposition of $In_xGa_yZn_zO$ for thin film transistors. *ACS Applied Materials and Interfaces* 7: 3671–3675.

Jeffrey, W.E., Martinson, A.B.F., Pellin, M.J., and Hupp, J.T. 2006. Atomic layer deposition of In_2O_3 using cyclopentadienyl indium: A new synthetic route to transparent conducting oxide films. *Chemistry of Materials* 18: 3571–3578.

Sarma, M.P., Kalita, J.M., and Wary, G. 2017b. Chemical bath deposited nanocrystalline TiO_2 thin film as X-ray radiation sensor. *Materials Research Express* 4: 045005.

Kang, S.J., and Joung, Y.H. 2007. Influence of substrate temperature on the optical and piezoelectric properties of ZnO thin films deposited by RF magnetron sputtering. *Applied Surface Science* 253: 7330–7335.

Kaupo, K., Marianna, K., Marko, V., Mikko, J.H., Kenichiro, M., Kristjan, K., Mikko, R., Markku, L., Ivan, K., and Karol, F. 2017. Atomic layer deposition and properties of mixed Ta_2O_5 and ZrO_2 films. *AIP Advances* 7. doi: 10.1063/1.4975928.

Kaupo, K., Mikko, R., Markku, L., Timo, S., Juhani, K., David, G., Sandeep, B., and Lata, P. 2002. Atomic layer deposition of Al_2O_3, ZrO_2, Ta_2O_5, and Nb_2O_5 based nanolayered dielectrics. *Journal of Non-Crystalline Solids* 303: 35–39.

Kerimova, A., Bagiyev, E., Aliyeva, E., and Bayramov, A. 2017. Nanostructured CdS thin films deposited by spray pyrolysis method. *Physica Status Solidi C* 14. doi: 10.1002/pssc.201600144.

Kim, J.Y. and George, S. 2010. Tin monosulfide thin films grown by atomic layer deposition using tin 2,4-pentanedionate and hydrogen sulfide. *The Journal of Physical Chemistry C* 114: 17597–17603.

Kim, S., Park, S., Son, M., and Kim, H. 2015. Ammonia treated ZnO nanoflowers based CdS/CdSe quantum dot sensitized solar cell. *Electrochimica Acta* 151: 531–536.

Korotcenkov, G., and Cho, B.K. 2017. Spray pyrolysis deposition of undoped SnO_2 and In_2O_3 films and their structural properties. *Progress in Crystal Growth and Characterization of Materials* 63: 1–47.

Kozhukharou, S., and Tchaoushev, S. 2013. Synthesis and characterization of multilayer thin films using spray pyrolysis technique. *Journal of Chemical Metallurgy* 4: 111–118.

Lee, G., Lai, B., and Phatak, C. 2013. A-thin high-dielectric constant nanolaminates for nanoelectronics. *Applied Physics Letters* 102. doi: 10.1063/1.4790838.

Li, E., Zhuo, H., He, H., Wang, N., and Liu, T. 2016. Structural, optical, and electrical properties of low-concentration Ga-doped CdO thin films by pulsed laser deposition. *Journal of Materials Science* 51: 7179–7185.

Li, X., and Nandhakumar, I.S. 2008. Direct electrodeposition of PbTe thin films on n-type silicon. *Electrochemistry Communications* 10: 363–366.

Lim, K.S., Ho, S.M., Anuar, K., and Saravanan, N. 2011. SEM, EDAX and UV-visible studies on the properties of Cu_2S thin films. *Chalcogenide Letters* 8: 405–410.

Lokhande, C.D., Sankapal, B.R., Pathan, H.M., Muller, M., Giersig, M., and Tributsch, H. 2001. Some structural studies on successive ionic layer adsorption and reaction (SILAR) deposited CdS thin films. *Applied Surface Science* 181: 277–282.

Mane, R.S., Sankapal, B.R., and Lokhande, C.D. 2000. Deposition of CdS thin films by the successive ionic layer adsorption and reaction (SILAR) method. *Materials Research Bulletin* 35: 177–184.

Martin, M.F., Glen, D.W., Dmitri, S., Torgny, G., Eric, G., Yves, J.C., John, G., and David, A.M. 2005. HfO_2 and Al_2O_3 gate dielectrics on GaAs grown by atomic layer deposition. *Applied Physics Letters* 86. doi: 10.1063/1.1899745.

Massaccesi, S., Sanchez, S., and Vedel, J. 1993. Cathodic deposition of copper selenide films on tin oxide in sulfate solutions. *Journal of the Electrochemical Society* 140: 2540–2546.

Minami, T., Oda, J., Nomoto, J., and Miyata, T. 2010. Effect of target on transparent conducting impurity doped ZnO thin films deposited by DC magnetron sputtering. *Thin Solid Films* 519: 385–390.

Mondal, S., and Mitra, P. 2008. Preparation of nanocrystalline SnS thin film by SILAR. *Material Science Research India* 5: 67–74.

Mukherjee, A., and Mitra, P. 2015. Structural and optical characteristics of SnS thin film prepared by SILAR. *Materials Science Poland* 33: 847–851.

Mukherjee, A., Satpati, B., Bhattacharyya, S.R., Ghosh, R., and Mitra, P. 2015. Synthesis of nanocrystalline CdS thin film by SILAR and their characterization. *Physica E: Low Dimensional Systems and Nanostructures* 65: 51–55.

Neil, P.D., Meng, X., Elam, J., and Martinson, A. 2015. Atomic layer deposition of metal sulfide materials. *Accounts of Chemical Research* 48: 341–348.

Ngai, C.F., Anuar, K., Ho, S.M., and Tan, W.T. 2011. Influence of triethanolamine on the chemical bath deposited NiS thin films. *American Journal of Applied Sciences* 8: 359–361.

Nicolau, Y.F. 1985. Solution deposition of thin solid compound films by a successive ionic layer adsorption and reaction process. *Applications of Surface Science* 22–23: 1061–1074.

Noraini, K., Anuar, K., Saravanan, N., and Ho, S.M. 2010. XRD and AFM studies of ZnS thin films produced by electrodeposition method. *Arabian Journal of Chemistry* 3: 243–249.

Otto, K. 2012. *Deposition of Indium Sulfide (In_2S_3) Thin Films by Chemical Spray Pyrolysis*. PhD Dissertation. Department of Materials Science, Tallinn University of Technology.

Pathak, T.K., Kumar, R., and Purohit, L.P. 2015. Preparation and optical properties of undoped and nitrogen doped ZnO thin films by RF sputtering. *International Journal of ChemTech Research* 7: 987–993.

Pathan, H.M., and Lokhande, C.D. 2004. Deposition of metal chalcogenide thin films by successive ionic layer adsorption and reaction (SILAR) method. *Bulletin of Materials Science* 27: 85–111.

Patil, P.S. 1999. Versatility of spray pyrolysis techniques. *Materials Chemistry and Physics* 59: 185–198.

Perendnis, D. 2003. *The Thin Film Deposition by Spray Pyrolysis and the Applications in Oxide Fuel Cell*. Doctoral Thesis. Swiss Federal Institute of Technology, ETH, Zurich.

Perendnis, D., and Gauckler, LJ. 2005. Thin film deposition using spray pyrolysis. *Journal of Electrochromic* 14: 103–111.

Ramanathan, K., Teeter, G., Keane, J.C., and Noufi, R. 2005. Properties of high efficiency $CuInGaSe_2$ thin film solar cells. *Thin Solid Films* 480–481: 499–502.

Riikka, L.P. 2005. Surface chemistry of atomic layer deposition: A case study for the trimethylaluminium/water process. *Journal of Applied Physics* 97. doi: 10.1063/1.1940727.

Riveros, R., Romero, E., and Gordillo, G. 2006. Synthesis and characterization of highly transparent and conductive SnO_2:F and In_2O_3:Sn thin films deposited by spray pyrolysis. *Brazilian Journal of Physics* 36: 1042–1045.

Rowlette, P.C., Allen, C., Bromley, O., Dubetz, A., and Wolden, C. 2009. Plasma-enhanced atomic layer deposition of semiconductor grade ZnO using dimethyl zinc. *Chemical Vapor Deposition* 15: 15–20.

Safonova, M., Nair, P.K., Mellikov, E., Garcia, A.R., Kerm, K., Revathi, N., Romann, T., Mikli, V., and Volobujeva, O. 2014. Chemical bath deposition of SnS thin films on ZnS and CdS substrates. *Journal of Materials Science: Materials in Electronics* 25: 3160–3165.

Saha, D., Amit, K.D., Ajimsha, R.S., Misra, P., and Kukreja, L.M. 2013. Effect of disorder on carrier transport in ZnO thin films grown by atomic layer deposition at different temperatures. *AIP Journal of Applied Physics* 114. doi: 10.1063/1.4815941.

Saha, D., Misra, P., Ajimsha, R., Joshi, M., and Kukreja, L. 2014. Phase-coherent electron transport in (Zn, Al)O_x thin films grown by atomic layer deposition. *Applied Physics Letters* 105. doi: 10.1063/1.4902513.

Saha, D., Misra, P., Das, G., Joshi, M., and Kukreja, L. 2016. Observation of dopant-profile independent electron transport in sub-monolayer TiO_x stacked ZnO thin films grown by atomic layer deposition. *Applied Physics Letters* 108. doi: 10.1063/1.4939926.

Sahu, S.R. 2015. Synthesis and characterization of multilayer thin films using spray pyrolysis. M. Sc Project Report. Department of Physics and Astronomy, National Institute of Technology, Rowrkds, India.

Sandy, X.L., Ryan, M.M., and Jane, P.C. 2005. Plasma enhanced atomic layer deposition of HfO_2 and ZrO_2 high-k thin films. *Journal of Vacuum Science and Technology A: Vacuum, Surfaces, and Films* 23. doi: 10.1116/1.1894666.

Sankapal, B.R., Mane, R.S., and Lokhande, C.D. 2000a. Deposition of CdS thin films by the successive ionic layer adsorption and reaction (SILAR) method. *Materials Research Bulletin* 35: 177–184.

Sankapal, B.R., Mane, R.S., and Lokhande, C.D. 2000b. Successive ionic layer adsorption and reaction (SILAR) method for the deposition of large area (10 cm^2) tin disulfide (SnS_2) thin films. *Materials Research Bulletin* 35: 2027–2035.

Sarma, M.P., Kalita, J.M., and Wary, G. 2017a. Chemically deposited ZnS thin film as potential X-ray radiation sensor. *Materials Science in Semiconductor Processing* 61: 131–136.

Schwierz, F., Pezoldt, J., and Granzner, R. 2015. Two-dimensional materials and their prospects in transistor electronics. *Nanoscale* 7: 8261–8283.

Shannon, C.R., Koegel, A., Emery, J., Pellin, M., and Martinson, A. 2017. Low-temperature atomic layer deposition of $CuSbS_2$ for thin-film photovoltaics. *ACS Applied Materials and Interfaces* 9: 4667–4673.

Sharon, M., Ramaiah, K.S., Kumar, M., Neumann-Spallart, M., and Levy-Clement, C. 1997. Electrodeposition of lead sulfide in acidic medium. *Journal of Electroanalytical Chemistry* 436: 49–52.

Sheng, J., Lee, H., Oh, S., and Park, J. 2016. Flexible and high-performance amorphous indium zinc oxide thin-film transistor using low-temperature atomic layer deposition. *ACS Applied Materials and Interfaces* 8: 33821–33828.

Shrestha, P., Gu, D., Tran, N.H., Tapily, K., Baumgart, H., and Namkoong, G. 2010. Investigation of Volmer-Weber growth during the nucleation phase of ALD platinum thin films and template based platinum nanotubes. *ECS Transactions* 33: 127–134.

Sriram, S., and Thayumanavan, A. 2013. Structural, optical and electrical properties of NiO thin films prepared by low cost spray pyrolysis technique. *International Journal of Materials Science and Engineering* 1: 118–121.

Subramanyam, T.K., Srinivasulu, B., and Uthanna, S. 2001. Studies on dc magnetron sputtered cadmium oxide films. *Applied Surface Science* 169–170: 529–534.

Susumu, N., and Adachi, S. 1995. Optical properties of wurtzite CdS. *Journal of Applied Physics* 78. doi: 10.1063/1.360355.

Takahashi, M., Ohshima, Y., Nagata, K., and Furuta, S. 1993. Electrodeposition of PbS films from acidic solution. *Journal of Electroanalytical Chemistry* 359: 281–286.

Teo, D., Anuar, K., Saravanan, N., Tan, W.T., and Ho, S.M. 2010. Chemical bath deposition of nickel sulfide (Ni_4S_3) thin films. *Leonardo Journal of Sciences* 16: 1–12.

Tewari, S., and Bhattacharjee, A. 2009. Synthesis and characterization of cadmium chalcogenide CdX (X = S, Te) thin films. *International Journal of Chemical Sciences* 7: 105–115.

Trilok, K.P., Vinod, K., Swart, H.C., and Purohit, L.P. 2015. P-type conductivity in doped and codoped ZnO thin films synthesized by RF magnetron sputtering. *Journal of Modern Optics* 62: 1368–1373.

Thomas, H., Martinson, A., Elam, J., Pellin, M., and Hupp, J. 2008. Aerogel templated ZnO dye-sensitized solar cells. *Advanced Materials* 20: 1560–1564.

Tomkiewicz, M., Ling, I., and Parsons, W.S. 1982. Morphology, properties, and performance of electrodeposited n-CdSe in liquid junction solar cells. *Journal of the Electrochemical Society* 129: 2016–2022.

Venables, J.A., Spiller, G.D.T., and Hanbucken, M. 1984. Nucleation and growth of thin films. *Reports on Progress in Physics* 47. doi: 10.1088/0034-4885/47/4/002.

Viljami, P., Hatanpaa, T., Ritala, M., and Leskela, M. 2009. Atomic layer deposition of metal tellurides and selenides using alkylsilyl compounds of tellurium and selenium. *Journal of the American Chemical Society* 131: 3478–3480.

Wu, X. 2004. High efficiency polycrystalline CdTe thin film solar cells. *Solar Energy* 77: 803–814.

Wu, Y., Potts, S., Hermkens, P., Knoops, H., Roozeboom, F., and Kessels, W. 2013. Enhanced doping efficiency of Al-doped ZnO by atomic layer deposition using dimethylaluminum isopropoxide as an alternative aluminum precursor. *Chemistry of Materials* 25: 4619–4622.

Yamaguchi, K., Yoshida, T., Sugiura, T., and Minoura, H. 1998. A novel approach for CdS thin-film deposition: Electrochemically induced atom-by-atom growth of CdS thin films from acidic chemical bath. *Journal of Physical Chemistry B* 102: 9677–9686.

Yang, P., and Lieber, C.M. 1996. Nanorod-superconductor composites: A pathway to materials with high critical current densities. *Science* 273: 1836–1840.

Yoffe, A.D. 1993. Low dimensional systems: Quantum size effects and electronic properties of semiconductor microcrystallites (zero dimensional systems) and some quasi-two-dimensional systems. *Advances in Physics* 42: 173–262.

Zaien, M., Ahmed, N.M., and Hassan, Z. 2013. Effects of annealing on the optical and electrical properties of CdO thin films prepared by thermal evaporation. *Materials Letters* 105: 84–86.

Zhang, K., Pillai, A., Bollenbach, K., Nminibapiel, D., Cao, W., and Baumgart, H. 2014. Atomic layer deposition of nanolaminate structures of alternating PbTe and PbSe thermoelectric films. *ECS Journal of Solid State Science and Technology* 3: P207–P212.

6

Carbon Nanotubes: Preparation and Surface Modification for Multifunctional Applications

Jingyao Sun, Jing Zhu, Merideth A. Cooper, Daming Wu, and Zhaogang Yang

CONTENTS

6.1 Introduction

With the development of nanotechnology, controlling and manipulating materials at macromolecular, molecular, and even atomic scales became possible. This enabled researchers to discover and prepare/fabricate new nanomaterials, improve their properties, and expand the range of applications (Yang, Xie, et al., 2016; Chen et al., 2016; Lee et al., 2016; Sahoo et al., 2011; Sun, Zhuang, et al., 2017; Sun, Wu, et al., 2017). As one of the most representative nanomaterials, carbon nanotubes (CNTs) are built with a one-dimensional (1D) cylindrical graphitic structure containing one or more graphitic layers. Arising from the all-carbon sp^2 hybrid structures with abundant delocalized π-electrons (Avouris, 2010; Gupta, Dharamvir, and Jindal, 2005; Castro Neto et al., 2009), CNTs present superior structural, mechanical, thermal, and electrical characteristics. These properties, together with their atomic scale dimension, give CNTs a broad range of application potential in organic optoelectronic, energy storage (Lee, Yabuuchi, et al., 2010; Lee, Kim, et al., 2010; Reddy et al., 2010; El-Kady et al., 2012), electrochemical and biological sensing (Zhang, 2005; Cao et al., 2008; Huang et al., 2010; Kim et al., 2009; Georgiou et al., 2012; Hwang et al., 2012), mechanically reinforced composite (Ajayan et al., 1994; Dalton et al., 2003; Zou et al., 2011; Lee, Lee, Kochuveedu, et al., 2011; Sun, Shen, et al., 2018), heterogeneous catalysis (Gong et al., 2009; Lee, Lee, Park, et al., 2011; Lee, Lee, Lee, et al., 2011), and many more areas.

CNTs have been an extremely popular research area in recent years; new research papers focusing on CNT-related topics are published every day. CNTs, as the name suggests, are cylindrical carbon tubes which are formed by the wrapping of graphene sheets. These cylindrical carbon tubes are composed of layers of all-carbon sp^2 hybrid structures, where every carbon atom is connected to three other atoms in the x–y plane. Based on geometry parameters (especially the number of cylindrical graphene sheets), CNTs can be divided into three broad classes: single-walled carbon nanotubes (SWCNTs), double-walled carbon nanotubes (DWCNTs), which have two cylindrical carbon tubes, and multi-walled carbon nanotubes (MWCNTs), which have several cylindrical carbon tubes. Among these three types of CNTs, SWCNTs and MWCNTs are the most commonly used (as shown in Figure 6.1).

In many cases, researchers regard nanotube structures that contains either two or more concentric graphene cylinders as MWCNTs. MWCNTs, which consist of several concentric graphene sheets, were first observed by Sumio Iijima of the NEC Corporation (Iijima, 1991) in 1991. Then, in 1993, a SWCNT that contains a single layer of cylindrical graphite was successfully

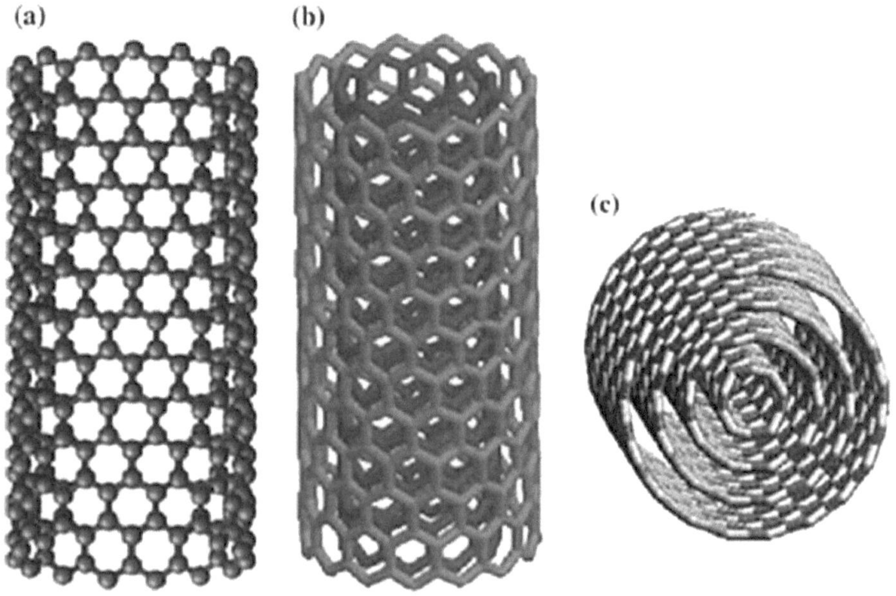

FIGURE 6.1 Schematic image of: (a) SWCNT, (b) DWCNT, and (c) MWCNT (Reproduced with permission (Eatemadi et al., 2014)).

synthesized by Bethune et al. of IBM and Iijima et al., independently. (Iijima and Ichihashi, 1993; Bethune et al., 1993).

The unique nanostructure of CNTs, only several nanometers in diameter and several hundred nanometers in length, ensures the outstanding properties and extensive applications of CNTs. The preparation of CNTs can be achieved using a variety of technologies, including arc discharge (Arora and Sharma, 2014), laser ablation (Chrzanowska et al., 2015), chemical vapor deposition (CVD) (Kong, Cassell, and Dai, 1998), glow discharge (Nozaki and Okazaki, 2008), solid-state pyrolysis (Shanmugam and Gedanken, 2006), and high-pressure carbon monoxide (HiPCO) (Nikolaev et al., 1999). As mentioned above, CNTs present superior mechanical, thermal, electrical, and magnetic characteristics. However, the exact magnitude of these properties is determined by the geometric parameters and chirality of the nanotubes. Besides, the type of CNTs, whether they belong to the single-walled, double-walled, or multi-walled form, also has a significant effect on the final properties. Table 6.1 shows the typical properties of carbon nanotubes.

SWCNTs can be divided into three types based on the wrapping of the cylinder: armchair, chiral, and zigzag. The structure of SWCNTs is characterized by a pair of indices (n, m) that describes the chiral vector. The number of unit vectors in the honeycomb crystal lattice of graphene along two directions is determined by the integers n and m. Generally speaking, when m = 0, CNTs are referred to as zigzag nanotubes; when m = n, CNTs are referred to as armchair nanotubes, and CNTs under other states are named chiral nanotubes. Figure 6.2 demonstrates the differences between the three forms of SWCNTs.

MWCNTs also can be divided into two models, the Russian doll model and the parchment model. The Russian doll model contains MWCNTs in which the carbon nanotube with a larger diameter contains another thinner nanotube inside it. On the other hand, the parchment model stands for a single graphene sheet that is wrapped around itself many times like a rolled-up paper scroll.

In order to improve the specific properties of CNTs and broaden their applications, it is necessary to tailor the chemical nature of nanotube's surface. For example, the surface modification can improve the poor dispersion in solvents that CNTs usually have and enhance their interactions with other materials.

The surface modification of CNTs can also be divided into two types. The first is the covalent modification and the second is the non-covalent modification of molecules. The covalent modification connects functional groups to the surface of nanotubes. However, the functional groups that are added to CNTs

TABLE 6.1

Typical Properties of Carbon Nanotubes

Physical properties	Classification	Values
Structure	Average diameter of SWCNT	1.2–1.4 nm
	Carbon bond length	1.42 Å
	Overlap energy of C–C bonding	~2.5 eV
Density	(17, 0) Zigzag	1.34 g/cm^3
	(10, 10) Armchair	1.33 g/cm^3
	(12, 6) Chiral	1.40 g/cm^3
Interlayer spacing	(n, 0) Zigzag	3.41 Å
	(n, n) Armchair	3.38 Å
	(2n, n) Chiral	3.39 Å
Lattice parameter	(17, 0) Zigzag	16.52 Å
	(10, 10) Armchair	16.78 Å
	(12, 6) Chiral	16.52 Å
Elastic behavior	Young's modulus of SWCNT	~1 TPa
	Young's modulus of MWCNT	~1.28 TPa
	Maximum tensile strength	~100 GPa
Electrical transport	Resistivity	10^{-4} Ω/cm
	Maximum current density	1013 A/m^2
Thermal transport	Thermal conductivity	~2000 W/m/K
	Relaxation time	~10 to 11 seconds

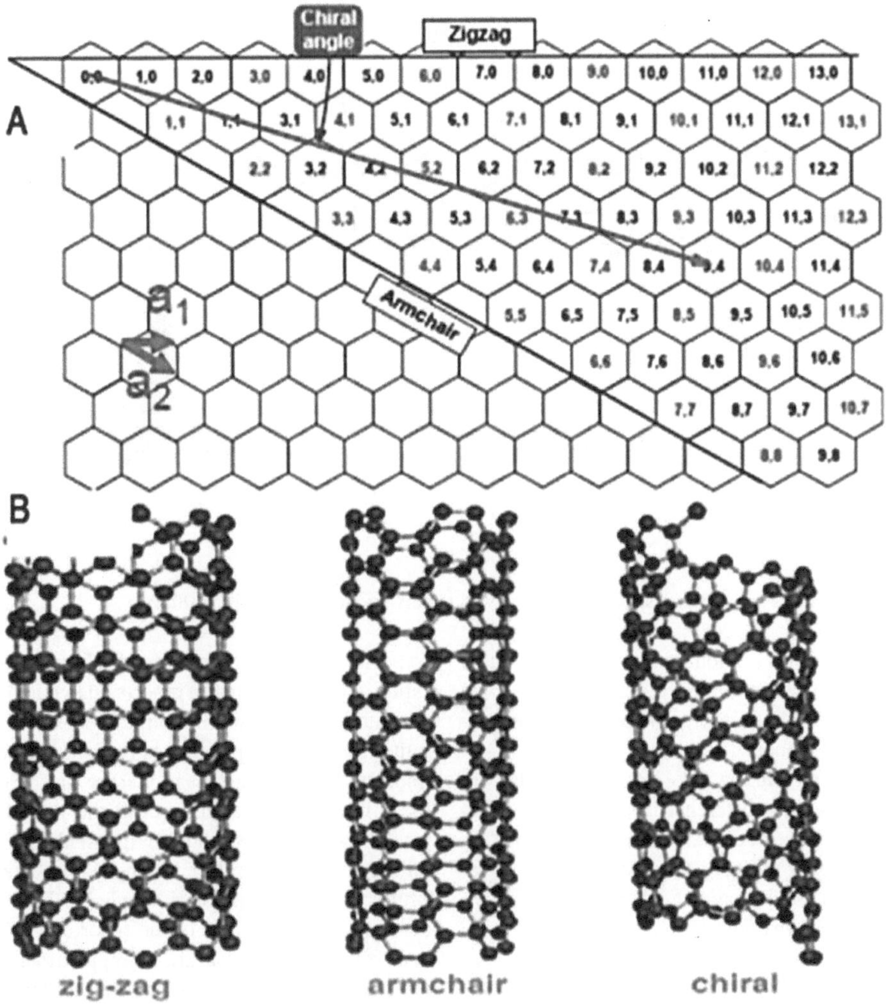

FIGURE 6.2 The directions of rolling a graphene sheet (a), Three different forms of SWCNTs (b). (Reproduced with permission (Eatemadi et al., 2014).).

are double-edged swords. On the one hand, the functional groups can enhance the properties of CNTs and even provide new functions that cannot be found in pristine CNTs. On the other hand, these functional groups may lead to defects because they destroy the perfect structure of the nanotube walls. Among the different modification methods, chemical functionalization is an attractive target as it can improve the machinability and solubility of CNTs. It will not only broaden the application area of CNTs, but also make it easier for manipulation. The non-covalent modification, controlled by thermodynamic criteria, changes the nature of a CNT surface. The advantage of non-covalent modification is that the structure of CNTs is not altered, hence the mechanical property can be retained and made more compatible with solution and polymer matrices. The potential disadvantage of non-covalent modification is that the interactions between CNTs and wrapping molecules may be very weak.

6.2 Preparation of Carbon Nanotubes

The laboratory-scale preparation of a CNT was started by Iijima in 1991 using the method of arc discharge. From then on, CNTs have become one of the most researched topics around the world. In order to make CNT preparation cost-effective and obtain a method for large-scale production, many other preparation methods have already been developed, one after the other, during the last two decades. However, all the commonly used CNT preparation methods can be roughly classified into three kinds: arc discharge, laser ablation, and chemical vapor deposition. Most of the other techniques are an optimized, developed, or complementary variation of the three basic methods. In the following section, we will give a brief introduction and discussion on these three preparation methods one by one.

6.2.1 Arc Discharge

Arc discharge was the first preparation method of CNTs used by Iijima in 1991. The breakthrough in MWCNT preparation by arc discharge was first achieved by Ebbesen and Ajayan in 1992 (Ebbesen and Ajayan, 1992). The preparation and purification of high-quality MWCNTs at the gram level was achieved with their efforts. Then in 1993, substantial amounts of SWCNTs prepared by the arc discharge method were achieved by Bethune et al. with the help of metal catalysts (Bethune et al., 1993). This

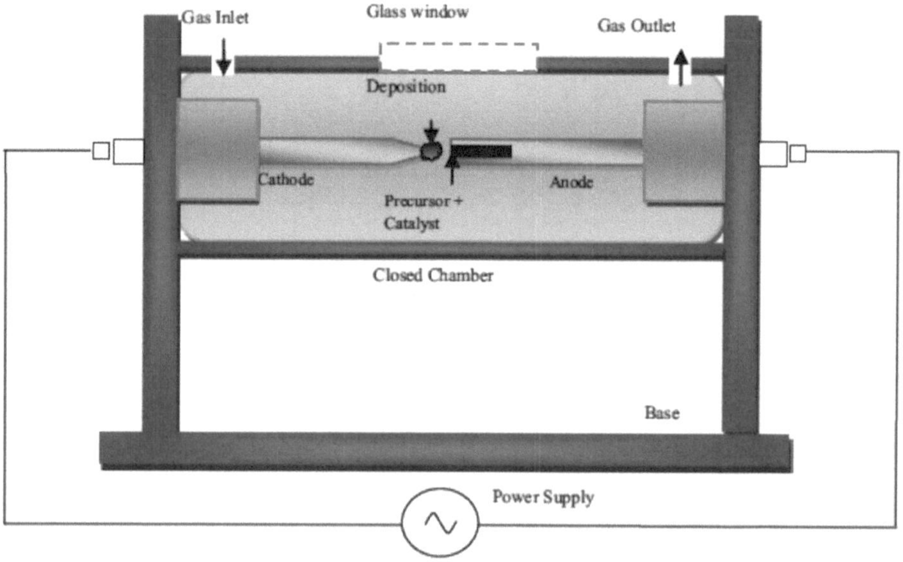

FIGURE 6.3 Schematic diagram of arc discharge setup (Reproduced with permission (Arora and Sharma, 2014).).

method was employed in a 100 Torr helium atmosphere with two graphite electrodes, with a 2 mm distance between them. Arc discharge is a physical method of CNT preparation and the process occurs in an inert atmosphere.

Arc plasma was used to vaporize the material in the arc discharge process. A potential difference should be provided between the two graphite electrodes to make the inner clipping gas electrically conductive for plasma generation. Arc discharge was the most convenient method to generate thermal plasma with the characteristics of high energy content and local thermal equilibrium state.

During CNT preparation using the arc discharge method, carbon vaporization under reduced gas atmosphere should be performed in the presence of catalysts (cobalt, boron, yttrium, iron, nickel, gadolinium, and so on). The vaporization is a result of the energy transfer from the arc to the anode made of graphite doped with catalyst. Arc current is another important variable in this process, in which low currents improve the quality of MWCNTs and high currents favor the production of SWCNTs.

The schematic diagram of the arc discharge setup is presented in Figure 6.3. As seen in Figure 6.3, this setup should be connected to both a gas supply and a vacuum line with a diffusion pump. Typically, the anode is a long rod with a diameter of 6 mm, while the cathode is a much shorter rod with a diameter of 9 mm. Researchers found that the efficient cooling of the cathode can be helpful for the quality improvement of the CNTs obtained. Normally, a DC bias of 20–30 V is applied between the two graphite electrodes in the arc discharge method. Carbon atoms are ejected from the anode and accumulate on the cathode in the form of nanotubes. It should be pointed out that the quantity and quality of obtained CNTs is greatly influenced by various parameters, including the kind of gas, inert gas pressure, metal concentration, system geometry, and current. The effects of various parameters on the size and yield of CNTs prepared using the arc discharge method are listed in Table 6.2.

After reviewing and comparing the results listed in Table 6.2, some observations can be made. First, current is the most important parameter of the arc discharge method in CNT preparation. However, different reports related to the large variations in current during the arc discharge process have led to some ambiguity. Therefore, further investigations are needed to deepen the understanding of current in the arc discharge method. Second, the influence of temperature on the growth and nucleation of CNTs is largely unexplored in the method of arc discharge. The vital role of temperature should also be researched carefully. Third, direct-current power supply has been widely used for arc discharge. Fourth, most of the reported CNT preparations have been performed with a graphite precursor, while other carbon materials (e.g., carbon black) have not been commonly utilized yet and are a potential research area.

6.2.2 Laser Ablation (Also Called Laser Vaporization)

Laser ablation (also called laser vaporization or pulsed laser vaporization) is currently being developed as a method for CNT preparation. Similarly to the arc discharge method, carbon is vaporized and later deposited onto a substrate. In 1995, laser ablation was first used for the preparation of CNTs (Guo et al., 1995). It was found that the laser ablation method has much better controllability over the growth of CNTs than other preparation methods such as the previously mentioned arc discharge method. A piece of graphite is used as the carbon source in this method and is vaporized by laser irradiation in an inert gas atmosphere (at 500 Torr pressure) under a very high temperature (between 800 and 1500 °C) (Liu et al., 2014; Mittal et al., 2015). The CNTs prepared with a pure graphite target in laser ablation were found to be MWCNTs (Guo et al., 1995). Furthermore, it was found that the yield and quality of CNT products rely on the reaction temperature. The best product quality can be obtained at the reaction temperature of 1200 °C. When the temperature goes lower, the quality of the CNT structure starts decreasing and presenting many defects. Moreover, catalysts play an important

TABLE 6.2

Effect of Different Parameters on the Size and Yield of CNTs Prepared Using the Arc Discharge Method.

Parameter	Effect on the size of CNTs	Effect on the yield of CNTs
Temperature	Higher temperature leads to a larger diameter in MWCNTs (Song, Liu, and Zhu, 2007)	/
	An increase in temperature reduces the diameter (Zhao, Liu, and Zhu, 2005; Zhao and Liu, 2004)	/
Atmosphere	Inserting gas leads to diameter changing (Su et al., 2012)	Krypton increases the yield (Sun et al., 2007)
	Changing from H_2/Ar to H_2/He results in a smaller diameter (Zhang, 2012)	/
	A larger diameter of CNT is formed in He than in air (Kim and Kim, 2006)	/
Catalyst	Cocatalyst increases the diameter (Yoshida, Sugai, and Shinohara, 2008)	The catalyst results in yield increase (Li et al., 2005)
	Ca-Ni catalyst decreases the diameter (Shi et al., 1999)	/
	Sulphur catalyst increases the diameter (Park et al., 2002; Saito, Nakahira, and Uemura, 2003)	/
Pressure	CO pressure higher than 4 kPa suppresses the growth of SWCNTs with smaller diameters (Su et al., 2011)	The yield increases under higher pressure (Mohammad et al., 2013)
	/	Low yield for pressure below 100 Torr (Chaudhary, Ali, and Yupapin, 2014)
Electrode	Significantly influences the diameter of SWCNTs (Afkhami et al., 2014)	A rotating electrode increases the yield (Zhao et al., 2012; Joshi et al., 2008)
	/	A smaller anode increases the yield (Fetterman, Raitses, and Keidar, 2008)

role in laser ablation for CNT preparation. After adding small quantities (several percent or less) of transition metals, such as nickel or cobalt, the CNT product will be significantly modified and SWCNTs can be obtained instead of MWCNTs under their corporation with the graphite pellet. Using a double-pulsed laser was also found to be helpful for increasing the production yield of SWCNTs (Goddard, 2007).

The schematic diagram of a laser ablation setup is shown in Figure 6.4. It is composed of a furnace, a water-cooled trap, a quartz tube with a window, a target carbon composite doped with catalytic particles, and flow controller systems for the buffer gas to maintain constant flow rates and pressures (Thess et al., 1996; Guo et al., 1992). During the laser ablation process, the laser beam (typically yttrium aluminum garnet or CO_2 laser) is introduced through lenses and windows then focused onto the target rod in the furnace. The target carbon source will be vaporized under a high temperature and conveyed by the Ar buffer gas to form SWCNTs on the trap at the end of the quartz glass tube. Typically, the flow rate and pressure of Ar buffer gas are 500 Torr and 1 cm/s. In order to keep the freshness of the

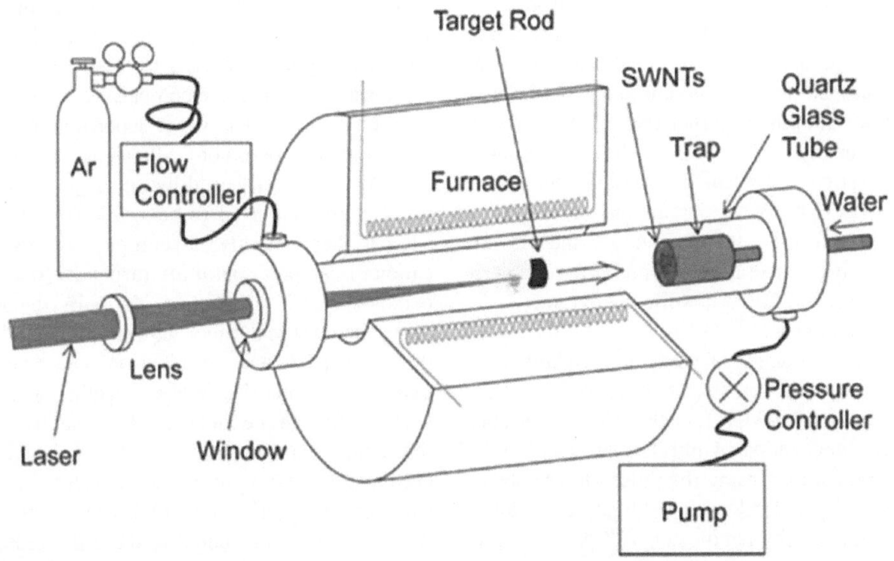

FIGURE 6.4 Schematic diagram of laser ablation setup for CNT preparation (Reproduced with permission (Ando et al., 2004).).

vaporization surface, one of the focus points of the laser beam and the target carbon source are designed to be moveable inside the quartz glass tube.

The laser ablation method presents several advantages, such as diameter control, new material production, and the ability to produce high-quality SWCNTs. With the development of the laser ablation method together with the purification process, researchers have produced high-quality SWCNTs with minimal defects and contaminants (e.g., amorphous carbon and catalytic metals) (Bandow et al., 1997; Ishii et al., 2003; Chiang et al., 2001). A high-power laser beam, a pure target carbon source, and homogeneous annealing conditions have been recognized as proper conditions for the generation of high-crystallinity products (Shinohara, 2000; Yudasaka et al., 2002). The laser ablation method is one of the most expensive among the different methods for CNT preparation. However, unlike in the arc discharge method, which can lead to a mixture of MWCNTs and SWCNTs, SWCNTs can be obtained without the presence of MWCNTs using the laser ablation method. Thus, laser ablation can be applied to many situations which would only need SWCNTs (Mittal et al., 2015).

Masera et al. (Maser et al., 1998) have reported a simple laser ablation method for the production of high-density SWCNTs. In this method, a continuous-wave CO_2 laser with a power of 250 W has been applied for the evaporation of graphite/bimetal targets in a vertical chamber. The system allows effective interaction between laser target and gas plasma, so the local conditions near the evaporation zone play an important role in the formation of SWCNTs (the SWCNT diameter is 1.1 to 1.6 nm). Muñoz et al. (Muñoz et al., 2000) also studied the preparation of SWCNTs using the same continuous wave 10.6 µm CO_2 laser under nitrogen, argon, and helium atmospheres in the pressure range of 50 to 500 Torr. Both nitrogen and argon can be used as buffer gases to prepare SWCNTs with a high yield from 200 to 400 Torr. The amount of SWCNTs in the soot will be drastically decreased below 200 Torr and the material will be dominated by amorphous carbon. Negligible amounts of SWCNTs were formed under a helium atmosphere. Yudasaka et al. (Yudasaka et al., 1999) investigated the formation of SWCNTs from a target composed of graphite, nickel, and cobalt. They performed a comparison between the use of a pulsed CO_2 laser (0.1 MW/cm^2 laser-power density, 20 ms pulse width) and a pulsed neodymium-doped yttrium aluminum garnet (Nd : YAG) laser (4 GW/cm^2 laser-power density, 5–7 ns pulse width). It was found that the CO_2 laser ablation (2 kW/cm^2 irradiation energy) and Nd : YAG laser ablation (2.4 kW/cm^2 irradiation energy) led to significant differences in forming the temperature and pressure for SWCNTs. The forming temperature and pressure of the CO_2 laser ablation were 300 K and 50 Torr, while the parameters for the Nd : YAG laser ablation were 1170 K and 200 Torr. They further noticed that, when compared with the standard metal carbon target, porous target carbon sources of graphite–cobalt–nickel yielded twice as much SWCNT in the whole laser ablation process. The first application of ultrafast laser pulses for large-scale SWCNT preparation using the pulsed-laser ablation technique was reported by Eklund et al. (Eklund et al., 2002). Carbon soot rich in high-quality bundles of SWCNTs was produced at a very high production rate of 1.5 g/h using only 20 % of the nominal average power of the 1 kW free electron laser. Besides the examples listed above, many researchers have prepared SWCNTs with high yield at °C in the presence of cobalt/nickel (Saghafi et al., 2014; Flahaut et al., 1999), graphite/bimetal targets (Mubarak et al., 2014), or rhodium/palladium graphite composite rod (Lebedkin et al., 2002).

The laser ablation method can be applied for the preparation of high-quality CNTs, but it is limited as a large-scale industrial technology because it suffers from the following disadvantages.

(1) The laser ablation technique is an energy-extensive process that requires a large amount of energy for the generation of laser beams. There is no doubt that such a large amount of energy is not economical for large-scale industrial production.
(2) CNTs prepared by the laser ablation method are in highly tangled form, and carbon products in unwanted forms or catalysts will be mixed together with CNTs. Thus, further purification processes to obtain purified products are required for CNTs prepared using this method. However, the additional design of such purification process is difficult and expensive.
(3) Large solid target carbon sources, which are very difficult to get for laser evaporation, are required in a large-scale production process. This significantly limits the possibility to use the laser ablation method as a large-scale process.

6.2.3 Chemical Vapor Deposition

Chemical vapor deposition was first developed for the preparation of carbon fibers in the 1900s. Then, in 1996, the chemical vapor deposition method emerged as a potential option for the preparation of CNTs (especially for large-scale production). Although there are many other CNT preparation methods, such as laser ablation and arc discharge as mentioned above, chemical vapor deposition is the most promising and versatile method due to its simplicity, low-temperature operation, bulk, and low-cost production (Shah and Tali, 2016). This technology requires solid, liquid, or gaseous carbon sources, nano-sized catalysts, and substrates as support materials for nanoparticle catalysts. Table 6.3 reveals a comprehensive comparison among arc discharge, laser ablation, and chemical vapor deposition for CNT preparation.

For the preparation of CNTs using the chemical vapor deposition method, fossil-based hydrocarbons and plant-based hydrocarbons are perceived as being the two most employed carbon sources. There are also several publications reporting the application of camphor, palm oil, turpentine oil, etc., as natural precursors for CNT preparation. Normally, the preparation of CNTs using the chemical vapor deposition method requires catalysts, metal particles, and an adhesion buffer layer (e.g., aluminum oxide) that binds the catalyst together with the targeting substrate. Although the metal catalysts can be selected from a wide range of pure metals, iron, cobalt, nickel, and their derived alloys are most commonly used and considered to be the most effective ones (Ishigami et al., 2008). The effectiveness of these metal catalysts can be attributed to the high solubility of carbon and the high carbon diffusivity under high temperatures (Kumar and Ando, 2010).

TABLE 6.3
Comprehensive Comparison among Arc Discharge, Laser Ablation, and Chemical Vapor Deposition Methods for CNT Preparation

Parameter	Arc discharge	Laser ablation	Chemical vapor deposition	Ref.
Carbon source	Graphite	Graphite	Fossil-based hydrocarbon and botanical hydrocarbon	(Nessim, 2010)
Yield	Low	Low	High	(Takagi et al., 2006)
Purity	Medium	Low to medium	Medium to high	(Takagi et al., 2006)
Cost	Medium to high	High	Low	(Takagi et al., 2006)
Temperature	~4000 °C	T_R –1000°C	500–1200 °C	(Takagi et al., 2006)
Atmosphere	500 Torr (argon or nitrogen)	Atmospheric pressure	Low-pressure (argon)	(Nessim, 2010)
Advantage	Open-air synthesis possible, no catalyst needed	Narrower distribution of SWCNT obtained	Simple process, large-scale production available	(Saifuddin, Raziah, and Junizah, 2013)

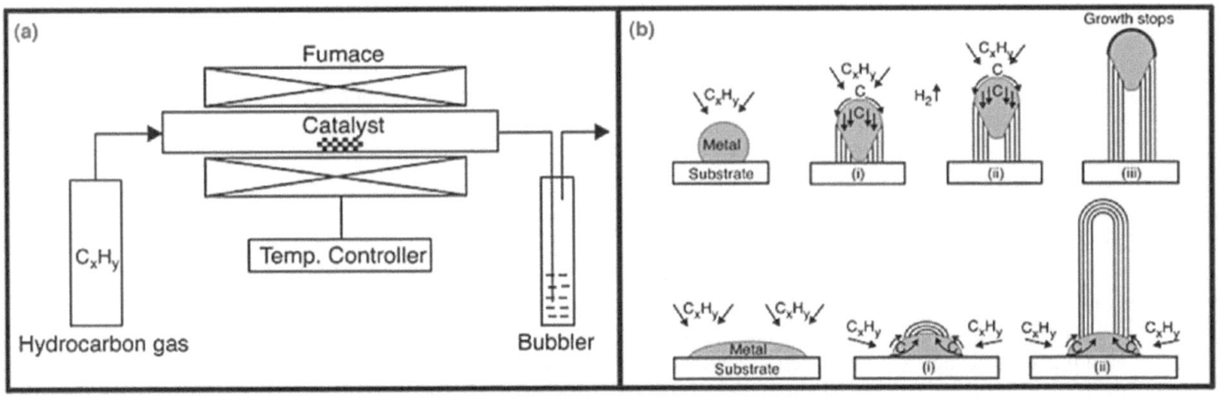

FIGURE 6.5 (a) Schematic diagram of a simple chemical vapor deposition setup. (b) Two proposed growth mechanisms of CNTs, tip-growth model (top) and base-growth model (bottom), using the chemical vapor deposition method (Reproduced with permission (Wang, Vinodgopal, and Dai, 2019).).

Figure 6.5a shows a basic chemical vapor deposition setup. In order to enhance the function of the chemical vapor deposition method, various additional pieces of equipment could be added to this basic setup. There are various kinds of chemical vapor deposition methods for the preparation of CNTs, each with specific add-ons and equipment. For example, a plasma generator is needed for the plasma-enhanced chemical vapor deposition method and a pressure controller is necessary for low-pressure chemical vapor deposition. In the general procedure, hydrocarbon sources, such as acetylene, ethylene, benzene, methane, etc., are generally applied in the chamber at high temperatures to grow CNTs. The decomposition of the hydrocarbon sources, known as pyrolysis, occurs under high temperature. After pyrolysis happens in the chamber, the carbon cylinder (i.e., CNT) will be formed by the carbon atoms with the help of catalyst particles. Many parameters such as temperature, pressure, growth time, flow rate/vapor pressure/concentration of carbon sources, catalyst materials, etc., will significantly influence the final length and quality of the CNTs. The contributions of the aforementioned parameters to the growth of CNTs are inter-related and can weigh differently under different circumstances.

The growth mechanism of CNTs can be described as follows. When the carbon species dissolve into the metal nanoparticles, the dissolved carbon will precipitate out as a tubular network after reaching the dissolution limit of carbon. As shown in Figure 6.5b, there are two proposed cases: (i) tip-growth model, and (ii) base-growth model. In the tip-growth model, the interaction between catalyst and substrate is quite weak, thus the precipitation of the carbon species from the metal nanoparticles can start from the bottom, and the catalyst would be lifted up during the growth process. On the contrary, the interaction between catalyst and substrate in the base-growth model is strong, thus the carbon species from the metal nanoparticles will be unable to lift the catalyst up and form tubular networks, which makes the precipitation happen from the tip of the catalyst. Therefore, in the base-growth model, the catalyst nanoparticles will anchor on the substrate, while the carbon species diffuse upwards to form carbon cylinders.

There are several ways to use other chemical vapor deposition methods such as water-assisted chemical vapor deposition (Hata, 2004; Ran et al., 2013), oxygen-assisted chemical vapor deposition (Chou, Wu, and Hsieh, 2016), hot-filament chemical vapor deposition (Piazza et al., 2014; Huang et al., 1998), microwave plasma chemical vapor deposition (Bower et al., 2000; Kar et al., 2016), radio frequency chemical vapor deposition (Wang et al., 2004), or plasma-enhanced chemical vapor deposition (Chhowalla et al., 2001; Gautier, Le Borgne, and El Khakani, 2016). Here we take the high-pressure carbon monoxide and carbon monoxide chemical vapor deposition, which are two important branches of the chemical vapor deposition technique, as typical examples. High-pressure carbon monoxide chemical vapor deposition, which stands for high-pressure catalytic decomposition of carbon monoxide, is a method for the preparation of SWCNTs using high-pressure carbon monoxide as the carbon source. The high-pressure carbon monoxide chemical vapor deposition technique was first developed by Nikolaev et al. from Rice University (Nikolaev et al., 1999). In the high-pressure carbon monoxide chemical vapor deposition process, volatile organometallic catalyst precursors are introduced into the reactor to form the used catalysts in this gas phase. The organometallic species decompose in the reactor under high temperature and then form metal clusters on which SWCNTs nucleate and grow. The yield of SWCNTs prepared by the high-pressure carbon monoxide chemical vapor deposition method can be as high as a kilogram per day, which is such a high level that no other methods can go beyond ig for now. Actually, carbon monoxide was the first feed gas used for the preparation of SWCNTs. In 1996, molybdenum-catalyzed disproportionation of carbon monoxide at 1200 °C was first applied for the preparation of SWCNTs by Dai et al. (Dai et al., 1996). It was reported that most of the resulting SWCNTs had catalytic particles attached to the ends, indicating that the growth of SWCNTs was catalyzed by preformed nanoparticles. Compared with the SWCNTs prepared using the same methane and catalyst, the application of carbon monoxide as a feed gas shows certain advantages on amorphous carbon reduction overhydrocarbons. Resasco's group in the University of Oklahoma (Kitiyanan et al., 2000) further developed the cobalt–molybdenum–catalyzed process in the carbon monoxide chemical vapor deposition method, which made it much more suitable for commercial applications. The most important advance in the large-scale preparation of SWCNTs was realized with the help of cobalt–molybdenum bimetallic catalysts and a fluidized-bed chemical vapor deposition reactor.

The advantages of chemical vapor deposition methods can be described as follows:

(1) The chemical vapor deposition method is a simple, scalable, and economical technique for the large-scale production of CNTs.
(2) Various kinds of substrates and hydrocarbons (no matter whether in solid, liquid, or gas form) can be utilized in CNT preparation using chemical the vapor deposition method.
(3) CNTs in different forms, including straight, coiled, aligned, entangled, etc., can be prepared using the chemical vapor deposition method. Furthermore, CNTs with the desired architecture also can be obtained with pre-defined cities of patterned substrates.

6.3 Carbon Nanotube Modification

CNTs prepared by different methods lead to different CNT structures, which can result in CNT products with different properties (such as mechanical, electrical, thermal, etc.) (Wang and Arash, 2014; Lee, 2016; Zhuang et al., 2018). In actuality, all the factors, such as the preparation method, purification treatment, and surface modification of CNTs have significant influence on the magnitude of these properties (Cividanes et al., 2017). CNTs have the tendency to aggregate and form bundles due to the high van der Waals attractions between the nanotubes, which limits the superior properties of CNTs and causes poor solubility in both aqueous and non-aqueous mediums (Khan, Gomes, and Altarawneh, 2010; Alpatova et al., 2010). CNT aggregates or bundles caused by poor solubility will further influence the targeting performances of CNT composites and restrict their application potentials. For example, the dispersion of CNTs in a polymer matrix has a strong effect on the performance (e.g. stretchability, conductivity, durability, etc.) of the CNT/polymer nanocomposite. While the carbon atoms on the CNT surface are chemically stable due to the aromatic nature of the bonds, the inert CNT can interact with the surrounding matrix only through van der Waals force, and cannot establish effective load transfer across the interface.

In order to modify the surface properties of CNTs, a variety of strategies and efforts in the development of CNT modification methods (or CNT functionalization methods) have been performed over the past decade. In common modification treatments, the CNT aggregates are broken up by ultrasonication. Then, specific chemical agents are added across the CNTs, which can change the CNT surface, prevent reaggregation, and improve the properties of CNTs (Husanu, Baibarac, and Baltog, 2008). The chemical properties of functionalized CNTs and the reaction mechanism between carbon nanotubes and various functional groups have been reviewed in several comprehensive papers (Ma et al., 2010; Gao et al., 2012; Balasubramanian and Burghard, 2005). The poor solubility of CNTs, the weak interactions with other materials, and their technological applications can be improved using chemical modification methods (Ferreira et al., 2015). These methods can be conveniently labeled as covalent modification (or covalent functionalization) and non-covalent modification (or non-covalent functionalization) based on the interactions between carbon atoms on the CNT surface and the active molecules. The covalent modification is based on the covalent bonding between the functionalization agent and the CNT surface, which is formed by chemical treatments. The non-covalent modification is built on the intermolecular interaction by van der Waals forces between chemical groups and the CNT surface (Figure 6.6).

Although much research has already been done on the topic of CNT modification based on the covalent and non-covalent methods, there exist many processes and material variables that have not been fully optimized. Moreover, structural changes caused by CNT modification and damages resulting from commonly applied ultrasonication treatments and other dispersion processes may result in inevitable adverse effects.

6.3.1 Covalent Modification

Covalent modification is based on the covalent linkage of functional groups on the carbon scaffold of CNTs. The functional groups can react with carboxylic groups (–COOH) of CNTs and other oxygen-containing groups formed during oxidation. The covalent modification process can be performed both at the sidewall and the end-tip of CNTs. The covalent modification of a CNT is associated with a change in the carbon hybridization (from sp^2 to sp^3) and a simultaneous loss of the local conjugation (Figure 6.7, Strategy A). This type of CNT modification can be performed by a chemical reaction with some highly reactive molecules and usually has a high modification yield (Balasubramanian and Burghard, 2005). CNTs can be functionalized by different kinds of functional groups, such as amino,

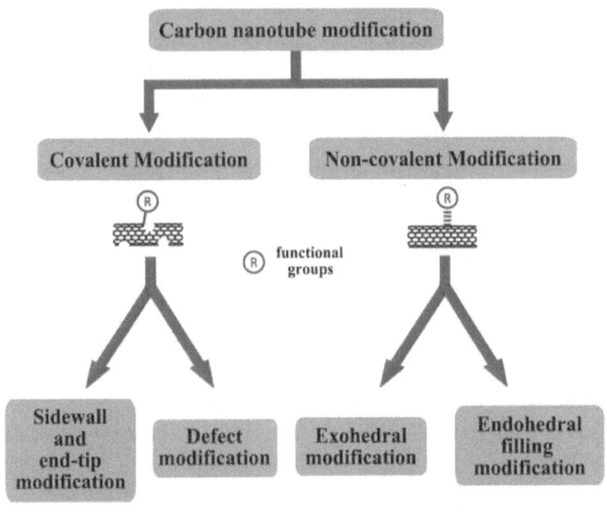

FIGURE 6.6 Overview scheme of CNT modification methods.

carboxylic, alkyl, and hydroxyl groups (Cividanes et al., 2014; Mohamed et al., 2015; Ferreira et al., 2015; Dinesh, Bianco, and Ménard-Moyon, 2016). Look at fluorine as an example. It is found that the fluorination process of pure SWCNTs, which occurs at temperatures up to 325 ° C, is reversible (the fluorine can be removed using anhydrous hydrazine) (Mickelson et al., 1998). As the C–F bonds in alkyl fluorides are stronger than those in fluorinated CNTs, the fluorinated CNTs provide substitution sites for additional modification (Kelly et al., 1999). Besides CNT fluorination on the sidewall, other similar methods, such as cycloaddition reaction, halogenation, nucleophilic addition, etc., have also been successfully applied and developed for sidewall and tip functionalization in recent years (Karousis, Tagmatarchis, and Tasis, 2010).

Another type of covalent modification method of CNTs is defect modification (Strategy B in Figure 6.7). Chemical transformation of defect sites is utilized in this process for CNT modification. Here, the defect sites can be the holes in the sidewall or end-tips of CNTs, oxygenated sites, and pentagonal or heptagonal irregularities in hexagonal graphene frames (e.g., the Stone–Wales defect, also known as 7-5-5-7 defect, as seen in Figure 6.8).

In the defect modification treatment, defects can be established both on the sidewalls and on the end-tips of CNTs using oxidation reactions with strong acids and oxidants (Yu et al., 1998), esterification and amidation reactions (Hamon et al., 2002), plasma activations (Wang, Chang, and Yuan, 2009; Ávila-Orta et al., 2009), etc. The oxidation defects on CNTs are stabilized by bonding with additional functional groups, including carboxylic acid (–COOH) and hydroxyl (–OH) groups. The rich chemistry of these functional groups together with CNTs can be utilized as precursors for further chemical modifications. The CNTs treated with defect modification methods can be soluble in many organic solvents due to the change in the nature of CNTs, from hydrophobic to hydrophilic, with the attachment of polar groups. Some of the most frequently reported methods for the covalent modification of CNTs in recent years are categorically listed below (see Table 6.4).

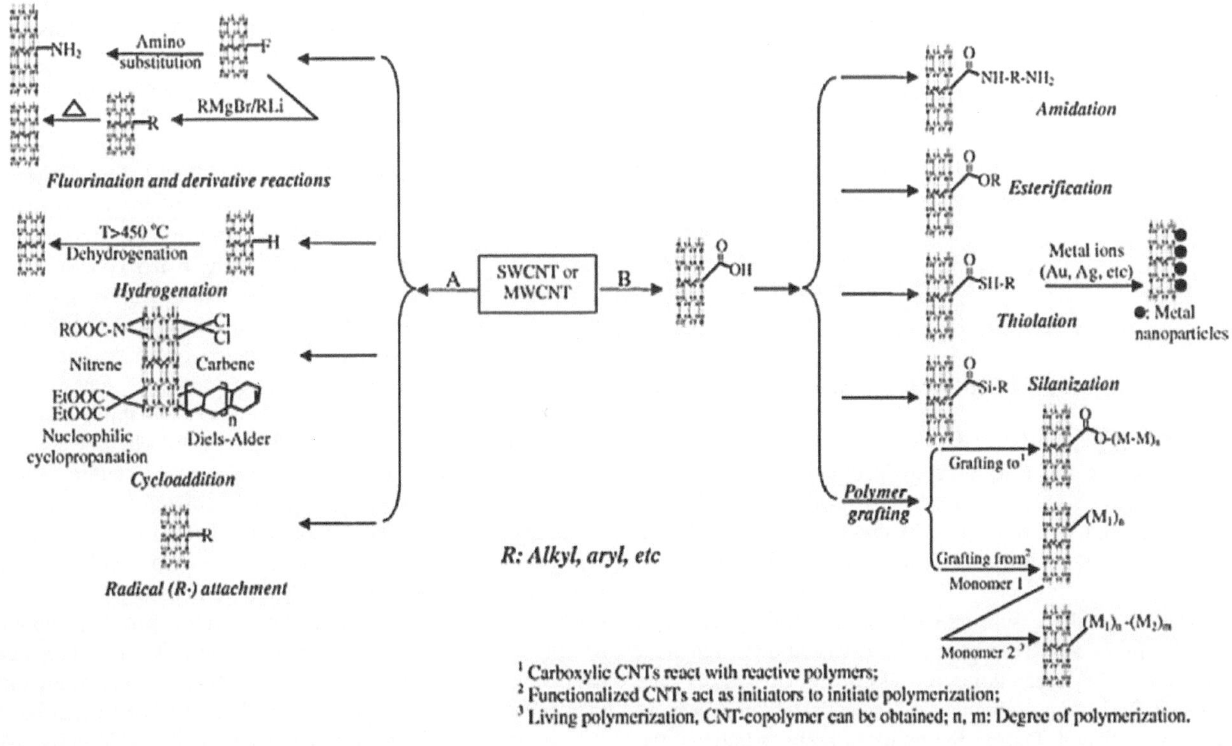

FIGURE 6.7 Covalent modification of CNTs (Strategy A: direct sidewall modification; Strategy B: defect modification) (Reproduced with permission (Ma et al., 2010).).

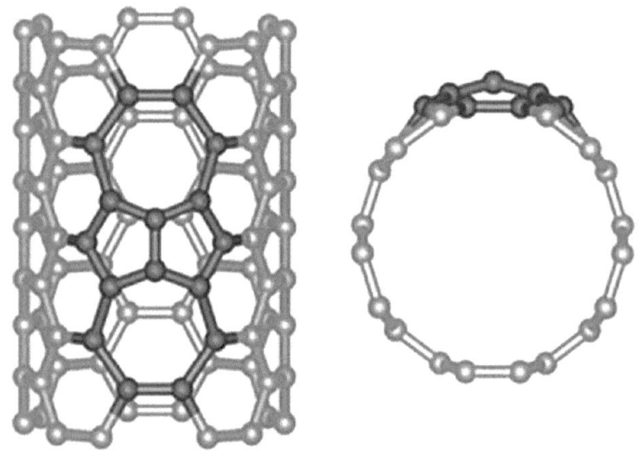

FIGURE 6.8 Stone–Wales defect on the sidewall of a CNT: (–) Thirteen-layer defect-free tube model and (α) Thirteen-layer defective tube model (Reproduced with permission (Lu, Chen, and Schleyer, 2005).).

TABLE 6.4

Studies Reported on the Covalent Modification of CNTs Using Different Approaches

Modification approach	Type of CNTs	Ref.
Amidation and cyclopropanation	MWCNT	(Roy et al., 2014)
	SWCNT	(Umeyama et al., 2007)
Esterification and Diels–Alder reaction	SWCNT	(Delgado et al., 2004)
Double polymerization	MWCNT	(Ozden et al., 2014)
Double arylation	SWCNT	(González-Domínguez et al., 2015; Wang et al., 2012)
1,3-Dipolar cycloaddition and esterification	MWCNT	(Samorì et al., 2010)
	SWCNT	(Alvaro et al., 2004; Ondera and Hamme Ii, 2014)
Double esterification	MWCNT	(Anbarasan and Peng, 2010)
	SWCNT	(Wei et al., 2006)
Triple arylation	MWCNT	(Ménard-Moyon et al., 2015)
	SWCNT	(Ménard-Moyon et al., 2011)
Radiation	SWCNT/ MWCNT	(Vanhorenbeke et al., 2013)

6.3.1.1 Sidewall and End-T Modification

6.3.1.1.1 Cycloaddition Reaction

Cycloaddition reactions can be divided into several classes, including nitrene cycloaddition, Diels–Alder cycloaddition, 1,3-cycloaddition of azomethine ylides, etc. (Dinesh, Bianco, and Ménard-Moyon, 2016; Karousis, Tagmatarchis, and Tasis, 2010).

Zhu et al. fixed the substituted C_2B_{10} carborane cages onto the sidewalls of SWCNTs by nitrene cycloaddition (Yinghuai et al., 2005). The decapitation process of the C_2B_{10} carborane cages was carried out through a reaction with sodium hydroxide in refluxing ethanol. During the reflux in basic conditions, the aziridine ring formed by the SWCNT and nitrene was opened to produce water-soluble SWCNTs in which the sidewalls are modified by both ethoxide moieties and substituted nido-C_2B_9 carborane units. The distribution of selected tissues showed that boron atoms tended to concentrate on cancer cells rather than on blood or other organs. This made boron atoms an attractive nano-vehicle for the delivery of effective boron neutron capture therapy to tumors in cancer treatment (Xie et al., 2016; Seno et al., 2015). Pastine at al. reported another nitrene cycloaddition reaction for the modification of vertically aligned CNTs. Covalent modification of the surface of CNT forests was performed using UV-triggered attachment of perfluoroarylazides with hydrophilic, hydrophobic, and polymerizable groups to prepare superhydrophobic surfaces containing superhydrophilic regions.

The Diels–Alder cycloaddition is also known to occur on the sidewalls of CNTs. Barron's group (Zhang et al., 2005) found that the fluorinated SWCNTs underwent a facile Diels–Alder cycloaddition with a range of dienes, including anthracene, 2,3-dimethyl-1,3-butadiene, and 2-trimethylsiloxyl-1,3-butadiene. This resulted in highly modified SWCNTs with a high C : substituent ratio between 20 : 1 and 32 : 1. The existence of electron-withdrawing fluorine atoms on SWCNT surfaces tended to raise the activation of the double bonds on the CNT sidewall, enhancing the reactive rate of the cycloaddition reaction. Mioskowski and co-workers further developed an original method to induce Diels–Alder cycloadditions on the surface of CNTs (Ménard-Moyon et al., 2006). The method is based on the double activation of the Diels–Alder cycloaddition reaction using a combination of high pressure (1.3 GPa) and hexacarbonyl chromium (transition metal complex). The experiment also showed that a synergistic effect of pressure and chromium complex existed for sidewall CNT modification.

The 1,3-cycloaddition of azomethine ylides on CNT sidewalls has also attracted the interest of researchers all around the world. The CNT surface was modified by a pyrrolidine ring in this situation. The ring carries a great number of functional groups for CNT modification, such as phthalocyanine addends (Ballesteros et al., 2007), amino ethylene glycol groups (Fabre et al., 2008), perfluoroalkylsilane groups (Georgakilas et al., 2008), etc. Ryu's group reported a surface modification of SWCNTs performed by introducing ylides groups containing anchored phenol structures (Bae et al., 2007). The successful modification of azomethine ylides was proven using elemental analysis and Fourier transform infrared spectroscopy. With the help of Raman spectroscopy, researchers also found that the surface modification had no effect on the basic crystal domain size of CNTs. The CNTs after modification presented higher zeta potential values, which meant the modified CNTs had better dispersant ability in water and acetone solvent compared with pure CNTs. In recent years, other novel methods of 1,3-cycloaddition of azomethine ylides have been developed. Montanari and co-workers reported a covalent double functionalization of CNTs by 1,3-dipolar cycloaddition of azomethine ylides. Differently from the old method, in which two different moieties were grafted on the surface of CNTs via cycloaddition reactions at one time using contemporaneously equimolar amounts of azomethine ylides, they prepared the double-modified CNTs by two successive 1,3-cycloadditions of azomethine ylides. The first cycloaddition process occurred in the presence of a phthalimide-protected amino acid derivative and paraformaldehyde, while the second cycloaddition process happened using a Boc-protected amino acid compound.

6.3.1.1.2 Halogenation

In the halogenation reactions for CNT modification, one or more halogens will be added to the surface of CNTs. The pathways of halogenation reactions vary with the applied functional groups and the structural features of the CNT sidewalls.

Janas et al. reported an innovative halogenation reaction for CNT modification (Janas, Boncel, and Koziol, 2014). Bias voltage was applied to the horizontally aligned CNT films to generate an electrothermal effect, which provided an entirely controlled temperature, up to 300 °C, for the chemical modification of the CNT film. They exposed the heated CNTs to gaseous halogens (Cl_2, Br_2, I_2) and kept monitoring the electrical properties. The conductivity of CNT films kept changing with the variety of experimental temperatures (100–300 °C). The maximum levels of Cl, Br, and I that can be successfully introduced into the framework of CNTs within one-minute reactions were found to be 6.7 %, 6.0 %, and 1.5 %, respectively. The conductivity of the CNT films was permanently increased and mildly purified compared with the starting CNT films because of the removal of various carbon–oxygen functional groups. The bromination of MWCNTs through both new and convenient methods was studied by Leila Moradi and Iman Etesami (Moradi and Etesami, 2016). Performing electrophilic addition and radical reactions using Br_2, N-Bromosuccinimide (NBS), and NH4NO3/NBS under UV and thermal conditions allowed bromine to be successfully attached onto the surfaces of MWCNTs.

Obraztsova's group introduced halogenation reactions induced by iodination from the gaseous phase to modify optical and electrical properties of extended SWCNT films (Tonkikh et al., 2015). The formation of different types of one-dimensional iodine crystals and polyiodide species inside CNTs was observed with Raman analysis and high-resolution transmission electron microscopy. A clear suppression of the optical absorption band of iodinated CNTs was demonstrated via the result of UV–vis–NIR optical absorption spectra. According to the formation of a charge transfer from the CNT to the polyiodide species, the conductivity type of CNTs was changed. In the case of filled CNTs, the metallic type of temperature-dependent conductivity behavior was observed at elevated temperatures. The conductivity of iodinated CNTs is comparable with that of indium tin oxide, which is the most popular material for the application of transparent conductive electrodes. Santana and co-workers (Santana et al., 2017) discussed a new method of CNT modification via a halogenation reaction using the compound α-bromoacid and the organic compound 2-(methacryloyloxy) ethyl phosphorylcholine (MPC). The modified CNTs with specific chemical and surface properties were further applied for the medical field in the cardiovascular area (Sun, Kormakov, et al., 2018). It was found that CNTs present a repellent behavior in contact with oxidized low-density lipoproteins (ox-LDL), which was suggested as a candidate to repel ox-LDL for the prevention of restenosis. The toxicity of CNTs and modified CNTs on the HepG2 cell line was analyzed and no damage to the cell membrane of HepG2 cells was found at doses below 1 mg/ml.

6.3.1.1.3 Nucleophilic Addition

Syrgiannis et al. (Syrgiannis et al., 2008) reported the sidewall modification of $(nPrNH)_n$-SWCNT, which is a SWCNT derivative containing n-propylamine addends. The in situ nucleophilic lithium n-propylamide was added to the sidewall of SWCNT, and then the $(nPrNH)_n$-$SWNT^{n-}$ was reoxidized to obtain CNT derivatives with covalently attached amino groups. Because of the electrostatic repulsion of negative charge intermediates, CNTs can be uniformly dispersed according to the reaction sequence. A drastic increase in the solubility of the resulting propylamine-modified $(nPrNH)_n$-SWCNTs in organic solvents was also observed. Another group from Iran developed a new and efficient method for the thioamidation of SWCNTs (Darabi, Roozkhosh, and Aghapoor, 2016). The intermediate-generated nitrogen-based nucleophile was added directly in their approach. The synergistic effect between CaH_2, thioacetic acid, and benzonitrile results in the formation of calcium thiolate, which was added to the sidewalls of CNTs subsequently. Raman, IR, X-ray photoelectron spectroscopy, and thermogravimetric analysis techniques were utilized to prove the resulting covalent modification of SWCNTs.

6.3.1.2 Defect Modification

6.3.1.2.1 Oxidation Reaction

Oxidation reaction for CNT modification is probably the most commonly used method among the various CNT surface modification techniques. In early work, oxidation reactions were normally carried out in air and oxidative plasmas. Then in 2009, Muhler and co-workers developed a much simpler and efficient technique for CNT modification using the vapor of HNO_3 (Xia et al., 2009). This gas-phase HNO_3 treatment significantly simplified the oxidation treatment by avoiding filtration, washing, and drying procedures. Compared with the conventional wet HNO_3 treatment, the gas-phase route also resulted in higher amounts of oxygen species. However, further studies found that purification and modification treatments using nitric acid led to different physical properties and the loss of SWCNT metallic character for multiple CNTs. Furthermore, the overoxidation of CNTs by nitric acid and the subsequent washing process using alkali liquor (e.g., sodium hydroxide) resulted in exfoliating the carbonaceous fragments (Bergeret et al., 2008; Yu et al., 2008; Pumera, 2009). The effect of MWCNT modification using dilute nitric acid under supercritical water conditions was investigated. Results showed that CNT modification with functional groups tended to proceed from the outer surface to the inner graphitic layers, while the nitric acid concentration and the reaction time increased (Park et al., 2005).

Hull's group (Xing et al., 2005) reported an easier approach for CNT oxidization, which applied a sonochemical treatment for density promotion of functional groups on CNT surfaces. In the process of the sonication treatment, some carbonyl groups were formed together with the carboxy groups, with the latter taking up the greater population. The structural damage of the surface of MWCNTs increased with the time of sonochemical treatment, which significantly altered the electronic properties. Avilés et al. (Avilés et al., 2009) evaluated five different kinds of mild acid oxidation treatments for MWCNT modification using H_2SO_4, HNO_3, and H_2O^2 at relatively low concentrations. The assistant sonication treatments of MWCNTs in acids were also set at low sonication power and short treatment times. These attempts were performed to find the most effective method for the oxidation of MWCNTs while minimizing nanotube damages. According to their results, the combination of low-power sonochemical treatment with 3.0 M HNO_3 for 2 h and the subsequent identical treatment with H_2O_2 for another 2 h were proven to be the best choice in this case.

In the same period, Galiotis and co-workers (Datsyuk et al., 2008) also discussed the possibility of controlling the amount of functionality, carboxylic groups, and hydroxyl groups for minimum damage on the structural integrity. The highest degree of degradation of acid-treated CNTs, including nanotube shortening and additional defect generation in graphitic networks, was clearly observed under reflux conditions via thermogravimetric analysis and electron microscopy. It was found that amorphous carbon and metal oxide impurities can be completely removed via basic oxidative treatment, while the structural integrity of CNTs stays intact. Bergmann's group (Osorio et al., 2008) demonstrated that the adsorption of functional groups showed higher efficiency on the modification process using the hybrids of sulfuric, nitric, and chloridric acids.

6.3.1.2.2 Esterification and Amidation Reactions

In esterification and amidation reactions, carboxylated CNTs were commonly applied as precursors for the covalent modification of the sidewalls and tips of CNTs. Through reactions with thionyl or oxalyl chloride, carboxylic groups of CNTs can be converted to acyl chloride groups, followed by the addition of the appropriate amine or alcohol. With the help of this procedure, many kinds of substituents including biomolecules, polymers, organic substances, and even several photosensitive compounds were able to interact with CNTs (Wu et al., 2007; Malarkey et al., 2008; Fagnoni et al., 2009; Deng et al., 2007).

The first publication mentioning the SWCNT modification with enzymes via the formation of covalent bonds, confirmed using Fourier transform infrared spectroscopy, was published in 2005 by Malhotra's group (Wang, Iqbal, and Malhotra, 2005). The first step of CNT modification with enzymes was the acylation of the nanotubes followed by an amidation reaction as the second step. With the help of this method, many other kinds of primary and secondary achiral and chiral amines were tethered to sidewalls and tips of SWNTs. A three-dimensional structured enzyme array, which was a potential material for biosensor applications, was obtained through this method of covalent modification. In 2007, Langa and co-workers (Delgado et al., 2007) reported the first preparation of the SWCNT-fullerene conjugated hybrid. This new kind of hybrid nanomaterial was created through the amidation of acid-modified SWCNTs by an amine-modified fullerene derivative (as shown in Figure 6.9). In order to enhance the solubility of the conjugated hybrid, a subsequent esterification reaction was carried out with n-pentanol on existing acyl chloride groups. The formation evidence of this hybridization of SWCNTs and addition of fullerene was demonstrated via high-resolution transmission electron microscopy and Raman spectra. Researchers believe that this new SWCNT–fullerene conjugated hybrid material has the potential for optoelectronic applications.

SWCNT functionalized by short acyl chloride with sterically hindered amines (e.g., 8-aminopentadecane) were further studied as a different field of nanotechnology. The resulting material in this reaction can be applied as a scaffold for the self-organization of C_{60} molecules on the SWCNT sidewall (Umeyama et al., 2008). The deposition processes of these composites onto nanostructured SnO_2 electrodes were performed using electrophoretical methods. A series of experiments on their photoelectrochemical and microscopic properties were performed to build the relationship between the spatial structure and their photoelectrochemical properties. In the same period, Prato's group (Bonifazi et al., 2006) presented a comprehensive spectroscopic and microscopic performance of chemically modified CNTs. Scanning tunneling microscopy (STM) investigation was employed to show that the covalent modification of short oxidized SWCNTs through amide reaction with aliphatic chains happened both on the sidewalls and on the tips of CNTs. In order to confirm the result of STM imaging, several other testing methods, including steady-state electronic absorption, thermogravimetric analysis, and Raman spectroscopy were also performed. The same research group also presented a series of donor–acceptor SWCNT nanoconjugates, in which tetrathiafulvalene analogues were grafted as electron donors (Herranz et al., 2006).

More recently, Battigelli et al. described a series of MWCNT conjugates, which were modified with different dendrons bearing positive charges such as ammonium and guanidinium groups at the tips of MWCNTs (Battigelli et al., 2013). Amidation and click reactions were utilized to anchor the dendrimer units onto the surfaces of CNTs. The performances of the final hybrids were characterized through complementary analytical techniques. Particularly, their interaction ability with siRNA was studied using agarose gel electrophoresis. Based on the results of comprehensive detections, the characterizations of final hybrids, including cell uptake capacity, low cytotoxicity, and the ability to silence cytotoxic genes, demonstrated that the MWCNT conjugates were promising carriers for the applications of genetic material. Bayazit and Coleman (Bayazit and Coleman, 2014) reported a versatile method to anchor ester functional groups onto the surface of SWCNTs. Reduced SWCNTs achieved by n-butyl lithium were directly reacted with haloformates for the preparation of ester-modified SWCNTs. When compared with pure ones, the solubility of successfully modified SWCNTs was significantly enhanced. The evidence of ester functional groups was further confirmed via gold tagging of positively charged SWCNTs, which shows the distribution of functional groups.

6.3.1.2.3 Plasma Activation

Plasma activation is a commonly used method for surface modification via plasma processing. There are many kinds of plasma that can be used in a plasma activation processes. Atmospheric pressure plasmas, including corona discharge, arc discharge, dielectric barrier discharge, etc., have found the most application because no expensive vacuum equipment or wet chemistry situations were needed during the modification process. Many kinds of industrial gases, such as oxygen, hydrogen, and nitrogen can be applied for CNT modification at atmospheric pressure (Kalita et al., 2009; Okpalugo et al., 2005).

A comparison between the nitrogen-doped CNTs (N-CNTs) obtained by plasma treatment and the conventional acid-treated CNTs (O-CNTs) were performed by Muhler's group (Chetty et

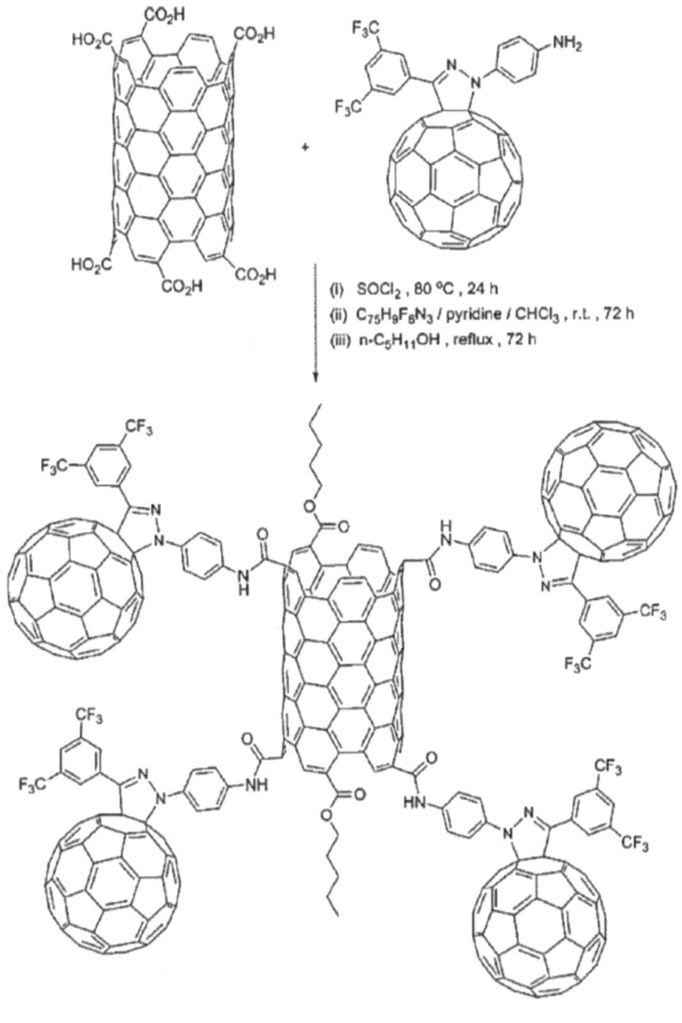

FIGURE 6.9 The preparation of SWCNT-fullerene conjugated hybrid nanomaterial (Reproduced with permission (Karousis, Tagmatarchis, and Tasis, 2010).).

al., 2009). During the anodic oxidation of methanol in direct methanol fuel cells, O-CNT was normally used as a catalyst carrier for platinum–ruthenium (PtRu) nanoparticles, which were prepared by impregnation reduction of chloride precursors.

Different kinds of testing methods, including X-ray photoelectron spectroscopy, X-ray diffraction, scanning electron microscopy, and transmission electron microscopy, were used for property characterization. Electrochemical testing results proved that the electrical property of PtRu/N-CNT was significantly higher than those of PtRu/O-CNT and the commercial E-TEK PtRu/C catalyst. Pyrrollic and pyridinic species on the surface of CNTs, which were produced by the nitrogen plasma treatment, provided a great deal of anchoring sites for PtRu particles and made N-CNTs an appropriate material for fuel cell electrocatalyst.

Localized surface modification using a plasma jet with an ultrafine atmospheric pressure was developed by Nagatsu and co-workers (Abuzairi et al., 2015; Abuzairi et al., 2016). The whole plasma surface modification process for vertically aligned CNTs can be divided into two procedures: pre-treatment for CNT surface activation and post-treatment for CNT surface modification. The result of chemical derivatization with the fluorescent dye showed that the vertically aligned CNTs were successfully modified with both amino and carboxyl groups. No interference between the two functional groups was observed after modification. This maskless modification technique provided an optional method for the preparation of multi-modified CNTs and has potential application in the field of microarray biosensors. Yang et al. (Yang, Shi, and Yang, 2015) reported an air-plasma activation method for the preparation of polypyrrole (PPy)-bonded CNTs using in situ chemical oxidative polymerization. According to the testing results of electrochemical properties via cyclic voltammetry and electrochemical impedance spectroscopy, the PPy was uniformly coated on the surface of plasma-activated CNTs (P-CNT). Both the conductivity and thermal stability of the P-CNT/PPy composite showed significant improvement. Furthermore, the specific capacitance and repeatability of the P-CNT/PPy composite were much higher than those of the CNT/PPy composite (188 and 148 F/g in 1 M KCl and 264, 210 F/g in 1 M H_2SO_4 for specific capacitance, and 89 % and 76 % of initial capacitance retained after charge–discharge tests for 1000 cycles). The superior properties of the

P-CNT/PPy composite made it a promising electrode material for high-performance supercapacitors.

6.3.2 Non-Covalent Modification

6.3.2.1 Exohedral Modification

Non-covalent aggregation of surfactants and wrapping polymers with CNTs are two typical species of exohedral modification. It was found that the nanotubes can be transferred to the aqueous phase with the reaction of surface-active, including benzalkonium chloride and sodium dodecyl sulfate (SDS) (Bandow et al., 1997; Krstic et al., 1998). Researchers believed that CNTs were in the hydrophobic interiors of the corresponding micelles, giving rise to stable dispersions. When the aromatics were contained in the hydrophobic part of the amphiphile, the effective π-π-stacking interactions could form a particularly strong interaction with the graphite sidewalls of CNTs. As the electronic properties of CNTs can be combined with the specific properties of the immobilized biosystems, this effect can be applied to the field of biosensors.

Song et al. (Ma et al., 2013) applied a covalently tethered polyoxometalate (POM)-pyrene hybrid (Py–SiW_{11}) for the non-covalent modification of SWCNTs. Perfect interaction between the nanoparticles and SWCNTs with no chemical decompositions was achieved using both pyrene and SiW_{11} moieties. After non-covalent modification, the electrical properties (e.g., discharge capacities, cycling stability, rate capacity, etc.) of the SWCNTs/Py–SiW_{11} nanocomposite were much superior to those of the individual components for the application of anode material in lithium-ion batteries.

Polymers are also used to form the supramolecular complexes of SWCNTs through non-covalent methods. Polymers wrapped around the nanotubes (as shown in Figure 10b) can be obtained with CNT suspension together with polymers including poly(arylene ethynylene)s (PPEs) and poly(mphenylene-co-2,5-dioctoxy-p-phenylenevinylene) (PmPV), in organic solvents such as toluene, THF, CH_3Cl, etc. (Bartelmess et al., 2011; Di Crescenzo, Ettorre, and Fontana, 2014). A brief list of specific molecules used for the dispersion of CNTs in different solvents is shown in Table 6.5.

Adronov and co-workers (Liang, Zhao, and Adronov, 2014) reported a non-covalent modified SWCNT with a conformationally switchable conjugated copolymer, which was prepared from fluorene and the monomer of vinylogous tetrathiafulvalene (TTFV). The conformational change of this copolymer was available upon the protonation with trifluoroacetic acid (TFA). Researchers found that the TTFV–fluorene copolymer formed

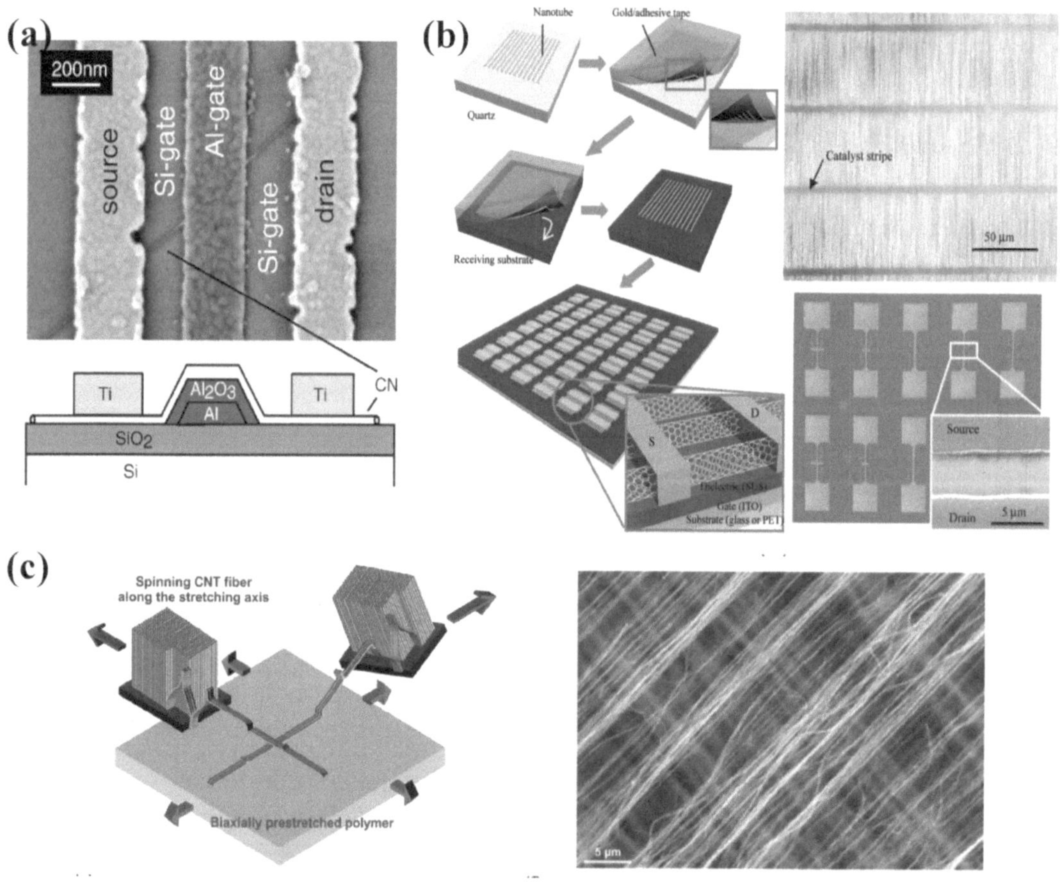

FIGURE 6.10 Selected CNT applications in electronics. (a) SEM (top) and schematic (bottom) images of the first TFET device. The schematic image shows the vertical arrangement of gates, contacts, and nanotube (Reproduced with permission (Appenzeller et al., 2004).). (b) Schematic and SEM images of the process for biaxial strain sensors fabrication (Reproduced with permission (Ryu et al., 2015).). (c) Schematic diagram of aligned SWCNTs transfer from quartz to transparent or/and flexible substrates (e.g., PDMS, PET, glass, etc.) (Reproduced with permission (Ishikawa et al., 2009).).

TABLE 6.5

Some Molecules Used for the Dispersion of CNTs in Different Solvents

Dispersant	Chemical property	Acronym	Solvent	Ref.
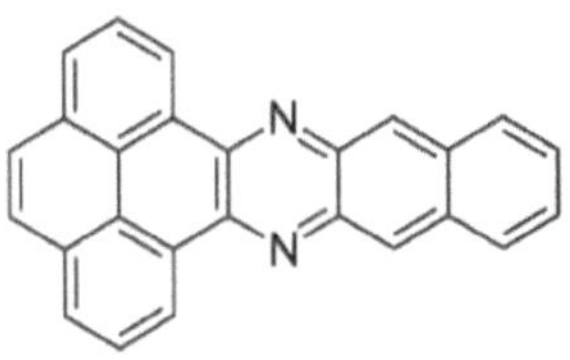	Lipophilic	Diazapentacene derivative	THF	(Mateo-Alonso et al., 2007)
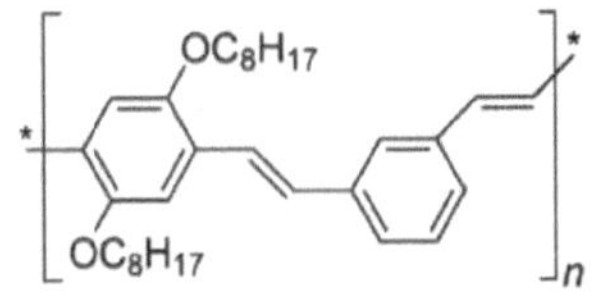	Lipophilic	PmPV	Toluene	(Murphy et al., 2002)
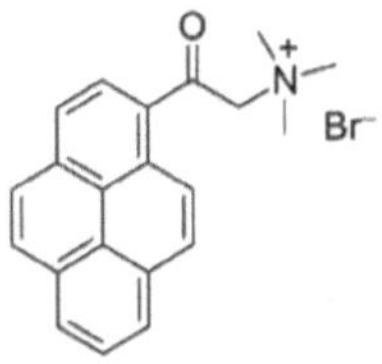	Amphiphilic	/	Water	(Toma et al., 2011)
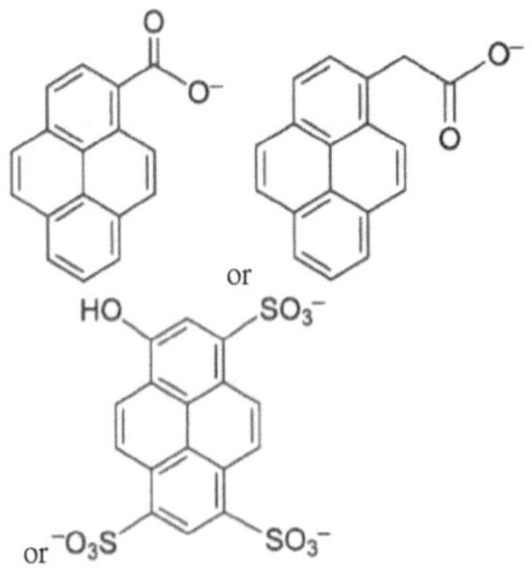	Amphiphilic	/	Water	(Guldi et al., 2005)
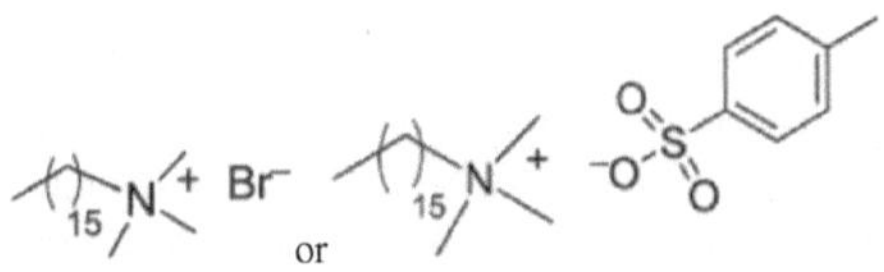	Amphiphilic	CTAB, CTAT	Water	(Attal, Thiruvengadathan, and Regev, 2006)
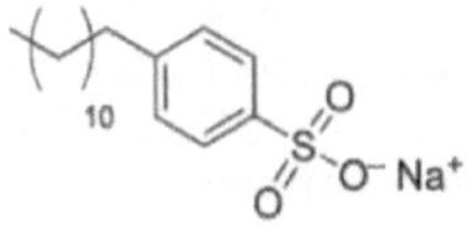	Amphiphilic	SDBS	Water	(Oh, Sim, and Ju, 2013; Zhong and Claverie, 2013)
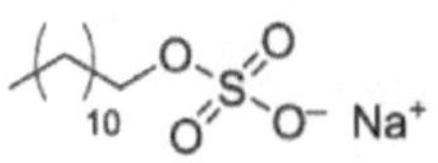	Amphiphilic	SDS	Water	(Richard, 2003)
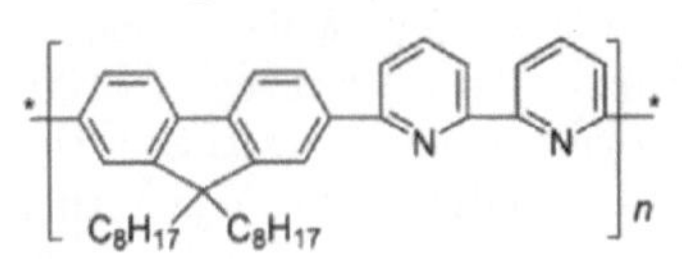	Lipophilic	PFO-BPy	THF, toluene	(Joo et al., 2014)

(Continued)

TABLE 6.5 (CONTINUED)

Some Molecules Used for the Dispersion of CNTs in Different Solvents

Dispersant	Chemical property	Acronym	Solvent	Ref.
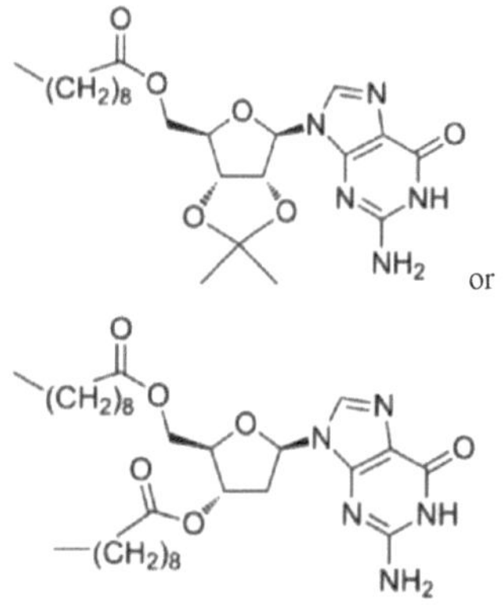	Lipophilic	LipoG1, LipoG2	$CHCl_3$	(Di Crescenzo et al., 2012)
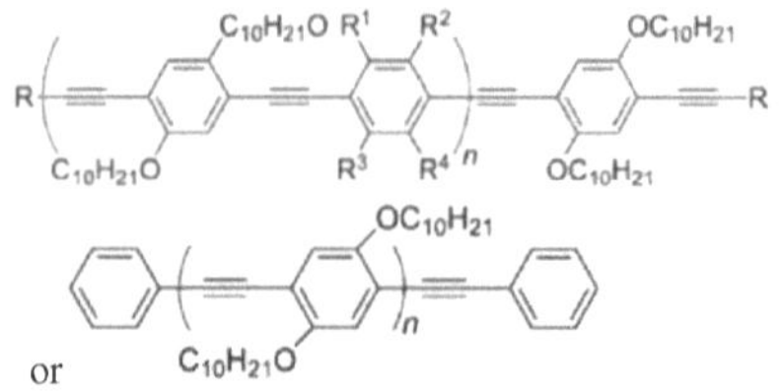	Lipophilic	PPE 1, PPE 2	CH_3Cl	(Chen et al., 2002)
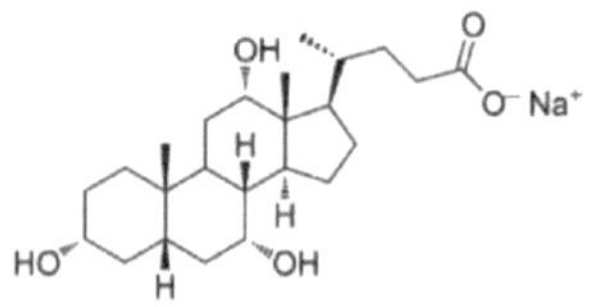	Amphiphilic	SC	Water	(Oh, Sim, and Ju, 2013; Cambré and Wenseleers, 2011)
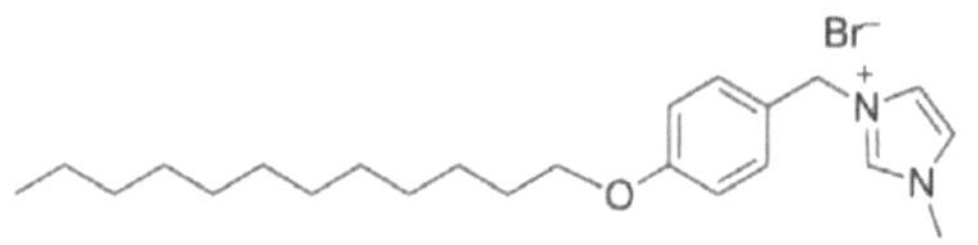	Amphiphilic	T3	Water	(Di Crescenzo et al., 2014)

strong interactions with the surface of SWCNTs with a simple mixing process, leading to stable, concentrated dispersions of nanotubes in toluene. The copolymer was further found selectively dispersed around low-diameter SWCNTs based on the results of photoluminescence excitation mapping. A rapid conformational change and desorption of the polymer from the surfaces of SWCNTs, which can be applied as purification process of SWCNTs, was carried out after the addition of TFA to the copolymer-SWCNT system.

More recently, the first experimental realization of exohedral non-covalent hybridization of unmodified CNTs and C_{60} was reported by Yekymov et al. (Yekymov et al., 2017). These exohedral van der Waals hybrids are expected to enable the engineering of carbonaceous materials with tunable chemical and physical properties owing to the unperturbed sp^2 hybridization of comprising allotropes. A two-step process was developed by these researchers for the preparation of CNT-C_{60} hybrids. In the first step, the three-dimensional network of unmodified CNTs was established based on random solution assembly and subsequent fast solvent quenching. Then, in the second step, the structural defects on the pre-assembled networks were applied as nucleation points for the growth of nanocrystals in the atmosphere of sublimated C_{60}s. The core-shell structure of CNT-C_{60} hybrids was formed with the reorganization of C_{60} nanocrystals after annealing at low temperature (80 °C). The fact that exohedral CNT-C_{60} hybrids were maintained with van der Waals force was confirmed based on Raman analyzing and the high mobility of the C_{60} on the CNT surface. Using this technique, researchers were able to prepare exohedral hybrids with controlled nanostructures. The formation of mono- or multi-layers of C_{60} shell with non-covalent bonds is also helpful for shaping the nano-morphology of the active layer in polymer-C_{60} hybrids, which can be applied as bulk-heterojunction solar cells.

6.3.2.2 *Endohedral Filling Modification*

As shown in Figure 10c, endohedral filling is another option for non-covalently modifying CNTs, as the inner cavity of CNTs can be used as storage space for guest molecules (e.g., C_{60}). The uptake process and the endohedral chemistry of metals and metal salts have been intensely studied for the last two decades (Han, 1997; Dujardin et al., 1998). Several kinds of metal nanothreads (e.g., gold, platinum, etc.) have already been created in the capillaries of the SWCNTs. Typically, a perchloromethane acid treatment at high temperature is needed for the metal endohedral filling modification of SWCNTs.

For the analysis of endohedral chemistry of SWNTs, the incorporation of Sm metallofullerenes (Sm@C82) or C_{60} (Smith and Luzzi, 2000) can be treated as especially impressive examples. In 2001, Shinohara's group (Okazaki et al., 2001) reported the first real-time observation of elongated nanocapsules formation using the incorporation of Sm@C82. They found that the inner hollow spaces of SWCNTs were ideal chemical reaction zones. Analytical methods, such as electron energy loss spectroscopy and high-resolution transmission electron microscopy, were applied to directly observe the reaction inside SWCNTs. The incorporation occurred both on the sidewalls and on the ends of SWCNTs. For the incorporation of C_{60} (as shown in Figure 6.10c), the encapsulated C_{60}s tends to form chains with the bonding forces (van der Waals forces). These kinds of arrayed nanoparticles are also known as "bucky peapods". More recently, Tobias and co-workers (Kierkowicz et al., 2017) reported a straightforward approach to remove the non-encapsulated compounds in an environmentally friendly and highly efficient way. SWCNTs were filled with lutetium chloride using water as a "green" solvent to minimize the residual waste of the removal process. Their method provided a possible pathway toward the mass production of high-quality closed-ended filled nanotubes or carbon nanocapsules using an endohedral filling modification technique.

6.4 Application

The explosive growth of the production capacity of CNTs in the last few years (at least a 10-fold increase since 2006) (De Volder et al., 2013) has reflected the worldwide commercial interest in this outstanding nanomaterial. Currently, most CNT production is used in bulk nanocomposites and thin films, including automotive parts, rechargeable batteries, etc. In these applications, the properties are limited and significantly lower than those of pure CNTs because of the unorganized CNT architectures. However, organized CNT architectures (e.g., CNT sheet, yarn, vertically aligned CNT forest, etc.) provide a potential way to obtain the properties of individual CNTs at a macro scale, which can help realize new functionalities, including high damping (Xu et al., 2010), terahertz polarization (Ren et al., 2009), shape recovery (Cao, 2005), large-stroke actuation (Aliev et al., 2009; Lima et al., 2012), etc. As the application field of pure and modified CNTs is very extensive, it is very challenging to describe all of them in one article. Herein, we mainly focus on three important branches of CNT application: functional polymer nanocomposites, electronics, and biotechnological applications.

6.4.1 Functional Nanocomposite Materials

Benefiting from the high aspect ratio of CNTs, they were first used as electrical conductive fillers in polymers to form a conductive network at a relatively low carbon content (as low as 0.1 wt %) (Bauhofer and Kovacs, 2009). The CNTs in these composite materials are normally disordered, since higher loadings lead to higher conductivities (Wu, Gao, et al., 2017). When carbon contents increase to 10 wt %, the conductivities of CNT/polymer nanocomposites can reach up to 10,000 S m^{-1}. Besides the excellent electrical properties, the outstanding mechanical and thermal properties of CNTs are also being used together with polymers for a vast range of potential industrial applications (Wu, Sun, et al., 2017; Jingyao et al., 2017; Zhuang et al., 2019; Sun, Li, et al., 2019). Researchers around the world have done a series of studies on CNT/polymer nanocomposites properties for industrial applications including sporting goods, automotive parts, filters, etc.

The field emission (FE) properties of modified-CNT/polymer nanocomposites, which were produced by solution processing method, were studied by Carey's group (Connolly et al., 2009). Based on the results, they reported that excellent electron emission can be realized at 0.7 % volume fraction loading of SWCNTs in the system. The charge transfer through the nanocomposites can be further improved by changing the CNT concentration and the type of polymer matrix. The solution processing method for the preparation of well-dispersed CNT/polymer nanocomposites made possible the scalable production of large area cathodes. More recently, Chattopadhyay and co-workers (Gupta, Maity, and Chattopadhyay, 2014) reported a simple in situ chemical polymerization method to avoid the influence of aggregation of MWCNTs, which provided immeasurable assistance for field emission enhancement of polypyrrole-multiwalled carbon nanotube (PPy-MWCNT) nanocomposites.

CNT/polymer nanocomposites can also be applied in the aerospace industry as anti-lightning, paints, anti-radar protectors, etc. (Cristina et al., 2012). Kim et al. (Kim et al., 2012) reported a PPy-coated MWCNT composite for the application of electromagnetic wave absorption materials. The thickness of PPy on the MWCNT surface decreased with the increase of hydrophilic groups, added by the oxyfluorination treatment. The composites presented noticeable enhancements in permeability, permittivity, and EMI shielding efficiency. Sundararaj and Al-Saleh (Al-Saleh and Sundararaj, 2009) analyzed the EMI shielding mechanisms of MWCNT/polypropylene (PP) nanocomposites both theoretically and experimentally. The EMI shielding effectiveness of the nanocomposites at four different loading contents and three different thicknesses was systematically measured. Absorption and reflection were found to be the primary and secondary shielding mechanisms based on the experimental results. Higher MWCNT content and larger shielding plate thickness of the MWCNT/PP nanocomposites led to greater EMI shielding effectiveness. The mechanism was further researched using a theoretical model based on the shielding of electromagnetic plane waves, which indicated that the overall EMI shielding effectiveness was significantly influenced by the multiple reflection within and between MWCNT internal and external surfaces.

The addition of CNTs into polymers could enhance the mechanical properties (e.g., stiffness, strength, durability, etc.) of the original polymer materials, making it an attractive option for sporting good applications, such as golf, bicycles, hockey sticks, ski poles, etc. Kumar and colleagues (Chae et al., 2009) reported a polyacrylonitrile (PAN)/carbon nanotube (CNT) nanocomposite-based carbon fiber fabrication using a gel spinning method. The spatial arrangement of carbons in the pyrolyzed fiber was significantly influenced by the added CNTs. This CNT-enhanced nanocomposite (1 μm diameter carbon fibers) obtained a 35 % promotion in strength (4.5 GPa) and stiffness (463 GPa) compared with control samples without CNTs. Moreover, while CNTs influence the rheology of a composite system, MWCNTs are also commercially applied as a flame-retardant additive to the polymer matrix (Kashiwagi et al., 2005). These CNT additives are commercially attractive as alternatives for halogenated flame retardants because of the use limitations caused by environmental regulations.

CNT/polymer nanocomposites can also be used as body parts for yachts and automobiles. For example, Shon et al. (Jeon, Park, and Shon, 2013) prepared a MWCNT-contained epoxy coating to increase the adhesion strength of the polymer matrix. With the help of hygrothermal cyclic tests and electrochemical impedance spectroscopy, the effect of the MWCNTs on the hydrophobicity, water transport behavior, and accompanying corrosion resistance performance were examined. The corrosion protection of a carbon steel coated with epoxy coating containing MWCNTs correlated well with water transport behavior and hydrophobicity. Similarly, Jin et al. (Jin et al., 2005) reported a superhydrophobic aligned polystyrene nanotube film with strong adhesion to the water. An alumina membrane template was utilized to fabricate the polystyrene nanotube, which perfectly mimics the keratinous hairs in geckos' feet. Their results revealed that aligned CNT structures can simultaneously enhance the hydrophobicity and adhesion forces with water through van der Waals forces. This characteristic of CNT/polymer nanocomposites showed potential for the application in coatings, paints, textiles, transports, etc. These CNT/polymer nanocomposites are commercialized by many companies around the world, including Hyperion, Toyota/Ube, Babolat, etc. (Mittal et al., 2015).

Beside CNT/polymer nanocomposites, specific amounts of CNTs are also added to metals (e.g., aluminum) for the enhancement of tensile strength and modulus (Bakshi and Agarwal, 2011). The strengths of commercial Al-MWCNT nanocomposites are comparable to those of stainless steels (0.7 to 1 GPa) at one-third of the density (2.6 g cm^{-3}). This strength is also comparable to Al-Li alloys, while the Al-MWNT nanocomposites are far more affordable.

6.4.2 Electronics

Electronics are one of the most important branches in the applications of pure and modified CNTs, including flexible transistors, chemical and biological sensors, batteries, supercapacitors, actuators, etc. (Park, Vosguerichian, and Bao, 2013; Sun, Zhuang, et al., 2017). In this section, we will mainly focus on CNT-based transistors and sensors (Sun, Zhao, et al., 2018; He et al., 2018; Huang et al., 2019). Figure 6.10 shows selected CNT applications in electronics.

CNTs are attractive materials for use in tunnel field effect transistors (TFETs). The low electron scattering, small and direct bandgap, light effective mass, and superior electrostatic control derived from the ultrathin body make CNTs one of the best choices in both materials and device geometry (Ionescu and Riel, 2011). The first TFET was reported by Avouris et al. in 2004 (as shown in Figure 6.10a) (Appenzeller et al., 2004). Differing from the gate voltage change limited by the Fermi distribution for an on/off switch in conventional devices, the electrostatics in the CNT were controlled by two independent gates in their TFET device. The structure of the nanotubes, which is the key component of the one-dimensional tunneling effect in their TFET device, was also discussed. Another kind of transistors made from CNTs, thin-film transistors (TFTs), are very suitable for the application of organic light-emitting diode (OLED) displays. The CNT TFTs show much higher mobility than that of amorphous silicon (~1 $cm^2\ V^{-1}\ s^{-1}$).

Recently, Ohno and co-workers (Sun et al., 2011) reported a simple fabrication method of high-performance integrated circuits and TFTs on flexible and transparent substrates. In their technique, floating-catalyst CVD was used for CNT deposition under low temperature without a vacuum environment. The CNT network consisted of long nanotubes (~10 mm) connected by Y-shaped junctions with low resistance, and the density was well controlled by its unique morphology. The flexible and transparent transistor devices simultaneously presented an on/off ratio of 6×10^6 and a mobility of 35 $cm^2\ V^{-1}s^{-1}$. Their fabrication method also showed potential for large-scale production using high-throughput printing techniques. Similarly, Rinzler et al. (McCarthy et al., 2011) fabricated a vertical SWCNT field effect transistor (FET) for OLED applications. The CNT FET presented sufficient current output for OLED operation at low-power dissipation, low-power and high-aperture ratio. The emission of three primary colors (i.e., red, green, blue) using the OLED device was available through this transparent CNT network. They further demonstrated that the performance of their device was at the same level with the display pixels driven by polycrystalline silicon-based transistors. Many other studies in this field, using low-cost techniques (e.g., printing, spin coating, etc.) for potential commercialization, have also been reported by researchers around the world in recent years (Lau et al., 2013; Sun et al., 2013; Homenick et al., 2016). To supplement the commonly used flexible CNT-based electronics mentioned above, a typical fabrication process of flexible electronics is shown in Figure 6.10b.

As one of the hottest topics, CNT-based sensors for chemical and biological applications have already been widely studied during the last two decades (Barsan, Ghica, and Brett, 2015; Kanoun et al., 2014; Kormakov et al., 2018). The successful realization of CNT-based sensors showed proper control of the physical and chemical characteristics of pure and modified CNTs. Up to now, both aligned and randomly dispersed CNTs have been used as nanostructured electrodes for the applications of CNT-based sensors (Gooding, 2005). Among different sensors, the elastic strain sensors for wearable devices (as shown in Figure 6.10c) that can monitor the physical movements of humans have become an essential factor for the realization of next-generation electronics. Ryu et al. (Ryu et al., 2015) reported a highly stretchable strain sensor for the preparation of wearable devices, which were made from dry-spun CNT fibers. Highly oriented CNT fibers were stretched and patterned on a flexible Ecoflex substrate. The device obtained can maintain high sensitivity, durability, and responsiveness at high strains (up to 900 %). They also developed another device with independent cross-sensitivity, which was helpful for the simultaneous measurement of strains along multiple axes, using biaxially oriented CNT fiber arrays (Figure 6.10c). Based on their superior electrical properties, these stretchable devices can be applied in many measuring areas, including strain gauge, single and multiaxial detecting motion sensors, etc.

Besides strain sensors (Sun, Zhuang, et al., 2019), CNT-based sensors for gases (Liu, Moh, and Swager, 2015), volatile organic compounds (Badhulika, Myung, and Mulchandani, 2014), glucose (Hyun et al., 2015), and living cells (Sun et al., 2015), biomacromolecules (e.g., DNAs, proteins, etc.) (Liu et al., 2016; Landry et al., 2017) have also been developed in recent years. Strano and co-workers (Landry et al., 2017) demonstrated real-time, label-free detection of individual proteins from yeast and bacteria using nanosensor arrays in a microfluidic chamber (Wu et al., 2018). Non-covalent modified SWCNTs were used for the fabrication of these nanosensor arrays and varying chemical spacers were applied for the optimization of sensor response. The detection of a unique protein produced from T7 bacteriophage

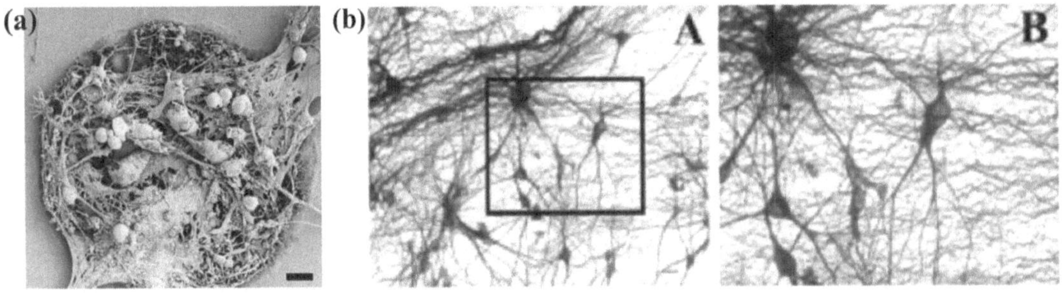

FIGURE 6.11 Selected biotechnological applications of pure and modified CNTs. (a) The growth of neuronal circuits on a CNT grid (Reproduced with permission (Harrison and Atala, 2007).). (b) Immunocytochemistry of hippocampal cultures grown on CNT (B is a higher magnification of the region in the inset in A). (Reproduced with permission (Lovat et al., 2005).).

infection of Escherichia coli was researched further to reveal the sensing ability of their fluorescent SWCNT sensor arrays in real-time, single-cell analysis of different kinds of protein products from various cell types.

6.4.3 Biotechnological Applications

Except for the application of biosensors discussed in Section 4.2, which is influenced by the size and chemical compatibility of CNTs with biomolecules (e.g., DNAs and proteins), pure and modified CNTs also find biotechnological applications in many areas (Marega and Bonifazi, 2014), such as tissue engineering (Harrison and Atala, 2007), drug and gene delivery (Pantarotto et al., 2004; Yang, Wang, et al., 2016; Yang et al., 2015), fluorescent and photoacoustic imaging (De La Zerda et al., 2008), etc. Selected nanobiotechnological applications of CNTs are shown in Figure 6.11.

CNTs are found to be an important material for tissue engineering to make possible and improve functions including tracking cells, delivering transfection agents, and serving as a scaffold for incorporation within the host's body, as shown in Figure 6.11a (Harrison and Atala, 2007). Ballerini's group (Lovat et al., 2005) found that CNTs have the potential to improve neural signal transfer, because of neurons adhesion and dendrite elongation support. Based on experiment results and SEM images, the growth of neuronal circuits on a CNT grid presented a significant increase with the CNT network (as shown in Figure 6.11b). The neural signal transmission of neuronal circuits on CNT grids simultaneously showed higher efficacy, which may be attributed to the high electrical conductivity of CNTs.

In the applications of drug and gene delivery (in vivo applications), CNTs are used as transfer vehicles because they can interact with many kinds of cells and are taken up by endocytosis (Shi et al., 2011). The internalized process of CNTs by cells begins with the binding of CNT tips and cell membranes, which enables the encapsulation of molecules inside the nanotubes and the transfection of those attached on the CNT sidewalls (Hong et al., 2010). For example, anti-cancer drugs (i.e., doxorubicin, methotrexate, taxol, etc.) are directly delivered. Some drugs, such as erythropoietin administered orally, are delivered with the assistance of CNTs. With the help of CNT vehicles, researchers have successfully delivered many kinds of biomacromolecules, including small interfering RNA, micro-RNA, and plasmid DNA (Karimi et al., 2015; Chen et al., 2017). Table 6.6 presents a brief summary of drug delivery using CNTs.

In order to enhance the ability to deliver drugs and genes, different modification treatments have been applied (Wu et al., 2014; Siu et al., 2014). Researchers still have a significant concern over the potential toxicity of CNTs, as their geometry and surface chemistry can influence biocompatibility. Furthermore, CNTs may also cause damage to the environment (Bianco, Kostarelos, and Prato, 2011). Thus, future medical acceptance of CNTs requires a deeper understanding of the immune response and the definition of exposure criteria for different uses (including injection, ingestion, inhalation, and skin contact) (De Volder et al., 2013).

Compared with most optical imaging techniques, the photoacoustic imaging of living subjects associated with CNTs can provide higher spatial resolution and deeper imaging of tissues (Zhang et al., 2006). It's necessary to administer a photoacoustic contrast agent for the detection of diseases in an early stage with no natural photoacoustic contrast. However, most of the existing contrast agents for photoacoustic imaging are helpless in disease targeting of living subjects (Eghtedari et al., 2007; Yang et al., 2007). Gambhir and co-workers (De La Zerda et al., 2008) reported a SWCNT-based contrast agent for the photoacoustic imaging of tumors. Cyclic Arg-Gly-Asp (RGD) peptides were used for the modification of SWCNTs in their study. Compared with mice injected with non-targeted CNTs, targeted CNTs showed a photoacoustic signal in the tumor that was eight

TABLE 6.6

Examples of Drug Delivery with the Help of CNTs

Drug	Type of CNTs	Target cell or tissue	Ref.
Cisplatin	SWCNTs	Squamous carcinoma	(Bhirde et al., 2009)
Doxorubicin	SWCNTs	Colon cancer	(Jabr-Milane et al., 2008)
		Breast cancer glioblastoma	(Liu et al., 2007)
		Cervical carcinoma	(Zhang et al., 2009)
Toxic siRNA sequence	Modified MWCNTs	Human lung xenograft model	(Podesta et al., 2009)
Paclitaxel	SWCNTs	Breast cancer	(Liu et al., 2008)
Taxoid	SWCNTs	Leukemia	(Chen et al., 2008)
Radionuclide	SWCNTs	Burkitt lymphoma	(McDevitt et al., 2007)

times higher. Raman microscopy was utilized to confirm the ex vivo results. This photoacoustic imaging technique has potential application in non-invasive cancer imaging and nanotherapeutic monitoring in living subjects.

Conclusion

The advances in CNT industries, including the preparation mechanism, mass-production methods, covalent and non-covalent modification, and wide range of applications of CNTs during the last two decades, were reviewed. A great number of techniques for CNT preparation with tunable wall number, high aspect ratio, and a relatively low cost have been successfully developed and optimized by many researchers around the world. Large-scale applications have also been widely explored for developing next-generation CNT-based products. In practice, both purified and modified CNTs have a high potential for unique applications in wide areas, including traditional industry, electronic, medicine, etc., due to their attractive physical and chemical properties. According to many business reports and research papers, there are a great number of companies focusing on the development of CNT-related techniques. However, the specific marketability level of the CNT industry is quite difficult to predict because companies tend to keep technical details hidden for a very long time before releasing them to make money. Thus, the comparison and discussions on CNTs in this review are based on the reports and papers that could be obtained.

Despite the tremendous contributions made in the CNT industry, there are still more formidable challenges and huge opportunities. For example, the realization of predictable and affordable mass production of SWCNTs with perfect physical and chemical properties. Compared with traditional commercialized chemical products, the controllable preparation and large-scale applications of CNTs still has a long way to go due to difficulties in both preparation and subsequent modification processes. The preparation of specific CNTs with controllable chirality, and predetermined length, size, and organization is even more challenging. For the applications of functional nanocomposite materials, most studies stay in the experimental stage only and have not been industrialized on a large scale. For the applications of electronics, the understanding of why the properties (e.g., mechanical strength, thermal and electrical conductivity, etc.) of organized CNTs, such as CNT sheets, forests, and yarns, remain much lower than those of individual CNTs, which greatly limits the development of next-generation electronics, is even more urgent. Most of the challenges that exist in CNT-based electronics also are challenges in biotechnological applications where CNTs are integrated with living biological systems. Furthermore, the toxicological impact and compatible degree of CNTs on living biological systems and the uptake mechanism for cells in the applications of CNT-based drug and gene delivery also need to be further studied.

With the rapid development of the CNT industry, larger quantities of purified and modified CNTs are being used every day, hence the establishment of CNT disposal and techniques to reuse CNTs are greatly needed. Academia, industry, and government should work together to reveal the environmental influence of CNTs throughout their whole life cycle, while cross-contamination may occur during the preparation, modification, storage, usage, and recycling processes of CNTs.

Lastly, there is still a huge gap between the basic scientific research and large-scale commercialization of CNT preparation, modification, and multifunctional applications. CNTs are an extraordinary structural platform to demonstrate the power of nanotechnology. With the unremitting efforts of worldwide researchers, the applications of specific CNT products may be realized in the coming decades and will definitely help us to develop a sustainable and wonderful world.

REFERENCES

Abuzairi, T., M. Okada, Y. Mochizuki, N.R. Poespawati, R.W. Purnamaningsih, and M. Nagatsu. 2015. "Maskless functionalization of a carbon nanotube dot array biosensor using an ultrafine atmospheric pressure plasma jet." *Carbon* no. 89:208–216. doi: 10.1016/j.carbon.2015.03.015.

Abuzairi, T., M. Okada, R.W. Purnamaningsih, N.R. Poespawati, F. Iwata, and M. Nagatsu. 2016. "Maskless localized patterning of biomolecules on carbon nanotube microarray functionalized by ultrafine atmospheric pressure plasma jet using biotin-avidin system." *Applied Physics Letters* no. 109 (2):023701. doi: 10.1063/1.4958988.

Afkhami, A., H. Khoshsafar, H. Bagheri, and T. Madrakian. 2014. "Construction of a carbon ionic liquid paste electrode based on multi-walled carbon nanotubes-synthesized Schiff base composite for trace electrochemical detection of cadmium." *Materials Science and Engineering: C* no. 35:8–14. doi: 10.1016/j.msec.2013.10.025.

Ajayan, P.M., O. Stephan, C. Colliex, and D. Trauth. 1994. "Aligned carbon nanotube arrays formed by cutting a polymer resin-nanotube composite." *Science* no. 265 (5176):1212–4. doi: 10.1126/science.265.5176.1212.

Al-Saleh, M.H., and U. Sundararaj. 2009. "Electromagnetic interference shielding mechanisms of CNT/polymer composites." *Carbon* no. 47 (7):1738–1746. doi: 10.1016/j.carbon.2009.02.030.

Aliev, A.E., J. Oh, M.E. Kozlov, A.A. Kuznetsov, S. Fang, A.F. Fonseca, R. Ovalle, M.D. Lima, M.H. Haque, Y.N. Gartstein, M. Zhang, A.A. Zakhidov, and R.H. Baughman. 2009. "Giant-stroke, superelastic carbon nanotube aerogel muscles." *Science* no. 323 (5921):1575–1578. doi: 10.1126/science.1168312.

Alpatova, A.L., W. Shan, P. Babica, B.L. Upham, A.R. Rogensues, S.J. Masten, E. Drown, A.K. Mohanty, E.C. Alocilja, and V.V. Tarabara. 2010. "Single-walled carbon nanotubes dispersed in aqueous media via non-covalent functionalization: Effect of dispersant on the stability, cytotoxicity, and epigenetic toxicity of nanotube suspensions." *Water Research* no. 44 (2):505–520. doi: 10.1016/j.watres.2009.09.042.

Alvaro, M., P. Atienzar, P. de la Cruz, J.L. Delgado, H. Garcia, and F. Langa. 2004. "Sidewall functionalization of single-walled carbon nanotubes with nitrile imines. Electron transfer from the substituent to the carbon nanotube." *The Journal of Physical Chemistry B* no. 108 (34):12691–12697. doi: 10.1021/jp0480044.

Anbarasan, R. and C.A. Peng. 2010. "Synthesis and characterization of Rosebengal/folicacid-functionalized multiwall carbon nanotubes." *Journal of Materials Science* no. 46 (4):992–998. doi: 10.1007/s10853-010-4857-x.

Ando, Y., X. Zhao, T. Sugai, and M. Kumar. 2004. "Growing carbon nanotubes." *Materials Today* no. 7 (10):22–29. doi: 10.1016/s1369-7021(04)00446-8.

Appenzeller, J., Y.M. Lin, J. Knoch, and Ph. Avouris. 2004. "Band-to-band tunneling in carbon nanotube field-effect transistors." *Physical Review Letters* no. 93 (19). doi: 10.1103/PhysRevLett.93.196805.

Arora, N. and N.N. Sharma. 2014. "Arc discharge synthesis of carbon nanotubes: Comprehensive review." *Diamond and Related Materials* no. 50:135–150. doi: 10.1016/j.diamond.2014.10.001.

Attal, S., R. Thiruvengadathan, and O. Regev. 2006. "Determination of the concentration of single-walled carbon nanotubes in aqueous dispersions using UV-visible absorption spectroscopy." *Analytical Chemistry* no. 78 (23):8098–8104. doi: 10.1021/ac060990s.

Ávila-Orta, C.A., V.J. Cruz-Delgado, M.G. Neira-Velázquez, E. Hernández-Hernández, M.G. Méndez-Padilla, and F.J. Medellín-Rodríguez. 2009. "Surface modification of carbon nanotubes with ethylene glycol plasma." *Carbon* no. 47 (8):1916–1921. doi: 10.1016/j.carbon.2009.02.033.

Avilés, F., J.V. Cauich-Rodríguez, L. Moo-Tah, A. May-Pat, and R. Vargas-Coronado. 2009. "Evaluation of mild acid oxidation treatments for MWCNT functionalization." *Carbon* no. 47 (13):2970–2975. doi: 10.1016/j.carbon.2009.06.044.

Avouris, P. 2010. "Graphene: Electronic and photonic properties and devices." *Nano Letters* no. 10 (11):4285–4294. doi: 10.1021/nl102824h.

Badhulika, S., N.V. Myung, and A. Mulchandani. 2014. "Conducting polymer coated single-walled carbon nanotube gas sensors for the detection of volatile organic compounds." *Talanta* no. 123:109–114. doi: 10.1016/j.talanta.2014.02.005.

Bae, J.H., A.M. Shanmugharaj, W.H. Noh, W.S. Choi, and S.H. Ryu. 2007. "Surface chemical functionalized single-walled carbon nanotube with anchored phenol structures: Physical and chemical characterization." *Applied Surface Science* no. 253 (9):4150–4155. doi: 10.1016/j.apsusc.2006.09.012.

Bakshi, S.R. and A. Agarwal. 2011. "An analysis of the factors affecting strengthening in carbon nanotube reinforced aluminum composites." *Carbon* no. 49 (2):533–544. doi: 10.1016/j.carbon.2010.09.054.

Balasubramanian, K. and M. Burghard. 2005. "Chemically functionalized carbon nanotubes." *Small* no. 1 (2):180–192. doi: 10.1002/smll.200400118.

Ballesteros, B., G. de la Torre, C. Ehli, G.M. Aminur Rahman, F. Agulló-Rueda, D.M. Guldi, and T. Torres. 2007. "Single-wall carbon nanotubes bearing covalently linked phthalocyanines – all carbon nanelectron transfer." *Journal of the American Chemical Society* no. 129 (16):5061–5068. doi: 10.1021/ja068240n.

Bandow, S., A.M. Rao, K.A. Williams, A. Thess, R.E. Smalley, and P.C. Eklund. 1997. "Purification of single-wall carbon nanotubes by microfiltration." *The Journal of Physical Chemistry B* no. 101 (44):8839–8842. doi: 10.1021/jp972026r.

Barsan, M.M., M. Emilia Ghica, and C.M.A. Brett. 2015. "Electrochemical sensors and biosensors based on redox polymer/carbon nanotube modified electrodes: A review." *Analytica Chimica Acta* no. 881:1–23. doi: 10.1016/j.aca.2015.02.059.

Bartelmess, J., C. Ehli, J.-J. Cid, M. García-Iglesias, P. Vázquez, T. Torres, and D.M. Guldi. 2011. "Tuning and optimizing the intrinsic interactions between phthalocyanine-based PPV oligomers and single-wall carbon nanotubes toward n-type/p-type." *Chemical Science* no. 2 (4):652–660. doi: 10.1039/c0sc00364f.

Battigelli, A., J.T.-W. Wang, J. Russier, T. Da Ros, K. Kostarelos, K.T. Al-Jamal, M. Prato, and A. Bianco. 2013. "Ammonium and guanidinium dendron-carbon nanotubes by amidation and click chemistry and their use for siRNA delivery." *Small* no. 9 (21):3610–3619. doi: 10.1002/smll.201300264.

Bauhofer, W. and J.Z. Kovacs. 2009. "A review and analysis of electrical percolation in carbon nanotube polymer composites." *Composites Science and Technology* no. 69 (10):1486–1498. doi: 10.1016/j.compscitech.2008.06.018.

Bayazit, M.K. and K.S. Coleman. 2014. "Ester-functionalized single-walled carbon nanotubes via addition of haloformates." *Journal of Materials Science* no. 49 (14):5190–5198. doi: 10.1007/s10853-014-8227-y.

Bergeret, C., J. Cousseau, V. Fernandez, J.-Y. Mevellec, and S. Lefrant. 2008. "Spectroscopic evidence of carbon nanotubes' metallic character loss induced by covalent functionalization via nitric acid purification." *The Journal of Physical Chemistry C* no. 112 (42):16411–16416. doi: 10.1021/jp806602t.

Bethune, D.S., C.H. Kiang, M.S. de Vries, G. Gorman, R. Savoy, J. Vazquez, and R. Beyers. 1993. "Cobalt-catalysed growth of carbon nanotubes with single-atomic-layer walls." *Nature* no. 363 (6430):605–607. doi: 10.1038/363605a0.

Bhirde, A.A., V. Patel, J. Gavard, G. Zhang, A.A. Sousa, A. Masedunskas, R.D. Leapman, R. Weigert, J.S. Gutkind, and J.F. Rusling. 2009. "Targeted killing of cancer cells in vivo and in vitro with EGF-directed carbon nanotube-based drug delivery." *ACS Nano* no. 3 (2):307–316. doi: 10.1021/nn800551s.

Bianco, A., K. Kostarelos, and M. Prato. 2011. "Making carbon nanotubes biocompatible and biodegradable." *Chemical Communications* no. 47 (37):10182. doi: 10.1039/c1cc13011k.

Bonifazi, D., C. Nacci, R. Marega, S. Campidelli, G. Ceballos, S. Modesti, M. Meneghetti, and M. Prato. 2006. "Microscopic and spectroscopic characterization of paintbrush-like single-walled carbon nanotubes." *Nano Letters* no. 6 (7):1408–1414. doi: 10.1021/nl060394d.

Bower, C., O. Zhou, W. Zhu, D.J. Werder, and S. Jin. 2000. "Nucleation and growth of carbon nanotubes by microwave plasma chemical vapor deposition." *Applied Physics Letters* no. 77 (17):2767–2769. doi: 10.1063/1.1319529.

Cambré, S. and W. Wenseleers. 2011. "Separation and diameter-sorting of empty (end-capped) and water-filled (open) carbon nanotubes by density gradient ultracentrifugation." *Angewandte Chemie International Edition* no. 50 (12):2764–2768. doi: 10.1002/anie.201007324.

Cao, A. 2005. "Super-compressible foamlike carbon nanotube films." *Science* no. 310 (5752):1307–1310. doi: 10.1126/science.1118957.

Cao, Q., H-S. Kim, N. Pimparkar, J.P. Kulkarni, C. Wang, M. Shim, K. Roy, M.A. Alam, and J.A. Rogers. 2008. "Medium-scale carbon nanotube thin-film integrated circuits on flexible plastic

substrates." *Nature* no. 454 (7203):495–500. doi: 10.1038/nature07110.
Castro Neto, A.H., F. Guinea, N.M.R. Peres, K.S. Novoselov, and A.K. Geim. 2009. "The electronic properties of graphene." *Reviews of Modern Physics* no. 81 (1):109–162. doi: 10.1103/RevModPhys.81.109.
Chae, H.G., Y.H. Choi, M.L. Minus, and S. Kumar. 2009. "Carbon nanotube reinforced small diameter polyacrylonitrile based carbon fiber." *Composites Science and Technology* no. 69 (3–4):406–413. doi: 10.1016/j.compscitech.2008.11.008.
Chaudhary, K.T., J. Ali, and P.P. Yupapin. 2014. "Growth of small diameter multi-walled carbon nanotubes by arc discharge process." *Chinese Physics B* no. 23 (3):035203. doi: 10.1088/1674-1056/23/3/035203.
Chen, J., H. Liu, W.A. Weimer, M.D. Halls, D.H. Waldeck, and G.C. Walker. 2002. "Noncovalent engineering of carbon nanotube surfaces by rigid, functional conjugated polymers." *Journal of the American Chemical Society* no. 124 (31):9034–9035. doi: 10.1021/ja026104m.
Chen, J., S. Chen, X. Zhao, L.V. Kuznetsova, S.S. Wong, and I. Ojima. 2008. "Functionalized single-walled carbon nanotubes as rationally designed vehicles for tumor-targeted drug delivery." *Journal of the American Chemical Society* no. 130 (49):16778–16785. doi: 10.1021/ja805570f.
Chen, Z., Z. Chen, A. Zhang, J. Hu, X. Wang, and Z. Yang. 2016. "Electrospun nanofibers for cancer diagnosis and therapy." *Biomaterials Science* no. 4 (6):922–32. doi: 10.1039/c6bm00070c.
Chen, Z., A. Zhang, X. Wang, J. Zhu, Y. Fan, H. Yu, and Z. Yang. 2017. "The advances of carbon nanotubes in cancer diagnostics and therapeutics." *Journal of Nanomaterials* no. 2017:1–13. doi: 10.1155/2017/3418932.
Chetty, R., S. Kundu, W. Xia, M. Bron, W. Schuhmann, V. Chirila, W. Brandl, T. Reinecke, and M. Muhler. 2009. "PtRu nanoparticles supported on nitrogen-doped multiwalled carbon nanotubes as catalyst for methanol electrooxidation." *Electrochimica Acta* no. 54 (17):4208–4215. doi: 10.1016/j.electacta.2009.02.073.
Chhowalla, M., K.B.K. Teo, C. Ducati, N.L. Rupesinghe, G.A.J. Amaratunga, A.C. Ferrari, D. Roy, J. Robertson, and W.I. Milne. 2001. "Growth process conditions of vertically aligned carbon nanotubes using plasma enhanced chemical vapor deposition." *Journal of Applied Physics* no. 90 (10):5308–5317. doi: 10.1063/1.1410322.
Chiang, I.W., B.E. Brinson, A.Y. Huang, P.A. Willis, M.J. Bronikowski, J.L. Margrave, R.E. Smalley, and R.H. Hauge. 2001. "Purification and characterization of single-wall carbon nanotubes (SWNTs) obtained from the gas-phase decomposition of CO (HiPco Process)." *The Journal of Physical Chemistry B* no. 105 (35):8297–8301. doi: 10.1021/jp0114891.
Chou, Y.-C., H.-C. Wu, and C.-K. Hsieh. 2016. "From graphene to carbon nanotube: The oxygen effect on the synthesis of carbon nanomaterials on nickel foil during CVD process." *Japanese Journal of Applied Physics* no. 55 (1S):01AE12. doi: 10.7567/jjap.55.01ae12.
Chrzanowska, J., J. Hoffman, A. Małolepszy, M. Mazurkiewicz, T.A. Kowalewski, Z. Szymanski, and L. Stobinski. 2015. "Synthesis of carbon nanotubes by the laser ablation method: Effect of laser wavelength." *Physica Status Solidi (b)* no. 252 (8):1860–1867. doi: 10.1002/pssb.201451614.
Cividanes, L.D.S., E.A.N. Simonetti, J.I.S. de Oliveira, A.A. Serra, J.C. de Souza Barboza, and G.P. Thim. 2017. "The sonication effect on CNT-epoxy composites finally clarified." *Polymer Composites* no. 38 (9):1964–1973. doi: 10.1002/pc.23767.
Cividanes, L.S., E.A.N. Simonetti, M.B. Moraes, F.W. Fernandes, and G.P. Thim. 2014. "Influence of carbon nanotubes on epoxy resin cure reaction using different techniques: A comprehensive review." *Polymer Engineering and Science* no. 54 (11):2461–2469. doi: 10.1002/pen.23775.
Connolly, T., R.C. Smith, Y. Hernandez, Y. Gun'ko, J.N. Coleman, and J.D. Carey. 2009. "Carbon-nanotube-polymer nanocomposites for field-emission cathodes." *Small* no. 5 (7):826–831. doi: 10.1002/smll.200801094.
Cristina, B.A.N., D. Ion, S. Adriana, and P. George. 2012. "Nanocomposites as advanced materials for aerospace industry." *Incas Bulletin* no. 4 (4):57–72. doi: 10.13111/2066-8201.2012.4.4.6.
Dai, H., A.G. Rinzler, P. Nikolaev, A. Thess, D.T. Colbert, and R.E. Smalley. 1996. "Single-wall nanotubes produced by metal-catalyzed disproportionation of carbon monoxide." *Chemical Physics Letters* no. 260 (3–4):471–475. doi: 10.1016/0009-2614(96)00862-7.
Dalton, A.B., S. Collins, E. Muñoz, J.M. Razal, V.H. Ebron, J.P. Ferraris, J.N. Coleman, B.G. Kim, and R.H. Baughman. 2003. "Super-tough carbon-nanotube fibres." *Nature* no. 423 (6941):703–703. doi: 10.1038/423703a.
Darabi, H.R., A. Roozkhosh, and K. Aghapoor. 2016. "The nucleophilic addition of in situ generated calcium thiolate of benzonitrile to the sidewall of single-walled carbon nanotubes: A new and direct approach for thioamidation." *Australian Journal of Chemistry* no. 69 (2):198. doi: 10.1071/ch15286.
Datsyuk, V., M. Kalyva, K. Papagelis, J. Parthenios, D. Tasis, A. Siokou, I. Kallitsis, and C. Galiotis. 2008. "Chemical oxidation of multiwalled carbon nanotubes." *Carbon* no. 46 (6):833–840. doi: 10.1016/j.carbon.2008.02.012.
De La Zerda, A., C. Zavaleta, S. Keren, S. Vaithilingam, S. Bodapati, Z. Liu, J. Levi, B.R. Smith, T.-J. Ma, O. Oralkan, Z. Cheng, X. Chen, H. Dai, B.T. Khuri-Yakub, and S.S. Gambhir. 2008. "Carbon nanotubes as photoacoustic molecular imaging agents in living mice." *Nature Nanotechnology* no. 3 (9):557–562. doi: 10.1038/nnano.2008.231.
De Volder, M.F.L., S.H. Tawfick, R.H. Baughman, and A.J. Hart. 2013. "Carbon nanotubes: Present and future commercial applications." *Science* no. 339 (6119):535–539. doi: 10.1126/science.1222453.
Delgado, J.L., P. de la Cruz, F. Langa, A. Urbina, J. Casado, and J.T. López Navarrete. 2004. "Microwave-assisted sidewall functionalization of single-wall carbon nanotubes by Diels–Alder cycloaddition." *Chemical Communications* (15):1734–1735. doi: 10.1039/b402375g.
Delgado, J.L., P. de la Cruz, A. Urbina, J.T. López Navarrete, J. Casado, and F. Langa. 2007. "The first synthesis of a conjugated hybrid of C60–fullerene and a single-wall carbon nanotube." *Carbon* no. 45 (11):2250–2252. doi: 10.1016/j.carbon.2007.06.019.
Deng, X., G. Jia, H. Wang, H. Sun, X. Wang, S. Yang, T. Wang, and Y. Liu. 2007. "Translocation and fate of multi-walled carbon nanotubes in vivo." *Carbon* no. 45 (7):1419–1424. doi: 10.1016/j.carbon.2007.03.035.

Di Crescenzo, A., S. Cambré, R. Germani, P. Di Profio, and A. Fontana. 2014. "Dispersion of SWCNTs with imidazolium-rich surfactants." *Langmuir* no. 30 (14):3979–3987. doi: 10.1021/la500151j.

Di Crescenzo, A., V. Ettorre, and A. Fontana. 2014. "Non-covalent and reversible functionalization of carbon nanotubes." *Beilstein Journal of Nanotechnology* no. 5:1675–1690. doi: 10.3762/bjnano.5.178.

Di Crescenzo, A., I. Kopf, S. Pieraccini, S. Masiero, E. Del Canto, G.P. Spada, S. Giordani, and A. Fontana. 2012. "Lipophilic guanosine derivatives as carbon nanotube dispersing agents." *Carbon* no. 50 (12):4663–4672. doi: 10.1016/j.carbon.2012.05.056.

Dinesh, B., A. Bianco, and C. Ménard-Moyon. 2016. "Designing multimodal carbon nanotubes by covalent multi-functionalization." *Nanoscale* no. 8 (44):18596–18611. doi: 10.1039/c6nr06728j.

Dujardin, E., T.W. Ebbesen, A. Krishnan, and M.M.J. Treacy. 1998. "Wetting of single shell carbon nanotubes." *Advanced Materials* no. 10 (17):1472–1475. doi: 10.1002/(sici)1521-4095(199812)10:17<1472::aid-adma1472>3.0.co;2-r.

Eatemadi, A., H. Daraee, H. Karimkhanloo, M. Kouhi, N. Zarghami, A. Akbarzadeh, M. Abasi, Y. Hanifehpour, and S. Joo. 2014. "Carbon nanotubes: Properties, synthesis, purification, and medical applications." *Nanoscale Research Letters* no. 9 (1):393. doi: 10.1186/1556-276x-9-393.

Ebbesen, T.W. and P.M. Ajayan. 1992. "Large-scale synthesis of carbon nanotubes." *Nature* no. 358 (6383):220–222. doi: 10.1038/358220a0.

Eghtedari, M., A. Oraevsky, J.A. Copland, N.A. Kotov, A. Conjusteau, and M. Motamedi. 2007. "High sensitivity of in vivo detection of gold nanorods using a laser optoacoustic imaging system." *Nano Letters* no. 7 (7):1914–1918. doi: 10.1021/nl070557d.

Eklund, P.C., B.K. Pradhan, U.J. Kim, Q. Xiong, J.E. Fischer, A.D. Friedman, B.C. Holloway, K. Jordan, and M.W. Smith. 2002. "Large-scale production of single-walled carbon nanotubes using ultrafast pulses from a free electron laser." *Nano Letters* no. 2 (6):561–566. doi: 10.1021/nl025515y.

El-Kady, M.F., V. Strong, S. Dubin, and R.B. Kaner. 2012. "Laser scribing of high-performance and flexible graphene-based electrochemical capacitors." *Science* no. 335 (6074):1326–1330. doi: 10.1126/science.1216744.

Fabre, B., F. Hauquier, C. Herrier, G. Pastorin, W. Wu, A. Bianco, M. Prato, P. Hapiot, D. Zigah, M. Prasciolu, and L. Vaccari. 2008. "Covalent assembly and micropatterning of functionalized multiwalled carbon nanotubes to monolayer-modified Si(111) surfaces." *Langmuir* no. 24 (13):6595–6602. doi: 10.1021/la800358w.

Fagnoni, M., A. Profumo, D. Merli, D. Dondi, P. Mustarelli, and Eliana Quartarone. 2009. "Water-miscible liquid multiwalled carbon nanotubes." *Advanced Materials* no. 21 (17):1761–1765. doi: 10.1002/adma.200801994.

Ferreira, F.V., W. Francisco, B.R.C. de Menezes, L. De Simone Cividanes, A. dos Reis Coutinho, and G.P. Thim. 2015. "Carbon nanotube functionalized with dodecylamine for the effective dispersion in solvents." *Applied Surface Science* no. 357:2154–2159. doi: 10.1016/j.apsusc.2015.09.202.

Fetterman, A.J., Y. Raitses, and M. Keidar. 2008. "Enhanced ablation of small anodes in a carbon nanotube arc plasma." *Carbon* no. 46 (10):1322–1326. doi: 10.1016/j.carbon.2008.05.018.

Flahaut, E., A. Govindaraj, A. Peigney, Ch. Laurent, A. Rousset, and C.N.R. Rao. 1999. "Synthesis of single-walled carbon nanotubes using binary (Fe, Co, Ni) alloy nanoparticles prepared in situ by the reduction of oxide solid solutions." *Chemical Physics Letters* no. 300 (1–2):236–242. doi: 10.1016/s0009-2614(98)01304-9.

Gao, C., Z. Guo, J.-H. Liu, and X.-J. Huang. 2012. "The new age of carbon nanotubes: An updated review of functionalized carbon nanotubes in electrochemical sensors." *Nanoscale* no. 4 (6):1948. doi: 10.1039/c2nr11757f.

Gautier, L.-A., V. Le Borgne, and M. Ali El Khakani. 2016. "Field emission properties of graphenated multi-wall carbon nanotubes grown by plasma enhanced chemical vapour deposition." *Carbon* no. 98:259–266. doi: 10.1016/j.carbon.2015.11.006.

Georgakilas, V., A.B. Bourlinos, R. Zboril, and C. Trapalis. 2008. "Synthesis, characterization and aspects of superhydrophobic functionalized carbon nanotubes." *Chemistry of Materials* no. 20 (9):2884–2886. doi: 10.1021/cm7034079.

Georgiou, T., T. Jalil, B.D. Belle, L. Britnell, R.V. Gorbachev, S.V. Morozov, Y.-J. Kim, A. Gholinia, S.J. Haigh, O. Makarovsky, L. Eaves, L.A. Ponomarenko, A.K. Geim, K.S. Novoselov, and A. Mishchenko. 2012. "Vertical field-effect transistor based on graphene–WS2 heterostructures for flexible and transparent electronics." *Nature Nanotechnology* no. 8 (2):100–103. doi: 10.1038/nnano.2012.224.

Goddard, W.A. 2007. *Handbook of Nanoscience, Engineering, and Technology*. 2nd ed. 1 vols, *The Electrical Engineering Handbook Series*. Boca Raton, FL: CRC Press.

Gong, K., F. Du, Z. Xia, M. Durstock, and L. Dai. 2009. "Nitrogen-doped carbon nanotube arrays with high electrocatalytic activity for oxygen reduction." *Science* no. 323 (5915):760–764. doi: 10.1126/science.1168049.

González-Domínguez, J.M., A. Santidrián, A. Criado, C. Hadad, M. Kalbáč, and T. Da Ros. 2015. "Multipurpose nature of rapid covalent functionalization on carbon nanotubes." *Chemistry – A European Journal* no. 21 (51):18631–18641. doi: 10.1002/chem.201503085.

Gooding, J.J. 2005. "Nanostructuring electrodes with carbon nanotubes: A review on electrochemistry and applications for sensing." *Electrochimica Acta* no. 50 (15):3049–3060. doi: 10.1016/j.electacta.2004.08.052.

Guldi, D.M., G.M. Aminur Rahman, N. Jux, D. Balbinot, U. Hartnagel, N. Tagmatarchis, and M. Prato. 2005. "Functional single-wall carbon nanotube nanohybrids – Associating SWNTs with water-soluble enzyme model systems." *Journal of the American Chemical Society* no. 127 (27):9830–9838. doi: 10.1021/ja050930o.

Guo, T., M.D. Diener, Y. Chai, M.J. Alford, R.E. Haufler, S.M. McClure, T. Ohno, J.H. Weaver, G.E. Scuseria, and R.E. Smalley. 1992. "Uranium stabilization of c28: A tetravalent fullerene." *Science* no. 257 (5077):1661–4. doi: 10.1126/science.257.5077.1661.

Guo, T., P. Nikolaev, A. Thess, D. T. Colbert, and R. E. Smalley. 1995. "Catalytic growth of single-walled nanotubes by laser vaporization." *Chemical Physics Letters* no. 243 (1–2):49–54. doi: 10.1016/0009-2614(95)00825-o.

Gupta, N.D., S. Maity, and K.K. Chattopadhyay. 2014. "Field emission enhancement of polypyrrole due to band bending induced tunnelling in polypyrrole-carbon nanotubes nanocomposite." *Journal of Industrial and Engineering Chemistry* no. 20 (5):3208–3213. doi: 10.1016/j.jiec.2013.11.067.

Gupta, S., K. Dharamvir, and V.K. Jindal. 2005. "Elastic moduli of single-walled carbon nanotubes and their ropes." *Physical Review B* no. 72 (16). doi: 10.1103/PhysRevB.72.165428.

Hamon, M.A., H. Hui, P. Bhowmik, H.M.E. Itkis, and R.C. Haddon. 2002. "Ester-functionalized soluble single-walled carbon nanotubes." *Applied Physics A: Materials Science and Processing* no. 74 (3):333–338. doi: 10.1007/s003390201281.

Han, W. 1997. "Synthesis of gallium nitride nanorods through a carbon nanotube-confined reaction." *Science* no. 277 (5330):1287–1289. doi: 10.1126/science.277.5330.1287.

Harrison, B.S. and A. Atala. 2007. "Carbon nanotube applications for tissue engineering." *Biomaterials* no. 28 (2):344–353. doi: 10.1016/j.biomaterials.2006.07.044.

Hata, K. 2004. "Water-assisted highly efficient synthesis of impurity-free single-walled carbon nanotubes." *Science* no. 306 (5700):1362–1364. doi: 10.1126/science.1104962.

He, X., Y. Huang, Y. Liu, X. Zheng, S. Kormakov, J. Sun, J. Zhuang, X. Gao, and D. Wu. 2018. "Improved thermal conductivity of polydimethylsiloxane/short carbon fiber composites prepared by spatial confining forced network assembly." *Journal of Materials Science* no. 53 (20):14299–14310. doi: 10.1007/s10853-018-2618-4.

Herranz, M.A., N. Martín, S. Campidelli, M. Prato, G. Brehm, and D.M. Guldi. 2006. "Control over electron transfer in tetrathiafulvalene-modified single-walled carbon nanotubes." *Angewandte Chemie International Edition* no. 45 (27):4478–4482. doi: 10.1002/anie.200504354.

Homenick, C.M., R. James, G.P. Lopinski, J. Dunford, J. Sun, H. Park, Y. Jung, G. Cho, and P.R.L. Malenfant. 2016. "Fully printed and encapsulated SWCNT-based thin film transistors via a combination of R2R gravure and inkjet printing." *ACS Applied Materials and Interfaces* no. 8 (41):27900–27910. doi: 10.1021/acsami.6b06838.

Hong, S.Y., G. Tobias, K.T. Al-Jamal, B. Ballesteros, H. Ali-Boucetta, S. Lozano-Perez, P.D. Nellist, R.B. Sim, C. Finucane, S.J. Mather, M.L H. Green, K. Kostarelos, and B.G. Davis. 2010. "Filled and glycosylated carbon nanotubes for in vivo radioemitter localization and imaging." *Nature Materials* no. 9 (6):485–490. doi: 10.1038/nmat2766.

Huang, S.-C.J., A.B. Artyukhin, N. Misra, J.A. Martinez, P.A. Stroeve, C.P. Grigoropoulos, J.-W.W. Ju, and A. Noy. 2010. "Carbon nanotube transistor controlled by a biological ion pump gate." *Nano Letters* no. 10 (5):1812–1816. doi: 10.1021/nl100499x.

Huang, Y, S. Kormakov, X. He, X. Gao, X. Zheng, Y. Liu, J. Sun, and D. Wu. 2019. "Conductive polymer composites from renewable resources: An overview of preparation, properties, and applications." *Polymers* no. 11 (2):187. doi: 10.3390/polym11020187.

Huang, Z.P., J.W. Xu, Z.F. Ren, J.H. Wang, M.P. Siegal, and P.N. Provencio. 1998. "Growth of highly oriented carbon nanotubes by plasma-enhanced hot filament chemical vapor deposition." *Applied Physics Letters* no. 73 (26):3845–3847. doi: 10.1063/1.122912.

Husanu, M., M. Baibarac, and I. Baltog. 2008. "Non-covalent functionalization of carbon nanotubes: Experimental evidence for isolated and bundled tubes." *Physica E: Low-Dimensional Systems and Nanostructures* no. 41 (1):66–69. doi: 10.1016/j.physe.2008.06.001.

Hwang, S.K., J.M. Lee, S. Kim, J.S. Park, H. Il Park, C.W. Ahn, K.J. Lee, T. Lee, and S. Ouk Kim. 2012. "Flexible multilevel resistive memory with controlled charge trap B- and N-doped carbon nanotubes." *Nano Letters* no. 12 (5):2217–2221. doi: 10.1021/nl204039q.

Hyun, K., S.W. Han, W.-G. Koh, and Y. Kwon. 2015. "Direct electrochemistry of glucose oxidase immobilized on carbon nanotube for improving glucose sensing." *International Journal of Hydrogen Energy* no. 40 (5):2199–2206. doi: 10.1016/j.ijhydene.2014.12.019.

Iijima, S. 1991. "Helical microtubules of graphitic carbon." *Nature* no. 354 (6348):56–58. doi: 10.1038/354056a0.

Iijima, S., and T. Ichihashi. 1993. "Single-shell carbon nanotubes of 1-nm diameter." *Nature* no. 363 (6430):603–605. doi: 10.1038/363603a0.

Ionescu, A.M., and H. Riel. 2011. "Tunnel field-effect transistors as energy-efficient electronic switches." *Nature* no. 479 (7373):329–337. doi: 10.1038/nature10679.

Ishigami, N., H. Ago, K. Imamoto, M. Tsuji, K. Iakoubovskii, and N. Minami. 2008. "Crystal plane dependent growth of aligned single-walled carbon nanotubes on sapphire." *Journal of the American Chemical Society* no. 130 (30):9918–9924. doi: 10.1021/ja8024752.

Ishii, H., H. Kataura, H. Shiozawa, H. Yoshioka, H. Otsubo, Y. Takayama, T. Miyahara, S. Suzuki, Y. Achiba, M. Nakatake, T. Narimura, M. Higashiguchi, K. Shimada, H. Namatame, and M. Taniguchi. 2003. "Direct observation of Tomonaga–Luttinger-liquid state in carbon nanotubes at low temperatures." *Nature* no. 426 (6966):540–544. doi: 10.1038/nature02074.

Ishikawa, F.N., H.-K. Chang, K. Ryu, P.-C. Chen, A. Badmaev, L. Gomez De Arco, G. Shen, and C. Zhou. 2009. "Transparent electronics based on transfer printed aligned carbon nanotubes on rigid and flexible substrates." *ACS Nano* no. 3 (1):73–79. doi: 10.1021/nn800434d.

Jabr-Milane, L.S., L.E. van Vlerken, S. Yadav, and M.M. Amiji. 2008. "Multi-functional nanocarriers to overcome tumor drug resistance." *Cancer Treatment Reviews* no. 34 (7):592–602. doi: 10.1016/j.ctrv.2008.04.003.

Janas, D., S. Boncel, and K.K.K. Koziol. 2014. "Electrothermal halogenation of carbon nanotube films." *Carbon* no. 73:259–266. doi: 10.1016/j.carbon.2014.02.062.

Jeon, H.R., J.H. Park, and M.Y. Shon. 2013. "Corrosion protection by epoxy coating containing multi-walled carbon nanotubes." *Journal of Industrial and Engineering Chemistry* no. 19 (3):849–853. doi: 10.1016/j.jiec.2012.10.030.

Jin, M., X. Feng, L. Feng, T. Sun, J. Zhai, T. Li, and L. Jiang. 2005. "Superhydrophobic aligned polystyrene nanotube films with high adhesive force." *Advanced Materials* no. 17 (16):1977–1981. doi: 10.1002/adma.200401726.

Jingyao, S., W. Daming, L. Ying, Y. Zhenzhou, and G. Pengsheng. 2017. "Rapid fabrication of micro structure on polypropylene by plate to plate isothermal hot embossing method." *Polymer Engineering and Science*. doi: 10.1002/pen.24651.

Joo, Y., G.J. Brady, M.S. Arnold, and P. Gopalan. 2014. "Dose-controlled, floating evaporative self-assembly and alignment of semiconducting carbon nanotubes from organic solvents." *Langmuir* no. 30 (12):3460–3466. doi: 10.1021/la500162x.

Joshi, R., J. Engstler, P. Kesavan Nair, P. Haridoss, and J.J. Schneider. 2008. "High yield formation of carbon nanotubes using a rotating cathode in open air." *Diamond and Related Materials* no. 17 (6):913–919. doi: 10.1016/j.diamond.2008.01.004.

Kalita, G., S. Adhikari, H.R. Aryal, R. Afre, T. Soga, M. Sharon, and M. Umeno. 2009. "Functionalization of multi-walled carbon nanotubes (MWCNTs) with nitrogen plasma for photovoltaic device application." *Current Applied Physics* no. 9 (2):346–351. doi: 10.1016/j.cap.2008.03.007.

Kanoun, O., C. Müller, A. Benchirouf, A. Sanli, T. Dinh, A. Al-Hamry, L. Bu, C. Gerlach, and A. Bouhamed. 2014. "Flexible carbon nanotube films for high performance strain sensors." *Sensors* no. 14 (6):10042–10071. doi: 10.3390/s140610042.

Kar, R., N.N. Patel, N. Chand, R.K. Shilpa, R.O. Dusane, D.S. Patil, and S. Sinha. 2016. "Detailed investigation on the mechanism of co-deposition of different carbon nanostructures by microwave plasma CVD." *Carbon* no. 106:233–242. doi: 10.1016/j.carbon.2016.05.027.

Karimi, M., N. Solati, A. Ghasemi, M.A. Estiar, M. Hashemkhani, P. Kiani, E. Mohamed, A. Saeidi, M. Taheri, P. Avci, A.R. Aref, M. Amiri, F. Baniasadi, and M.R. Hamblin. 2015. "Carbon nanotubes part II: A remarkable carrier for drug and gene delivery." *Expert Opinion on Drug Delivery* no. 12 (7):1089–1105. doi: 10.1517/17425247.2015.1004309.

Karousis, N., N. Tagmatarchis, and D. Tasis. 2010. "Current progress on the chemical modification of carbon nanotubes." *Chemical Reviews* no. 110 (9):5366–5397. doi: 10.1021/cr100018g.

Kashiwagi, T., F. Du, J.F. Douglas, K.I. Winey, R.H. Harris, and J.R. Shields. 2005. "Nanoparticle networks reduce the flammability of polymer nanocomposites." *Nature Materials* no. 4 (12):928–933. doi: 10.1038/nmat1502.

Kelly, K.F., I.W. Chiang, E.T. Mickelson, R.H. Hauge, J.L. Margrave, X. Wang, G.E. Scuseria, C. Radloff, and N.J. Halas. 1999. "Insight into the mechanism of sidewall functionalization of single-walled nanotubes: An STM study." *Chemical Physics Letters* no. 313 (3–4):445–450. doi: 10.1016/s0009-2614(99)00973-2.

Khan, M.U., V.G. Gomes, and I.S. Altarawneh. 2010. "Synthesizing polystyrene/carbon nanotube composites by emulsion polymerization with non-covalent and covalent functionalization." *Carbon* no. 48 (10):2925–2933. doi: 10.1016/j.carbon.2010.04.029.

Kierkowicz, M., J.M. González-Domínguez, E. Pach, S. Sandoval, B. Ballesteros, T. Da Ros, and G. Tobias. 2017. "Filling single-walled carbon nanotubes with lutetium chloride: A sustainable production of nanocapsules free of nonencapsulated material." *ACS Sustainable Chemistry and Engineering* no. 5 (3):2501–2508. doi: 10.1021/acssuschemeng.6b02850.

Kim, H.H., and H.J. Kim. 2006. "Preparation of carbon nanotubes by DC arc discharge process under reduced pressure in an air atmosphere." *Materials Science and Engineering: B* no. 133 (1–3):241–244. doi: 10.1016/j.mseb.2006.06.017.

Kim, K.S., Y. Zhao, H. Jang, S.Y. Lee, J.M. Kim, K.S. Kim, J.-H. Ahn, P. Kim, J.-Y. Choi, and B.H. Hong. 2009. "Large-scale pattern growth of graphene films for stretchable transparent electrodes." *Nature* no. 457 (7230):706–710. doi: 10.1038/nature07719.

Kim, Y.-Y., J. Yun, H.-I. Kim, and Y.-S. Lee. 2012. "Effect of oxy-fluorination on electromagnetic interference shielding of polypyrrole-coated multi-walled carbon nanotubes." *Journal of Industrial and Engineering Chemistry* no. 18 (1):392–398. doi: 10.1016/j.jiec.2011.11.103.

Kitiyanan, B., W.E. Alvarez, J.H. Harwell, and D.E. Resasco. 2000. "Controlled production of single-wall carbon nanotubes by catalytic decomposition of CO on bimetallic Co–Mo catalysts." *Chemical Physics Letters* no. 317 (3–5):497–503. doi: 10.1016/s0009-2614(99)01379-2.

Kong, J., A.M. Cassell, and H. Dai. 1998. "Chemical vapor deposition of methane for single-walled carbon nanotubes." *Chemical Physics Letters* no. 292 (4–6):567–574. doi: 10.1016/s0009-2614(98)00745-3.

Kormakov, S., Z. He, Y. Huang, Y. Liu, J. Sun, X. Zheng, I. Skopincev, X. Gao, and D. Wu. 2018. "A mathematical model for predicting conductivity of polymer composites with a forced assembly network obtained by SCFNA method." *Polymer Composites* no. 40 (5):1819–1827. doi: 10.1002/pc.24942.

Krstic, V., G.S. Duesberg, J. Muster, M. Burghard, and S. Roth. 1998. "Langmuir–Blodgett films of matrix-diluted single-walled carbon nanotubes." *Chemistry of Materials* no. 10 (9):2338–2340. doi: 10.1021/cm980207f.

Kumar, M., and Y. Ando. 2010. "Chemical vapor deposition of carbon nanotubes: A review on growth mechanism and mass production." *Journal of Nanoscience and Nanotechnology* no. 10 (6):3739–3758. doi: 10.1166/jnn.2010.2939.

Landry, M.P., H. Ando, A.Y. Chen, J. Cao, V.I. Kottadiel, L. Chio, D. Yang, J. Dong, T.K. Lu, and M.S. Strano. 2017. "Single-molecule detection of protein efflux from microorganisms using fluorescent single-walled carbon nanotube sensor arrays." *Nature Nanotechnology* no. 12 (4):368–377. doi: 10.1038/nnano.2016.284.

Lau, P.H., K. Takei, C. Wang, Y. Ju, J. Kim, Zhibin Y., Toshitake Takahashi, G. Cho, and A. Javey. 2013. "Fully printed, high performance carbon nanotube thin-film transistors on flexible substrates." *Nano Letters* no. 13 (8):3864–3869. doi: 10.1021/nl401934a.

Lebedkin, S., P. Schweiss, B. Renker, S. Malik, F. Hennrich, M. Neumaier, C. Stoermer, and M.M. Kappes. 2002. "Single-wall carbon nanotubes with diameters approaching 6 nm obtained by laser vaporization." *Carbon* no. 40 (3):417–423. doi: 10.1016/s0008-6223(01)00119-1.

Lee, D.H., W.J. Lee, W.J. Lee, S. Ouk Kim, and Y.-H. Kim. 2011. "Theory, synthesis, and oxygen reduction catalysis of Fe-porphyrin-like carbon nanotube." *Physical Review Letters* no. 106 (17). doi: 10.1103/PhysRevLett.106.175502.

Lee, K.S., W.J. Lee, N.-G. Park, S. Ouk Kim, and J.H. Park. 2011. "Transferred vertically aligned N-doped carbon nanotube arrays: Use in dye-sensitized solar cells as counter electrodes." *Chemical Communications* no. 47 (14):4264. doi: 10.1039/c1cc10471c.

Lee, L.J., Z. Yang, M. Rahman, J. Ma, K.J. Kwak, J. McElroy, K. Shilo, C. Goparaju, L. Yu, W. Rom, T.K. Kim, X. Wu, Y. He, K. Wang, H.I. Pass, and S. P. Nana-Sinkam. 2016. "Extracellular mRNA detected by tethered lipoplex nanoparticle biochip for lung adenocarcinoma detection." *American Journal of Respiratory and Critical Care Medicine* no. 193 (12):1431–3. doi: 10.1164/rccm.201511-2129LE.

Lee, S.W. 2016. "Mechanical properties of suspended individual carbon nanotube studied by atomic force microscope." *Synthetic Metals* no. 216:88–92. doi: 10.1016/j.synthmet.2015.09.014.

Lee, S.W., N. Yabuuchi, B.M. Gallant, S. Chen, B.-S. Kim, P.T. Hammond, and Y. Shao-Horn. 2010. "High-power lithium batteries from functionalized carbon-nanotube electrodes." *Nature Nanotechnology* no. 5 (7):531–537. doi: 10.1038/nnano.2010.116.

Lee, S.H., H.W. Kim, J.O. Hwang, W.J. Lee, J. Kwon, C. W. Bielawski, R.S. Ruoff, and S. Ouk Kim. 2010. "Three-dimensional self-assembly of graphene oxide platelets into mechanically flexible macroporous carbon films." *Angewandte Chemie* no. 122 (52):10282–10286. doi: 10.1002/ange.201006240.

Lee, W.J., J.M. Lee, S.T. Kochuveedu, T.H. Han, H.Y. Jeong, M. Park, J.M. Yun, J. Kwon, K. No, D. Ha Kim, and S. Ouk Kim. 2011. "Biomineralized N-doped CNT/TiO2 core/shell nanowires for visible light photocatalysis." *ACS Nano* no. 6 (1):935–943. doi: 10.1021/nn204504h.

Li, X., F. Li, C. Liu, and H.-M. Cheng. 2005. "Synthesis and characterization of double-walled carbon nanotubes from multi-walled carbon nanotubes by hydrogen-arc discharge." *Carbon* no. 43 (3):623–629. doi: 10.1016/j.carbon.2004.10.028.

Liang, S., Y. Zhao, and A. Adronov. 2014. "Selective and reversible noncovalent functionalization of single-walled carbon nanotubes by a pH-responsive vinylogous tetrathiafulvalene–fluorene copolymer." *Journal of the American Chemical Society* no. 136 (3):970–977. doi: 10.1021/ja409918n.

Lima, M.D., N. Li, M. Jung de Andrade, S. Fang, J. Oh, G.M. Spinks, M.E. Kozlov, C.S. Haines, D. Suh, J. Foroughi, S.J. Kim, Y. Chen, T. Ware, M.K. Shin, L.D. Machado, A.F. Fonseca, J.D.W. Madden, W.E. Voit, D.S. Galvao, and R.H. Baughman. 2012. "Electrically, chemically, and photonically powered torsional and tensile actuation of hybrid carbon nanotube yarn muscles." *Science* no. 338 (6109):928–932. doi: 10.1126/science.1226762.

Liu, S.F., L.C.H. Moh, and T.M. Swager. 2015. "Single-walled carbon nanotube–metalloporphyrin chemiresistive gas sensor arrays for volatile organic compounds." *Chemistry of Materials* no. 27 (10):3560–3563. doi: 10.1021/acs.chemmater.5b00153.

Liu, W.-W., S.-P. Chai, A.R. Mohamed, and U. Hashim. 2014. "Synthesis and characterization of graphene and carbon nanotubes: A review on the past and recent developments." *Journal of Industrial and Engineering Chemistry* no. 20 (4):1171–1185. doi: 10.1016/j.jiec.2013.08.028.

Liu, X., H.-L. Shuai, Y.-J. Liu, and K.-J. Huang. 2016. "An electrochemical biosensor for DNA detection based on tungsten disulfide/multi-walled carbon nanotube composites and hybridization chain reaction amplification." *Sensors and Actuators B: Chemical* no. 235:603–613. doi: 10.1016/j.snb.2016.05.132.

Liu, Z., K. Chen, C. Davis, S. Sherlock, Q. Cao, X. Chen, and H. Dai. 2008. "Drug delivery with carbon nanotubes for in vivo cancer treatment." *Cancer Research* no. 68 (16):6652–6660. doi: 10.1158/0008-5472.can-08-1468.

Liu, Z., X. Sun, N. Nakayama-Ratchford, and H. Dai. 2007. "Supramolecular chemistry on water-soluble carbon nanotubes for drug loading and delivery." *ACS Nano* no. 1 (1):50–56. doi: 10.1021/nn700040t.

Lovat, V., D. Pantarotto, L. Lagostena, B. Cacciari, M. Grandolfo, M. Righi, G. Spalluto, M. Prato, and L. Ballerini. 2005. "Carbon nanotube substrates boost neuronal electrical signaling." *Nano Letters* no. 5 (6):1107–1110. doi: 10.1021/nl050637m.

Lu, X., Z. Chen, and P.v.R. Schleyer. 2005. "Are Stone–Wales defect sites always more reactive than perfect sites in the sidewalls of single-wall carbon nanotubes?" *Journal of the American Chemical Society* no. 127 (1):20–21. doi: 10.1021/ja0447053.

Ménard-Moyon, C., H. Ali-Boucetta, C. Fabbro, O. Chaloin, K. Kostarelos, and A. Bianco. 2015. "Controlled chemical derivatisation of carbon nanotubes with imaging, targeting, and therapeutic capabilities." *Chemistry – A European Journal* no. 21 (42):14886–14892. doi: 10.1002/chem.201501993.

Ménard-Moyon, C., F. Dumas, E. Doris, and C. Mioskowski. 2006. "Functionalization of single-wall carbon nanotubes by tandem high-pressure/Cr(CO)6 activation of Diels–Alder cycloaddition." *Journal of the American Chemical Society* no. 128 (46):14764–14765. doi: 10.1021/ja065698g.

Ménard-Moyon, C., C. Fabbro, M. Prato, and A. Bianco. 2011. "One-pot triple functionalization of carbon nanotubes." *Chemistry – A European Journal* no. 17 (11):3222–3227. doi: 10.1002/chem.201003050.

Ma, D., Liying L., W. Chen, H. Liu, and Y.-F. Song. 2013. "Covalently tethered polyoxometalate-pyrene hybrids for noncovalent sidewall functionalization of single-walled carbon nanotubes as high-performance anode material." *Advanced Functional Materials* no. 23 (48):6100–6105. doi: 10.1002/adfm.201301624.

Ma, P.-C., N.A. Siddiqui, G. Marom, and J.-K. Kim. 2010. "Dispersion and functionalization of carbon nanotubes for polymer-based nanocomposites: A review." *Composites Part A: Applied Science and Manufacturing* no. 41 (10):1345–1367. doi: 10.1016/j.compositesa.2010.07.003.

Malarkey, E.B., R.C. Reyes, B. Zhao, R.C. Haddon, and V. Parpura. 2008. "Water soluble single-walled carbon nanotubes inhibit stimulated endocytosis in neurons." *Nano Letters* no. 8 (10):3538–3542. doi: 10.1021/nl8017912.

Marega, R., and D. Bonifazi. 2014. "Filling carbon nanotubes for nanobiotechnological applications." *New Journal of Chemistry* no. 38 (1):22–27. doi: 10.1039/c3nj01008b.

Maser, W.K., E. Muñoz, A.M. Benito, M.T. Martínez, G.F. de la Fuente, Y. Maniette, E. Anglaret, and J.L. Sauvajol. 1998. "Production of high-density single-walled nanotube material by a simple laser-ablation method." *Chemical Physics Letters* no. 292 (4–6):587–593. doi: 10.1016/s0009-2614(98)00776-3.

Mateo-Alonso, A., C. Ehli, K.H. Chen, D.M. Guldi, and M. Prato. 2007. "Dispersion of single-walled carbon nanotubes with an extended diazapentacene derivative." *The Journal of Physical Chemistry A* no. 111 (49):12669–12673. doi: 10.1021/jp0765648.

McCarthy, M.A., B. Liu, E.P. Donoghue, I. Kravchenko, D.Y. Kim, F. So, and A.G. Rinzler. 2011. "Low-voltage, low-power, organic light-emitting transistors for active matrix displays." *Science* no. 332 (6029):570–573. doi: 10.1126/science.1203052.

McDevitt, M.R., D. Chattopadhyay, B.J. Kappel, J.S. Jaggi, S.R. Schiffman, C. Antczak, J.T. Njardarson, R. Brentjens, and D.A. Scheinberg. 2007. "Tumor targeting with antibody-functionalized, radiolabeled carbon nanotubes." *Journal of Nuclear Medicine* no. 48 (7):1180–1189. doi: 10.2967/jnumed.106.039131.

Mickelson, E.T., C.B. Huffman, A.G. Rinzler, R.E. Smalley, R.H. Hauge, and J.L. Margrave. 1998. "Fluorination of single-wall carbon nanotubes." *Chemical Physics Letters* no. 296 (1–2):188–194. doi: 10.1016/s0009-2614(98)01026-4.

Mittal, G., V. Dhand, K.Y. Rhee, S.-J. Park, and W.R. Lee. 2015. "A review on carbon nanotubes and graphene as fillers in reinforced polymer nanocomposites." *Journal of Industrial and Engineering Chemistry* no. 21:11–25. doi: 10.1016/j.jiec.2014.03.022.

Mohamed, A., A.K. Anas, S. Abu Bakar, T. Ardyani, W.M.W. Zin, S. Ibrahim, M. Sagisaka, P. Brown, and J. Eastoe. 2015. "Enhanced dispersion of multiwall carbon nanotubes in natural rubber latex nanocomposites by surfactants bearing phenyl groups." *Journal of Colloid and Interface Science* no. 455:179–187. doi: 10.1016/j.jcis.2015.05.054.

Mohammad, M.I., A.A. Moosa, J.H. Potgieter, and M.K. Ismael. 2013. "Carbon nanotubes synthesis via Arc discharge with a Yttria catalyst." *ISRN Nanomaterials* no. 2013:1–7. doi: 10.1155/2013/785160.

Moradi, L., and I. Etesami. 2016. "New route for bromination of multiwalled carbon nanotubes under mild and efficient conditions." *Fullerenes, Nanotubes and Carbon Nanostructures* no. 24 (3):213–218. doi: 10.1080/1536383x.2015.1136820.

Mubarak, N.M., E.C. Abdullah, N.S. Jayakumar, and J.N. Sahu. 2014. "An overview on methods for the production of carbon nanotubes." *Journal of Industrial and Engineering Chemistry* no. 20 (4):1186–1197. doi: 10.1016/j.jiec.2013.09.001.

Muñoz, E., W.K. Maser, A.M. Benito, M.T. Martínez, G.F. de la Fuente, Y. Maniette, A. Righi, E. Anglaret, and J.L. Sauvajol. 2000. "Gas and pressure effects on the production of single-walled carbon nanotubes by laser ablation." *Carbon* no. 38 (10):1445–1451. doi: 10.1016/s0008-6223(99)00277-8.

Murphy, R., J.N. Coleman, M. Cadek, B. McCarthy, M. Bent, A. Drury, R.C. Barklie, and W.J. Blau. 2002. "High-yield, nondestructive purification and quantification method for multiwalled carbon nanotubes." *The Journal of Physical Chemistry B* no. 106 (12):3087–3091. doi: 10.1021/jp0132836.

Nessim, G.D. 2010. "Properties, synthesis, and growth mechanisms of carbon nanotubes with special focus on thermal chemical vapor deposition." *Nanoscale* no. 2 (8):1306. doi: 10.1039/b9nr00427k.

Nikolaev, P., M.J. Bronikowski, R.K. Bradley, F. Rohmund, D.T. Colbert, K.A. Smith, and R.E. Smalley. 1999. "Gas-phase catalytic growth of single-walled carbon nanotubes from carbon monoxide." *Chemical Physics Letters* no. 313 (1–2):91–97. doi: 10.1016/s0009-2614(99)01029-5.

Nozaki, T. and K. Okazaki. 2008. "Carbon nanotube synthesis in atmospheric pressure glow discharge: A review." *Plasma Processes and Polymers* no. 5 (4):300–321. doi: 10.1002/ppap.200700141.

Oh, H., Jinsook S., and S.-Y. Ju. 2013. "Binding affinities and thermodynamics of noncovalent functionalization of carbon nanotubes with surfactants." *Langmuir* no. 29 (35):11154–11162. doi: 10.1021/la4022933.

Okazaki, T., K. Suenaga, K. Hirahara, S. Bandow, S. Iijima, and H. Shinohara. 2001. "Real time reaction dynamics in carbon nanotubes." *Journal of the American Chemical Society* no. 123 (39):9673–9674. doi: 10.1021/ja016415h.

Okpalugo, T.I.T., P. Papakonstantinou, H. Murphy, J. McLaughlin, and N.M.D. Brown. 2005. "Oxidative functionalization of carbon nanotubes in atmospheric pressure filamentary dielectric barrier discharge (APDBD)." *Carbon* no. 43 (14):2951–2959. doi: 10.1016/j.carbon.2005.06.033.

Ondera, T.J. and A.T. Hamme Ii. 2014. "A gold nanopopcorn attached single-walled carbon nanotube hybrid for rapid detection and killing of bacteria." *Journal of Materials Chemistry B* no. 2 (43):7534–7543. doi: 10.1039/c4tb01195c.

Osorio, A.G., I.C.L. Silveira, V.L. Bueno, and C.P. Bergmann. 2008. "H2SO4/HNO3/HCl—Functionalization and its effect on dispersion of carbon nanotubes in aqueous media." *Applied Surface Science* no. 255 (5):2485–2489. doi: 10.1016/j.apsusc.2008.07.144.

Ozden, S., L. Ge, T.N. Narayanan, A.H.C. Hart, H. Yang, S. Sridhar, R. Vajtai, and P.M. Ajayan. 2014. "Anisotropically functionalized carbon nanotube array based hygroscopic scaffolds." *ACS Applied Materials and Interfaces* no. 6 (13):10608–10613. doi: 10.1021/am5022717.

Pantarotto, D., R. Singh, D. McCarthy, M. Erhardt, J.-P. Briand, M. Prato, K. Kostarelos, and A. Bianco. 2004. "Functionalized carbon nanotubes for plasmid DNA gene delivery." *Angewandte Chemie* no. 116 (39):5354–5358. doi: 10.1002/ange.200460437.

Park, K.C., T. Hayashi, H. Tomiyasu, M. Endo, and M.S. Dresselhaus. 2005. "Progressive and invasive functionalization of carbon nanotube sidewalls by diluted nitric acid under supercritical conditions." *Journal of Materials Chemistry* no. 15 (3):407. doi: 10.1039/b411221k.

Park, S., M. Vosguerichian, and Z. Bao. 2013. "A review of fabrication and applications of carbon nanotube film-based flexible electronics." *Nanoscale* no. 5 (5):1727. doi: 10.1039/c3nr33560g.

Park, Y.S., K.S. Kim, H.J. Jeong, W.S. Kim, J.M. Moon, K.H. An, D.J. Bae, Y.S. Lee, G.-S. Park, and Y.H. Lee. 2002. "Low pressure synthesis of single-walled carbon nanotubes by arc discharge." *Synthetic Metals* no. 126 (2–3):245–251. doi: 10.1016/s0379-6779(01)00563-x.

Piazza, F., G. Morell, J. Beltran-Huarac, G. Paredes, M. Ahmadi, and M. Guinel. 2014. "Carbon nanotubes coated with diamond nanocrystals and silicon carbide by hot-filament chemical vapor deposition below 200°C substrate temperature." *Carbon* no. 75:113–123. doi: 10.1016/j.carbon.2014.03.043.

Podesta, J.E., K.T. Al-Jamal, M.A. Herrero, B. Tian, H. Ali-Boucetta, V. Hegde, A. Bianco, M. Prato, and K. Kostarelos. 2009. "Antitumor activity and prolonged survival by carbon-nanotube-mediated therapeutic siRNA silencing in a human lung xenograft model." *Small*. doi: 10.1002/smll.200801572.

Pumera, M. 2009. "The electrochemistry of carbon nanotubes: Fundamentals and applications." *Chemistry – A European Journal* no. 15 (20):4970–4978. doi: 10.1002/chem.200900421.

Ran, M., W. Sun, Y. Liu, W. Chu, and C. Jiang. 2013. "Functionalization of multi-walled carbon nanotubes using water-assisted chemical vapor deposition." *Journal of Solid State Chemistry* no. 197:517–522. doi: 10.1016/j.jssc.2012.08.014.

Reddy, A.L.M., A. Srivastava, S.R. Gowda, H. Gullapalli, M. Dubey, and P.M. Ajayan. 2010. "Synthesis of nitrogen-doped graphene films for lithium battery application." *ACS Nano* no. 4 (11):6337–6342. doi: 10.1021/nn101926g.

Ren, L., C.L. Pint, L.G. Booshehri, W.D. Rice, X. Wang, D.J. Hilton, K. Takeya, I. Kawayama, M. Tonouchi, R.H. Hauge, and J. Kono. 2009. "Carbon nanotube terahertz polarizer." *Nano Letters* no. 9 (7):2610–2613. doi: 10.1021/nl900815s.

Richard, C. 2003. "Supramolecular self-assembly of lipid derivatives on carbon nanotubes." *Science* no. 300 (5620):775–778. doi: 10.1126/science.1080848.

Roy, D., N. Tiwari, K. Mukhopadhyay, T. Shripathi, and A.K. Saxena. 2014. "The role of functional moieties on carbon nanotube surfaces in solar energy conversion." *ChemPhysChem* no. 15 (17):3839–3847. doi: 10.1002/cphc.201402268.

Ryu, S., P. Lee, J.B. Chou, R. Xu, R. Zhao, A.J. Hart, and S.-G. Kim. 2015. "Extremely elastic wearable carbon nanotube fiber strain sensor for monitoring of human motion." *ACS Nano* no. 9 (6):5929–5936. doi: 10.1021/acsnano.5b00599.

Saghafi, M., F. Mahboubi, S. Mohajerzadeh, and R. Holze. 2014. "Preparation of co-ni oxide/vertically aligned carbon nanotube and their electrochemical performance in supercapacitors." *Materials and Manufacturing Processes* no. 30 (1):70–78. doi: 10.1080/10426914.2014.952026.

Sahoo, N.G., H. Bao, Y. Pan, M. Pal, M. Kakran, H.K. Cheng, L. Li, and L.P. Tan. 2011. "Functionalized carbon nanomaterials as nanocarriers for loading and delivery of a poorly water-soluble anticancer drug: A comparative study." *Chemical Communications (Camb)* no. 47 (18):5235–7. doi: 10.1039/c1cc00075f.

Saifuddin, N., A.Z. Raziah, and A.R. Junizah. 2013. "Carbon nanotubes: A review on structure and their interaction with proteins." *Journal of Chemistry* no. 2013:1–18. doi: 10.1155/2013/676815.

Saito, Y., T. Nakahira, and S. Uemura. 2003. "Growth conditions of double-walled carbon nanotubes in arc discharge." *The Journal of Physical Chemistry B* no. 107 (4):931–934. doi: 10.1021/jp021367o.

Samorì, C., H. Ali-Boucetta, R. Sainz, C. Guo, F.M. Toma, C. Fabbro, T. da Ros, M. Prato, K. Kostarelos, and A. Bianco. 2010. "Enhanced anticancer activity of multi-walled carbon nanotube–methotrexate conjugates using cleavable linkers." *Chemical Communications* no. 46 (9):1494–1496. doi: 10.1039/b923560d.

Santana, C.I., L.M. Hoyos, J.F. Pérez, J. Bustamante, and A.G. García. 2017. "A novel functionalization method for carbon nanotubes to repel ox-LDL in treatments after stent placement." *Materials Science and Engineering: C* no. 79:30–36. doi: 10.1016/j.msec.2017.05.002.

Seno, M., X. Yang, S. Yang, H. Chai, Z. Yang, R.J. Lee, W. Liao, and L. Teng. 2015. "A novel isoquinoline derivative anticancer agent and its targeted delivery to tumor cells using transferrin-conjugated liposomes." *Plos One* no. 10 (8):e0136649. doi: 10.1371/journal.pone.0136649.

Shah, K.A. and B.A. Tali. 2016. "Synthesis of carbon nanotubes by catalytic chemical vapour deposition: A review on carbon sources, catalysts and substrates." *Materials Science in Semiconductor Processing* no. 41:67–82. doi: 10.1016/j.mssp.2015.08.013.

Shanmugam, S. and A. Gedanken. 2006. "Generation of hydrophilic, bamboo-shaped multiwalled carbon nanotubes by solid-state pyrolysis and its electrochemical studies." *The Journal of Physical Chemistry B* no. 110 (5):2037–2044. doi: 10.1021/jp055749g.

Shi, X., A. von dem Bussche, R.H. Hurt, A.B. Kane, and H. Gao. 2011. "Cell entry of one-dimensional nanomaterials occurs by tip recognition and rotation." *Nature Nanotechnology* no. 6 (11):714–719. doi: 10.1038/nnano.2011.151.

Shi, Z., Y. Lian, X. Zhou, Z. Gu, Y. Zhang, S. Iijima, H. Li, K. To Yue, and S.-L. Zhang. 1999. "Production of single-wall carbon nanotubes at high pressure." *The Journal of Physical Chemistry B* no. 103 (41):8698–8701. doi: 10.1021/jp991531g.

Shinohara, H. 2000. "Endohedral metallofullerenes." *Reports on Progress in Physics* no. 63 (6):843–892. doi: 10.1088/0034-4885/63/6/201.

Siu, K.S., D. Chen, X. Zheng, X. Zhang, N. Johnston, Y. Liu, K. Yuan, J. Koropatnick, E.R. Gillies, and W.-P. Min. 2014. "Non-covalently functionalized single-walled carbon nanotube for topical siRNA delivery into melanoma." *Biomaterials* no. 35 (10):3435–3442. doi: 10.1016/j.biomaterials.2013.12.079.

Smith, B.W. and D.E. Luzzi. 2000. "Formation mechanism of fullerene peapods and coaxial tubes: A path to large scale synthesis." *Chemical Physics Letters* no. 321 (1–2):169–174. doi: 10.1016/s0009-2614(00)00307-9.

Song, X., Y. Liu, and J. Zhu. 2007. "Multi-walled carbon nanotubes produced by hydrogen DC arc discharge at elevated environment temperature." *Materials Letters* no. 61 (2):389–391. doi: 10.1016/j.matlet.2006.04.068.

Su, Y., Z. Yang, H. Wei, E.S.-W. Kong, and Y. Zhang. 2011. "Synthesis of single-walled carbon nanotubes with selective diameter distributions using DC arc discharge under CO mixed atmosphere." *Applied Surface Science* no. 257 (7):3123–3127. doi: 10.1016/j.apsusc.2010.10.127.

Su, Y., Y. Zhang, H. Wei, B. Qian, Z. Yang, and Y. Zhang. 2012. "Length-controlled synthesis of single-walled carbon nanotubes by arc discharge with variable cathode diameters." *Physica E: Low-dimensional Systems and Nanostructures* no. 44 (7–8):1548–1551. doi: 10.1016/j.physe.2012.03.025.

Sun, D.-M., C. Liu, W.-C. Ren, and H.-M. Cheng. 2013. "A review of carbon nanotube- and graphene-based flexible thin-film transistors." *Small* no. 9 (8):1188–1205. doi: 10.1002/smll.201203154.

Sun, D.-M., M.Y. Timmermans, Y. Tian, A.G. Nasibulin, E.I. Kauppinen, S. Kishimoto, T. Mizutani, and Y. Ohno. 2011. "Flexible high-performance carbon nanotube integrated circuits." *Nature Nanotechnology* no. 6 (3):156–161. doi: 10.1038/nnano.2011.1.

Sun, J., S. Kormakov, Y. Liu, Y. Huang, D. Wu, and Z. Yang. 2018. "Recent progress in metal-based nanoparticles mediated photodynamic therapy." *Molecules* no. 23 (7):1704. doi: 10.3390/molecules23071704.

Sun, J., H. Li, Y. Huang, X. Zheng, Y. Liu, J. Zhuang, and D. Wu. 2019. "Simple and affordable way to achieve polymeric superhydrophobic surfaces with biomimetic hierarchical roughness." *ACS Omega* no. 4 (2):2750–2757. doi: 10.1021/acsomega.8b03138.

Sun, J., J. Shen, S. Chen, M. Cooper, H. Fu, D. Wu, and Z. Yang. 2018. "Nanofiller reinforced biodegradable PLA/PHA composites: Current status and future trends." *Polymers* no. 10 (5):505. doi: 10.3390/polym10050505.

Sun, J., D. Wu, Y. Liu, L. Dai, and C. Jiang. 2017. "Numerical simulation and experimental study of filling process of micro prism by isothermal hot embossing in solid-like state." *Advances in Polymer Technology*, no. 37 (6):1581–1591. doi: 10.1002/adv.21815.

Sun, J., Y. Zhao, Z. Yang, J. Shen, E. Cabrera, M.J. Lertola, W. Yang, D. Zhang, A. Benatar, J.M. Castro, D. Wu, and L.J. Lee. 2018. "Highly stretchable and ultrathin nanopaper composites for epidermal strain sensors." *Nanotechnology* no. 29 (35):355304. doi: 10.1088/1361-6528/aacc59.

Sun, J., J. Zhuang, H. Jiang, Y. Huang, X. Zheng, Y. Liu, and D. Wu. 2017. "Thermal dissipation performance of metal-polymer composite heat exchanger with V-shape microgrooves: A numerical and experimental study." *Applied Thermal Engineering* no. 121:492–500. doi: 10.1016/j.applthermaleng.2017.04.104.

Sun, J., J. Zhuang, J. Shi, S. Kormakov, Y. Liu, Z. Yang, and D. Wu. 2019. "Highly elastic and ultrathin nanopaper-based nanocomposites with superior electric and thermal characteristics." *Journal of Materials Science* no. 54 (11):8436–8449. doi: 10.1007/s10853-019-03472-1.

Sun, X., W. Bao, Y. Lv, J. Deng, and X. Wang. 2007. "Synthesis of high quality single-walled carbon nanotubes by arc discharge method in large scale." *Materials Letters* no. 61 (18):3956–3958. doi: 10.1016/j.matlet.2006.12.070.

Sun, Y., K. He, Z. Zhang, A. Zhou, and H. Duan. 2015. "Real-time electrochemical detection of hydrogen peroxide secretion in live cells by Pt nanoparticles decorated graphene–carbon nanotube hybrid paper electrode." *Biosensors and Bioelectronics* no. 68:358–364. doi: 10.1016/j.bios.2015.01.017.

Syrgiannis, Z., F. Hauke, J. Röhrl, M. Hundhausen, R. Graupner, Y. Elemes, and A. Hirsch. 2008. "Covalent sidewall functionalization of SWNTs by nucleophilic addition of lithium amides." *European Journal of Organic Chemistry* no. 2008 (15):2544–2550. doi: 10.1002/ejoc.200800005.

Takagi, D., Y. Homma, H. Hibino, S. Suzuki, and Y. Kobayashi. 2006. "Single-walled carbon nanotube growth from highly activated metal nanoparticles." *Nano Letters* no. 6 (12):2642–2645. doi: 10.1021/nl061797g.

Thess, A., R. Lee, P. Nikolaev, H. Dai, P. Petit, J. Robert, C. Xu, Y.H. Lee, S.G. Kim, A.G. Rinzler, D.T. Colbert, G.E. Scuseria, D. Tomanek, J.E. Fischer, and R.E. Smalley. 1996. "Crystalline ropes of metallic carbon nanotubes." *Science* no. 273 (5274):483–7.

Toma, F.M., A. Sartorel, M. Iurlo, M. Carraro, S. Rapino, L. Hoober-Burkhardt, T. Da Ros, M. Marcaccio, G. Scorrano, F. Paolucci, M. Bonchio, and M. Prato. 2011. "Tailored functionalization of carbon nanotubes for electrocatalytic water splitting and sustainable energy applications." *ChemSusChem* no. 4 (10):1447–1451. doi: 10.1002/cssc.201100089.

Tonkikh, A.A., V.I. Tsebro, E.A. Obraztsova, K. Suenaga, H. Kataura, A.G. Nasibulin, E.I. Kauppinen, and E.D. Obraztsova. 2015. "Metallization of single-wall carbon nanotube thin films induced by gas phase iodination." *Carbon* no. 94:768–774. doi: 10.1016/j.carbon.2015.07.062.

Umeyama, T., N. Tezuka, M. Fujita, S. Hayashi, N. Kadota, Y. Matano, and H. Imahori. 2008. "Clusterization, electrophoretic deposition, and photoelectrochemical properties of fullerene-functionalized carbon nanotube composites." *Chemistry – A European Journal* no. 14 (16):4875–4885. doi: 10.1002/chem.200702053.

Umeyama, T., N. Tezuka, M. Fujita, Y. Matano, N. Takeda, K. Murakoshi, K. Yoshida, S. Isoda, and H. Imahori. 2007. "Retention of intrinsic electronic properties of soluble single-walled carbon nanotubes after a significant degree of sidewall functionalization by the Bingel reaction." *The Journal of Physical Chemistry C* no. 111 (27):9734–9741. doi: 10.1021/jp071604t.

Vanhorenbeke, B., C. Vriamont, F. Pennetreau, M. Devillers, O. Riant, and S. Hermans. 2013. "Radical addition of xanthates on carbon nanotubes as an efficient covalent functionalization method." *Chemistry – A European Journal* no. 19 (3):852–856. doi: 10.1002/chem.201203207.

Wang, F., K.-J. Deng, L. Zhou, J.-B. Zhao, X.-H. Ke, and L.-L. Wen. 2012. "Improving the degree of functionalization and solubility of single-walled carbon nanotubes via covalent multiple functionalization." *Journal of Inorganic and Organometallic Polymers and Materials* no. 22 (5):1182–1188. doi: 10.1007/s10904-012-9689-5.

Wang, J., M. Zhu, R.A. Outlaw, X. Zhao, D.M. Manos, and B.C. Holloway. 2004. "Synthesis of carbon nanosheets by inductively coupled radio-frequency plasma enhanced chemical vapor deposition." *Carbon* no. 42 (14):2867–2872. doi: 10.1016/j.carbon.2004.06.035.

Wang, Q., and B. Arash. 2014. "A review on applications of carbon nanotubes and graphenes as nano-resonator sensors." *Computational Materials Science* no. 82:350–360. doi: 10.1016/j.commatsci.2013.10.010.

Wang, S.C., K.S. Chang, and C.J. Yuan. 2009. "Enhancement of electrochemical properties of screen-printed carbon electrodes by oxygen plasma treatment." *Electrochimica Acta* no. 54 (21):4937–4943. doi: 10.1016/j.electacta.2009.04.006.

Wang, X.-D., K. Vinodgopal, and G.-P. Dai. 2019. "Synthesis of carbon nanotubes by catalytic chemical vapor deposition." doi: 10.5772/intechopen.86995.

Wang, Y., Z. Iqbal, and S.V. Malhotra. 2005. "Functionalization of carbon nanotubes with amines and enzymes." *Chemical Physics Letters* no. 402 (1–3):96–101. doi: 10.1016/j.cplett.2004.11.099.

Wei, Z., M. Kondratenko, L.H. Dao, and D.F. Perepichka. 2006. "Rectifying diodes from asymmetrically functionalized single-wall carbon nanotubes." *Journal of the American Chemical Society* no. 128 (10):3134–3135. doi: 10.1021/ja053950z.

Wu, D., X. Gao, J. Sun, D. Wu, Y. Liu, S. Kormakov, X. Zheng, L. Wu, Y. Huang, and Z. Guo. 2017. "Spatial confining forced network assembly for preparation of high-performance conductive polymeric composites." *Composites Part A: Applied Science and Manufacturing* no. 102:88–95. doi: 10.1016/j.compositesa.2017.07.027.

Wu, D., J. Sun, Y. Liu, Z. Yang, H. Xu, X. Zheng, and P. Gou. 2017. "Rapid fabrication of microstructure on PMMA substrate by the plate to plate transition-spanning isothermal hot embossing method nearby glass transition temperature." *Polymer Engineering and Science* no. 57 (3):268–274. doi: 10.1002/pen.24408.

Wu, H., J. Zhu, Y. Huang, D. Wu, and J. Sun. 2018. "Microfluidic-based single-cell study: Current status and future perspective." *Molecules* no. 23 (9):2347. doi: 10.3390/molecules23092347.

Wu, X., H. Shi, H. Zhang, X. Wang, Y. Yang, C. Yu, C. Hao, J. Du, H. Hu, and S. Yang. 2014. "Prostate stem cell antigen antibody-conjugated multiwalled carbon nanotubes for targeted ultrasound imaging and drug delivery." *Biomaterials* no. 35 (20):5369–5380. doi: 10.1016/j.biomaterials.2014.03.038.

Wu, Z., W. Feng, Y. Feng, Q. Liu, X. Xu, T. Sekino, A. Fujii, and M. Ozaki. 2007. "Preparation and characterization of chitosan-grafted multiwalled carbon nanotubes and their electrochemical properties." *Carbon* no. 45 (6):1212–1218. doi: 10.1016/j.carbon.2007.02.013.

Xia, W., C. Jin, S. Kundu, and M. Muhler. 2009. "A highly efficient gas-phase route for the oxygen functionalization of carbon nanotubes based on nitric acid vapor." *Carbon* no. 47 (3):919–922. doi: 10.1016/j.carbon.2008.12.026.

Xie, J., Z. Yang, C. Zhou, J. Zhu, R.J. Lee, and L. Teng. 2016. "Nanotechnology for the delivery of phytochemicals in cancer therapy." *Biotechnology Advances* no. 34 (4):343–353. doi: 10.1016/j.biotechadv.2016.04.002.

Xing, Y., L. Li, C.C. Chusuei, and R.V. Hull. 2005. "Sonochemical oxidation of multiwalled carbon nanotubes." *Langmuir* no. 21 (9):4185–4190. doi: 10.1021/la047268e.

Xu, M., D.N. Futaba, T. Yamada, M. Yumura, and K. Hata. 2010. "Carbon nanotubes with temperature-invariant viscoelasticity from –196 to 1000 C." *Science* no. 330 (6009):1364–1368. doi: 10.1126/science.1194865.

Yang, L., Z. Shi, and W. Yang. 2015. "Polypyrrole directly bonded to air-plasma activated carbon nanotube as electrode materials for high-performance supercapacitor." *Electrochimica Acta* no. 153:76–82. doi: 10.1016/j.electacta.2014.11.146.

Yang, X., S.E. Skrabalak, Z.-Y. Li, Y. Xia, and L.V. Wang. 2007. "Photoacoustic tomography of a rat cerebral cortex in vivo with Au nanocages as an optical contrast agent." *Nano Letters* no. 7 (12):3798–3802. doi: 10.1021/nl072349r.

Yang, Z., J. Xie, J. Zhu, C. Kang, C. Chiang, X. Wang, X. Wang, T. Kuang, F. Chen, Z. Chen, A. Zhang, B. Yu, R.J. Lee, L. Teng, and L.J. Lee. 2016. "Functional exosome-mimic for delivery of siRNA to cancer: In vitro and in vivo evaluation." *Journal of Controlled Release* no. 243:160–171. doi: 10.1016/j.jconrel.2016.10.008.

Yang, Z., L. Chang, W. Li, and J. Xie. 2015. "Novel biomaterials and biotechnology for nanomedicine." *European Journal of BioMedical Research* no. 1 (3):1. doi: 10.18088/ejbmr.1.3.2015.pp1-2.

Yang, Z., X. Wang, X. Huang, J. Xie, and C. Zhou. 2016. "Nanotechnology in gene delivery: Pharmacokinetic and pharmacodynamic perspectives." no. 1:295–326. doi: 10.1142/9789813202528_0008.

Yekymov, E., C. Bounioux, R. Itzhak-Cohen, L. Zeiri, E. Shahnazaryan, E.A. Katz, and R. Yerushalmi-Rozen. 2017. "All carbon non-covalent exohedral hybrids: C 60 aggregates on nanotube networks." *Journal of Energy Chemistry*. doi: 10.1016/j.jechem.2017.10.035.

Yinghuai, Z., A.T. Peng, K. Carpenter, J.A. Maguire, N.S. Hosmane, and M. Takagaki. 2005. "Substituted carborane-appended water-soluble single-wall carbon nanotubes: New approach to boron neutron capture therapy drug delivery." *Journal of the American Chemical Society* no. 127 (27):9875–9880. doi: 10.1021/ja0517116.

Yoshida, H., T. Sugai, and H. Shinohara. 2008. "Fabrication, purification, and characterization of double-wall carbon nanotubes via pulsed arc discharge." *The Journal of Physical Chemistry C* no. 112 (50):19908–19915. doi: 10.1021/jp806529v.

Yu, H., Y. Jin, F. Peng, H. Wang, and J. Yang. 2008. "Kinetically controlled side-wall functionalization of carbon nanotubes by nitric acid oxidation." *The Journal of Physical Chemistry C* no. 112 (17):6758–6763. doi: 10.1021/jp711975a.

Yu, R., L. Chen, Q. Liu, J. Lin, K.-L. Tan, S.C. Ng, H.S.O. Chan, G.-Q. Xu, and T.S.A. Hor. 1998. "Platinum deposition on carbon nanotubes via chemical modification." *Chemistry of Materials* no. 10 (3):718–722. doi: 10.1021/cm970364z.

Yudasaka, M., Y. Kasuya, F. Kokai, K. Takahashi, M. Takizawa, S. Bandow, and S. Iijima. 2002. "Causes of different catalytic activities of metals in formation of single-wall carbon nanotubes." *Applied Physics A: Materials Science & Processing* no. 74 (3):377–385. doi: 10.1007/s003390101070.

Yudasaka, M., F. Kokai, K. Takahashi, R. Yamada, N. Sensui, T. Ichihashi, and S. Iijima. 1999. "Formation of single-wall carbon nanotubes: Comparison of CO_2 laser ablation and Nd:YAG laser ablation." *The Journal of Physical Chemistry B* no. 103 (18):3576–3581. doi: 10.1021/jp990072g.

Zhang, H.F., K. Maslov, G. Stoica, and L.V. Wang. 2006. "Functional photoacoustic microscopy for high-resolution and noninvasive in vivo imaging." *Nature Biotechnology* no. 24 (7):848–851. doi: 10.1038/nbt1220.

Zhang, L., J. Yang, C.L. Edwards, L.B. Alemany, V.N. Khabashesku, and A.R. Barron. 2005. "Diels–Alder addition to fluorinated single walled carbon nanotubes." *Chemical Communications* (26):3265. doi: 10.1039/b500125k.

Zhang, M. 2005. "Strong, transparent, multifunctional, carbon nanotube sheets." *Science* no. 309 (5738):1215–1219. doi: 10.1126/science.1115311.

Zhang, X., L. Meng, Q. Lu, Z. Fei, and P.J. Dyson. 2009. "Targeted delivery and controlled release of doxorubicin to cancer cells using modified single wall carbon nanotubes." *Biomaterials* no. 30 (30):6041–6047. doi: 10.1016/j.biomaterials.2009.07.025.

Zhang, Y. 2012. "Synthesis of few-walled carbon nanotube–Rh nanoparticles by arc discharge: Effect of selective oxidation." *Materials Characterization* no. 68:102–109. doi: 10.1016/j.matchar.2012.03.013.

Zhao, J., J. Zhang, Y. Su, Z. Yang, L. Wei, and Y. Zhang. 2012. "Synthesis of straight multi-walled carbon nanotubes by arc discharge in air and their field emission properties." *Journal of Materials Science* no. 47 (18):6535–6541. doi: 10.1007/s10853-012-6583-z.

Zhao, T. and Y. Liu. 2004. "Large scale and high purity synthesis of single-walled carbon nanotubes by arc discharge at controlled temperatures." *Carbon* no. 42 (12–13):2765–2768. doi: 10.1016/j.carbon.2004.05.033.

Zhao, T., Y. Liu, and J. Zhu. 2005. "Temperature and catalyst effects on the production of amorphous carbon nanotubes by a modified arc discharge." *Carbon* no. 43 (14):2907–2912. doi: 10.1016/j.carbon.2005.06.005.

Zhong, W. and J.P. Claverie. 2013. "Probing the carbon nanotube-surfactant interaction for the preparation of composites." *Carbon* no. 51:72–84. doi: 10.1016/j.carbon.2012.08.014.

Zhuang, J., D.M. Wu, H. Xu, Y. Huang, Y. Liu, and J.Y. Sun. 2019. "Edge effect in hot embossing and its influence on global pattern replication of polymer-based microneedles." *International Polymer Processing* no. 34 (2):231–238. doi: 10.3139/217.3726.

Zhuang, J., W. Hu, Y. Fan, J. Sun, X. He, H. Xu, Y. Huang, and D. Wu. 2018. "Fabrication and testing of metal/polymer microstructure heat exchangers based on micro embossed molding method." *Microsystem Technologies* no. 25 (2):381–388. doi: 10.1007/s00542-018-3988-x.

Zou, G.F., H.M. Luo, S. Baily, Y.Y. Zhang, N.F. Haberkorn, J. Xiong, E. Bauer, T.M. McCleskey, A.K. Burrell, L. Civale, Y.T. Zhu, J.L. MacManus-Driscoll, and Q.X. Jia. 2011. "Highly aligned carbon nanotube forests coated by superconducting NbC." *Nature Communications* no. 2:428. doi: 10.1038/ncomms1438.

7

Carbon Dots: Scalable Synthesis, Physicochemical Properties, and Biomedical Application

Savita Chaudhary and Pooja Chauhan

CONTENTS

7.1 Introduction

Carbon is one of the most significant elements that constitute living beings in all types of organic life (Zheng, 2018). In addition, carbon-based particles in the nanometric range play a central role in the development of nanomaterials in three-dimensional materials to one-dimensional carbon dots (Das, 2018). Carbon-based nanomaterials have been broadly investigated due to their exceptionally tiny size with large surface-to-volume ratio. These physical aspects have mainly contributed to the strengthening of the physical and chemical properties of carbon-based nanoparticles (Isnaeni, 2018). Carbon-based nanomaterials have set a new pathway, encouraging scientists to investigate their potential role in overcoming the drawbacks of conventional materials (Momper, 2018). Within the diverse range of carbon nanoparticles (Figure 7.1), carbon dots are considered as one of the potential candidates of the carbon family. Carbon dots are a class of zero-dimensional fluorescent discrete quasi-spherical nanoparticles having a size lower than 10 nm. Carbon dots are emerging as the rising star of carbon nanomaterials (Jun, 2018). They have displayed outstanding optical properties as conventional quantum dots such as superior photo stability, extensive excitation and emission spectra, and strokes shift. However, in comparison with quantum dots, they acquire minor or no toxicity and have enhanced composition; moreover, their admirable biocompatibility excludes lethal heavy-metal elements (Siddique, 2018).

Carbon dots are mainly defined as particles smaller than 10 nm with excellent qualities including superior conductivity, higher chemical permanence, environmental compatibility, diverse range of photoluminescence activity, and easy preparation methodologies, which make them an effective material in several classes of applications (Tuerhong, 2017). Hence, it can be said that carbon dots are evolving as a persuasive substitute to conventional metal-based semiconductor quantum dots as they possess various advantages over conventional materials, such as prominent biocompatibility, colourful photoluminescence, and, most importantly, low cost of fabrication. Their applications have been decided by their various properties. They are also being used in the field of super capacitors. Moreover, their unique ability for storing and transporting electrons has been utilized in making solar cells (Sun, 2017). Carbon dots were unintentionally exposed by Xu et al. in 2004 throughout the fine-tuning of single-walled carbon nanotubes. This discovery has promoted the utilization of carbon dots in diverse fields. Carbon dots were found to be different from semiconducting quantum dots (SQDs) and graphene-based quantum dots (GQDs) (Xu, 2004). In the case of carbon dots, there was a mixture of crystalline structure along with amorphous carbon (Kuai, 2018). Therefore, it is well documented in the literature that the nanoparticles of carbon with crystalline core which have sp^2 and sp^3 hybridization are called carbon quantum dots (CQDs) (Sarkar, 2016). On the other hand, the nanostructure materials made up of a disordered structure mixture of sp^3 carbons are termed carbon nanodots (CNDs) (Omer, 2018). Carbon dots have displayed two types of lattice spacing in their crystalline structure, i.e., 0.34 nm and 0.21 nm related to (002) and (100) facets in graphite, respectively (Yang, 2011). The comparison between carbon dots and other carbon-based nanoparticles is illustrated in Table 7.1. The size restriction within the range of 2–10 nm has appreciably influenced the emission intensity of the carbon dots and made them effective for fluorescence-sensing applications. Due to their exceptional physicochemical and electronic properties, carbon dots were explored as biosensors, gene transmitters, drug carriers, and bioimaging probes (Zhu, 2013). As a result, it is right to say that carbon dots are one of the most effective members of carbon-based materials and possess a remarkable position in the field of remediation (Singh, 2018). The excitation-dependent fluorescence properties in carbon dots have further enhanced the scope of their utilization in the field of optical imaging (Sie, 2018).

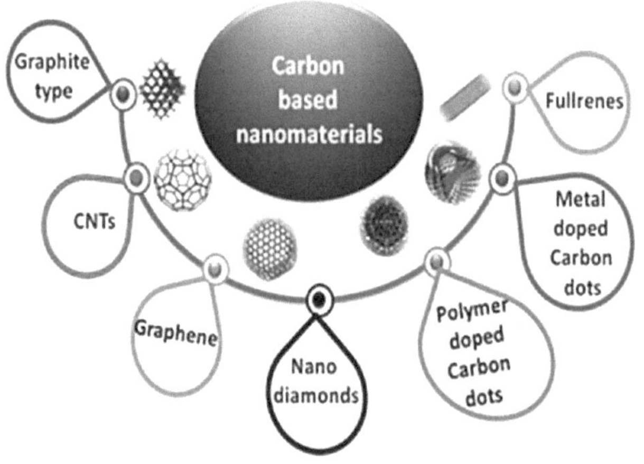

FIGURE 7.1 Various types of carbon-based nanostructures.

7.2 Characteristic Properties of Carbon Dots

The fluorescent characteristics of carbon dots are mainly dependent on the size of the formed particles. There is a significant enhancement in the emission peak when increasing the size of carbon dots (Athar, 1999; Jana, 2018). Along with the size, the local environment around the formed particles, solvent media, and mode of preparation also affect the PL intensities of carbon dots (Firdaus, 2018). The origin of the PL emission in the formed particles was explained in several ways by different researchers in the literature (Kejik, 2016). Some explanations have attributed it to the quantum confinement effect in carbon dots (Zhu, 2015). In addition, the existing valence states, the nature of the functional groups, and π conjugation also influenced the photoluminescence properties of Carbon dots (Jin, 2013). Carbon dots have displayed strong optical absorption in the ultra violet region between 280 and 360 nm. The optical properties of the particles are significantly optimized by using the surface modification of the prepared carbon dots (Zhao, 2013).

For instance, when polymer molecules are used, the surface modification has a tendency to make the particles absorb in a visible region, i.e., 460–495 nm. This variation was associated with the non-uniform passivation of carbon dots by using polymer moieties. As a result, the particles were excited at a longer wavelength as compared to non-functionalized carbon dots (Huang, 2014). The mode of synthesis has also affected the absorption wavelength of formed nanoparticles. In addition to optimized absorption properties, these carbon dots have also displayed a phosphorescence behavior in aqueous media (Sun, 2013). For instance, PVA-functionalized carbon dots have shown phosphorescence in the presence of UV light. The existence of aromatic carbonyl groups in PVA has provided a triplet excited state for the phosphorescence (Wang, 2014). The main emission peaks associated with carbon dots are in the region of blue and green in the presence of UV light. The presence of a high dosage of UV and NIR radiations has the tendency to produce full-color light emission in carbon dots (Jiang, 2015). The estimation of radioactive materials was also performed by using the chemiluminescence properties of carbon dots (Tao, 2012). The emission intensity was easily optimized as a function of concentration of carbon dots within the limit of the absorption range of the particles. In certain cases, carbon dots have also possessed up-converted PL properties (Wang, 2011). In this case, particles absorbed light in the longer wavelength and had the tendency to emit in a shorter-wavelength region. This process possess mainly two or multiphoton active behavior and acts as better material in optics (Li, 2014). The PL spectra of carbon dots are mainly affected by the nature of the solvent media. The favorable fluorescence properties of carbon dots have displayed enormous potential for their use in analytical chemistry, especially in the environment and in biological-contaminant sensing. Khan et al. found that the photoluminescence spectra of carbon dots were inconsistently widened because of the low relaxation rate of the solvent molecules around the formed carbon dots (Khan et al., 2015). The excitation-dependent emission spectra violate the Kasha–Vavilov rule. The results were further verified with the help of time-resolved measurements. It was found that the main factors, such as energy redeployment, relaxation speed of emitting states, and tendency of spectral migration of spectra with time are mainly responsible for the widening of the spectra (Figure 7.2). The colour emission was generated due to the relative amount of particles in excited states.

7.3 Synthesis and Application of Carbon Dots

Lin et al. synthesized carbon dots by using ash from burning paper and employed them to detect pesticide residues (Lin et al., 2018). This strategy was quite useful for the estimation of pesticides and highly effective for their detection in nano molar quantities. The fluorescence of carbon dots could be turned off in the presence of Fe^{3+}, which was obtained from the oxidization of Fe^{2+} in the presence of peroxide. The presence of pesticides could efficiently hinder the production of peroxide by damaging the activity of the acetyl cholinesterase enzymes. As a result, the fluorescence intensity of carbon dots turn on in the presence of pesticides. Currently, carbon dots possess significant importance in the area of clinical medicine and diagnostics. They might help in the treatment of cancer. Drug delivery vehicles are also prepared by using carbon dots (Karthik, 2013). Feng et al. developed PEG-coated carbon dots and used them in diagnostics. The prepared particles displayed pH-dependent dual response in the extracellular environment of tumors. The variations in the pH of the media triggered the release of anticancer drug from the carbon-dots-based carriers (Figure 7.3). The external coating of carbon dots was exposed while encountering the extracellular microenvironment of the tumor cell. The pH of the environment was around pH 6.5–6.8 as compared to the neutral pH of the particles. Drug imaging was also possible due to the multicolor emission of the carbon dots. The external coating provides a tumor-triggered targeting property to the particles at pH 6.8 and assists the internalization of cis-platin in tumor cells (Feng et al., 2016).

Carbon dots possess a prime position in regenerative tissue engineering. Their high biocompatibility as well as their association with biologically important materials is the major cause for their use in regenerative-tissue engineering (Harrison, 2007). Currently, such materials are emerging as competent, better-quality fluorophores for sensing harmful toxins (Wee, 2013). The employment of carbon dots in the environment science is well known in the literature (Yang et al., 2009). Therefore, the

TABLE 7.1
Contrasting Features of Carbon Dots and Other Carbon-Based Particles

Carbon-based nanomaterials	Distinctive features	Functions	Constraints	Ref
Carbon nanotubes 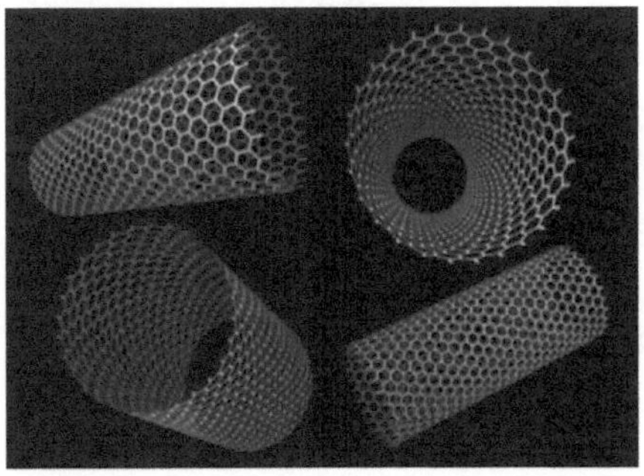	High strength, stiff nature, thermal stability, high electrical conductance, fibrous material	In the biomedical field, effectively combine with bioactive materials, effective therapeutic agent	Display toxicity to human keratinocytes, display cellular oxidative stress, mainly insoluble	(Meena, 2018; Zaporotskova, 2016)
Graphene 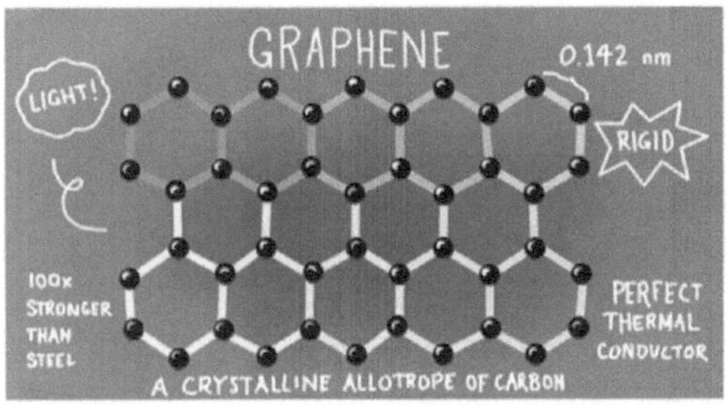 	Possess one atomic thick layer of sp^2-hybridized carbon sheet	Useful for the fabrication of materials with high mechanical strength and elasticity	The commercial availability of graphene is limited. Sensing efficiency is lower	(Ren, 2018; Park, 2018)
Carbon dots  	Have particles in zero dimensions with size under 10 nm	Helpful in fluorescence sensing, targeted drug delivery, bioimaging, and diagnostics	The long-term stability of particles is lower. Need extra functionalization for stabilization	(Aji, 2018; Zhang et al., 2018)

last decade has witnessed a great number of synthetic methodologies for the fabrication of carbon dots (Deng, 2014). Various methods have been employed for their synthesis. Out of all the approaches, the green method for the fabrication of carbon dots has received the greatest attention. This method reduces costs, time, and results in the high-yield formation of environment-friendly materials. Carbon dots are also relevant in food examination, which includes the detection of organic pollutants, the analysis of the nutritional components of food, and the detection of heavy metal ions. There exists a vast amount of literature on how to fabricate carbon dots by utilizing cost-effective and easy strategies. These methods have provided good control over the optical and fluorescence properties of particles (Namdari, 2017). In general, these synthetic strategies are divided into two main groups: top-down and bottom-up approaches, depending upon the source and the formed products. In the case of a top-down approach, the respective sources of carbonaceous materials are converted into carbon dots by the application of physicochemical methods involving hydrothermal, CVD, oxidation, and solvothermal techniques, beside arc discharge, ultrasonic synthesis, and laser ablations (Table 7.2). Therefore, carbon dots can also be signified as graphene nanodots and carbon nanodots, respectively, whereas the bottom-up approach mainly utilizes the application of molecular precursors for the fabrication of carbon dots by using glucose and sucrose through microwave synthesis (Ye, 2017; Peng, 2017; Jaleel et al., 2018). The top-down approach recurs to slicing or cutting a bulk substance to obtain nanosized particles, whereas the bottom-up approach is a method in which devices create themselves by self-assembly. Chemical synthesis is a good example of a bottom-up approach. The main techniques

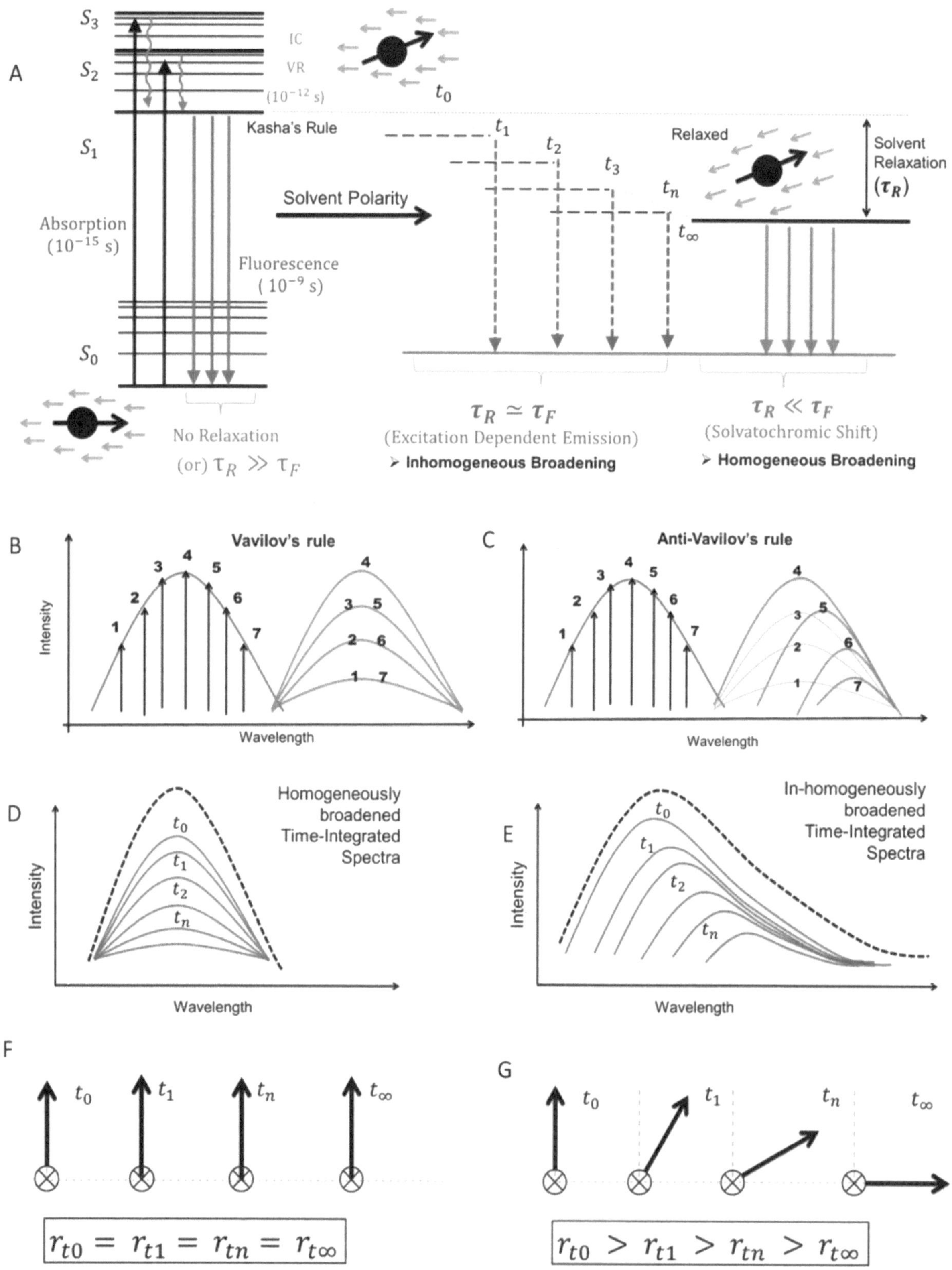

FIGURE 7.2 Schematic representation of different fluorescence stages: (A) Solvent relaxation study with fluorescence decay time disobeying the Kasha–Vavilov rule. (B) Process of emission through the Vavilov rule. (C) Disobedience to the Vavilov rule. (D) Time-incorporated spectra with homogenous broadening. (E) Migration of emission spectra towards larger wavelengths, resulting in homogenous broadening. (F, G) Emergence of depolarization or energy relaxation due to dipolar rotation. A higher value of depolarization leads to lower anisotropy, which results in red shift. Figure adapted from Khan, 2015 with permission from copyright (2015), American Chemical Society (Washington, DC, USA).

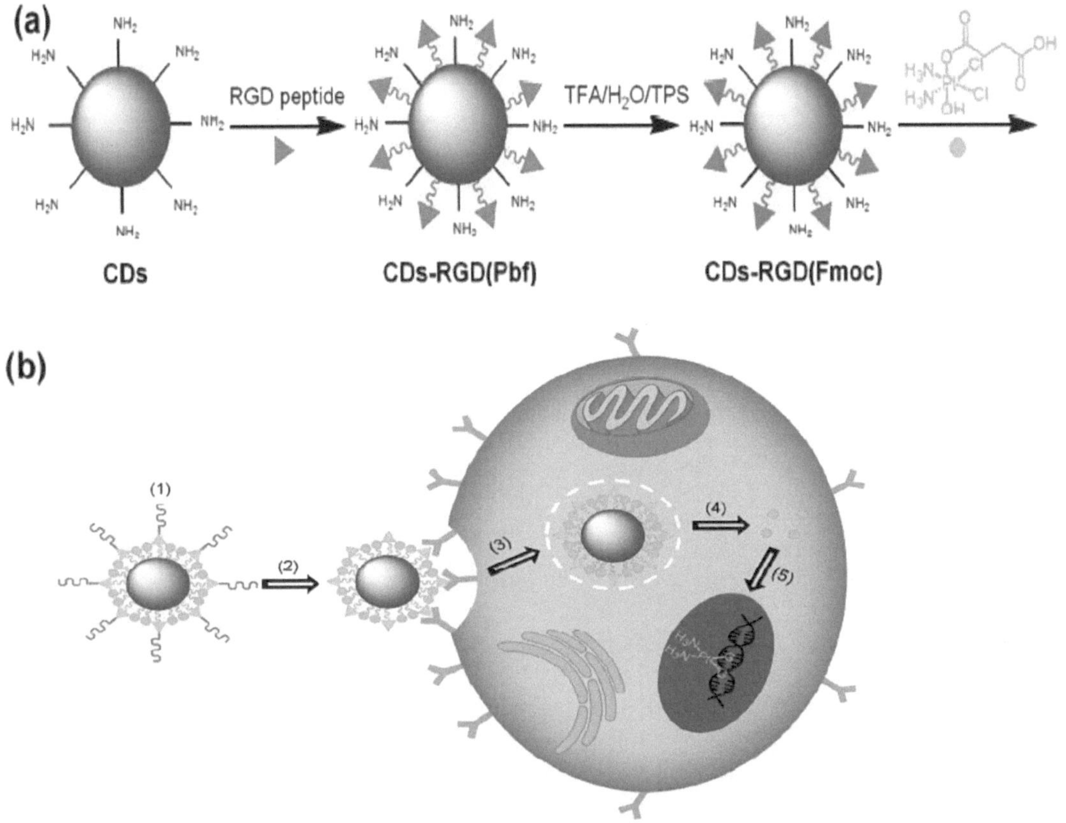

FIGURE 7.3 (a) Pictorial representation of the synthesis of carbon dot–based drug nanocarriers with tumor-triggered targeting properties. (b) Process of drug delivery by using drug nanocarriers. (1) Pathway of drug nanocarriers by multicolor fluorescent properties of carbon dots under regular conditions. (2) Tumor-targeted ligand with pH 6.8. (3) Ligand receptor binding for uptake activity by cancer cells. (4) Discharge of cis-platin (IV) from cis-platin. (5) Requisites of cis-platin with DNA. Figure adapted from Feng, 2016 with permission from copyright (2016), American Chemical Society (Washington, DC, USA).

employed for the bottom-up approach are microwaving-mediated synthesis and pyrolysis of starting materials. Electrochemical techniques were also utilized for the production of carbon dots with controlled size and shape. To enhance the scope of carbon dots in the biomedicinal field, surface-functionalized carbon dots were also prepared in large scale (Ng, 2018). This functionalization not only modulated the surface but also affected the solubilization rate of prepared carbon dots in aqueous media. The emission properties of prepared particles were also affected in the presence of external templating agents. The surface modification further affected the photostability of prepared carbon dots. The rate of photo-bleaching was controlled in the presence of external stabilising agents. The nature of bonding between the carbon dots and the templating agent further affects the stability rate of formed particles in aqueous media (Lin, 2018). The surface modification of carbon dots directly influences their potential role in drug delivery, bio-imaging, and target-based sensing of harmful analytes. The biocompatible covering over the surface of carbon dots has minimized their toxicity and enhanced their use in the medicinal field. The higher rate of water solubility of prepared particles has further affected their elimination rate from the living system. For instance, surface modification by using polymers has modulated their optical properties. Such kind of amendments in reaction to external chemical or mechanical agents have shown various sensing applications of carbon dots, especially if an external stimulus reaction happens in aqueous media. The doping of carbon dots has also produced white luminescence by regulating the equilibrium of blue-light-emitting carbon dots with green- and red-light-emitting lanthanide ions as dopants. (Figure 7.4). These doped carbon dots have showed chromic receptiveness towards numerous external agents, such as pH, organic moieties, heavy transition metal ions, and variations in temperature (Ding, 2016).

Zhi, et al. prepared malic-acid-functionalized Carbon dots. The prepared particles displayed photo blinking activities which made them superior as compared to commercially available organic dyes. The formed particles were considered as potential candidates for super-resolution fluorescence localization microscopy. At diverse ranges of external conditions, the particles have shown effective imaging of fixed and live trout-gill epithelial cells at high resolution (Zhi et al., 2018). The other major advantage of malic-acid functionalization is the non-observable "excitation wavelength-dependent" emission in carbon dots. This makes the particles act as a time-saving bio-imaging agent. The additional benefit is the easy separation of malic-acid-functionalized carbon dots by employing the reversed-phase silica gel column chromatographic technique. Such kind of separation has provided a clear view of the size effect over the optical properties of nanoparticles, including the band gap and photoluminescence activities of carbon dots (Figure 7.5).

TABLE 7.2

Different Schemes for the Separation of Carbon Dots

Fabrication process	Size	Color	Nature	Quantum yield	Advantages	Ref
Electrochemical process	3–5 nm	Green luminescence	Monodisperse. Shelve life 3 months without variations in physicochemical properties	26 %	Easy optimization of size and shape of particles. Good stability and higher dispersion ability in aqueous media	(Ming et al., 2012; Bao et al., 2011)
Chemical ablation process	>10 nm	Green luminescence	Polydisperse, show agglomeration after a few days	5–6 %	The precursors employed for the synthesis are easily accessible	(Shen, 2013; Bao, 2011)
Laser ablation	~10 nm	Blue luminescence	Polydisperse, quantum yield can be improved by using external dopant	4–10 %	The synthesis process is fast. Easily amendable photoluminescence properties	(Li, 2010)
Solvothermal method	10–15 nm	Green luminescence	Monodisperse	>15 %	Economical method of preparation with good control over size and shape; eco-friendly process and biocompatible particles	(Bhunia, 2013; Yang et al., 2012)
Microwave-assisted fabrication of carbon dots	>5 nm	Greenish-yellow luminescent	Monodisperse	11–25 %	The synthesis process is quite fast, scalable to a large extent, economical, and environmentally friendly	(Liu et al., 2014)
Plasma-induced pyrolysis	10–15 nm	----	Polydisperse	>15 %	Easily accessible precursors	(Wang, 2012)
Using natural bio-sources, such as bee pollen, biomass waste, and eutrophic algal blooms	5–10 nm >10 nm 5–10 nm	Bluish green, green-blue	Controlled size. Amendable size and controlled physicochemical properties. Polydisperse	------ 15–20 % >10 %	Easily accessible starting material. Utilizes the application of waste biomass. Non-toxic particles	(Zhang, 2016; Sahu, 2012; Ramanan, 2016)
Arc discharge method	>5 nm	Bluish green	Controlled size	>5 %	Optimized physicochemical properties	(Xu, 2004; Pirsaheb, 2018; Ravi and Vadukumpully, 2016; Sharma, 2018; Hoang, 2018)

Surface-functionalized carbon dots are also created to make the particles dispersible in aqueous media. The surface coating is mainly categorized into two types: soft and hard template covering over the surface of nanoparticles (Krueger, 2008). In the case of the soft-coating process, the formed nanostructures were functionalized with the mutual assemblage of monomer moieties of surfactant molecules (Zhang et al., 2017). The presence of precursor molecules is also not ruled out during the synthesis. The removal of surfactant molecules has the tendency to produce mesoporosity in the formed particles. The nanocasting methodology comes mainly under this category. These particles were prepared by using both natural and synthetic precursor during the synthesis. In the case of natural precursor strategies, mainly natural resources involving wheat, bajra, ragi, grass, sugar, jaggery, eggs, soya milk, juices, vegetable waste, bread, beverages, coffee seeds, paper, wood, biomass, bee pollens, etc., were employed as starting materials (Chaudhary, 2016; Ansi, 2018; Miao, 2016; Wang et al., 2018; Kim, 2018). The surface-engineered particles of carbon dots were also prepared with optimized physicochemical properties. Post-synthetic techniques such as dialysis, ultracentrifugation, filtration, and chromatography were also reported in the literature as getting the purest form of carbon dots (Zheng et al., 2018). The application of green methodologies for the production of carbon dots is highly recommended in the current scenario as compared to the commercial physical and chemical treatment processes (Tabaraki, 2018). These natural bio-sources have significantly reduced the usage of toxic precursor materials, and the formed carbon dots possess a higher order of biocompatibility with living beings as compared to other methods (Ensafi, 2017). For instance, Zhou et al. used watermelon peels for the synthesis of highly fluorescent particles of carbon dots with size ranges between 1 and 2 nm. The synthesis was done at a relatively low temperature and the

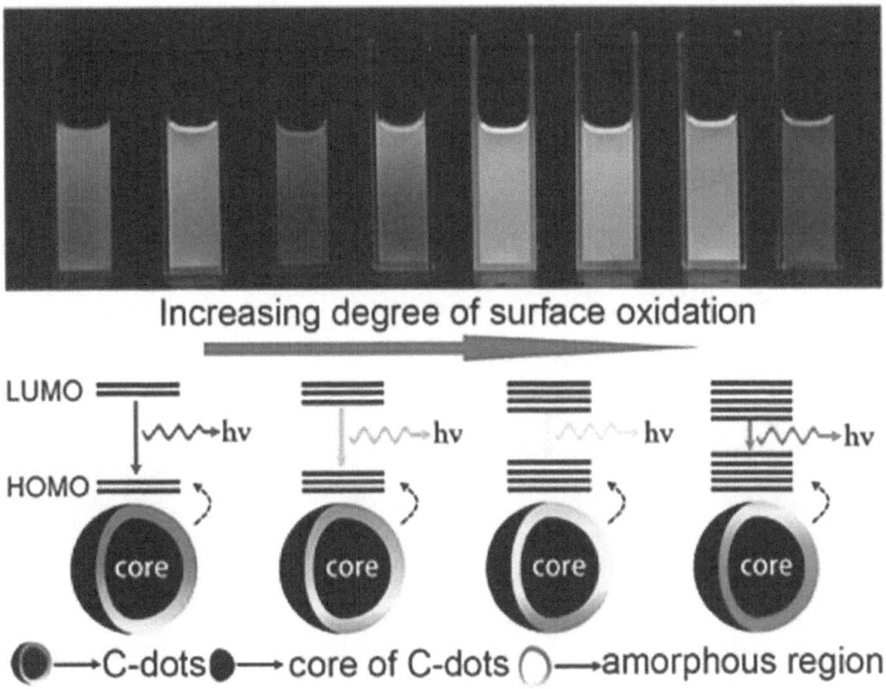

FIGURE 7.4 Different carbon dot samples under 365 nm UV light and model for photoluminescence of carbon dots with different degrees of oxidation. Figure adapted from Ding, 2016 with permission from copyright (2016), American Chemical Society (Washington, DC, USA).

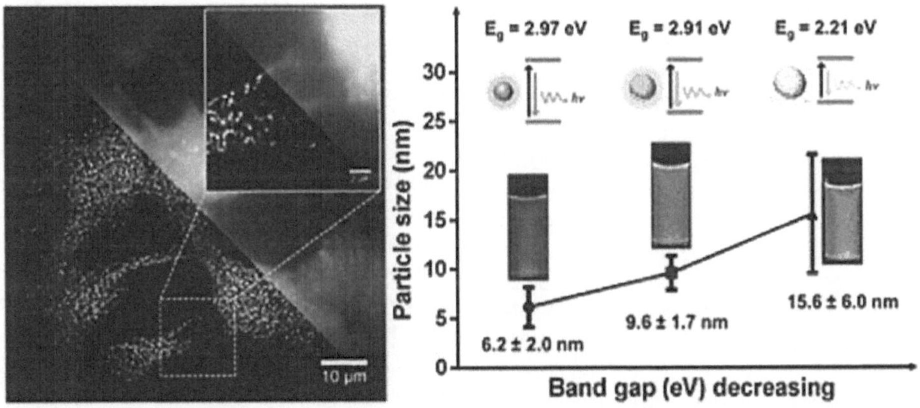

FIGURE 7.5 The band gap variation of malic acid–functionalized carbon dots. Figure adapted from Zhi et al., 2018 with permission from copyright (2018), American Chemical Society (Washington, DC, USA).

process was eco-friendly. The prepared particles displayed a significant role in the bioimaging of HeLa cells. The formed particles displayed bright luminescence at excitation wavelength of 488 nm and acted as efficient bio probes for tumor cells (Zhou et al., 2012).

The reactivity of carbon dots in the reaction media is a major issue that needs extensive investigation. The applications of carbon dots in medicines and bioimaging as well as in sensing have made researchers eager to understand their behaviour in the colloidal phase (Pirsaheb, 2018). It is necessary to understand the stability of particles in solvent media in order to use them while respecting the ecosystem. The toxicity caused by these particles for living beings is another sector which needs thorough investigation. It is necessary to revise the currently available knowledge of carbon dots and their environmental performance for the commercialization of these products (Ravi, 2016). Carbon dots are described commonly in terms of a cacogenic core with surface functional groups. The best properties associated with carbon dots are high water solubility, good conductivity, high stability, and optical properties comparable to those of quantum dots. They also show excellent biocompatibility. The presence of different functional groups associated with carbon dots affects the biocompatible nature and toxicity of the system. In order to fabricate biocompatible nanostructures of carbon dots which are effectively usable in medical applications and utilize less harmful chemical compounds, biocompatible agents such as fructose, glucose, and ascorbic acid might be used as a starting material for the synthesis of carbon dots. This methodology does not require any kind of external templating agent for the stabilization of prepared carbon dots. The prepared particles have also been used as effective drug nanocarriers for the drug DOx (Sharma, 2018).

Carbon dots have also affected the biological nitrogen fixation rate in living media. These carbon dots have a tendency

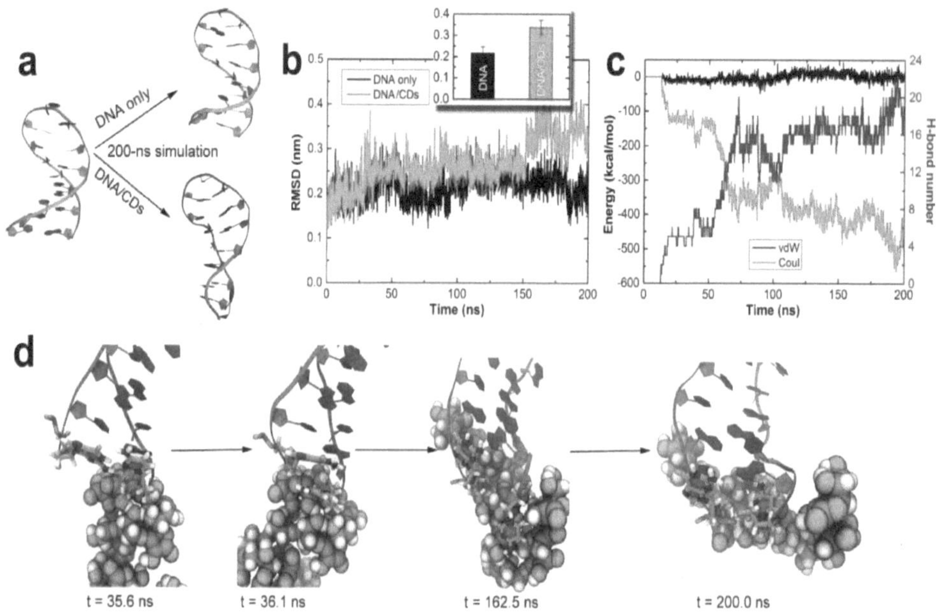

FIGURE 7.6 (a) The different configurations of DNA hairpin in control and carbon dots. (b) Time study of DNA in control and carbon dot system by utilizing the root mean square deviation method. (c) Representation of non-bonding interaction (black curve), coulomb interaction energy (light gray graph), and hydrogen bond (dark gray curve) established in carbon dots and DNA hairpin as a function of time. (d) Formation of some configurations from key time points for the representation of the denaturation process. Figure adapted from Li et al., 2018 with permission from copyright (2018), American Chemical Society (Washington, DC, USA).

to improve the nitrogen-fixing action of azotobacter chroococcum. The data has shown that the presence of 4 μg/mL of carbon dots can increase the activity to around 158 % as compared to the control. Carbon dots combine with the nitrogenise enzyme and influence the secondary structure of nitrogenise. This behavioural variation influences the electron transfer rate. The introduction of carbon dots has provided an effective and economically sound method for convalescing the biological nitrogen fixation in soil media and solved the problem of nitrogen deficiency on Earth. The determination of the pH in living cells is very useful to understand cellular functioning. Such kinds of measurements provide information about the early identification of diseases in living beings. In this regard, carbon dots have a significant role in the intracellular ratiometric sensing of pH variations. The occurrence of dual emission at 475 and 545 nm with one excitation wavelength in carbon dots has made them an effective agent for the preparation of a pH-sensitive sensor. N- or B-doped carbon dots have the potential to act as a pH and fluorescence sensor for a diverse range of analytes. The high range of quantum yield makes them an effective material for dual-sensing agents (Hoang, 2018). Their low-toxicity range and good stability make them effective in bioimaging applications.

Li et al. have used a single-step electrochemical process during the manufacturing of biocompatible carbon dots with low toxicity in addition to the degradable nature of vitamin C as starting material. The prepared carbon dots have displayed superior activities and wide-ranging antimicrobial and antimold action. The prepared particles were able to restrain the augmentation procedure of *B. subtilis*, S. *aureus*, *WL-6, Bacillus sp., E. coli*, ampicillin-resistant *E. coli*, *R. solani*, and *P. grisea* (Li et al., 2018). The retardation rate was mainly explained by the entrance of carbon dots inside the cell by diffusion phenomena (Figure 7.6).

7.4 Future Prospects of Carbon Dots

Carbon dots have the ability to annihilate the bacterial wall and combine to the DNA and RNA of the bacteria. The genetic processing of bacteria and fungi is affected, and finally the bacteria and fungi are exterminated at minute concentrations. In addition, carbon dots have the ability to cover the exterior cell wall, impose toxicity on the bacterial cells, and isolate the cells from the growing media. As a result, bacterial cells die due to the non-consumption of nutrients. The wall of the bacterial cell is also affected in the presence of particles, and cell growth is damaged. There has been a lot of research on the fabrication, properties, and application of carbon dots because of their exceptional optical and electronic properties. It is encouraging to witness the diversion of research interest in carbon dots away from traditional fields such as bioimaging, drug delivery, and biosensing. To date, numerous cost-effective, easy, large-scale, and size-controllable schemes have been reported for the production of carbon dots with various structures and compositions, despite the fact that some issues such as the defects of carbon dots have an important effect on their optical and electronic properties. Carbon dots are still up in the air and numerous stimulating challenges are still waiting to discover their enormous potential.

Conclusion

Carbon dots are considered as the rising star of the carbon nano world. They have received a lot of attention due to their excellent optical and physicochemical properties Various methods have been reported for the fabrication of carbon dots, out of which the green strategy is considered as the easiest and cheapest method. The emission properties of carbon dots can be augmented by using different modification methods, capping-agents, and dopants. Carbon dots are highly biocompatible and non-toxic in nature; for this reason, they have been further utilized in the fields of bioimaging and drug delivery, as well as in the area of toxin and chemical pollutant sensing. The enormous potential of carbon dots is still to be uncovered.

REFERENCES

Aji, M. P., Wati, A. L., Priyanto, A., et al. 2018. Polymer carbon dots from plastic waste upcycling. *Environ. Nanotech. Monit. Manage.* 9:136–40.

Ansi, V. A. and N. K. Renuka. 2018. Table sugar derived carbon dot – A naked eye sensor for toxic Pb^{2+} ions. *Sens. Actuators B: Chem.* 264:67–75.

Athar, H. 1999. Use of fluorescence enhancement technique to study bilirubin-albumin interaction. *Int. J. Biol. Macromol.* 25:353–58.

Bao, L., Zhang, Z. L., Tian, Z. Q., et al. 2011. Electrochemical tuning of luminescent carbon nanodots: From preparation to luminescence mechanism. *Adv. Mater.* 23:5801–06.

Bhunia, S. K. 2013. Carbon nanoparticle-based fluorescent bioimaging probes. *Sci. Rep.* 3:1–7.

Chaudhary, S. 2016. Potential prospects for carbon dots as a fluorescence sensing probe for metal ions. *RSC Adv.* 6:90526–36.

Das, R. 2018. Carbon quantum dots from natural resource: A review. *Mater. Today Chem.* 8:96–109.

Deng, Y. 2014. Environment-dependent photon emission from solid state carbon dots and its mechanism. *Nanoscale* 6:10388–93.

Ding, H. 2016. Full-color light-emitting carbon dots with a surface-state-controlled luminescence mechanism. *ACS Nano* 10:484–91.

Ensafi, A. A. 2017. A novel one-step and green synthesis of highly fluorescent carbon dots from saffron for cell imaging and sensing of prilocaine. *Sens. Actuators B: Chem.* 253:451–60.

Feng, T. 2016. Dual-responsive carbon dots for tumor extracellular microenvironment triggered targeting and enhanced anticancer drug. *ACS Appl. Mater. Interfaces* 8:18732–40.

Firdaus, F. 2018. Benzidine based fluorescent probe for the sensitive detection of heavy metal ions via chelation enhanced fluorescence mechanism—A multiplexed sensing platform. *J. Lumin.* 199:475–82.

Harrison, B. S., and A. Atala. 2007. Carbon nanotube applications for tissue engineering. *Biomaterials* 28:344–53.

Hoang, V. C. 2018. Coal derived carbon nanomaterials – Recent advances in synthesis and applications. *Appl. Mater. Today* 12:342–58.

Huang, J. J. 2014. An easy approach of preparing strongly luminescent carbon dots and their polymer based composites for enhancing solar cell efficiency. *Carbon* 70:190–98.

Isnaeni. 2018. Concentration effect on optical properties of carbon dots at room temperature. *J. Lumin.* 198:215–19.

Jaleel, A. J., and K. Pramod. 2018. Artful and multifaceted applications of carbon dot in biomedicine. *J. Control. Release* 269:302–21.

Jana, J. 2018. Fluorescence enhancement via varied long chain thiol stabilized gold nanoparticles: A study of far field effect. *Spectrochim. Acta, Part A: Mol. Biomol. Spectrosc.* 188:551–60.

Jiang, K., Sun, S., Zhang, L., et al. 2015. Red, green, and blue luminescence by carbon dots: Full-color emission tuning and multicolor cellular imaging. *Angew. Chem.* 54:5360–63.

Jin, S. H. 2013. Tuning the photoluminescence of graphene quantum dots through the charge transfer effect of functional groups. *ACS Nano* 7:1239–45.

Jun, L. Y., Mubarak, N. M., Yee, M. J., et al. 2018. An overview of functionalised carbon nanomaterials for organic pollutant removal. *J. Indus. Eng. Chem.* 67:175–86.

Karthik, S. 2013. Photoresponsive quinoline tethered fluorescent carbon dots for regulated anticancer drug delivery. *Chem. Comm.* 49:10471–73.

Kejik, Z., Kaplanek, R., Havlik, M., et al. 2016. Aluminium(III) sensing by pyridoxal hydrazone utilising the chelation enhanced fluorescence effect. *J. Lumin.* 180:269–72.

Khan, S. 2015. Time-resolved emission reveals ensemble of emissive states as the origin of multicolor fluorescence in carbon dots. *Nano Lett.* 15:8300–05.

Kim, M., Park, S. Y., Park, S. K., et al. 2018. Label-free fluorescent detection of alkaline phosphatase with vegetable waste-derived green carbon probes. *Sens. Actuators B: Chem.* 262:469–76.

Krueger, A. 2008. New carbon materials: Biological applications of functionalized nanodiamond materials. *Chem. Euro. J.* 14:1382–90.

Kuai, Y. 2018. A convenient way of activating carbon quantum dots and the efficient isolation. *Mater. Res. Bull.* 104:119–23.

Li, H., Huang, J., Song, Y., Zhang, M., et al. 2018. Degradable carbon dots with broad-spectrum antibacterial activity. *ACS Appl. Mater. Interfaces* 32:26936–46.

Li, X. 2010. Preparation of carbon quantum dots with tunable photoluminescence by rapid laser passivation in ordinary organic solvents. *Chem. Comm.* 47:932–34.

Li, X. 2014. Engineering surface states of carbon dots to achieve controllable luminescence for solid-luminescent composites and sensitive Be^{2+} detection. *Sci. Rep.* 4:1–8.

Lin, B. 2018. Modification free carbon dots as turn on fluorescence probe for the detection of organophosphorous pesticides. *Food Chem.* 245:1176–82

Lin, Y., Zheng, Y., Guo, Y., et al. 2018. Peptide-functionalized carbon dots for sensitive and selective Ca^{2+} detection. *Sens. Actuators B: Chem.* 273:1654–59.

Liu, Y., Xiao, N., Gong, N., et al. 2014. One-step microwave- assisted polyol synthesis of green luminescent carbon dots as optical nanoprobes. *Carbon* 68:258–64.

Meena, S. 2018. Effect of functionalisation of carbon nanotubes on its spin transport properties. *Mater. Chem. Phys.* 217:175–81.

Miao, H. 2016. Label-free fluorimetric detection of CEA using carbon dots derived from tomato juice. *Biosens. Bioelectron.* 86:83–9.

Ming, H., Ma, Z., Liu, Y., et al. 2012. Large scale electrochemical synthesis of high-quality carbon nanodots and their photocatalytic property. *Dalton Trans.* 41:9526–31.

Momper, R., Steinbrecher, J., Dorn, M., et al. 2018. Enhanced photoluminescence of a carbon dots system through surface interaction with polymeric nanoparticles. *J. Colloid Interface Sci.* 518:11–20.

Namdari, P. 2017. Synthesis, properties and biomedical applications of carbon-based quantum dots: An updated review. *Biomed. Pharmacother.* 87:209–22.

Ng, Y. H. 2018. Utilising the interface interaction on tris(hydroxymethy)aminomethane-capped carbon dots to enhance the sensitivity and selectivity towards the detection of Co(II) ions. *Sens. Actuators B: Chem.* 273:83–92.

Omer, K. M. 2018. Carbon nanodots as efficient photo sensitizer to enhance visible driven photocatalytic activity. *J. Photochem. Photobiol. A: Chem.* 364:53–8.

Park, J., Cho, Y. S., Sung, S. J., et al. 2018. Characteristics tuning of graphene oxide based to various end uses. *Energy Storage Mater.* 14:8–21.

Peng, Z., Han, X., Li, S., et al. 2017. Carbon dots: Biomacromolecule interaction, bioimaging and nanomedicine. *Coord. Chem. Rev.* 343:256–77.

Pirsaheb, M. 2018. Application of carbon dots as efficient catalyst for the green oxidation of phenol: Kinetic study of the degradation and optimization using response surface methodology. *J. Hazard. Mater.* 353:444–53.

Ramanan, V. 2016. Outright green synthesis of fluorescent carbon dots from eutrophic algal blooms for in vitro imaging. *ACS Sustain. Chem. Eng.* 4:4724–31.

Ravi, S., and S. Vadukumpully. 2016. Sustainable carbon nanomaterials: Recent advances and its applications in energy and environmental remediation. *J. Environ. Chem. Eng.* 4:835–56.

Ren, S. 2018. Preparation, properties and applications of functional graphene devices: A concise review. *Ceram. Int.* 44:11944–50.

Sahu, S. 2012. Simple one-step synthesis of highly luminescent carbon dots from orange juice: Application as excellent bio-imaging agents. *Chem. Comm.* 48:8835–37.

Sarkar, S. 2016. Size dependent photoluminescence property of hydrothermally synthesized crystalline carbon quantum dots. *J. Lumin.* 178:314–23.

Sharma, S. 2018. Photoluminescent C-dots: An overview on the recent development in the synthesis, physiochemical properties and potential applications. *J. Alloys Compd.* 748:818–53.

Shen, L. 2013. The production of pH-sensitive photoluminescent carbon nanoparticles by the carbonization of polyethylenimine and their use for bioimaging. *Carbon* 55:343–49.

Siddique, A. B. 2018. Facile synthesis and versatile application of amorphous carbon dots. *Mater. Today Proceedings* 5:10077–83.

Sie, Y. W. 2018. A novel fluorescence sensor for dual sensing of Hg^{2+} and Cu^{2+} ions. *J. Photochem. Photobiol. A: Chem.* 353:19–25.

Singh, R. 2018. Flourogenic detection of Hg^{2+} and Ag^{+} ions via two mechanistically discrete signal genres: A paradigm of differently responsive metal ion sensing. *Sens. Actuators B: Chem.* 258:478–83.

Sun, D. 2013. Hair fiber as a precursor for synthesizing of sulfur- and nitrogen-co-doped carbon dots with tunable luminescence properties. *Carbon* 64:424–34.

Sun, X. 2017. Fluorescent carbon dots and their sensing ability. *TrAC Trends Anal. Chem.* 89:163–80.

Tabaraki, R., and N. Sadeghinejad. 2018. Microwave assisted synthesis of doped carbon dots and their application as green and simple turn off–on fluorescent sensor for mercury (II) and iodide in environmental samples. *Ecotoxicol. Environ. Saf.* 153:101–6.

Tao, H., Yang, K., Ma, Z., et al. 2012. In vivo NIR fluorescence imaging, biodistribution, and toxicology of photoluminescent carbon dots produced from carbon nanotubes and graphite. *Nano Micro Small* 8:281–90.

Tuerhong, M. 2017. Review on carbon dots and their applications. *Chinese J. Anal. Chem.* 45:139–50.

Wang, J. 2012. Amphiphilic egg-derived carbon dots: Rapid plasma fabrication, pyrolysis process, and multicolor printing patterns. *Angew. Chem.* 51:9297–9301.

Wang, S., Wang, H., Zhang, R., et al. 2018. Egg yolk-derived carbon: Achieving excellent fluorescent carbon dots and high performance lithium-ion batteries. *J. Alloys Compd.* 746:567–75.

Wang, X. 2011. Microwave assisted one-step green synthesis of cell-permeable multicolor photoluminescent carbon dots without surface passivation reagents. *J. Mater. Chem.* 21:2445–50.

Wang, Y. 2014. Carbon quantum dots: Synthesis, properties and applications. *J. Phys. Chem. C* 2:6921–39.

Wee, S. S. 2013. Synthesis of fluorescent carbon dots via simple acid hydrolysis of bovine serum albumin and its potential as sensitive sensing probe for lead (II) ions. *Talanta* 116:71–6.

Xu, X., Ray, R., Gu, Y., et al. 2004. Electrophoretic analysis and purification of fluorescent single-walled carbon nanotube fragments. *J. Am. Chem. Soc.* 126:12736–37.

Yang, S. T., Cao, L., Luo, P. G., et al. 2009. Carbon dots for optical imaging in vivo. *J. Am. Chem. Soc.* 131:11308–9.

Yang, Y., Cui, J., Zheng, M., et al. 2012. One-step synthesis of amino-functionalized fluorescent carbon nanoparticles by hydrothermal carbonization of chitosan. *Chem. Comm.* 48:380–82.

Yang, Z. C., Wang, M., Yong, A. M., et al. 2011. Intrinsically fluorescent carbon dots with tunable emission derived from hydrothermal treatment of glucose in the presence monopotassium phosphate. *Chem. Comm.* 47:11615–17.

Ye, S. L., Huang, J. J., Luo, L., et al. 2017. Preparation of carbon dots and their application in food analysis as signal probe. *Chin. J. Anal. Chem.* 45:1571–81.

Zaporotskova, I. V. 2016. Carbon nanotubes: Sensor properties. A review. *Modern Electron. Mater.* 2:95–105.

Zhang, J., and S. H. Yu. 2016. Carbon dots: Large-scale synthesis, sensing and bioimaging. *Mater. Today* 19:382–393.

Zhang, X., Wang, Y., Liu, W., et al. 2017. Facile preparation of surface functional carbon dots and their application in doxorubicin hydrochloride delivery. *Mater. Lett.* 209:360–64.

Zhang, Y., Gao, Z., Zhang, W., et al. 2018. Fluorescent carbon nanodots as nanoprobe for determination of lidocaine hydrochloride. *Sens. Actuators B: Chem.* 262:928–37.

Zhao, L., Di, F., Wang, D., et al. 2013. Chemiluminescence of carbon dots under strong alkaline solutions: A novel insight into carbon dot optical properties. *Nanoscale* 5:2655–58.

Zheng, A. Q. 2018. Probing pH variation in living cells and assaying hemoglobin in blood with nitrogen enriched carbon dots. *Talanta* 188:788–94.

Zheng, M., Wang, C., Wang, Y., et al. 2018. Green synthesis of carbon dots functionalized silver nanoparticles for the colorimetric detection of phoxim. *Talanta* 185:309–15.

Zhi, B., Cui, Y., Wang, S., et al. 2018. Malic acid carbon dots: From superresolution live-cell imaging to highly efficient separation. *ACS Nano* 12:5741–5752.

Zhou, J. 2012. Facile synthesis of fluorescent carbon dots using watermelon peel as a carbon source. *Mater. Lett.* 66:222–224.

Zhu, S. 2015. The photoluminescence mechanism in carbon dots (graphene quantum dots, carbon nanodots, and polymer dots): Current state and future perspective. Nano Res. 8:355–81.

Zhu, S., Meng, Q., Weng, L., et al. 2013. Highly photoluminescent carbon dots for multicolour patterning sensors, and bioimaging. *Angew. Chem.* 125:4045–49.

8

Investigations on Exotic Forms of Carbon: Nanotubes, Graphene, Fullerene, and Quantum Dots

Mahe Talat, Kalpana Awasthi, Vikas Kumar Singh, and O.N. Srivastava

CONTENTS

8.1 Introduction

The epoch of carbon nanomaterials (CNMs) started with the initial reports on fullerenes and related compounds around the mid-eighties, and a remarkable increase in the research activity in the field has been observed ever since. A breakthrough took place after the finding of graphene as a material available for detailed investigations. The progress in understanding the properties of CNMs created new avenues for universal applications of nanomaterials and motivated the community of scientists to work in an interdisciplinary area involving materials science, organic chemistry, and physics. The sp^3 to sp^2-hybridized CNMs combine the quantum effects at the nanoscale, as well as the stability of resonance structures, and have distinctive tunable physicochemical properties (Mauter et al., 2008). There are several variants in the family of CNMs including fullerenes, CNTs, graphene, etc. (Georgakilas et al., 2015). The structures of different CNMs are shown in Figure 8.1. Fullerenes were discovered in 1985; they are a closed-cage form of carbon atoms (diameter $\leq$1 nm) (Kroto et al., 1985). The most common form of fullerene is Buckminster fullerene (C_{60}), in which 60 carbon atoms are arranged in a spherical structure. Another carbon nanostructure is graphene, a one-atom-thick planar sheet of sp^2-bonded carbon atoms tightly packed into a two-dimensional honeycomb lattice (Geim et al., 2007). CNTs are cylindrical molecules that contain rolled-up sheets of single- or multiple-sheet carbon atoms. They can be categorized into single-walled (SWCNTs) and multi-walled nanotubes (MWCNTs) depending on the number of concentric layers of rolled graphene sheets. These CNMs have attracted a lot of attention due to their outstanding thermal, electrical, optical, and mechanical properties, as well as to their promising applications in different fields such as field emission transistors, sensors, super capacitors, composites, biomaterials, and environment remediation (Xia et al., 2010; Dasari et al., 2017; Lota et al., 2011; Jiang et al., 2013; Wang et al., 2015; Moyon, 2018; Singh et al., 2018). Carbon quantum dots (CQDs) are a new class of small, fluorescent carbon nanoparticles with particle size under 10 nm. They were first discovered in 2004 during the purification of single-walled CNTs (Xu et al., 2004). Due to their strong and tunable fluorescence properties, CQDs have been used in biomedicines, optronics, sensors, and catalytic applications (Limet al., 2015). The presence of carboxyl groups on the surface of the CQD gives good solubility in water and biocompatibility (Namdari et al., 2017). The present chapter discusses the development of the research in the areas of synthesis and characterization of carbon nanomaterials (fullerene, graphene, CNTs, and CQDs). It also presents our R and D efforts towards the formation of these CNMs.

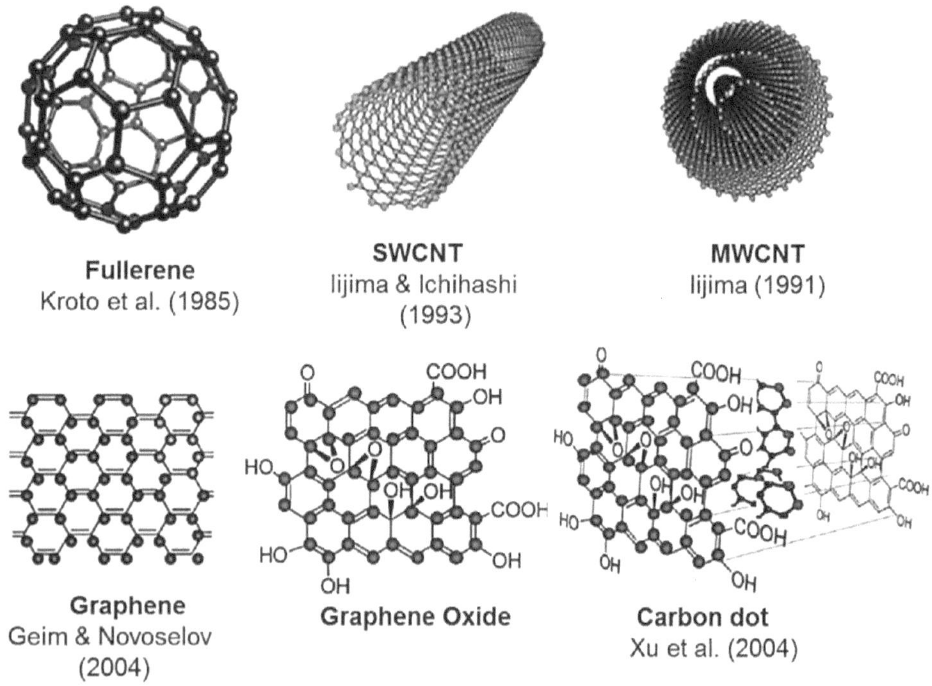

FIGURE 8.1 Different forms of carbon nanostructures (source: Google images).

8.2 Synthesis Methods of Different Carbon Nanomaterials

8.2.1 Fullerene

Fullerene has potential applications in different areas such as electronic devices, superconductors, sensors, and catalysts, as well as in the biomedical field (Thakral et al., 2006; Coro et al., 2016). Fullerene was firstly synthesized by vaporization of graphite rod using an Nd:YAG laser in a low-pressure environment (Kroto et al., 1985). In 1990, the Krätschmer and Huffman group prepared mg quantities of C_{60} with higher fullerenes (e.g., C_{70}, C_{76}, etc.) by a.c. arc discharge of graphite rods under a helium atmosphere (Krätschmer et al., 1990). Conventional chromatography and NMR techniques were used to separate C_{60} and C_{70} and for their characterization. In the arc discharge method, graphite was vaporized under a helium atmosphere at 100 Torr pressure and condensed soot contains fullerene is most popular method. Above 100 Torr helium pressure, CNT formation was observed. There are several reports on the formation of fullerene using the arc discharge method (Haufler et al., 1991; Diederich et al., 1992; Churilov, 2008). Recently, environment-friendly fullerene separation methods were reported by the Zeng group (Zeng et al., 2017). Fullerenes were prepared using premixed hydrocarbon flames under reduced-pressure and fuel-rich conditions (Gerhardt et al., 1987; Howard et al., 1991; Howard et al., 1992). C_{60} was synthesized in a low-pressure flame of benzene and acetylene (Mckinnon et al., 1992; Ozawa et al., 1999). In 2000, Hebgen et al. synthesized fullerene by the low-pressure diffusion flames method using benzene/argon/oxygen. The yield of fullerene in collected soot was determined by the high-performance liquid chromatography method. The concentration of fullerenes was well above the visible stoichiometric surface of a flame. The fullerenic nanostructures, e.g., spheroids including highly ordered multi-layered or onion-like structures are shown in Figure 8.2. Fullerenes (i.e., C_{60} and C_{70}) were also prepared by the microwave synthesis method using graphite and fluorinated graphite (Hetzel et al., 2012). Using this synthesis method, carbon allotropes are produced quickly. There are various types of fullerenes, such as are alkali-doped, endohedral, exohedral, and hetero fullerenes. In alkali-doped fullerenes, alkali metal atoms fill in the space between fullerenes and donate valence electrons to the neighbouring fullerenes. It is also possible to enclose another atom inside the fullerene because of its hollow cage. This type of fullerene is known as endohedral fullerene. If the metal atom is trapped inside the fullerene, metallofullerenes are obtained. Due to the small size of C_{60} fullerene, endohedral fullerene was prepared using C_{82}, C_{84}, or higher fullerene. Exohedral fullerenes were synthesized by a chemical reaction between fullerenes and other chemical groups. In heterofullerenes, one or more carbon atoms of the cage are substituted by hetero-atoms, e.g., nitrogen or boron atoms (Coro et al., 2016).

8.2.2 Carbon Nanotubes (CNTs)

CNTs are prepared by arc discharge, laser vaporization, and chemical vapor deposition methods, as described in our previously published chapters (Awasthi et al., 2008; Awasthi et al., 2012). These synthesis methods of CNTs are summarized in Figure 8.3 and outlined in the following sections.

8.2.2.1 Arc Discharge

In 1991, S. Iijima used this method for the synthesis of CNTs (Iijima, 1991). The experimental setup and conditions are the

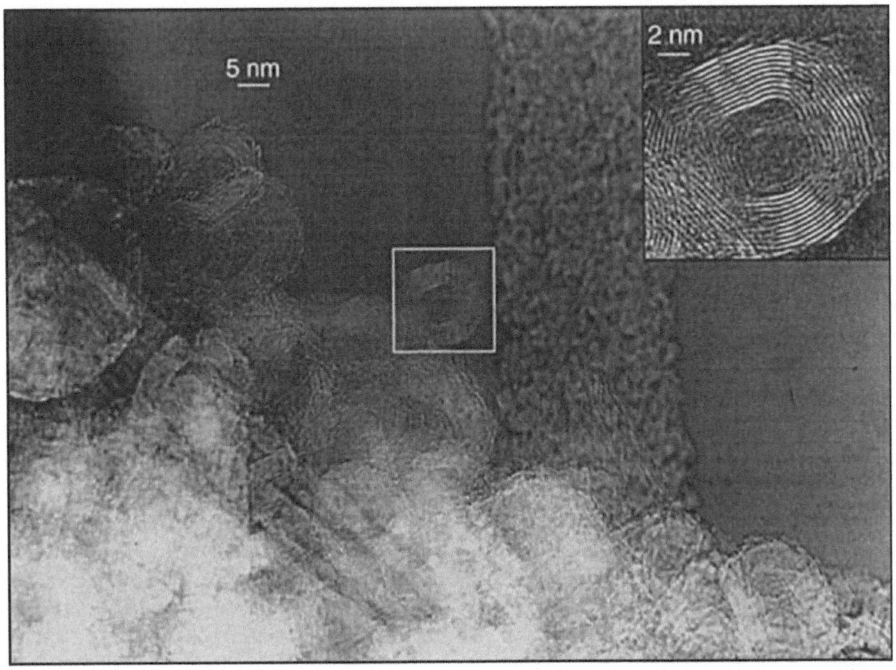

FIGURE 8.2 HRTEM image of fullerenic nanostructures incondensable soot. The inset shows onion-like structures (Hebgen et al., 2000).

same as those used for the formation of fullerenes. In this method, the arc is generated between two graphite electrodes (separated by ~1 mm) in a chamber filled with inert gas (e.g., helium and argon) at a pressure above 200 Torr. The temperature in the arc chamber is so high (~4000–6000 K) that it evaporates carbon from the anode graphite electrodes. The evaporated carbon re-condenses on the cathode electrode, and the deposit contains CNTs. The yield of CNTs depends on parameters such as current density, uniformity of plasma, chamber pressure, temperature, and cooling of electrodes and chamber (Arora et al., 2014). The high yield of multi-walled CNTs (MWCNTs) was prepared by the plasma rotating arc discharge method (Lee et al., 2002). The diameters of MWCNTs varies from 2 to 25 nm, the tube length does not exceed 1 μm, and the tubes have closed tips. MWCNTs were also synthesized by arc discharge under an NH_3, ethanol, acetone, and hexane atmosphere (Jiang et al., 2009 and Shimotani et al., 2001). The electric arc discharge was conducted in NaCl solution, liquid N_2, and water and air without evacuation of the chamber (Wang and Chang, 2005; Jung and Kim, 2003; Guo and Wang, 2007; Parkansky and Boxman, 2004). Single-walled CNTs (SWCNTs) were produced using metal-filled graphite anodes. The catalysts used transition metals such as Fe, Co, Ni, and rare earth metals such as Y and Gd. Better results were obtained using bimetallic catalysts such as Fe-Ni, Co-Ni, and Y-Ni (Iijima et al., 1993; Bethune et al., 1993; Seraphin et al., 1994; Shi et al., 2000). Most of the SWCNTs

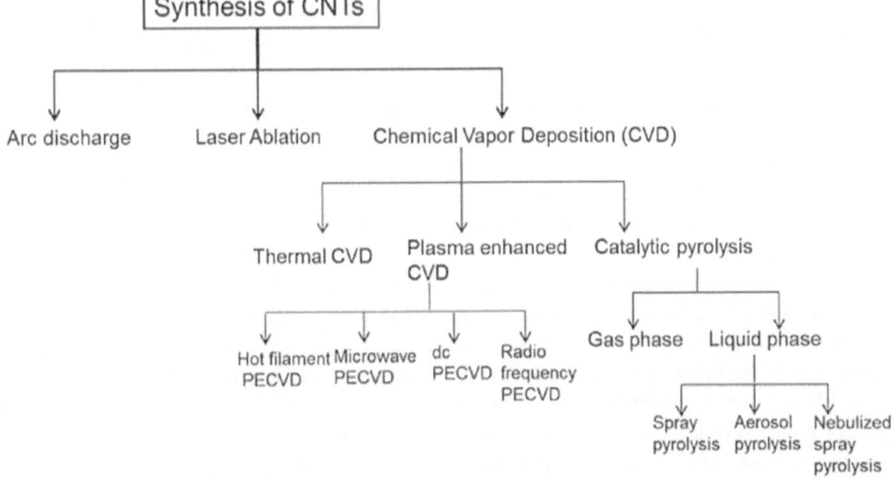

FIGURE 8.3 Synthesis methods of CNTs.

have a diameter of ~1.1–1.5 nm and are several microns in length. The microscopic-oriented web of SWCNTs was prepared by arc discharge of Fe-filled graphite electrodes in Ar-H_2 mixture (Zhao et al., 2003). Few-walled CNTs (FWCNTs) have attracted particular attention due to their unique structures, being quite different from SWCNTs and MWCNTs. FWCNTs have been synthesized using Fe as a catalyst and sulphur as a promoter (Su et al., 2013). Synthesized FWCNTs have a diameter ~1.6–6.0 nm and consist of single, double, and triple-walled CNTs. CNTs were also prepared using coal rod in place of graphite electrodes. A catalyst-doped coal rod was used as anode while a high-purity graphite rod was used as cathode (Qiu et al., 2003; Qiu et al., 2004; Mathur et al., 2007; Qiu et al., 2007; Awasthi et al., 2015). The disadvantage of this technique is that carbon impurities such as polyhedral graphite particles, amorphous carbon, and encapsulated nanoparticles are usually produced beside CNTs. Different purification methods such as oxidation, acidic and thermal treatments, etc., were employed to purify the synthesized CNTs (Aqel et al., 2012; Cho et al., 2009).

8.2.2.2 Laser Ablation

In this method, the graphite target is placed in a quartz tube surrounded by a furnace heated at 800–1500 °C under inert gas. The soot was deposited on to a water-cooled Cu collector. The arc discharge and this method are based on the condensation of carbon from the vaporization of graphite. In 1996, Thess et al. produced SWCNTs by vaporization of the graphite target with the mixture of Co and Ni powders (50 : 50) at 1200 °C in an argon (Ar) atmosphere. These SWCNTs were self-organized into bundles of 5–20 nm and have remarkably uniform diameters. The quantity and quality of produced CNT materials were found to depend on the reaction temperature, catalyst composition, gas ambient, and pressure and laser parameters. It was suggested that the continuous mode is more efficient for the high-yield synthesis of CNTs. In addition to the commonly used Nd:YAG, CO_2 lasers for SWCNT synthesis, KrF and XeCl excimer lasers were also used (Lebel et al., 2010; Kusaba et al., 2006). The CNTs produced by this method has are of high quality, but the production rate is low.

8.2.2.3 Chemical Vapor Deposition

Arc discharge and laser ablation synthesis methods of CNTs are consuming more energy. The main technological drawback of the above-described methods is that the CNTs stand on their own. A significant effort is being made towards production processes that offer more controllable routes to CNT synthesis. Chemical vapor deposition (CVD) is considered for economical large-scale CNT production and for the integration of CNTs into various devices. High-quality CNTs can be prepared in high yield as a raw material and grown onto the substrate. This method is based on the catalytic decomposition of hydrocarbon or carbon monoxide feedstock in the presence of transition metal catalysts. A supported catalyst is heated in a furnace to 600–1200 °C together with hydrocarbon gas for a period of time and the as-deposited carbon sample is allowed to cool down in an inert gas ambient. Many CVD methods were reported in the literature: these include thermal CVD, plasma-enhanced CVD (PECVD), and catalytic pyrolysis of hydrocarbon, all of which are capable of producing large-scale CNTs (Awasthi et al., 2008). Several carbon sources such as methane, benzene, ethanol, ethane, alcohol, carbon monoxide, hexane, toluene, xylene cyclohexane, naphthalene, anthracene, turpentine oil, camphor, eucalyptus oil, and others were used to produce CNTs (Awasthi et al., 2008; Awasthi et al., 2012; Awasthi et al., 2010). Carbon monoxide and methane have been found to be effective for the synthesis of SWCNTs. Other metals such as iron, nickel, cobalt, and molybdenum were used for the formation of CNTs. Szaboet al. (2010) studied the composition and the morphology of the catalyst nanoparticles on the growth of CNTs. The catalyst was deposited on the following substrates: Ni, Si, SiO_2, Cu, stainless steel or glass, and, rarely, Al_2O_3, $CaCO_3$, zeolites, mesoporous silica, graphite, and tungsten foil. The solution, electron beam evaporation, and physical sputtering methods were used for the deposition of the catalyst onto the substrates. Thermal CVD and plasma-enhanced CVD are commonly used to grow aligned MWCNTs and SWCNTs on various substrates such as Ni, Si, SiO_2, Cu/Ti/Si, stainless steel, and glass. The CNTs were deposited from various plasma techniques such as hot-filament PECVD, microwave PECVD, dc PECVD, and radio frequency PECVD. Another CVD method is the pyrolysis method, in which catalyst nanoparticles (e.g., organometallic precursors) with a carbon feedstock are injected into the CVD chamber. Organometallic compounds (e.g., metallocenes, iron pentacarbonyl, and iron (II) phthalocyanine) are used as precursors for the catalyst. The CNTs were grown on quartz (SiO_2), in the form of either a specific substrate or the reactor wall. The pyrolysis of metallocenes such as ferrocene, cobaltocene, and nickelocene produced the transition metal catalyst particles (Fe, Co, and Ni) (Sen et al., 1997). The MWCNTs were synthesized by the pyrolysis of ferrocene in an Ar-H_2 mixture at 900 °C. The pyrolysis method is divided into gas phase and liquid phase (Figure 8.3). In gas phase pyrolysis, hydrocarbon gas and vapor of metallocene were passed into the reaction zone of a quartz tube under inert atmosphere, e.g., Ar or Ar and H_2. In liquid phase pyrolysis, liquid hydrocarbon was passed into the pyrolysis furnace either as a vapor in a gas stream (Ar or Ar/H_2 mixtures) or by liquid injection (Awasthi et al., 2012; Kamalakaran et al., 2000). The CNTs were deposited into the inner wall of a quartz tube and on a quartz substrate. The main advantage of the liquid pyrolysis technique is that the catalyst particles and carbon precursor are continuously supplied, thus a continuous growth of nanotubes occurs. The growth of CNTs (yield, length, diameter, etc.) can be controlled.

8.2.3 Preparation of Graphene

The simplest way to prepare graphene is the scotch tape technique, where the tape is stuck to the graphite and peeled off, and a layer of graphite made of carbon atoms is obtained. This process is repeated again and again so that in the end a thin sheet of carbon atoms is obtained. This crude method of peeling single sheets of carbon atoms is known as mechanical exfoliation. Exfoliation includes graphite oxide reduction, shearing, sonication, electrochemical synthesis, etc. Graphene flakes were prepared by rapid heating of graphite oxide. Several methods have been reported for this kind of exfoliation of graphite oxide. The most commonly used are the Hummers' method and

TABLE 8.1
Summary of Experiment for the Synthesis of CNTs

Synthesis ethod	Carbon source	Catalyst	Synthesis condition	Atmosphere	Product	Ref.
Arc discharge	Bituminous coal: Anode electrode Graphite: cathode	Not required	d.c. voltage: 20–30 V Current: 100 Amp.	Ar (200 Torr)	MWCNTs (diameter ~8–20 nm)	Awasthi et al., 2015
		Fe		Ar (200 Torr)	SWCNTs (diameter ~1.7 nm)	
		Ni–Y (3 : 1)			SWCNTs (diameter ~1.2 nm)	
		No		Ar (250 Torr) H_2 (100 Torr)	Graphene nanosheet	
Arc discharge	Graphite: anode & cathode	Fe	d.c. voltage: 22–30 V Current: 100–150 Amp.	Ar (200 Torr)	SWCNTs web (~1.8 nm)	Awasthi et al., 2018
		No		Ar (350 Torr)	A few layers (~4) of graphene nanosheets	
Spray pyrolysis-assisted CVD	Turpentine oil	Ferrocene	Reaction temp.: 800 °C	Ar ~100 sccm sccm: standard cubic centimeter per minute	Bundles of CNTs (length: ~70–130 μm) CNTs (diameter ~15–40 nm)	Awasthi et al., 2010
Spray pyrolysis-assisted CVD	Biodiesel	Ferrocene	850 °C	Ar ~100 sccm	CNTs (diameter ~20–50 nm)	Kumar et al., 2013
	Biodiesel	Ferrocene-acetonitrile	800 °C	Ar ~100 sccm	C–N nanotubes (diameter ~30–60 nm) bamboo-shaped morphology	Kumar et al., 2013
Low-pressure CVD	Methane, ethylene	Ferritin	800 °C	Ar ~500 sccm Hydrogen (500 sccm) Methane (1400 sccm) Ethylene (30 sccm)	SWCNTs (~5 nm)	Awasthi et al., 2013

the Staudenmaier method (Hummers et al. 1958; Staudenmaier et al., 1998). However, both these methods include the use of toxic chemicals. Later, improved versions of such methods were reported in which the use of toxic chemicals was avoided to an extent, and were mentioned as an improved Hummers' method (Marcano et al., 2010). Techniques such as this are complex and intricate and elucidate why graphene is currently the most costly material on the globe. These methods are fine to use at a laboratory scale, but synthesizing at an industrial scale in bulk quantity is difficult. Thus, some other methods have been employed, such as the CVD method. In this method, graphene is formed by depositing chemicals (carbon source) from a gas.

8.2.4 Synthesis of CQDs

Quantum dots are very small semiconductor particles, only a few nanometres in size (2–10 nm), so small that their optical and electronic properties differ from those of larger particles and glow into a particular color after being illuminated by light. The color of the glow depends on the size of the quantum dot. After being illuminated by UV light, some of the electrons free themselves from the atoms. This property allows them to move around the quantum dots creating a conductance band where electrons can freely move and conduct electricity. The energy difference between the conductance band and the valence band is responsible for the color of that light. There are various sources and methods for synthesizing CQDs. One such method is "cutting" or fragmenting the graphene sheets into particles smaller than the radius (typically below 15 nm) which will result in graphene quantum dots (GQD). GQDs show fascinating physical and chemical properties such as high stability and luminescence on excitation, which are assigned to pronounced quantum confinement and edge effects.

8.3 Our Group's R and D Efforts towards Synthesis and Characterization of CNTs, Graphene, Fullerene, and Quantum Dots

8.3.1 Synthesis of CNTs and Fullerene

In our laboratory, fullerenes were synthesized by a.c. electric arc discharge of high-purity graphite rods under a helium ambient (120 Torr) (Awasthi et al., 1997). The as-deposited carbon soot was collected and dissolved in benzene. The red-to-brown-colour liquid confirms the formation of C_{60} in as-deposited soot. Fullerenes solids (black residue) were extracted from this solution using the solvent extraction method and washed with either to remove hydrocarbon contaminants.

CNTs (both MWCNTs and SWCNTs) were synthesized by the following methods:

- **Electric arc discharge**
  MWCNTs using coal rod in an argon atmosphere

SWCNTs using catalyst coal rod filled with Fe–Ni–Y in an argon atmosphere

SWCNTs web using Fe-filled graphite electrodes in an argon atmosphere

- **Chemical vapor deposition**

  By spray pyrolysis of ferrocene and hydrocarbon mixture:

  Bundles of aligned CNTs (ferrocene and turpentine oil)

  CNTs (ferrocene and biodiesel)

  Carbon–nitrogen (C–N) CNTs (biodiesel and ferrocene–acetonitrile)

  CNTs by low pressure CVD using ferritin-based Fe catalyst substrate

The details of these CNT synthesis techniques are described in our previously published chapter (Awasthi et al., 2012). The experimental conditions for the synthesis of CNTs in our group are summarized in Table 8.1 (Awasthi et al., 2015; Awasthi et al., 2010; Kumar et al., 2013; Awasthi et al., 2013). Recently, we have synthesized SWCNTs web and buckybook by d.c. electric arc discharge of high-purity graphite electrodes in an Ar atmosphere using Fe as catalyst. SEM analysis revealed that webs contain SWCNTs and are ~75 to 100 μm thick. The diameter of individual SWCNTs as calculated using TEM was found to be 1.8 nm (Awasthi et al., 2018).

8.3.2 Synthesis of Graphene

In our laboratory, different methods such as thermal exfoliation of graphite oxide, CVD, and arc discharge have been used for the preparation of graphene. A detailed discussion is to be found in our book chapter (Talat et al., 2014). Another energy-saving and quick method which is used to prepare graphene is microwave irradiation. We have exfoliated a few layers of graphene oxide using microwave irradiation, which is convenient and less energy-consuming (Talat et al. 2016). A synthesized graphene sheet was further functionalized and used for various applications including enzyme immobilization. The wrinkled and folded regions can be seen in the prepared graphene oxide [Figure 8.4(a)]. The magnified view of the transparent graphene sheet is shown in Figure 8.4(b). The SEM image of graphene oxide [Figure 8.4(c)] shows the agglomerated sheets of graphene oxide.

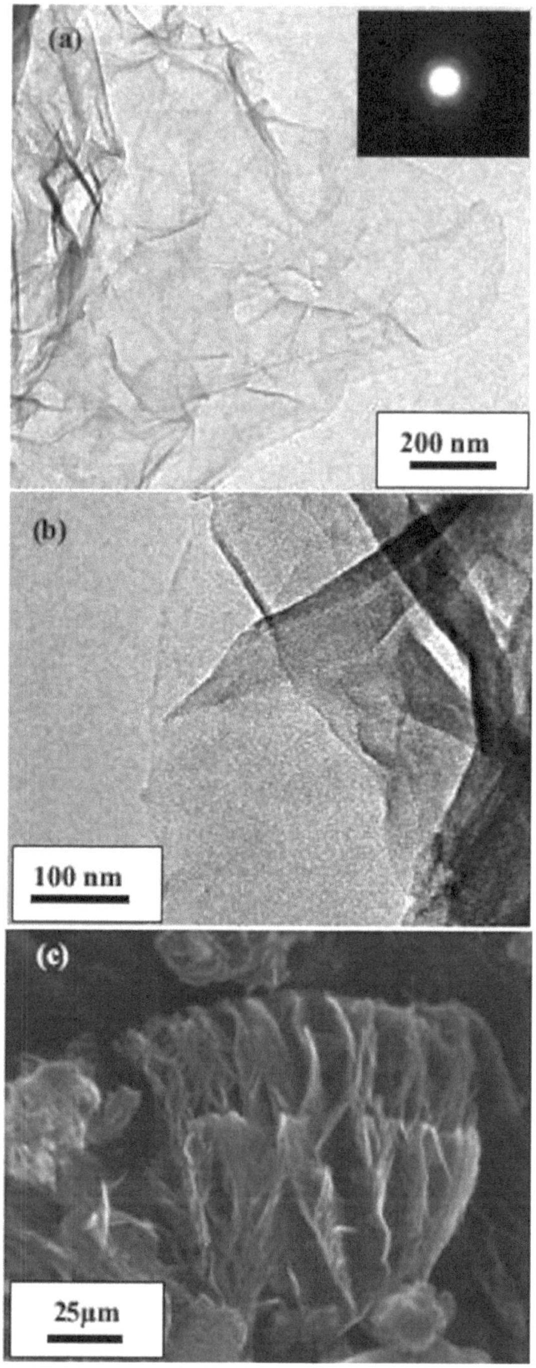

FIGURE 8.4 (a) and (b) TEM images of as-prepared graphene. The inset of (a) is a SAED pattern of a graphene sheet, (c) SEM image of a graphene sample (Talat et al., 2016).

8.3.3 Synthesis of CQDs

We have also attempted to synthesize highly fluorescent nitrogen and sulphur dual-doped carbon quantum dots (NS-CQDs) through an eco-friendly and simple hydrothermal method employing histidine and thiourea as nitrogen and sulphur sources (Singh et al., 2018). Initially, histidine and thiourea were dissolved in distilled water separately at room temperature by the sonication method. Then, the solution was transferred to a Teflon-line stainless-steel autoclave and kept in the oven. After the completion of the reaction, the autoclave was cooled to room temperature naturally and a dark-brown solution was obtained. The solution was filtered to remove the larger unreacted particles, and the filtrate was centrifuged at (15000 rpm) to remove the large and agglomerate particles. The resulting supernatant was dialyzed against distilled water to remove impurities. Finally, light-brown pure NS-CQDs were obtained. Furthermore, the quantum dots obtained are thoroughly characterised by TEM, Fourier-transformed infrared, and X-ray photon spectroscopy (FTIR and XPS). These fluorescent NS-CQDs are ideal and convenient for the colorimetric

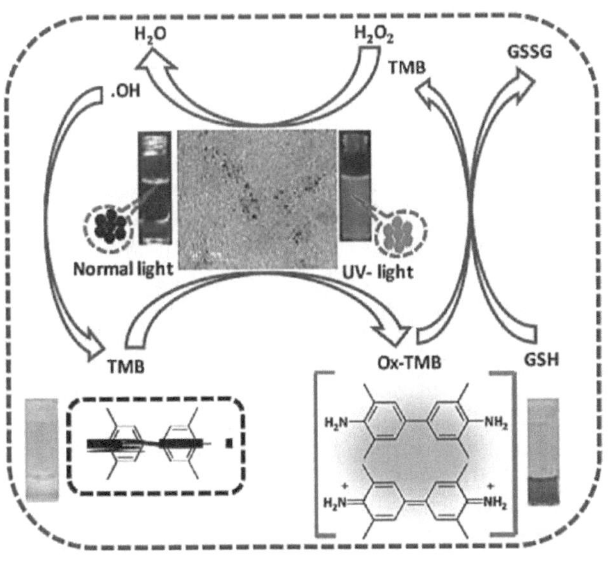

FIGURE 8.5 Schematic illustration of peroxidase mimetic activity of NS-CQDs and colorimetric detection of H_2O_2 and GSH (Singh et al., 2018b).

detection of H_2O_2 and glutathione in human blood serum with a limit of detection of 0.004 mM Figure 8.5.

Highly fluorescent nitrogen and phosphorus-doped carbon quantum dots (N, P-CQDs) were synthesized through the one-step hydrothermal method (Singh et al., 2018b). Precursors such as aspartic acid, diethylenetriamine, and phosphoric acid were used and nitrogen and phosphorous doping were attained. Due to their high quantum yield and bright-blue colour, N, P-CQDs could act as an on –off fluorescent nanoprobe which was used further for the selective and sensitive detection of highly toxic Cr (VI) in a pure aqueous medium at a physiological pH of 7.1 Figure 8.6. The calculated limit of detection was found to be 24 mg L^{-1}, which was below the permissible limit (50 mg L^{-1}) in drinking water.

8.4 Conclusions

In the past two decades, carbon nano allotropes were intensively investigated owing to their unique hybridization properties and sensitivity to change while synthesizing, allowing for fine manoeuvring of the material characteristics. Hence, entirely new carbon nanostructures (fullerenes, nanotubes, graphene, quantum dots, etc.) have been synthesized. Indeed, we might look forward to generating many breakthroughs and a new scenario for the world's economy from advances in nanotechnology. The cost effectiveness of carbon makes it a smart choice to replace conventional materials for applications. Given the potentially extensive application of carbon nanomaterials in the future, these may be comprehensively used in various fields, such as medicine, energy, biomedical, bio-imaging, solar cells, biosensors, etc. Due to their biocompatibility, stability, and surface chemistry, their

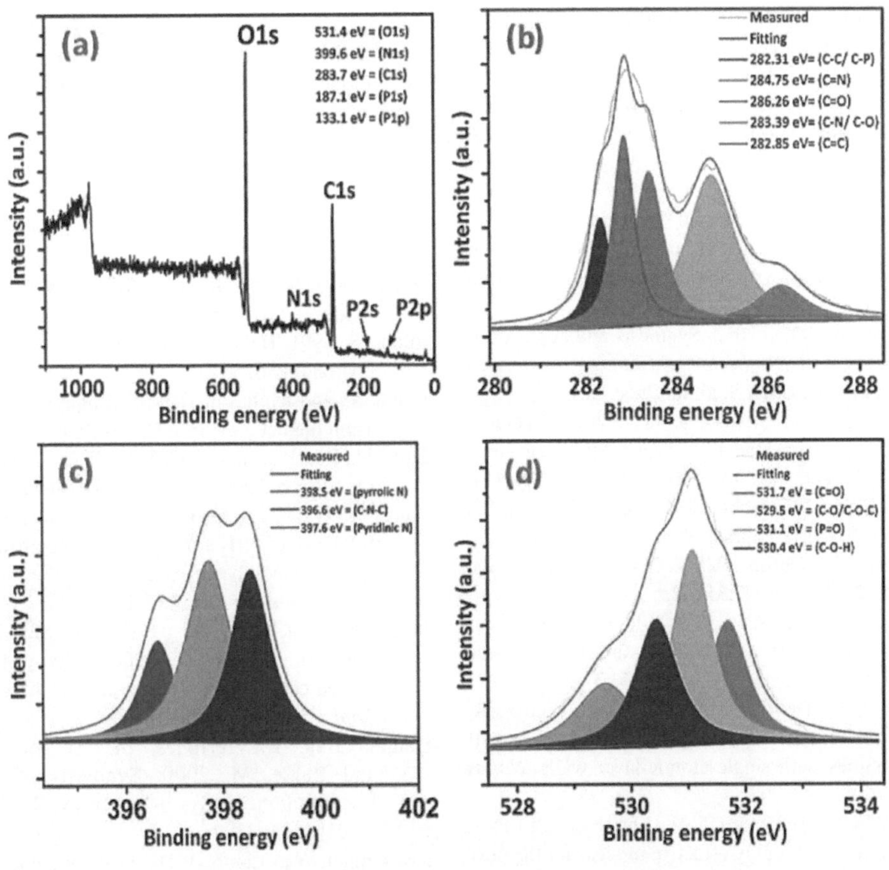

FIGURE 8.6 (a) XPS spectrum, (b) C 1s XPS spectrum, (c) N 1s XPS spectrum, and (d) O 1s XPS spectrum of the as-prepared N, P-CQDs (Singh et al.,2018).

use is highly encouraged in clinical diagnosis, tumour therapy, and radio-diagnostics. In the present chapter, different methods to synthesize these CNMs for various applications have been explored. **Achieving a high-quality material on a large scale is still a concern in order to compete with the conventional materials available in the market**. Apart from having high yield and additional variants of fullerene derivatives, CNTs and graphene oxide could help to **boost the performance and** properties of these variants. Yet, a lot more needs to be improved and explored to achieve good quality and high yield with a view to bringing these wonder materials at commercial scale.

Acknowledgments

MT greatly acknowledge the financial support from the WOS-A (DST) projects. We would also like to acknowledge the IIT (BHU) and MHRD, New Delhi, India for their financial support.

REFERENCES

Aqel, A. Abou, El-Nour, K.M.M., Ammar, R.A.A., and Al-Warthan, A. 2012. Carbon nanotubes, science and technology part (I) structure, synthesis and characterisation. *Arab. J. Chem.* 5(1):1–23.

Arora, N., and Sharma, N.N. 2014. Arc discharge synthesis of carbon nanotubes: Comprehensive review. *Dia. Relat. Mat.* 50:135–150.

Awasthi, K., and Srivastava, O.N. 1997. Synthesis and studies of new carbon variants fullerene (C60) and tubulenes. M.Sc. dissertation, Banaras Hindu University.

Awasthi, K., and Srivastava, O.N. 2008. Synthesis of carbon nanotubes. In: *Chemistry of Carbon Nanotubes*, eds. V.A. Basiuk and E.V.Basiuk, pp. 1–26. USA: American Scientific Publishers.

Awasthi, K., Kumar, R., Tiwari, R.S., and Srivastava, O.N. 2010. Large scale synthesis of bundles of aligned carbon nanotubes using a natural precursor: Turpentine oil. *J. Expt. Nanosci.* 5(6):498–508.

Awasthi, K., and Srivastava, O.N. 2012. Synthesis and applications of carbon nanotubes. In: *Synthesis and Characterization*, ed. J.N. Govil, pp. 121–171. USA: Studium Press LLC.

Awasthi, K., Awasthi, S., and Srivastava, O.N. 2013. Synthesis of CNTs by low pressure chemical vapor deposition using ferritin based Fe catalyst. In: *Emerging Paradigms in Nanotechnology*, eds. R.C. Sobti, A. Kaushik, B. Singh and S.K. Tripathi pp. 104–107. Pearson Press.

Awasthi, S., Awasthi, K., Ghosh, A.K., Srivastava, S.K., and Srivastava, O.N. 2015. *Fuel* 147:35–42.

Awasthi, S., Awasthi, K., and Srivastava, O.N. 2018. Formation of single-walled carbon nanotube buckybooks, graphene nanosheets and metal decorated graphene. *J. Nano Res.* 53:37–53.

Bethune, D.S., Kiang, C.H., Devries, M.S., Gorman, G., Savoy, R., Vazquez, J., and Beyers, R. 1993. Cobalt-catalysed growth of carbon nanotubes with single-atomic-layer walls. *Nature* 363(6430):605–606.

Cho, H.G., Kim, S.W., Lim, H.J., Yun, C.H., Lee, H.S., and Park, C.R. 2009. A simple and highly effective process for the purification of single-walled carbon nanotubes synthesized with arc-discharge. *Carbon* 47(15):3544–3549.

Churilov, G.N. 2008. Synthesis of fullerenes and other nanomaterials in arc discharge. *Fuller. Nano. Carbon Nanost.* 16(5–6):395–403.

Coro, J., Suárez, M., Silva, L.S.R., Eguiluz, K.I.B., and Salazar-Banda, G.R. 2016. Fullerene applications in fuel cells: A review. *Int. Jr. Hyd. Energy* 41(40):17944–17959.

Dasari, B.L., Nouri, J.M., Brabazon, D., and Naher, S. 2017. Graphene and derivatives – Synthesis techniques, properties and their energy applications. *Energy* 1401:766–778.

Diederich, F., and Whetten, R.L. 1992. Beyond C_{60}: The higher fullerenes. *Acc. Chem. Res.* 25(3):119–126.

Geim, A.K., and Novoselov, A.K. 2007. *Nat. Mater.* 6(3):183–191.

Georgakilas, V., Perman, J.A., Tucek, J., and Zboril, R. 2015. Broad family of carbon nanoallotropes: Classification, chemistry, and applications of fullerenes, carbon dots, nanotubes, graphene, nanodiamonds, and combined superstructures. *Chem. Rev.* 115(11):4744–4822.

Gerhardt, P., Löffler, S., and Homann, K.-H. 1987. Polyhedral carbon ions in hydrocarbon flame. *Chem. Phys. Lett.* 137(4):306–310.

Guo, J.J., Wang, X.M., Yao, Y.L., Yang, X.W., Liu, X.G., and Xu, B.S. 2007. Structure of nanocarbons prepared by arc discharge in water. *Mater. Chem. Phys.* 105(2–3):175–178.

Haufler, R.E., Chai, Y., Chibante, L.P.F., Conceicao, J., Jin, C., Wang, L.S., Maruyama, S., and Smalley, R.E. 1991. Carbon arc generation of C_{60}. *Mat. Res. Soc. Symp. Proc.* 206:627–637.

Hebgen, P., Goel, A., Howard, J.B., Rainey, L.C., and Sande, J.B.V. 2000. Synthesis of fullerenes and fullerenic nanostructures in a low-pressure benzene/oxygen diffusion flame. *Proc. Combust. Inst.* 28(1):1397–1404.

Hetzel, R., Manning, T., Lovingood, D., Strouse, G., and Phillips, D. 2012. Production of fullerenes by microwave synthesis. *Fullerenes Nanotubes Carbon Nanost* 20(2):99–108.

Howard, J.B., McKinnon, J.T., Makarovsky, Y., Lafleur, A.L., and Johnson, M.E. 1991. Fullerenes C_{60} and C_{70} in flames. *Nature* 352(6331):139–141.

Howard, J.B., Lafleur, A.L., Makarovsky, Y., Mitra, S., Pope, C.J., and Yadav, T.K. 1992. Fullerenes synthesis in combustion. *Carbon* 30(8):1183.

Hummers, W.S., and Offeman, R.E. 1958. Preparation of graphitic oxide. *J. Am. Chem. Soc.* 80(6):1339.

Iijima, S. 1991. Helical microtubules of graphitic carbon. *Nature* 354(6348):56–58.

Iijima, S., and Ichihashi, T. 1993. Single-shell carbon nanotubes of 1-nm diameter. *Nature* 363(6430):603–604.

Jiang, H., Lee, P.S., and Li, C. 2013. 3D carbon based nanostructures for advanced supercapacitors. *Energy Env. Sci.* 6(1):41–53.

Jiang, Y., Wang, H., Shang, X.F., Li, Z.H., and Wang, M. 2009. Influence of NH_3 atmosphere on the growth and structures of carbon nanotubes synthesized by the arc-discharge method. *Inorg. Mater.* 45(11):1237–1239.

Jung, S.H., Kim, M.R., Jeong, S.H., Kim, S.U., Lee, O.J., Lee, K.H., Suh, J.H., and Park, C.K. 2003. High-yield synthesis of multiwalled carbon nanotubes by arc discharge in liquid nitrogen. *Appl. Phys. A* 76(2):285–286.

Kamalakaran, R., Terrones, M., Seeger, T., Redlich, P.K., and Ruhle, M. 2000. Synthesis of thick and crystalline nanotube arrays by spray pyrolysis. *Appl. Phys. Lett.* 77(21):3385–3387.

Krätschmer, W., Lamb, L.D., Fostiropoulos, K., and Huffman, D.R. 1990. Solid C_{60}: A new form of carbon. *Nature* 347(6291):354–358.

Kroto, H.W., Heath, J.R., O'Brien, S.C., Curl, R.E., and Smalley, R.E. 1985. C60: Buckminsterfullerene. *Nature* 318(6042):162–164.

Kumar, R., Yadav, R.M., and Awasthi, K. 2013. Synthesis of carbon and carbon-nitrogen nanotubes using green precursor: Jatropha-derived biodiesel. *J. Expt. Nanosci.* 8(4):606–620.

Kusaba, M., and Tsunawaki, Y. 2006. Production of single-wall carbon nanotubes by a XeClexcimer laser ablation. *Thin Solid Films* 506:255–258.

Lebel, L.L., Aissa, B., ElKhakani, M.A., and Therriault, D. 2010. Preparation and mechanical characterization of laser ablated single-walled carbon-nanotubes/polyurethane nanocomposite microbeams. *Comp. Sci. Technol.* 70(3):518–524.

Lee, S.J., Baik, H.K., Yoo, J.-E., and Han, J.H. 2002. Large scale synthesis of carbon nanotubes by plasma rotating arc discharge technique. *Dia Relat. Mat.* 11:914–917.

Lim, S.Y., Shen, W., and Gao, Z. 2015. Carbon quantum dots and their applications. *Chem. Soc. Rev.* 44(1):362–381.

Lota, G., Fic, K., and Frackowiak, E. 2011. Carbon nanotubes and their composites in electrochemical applications. *Energy Env. Sci.* 4(5):1592–1605.

Marcano, D.C., Kosynkin, D.V., and Berlin, J.M. 2010. Improved synthesis of graphene oxide. *ACS Nano* 4(8):4806–4814.

Mauter, M.S., and Elimelech, M. 2008. Environmental applications of carbon-based nanomaterials. *Env. Sci. Tech.* 42(16):5843–5859.

Mathur, R.B., Lal, C., and Sharma, D.K. 2007. Catalyst-free carbon nanotubes from coal-based material. *Energy Sources A* 29(1):21–27.

Mckinnon, J.T., and William, L.B. 1992. Combustion synthesis of fullerenes. *Combust. Flame* 88(1):102–112.

Moyon, C.M. 2018. Applications of carbon nanotubes in the biomedical field. *Smart Nanopart. Biomed.* :83–101.

Namdari, P., Negahdari, B., and Eatemadi, A. 2017. Synthesis, properties and biomedical applications of carbon-based quantum dots: An updated review. *Biomed. Pharma.* 87:209–222.

Ozawa, M., Deota, P., and Osawa, E. 1999. Production of fullerenes by combustion. *Fullerene Sci. Technol.* 7(3):387–409.

Parkansky, N., Boxman, R.L., Alterkop, B., Zontag, I., Lereah, Y., and Barkay, Z. 2004. Single-pulse arc production of carbon nanotubes in ambient air. *J. Phys. D: Appl. Phys.* 37(19):2715–2719.

Qiu, J., Li, Y., Wang, Y., Wang, T., Zhao, Z., Zhou, Y., Li, F., and Cheng, H. 2003. High purity single-wall carbon nanotubes synthesized from coal by arc discharge. *Carbon* 41(11):2159–2179.

Qiu, J., Li, Y., Wang, Y., and Li, W. 2004. Production of carbon nanotubes from coal. *Fuel Process. Technol.* 85(15):1663–1670.

Qiu, J., Wang, Z., Zhao, Z., and Wang, T. 2007. Synthesis of double-walled carbon nanotubes from coal in hydrogen-free atmosphere. *Fuel* 86(1–2):282–286.

Sen, R., Govindaraj, A., and Rao, C.N.R. 1997. Carbon nanotubes by the metallocene route. *Chem. Phys. Lett.* 267(3–4):276–280.

Seraphin, S., and Zhou, D. 1994. Single-walled carbon nanotubes produced at high yield by mixed catalysts. *Appl. Phys. Lett.* 64(16):2087–2089.

Shi, Z., Lian, Y., Liao, F.H., Zhou, X., Gu, Z., Zhang, Y., Iijima, S., Li, H., Tue, K.T., and Zhang, S.L. 2000. Large scale synthesis of single-wall carbon nanotubes by arc-discharge method. *J. Phys. Chem. Solids* 61(7):1031–1036.

Shimotani, K., Anazawa, K., Watanabe, H., and Shimizu, M. 2001. New synthesis of multi-walled carbon nanotubes using an arc discharge technique under organic molecular atmospheres. *Appl. Phys. A* 73(4):451–454.

Singh, D.P., Herrera, C.E., Singh, B., Singh, S., Singh, R.K., and Kumar, R. 2018. Graphene oxide: An efficient material and recent approach for biotechnological and biomedical applications. *Mat. Sci. Eng.* 86:173–197.

Singh, V.K., Singh, V., and Yadav, P.K. 2018a. Bright-blue-emission nitrogen and phosphorus doped carbon quantum dots as a promising nanoprobe for detection of Cr (VI) and ascorbic acid in pure aqueous solution and in living cells. *New J. Chem.* 42(15):12990–12997.

Singh, V.K., Yadav, P.K., Chandra, S., Bano, D., Talat, M., and Hasan, S.H. 2018b. Peroxidase mimetic activity of fluorescent NS-carbon quantum dots and its application for colorimetric detection of H_2O_2 and glutathione in human blood serum. *J. Mater. Chem. B.*

Staudenmaier, L., and der Graphitsaure, V. D. 1998. *Ber. Deut Chem. Ges.* 31:1481.

Su, Y., Zhou, P., Zhao, J., Yang, Z., and Zhang, Y. 2013. Large-scale synthesis of few-walled carbon nanotubes by DC arc discharge in low-pressure flowing air. *Mat. Res. Bull.* 48(9): 3232–3235.

Szabo, A., Perri, C., Csato, A., Giordano, G., Vuono, D., and Nagy, J.B. 2010. Synthesis methods of carbon nanotubes and related materials. *Materials* 3(5):3092–3140.

Talat, M., and Srivastava, O.N. 2014. Synthesis, characterization and functionalization of carbon nanotubes and graphene: A glimpse of their application. In: *Advanced Carbon Materials and Technology.*

Talat, M., Awasthi, K., and Srivastava, O.N. 2016. Microwave assisted exfoliation of graphene and its application in enzyme, immobilization. *Curr. Nanomat.* 1:159–163.

Thakral, S., and Mehta, R.M. 2006. Fullerenes: An introduction and overview of their biological properties. *Indian Jr. Pharm. Sci.* :13–19.

Thess, A., Lee, R., and Nikolaev, P. 1996. Crystalline ropes of metallic carbon nanotubes. *Science* 273(5274):483–487.

Wang, S.D., Chang, M.H., Lan, K.M.D., Wu, C.C., Cheng, J.J., and Chang, H.K. 2005. Synthesis of carbon nanotubes by arc discharge in NaCl solution. *Carbon* 43(8):1778–1814.

Wang, Y., Wei, H., Lu, Y., Wei, S., Wujcik, E.K., and Guo, Z. 2015. Multifunctional carbon nanostructures for advanced energy storage applications. *Nanomaterials* 5(2):755–777.

Xia, F., Farmer, D.B., Lin, Y., and Avouris, P. 2010. Graphene field-effect transistors with high on/off current ratio and large transport band gap at room temperature. *Nano Lett.* 10(2): 715–718.

Xu, X., Ray, R., Gu, Y., Ploehn, H.J., Gearheart, L., Raker, K., and Scrivens, W.A. 2004. Electrophoretic analysis and purification of fluorescent single-walled carbon nanotube fragments. *J. Am. Chem. Soc.* 126(40):12736–12737.

Zeng, H.Y.G., Lai, C., and Huang, D. 2017. Environment-friendly fullerene separation methods. *Chem. Eng. J.* 330: 134–145.

Zhao, X., Inoue, S., Jinno, M., Suzuki, T., and Ando, Y. 2003. Macroscopic oriented web of single-wall carbon nanotubes. *Chem. Phys. Lett.* 373(3–4):266–271.

9

Nanodiamonds and Other Organic Nanoparticles: Synthesis and Surface Modifications

Navneet Kaur, Chander Prakash, Aman Bhalla, and Ganga Ram Chaudhary

CONTENTS

9.1 Introduction

In the past century, nanocarbon research has witnessed significant developments. Many new members, such as carbon nanotubes and graphene, have been enrolled in the class of carbon-containing materials, which led to the discovery of many new properties apart from the long-known forms of carbon, i.e., diamond and graphite (Liu et al., 2014). Sparked by the discovery of fullerenes, which highlighted the carbon structures with specific covalent bonding, a lot of has developed among the scientific communities who aim to discover and synthesize the materials that show specific bonding patterns of carbon atoms directed to specific properties. Significant work was seen in this direction in the late 20th century, marking an era of vast research in the area of carbon nanostructures that have contributed enormously to the fields of chemical, physical, and biomedical sciences (Gogotsi and Presser, 2013; Wang and Astruc, 2017; Kaur et al., 2018). Nanodiamonds are one of the most recently discovered nanostructures belonging to the nanocarbon family. Though they were discovered in the early 60s by the Russians, the curtain raiser about their existence and discovery happened to be in late 80s (Danilenko, 2004). Nanodiamonds are zero-dimensional nanostructures made of sp^3-bonded carbon atoms, unlike other members of the nanocarbon family that comprise of sp^2-bonded carbon atoms, such as one-dimensional carbon nanotubes and two-dimensional graphene nanostructures. The sp^3-bonded structural arrangement of carbon atoms in nanodiamonds gives them exceptional chemical and mechanical properties (Iakoubovskii, 2000). These nanosized structures of carbon carry a large percentage of atoms on the surface, which imparts to them many exceptional mechanical properties, different from those of bulk diamonds. It is due to these properties that nanodiamonds have found large application as abrasives, hard protective coatings for aircrafts, cutting tools, and additives in lubricants to ensure a reduction in the friction of sliding parts and increased engine life (Ekimov et al., 2002). The introduction of nanodiamond suspensions into the aluminum alloys or carbon steel ensures frictionless and wear-resistant surfaces for a wide use (Chou and Lee, 2010). Lately, the chemical vapor deposition technique has been known to generate thin films of nanodiamonds which have found significant use in the fabrication of nano-mechanical and electromechanical resonant structures, cold cathodes, and field emission displays (Williams et al., 2008). Apart from their inherent catalytic properties, nanodiamonds are known to be the precursors of carbon onions, which have potential as energy storage materials (Butenko et al., 2005; Han et al., 2011). Having been popularized as excellent material owing to their mechanical properties, nanodiamonds have revolutionized the biomedical field with their chemical and optical properties. A significant success has been achieved in the biomedical research area with the advent of nanodiamonds. Properties such as biocompatibility, chemical inertness, and highly tunable surface chemistry have popularized them as potential tools in a number of biological applications, including protein extraction, gene and therapeutic drug delivery, etc. (Zhu et al., 2012). Certain structural variations induced in nanodiamonds through ion radiation techniques can induce useful optical properties in them. Optical properties such as photoluminescence allow the application of nanodiamonds in cell imaging, cell tracking, fluorescence resonance energy transfer, high resolution magnetic sensing, etc. (Vaijayanthimala and Chang, 2009). Nanodiamonds have also significantly contributed to the field of tissue engineering, where the amalgamation of the mechanical properties of nanodiamonds with the biocompatibility of polymers has produced mechanically robust composites of clinical relevance (Ullah et al., 2015).

Apart from nanodiamonds, there is another class of carbon-based nanomaterials, broadly known as organic nanoparticles. These nanoparticles are not whole-carbon nanomaterials; instead they comprise of organic molecules with carbon as their major component. These organic nanoparticles are of major interest in the field of soft materials and, especially, life sciences. The nano range of their size along with their biocompatibility puts organic nanoparticles on the forefront in attempts to develop new, efficient, and reliable technologies in medicine, clinical research, and biotechnologies (Qin et al., 2012). Due to the non-toxicity of organic nanoparticles, they exhibit notable competence with semiconductor quantum dots when applied in the fields of sensing and diagnostics. They find wide applicability in the development of biosensors and molecular switches, and in the targeted delivery of diagnostic and therapeutic agents. This is attributed to their biocompatibility and efficiency as scaffolds for encapsulating different materials (Jana et al., 2012). The molecules or structures that fall under the category of organic nanoparticles are micelles, vesicles, liposomes, polymerosomes, polymer conjugates, solid lipid nanoparticles, polymeric nanoparticles, and dendrimers (Figure 9.1). Whereas dendrimers, solid lipid nanoparticles, and polymeric nanoparticles comprise of a single molecule falling in the nanometric range and possessing different properties, the other forms of organic nanoparticles such as micelles, liposomes, vesicles, polymer conjugates, and polymerosomes are essentially self-assembled structures of a number of molecules bound together through different forces. These self-assembled structures are such that they have a particular size (in nm) and shape (Kumar and Lal, 2014). Also, these structures, based on the comprising molecules, are capable of encapsulating and carrying organic molecules such as drugs for therapeutics and dyes or contrast agents in diagnostics (Li and Liu, 2014). Whether it is through encapsulation of molecules or their attachment on the

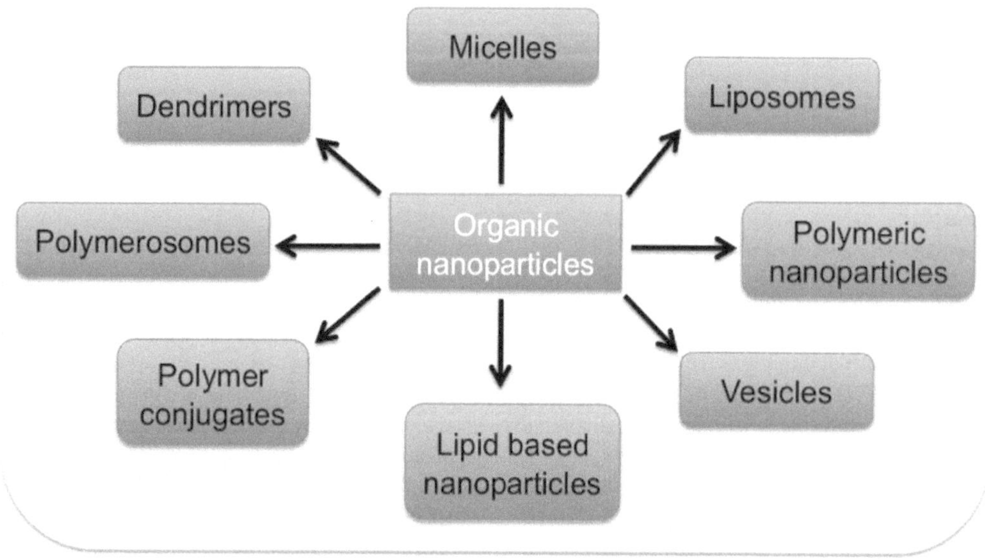

FIGURE 9.1 Classification of organic nanoparticles.

surface of these self-assembled structures, the thus-formed self-assemblies form an efficient carrier system that is stable in biological systems.

Apart from carrying important drugs and other molecules relevant to biological applications, the micellar solutions of amphiphilic molecules such as surfactants and nanoemulsions show remarkable encapsulating properties. This gives the possibility to employ them as nanoreactors for carrying out reactions between organic reactants that are otherwise immiscible with aqueous solvents (Kaur et al., 2019). Thus, these organic nanoparticles act as catalytic systems by localizing the reacting species in nano-sized regions, thereby making the reaction feasible (Shchukin and Sukhorukov, 2004). The functionalization of organic nanoparticles with biomolecules such as DNA has contributed to the development of processes such as molecular recognition and protein-based membrane fusion (Guerrero-Cázares et al., 2014).

Chemical inertness and a wide scope of surface modification in these carbon-based nanomaterials (nanodiamonds and organic nanoparticles) make them a potent tool in a number of significant applications in the fields of materials and biosciences. The aforementioned properties and applications of these carbon-based nanomaterials have made it essential to look in detail at the synthesis and properties of these nanomaterials. Detailed reports about synthetic procedures can give insight into the resulting structures and properties. This also enables us to discover the properties that can be incorporated into the structures right at the synthesis stage. This chapter comprises of two sections – one highlighting the importance, synthetic procedures, and possible modifications of nanodiamonds, and the other highlighting various classes of organic nanoparticles and their synthetic protocols.

9.2 Nanodiamonds

9.2.1 Structure of Nanodiamonds

Nanodiamonds, as the name suggests, are nanostructures similar to bulk diamonds having sp^3-bonded carbon in their structures. Due to their nano sizes, nanodiamonds are imparted with many properties that are unique with respect to the structures of bulk diamonds. The crystal of nanodiamonds comprises of two interchangeable closely packed *fcc* lattices that can be displaced along the direction of space diagonal up to one-fourth of its length (Figure 9.2) (Bhosale et al., 2013; Zhang et al., 2018).

Here, it is noteworthy that sp^3 hybridization of carbon in nanodiamonds leaves dangling bonds on the surface which may lead to the agglomeration and aggregation of individual nanodiamond structures. The surface atoms are therefore terminated either by graphitization, where sp^3 carbon is converted to sp^2, or by some functional groups. The majority of functional groups terminate the surface atoms of nanodiamonds through nitrogen or oxygen atoms (Figure 9.3). However, surface stabilization is also reported by hydrogen or hydrocarbon chains through termination. Leaving aside the graphitization that causes the formation of sp^2-carbon chains shell around the nanodiamond's core, the termination of sp^3-carbon surfaces through functional groups results in the versatility of nanodiamond applications. It is the presence of these surface functional groups that is instrumental in the attachment of many biomolecules to nanodiamonds, which opens up a plethora of applications of nanodiamonds in the biomedical field (Osswald et al., 2006; Ho, 2010).

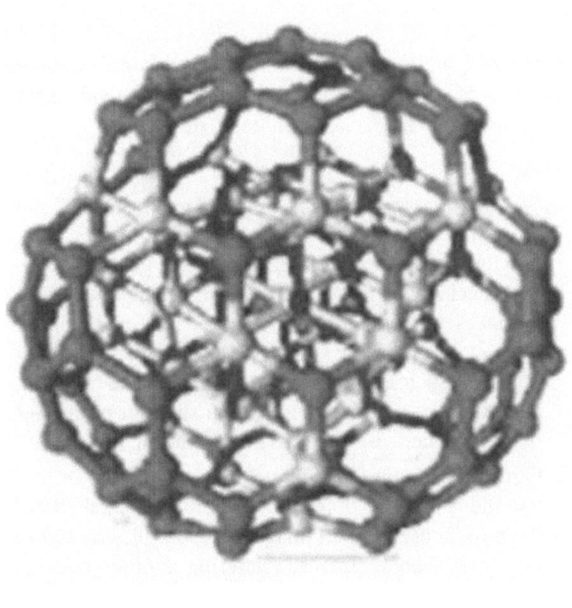

FIGURE 9.2 Crystalline structure of nanodiamonds.

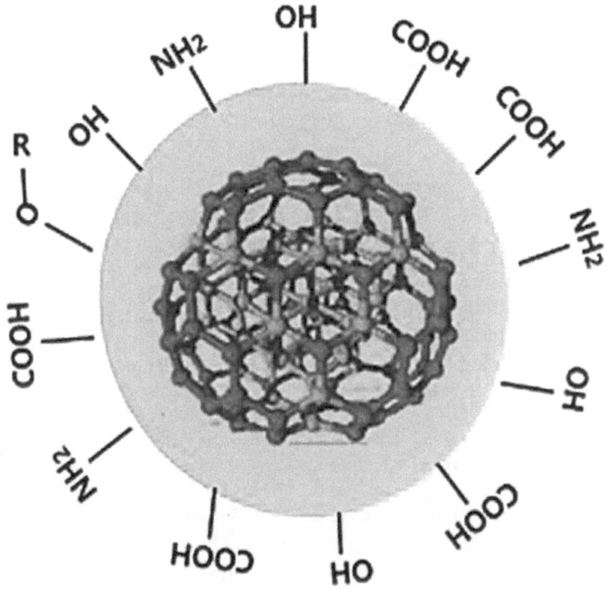

FIGURE 9.3 Structural features of nanodiamond surface.

Also, there is a large scope for the covalent functionalization of nanodiamond surfaces without affecting the inherent properties of core nanodiamond structures. The tunability of the surface not only helps introducing desired moieties on nanodiamond surfaces for specific applications but also plays an important role in size control and surface stabilization (Barnard et al., 2003).

9.2.2 Significant Properties of Nanodiamonds

A large number of applications of nanodiamonds are possible due to their unique qualities, which comprise of mechanical properties that are inherited from their bulk analogues and of optical and biological properties that are exclusively a result of the nano size and structural modifications of nanodiamonds. Below are some important properties of nanodiamonds that render them useful in diverse applications.

9.2.2.1 Physical Properties

i. Mechanical properties: Nanodiamonds possess excellent mechanical properties such as Young's modulus, hardness, bulk modulus, low coefficient of friction, etc. (Maitra et al., 2009). Compared to many known hard materials such as titanium, stainless steel, and titanates, nanodiamonds are found to exhibit superior properties that make them useful as abrasives, cutting tools, etc.
ii. Optical properties: Nanodiamonds have a high refractive index and show optical transparency which allows them to be used in cell imaging in both *in vivo* and *in vitro* experiments. Intense scattering is observed with high-resolution light microscopy in the case of cells that are cultured with nanodiamonds. Fluorescence properties can be induced in nanodiamonds by introducing nitrogen vacancies through ion beam irradiation. The defects and impurities in the crystal lattice of nanodiamonds give rise to bright and stable internal fluorescence, which qualifies them as fluorescent labels in cellular tracking and as biomarkers (Baranov et al., 2011).
iii. Thermal properties: Nanodiamonds have optimum values of thermal expansion coefficient and thermal conductivity, which helps them withstand steep temperature changes (Branson et al., 2013). This is quite helpful in using nanodiamonds as materials for designing autoclaves or nitrogen-storage cylinders as well as for tuning the thermal properties of materials.

9.2.2.2 Chemical Properties

The surfaces of nanodiamonds are chemically inert to corrosion or any pH change in their environment. This chemical inertness, along with high purity, helps nanodiamonds find utility as protective materials for the surface coating of surgical and dental implants (Aversa et al., 2017). The structural analysis of nanodiamonds suggests that the unsatisfied fourth covalency of sp^3 carbon on the nanodiamond surface can be either satisfied by conversion of surface sp^3 carbon to sp^2 carbon, thus forming a graphitic shell around nanodiamond cores. This graphitic shell facilitates the adsorption of hydrophobic biomolecules on the surfaces of nanodiamonds with large surface areas (Chang et al., 2009). This property finds application in radiation therapy, where surface-modified nanodiamonds absorb electromagnetic radiation to produce localized heating effects (Chen et al., 2018). The other possibility of surface stabilization of nanodiamond surface is through functional groups containing oxygen. The hydrophilic nature of these nanodiamond surfaces facilitates further functionalization to get the desired polymers, biomolecules, or drugs attached to the nanodiamond surface for various applications such as drug delivery, cellular tracking, and imaging (Say et al., 2011). Also, in electrochemical applications, nanodiamonds show high chemical sensitivity and electrode stability (Zang et al., 2007).

9.2.2.3 Biological Properties

Nanodiamonds show high biocompatibility and low cytotoxicity to a variety of cells. It is due to these properties that nanodiamonds largely find applications in the biomedical field (Yu et al., 2005). From dental and surgical implants to *in vivo* cellular tracking, nanodiamonds find application in many other diagnostic tools and targeted drug therapies where semiconductor quantum dots have limitations.

9.2.3 Synthesis of Nanodiamonds

After the Russians first introduced the world to the synthesis of nanodiamonds through the detonation of explosives, there has been a great advancement in the synthetic procedure for nanodiamonds. According to astronomical studies, naturally occurring nanodiamonds are found in the protoplanetary disks of stars which are not in the range of human exploitation (Hill et al., 1998). The alluring properties of nanodiamonds have propelled the scientific community to go in depth to understand the challenges in the synthesis of desired nanodiamond structures. Very fine and phase-pure nanodiamonds can now be synthesized through different techniques. The following section discusses different methods of synthesis of nanodiamonds with varied properties and purities, along with various steps involved in purification, surface stabilization, and functionalization.

9.2.3.1 Detonation Synthesis

This method of synthesis of nanodiamonds was accidentally observed by Russian scientists while they were working on explosives. Ever since the official declaration of the discovery of nanodiamond synthesis, this has been the most popular and commercial method(Aharonovich and Shenderova, 2012). The popularity and acknowledgment of the detonation synthesis method is evidenced from the fact that the Ig Nobel Peace Prize 2012 was awarded for the conversion of Russian ammunition into useful nanodiamonds by the SKN company. The primary advantage of this synthetic approach is that it ensures the bulk production of nanodiamonds withinunder 10 nm in size. The complete procedure of obtaining useful nanodiamonds from the detonation technique comprises of careful synthesis, purification, and functionalization of nanodiamonds. The detailed account of the aforementioned steps is given as follows.

9.2.3.1.1 Synthesis

The synthetic procedure involves the use of explosive molecules as precursors which are exploded in a non-oxidizing environment in a metallic chamber, resulting in the formation of nanodiamonds. It is the carbon present in the precursors that makes up the nanodiamond structure after detonation; therefore, the choice of starting explosive material is important as it decides the size of the resulting nanodiamonds. Generally, a starting explosive material is chosen that has negative oxygen balance, thereby making the system short of oxygen and rich in carbon. The most widely used explosive in detonation synthesis is a mixture of trinitrotoluene and hexogen that yields nanodiamonds of under10 nm in size (Kuznetsov et al., 1994). A typical procedure involves an explosive mixture comprising of trinitrotoluene (TNT) and hexogen in a 6 : 4 ratio, which is detonated in a metallic chamber in the presence of coolants in a non-oxidizing environment. The resultant product after the detonation of explosives is in the form of carbon soot, which comprises of both diamond and graphite phases along with some metal impurities from the detonation chamber (Dolmatov, 2001). The nanodiamond formation mechanism can be understood on the basis of a phase diagram of carbon highlighting temperature/pressure conditions in which different phases of carbon are stable (Figure 9.4).

When we look at the phase diagram, it is evident that the diamond phase of carbon is stable at very high temperature and pressure. The optimal conditions for the stability of nanodiamonds are indicated by the phase diagram to explain the principle of detonation synthesis. A similar phase diagram can be considered to understand the different phases of nanocarbon with an exception for the liquid phase, which appears above 4500 K for bulk carbon, and at a lower temperature for nanocarbon (Viecelli et al., 2001). A large amount of heat is generated as a result of the compression of explosive precursors on detonation. The release of enormous amounts of heat in a short fraction of time is a result of the chemical decomposition of carbon-containing precursors (Figure 9.5).

The shock resulting from the detonation propagates and creates high temperatures and pressures corresponding to point A (Figure 9.4), a region of liquid carbon which, when the temperature is lowered with coolants, generates nanocarbon clusters. This homogeneous nucleation of supersaturated free carbon into nanoclusters followed by coalescence and crystallization yields nanodiamonds of very small size (2–3 nm) (Shenderova and Nunn, 2017). It is noteworthy that a decrease in pressure inside the detonation chamber favors the formation of thermodynamically stable graphite, thus hampering the growth of

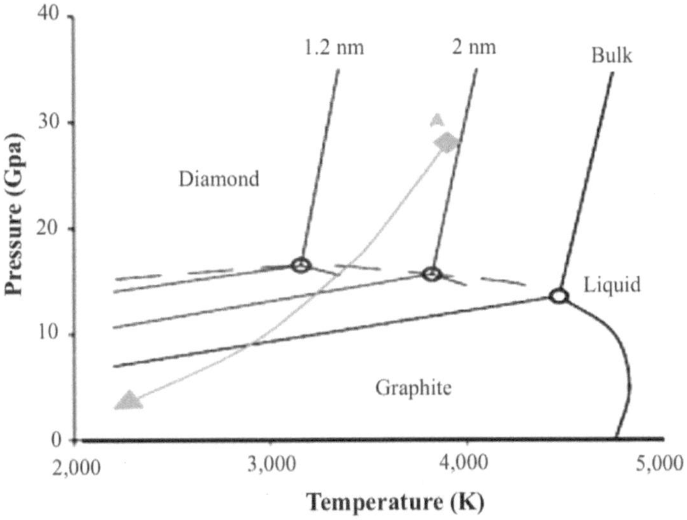

FIGURE 9.4 Phase diagram highlighting different phases of carbon and stability conditions.

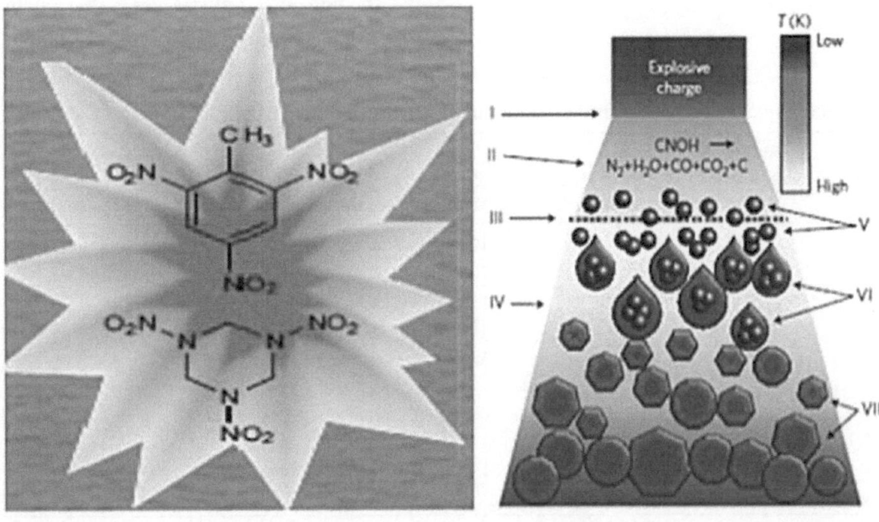

FIGURE 9.5 Synthetic procedure of detonation nanodiamonds.

nanodiamonds. Only 4–10 % of the initial explosive weight is obtained as detonation soot in the final product. The soot comprises majorly of a diamond phase (75 %), along with graphite carbon and metal impurities (Aleksenskiy et al., 2010).

The essential requirements for the detonation synthesis, which determine the quality and yield of final products, are listed below.

(a) Composition of precursor explosive molecules

It is very important to select such compositions of explosive mixtures that have negative oxygen balance, i.e., carbon to oxygen (C/O) is less than 1. This ensures that the oxygen constituting the molecule is not enough to oxidize the combustibles after the chemical decomposition of the explosive molecules. This ensures that elemental carbon will be obtained in detonation soot to form nanodiamonds, and their oxidation will be prevented (Shenderova et al., 2006).

(b) Cooling media

To achieve the desired pressure and temperature conditions corresponding to a stable diamond phase of carbon, it is important to employ a cooling medium in the detonation chamber along with explosives. Apart from acting as a cooling medium to bring down high temperatures, it also facilitates the travel of the detonation wave. Based on the type of cooling medium employed, the detonation synthesis of nanodiamonds is classified into dry (in case of pressurized N_2, Ar, CO_2) and wet synthesis (H_2O). The resulting phase of carbon largely depends on the cooling rate of the cooling media, which should be adjusted so as to prevent the transformation of initially formed nanodiamonds into the graphite phase. The ideal cooling rate to ensure a good percentage of nanodiamonds in the resulting detonation soot, as reported in the literature, is 3000 K/min. Moreover, the choice of cooling media is decisive in controlling the size and aggregation of the nanodiamonds produced. Reports suggest that, in comparison with wet synthesis, dry detonation synthesis yields basically smaller nanodiamonds which are less prone to aggregation (Dolmatov, 2001).

(c) Weight of the detonation charge

Detonation charge refers to the explosive mixture. A cooling medium is introduced into the detonation chamber before the explosion. The nanodiamond product yield is significantly influenced by the relative weight ratios of explosive mixture and cooling media, which control the cooling rate during the formation of nanodiamonds in detonation chambers (Dolmatov, 2001).

9.2.3.2 Laser-Based Synthesis

Laser-based synthetic approaches have been recently explored to generate nanodiamonds of high purity. Traditional approaches such as detonation synthesis result in a high-percentage yield of products, but the multi-step involving the purification and post-treatment of nanodiamond products does not ensure uniformity in sizes or the purity of the nanodiamonds obtained (Mitev et al., 2007). Also, the safety and environmental issues associated with the post-synthesis treatment processes of nanodiamonds have caused the need for cleaner and less complicated synthetic approaches. Laser-based synthetic methods are successful in the laboratory-scale synthesis of nanodiamonds as they ensure high-purity (>98 %) products resulting from a single step. Tuning certain parameters during the synthesis stage offers a good control over the sizes of the resulting nanodiamonds (Baidakova et al., 2013). The two synthetic approaches that involve lasers for nanodiamond synthesis are the pulsed laser ablation in liquids (PLAL) and the light hydro-dynamic pulse (LHDP) syntheses. Both of these methods require a pulsed beam of high density (Nd–YAG laser), a solid-target source containing graphitic carbon, and a liquid. The detailed account of these methods is given below.

9.2.3.2.1 Pulsed Laser Ablation in Liquids (PLAL)

This is a top-down approach of nanodiamond synthesis where a sp^2 carbon source such as graphite is acted upon by a high-energy pulsed laser beam. As the name suggests, the pulsed laser beam is focused on the carbon source placed in a particular liquid. When a high-energy pulsed beam of laser is focused on the target's surface, decomposition of the material takes place due to high temperatures generated on the surface, resulting

TABLE 9.1

Dependence of Various Parameters on the Resulting Nanodiamond Sizes Obtained from PLAL

S.No. S. No.	Wavelength (nm)	Radio Frequency (Hz)	Power Density (Wcm^{-2})	Pulse Width	Graphite Target	Liquid	ND size (nm)	Reference
1.	532	5	10^{11}	10 ns	Poly crystalline	Water	300–400	Yang et al., 1998
2.	532	5	10^{11}	10 ns	Poly crystalline	Acetone	30–40	Wang et al., 2002
3.	1064	20	4×10^6	1.2 ms	Graphite particles (<0.2 μm)	Water	3–6	Sun et al., 2006
4.	355	10	40–100 Spot: 0.1–0.5 mm	5 ns	Pyrolytic graphite	Water	5–15	Amans, 2008
5.	1064	20	4×10^6	0.4 ms	Graphite particles (<2.0 μm)	Water	2–7	Bai, 2010
6.	1064	20	4×10^6	1.2 ms	Graphite particles (<2.0 μm)	Water	2–13	Bai et al., 2010

in vaporization products called ablation plume. The ablation plume interacts with the surrounding liquid medium, giving rise to cavitation bubbles. The high-temperature and high-pressure conditions inside the collapsing bubbles lead to the nucleation and growth of the diamond structure, which is quickly solidified by fast quenching due to the surrounding liquid (Amans et al., 2009). The first synthesis of NDs by PLAL reported in 1998 involved the irradiation of graphite with pulsed YAG laser in water. Soon after the successful synthesis, there were many reports for PLAL synthesis of nanodiamonds with various laser parameters, target sources, and liquids. Table 9.1 illustrates the dependence of these parameters on the resulting nanodiamond sizes (Yang et al., 2007; Yang et al., 1998; Wang et al., 2002; Sun et al., 2006; Amans et al., 2009; Bai et al., 2010).

Despite its efficiency in yielding size-controlled nanodiamonds of high purity, this method, being expensive, cannot compete well in terms of wide applicability scope on the industrial scale.

9.2.3.2.2 Light Hydro-Dynamic Pulse (LHDP) Synthesis

The cost issues associated with the PLAL synthesis of nanodiamonds pushed the researchers to find a new laser-based synthetic approach which is essentially similar to PLAL, but with a little operational alteration that resolves the large-scale applicability issues of laser-based synthetic approaches. The LHDP synthesis of nanodiamonds, similar to PLAL, involves a high-power laser beam source, a fluid/liquid, and a carbon source. However, LHDP synthesis involves two important alterations in conventional conditions:

(i) The target source of carbon is now a mixture of carbon and a binder containing hydrocarbons.
(ii) The high energy pulsed laser beam does not actually hit the target. Instead, it is focused at a distance which is closer than the target position.

The second alteration in technical procedure accounts for the basic difference between PLAL and LHDP. In PLAL, the laser beam is focused on the source of carbon and it is the action of plasma, thus generated, on the target carbon source (mainly graphite), that results into the formation of nanodiamonds. However, LHDP involves the laser assisted-generation of a shock wave that travels through plasma and acts on the composite target. The mechanism involves a high laser beam (1010 W $cm^{-2)}$ which is focused at a predetermined distance from the carbon target placed in a liquid medium. At this point, the choice of the medium is important as the ratio of physical quantities such as refractive index and light flow intensity in the medium gives rise to a self-focusing effect. When a laser beam passes through a suitable medium, it causes a self-focusing effect and generates a large amount of heat in the medium, which results in an acoustic shock wave. This shock wave interacts with the target mixture containing carbon and generates very high temperature and pressure conditions in which the diamond form of carbon is thermodynamically stable. The resulting product contains nanodiamonds along with hydrocarbons that were present in the starting target mixture. The purification of the product is carried out by washing the hydrocarbons with organic solvents, then in deionized water, and finally drying (Chiao et al., 1964). Moreover, nanodiamonds have been synthesized through laser irradiation of ethanol. The solvent absorbs energy upon irradiating with a femtosecond laser (1025 nm), and, at energy higher than 300 μJ, the dissociation of ethanol into C, H, and O takes place. The nucleation of this formed carbon species in the confined plasma results in ultra-small nanodiamonds (<5 nm).

The LHDP method of nanodiamond synthesis ensures nanodiamond products of high purity ad free from metal impurities. The size of the products can be simply controlled on the basis of laser beam parameters (Zousman and Levinson, 2014). The purification process is simple and does not involve any costly or environment-harmful techniques. Since the synthesis can be carried out at normal room temperature and pressure conditions and involves the action of an acoustic shock wave on the target, it is feasible to scale up this method on industrial scale in a cost-effective and greener way (Williams, 2014).

9.2.3.3 High-Pressure High-Temperature Synthesis

The high-pressure high-temperature synthetic approach is one of the oldest and most widely used approaches for the production of man-made diamonds. The reproduction of temperature pressure conditions similar to those essential for natural diamonds has popularized this approach in industries and research. The technical process involves the application of high pressure using large press tools, accompanied by heating to high temperatures. The compression of the carbon source at high temperatures raises the pressure inside the capsule, creating the right conditions for stable nanodiamonds (Dubrovinskaia et al., 2005; Boudou et al., 2013).

The synthesis setup involves large press tools of tons of weight to produce high pressures up to 5 GPa through tungsten carbide anvils. Based on the design of anvils (two anvils – upper and lower – in case of belt press and six anvils in case of cubic press), different pressure conditions can be achieved (Bach, 2011; Sung, 2008). At the top of a cylindrical titanium capsule, a pure carbon source is present along with metal additives such as Ti (for N vacancy) and Cu (to inhibit TiC formation) below it. The bottom of the capsule contains synthetic nanodiamond crystal seeds. When the capsule is pressed through anvils, very high pressure is generated along with high temperature (1773 K) achieved by heaters. Upon reaching the ideal conditions of pressure and temperature, carbon from the source melts and moves through the metal slug. The upper and lower parts of the capsule are at a temperature difference of 293–323 K. This temperature difference causes the precipitation and crystallization of carbon on diamond seeds to form diamond crystals of nano size (Sumiya et al., 2005). Compared to belt presses containing two anvils, multi-anvil presses are more effective as they induce higher pressure/temperature conditions at a faster rate, leading to faster synthesis of nanodiamonds. Single crystalline nanodiamonds (50–100 nm) can be synthesized from mesoporous carbon source at 14 GPa and 1673 K (Mandal et al., 2014). A recent high-pressure, high-temperature-based synthesis of ultra-small nanodiamonds (3 nm) involved adamantane as carbon source subjected to pressures of 9 GPa and 1600 K (Ekimov et al., 2018). This opens up an opportunity to focus on carbon sources other than graphite powder, such as functionalized adamantine mixtures, to get nanodiamond products with the desired functionalities.

9.2.3.4 Ultrasonic Cavitation

The ultrasonic cavitation method is an innovative technique that mimics the necessary conditions for nanodiamond synthesis under ambient conditions. Ultrasonication generates sound waves in liquids, the strength of which depends on the frequencies of the probe used. The propagation of sound waves through the medium generates alternating cycles of high and low pressures. These alternating cycles give rise to cavitation effects in liquid medium, which involve small vacuum bubbles formation at low pressure. These subsequently collapse during the high-pressure cycle. During the cavitation process, very high temperatures (approx. 5000 K) and pressures (approx. 2000 atm) are reached locally. It is the localized generation of high pressure/temperature conditions that can be exploited for the generation of nanodiamonds using ultrasonic cavitation of liquids containing a carbon source. The conditions of very high pressure and temperature required for the conversion of carbon to graphite can be conveniently achieved under ambient conditions through ultrasonication. Galimov et al. performed the cavitation-assisted synthesis of nanodiamonds using benzene as a carbon source as well as a medium for the propagation of ultrasound waves. Nanodiamond aggregates of size 10–30 nm were formed along with graphite and a new carbon allotrope, n-diamond (Galimov et al., 2004). Khachatryan et al. (2008) employed the ultrasonication cavitation approach to convert graphite into highly pure diamond crystals of reduced sizes in micro range. Their report highlights the ultrasonication of graphite powder suspension in organic liquid comprising of aromatic oligomers. High boiling point and low vapor pressure of aromatic oligomers serve as an efficient cavitation medium, generating high-temperature/pressure conditions resulting in the conversion (10 %) of graphite to diamond. The resulting diamond structures are well crystalline, monodisperse in the size range of 6–9 μm. Factors such as cavitation fluid density (solid : fluid weight ratio), cavitation fluid, power, and frequencies of sonication are decisive in determining the size and morphology of the resultant diamond structures (Khachatryan et al., 2008). Tuning these parameters can optimize the system to produce better yield and make the ultrasonic cavitation-based synthetic approach for nanodiamonds an efficient alternative to costly high-pressure high-temperature synthesis.

9.2.3.5 Chemical Vapor Deposition

Chemical vapor deposition is the most popular technique for the synthesis of nanomaterial films. This technique is being widely exploited for the fabrication of nanocrystalline diamond films. It involves the deposition of active carbon species (generated from precursor gases) on different substrates (mainly Si).

In general, the chemical vapor deposition processes for nanodiamonds involve a carbon source (mainly methane) in hydrogen and an inert gas, argon. The primary step for nanodiamond film generation involves the generation of active growth species from precursor gases. The gas phase diffusion of the reactive radicals ($CH_2\cdot$, H·, C) onto the substrate surface depends on fed gas composition and chamber pressure (Stacey et al., 2009). Whereas the lifetime of reactive species is largely determined by pressure and gas flow rate conditions inside the chamber, the film growth on the substrate depends on its temperature, carbon diffusion coefficient, and seeding efficiency. Seeding refers to the introduction of pure nanodiamond crystal to the clean etched substrate's surface either by polishing or through the ultrasonic treatment of powdered diamond dispersion or detonation nanodiamond. The seeding process carries special significance in the generation of homogenous films as it increases the nucleation density on the substrate (Balmer et al., 2009).

Generally classified on the basis of the activation mechanism of precursor gases, widely used chemical vapor deposition techniques for nanodiamond synthesis are hot filament chemical vapor deposition and microwave plasma-enhanced chemical vapor deposition (Figures 9.6a and 9.6b, respectively). Whereas microwave plasma-enhanced chemical vapor deposition involves the use of microwaves, the hot filament chemical vapor deposition

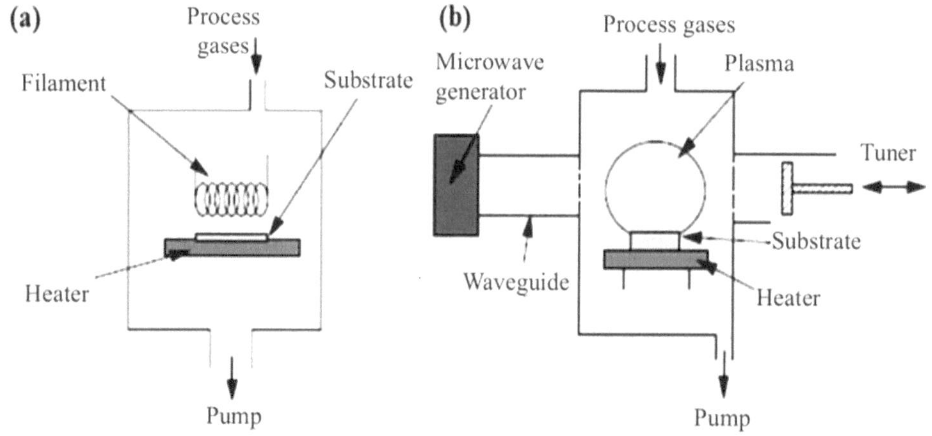

FIGURE 9.6 Experimental setup for (a) Hot filament chemical vapor deposition and (b) Microwave plasma-enhanced chemical vapor deposition synthesis of nanodiamonds.

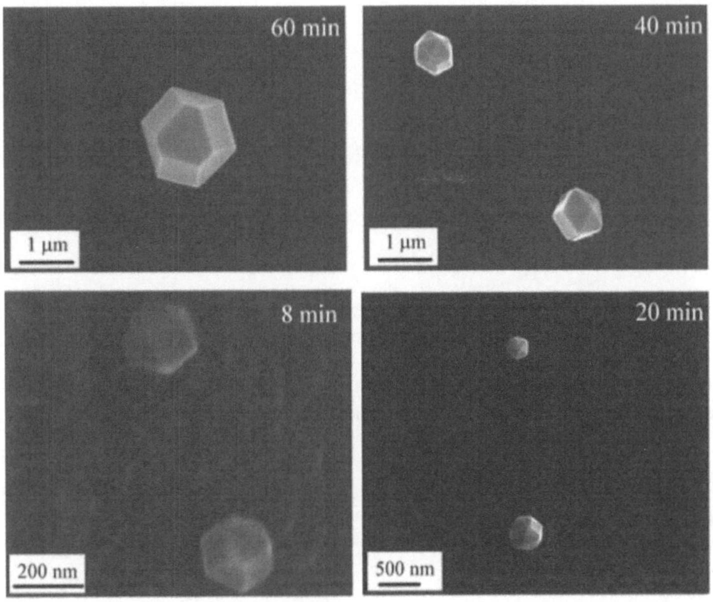

FIGURE 9.7 Scanning electron microscope images showing correlation of nanodiamond sizes with deposition times.

method involves heated filaments of metals such as tungsten or molybdenum for the activation of precursor gases (Butler and Sumant, 2008).

Additionally, on the basis of the nucleation and growth mechanisms, the resulting products can be classified as ultra-nanocrystalline diamond and nanocrystalline diamond. Argon-rich and hydrogen-deficient chemical vapor deposition conditions yield highly pure ultra-nanocrystalline nanodiamonds of grain size 2–5 nm containing >95 % of sp^3-bonded carbon. Hydrogen excess and carbon-deficient gases in the chemical vapor deposition process result in nanocrystalline diamonds of varied purity and grain size (Balmer et al., 2009).

Apart from the well-explored nanodiamond films, the microwave plasma-enhanced chemical vapor deposition technique is known to produce isolated nanocrystalline diamonds. Stacey et al. have grown nanocrystals of average size 200 nm on silicon and sapphire substrates, which were etched with oxygen plasma with 1 : 5 ratio of O_2 and Ar in H_2 before the introduction of methane gas (2 %). Crystals ranging from sizes 0.2 to 1.5 µm were obtained depending upon the deposition time (Stacey et al., 2009). Figure 9.7 shows the scanning electron microscope images of diamond crystals obtained as a result of increasing deposition times. The results lead to the inference that microwave plasma-enhanced chemical vapor deposition at higher microwave powers using nanodiamond seeds can generate well-faceted diamond crystals of tunable sizes depending upon the deposition time of vapors on the silicon substrate.

A recent study by Degutis and coworkers presents a nanocrystalline diamond growth from a nanodiamond seed buried under a thin Cr layer on the substrate (Si). The growth mechanism of diamond crystals is given in Figure 9.8, which shows that the Cr layer plays a sacrificial role by getting carburized by the action of carbon-containing plasma under high MW power. As the deposition time increases, nanocrystalline diamond grows on the surface similar to free Si substrate due to diffusion of carburized

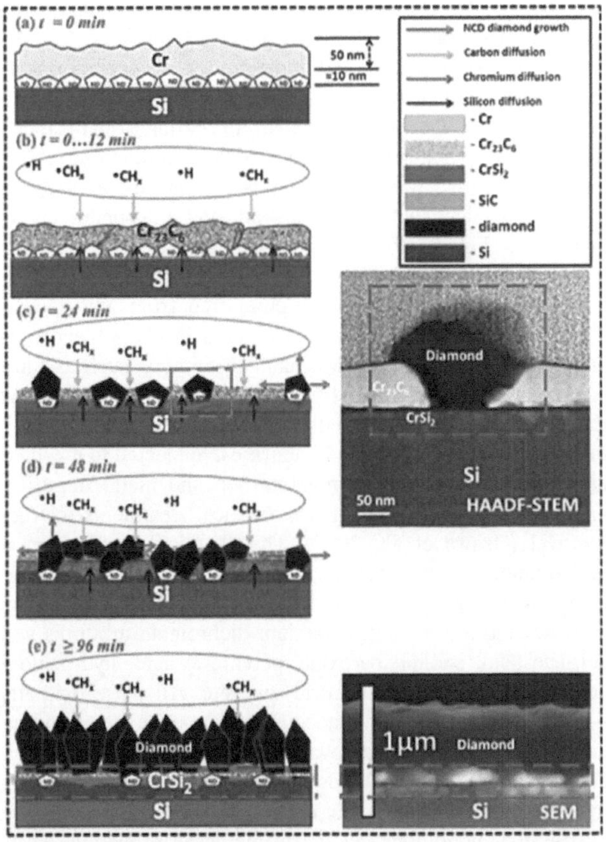

FIGURE 9.8 Time dependent growth of nanodiamond film on the Cr covered Si substrate.

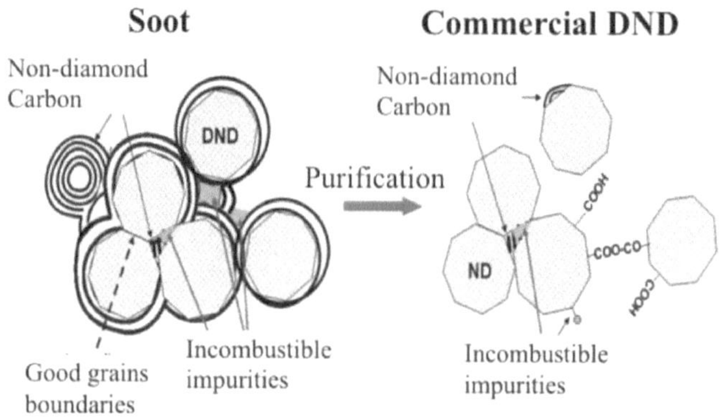

FIGURE 9.9 Components of detonation soot obtained during detonation nanodiamond synthesis and resulting purified nanodiamonds.

Cr to Si $CrSi_2$. This leads to exposure of nanodiamond seeds buried beneath to allow further growth of nanocrystalline films (Degutis et al., 2016).

9.2.4 Purification of Nanodiamonds

The large-scale synthesis of nanodiamonds, obtained by the detonation method, is accompanied by intense purification steps to obtain pure desirable products. It is noteworthy that, apart from nanodiamond content, the resultant powder obtained from the detonation chamber at the end of the process also contains some non-diamond impurities. These impurities comprise of incombustible metal impurities (1–8 weight %) coming from the metallic detonation chamber and combustible non-diamond carbon structures (25–45 weight %) such as carbon onions, graphitic carbon associated with nanodiamond, and amorphous carbon aggregates (Figure 9.9) (Osswald et al., 2006; Dolmatov, 2001).

Based on the applications, it is essential to ensure that the nanodiamond product is phase-pure, free from any undesirable impurities. The purification process involves mechanical and chemical procedures. Mechanical methods such as ultrasonication and centrifugal fractionation are required to break the clusters of primary nanodiamond structures holding tightly inside them the non-diamond impurities. Subjected to mechanical treatment, the non-diamond carbon and metal impurities are rendered exposed to further chemical treatments for their removal (Ozawa et al., 2007). The chemical purification of nanodiamonds is based majorly on the oxidation of impurities using oxidizing agents. The most used oxidizing agents are sulfuric acid, nitric acid, potassium dichromate mixtures with sulfuric acid, sodium peroxide, perchloric acid, hydrochloric acid, hydrofluoric acid in nitric acid, etc. All these oxidizing liquids result in the simultaneous oxidation of metal and sp^2 carbon impurities (Shenderova and Gruen, 2006). Though efficient in removing both kinds of impurities, the liquid oxidizing agents suffers from major drawbacks such as high-pressure/temperature conditions leading to increased processing costs, safety issues, and potential poisoning of the nanodiamond surface due to preferential adsorption of metal ions such as Cr^{3+} from oxidizers, which compromises their purity standards for biomedical applications (Dolmatov, 2003). As an alternative to liquid oxidizers, a thermal and gaseous treatment (using chlorine and ozone gas) of the detonation product is used. Thermal heating in a range of 673–873 K in air causes the oxidation of sp^2 carbon impurities. Oxidation using Cl_2 or ozone-rich air at elevated temperatures ensures the successful removal of non-diamond carbon (Shenderova et al., 2011). Although this method avoids any undue generation and pretreatment of waste liquid effluents (as in liquid oxidizers), it is expensive and involves the risk of explosion or ignition of the product. Moreover, it fails to eliminate the metal impurities. To specifically remediate these impurities without any undue toxification of the nanodiamond product, metal complexing agents such as hexamethylenetetramine, potassium thiocyanate, disodium dihydrogen ethylenediaminetetraacetate, thiourea, dicyandiamide, and sodium 2,3-dimercaptopropanesulfonate (Unithiol) are used to form water-removable complexes of impurities of metals such as Cr, Mn, Fe, Ni, and Cu. The detonation products, when subjected to the aqueous/organic solutions of complexing agents under conditions such as cavitation disintegration or high-temperature and pressure, undergo metal complexation, which leads to the formation of water-soluble complexes that are subsequently washed away with water to ensure a metal-free product (Dolmatova et al., 2013). This approach selectively removes metal impurities while leaving undisturbed the traces of graphitic carbon attached to the aggregates of nanodiamonds. This approach is helpful in fields such as medicinal biology and polymer chemistry where only metal-free products are desirable, and a non-diamond carbon content associated with nanodiamond clusters carries significance in various applications and modifications.

9.2.5 Functionalized Nanodiamonds

The nanodiamonds obtained through different synthesis methods undergo various purification steps that introduce a number of surface-terminating atoms or groups including hydrogen carboxyl and hydroxy groups. These surface groups on the nanodiamonds account for the versatile chemistry involved in the attempts to introduce a variety of desired functional groups (Krueger, 2008). There are numerous methods involving the gaseous treatment (at high temperatures) and wet chemical treatment (at ambient conditions) of nanodiamonds to functionalize them. The functionalization of nanodiamonds involves attaching different molecules to their surfaces or introducing new functionalities to them, thus imparting new properties (Liu et al., 2004). The hydrophilicity introduced to the nanodiamonds by the surface functional groups

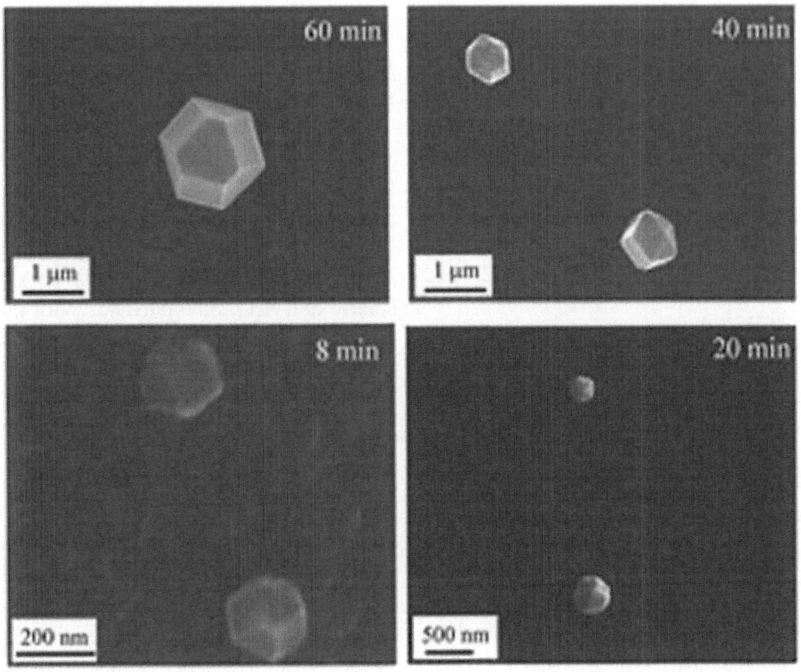

FIGURE 9.10 Surface functionalization of carboxylic-group-terminated nanodiamond.

has envisioned possible further modifications with different molecules. A versatile surface chemistry is known to be possessed by nanodiamonds carrying carboxyl terminating groups. To control any interference by other functional groups, nanodiamonds need to possess only a single starting functional group on the surface. The –COOH-terminated nanodiamonds are obtained by treating metal-free nanodiamonds with ozone/air, then with acid (HCl). The carboxyl group offers the introduction of many functional groups depending upon the reagents and conditions used (Liang et al., 2009; Zhang et al., 2018). Figure 9.10 depicts different reaction conditions and reagents used to obtain desired properties in nanodiamonds. The nanodiamond surfaces, rich with –OH functionalization, are obtained by reduction with diboranes/$LiAlH_4$ to remove any carboxyl functionalities. The –OH functionalities on nanodiamonds can be helpful in the attachment of important peptides through the silanization of –OH groups. The surface –OH groups can attach different alkyl chains through esterification reaction of long-chain acid chlorides (Arnault et al., 2011). A treatment with hydrogen gas at high temperature leads to the graphitization of nanodiamonds, converting them to carbon onions (Kuznetsov et al., 1994).

The sp^2 graphitic carbon is either already present as shell around the nanodiamond structure or created specifically through graphitization (under vacuum at high temperatures), and can attach different organic moieties through strong C–C bonds (Osswald et al., 2006). There are reports involving the functionalization of nanodiamonds via Diels–Alder reactions or Diazonium compounds (Jarre et al., 2011; Liang et al., 2011). The hydrogen-terminated nanodiamonds bearing hydrophobic surfaces can successfully attach biomolecules (DNA) and exhibit excellent substrates to attach organic moieties through C–C bonds (Purtov et al., 2008). The treatment of amines with carboxyl-rich nanodiamond surfaces can lead to fluorophore nanodiamond conjugates (Zhang et al., 2011).

There is a wide scope in the functionalization of nanodiamonds with various molecules such as drugs, DNA, proteins, polymers, etc. These assist in the application of nanodiamonds in the fields of medicine, drug delivery, biosensing, and bioimaging. Additionally, it is possible to dope the intrinsic structure of nanodiamonds for the purpose of modifying optical and electrical properties. The doping procedure involves the incorporation of an atom (other than carbon) in the nanodiamond lattice. Numerous reports indicate nitrogen (N) as being the mostly inserted atom in nanodiamond lattice. The N incorporation in nanodiamond films improves their electrical behavior, imparting semiconductor characteristics to them. The insertion of N atoms into the lattice takes place during the growth of films in chemical vapor deposition conditions with N_2 gas in the fed gaseous mixtures (Rabeau et al., 2007).

In isolated nanodiamond structures, N atoms in the lattices are responsible for the intrinsic fluorescence emission characteristics of nanodiamonds. This is due to the presence of nitrogen vacancy centers. When nanodiamonds are irradiated with high-energy alpha particles, certain vacancies are generated in the nanodiamond crystals and get trapped by N atoms (already present in nanodiamonds as impurities) during the annealing process (at 873–1073 K). It is due to these nitrogen vacancy centers that nanodiamonds exhibit a stable luminescence. Recent reports suggest the in situ generation of nitrogen vacancies in the detonation synthesis of nanodiamonds (Vlasov et al., 2010). The nitrogen content present in explosive precursors and the cooling conditions during the process account for the nitrogen vacancy centers in detonation nanodiamonds. However, the presence of surface defects along with other internal defects in nanodiamond crystals hampers the fluorescence intensity of the resulting nanocrystalline diamonds.

Thus, it is well illustrated that nanodiamonds possess a wide scope of surface functionalization to attach various biomolecules and drugs. Also, the nitrogen vacancy defects resulting in the optical properties of nanodiamonds have widened the possibilities of doping nanodiamond crystals with different metal atoms to achieve new possibilities in the fields of cell imaging and diagnostics.

9.3 Organic Nanoparticles

Organic nanoparticle is the term given to a class of carbon containing compounds to demarcate them from traditional inorganic nanoparticles. Whereas inorganic nanoparticles are known to be synthesized by precipitation reactions between precursors, organic nanoparticles are generally single-molecule entities of large molecules or their aggregated structures falling in nano-size ranges. The molecules comprising the organic nanoparticles are mostly natural polymers, proteins, lipids, and surfactants, along with drug or polymer molecules that are reduced to nano sizes. The structures that can be discussed under the category of organic nanoparticles are polymeric nanoparticles, dendrimers, solid lipid nanoparticles (independent nano-sized entities), micelles, liposomes, vesicles, polymer conjugates, polymerosomes, and nano capsules (aggregated structures of a number of molecules). The vast range of nanostructures collectively called organic nanoparticles greatly contributes to the fields of material and biomedical science. From single organic nanoparticles such as dendrimers and polymeric nanoparticles to aggregated nanostructures such as micelles, vesicles, liposomes, etc. – all these have led to enormous innovations to replace the existing technologies in the fields of diagnostic cell imaging, sustained drug delivery, materials for surgical implants, and fillers (Ng and Zheng, 2015). Organic nanoparticles are popularized as efficient and reliable materials for developing future technologies because of their remarkable properties, which make them outshine the inorganic nanoparticles used in preexisting technologies (Kaur et al., 2019). The high stability and surface reactivity of these organic nanoparticles combined with their non-toxicity and biodegradability means that they are safe and efficient materials in biomedical applications such as targeted drug and gene therapy (Singh et al., 2017).

9.3.1 General Synthetic Approaches for the Fabrication of Organic Nanoparticles

Organic nanoparticles can be perceived as engineered structures made up of either a single molecule or a number of molecules. The terms 'top-down' or 'bottom-up' can be used for the broad classification of the synthetic approaches for the fabrication of organic nanoparticles.

9.3.1.1 Top-Down Approaches

The essence of the top-down approaches for organic nanoparticles synthesis is the size reduction of macro and micro range source material to nano range. The techniques that fall under the top-down approach include mechanical milling, which involves the mechanical grinding of macro- or micro-sized molecules with the help of abrasives. The mechanical milling technique finds great application in the fabrication of nanoparticles of drugs that are poorly soluble in water from their crystals using pearl mills. Despite having the advantage of reducing sizes from macro/micro to nanoscale, the technique suffers with drawbacks such as inadvertent product contamination and a wide size distribution of the nanoparticles obtained. The drawbacks of mechanical milling are overcome by the microfluids and lithography techniques for organic nanoparticles fabrication. It is possible to synthesize polymeric and drug nanoparticles with good control over shape, size, and composition with the help of these advanced top-down approaches (Müller et al., 2011; Rolland et al., 2004; Rolland et al., 2005).

9.3.1.2 Bottom-Up Approaches

Unlike top-down approaches, which are mainly employed to synthesize polymeric or drug nanoparticles, bottom-up synthetic approaches do not require any complex setup. Essentially, they are based upon the building of structures in the nano range starting from a single precursor molecule by following the principles of synthetic chemistry and self-aggregation (Texter, 2001; Zhang et al., 2008; Euliss et al., 2006; Horn and Rieger, 2001). A careful choice of the precursor molecule and control of various reaction parameters such as concentration of precursors, solvent, and temperature can lead to the desired structures possessing particular shape, size, and properties. Bottom-up techniques can lead to the generation of a variety of nanostructures, e.g., micelles, vesicles, liposomes, dendrimers, polymeric nanoparticles, polymerosomes, polymer conjugates, and solid lipid nanoparticles, which can be collectively called organic nanoparticles. The following section gives a detailed account of various organic nanoparticles structures and provides an insight into the principles governing their synthetic processes via bottom-up approaches.

9.3.2 Synthesis of Organic Nanoparticles

9.3.2.1 Micelles

Micelles are the colloidal aggregates of amphiphilic molecules that assemble in a particular fashion to form structures in micro or nano range. Amphiphilic molecules consist of two parts, i.e., a non-polar hydrophobic tail and a polar hydrophilic head group. Depending on the nature of the solvent they are introduced to, the resulting aggregates are of two kinds, i.e., micelles and inverted micelles (Figure 9.11). When added to polar solvents such as water, the hydrophilic head groups of the molecules interact strongly with the solvent molecules, and the hydrophobicity of non-polar tails pushes them away to form spherical or cylindrical aggregates, known as micelles. The hydrophilic head groups of molecules comprising the micelles are in direct contact with the polar solvent; however, the tails are pointing inwards. In the case of inverted micelles, where the aggregates are formed in non-polar solvents, the tails are pointing outwards and the head groups are pointing towards the core of the aggregated structure. The main driving force behind the self-assembly of amphiphilic molecules are the hydrophobic/hydrophilic interactions with the solvent.

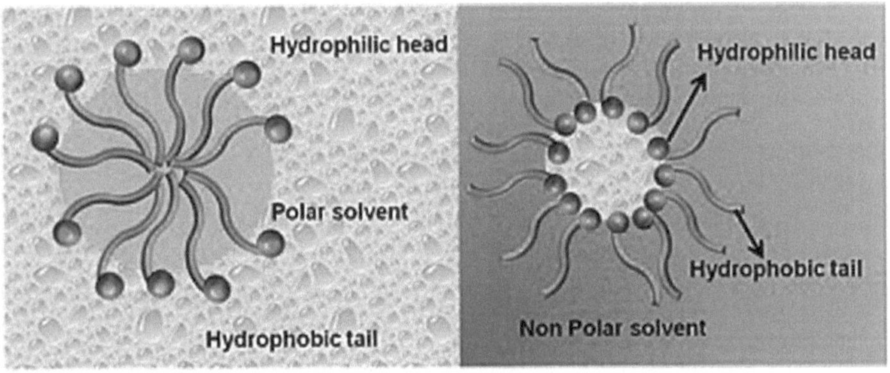

FIGURE 9.11 Thiol-ene reaction-based divergent synthesis of G2 dendrimer.

The essential conditions for the formation of micelles are as follows.

The self-assembly of molecules in the form of aggregates requires a particular concentration of amphiphilic molecules. This particular concentration, which marks the onset of aggregation of molecules in the form of micelles, is called critical micelle concentration (Lindman and Wennerstrom, 1980).

The parameters of the resulting solution, such as temperature, pH, and ionic strength, and the nature and size of the amphiphilic molecule, are decisive in controlling the shape and size of the resulting aggregated structures. The structures are observed to undergo a transition in shape from spherical to cylindrical depending upon the monomers added to the solution.

The amphiphiles that come across as constituents of micellar systems are surfactants such as CTAB, CTAC, SDS, AOT, etc. All these molecules have different hydrophilic head groups along with hydrophobic carbon chains. The size of the micellar aggregates formed by them depends on the relative sizes of both head groups and tail lengths. Apart from the long-known surfactants, lipids and proteins are also known to form micellar aggregated structures (Dubertret et al., 2002).

9.3.2.2 Vesicles and Liposomes

Vesicles and liposomes are three-dimensional structures comprising of amphiphilic molecules. The amphiphilic molecules are aggregated in the form of bilayers to form hollow spherical structures. The bilayer structures are formed as result of a strong interaction of the hydrophilic head groups of amphiphilic lipid molecules with water, thereby shielding the hydrophobic hydrocarbon chains (Davies et al., 2006).

Based on the nature of the solvent in which bilayer structures are formed, vesicles can be of two types: (i) normal vesicles and (ii) inverted vesicles. Normal vesicles are composed of a spherical bilayer of amphiphilic molecules containing an aqueous core, and the whole aggregated body is suspended in aqueous medium. On the other hand, inverted vesicles are composed of an inverted bilayer in which the hydrophobic portions of the precursor amphiphilic molecules are exposed to the solvent of low polarity, both in the core and in the exterior (Dill and Flory, 1981). Owing to the self-closed bilayer structures, these vesicular structures are capable of carrying various polar and non-polar substances and find potential applications in drug and gene delivery, molecular recognition, electronic and electrical devices, templating nanomaterials, and gelation; they are also employed as nanoreactors (Walde et al., 2010).

The amphiphilic molecules that are generally considered for vesicle formation are surfactants and lipids (glycolipids and phospholipids), whereas the vesicular structures comprising of phospholipids are known as liposomes (Bangham, 1961). The nature of the constituent amphiphiles largely affects properties such as permeability, degree of ionization, and surface potential of vesicular structures, whereas the factors such as the size, polydispersity, and physical stability of the resulting vesicles are decided by the adopted synthetic approach. Before discussing the various methods involved in the synthetic procedures for vesicles, it is important to distinguish between the two possible vesicular structures. Unilamellar vesicles are composed of only one amphiphilic bilayer, while multilamellar vesicles consist of multiple bilayers of amphiphiles (Hope et al., 1986).

Synthetic methods can be broadly classified into two classes: one does not involve any external supply of energy and is spontaneously formed on applying stress; the other involves induced vesicle formation by applying external energy in the form of sonication, extrusion, and other shearing methods (Courbin and Panizza, 2004). These are the film hydration and solvent spherule evaporation methods.

The film hydration method involves the dissolution of lipids in organic solvent followed by vacuum evaporation to form thin film (Kirby and Gregoriadis, 1984). This dry thin film is then hydrated with aqueous buffer solution along with gentle agitation resulting into multilamellar vesicles. The initial lipid concentration and intensity of shaking or agitation greatly influences the size distribution of the resulting multilamellar vesicles.

The second most followed method for the synthesis of multilamellar vesicles is reverse phase solvent spherule evaporation. This method involves the evaporation of the organic phase from the aqueous dispersion of amphiphilic lipids dissolved in volatile hydrophobic solvents (Kim et al., 1985).

The third method, known as detergent depletion technique, involves the initial dissolution of dry lipids into aqueous solutions of detergents which are later removed by different extraction techniques. The lipids adopt bilayer structures in detergents which result in unilamellar vesicles after the removal of the detergents (Schubert, 2003).

These methods can also be used to form unilamellar vesicles by controlling certain parameters, such as the concentration of amphiphile, the injection speed of the amphiphile solution of organic solvent into the aqueous phase, sonication, etc.

The most commonly employed methods for the formation of unilamellar vesicles from multilamellar vesicles are as follows.

1. Ultrasonication: This method involves the application of ultrasound waves to the solutions containing multilamellar vesicles that result in the disruption of the multilamellar lipid structures. The resulting bilayer fragments and smaller aggregates undergo rearrangement to form smaller and unilamellar structures (Kondo et al., 1995). Despite the advantages, to some extent, the method suffers from certain drawbacks such as non-uniform size distribution, instability of the resulting structures towards aggregation, metal contamination, lipid degradation.
2. Extrusion: This involves size reduction in structures by passing the suspensions through a small orifice of a defined size under pressure. This technique is known as French press and it generally involves polycarbonate filters with small pore sizes in order to achieve UV of sizes comparable to the pore size of filters (Hamilton et al., 1980).

Different techniques such as dynamic light scattering, static light scattering, atomic force microscopy and transmission electron microscopy are used to gain insight into the sizes and morphologies of vesicular structures. Also, with small-angle X-ray scattering and small-angle neutron scattering, it is possible to determine the thickness of bilayer structures (Hope et al., 1985).

9.3.2.3 Dendrimers

Dendrimers are highly branched monodispersed macromolecules with well-defined 3-dimensional structures in the nano range. The dendritic structure is divided into three regions (Figure 9.12):

(i) Core: It is generally a small molecule bearing functional groups and is decisive in the final branched structure of a dendrimer.
(ii) Inner shell: It refers to the interior layers comprising of repeating monomers radically attached to the central core. All the repeating molecules forming the inner shell are defined as the generation (G) in the dendritic structure.
(iii) Outer shell: It comprises of the surface groups present on the periphery of the dendrimer that depend largely on the number of generations present in the inner shells (Frechet and Tomalia, 2001).

A stepwise and repetitive synthetic sequence enables to control the size (number of branches) and chemistries of both inner shell and surface. Because of the chemical tunability of their surface and of their molecular architecture, dendrimers show unique chemical and physical properties as compared to linear polymers.

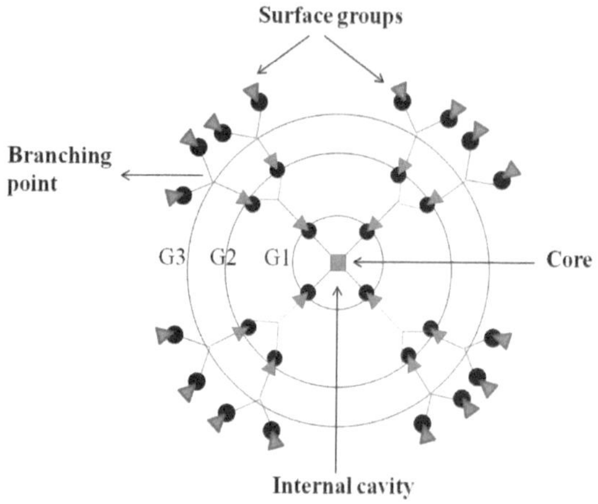

FIGURE 9.12 Structure of dendrimer.

The properties of dendrimers, such as monodispersity, water solubility, and encapsulation of important molecules in the internal cavities of dendrimer structures make them suitable for many biomedical and industrial applications. Dendrimers have been known to contribute significantly to industrial processes such as nanocatalysis. The biomedical applications based on the encapsulation tendency of the structures on account of the functional groups and the possible surface modifications have popularized dendrimers as drug and gene delivery systems, therapeutic agents, contrast agents in magnetic resonance imaging, and scaffolds in the field of regenerative medicine (Astruc et al., 2010).

The synthesis of dendrimers can be achieved in two possible ways, i.e., convergent and divergent methods (Figure 9.13)

1. Divergent method: As the name suggests, this method follows the outward growth of the molecule from the core. The monomer attaches to the core through a reactive group, while the other groups remain dormant. This marks the formation of a first generation of dendrimers. The other dormant groups present at the periphery are then activated to further attach to the next group of monomers. This repetitive activation of dormant groups and successive attachment of monomers through active groups gives rise to large dendrimer structures (Ihre et al., 2001).

The increasing number of generations in the dendrimer structure leads to a congestion of the groups at the periphery and to possible surface defects. The increasing possibility of structural defects at each additive step gives rise to difficulties in the purification and characterization of the resulting dendrimer structures. However, these drawbacks can be overcome by improving the efficiencies of synthetic methods. The classical method based on the divergent synthesis approach is a two-step process involving the coupling of monomers and subsequent activation of end groups. Highly monodisperse poly(amidoamine) (PAMAM) dendrimers up to seven generations have been prepared by following a classical two-step synthesis, i.e., Michael addition and amidation in the presence of a large amount of reagents (Tomalia et al., 1985). The other examples of classical divergent approaches include the synthesis of high-generation (G6) dendrimers based on organometallic chemistry and four-generation (G4) polyamide-based dendrimers with high yields (95 %), but involves purification using precipitation in alkaline water in each step of the synthesis. The drawbacks, such as the use of large

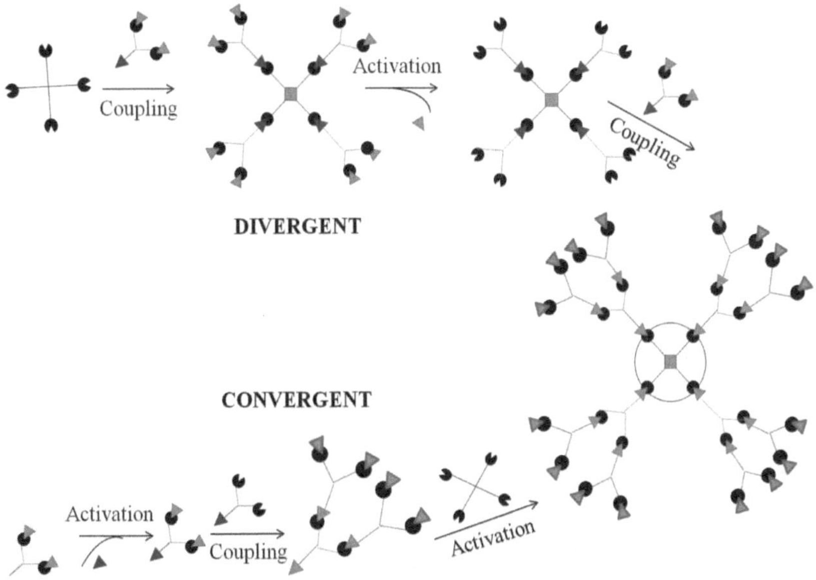

FIGURE 9.13 Divergent and convergent approaches for the synthesis of a dendrimer.

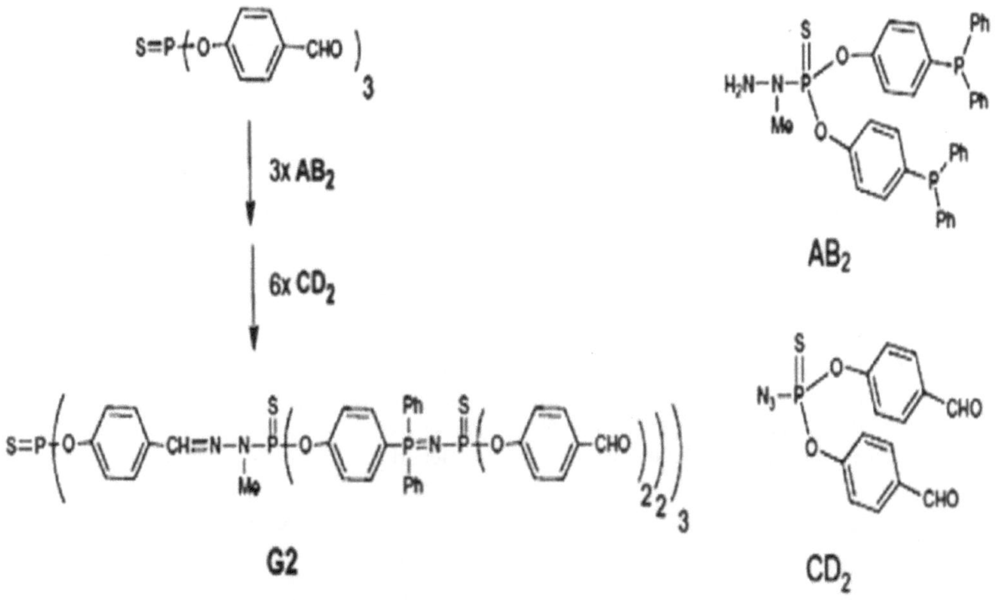

FIGURE 9.14 Orthogonal coupling strategy for the synthesis of a dendrimer.

amounts of reagents in the activation steps and the further purification of the product pave the way for advanced approaches that avoid any purification steps and give high yields of dendrimers in fewer steps as compared to classical approaches (Onitsuka et al., 2003; Washio et al., 2007).

The orthogonal strategy is an alternative to the classical approach for dendrimer synthesis (Brauge et al., 2001). It excludes the activation step of the classical approach, involving instead a careful selection of the monomers that couple selectively under particular conditions. This requires two monomers that carry orthogonal functional groups, which means that a particular functional group on the first monomer reacts only with a particular functional group under a given set of conditions. This can be exemplified with two monomers, AB_2 and CD_2, carrying orthogonal functional groups (A–D and B–C). Functional group A reacts only with functional group D on the other monomer and, similarly, functional group C reacts with B the first monomer. The typical work done by Caminade and his coworkers involved a condensation reaction between phosphorhydrazides (A) and aldehydes (D) and a Staudinger reaction between phosphines (B) and azides (C). The one-pot synthesis of G4 dendrimers was possible without any activating reagent or purification steps (Figure 9.14).

On similar lines, the synthesis of a G4 dendrimer based on poly(arylethers) was possible in a total of four steps, giving high yields. The classical synthetic method for the same dendrimer,

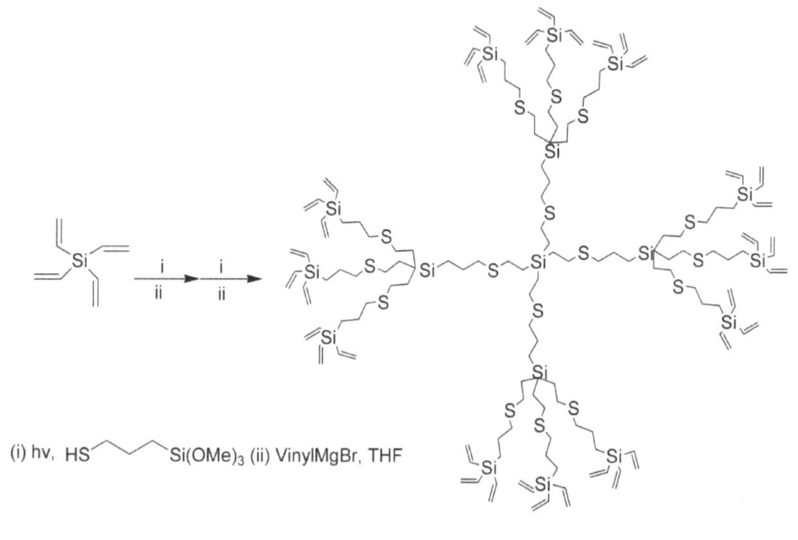

FIGURE 9.15 thiol-ene reaction based divergent synthesis of G2 dendrimer.

however, required eight steps and several purifications (Antoni et al., 2007).

The development of click chemistry reactions such as copper-catalyzed azide-alkyne 1,3 dipolar cycloaddition, Diels–Alder reactions, thiol-ene reactions, etc., has made possible the large-scale synthesis of dendrimers in environment-friendly and economic ways (Vieyres et al., 2010; Killops et al., 2008). The synthetic strategy based on the thiol-ene reaction for the preparation of G2 carbosilane-thioether dendrimers is given in Figure 9.15.

Divergent synthetic techniques offer great possibilities in tuning the dendrimer structure during its formation because of a stepwise increase in dendrimer size. These methods greatly contribute to the industrially adopted protocols for dendrimer synthesis.

2. Convergent method: This method involves the inward stepwise growth of dendrimer structures. The term convergent is used due to the fact that the total synthetic process starts from the synthesis of dendron structures constituting the periphery of the final dendrimer structure. It involves the initial synthesis of dendron structures comprising the dendrimer and then its attachment to the multifunctional core (Xu et al., 1994).

The strategy of joining a number of segments to the core at each step is advantageous because fewer bond formations occur. Therefore, the process does not require many reagents and purification steps. The single step involved in the growth of each generation assures the high monodispersity of the product. Figure 9.16 shows the convergent synthesis of a polyether dendrimer from the pre-synthesized high generation dendrons. G3 dendrons have been synthesized by a repetitive reaction between 3,5-dihydroxybenzyl alcohol and benzyl bromide. The attachment of these large dendrons to the core in a single step generates the resulting dendrimer (Hawker and Fre´chet, 1990).

The synthesis of a variety of dendrimers is possible through click-chemistry-based reactions between dendrons and multifunctional cores. Apart from the covalent interactions, non-covalent interactions between specific units are also known to result in dendrimer structures. Tomalia and coworkers synthesized

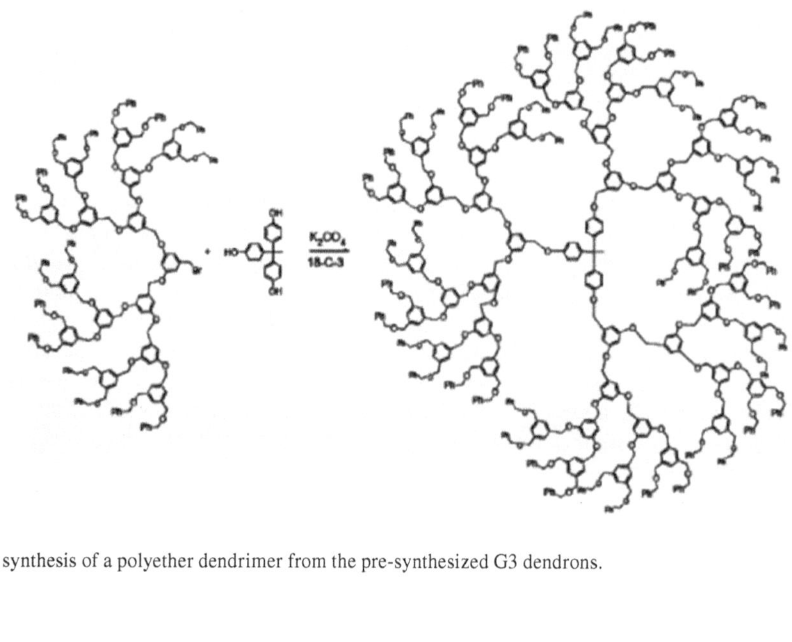

FIGURE 9.16 Convergent synthesis of a polyether dendrimer from the pre-synthesized G3 dendrons.

dendrimers starting from DNA-functionalized dendrons that were stabilized through non-covalent interactions (DeMattei et al., 2004).

It is also possible to attach different dendrons to the core, which results in unsymmetrical dendrimers possessing different properties as compared to their symmetrical counterparts. It is possible to combine the functionalities of different dendrons. The first report on an unsymmetrical dendrimer (Frechet, 1993) was based on two different dendrons bearing an electron-withdrawing and an electron-donating group, respectively (Wooley et al., 1993). Enhanced dipole moment was observed in the dendrimers resulting from these dendrons bearing different functional groups in comparison to the dendrimers resulting from symmetrical dendrons. The ease with which synthetic strategies prepare unsymmetrical dendrimers has enhanced the versatility of dendrimers as it is now possible to include a number of desired moieties in a single stable structure.

The other widely followed methodology for the efficient and accelerated synthesis of dendrimers is the double exponential growth strategy. Branched monomers carrying orthogonally protected functional groups ensure the accelerated and targeted growth of dendrons along the focal point.

The only disadvantage of the convergent methodology of dendrimer synthesis is the stearic hindrance between bulky dendrons, which results in reduced reactivity between the functional groups on the core and large dendrons.

Despite the difficulty and complexity of synthetic procedures for dendrimers, these monodispersed macromolecular structures find significant applications as biocompatible materials in the biomedical field (Svenson and Tomalia, 2012). The control of the tuning of functional groups and moieties carried by constituting dendrons in a dendrimer opens up a plethora of opportunities to design structures showing spectroscopic and other important properties.

9.3.2.4 Polymeric Nanoparticles

Polymeric nanoparticles are sub-micron-sized solid nanospheres containing polymer matrix, which, in addition to their inherent properties, are also capable of carrying a cargo molecule through encapsulation or surface adsorption. Polymeric nanoparticles are synthesized as stable colloidal dispersions known as latex. Latexes obtained from polymethacrylates, polyacrylamides, and polyalkylcyanoacrylates find application in biotechnology, molecular biology, and biomedicine for the analytical detection of soluble metal ions and as retention aids in the papermaking processes, as well as in protein separation and purification. Moreover, they are employed as substrates for living tissues, in surgical glues, and for drug delivery and diagnosis (Kumari et al., 2010). The synthesis of polymeric nanoparticles is achieved through either in situ polymerization techniques or direct synthesis from polymer molecules. The in situ polymerization techniques are as follows.

(i) Polymerization in microemulsions

In this method, the polymerization reaction occurs in the microemulsions, i.e., o/w and w/o. The monomers along with the reaction initiators are dissolved in the dispersed phase of the microemulsion. The nature of the monomers decides the choice of microemulsion type, i.e., w/o or o/w. The composition, size, and shape of the resulting polymeric nanoparticles is governed by the reaction conditions and other factors such as microemulsion composition, concentration, and solubility of monomer and initiator (Munshi et al., 1997). This method also allows the formation of polymeric nanospheres with incorporated biomolecules. The active biomolecule to be incorporated in the polymeric nanospheres is added to the microemulsion before the polymerization starts. Zhao and coworkers followed a similar methodology to synthesize polymeric nanoparticles of methylmethacrylate (40–80 nm) incorporated with fluorescent organic molecules (Zhao et al., 2006).

(ii) Direct solvent evaporation

This method utilizes dispersed-phase droplets of microemulsion formed from volatile organic solvent as templates for the preparation of nanoparticles. The solution of polymer in the volatile organic solvent is emulsified with water in presence of surfactants as emulsifiers, resulting in an o/w microemulsion (Margulis-Goshen et al., 2010). The direct evaporation of organic solvent and water through techniques such as spray drying, freeze drying, etc. results in the formation of solid polymeric nanoparticles (Figure 9.17).

A similar methodology can be adopted for the preparation of nanoparticles of important molecules such as drugs with low aqueous solubility, pesticides, and fluorescent materials.

(iii) Solvent displacement

The technique also known as nanoprecipitation causes the precipitation of polymer in the solvent present as a dispersed phase of microemulsion. Polymer dissolved in organic solvent is injected into the w/o-type microemulsion from where the polymer molecules diffuse into the aqueous cores in presence of mechanical stirring or ultrasonication. The hydrophobic polymer molecules are precipitated in the aqueous cores, forming the nuclei. The rapid collision of aqueous droplets allows the growth of the nuclei, resulting into polymeric nanoparticles (Destrée et al., 2010). The growth and size of the nanoparticles are determined by the concentration of active polymer molecules added into the microemulsion (Figure 9.18). The synthesis of polymeric nanoparticles of PLA, PLGA, PCL, and PMMA has been reported with this technique (Cirpanli et al., 2010; Schubert et al., 2011).

9.3.2.5 Polymer-Based Nanostructures

Polymer-based nanostructures comprise of polymeric aggregates or self-assemblies as well as individual polymeric nanoparticles. The most widely known polymeric self-assemblies are polymer micelles and polymerosomes. Polymer micelles consist of diblock copolymers as monomers which act as amphiphiles (Riess, 2003). One of the blocks of a diblock copolymer, being

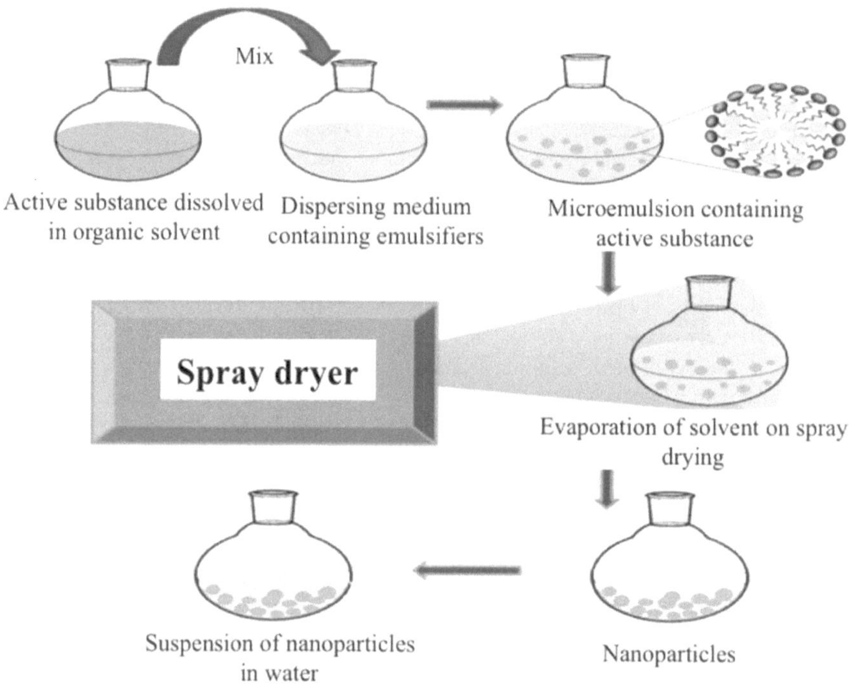

FIGURE 9.17 Direct solvent Evaporation technique.

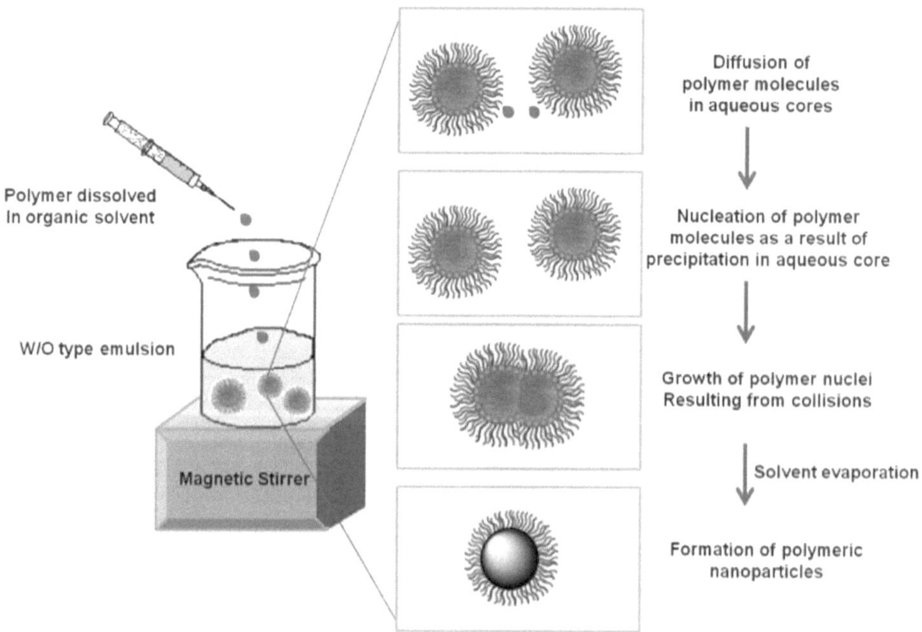

FIGURE 9.18 Solvent displacement technique for polymeric nanoparticles based on microemulsions.

soluble in the solvent, forms the shell, while the insoluble block constitutes the core of the resulting micellar structure in a given solvent. Depending on the type of block copolymer used, various structured micelles are possible (Figure 9.19).

The polymer micelles, on account of their nano range and biocompatibility, have a wide range of applications such as in emulsifiers, dispersants, wetting agents, drug carriers, nanoreactors for catalysis, etc. In general, the hydrophobic core of the micelle facilitates the encapsulation of water-insoluble drugs into the polymeric micelles. The stable micelle structure, on account of various chemical, physical, or electrostatic interactions, controls the release of the encapsulated drug. The possibility of tuning the micellar surface by attaching certain different molecules accounts for the success of polymer micelles in targeted therapy (Figure 9.20).

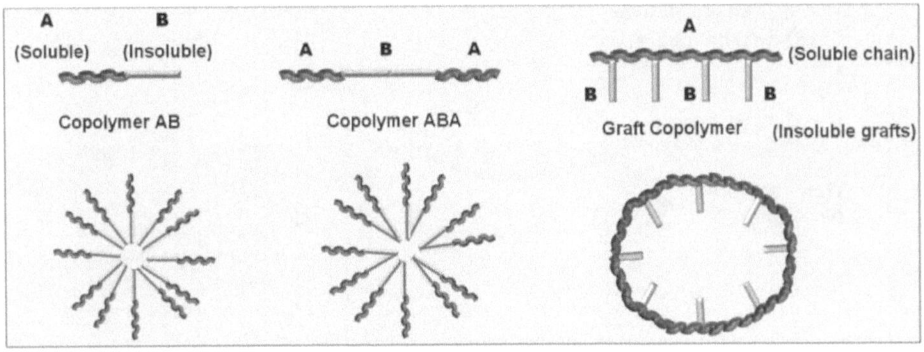

FIGURE 9.19 Various micellar structures resulting from different copolymers.

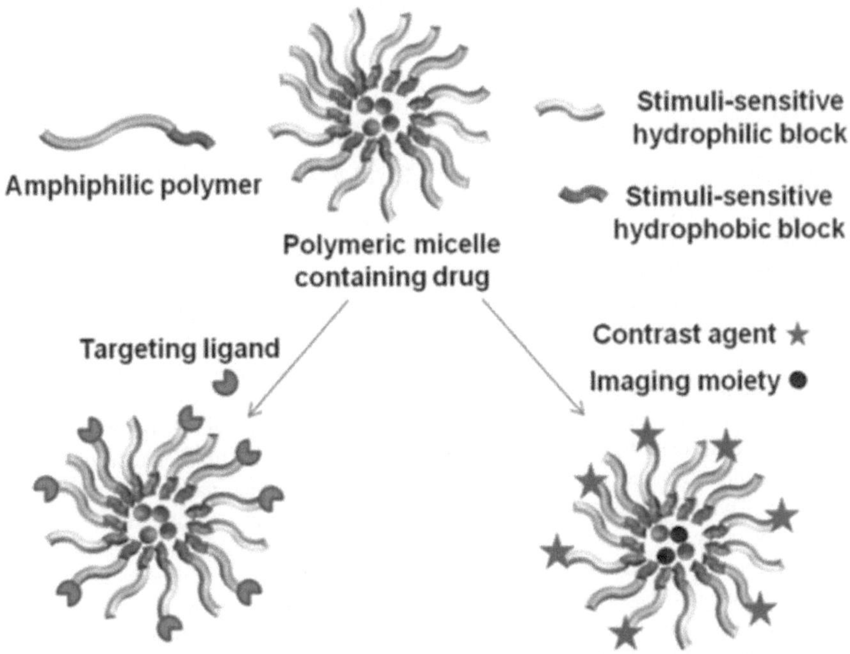

FIGURE 9.20 Micellar structure of block copolymers and possible modifications.

The safe transport of encapsulated therapeutic drugs through the body is possible due to a specific target ligand attached to the polymer block constituting the shell of the micelle (Li et al., 2012). It is also possible to incorporate the contrast agents into the micelle structure. This can be helpful in cellular imaging (Shiraishi et al., 2009). The release rate of the encapsulated drug is dependent on the stimuli-responsive groups/molecules that trigger the release of the drug in presence of stimuli which can be pH, temperature, ultrasonic wave, magnetic field, etc. Polymer micelles have been shown to be effective vehicles in targeted drug therapies involving poorly water-soluble drugs.

The next class of polymer-based aggregates is polymerosomes. As suggested by the name, polymerosomes are analogous to liposomes or vesicles with the only difference that, in polymerosomes, the amphiphilic molecule is a di- or tri-block copolymer that self-aggregates into lamellar bilayer structures, resulting into vesicular structures known as polymerosomes. These are three-dimensional spherical structures comprising of a core capable of encapsulating molecules or drugs.

Owing to their structural features, similar to those of liposomes, polymerosomes show effective drug-carrying capacities and stability towards unwanted stimuli as compared to polymer micelles (Catalin Balaure and Mihai Grumezescu, 2014). The polymerosomes constituted by block copolymers such as poly(ethylene oxide)-b-polycaprolactone and polyethyleneoxide-polyethylethylene have been reported to carry toxic chemotherapeutic drugs such as doxorubicin with tuned pharmacokinetics (Ghoroghchian et al., 2006). Apart from therapeutic drugs, polymersome loading with photoresponsive molecules or materials makes them viable as cellular labels and tracking agents in the field of diagnostics. Also, these vesicular structures are used for environment remediation by selective trapping of toxic metal ions from aqueous solutions into the cores of polymerosomes (Kapakoglou et al., 2008).

The synthetic methods for the preparation of polymerosomes involve traditional techniques such as solvent-free and solvent-displacement techniques. Solvent-free techniques include film hydration and electro formation, where the rehydration of block copolymers is assisted by the electric field, while the solvent

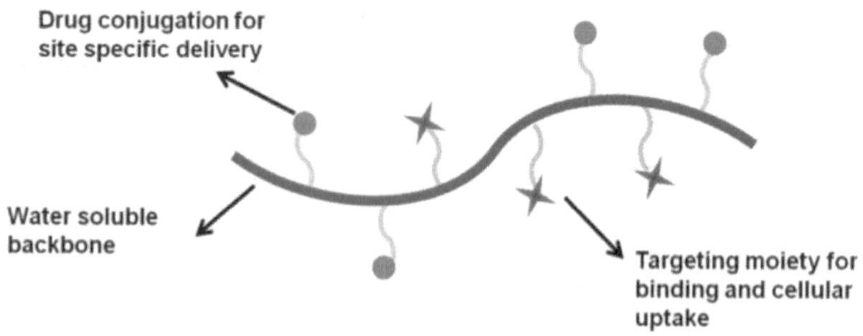

FIGURE 9.21 Model of polymer conjugate functionalized with target moiety.

displacement methods include the injection method, solvent evaporation, and detergent displacement (Discher et al., 1999). All these have been discussed in detail in the above section. Polymerosomes can also be prepared through the double-emulsion method. For example, the water in oil in water (w/o/w) double emulsion consists of water-carrying oil droplets in water with the oil phase constituted by hydrophobic organic solvents. Each block of a diblock copolymer is dissolved in respective phases of the double emulsion. The slow evaporation of organic solvents from the double emulsion causes the self-assembled structure of dissolved copolymer blocks into vesicular structures. This methodology is advantageous in terms of size control of polymerosomes and its encapsulation efficiencies (Shum et al., 2011).

Other than polymer-based aggregates structures, there are individual polymer-based nanoparticles called polymer conjugates. Polymer conjugates are water-soluble hydrid structures having well defined structures in nano sizes and narrow molecular weight distributions. The classic model of a polymer conjugate consists of a drug or bioactive molecule linked to a polymer chain through a biodegradable molecule that acts as a linker (Figure 9.21). The significance of the linker in the structure lies with it being the break point at the drug release site. The hybrid structures based on polymers show a high drug retention capacity and their bioavailability can be enhanced by modifying the structure of the macromolecules with target-specific receptor ligands (Alarcón et al., 2005).

The size of the resulting conjugate structures is decided by the length of the polymer chain. Thus, synthetic methods are mainly based on radical proliferation techniques such as atom transfer radical polymerization (ATRP) and reversible addition–fragmentation chain transfer (RAFT), which offer a high level of size control (Grover and Maynard, 2010). There are three different strategies that involve the introduction of drug molecules to the polymer at different stages of the preparation of the polymer–drug conjugate.

(i) Grafting through: It involves the molecule to be conjugated (drug or protein), which participates in the polymerization as a monomer.
(ii) Grafting from: It involves the polymerization of modified monomers carrying drug or protein molecules.
(iii) Grafting to: This requires the initial polymerization of the monomers to form a long chain, which is later functionalized with drug molecules through covalent interactions.

The different polymeric structures used for the binding of drug molecules include linear polymers (PEG, PVP, PVA, EI, N-(2hydroxypropyl methacrylamide, HPMA) and dendrimers (polyamidoamine, PMAM) (Haag and Kratz, 2006).

9.3.2.6 *Lipid-Based Nanoparticles*

The earliest lipid-based nanostructures were mainly liposomes, but technical drawbacks such as poor stability, low encapsulation efficiency, and poor cell interactions limited their use as pharmaceutical carriers. The latest lipid methodologies, including solid lipid nanoparticles and nanostructured lipid carriers, have proved to be more efficient in pharmaceutical applications such as drug delivery and release as compared to traditional liposomes. The term lipid-based nanoparticles is collectively used for both systems, i.e., solid lipid nanoparticles and nanostructured lipid carriers. Solid lipid nanoparticles and nanostructured lipid carriers are spherical structures with size range of 40–1000 nm and consist of lipids and surfactants which are dispersed in aqueous solvents. The major component is the lipid itself, whereas the surfactant, which is present in small concentrations (5–20 %), acts as a stabilizer (Müller et al., 2002). The physicochemical properties and sizes of nanoparticles are specific to the choice of lipid and surfactant as well. The most used lipids for the synthesis of solid lipid nanoparticles are fatty acids, (mono-, di-, and tri-) glycerides, steroids, and waxes.

The difference between solid lipid nanoparticles and nanostructured lipid carriers lies in the compositional differences in the lipid matrix. Solid lipid nanoparticles are made up of a lipid phase consisting of solid fat, which is stable at both body and ambient temperatures, whereas in nanostructured lipid carriers, the lipid phase comprises of both solid fats and liquid oils. Nanostructured lipid carriers are known to be modified and improved versions of solid lipid nanoparticles as they exhibit improved drug loading and stability (Naseri et al., 2015). This is attributed to various features possible in the lipid matrix of nanostructured lipid carriers, which can be divided into three classes, i.e., class I, class II, and class III. Class I refers to an imperfect type of lipid matrix which contains different crystal structures of lipid contents, i.e., fat and oils. Class II consists of lipid matrices that are amorphous in nature and formless. This involves the mixtures of certain lipids that prevent the crystallization of any component. The non-crystalline nature of the matrices in class II of nanostructured lipid carriers is

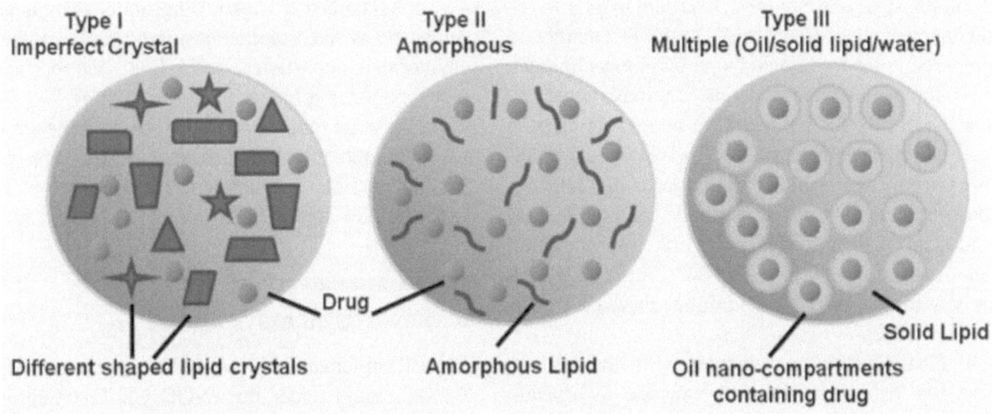

FIGURE 9.22 Various classes of nanostructured lipid carriers.

advantageous as it prevents the expulsion of loaded drug. Class III of nanostructured lipid carriers consists of a mixture of both solid (fat) and liquid (oil) phases (Jaiswal et al., 2014). In this case, the stability of the loaded drug is maintained by its higher solubility in the liquid phase compared to that of the solid phase (Figure 9.22).

9.3.2.6.1 Synthesis of Lipid-Based Nanoparticles

The incorporation of drugs the lipid nanoparticles is dependent on their synthetic method. The most commonly used methods for the synthesis of lipid nanoparticles are discussed below.

9.3.2.6.1.1 High-Pressure Homogenization The high-pressure homogenization method ensures a large-scale production of lipid nanoparticles through the homogenization of lipid-melt-containing drugs or pharmaceutical compounds under high pressure (100–2000 bar), which yield particles in the sub-micron size range (Jenning et al., 2002). Based on the temperature conditions during the process, the homogenization process can be classified into two classes:

(i) Hot homogenization: In this method, the lipid melt is loaded with a drug, at a temperature higher than the melting point of lipids, and dispersed in a hot aqueous solution containing emulsifiers. The isothermal mixing of two solutions followed by high-pressure homogenization results into lipid nanoparticles of reduced sizes. This could be attributed to reduced viscosity at higher temperatures (Parhi and Suresh, 2012). Despite the narrow-sized lipid nanoparticles, the homogenization at high temperature comes with some disadvantages, such as:
  (a) Drug degradation.
  (b) Expulsion of drug into aqueous phase.
  (c) Complex crystallization steps, formation of super-cooled melts.

(ii) Cold homogenization: This technique resolves the problems of the hot homogenization process as in this case the homogenization takes place at ambient temperature. The method involves, initially, the rapid cooling of drug-loaded lipid melt by liquid nitrogen or dry ice. The milling of this super-cooled drug-loaded lipid melt generates nanoparticles which form a suspension with a cold surfactant solution. The high-pressure homogenization of this cold emulsion leads to the breakage of nanoparticles to solid lipid nanoparticles.

Despite the fact that high-temperature/pressure conditions are involved in this process, the process is still considered to be the most exploited and viable method as it precludes the use of organic solvents and gives improved product stability and drug loading (Kamboj et al., 2010).

9.3.2.6.1.2 Solvent Emulsification This method requires an o/w emulsion containing lipid dissolved in the organic solvent and surfactant dissolved in the aqueous phase. The evaporation of solvent from the emulsion under reduced pressure leads to precipitation of lipids in aqueous phase. The resulting particle sizes of solid lipid nanoparticles vary with the choice of lipid and surfactant in the emulsion system (Trotta et al., 2003). The process does not induce any thermal stress during the formation of solid lipid nanoparticles, but the involvement of organic solvent is a disadvantage associated with the process.

9.3.2.6.1.3 Supercritical Fluid Extraction of Emulsions This method is an advancement from the solvent emulsification approach in which the solvent is removed from the emulsion via evaporation at reduced pressures. In this advanced approach, the solvent from the emulsion is extracted using supercritical carbon dioxide (Chattopadhyay et al., 2007; Campardelli et al., 2013).

9.3.2.6.1.4 Ultrasonication This technique is based on high-shear homogenization using ultrasonication. The dispersions containing lipid as oil phase and a surfactant-rich aqueous phase are homogenized to obtain lipid nanoparticles. In spite of the operational simplicity in comparison to the HPH method, homogenization under ultrasonication gives less stable nanoparticles with a wide range of size distribution (Silva et al., 2011).

9.3.2.6.1.5 Spray Drying Spray drying is a cost-effective technique in which drug-loaded solid lipid nanoparticles can be achieved. Spray drying causes the evaporation of both the organic and the aqueous phases of emulsion-carrying lipids and drug molecules in the oil phase of o/w emulsion, resulting in the formation of solid lipid nanoparticles (Zhang et al., 2008). The method is economic as compared to the traditional lyophilization techniques of extracting drug molecules from aqueous solutions. However, this method is suitable only for lipids with a melting point >70 ^{0}C, as the high temperature and shear forces involved during the process can cause particle aggregation (Mukherjee et al., 2009).

Depending on the variations in the synthetic methodologies, three different models of drug-loaded solid lipid nanoparticles are possible, including the solid solution model (Figure 9.23) and core-shell models with a drug present in either the core or the shell (Pardeshi et al., 2012).

The common disadvantages associated with solid lipid nanoparticles are low drug incorporation rate due to crystalline nature, gelation tendency, particle growth, and polymorphic transitions. All these problems are solved with nanostructured lipid carriers having enhanced stability and drug encapsulation properties owing to their unordered lipid matrix.

The advantages of lipid nanoparticles, such as economic and high-scale production, physical stability, chemical versatility, sustained drug release profiles, biocompatibility, and degradability have made them a popular subject in the field of pharmaceutical research and application (Mehnert and Mäder, 2012).

Conclusion

This chapter discusses in detail two important classes of nanostructures, i.e., nanodiamonds and organic nanoparticles. Nanodiamonds, owing to their unique structures and diverse surface chemistry, contribute significantly to the research in the fields of material science and biomedical science. These carbon-based nanostructures have unique mechanical, chemical, and optical properties, which widen their scope of application as cutting tools, surgical implants, biomarkers, and carriers of important biomolecules in targeted drug and gene therapy. In order to discover the properties of nanodiamonds and fully explore their potential applications, it is essential to develop methodologies that yield pure nanodiamonds on a large scale. The detonation synthesis of nanodiamonds is, to date, the most widely accepted method for the large-scale synthesis of nanodiamonds; however, several purification steps are involved. The other techniques for nanodiamond synthesis, such as chemical vapor deposition, high pressure, high tempertaure, laser ablation, ultrasonication, etc., produce pure nanodiamonds for biomedical applications and are also capable of introducing a few structural defects to impart new properties to nanodiamonds. Several important functional groups introduced to nanodiamond surfaces as a result of their synthetic procedures contribute to their rich and easily tunable surface chemistry, thereby allowing the attachment of active biomolecules to nanodiamonds. Similarly, organic nanoparticles comprise of a number of different nanostructures derived from organic molecules. The majority of them are based on drugs, lipids, and other biodegradable polymers. The versatility in their structures and their inherent ability to encapsulate or attach different molecules to their surfaces has put them on the forefront of many other carbon-based nanostructures that find extensive utility in pharmaceutical and biomedical applications. The chapter talks in detail about the synthetic procedure and the features of various nanostructures falling under the category of organic nanoparticles, such as micelles, vesicles, liposomes, dendrimers, polymeric nanoparticles, solid lipid nanoparticles, polymerosomes, polymer conjugates, etc.

It is believed that the in-depth study of the structural features and surface chemistries of these nanostructures gives an insight into the possible developments in their respective fields, in a search for new applications.

Acknowledgments

Ganga Ram Chaudhary would like to acknowledge the support of UGC, India under the INDO-US 21st Century Knowledge Initiative Project [F.No. 194-2/2016 (IC)]. Navneet Kaur is thankful to UGC, Delhi for the fellowships. This work is also supported by a PURSE II grant.

REFERENCES

Aharonovich, I. and Shenderova, O. 2012. Brilliant explosions. *Nat. Mater.* 11(12): 996.

Aleksenskiy, A., Baidakova, M., Osipov, V., and Vul', A. 2010. *Nanodiamonds: Applications in Biology and Nanoscale Medicine*. Springer, 55.

Amans, D., Chenus, A.C., Ledoux, G., Dujardin, C., Reynaud, C., Sublemontier, O., Masenelli-Varlot, K., and Guillois, O. 2009. Nanodiamond synthesis by pulsed laser ablation in liquids. *Diam. Relat. Mater.* 18(2–3): 177–180.

Antoni, P., Nyström, D., Hawker, C.J., Hult, A., and Malkoch, M. 2007. A chemoselective approach for the accelerated synthesis of well-defined dendritic architectures. *Chem. Comm.* 22(22): 2249–2251.

Arnault, J.C., Petit, T., Girard, H., Chavanne, A., Gesset, C., Sennour, M., and Chaigneau, M. 2011. Surface chemical modifications and surface reactivity of nanodiamonds hydrogenated by CVD plasma. *Phys. Chem. Chem. Phys.* 13(24): 11481–11487.

Astruc, D., Boisselier, E., and Ornelas, C. 2010. Dendrimers designed for functions: From physical, photophysical, and supramolecular properties to applications in sensing, catalysis, molecular electronics, photonics, and nanomedicine. *Chem. Rev.* 110(4): 1857–1959.

Aversa, R., Petrescu, R.V., Apicella, A., and Petrescu, F.I. 2016. Nano-diamond hybrid materials for structural biomedical application. *Am. J. Biochem. Biotechnol.* 13(1): 34–41.

Bach, K.C. and US Synthetic Corp. 2011. Cell assembly for use in a high-pressure cubic press. U.S. Patent 8,074,566.

Bai, P., Hu, S., Zhang, T., Sun, J., and Cao, S. 2010. Effect of laser pulse parameters on the size and fluorescence of nanodiamonds formed upon pulsed-laser irradiation. *Mater. Res. Bull.* 45(7): 826–829.

Baidakova, M.V., Kukushkina, Y.A., Sitnikova, A.A., Yagovkina, M.A., Kirilenko, D.A., Sokolov, V.V., Shestakov, M.S., Vul, A.Y., Zousman, B., and Levinson, O. 2013. Structure of nanodiamonds prepared by laser synthesis. *Phys. Solid State* 55(8): 1747–1753.

Balmer, R.S., Brandon, J.R., Clewes, S.L., Dhillon, H.K., Dodson, J.M., Friel, I., Inglis, P.N., Madgwick, T.D., Markham, M.L., Mollart, T.P., and Perkins, N. 2009. Chemical vapour deposition synthetic diamond: Materials, technology and applications. *J. Phys. Condens. Matter* 21(36): 364221.

Bangham, A.D. 1961. A correlation between surface charge and coagulant action of phospholipids. *Nature* 192(4808): 1197.

Baranov, P.G., Soltamova, A.A., Tolmachev, D.O., Romanov, N.G., Babunts, R.A., Shakhov, F.M., Kidalov, S.V., Vul', A.Y., Mamin, G.V., Orlinskii, S.B., and Silkin, N.I. 2011. Enormously high concentrations of fluorescent nitrogen-vacancy centers fabricated by sintering of detonation nanodiamonds. *Small* 7(11): 1533–1537.

Barnard, A.S., Russo, S.P. and Snook, I.K. 2003. Structural relaxation and relative stability of nanodiamond morphologies. *Diam. Relat. Mater.* 12(10–11): 1867–1872.

Bhosale, R.R., Osmani, R.A., Ghodake, P.P., Harkare, B.R., Shaikh, S.M., and Chavan, S.R. 2013. Nanodiamonds: A new-fangled drug delivery system. *J. Pharm. Res.* 3(12).

Boudou, J.P., Tisler, J., Reuter, R., Thorel, A., Curmi, P.A., Jelezko, F., and Wrachtrup, J. 2013. Fluorescent nanodiamonds derived from HPHT with a size of less than 10 nm. *Diam. Relat. Mater.* 37: 80–86.

Branson, B.T., Beauchamp, P.S., Beam, J.C., Lukehart, C.M., and Davidson, J.L. 2013. Nanodiamond nanofluids for enhanced thermal conductivity. *ACS Nano* 7(4): 3183–3189.

Brauge, L., Magro, G., Caminade, A.M., and Majoral, J.P. 2001. First divergent strategy using two AB2 unprotected monomers for the rapid synthesis of dendrimers. *J. Am. Chem. Soc.* 123(27): 6698–6699.

Butenko, Y.V., Krishnamurthy, S., Chakraborty, A.K., Kuznetsov, V.L., Dhanak, V.R., Hunt, M.R.C. and Šiller, L. 2005. Photoemission study of onionlike carbons produced by annealing nanodiamonds. *Phys. Rev. B Condens. Matter* 71(7): 075420.

Butler, J.E. and Sumant, A.V. 2008. The CVD of nanodiamond materials. *Chem. Vap. Depos.* 14(7–8): 145–160.

Campardelli, R., Cherain, M., Perfetti, C., Iorio, C., Scognamiglio, M., Reverchon, E., and Della Porta, G. 2013. Lipid nanoparticles production by supercritical fluid assisted emulsion–diffusion. *J. Supercrit. Fluids* 82: 34–40.

Catalin Balaure, P. and Mihai Grumezescu, A. 2014. Methods for synthesizing the macromolecular constituents of smart nanosized carriers for controlled drug delivery. *Curr. Med. Chem.* 21(29): 3333–3374.

Chang, I.P., Hwang, K.C., Ho, J.A.A., Lin, C.C., Hwu, R.J.R., and Horng, J.C. 2009. Facile surface functionalization of nanodiamonds. *Langmuir* 26(5): 3685–3689.

Chattopadhyay, P., Shekunov, B.Y., Yim, D., Cipolla, D., Boyd, B., and Farr, S. 2007. Production of solid lipid nanoparticle suspensions using supercritical fluid extraction of emulsions (SFEE) for pulmonary delivery using the AERx system. *Adv. Drug Deliv. Rev.* 59(6): 444–453.

Chen, X., Wang, H., Wang, H., Fu, Y., Liu, J., and Liu, R. 2018. The radiosensitizing effect of nanodiamonds (NDs) on HeLa cells under x-ray irradiation. *Phys. Status Solidi A* 215(6): 1700715.

Chiao, R.Y., Garmire, E., and Townes, C.H. 1964. Self-trapping of optical beams. *Phys. Rev. Lett.* 13(15): 479.

Chou, C.C. and Lee, S.H. 2010. Tribological behavior of nanodiamond-dispersed lubricants on carbon steels and aluminum alloy. *Wear* 269(11–12): 757–762.

Cirpanli, Y., Yerlikaya, F., Ozturk, K., Erdogar, N., Launay, M., Gegu, C., Leturgez, T., Bilensoy, E., Calis, S., and Capan, Y. 2010. Comparative evaluation of in vitro parameters of tamoxifen citrate loaded poly (lactide-co-glycolide), poly (ε-caprolactone) and chitosan nanoparticles. *Die Pharmazie-An Int. J. Pharm. Sci.* 65(12): 867–870.

Courbin, L. and Panizza, P. 2004. Shear-induced formation of vesicles in membrane phases: Kinetics and size selection mechanisms, elasticity versus surface tension. *Phys. Rev. E* 69(2): 021504.

Danilenko, V.V. 2004. On the history of the discovery of nanodiamond synthesis. *Phys. Solid State* 46(4): 595–599.

Davies, T.S., Ketner, A.M., and Raghavan, S.R. 2006. Self-assembly of surfactant vesicles that transform into viscoelastic wormlike micelles upon heating. *J. Am. Chem. Soc.* 128(20): 6669–6675.

De las Heras Alarcón, C., Pennadam, S., and Alexander, C. 2005. Stimuli responsive polymers for biomedical applications. *Chem. Soc. Rev.* 34(3): 276–285.

Degutis, G., Pobedinskas, P., Turner, S., Lu, Y.G., Al Riyami, S., Ruttens, B., Yoshitake, T., D'Haen, J., Haenen, K., Verbeeck, J., and Hardy, A. 2016. CVD diamond growth from nanodiamond seeds buried under a thin chromium layer. *Diam. Relat. Mater.* 64: 163–168.

DeMattei, C.R., Huang, B., and Tomalia, D.A. 2004. Designed dendrimer syntheses by self-assembly of single-site, ssDNA functionalized dendrons. *Nano Lett.* 4(5): 771–777.

Destree, C., Ghijsen, J., and Nagy, B.J. 2007. Preparation of organic nanoparticles using microemulsions: Their potential use in transdermal delivery. *Langmuir* 23(4): 1965–1973.

Dill, K.A. and Flory, P.J. 1981. Molecular organization in micelles and vesicles. *Proc. Natl Acad. Sci. U. S. A.* 78(2): 676–680.

Discher, B.M., Won, Y.Y., Ege, D.S., Lee, J.C., Bates, F.S., Discher, D.E., and Hammer, D.A. 1999. Polymersomes: Tough vesicles made from eblock copolymers. *Science* 284(5417): 1143–1146.

Dolmatov, V.Y. 2001. Detonation synthesis ultradispersed diamonds: Properties and applications. *Russ. Chem. Rev.* 70(7): 607–626.

Dolmatov, V.Y. 2003. *Ul'tradispersnye Almazy Detonatsionnogo Sinteza* (*Ultrasdispersed Diamonds of Detonation Synthesis*). St. Petersburg: Sankt-Peterb. Gos. Politekh. Univ.

Dolmatov, V.Y., Vehanen, A., Myllymäki, V., Rudometkin, K.A., Panova, A.N., Korolev, K.M., and Shpadkovskaya, T.A. 2013. Purification of detonation nanodiamond material using high-intensity processes. *Russ. J.Appl. Chem.* 86(7): 1036–1045.

Dubertret, B., Skourides, P., Norris, D.J., Noireaux, V., Brivanlou, A.H., and Libchaber, A. 2002. In vivo imaging of quantum dots encapsulated in phospholipid micelles. *Science* 298(5599): 1759–1762.

Dubrovinskaia, N., Dubrovinsky, L., Langenhorst, F., Jacobsen, S., and Liebske, C. 2005. Nanocrystalline diamond synthesized from C60. *Diam. Relat. Mater.* 14(1): 16–22.

Ekimov, E.A., Gromnitskaya, E.L., Gierlotka, S., Lojkowski, W., Palosz, B., Swiderska-Sroda, A., Kozubowski, J.A., and Naletov, A.M. 2002. Mechanical behavior and microstructure of nanodiamond-based composite materials. *J. Mater. Sci. Lett.* 21(21): 1699–1702.

Ekimov, E.A., Kudryavtsev, O.S., Mordvinova, N.E., Lebedev, O.I., and Vlasov, I.I. 2018. High-pressure synthesis of nanodiamonds from adamantane: Myth or reality? *ChemNanoMat* 4(3): 269–273.

Euliss, L.E., DuPont, J.A., Gratton, S., and DeSimone, J. 2006. Imparting size, shape, and composition control of materials for nanomedicine. *Chem. Soc. Rev.* 35(11): 1095–1104.

Frechet, J.M. and Tomalia, D.A. 2001. *Dendrimers and Other Dendritic Polymers*. John Wiley and Sons Inc.

Galimov, E.M., Kudin, A.M., Skorobogatskii, V.N., Plotnichenko, V.G., Bondarev, O.L., Zarubin, B.G., Strazdovskii, V.V., Aronin, A.S., Fisenko, A.V., Bykov, I.V., and Barinov, A.Y. 2004. Experimental corroboration of the synthesis of diamond in the cavitation process. *Dok. Phys.* 49(3): 150–153.

Ghoroghchian, P.P., Li, G., Levine, D.H., Davis, K.P., Bates, F.S., Hammer, D.A., and Therien, M.J. 2006. Bioresorbable vesicles formed through spontaneous self-assembly of amphiphilic poly (ethylene oxide)-block-polycaprolactone. *Macromolecules* 39(5): 1673–1675.

Gogotsi, Y. and Presser, V. 2013. *Carbon Nanomaterials*. CRC press.

Grover, G.N. and Maynard, H.D. 2010. Protein–polymer conjugates: Synthetic approaches by controlled radical polymerizations and interesting applications. *Curr. Opin. Chem. Biol.* 14(6): 818–827.

Guerrero-Cázares, H., Tzeng, S.Y., Young, N.P., Abutaleb, A.O., Quiñones-Hinojosa, A., and Green, J.J. 2014. Biodegradable polymeric nanoparticles show high efficacy and specificity at DNA delivery to human glioblastoma in vitro and in vivo. *ACS Nano* 8(5): 5141–5153.

Haag, R. and Kratz, F. 2006. Polymer therapeutics: Concepts and applications. *Angew. Chem. Int. Ed.* 45(8): 1198–1215.

Hamilton, R.L., Goerke, J., Guo, L.S., Williams, M.C., and Havel, R.J. 1980. Unilamellar liposomes made with the French pressure cell: A simple preparative and semiquantitative technique. *J. Lipid Res.* 21(8): 981–992.

Han, F.D., Yao, B., and Bai, Y.J. 2011. Preparation of carbon nano-onions and their application as anode materials for rechargeable lithium-ion batteries. *J. Phys. Chem. C* 115(18): 8923–8927.

Hawker, C.J. and Frechet, J.M. 1990. Preparation of polymers with controlled molecular architecture. A new convergent approach to dendritic macromolecules. *J. Am. Chem. Soc.* 112(21): 7638–7647.

Hill, H.G.M., Jones, A.P., and d'Hendecourt, L.B. 1998. Diamonds in carbon-rich proto-planetary nebulae. *Astron. Astro Phys.* 336: L41–L44.

Ho, D.N. 2010. Applications in Biology and Nanoscale Medicine. *Springer (US)* 10: 978–971.

Hope, M.J., Bally, M.B., Mayer, L.D., Janoff, A.S., and Cullis, P.R. 1986. Generation of multilamellar and unilamellar phospholipid vesicles. *Chem. Phys. Lipids* 40(2–4): 89–107.

Hope, M.J., Bally, M.B., Webb, G., and Cullis, P.R. 1985. Production of large unilamellar vesicles by a rapid extrusion procedure. Characterization of size distribution, trapped volume and ability to maintain a membrane potential. *Biochim. Biophys. Acta Biomembr.* 812(1): 55–65.

Horn, D. and Rieger, J. 2001. Organic nanoparticles in the aqueous phase-theory, experiment, and use. *Angew. Chem.Int. Ed.* 40(23): 4330–4361.

Iakoubovskii, K., Baidakova, M.V., Wouters, B.H., Stesmans, A., Adriaenssens, G.J., Vul, A.Y., and Grobet, P.J. 2000. Structure and defects of detonation synthesis nanodiamond. *Diam. Relat. Mater.* 9(3–6): 861–865.

Ihre, H., Padilla De Jesús, O.L., and Fréchet, J.M. 2001. Fast and convenient divergent synthesis of aliphatic ester dendrimers by anhydride coupling. *J. Am. Chem. Soc.* 123(25): 5908–5917.

Jaiswal, P., Gidwani, B., and Vyas, A. 2016. Nanostructured lipid carriers and their current application in targeted drug delivery. *Artif. Cell Nanomed. Biotechnol.* 44(1): 27–40.

Jana, A., Devi, K.S.P., Maiti, T.K., and Singh, N.P. 2012. Perylene-3-ylmethanol: Fluorescent organic nanoparticles as a single-component photoresponsive nanocarrier with real-time monitoring of anticancer drug release. *J. Am. Chem. Soc.* 134(18): 7656–7659.

Jarre, G., Liang, Y., Betz, P., Lang, D., and Krueger, A. 2011. Playing the surface game—Diels–Alder reactions on diamond nanoparticles. *Chem. Comm.* 47(1): 544–546.

Jenning, V., Lippacher, A., and Gohla, S.H. 2002. Medium scale production of solid lipid nanoparticles (SLN) by high pressure homogenization. *J. Microencapsul.* 19(1): 1–10.

Kamboj, S., Bala, S., and Nair, A.B. 2010. Solid lipid nanoparticles: An effective lipid based technology for poorly water soluble drugs. *Int. J. Pharm. Sci. Rev. Res.* 5(2): 78–90.

Kapakoglou, N.I., Giokas, D.L., Tsogas, G.Z., and Vlessidis, A.G. 2008. Coacervation of surface-functionalized polymerized vesicles derived from ammonium bromide surfactants. Application to the selective speciation of chromium in environmental samples. *Anal. Chem.* 80(24): 9787–9796.

Kaur, N., Kaur, G., Bhalla, A., Dhau, J.S., and Chaudhary, G.R. 2018. Metallosurfactant based Pd–Ni alloy nanoparticles as a proficient catalyst in the Mizoroki Heck coupling reaction. *Green Chem.* 20(7): 1506–1514.

Kaur, N., Kaur, S., Kaur, G., Bhalla, A., Srinivasan, S., and Chaudhary, G.R. 2019. Metallovesicles as smart nanoreactors for green catalytic synthesis of benzimidazole derivatives in water. *J. Mater. Chem. A* 7(29): 17306–17314.

Khachatryan, A.K., Aloyan, S.G., May, P.W., Sargsyan, R., Khachatryan, V.A. and Baghdasaryan, V.S. 2008. Graphite-to-diamond transformation induced by ultrasound cavitation. *Diam. Relat. Mater.* 17(6): 931–936.

Killops, K.L., Campos, L.M., and Hawker, C.J. 2008. Robust, efficient, and orthogonal synthesis of dendrimers via thiol-ene "click" chemistry. *J. Am. Chem. Soc.* 130(15): 5062–5064.

Kim, S., Jacobs, R.E., and White, S.H. 1985. Preparation of multilamellar vesicles of defined size-distribution by solvent-spherule evaporation. *Biochim. Biophys. Acta Biomembr.* 812(3): 793–801.

Kirby, C. and Gregoriadis, G. 1984. Dehydration-rehydration vesicles: A simple method for high yield drug entrapment in liposomes. *Biotechnology* 2(11): 979.

Kondo, Y., Uchiyama, H., Yoshino, N., Nishiyama, K., and Abe, M. 1995. Spontaneous vesicle formation from aqueous solutions of didodecyldimethylammonium bromide and sodium dodecyl sulfate mixtures. *Langmuir* 11(7): 2380–2384.

Krueger, A. 2008. The structure and reactivity of nanoscale diamond. *J. Mater. Chem.* 18(13): 1485–1492.

Kumar, R. and Lal, S. 2014. Synthesis of organic nanoparticles and their applications in drug delivery and food nanotechnology: A review. *J. Nanomater. Molec. Nanotech.* 3: 4.

Kumari, A., Yadav, S.K., and Yadav, S.C. 2010. Biodegradable polymeric nanoparticles based drug delivery systems. *Colloids Surf. B Biointerfaces* 75(1): 1–18.

Kuznetsov, V.L., Chuvilin, A.L., Butenko, Y.V., Mal'kov, I.Y., and Titov, V.M. 1994. Onion-like carbon from ultra-disperse diamond. *Chem. Phys. Lett.* 222(4): 343–348.

Kuznetsov, V.L., Chuvilin, A.L., Moroz, E.M., Kolomiichuk, V.N., Shaikhutdinov, S.K., Butenko, Y.V., and Mal'kov, I.Y. 1994. Effect of explosion conditions on the structure of detonation soots: Ultradisperse diamond and onion carbon. *Carbon* 32(5): 873–882.

Li, K. and Liu, B. 2014. Polymer-encapsulated organic nanoparticles for fluorescence and photoacoustic imaging. *Chem. Soc. Rev.* 43(18): 6570–6597.

Li, X., Zhang, Z., Li, J., Sun, S., Weng, Y., and Chen, H. 2012. Diclofenac/biodegradable polymer micelles for ocular applications. *Nanoscale* 4(15): 4667–4673.

Liang, Y., Meinhardt, T., Jarre, G., Ozawa, M., Vrdoljak, P., Schöll, A., Reinert, F., and Krueger, A. 2011. Deagglomeration and surface modification of thermally annealed nanoscale diamond. *J. Colloid Interface Sci.* 354(1): 23–30.

Liang, Y., Ozawa, M., and Krueger, A. 2009. A general procedure to functionalize agglomerating nanoparticles demonstrated on nanodiamond. *ACS Nano* 3(8): 2288–2296.

Lindman, B. and Wennerström, H. 1980. Micelles. In: *Micelles*, 1–83. Berlin, Heidelberg: Springer.

Liu, W.W., Chai, S.P., Mohamed, A.R., and Hashim, U. 2014. Synthesis and characterization of graphene and carbon nanotubes: A review on the past and recent developments. *J. Ind. Eng. Chem.* 20(4): 1171–1185.

Liu, Y., Gu, Z., Margrave, J.L., and Khabashesku, V.N. 2004. Functionalization of nanoscale diamond powder: Fluoro-, alkyl-, amino-, and amino acid-nanodiamond derivatives. *Chem. Mater.* 16(20): 3924–3930.

Maitra, U., Prasad, K.E., Ramamurty, U., and Rao, C.N.R. 2009. Mechanical properties of nanodiamond-reinforced polymer-matrix composites. *Solid State Commun.* 149(39–40): 1693–1697.

Mandal, M., Haso, F., Liu, T., Fei, Y., and Landskron, K. 2014. Size tunable synthesis of solution processable diamond nanocrystals. *Chem. Comm.* 50(77): 11307–11310.

Margulis-Goshen, K., Netivi, H.D., Major, D.T., Gradzielski, M., Raviv, U., and Magdassi, S. 2010. Formation of organic nanoparticles from volatile microemulsions. *J. Colloid Interface Sci.* 342(2): 283–292.

Mehnert, W. and Mäder, K. 2012. Solid lipid nanoparticles: Production, characterization, and applications. *Adv. Drug Deliv. Rev.* 64: 83–101.

Mitev, D., Dimitrova, R., Spassova, M., Minchev, C., and Stavrev, S. 2007. Surface peculiarities of detonation nanodiamonds in dependence of fabrication and purification methods. *Diam. Relat. Mater.* 16(4–7): 776–780.

Mukherjee, S., Ray, S., and Thakur, R.S. 2009. Solid lipid nanoparticles: A modern formulation approach in drug delivery system. *Indian J. Pharma. Sci.* 71(4): 349.

Müller, R.H., Gohla, S., and Keck, C.M. 2011. State of the art of nanocrystals–special features, production, nanotoxicology aspects and intracellular delivery. *Eur. J Pharma. Biopharma* 78(1): 1–9.

Müller, R.H., Radtke, M., and Wissing, S.A. 2002. Solid lipid nanoparticles (SLN) and nanostructured lipid carriers (NLC) in cosmetic and dermatological preparations. *Adv. Drug Deliv. Rev.* 54 Supplement 1: 131–155.

Munshi, N., De, T.K., and Maitra, A. 1997. Size modulation of polymeric nanoparticles under controlled dynamics of microemulsion droplets. *J. Colloid Interface Sci.* 190(2): 387–391.

Naseri, N., Valizadeh, H., and Zakeri-Milani, P. 2015. Solid lipid nanoparticles and nanostructured lipid carriers: Structure, preparation and application. *Adv. Pharma. Bull.* 5(3): 305.

Ng, K.K. and Zheng, G. 2015. Molecular interactions in organic nanoparticles for phototheranostic applications. *Chem. Rev.* 115(19): 11012–11042.

Onitsuka, K., Shimizu, A., and Takahashi, S. 2003. A divergent approach to the precise synthesis of giant organometallic dendrimers using platinum–acetylides as building blocks. *Chem. Comm.* 2(2): 280–281.

Osswald, S., Yushin, G., Mochalin, V., Kucheyev, S.O., and Gogotsi, Y. 2006. Control of sp2/sp3 carbon ratio and surface chemistry of nanodiamond powders by selective oxidation in air. *J. Am. Chem. Soc.* 128(35): 11635–11642.

Ozawa, M., Inaguma, M., Takahashi, M., Kataoka, F., Krueger, A., and Ōsawa, E. 2007. Preparation and behavior of brownish, clear nanodiamond colloids. *Adv. Mater.* 19(9): 1201–1206.

Pardeshi, C., Rajput, P., Belgamwar, V., Tekade, A., Patil, G., Chaudhary, K., and Sonje, A. 2012. Solid lipid based nanocarriers: An overview. *Acta Pharm.* 62(4): 433–472.

Parhi, R. and Suresh, P. 2012. Preparation and characterization of solid lipid nanoparticles – A review. *Curr. Drug Discov. Technol.* 9(1): 2–16.

Purtov, K.V., Burakova, L.P., Puzyr, A.P., and Bondar, V.S. 2008. The interaction of linear and ring forms of DNA molecules with nanodiamonds synthesized by detonation. *Nanotechnology* 19(32): 325101.

Qin, W., Ding, D., Liu, J., Yuan, W.Z., Hu, Y., Liu, B., and Tang, B.Z. 2012. Biocompatible nanoparticles with aggregation-induced emission characteristics as far-red/near-infrared fluorescent bioprobes for in vitro and in vivo imaging applications. *Adv. Funct. Mater.* 22(4): 771–779.

Rabeau, J.R., Stacey, A., Rabeau, A., Prawer, S., Jelezko, F., Mirza, I., and Wrachtrup, J. 2007. Single nitrogen vacancy centers in chemical vapor deposited diamond nanocrystals. *Nano Lett.* 7(11): 3433–3437.

Riess, G. 2003. Micellization of block copolymers. *Prog. Polym. Sci.* 28(7): 1107–1170.

Rolland, J.P., Hagberg, E.C., Denison, G.M., Carter, K.R., and De Simone, J.M. 2004. High-resolution soft lithography: Enabling materials for nanotechnologies. *Angew. Chem.Int. Ed.* 43(43): 5796–5799.

Rolland, J.P., Maynor, B.W., Euliss, L.E., Exner, A.E., Denison, G.M., and DeSimone, J.M. 2005. Direct fabrication and harvesting of monodisperse, shape-specific nanobiomaterials. *J. Am. Chem. Soc.* 127(28): 10096–10100.

Say, J.M., Bradac, C., Vreden, C.V., Hill, C., Reilly, D., King, N., Herbert, B., Brown, L., and Rabeau, J.R. 2011. Fluorescent nanodiamonds for biological applications. *Quantum Electronics Conference & Lasers and Electro-Optics (CLEO/IQEC/PACIFIC RIM)*, IEEE, pp. 1708–1710.

Schubert, R. 2003. Liposome preparation by detergent removal. *Methods Enzymol.* 367: 46–70.

Schubert, S., Delaney Jr, J.T., and Schubert, U.S. 2011. Nanoprecipitation and nanoformulation of polymers: From history to powerful possibilities beyond poly (lactic acid). *Soft Matter* 7(5): 1581–1588.

Shchukin, D.G. and Sukhorukov, G.B. 2004. Nanoparticle synthesis in engineered organic nanoscale reactors. *Adv. Mater.* 16(8): 671–682.

Shenderova, O., Koscheev, A., Zaripov, N., Petrov, I., Skryabin, Y., Detkov, P., Turner, S., and Van Tendeloo, G. 2011. Surface chemistry and properties of ozone-purified detonation nanodiamonds. *J. Phys. Chem. C* 115(20): 9827–9837.

Shenderova, O. and Nunn, N. 2017. Production and purification of nanodiamonds. In: *Nanodiamonds*. Elsevier.

Shenderova, O.A. and Gruen, D.M. 2012. *Ultrananocrystalline Diamond: Synthesis, Properties and Applications*. William Andrew.

Shiraishi, K., Kawano, K., Minowa, T., Maitani, Y., and Yokoyama, M. 2009. Preparation and in vivo imaging of PEG-poly (L-lysine)-based polymeric micelle MRI contrast agents. *J. Control. Release* 136(1): 14–20.

Shum, H.C., Zhao, Y.J., Kim, S.H., and Weitz, D.A. 2011. Multicompartment polymersomes from double emulsions. *Angew. Chem. Int. Ed. Engl.* 50(7): 1648–1651.

Silva, A.C., González-Mira, E., García, M.L., Egea, M.A., Fonseca, J., Silva, R., Santos, D., Souto, E.B., and Ferreira, D. 2011. Preparation, characterization and biocompatibility studies on risperidone-loaded solid lipid nanoparticles (SLN): High pressure homogenization versus ultrasound. *Colloids Surf. B Biointerfaces* 86(1): 158–165.

Singh, B.N., Gupta, V.K., Chen, J., and Atanasov, A.G. 2017. Organic nanoparticle-based combinatory approaches for gene therapy. *Trends Biotechnol.* 35(12): 1121–1124.

Stacey, A., Aharonovich, I., Prawer, S., and Butler, J.E. 2009. Controlled synthesis of high-quality micro/nano-diamonds by microwave plasma chemical vapor deposition. *Diam. Relat. Mater.* 18(1): 51–55.

Sumiya, H., Toda, N., and Satoh, S. 2005. Development of high-quality large-size synthetic diamond crystals. *SEI Tech. Rev.-English Edition* 60: 10.

Sun, J., Hu, S.L., Du, X.W., Lei, Y.W., and Jiang, L. 2006. Ultrafine diamond synthesized by long-pulse-width laser. *Appl. Phys. Lett.* 89(18): 183115.

Sung, C.M. 2008. Polycrystalline grits and associated methods. U.S. Patent 7,384,436.

Svenson, S. and Tomalia, D.A. 2012. Dendrimers in biomedical applications—Reflections on the field. *Adv. Drug Deliv. Rev.* 64: 102–115.

Texter, J. 2001. Precipitation and condensation of organic particles. *J. Dispers. Sci. Technol.* 22(6): 499–527.

Tomalia, D.A., Baker, H., Dewald, J., Hall, M., Kallos, G., Martin, S., Roeck, J., Ryder, J., and Smith, P. 1985. A new class of polymers: Starburst-dendritic macromolecules. *Pol. J. Chem.* 17(1): 117.

Trotta, M., Debernardi, F., and Caputo, O. 2003. Preparation of solid lipid nanoparticles by a solvent emulsification–diffusion technique. *Int. J. Pharm.* 257(1–2): 153–160.

Ullah, M., Kausar, A., Siddiq, M., Subhan, M., and Abid Zia, M. 2015. Reinforcing effects of modified nanodiamonds on the physical properties of polymer-based nanocomposites: A review. *Polym. Plast. Technol. Eng.* 54(8): 861–879.

Vaijayanthimala, V. and Chang, H.C. 2009. Functionalized fluorescent nanodiamonds for biomedical applications. *Nanomedicine* 4(1): 47–55.

Viecelli, J.A., Bastea, S., Glosli, J.N., and Ree, F.H. 2001. Phase transformations of nanometer size carbon particles in shocked hydrocarbons and explosives. *J. Chem. Phys.* 115(6): 2730–2736.

Vieyres, A., Lam, T., Gillet, R., Franc, G., Castonguay, A., and Kakkar, A. 2010. Combined CuI-catalysed alkyne–azide cycloaddition and furan–maleimide Diels–Alder "click" chemistry approach to thermoresponsive dendrimers. *Chem. Comm.* 46(11): 1875–1877.

Vlasov, I.I., Shenderova, O., Turner, S., Lebedev, O.I., Basov, A.A., Sildos, I., Rähn, M., Shiryaev, A.A., and Van Tendeloo, G. 2010. Nitrogen and luminescent nitrogen-vacancy defects in detonation nanodiamond. *Small* 6(5): 687–694.

Walde, P., Cosentino, K., Engel, H., and Stano, P. 2010. Giant vesicles: Preparations and applications. *ChemBioChem* 11(7): 848–865.

Wang, D. and Astruc, D. 2017. The recent development of efficient earth-abundant transition-metal nanocatalysts. *Chem. Soc. Rev.* 46(3): 816–854.

Wang, J.B., Zhang, C.Y., Zhong, X.L., and Yang, G.W. 2002. Cubic and hexagonal structures of diamond nanocrystals formed upon pulsed laser induced liquid–solid interfacial reaction. *Chem. Phys. Lett.* 361(1–2): 86–90.

Washio, I., Shibasaki, Y., and Ueda, M. 2007. Facile synthesis of amine-terminated aromatic polyamide dendrimers via a divergent method. *Org. Lett.* 9(7): 1363–1366.

Williams, O.A. ed. 2014. *Nanodiamond*. Royal Society of Chemistry.

Williams, O.A., Nesladek, M., Daenen, M., Michaelson, S., Hoffman, A., Osawa, E., Haenen, K., and Jackman, R.B. 2008. Growth, electronic properties and applications of nanodiamond. *Diam. Relat. Mater.* 17(7–10): 1080–1088.

Wooley, K.L., Hawker, C.J., and Frechet, J.M. 1993. Unsymmetrical three-dimensional macromolecules: Preparation and characterization of strongly dipolar dendritic macromolecules. *J. Am. Chem. Soc.* 115(24): 11496–11505.

Xu, Z., Kahr, M., Walker, K.L., Wilkins, C.L., and Moore, J.S. 1994. Phenylacetylene dendrimers by the divergent, convergent, and double-stage convergent methods. *J. Am. Chem. Soc.* 116(11): 4537–4550.

Yang, G.W., Wang, J.B., and Liu, Q.X. 1998. Preparation of nano-crystalline diamonds using pulsed laser induced reactive quenching. *J. Phys. Condens. Matter* 10(35): 7923.

Yang, L., May, P.W., Yin, L., Smith, J.A., and Rosser, K.N. 2007. Growth of diamond nanocrystals by pulsed laser ablation of graphite in liquid. *Diam. Relat. Mater.* 16(4–7): 725–729.

Yu, S.J., Kang, M.W., Chang, H.C., Chen, K.M., and Yu, Y.C. 2005. Bright fluorescent nanodiamonds: No photobleaching and low cytotoxicity. *J. Am. Chem. Soc.* 127(50): 17604–17605.

Zang, J.B., Wang, Y.H., Zhao, S.Z., Bian, L.Y., and Lu, J. 2007. Electrochemical properties of nanodiamond powder electrodes. *Diam. Relat. Mater.* 16(1): 16–20.

Zhang, H., Wang, D., Butler, R., Campbell, N.L., Long, J., Tan, B., Duncalf, D.J., Foster, A.J., Hopkinson, A., Taylor, D., and Angus, D. 2008. Formation and enhanced biocidal activity of water-dispersable organic nanoparticles. *Nat. Nanotech.* 3(8): 506.

Zhang, Q., Mochalin, V.N., Neitzel, I., Knoke, I.Y., Han, J., Klug, C.A., Zhou, J.G., Lelkes, P.I., and Gogotsi, Y. 2011. Fluorescent PLLA-nanodiamond composites for bone tissue engineering. *Biomaterials* 32(1): 87–94.

Zhang, X., Pan, W., Gan, L., Zhu, C., Gan, Y., and Nie, S. 2008. Preparation of a dispersible PEGylate nanostructured lipid carriers (NLC) loaded with 10-hydroxycamptothecin by spray-drying. *Chem. Pharm. Bull.* 56(12): 1645–1650.

Zhang, Y., Rhee, K.Y., Hui, D., and Park, S.J. 2018. A critical review of nanodiamond based nanocomposites: Synthesis, properties and applications. *Composites B* 143: 19–27.

Zhao, L., Lei, Z., Li, X., Li, S., Xu, J., Peng, B., and Huang, W. 2006. A novel approach of preparation and patterning of organic fluorescent nanomaterials. *Chem. Phys. Lett.* 420(4–6): 480–483.

Zhu, Y., Li, J., Li, W., Zhang, Y., Yang, X., Chen, N., Sun, Y., Zhao, Y., Fan, C., and Huang, Q. 2012. The biocompatibility of nanodiamonds and their application in drug delivery systems. *Theranostics* 2(3): 302.

Zousman, B. and Levinson, O. 2014. Pure nanodiamonds produced by laser-assisted technique. In: *Nanodiamond Royal Society of Chemistry*, pp. 112–127.

10

Polymeric Nanoparticles: Preparation and Surface Modification

A. Chander, R. Santhosh, S. Avinash, M. Priyanka, T. Guping, B. Murali, R. Karthik, and S. N. Rath

CONTENTS

10.1 Introduction

Globally, there is a lot of interest in the fields of nanotechnology and nanobiotechnology through the applications of polymeric nanoparticles (Wu et al., 2018; Wang et al., 2015; Sun et al., 2014; Puskas et al., 2011). Researchers designed and developed various nanomaterials such as carbon dots and nanotubes (CNTs) (Fadel et al., 2014), quantum dots (Bradburne et al., 2013), metallic nanoparticles (gold and silver nanoparticles) (Li et al., 2018), magnetic nanoparticles (iron oxide nanoparticles), SnO_2 nanospheres (Li et al., 2014), protein-SiO_2 nanoparticles (Argyo et al., 2014; Wang et al., 2013), etc. All the above-mentioned nanoparticles are non-polymeric. Apart from these nanomaterials, polymeric nanoparticles and nanocapsules such as dendrimers, core-shell nanoparticles, block copolymers, mesoporous silica nanoparticles (Lai et al., 2003), graphene nanosheets (Wang et al., 2011), hollow polymer nanoparticles (Li et al., 2014; Wang et al., 2005), hollow core-shell nanoparticles, polymer blends, polymer nanocomposites, polymeric micelles, lipid nanocapsules (Mona et al., 2011; Venturini et al., 2011), liposomes, and chitosan (Ortegaa et al., 2011) have also been reported in the literature. The interest in the synthesis of polymers followed by the development of polymeric nanoparticles is rising because of their facile synthesis, engineering them by significant functionalization and surface modifications based on nanotechnological applications. These polymeric nanoparticles are very useful in the fields of drug-, cell-, and gene-delivery systems. The porosity of polymer nanoparticles is a unique and novel characteristic property to load and release drug molecules with tumor and target specificity. Apart from the above-mentioned polymeric nanoparticles, there are amino acid–based [(L-GluA)-(PCL)] (PCL-polycaprolactone) nanoporous polymeric carrier particles/capsules which are much more suitable for biomedical and pharmaceutical applications. Polymeric microparticles (particle size <2 nm, in diameter), meso-particles (particle size 2–50 nm), and nanoparticles (particle size 50–200 nm) have been involved as effective and potential particles for many applications in various fields (Lai et al., 2014). Various biocompatible polymers with their chemical formula and structure are listed in Table 10.1. However, due to side effects and biocompatibility issues, quantum dots and carbon nanotubes are entirely banned or avoided in the fields of biomedicals and pharmaceuticals.

10.2 Polymers

The word polymer derives from the Greek with 'poly' meaning 'many and 'mer/s' means 'units or parts'. We aware of that unit cells are the smallest repeating units in the construction of crystal. Similarly, amino acids are the building block of proteins. Polymers are mainly prepared based on two polymerization reactions such as (1) addition polymerization and (2) condensation polymerization.

The 'n' in Table 10.1 represents the number of monomer units involved in the preparation and synthesis of polymer molecules. There are various modified chemical synthetic polymerization methods apart from these polymerization processes. Based on their occurrence, polymers are classified into three types: (1) natural polymers such as cellulose, chitosan, and starch, (2) synthetic polymers such as polymethyl methacrylate amide, nylon-6,6, and polystyrene, and (3) semi-synthetic polymers such as cellulose derivatives – cellulose nitrate, vulcanized rubber, rayon, and guncotton. There are some polymers that are extensively utilized in plastic industries, for example, polyvinyl chloride. These polymers are further categorized into low-density and high-density polymers based on their usage (Magnusson et al., 2011; Teixeira et al., 2005; Liu et al., 2014). Polymer tetrafluoroethane $(C_2F_4)_n$ has been utilized for teflon tape as a sealant for leakages. Further, isoprene used for natural rubber and vulcanized sulfer has been used for tyre and plastic industry. The combination of formaldehyde and benzophenol referred as bakalite and used for design and development of combs and other plastic wares (Bertram et

TABLE 10.1

List of the Different Biocompatible Polymers with Their Chemical Formula and Structure

S. No.	Polymer name	Polymer	Structure	Ref.
1	PCL (Polycaprolactone)	$(C_6H_{10}O_2)_n$	Fig 10.1(a)	(Baudry et al., 2003) (Wang et al., 2013)
2	PLGA [Poly(lactic-co-glycolic acid)]	$(C_4H_6O_2)_n$	Fig 10.1(b)	(Teixeira et al., 2005)
3	PNIPAM (Poly-N-isopropyl acrylamide)	$(C_6H_{11}NO)_n$	Fig 10.1(c)	(Li et al., 2014) (Guo et al., 2006)
4	PEG (Polyethylene glycol)	$(C_{2n}H_{4n+2}O_{n+1})_n$	Fig 10.1(d)	(Quadir et al., 2014)
5	PS (Polystyrene)	$(C_8H_8)_n$	Fig 10.1(e)	(Nuruzatulifah et al., 2016)
6	PMMA (Polymethyl methacrylate)	$(C_5O_2H_8)_n$	Fig 10.1(f)	(Yuan et al., 2013)

al., 2009; Chaubey et al., 2014; Li et al., 2014). Thechemical Structure of some polymers along with their active functional groups are shown in Figure 10.1.

10.3 Polymer Properties

The selection of polymers for the design and development of nanoparticles possessing specific applications depends on various factors such as synthetic methods, molecular weight, density, network forming nature, cross-linking nature, branching, surface tension, hydrophobic/hydrophilic nature, and interlocking nature, followed by the self-assembly among active functional groups present in the polymers (Figure 10.2). Based on their applications, polymeric nanoparticles should be biocompatible, bio-safe, and biodegradable; moreover, the viscosity and degradation temperature of polymers for biomedical applications should be reasonably low (Bergbreiter et al., 1998; Yin et al., 2018).

Besides, polymer particles have to show good cell viability for pharmaceutical usages. However, every nanoparticle system has limitations for their use in biotechnology and nanotechnology applications (Song and Kretzschmar, 2009; Ashjari et al., 2012). The development of surfactant-free and cost-effective nanoparticles without using any harmful catalysts and initiators is a novel approach to prepare green nanoparticles (~200 nm). The term green nanoparticles refers to nanoparticles that have no harmful side effects upon usage in biomedical therapies and treatments. Amino acid–based nanoparticles are termed green nanoparticles

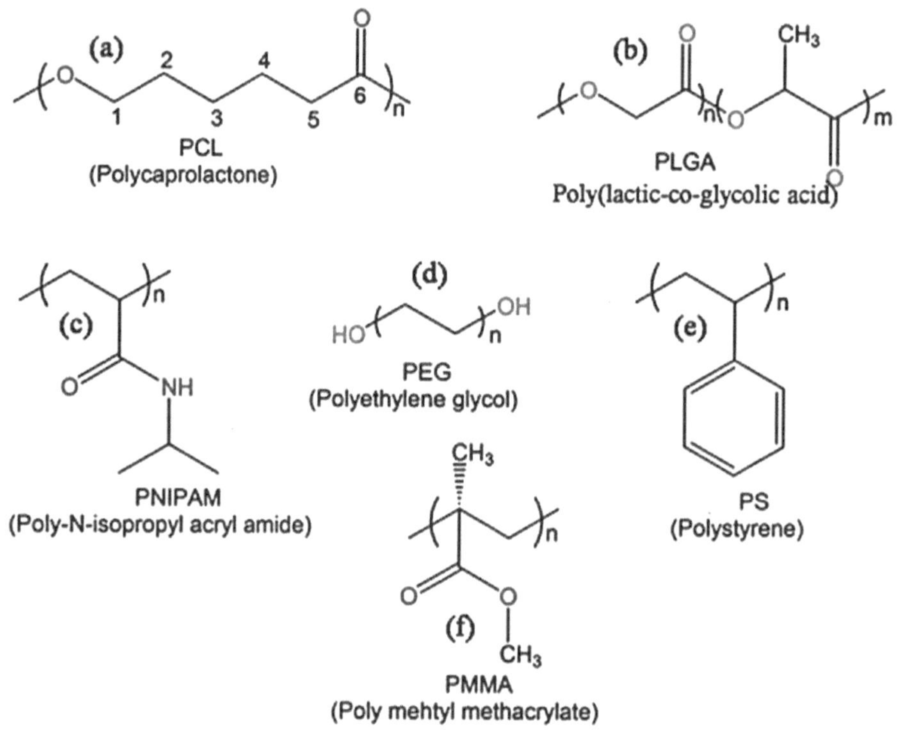

FIGURE 10.1 Chemical structure of some polymers along with their active functional groups involved in the self-assembly for the formulation of polymer-based nanoparticles (a-f) Polymers further described in Table 10.1.

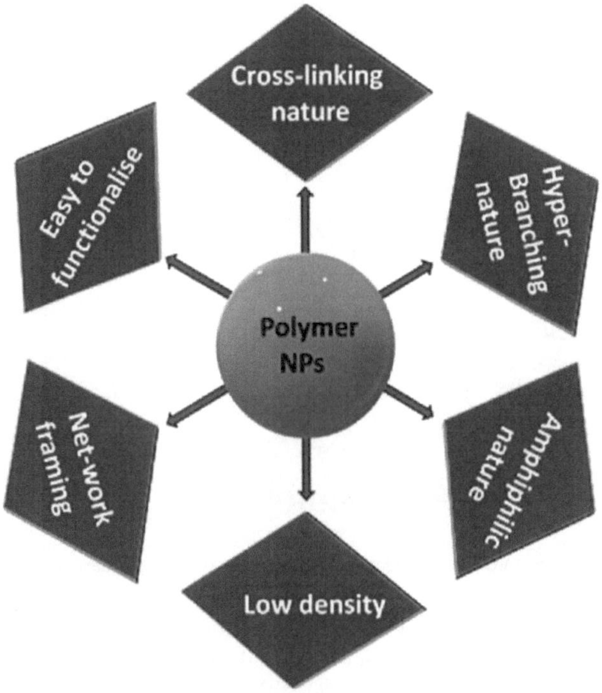

FIGURE 10.2 Schematic corroborates the factors and properties that need to be considered for the selection of polymers for the synthesis of surface-functionalized polymer-based nanoparticles.

because our body has various essential/non-essential amino acids, and particles developed from these amino acids can show great biocompatibility and cell survival rates (Park et al., 2012). The salient features and properties of the polymer nanoparticles required for usage in the biomedical field are given in Figure 10.3.

The morphology, size, and shape of nanoparticles can be tuned by switching the solvents and temperature as well. However, self-assembly between active functional groups of polymers leads to particles with a spherical shape, and internal as well as external

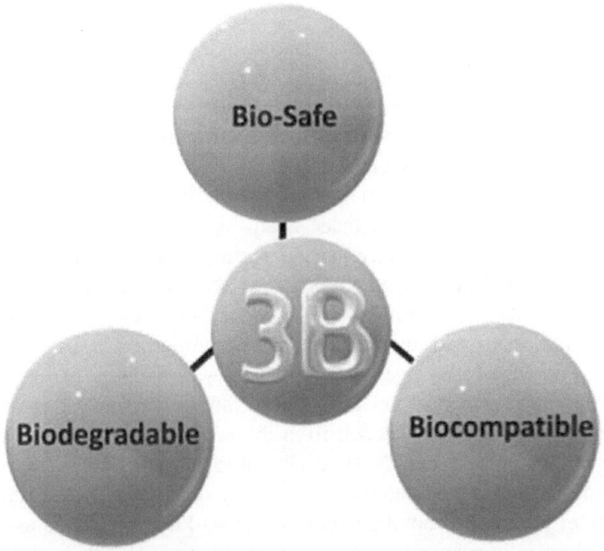

FIGURE 10.3 Schematic represents the characteristic properties of polymer nanoparticles for biomedical applications.

phase separation depends on the characteristic properties of polymers, i.e., hydrophilic (solvent-loving) and hydrophobic (solvent-hating) properties, which means that the effects of solvents need to be considered for the preparation of polymer nanoparticles (Branco and Schneider, 2009; Salmaso and Calicetti, 2013). The self-assembly of functional groups such as hydroxyl (–OH), amide (–NH), keto (–C = O), ester (R-COO-R), acid (–COOH), aldehyde (–CHO), and saturated and unsaturated hydrocarbons leads to the formation of spheroids (spherical-shaped nanoparticles) (Chen et al., 2011). The cleavage of intra- and intermolecular secondary bonding between amphiphilic polymers and polycaprolactone can allow to modify the surface of the nanoparticles.

10.4 Nanoparticles

Based on their constituent elements, nanoparticles can be broadly classified as polymeric and non-polymeric. The complete classification of nanoparticles is shown in Figure 10.4. Polymeric micro/nanomaterials or carriers are widely used in biotechnology and nanobiotechnology for drug- and gene-delivery applications. Block copolymers are temperature-sensitive (Delcea et al., 2011; Kassi et al., 2013), which means that their morphology, pore size, and shape change as the temperature varies. Ultimately, novel porous polymer capsules loaded with anti-cancer drugs have been used for cancer therapy. The drug loading and releasing efficiency of several nanocapsules have been investigated. The role of pores has been studied for the sustainable delivery of anti-cancer drugs. Subsequently, the efficiency of novel formulations for the treatment of cancer-specific tumor cells has been investigated *in vitro* through a series of studies. The synthetic approach of nanocarriers is very important in the drug-delivery field and the major synthetic approaches are ring-opening polymerization, polymer precipitation, self-assembly of polymer materials, etc.

Polymers have the ability to form network-like structures due to their entanglement and their chain cross-linking and branching nature (Chang and Lee, 2009). The synthesis of nanocarriers such as nanocapsules, nanospheres, and nanoparticles with characteristic features like mesopores (cavities or voids), core-shell (host-guest) (Figure 10.5) nature is an important concern to absorb, encapsulate, or entrap the drug molecules.

To overcome the limitations, block copolymer-based micro and nanocarriers for drug loading in the range of ~200 nm to 5 μm have been synthesized through the ring-opening polymerization method. The compounds can be characterized through the Fourier transform infrared (FTIR) spectrophotometer, scanning electron microscopy (SEM), Field emission scanning electron microscope (FE-SEM), high-resolution transmission electron microscope (HR-TEM), dynamic light scattering (DLS), circular dichroism (CD), laser scanning confocal microscope (LSCM), and MTT assay. Drug, gene, protein loading, and release profile studies have been examined through the UV-visible spectrophotometer.

10.5 Strategies to Functionalize Nanoparticles

The synthetic approach of polymeric nanoparticles is very important to tune their properties and monodispersed sizes to maintain the surface modifications based on our specifications. However,

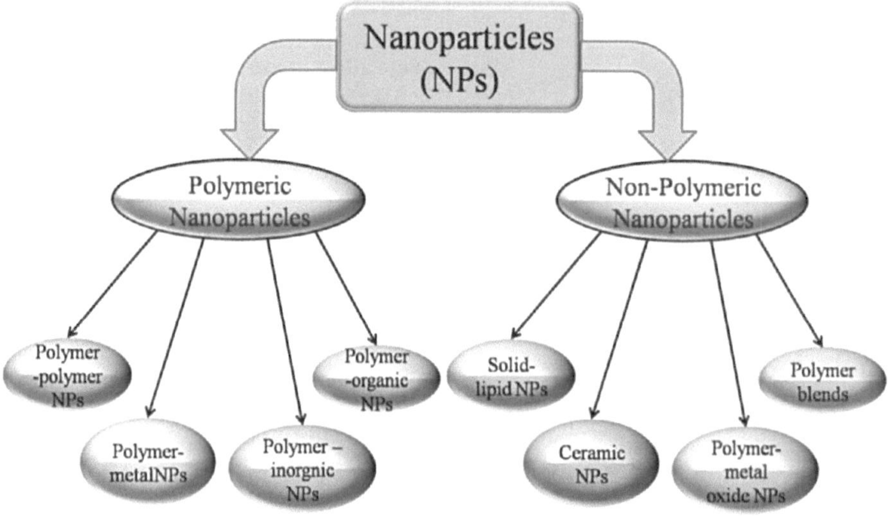

FIGURE 10.4 The schematic illustrates the classification of nanoparticles into polymer and non-polymer nanoparticles based on source materials.

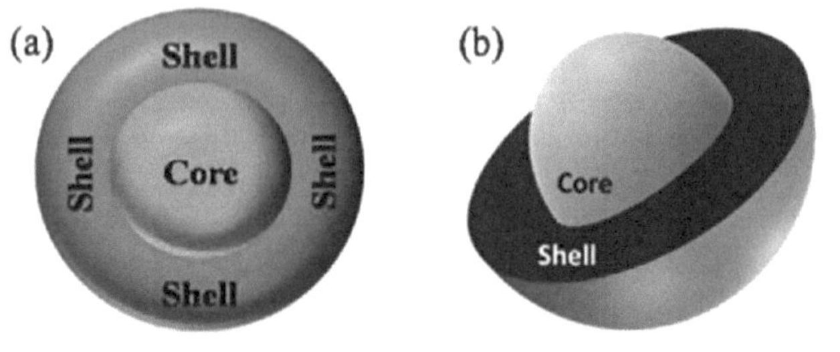

FIGURE 10.5 Core-shell concept of nanoparticles: (a) Core part (drug) surrounded by the polymer shell, and (b) core part partially surrounded by the shell material.

nanoparticles can be synthesized through various approaches, such as monomer (single-unit), polymer (formed with the help of monomers), and group-transfer polymerization (Shen et al., 2018), ring-opening polymerization (Chang and Lee, 2009), polymer re-precipitation, emulsion, micro and macro initiation, addition polymerization, atom transfer radical polymerization, condensation polymerization, oxidation, reduction followed by redox reactions, and self-assembly of polymer materials. Oxidation, reduction, and redox reactions are involved in the synthesis of polymers followed by the design and development of nanoparticles. The hydrophobic, hydrophilic, and amphiphilic nature of polymers needs to be considered with the solvent which is used in the synthesis of nanoparticles. Apart from the above-mentioned methods, several modified chemical synthetic routes can be followed to elucidate the nanoparticles. The formulation and development of novel nanoparticles with characteristic features in the field of nanotechnology is a unique approach to functionalize them for various applications (Shen et al., 2018; Gunti et al., 2017).

10.6 Characterizations of Polymeric Nanoparticles

Molecular structures and functional groups present in polymeric nanoparticles are studied through various spectroscopic techniques such as FTIR (Fourier transform infrared) and NMR (nuclear magnetic resonance; ^{1}H and ^{13}C) spectroscopy. SEM (scanning electron microscope), SEM-EDEX (energy dispersive X-ray analysis), and FE-SEM (field emission scanning electron microscope) are used to understand the topography, morphology, and compositional information of nanoparticles. XRD (X-ray diffraction), TEM (transmission electron microscope), and HR-TEM (high-resolution transmission electron microscope) are used to identify the solid states (amorphous, crystalline and semi-crystalline) and particle size of nanoparticles. The AFM (atomic force microscope) is used to study the texture of the surface of nanoparticles followed by the pore profile such as pore depth and width in case of porous capsules or particles. DLS (dynamic light scattering) is used for the particle size distribution of nanoparticles. DSC (differential scanning calorimetry) is used to determine the heat flow (endothermic; system which absorbs heat from surrounding and exothermic; a system which releases heat to the surroundings), fusion, and glass transition temperature of the polymeric nanoparticles. TGA (thermogravimetric analysis) is employed to identify the weight loss and degradation temperature of polymer nanoparticles. Polymeric micro nanocapsules and particles play an important role in biological applications such as as anticancer drugs and nanomedicines. The selection of polymeric materials for the design and development of nanoparticles needs to consider key surface properties such as surface tension, hydrophobic and hydrophilic nature, etc. The synthetic approaches and methods play a vital role in the engineering of polymeric nanocapsules (NCs) and nanoparticles with their unique surface properties to tune their morphology based on circumstances. However, Figure 10.6c corroborates the pores throughout the surface of the single capsule/particle which is developed from the combination of polymers such as [(MeO-PEG-NH)-(L-GluA)]-PCL. The main reason and mechanism behind the formation of such novel porous capsules are the hydrophilic nature of [(MeO-PEG-NH)-(L-GluA)] polymers imprinted with the hydrophobic nature of PCL (poly-Ɛ-caprolactone). Moreover, Figure 10.6a illustrates the smooth surface of the capsule/particle which is developed from the

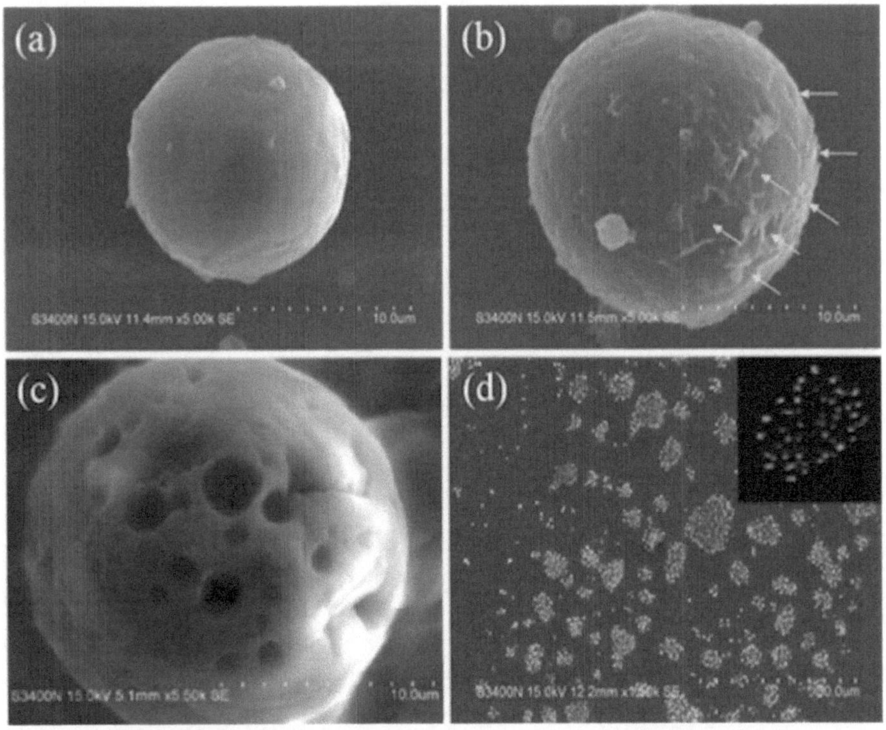

FIGURE 10.6 Scanning electron microscopic images of polymeric nanoparticles and capsules: image (a) elucidates the particle with a smooth surface, image (b) shows the particle with a rough surface, image (c) shows the particle with huge pores throughout its surface, and image (d) illustrates the aggregates of polymeric nanoparticles at lower magnification (Amgoth and Joshi, 2017).

complex [(MeO-PEG-NH)-(L-GluA)] (Amgoth and Joshi, 2017). The self-assembly nature of polymers such as α-methoxy-ω-amino polyethylene glycol (MeO-PEG-NH_2), and L-glutamic acid (L-GluA) leads to the formation of a spherical-shaped capsule with a smooth surface, but upon the addition of a hydrophobic solution of PCL, pore formation takes place and the smooth surface is modified into a porous surface.

Interestingly, Figure 10.6b illustrates the rough surface of the capsule/particle and this has been engineered through the ultrasonication of smooth surface particles at room temperature for 5 minutes in an ethanol solvent. Similarly, Figure 10.6c shows spherical-shaped nanocapsules with a huge number of pores throughout the surface of the particle. During sonication, some functional groups are detached, or bond cleavage/breakage takes place which results in the formation of a rough surface.

Loosely bound patches are separated from the top surface and become rough and porous, which is confirmed by AFM microscopic characterization. The yellow arrows represent the rough surface of the capsule/particle. Furthermore, Figure 10.6d corresponds to the polymer nanoparticles formed over the ultra-sonication of polymer material synthesized through the [(MeO-PEG-NH)-(L-GluA)] (Amgoth and Joshi, 2017; Gustafson et al., 2014). The scale bar for the images in Figure 10.6a–c is 10 μm and for Figure 10.6d is 50 μm. The figure shows the well-dispersed polymer nanoparticles without any pores throughout the surface. The inset represents the irregular-shaped polymeric nanoparticles. The monodispersion process is essential to maintain the particles size without any shape irregularities. The synthesis methods and solvents used for the dispersion play a crucial role in tuning and molding the spherical shape of particles/capsules and in changing the surface based on demands. Polymer synthesis and preparation methods such as ring-opening polymerization, polymer re-precipitation, emulsion polymerization, and atom transfer radical polymerization influence the self-assembly between polymer functional groups, and leads to the formation of spherical and sphere-shaped nanoparticles. Image (a) from Figure 10.7 corroborates the monodispersed silica (SiO_2) nanoparticles prepared by using a modified Stöber 'sol-gel' method. The formation of spherical-shaped SiO_2 nanoparticles depends on the stirring methods, solvent, rotations per minute (RPM), purity of chemicals used for the synthesis of SiO_2 nanoparticles, surfactants, and types of equipment employed for the experiments. Their surface properties were investigated by the morphological characterizations using various microscopic techniques such as SEM, FE-SEM, TEM, and AFM for the size, shape, physical appearance, and surface profile. Figure 10.7b shows silica (SiO_2) nanoparticles embedded inside the polymer matrix; Figure 10.7c shows the gold nanoparticles (Au nanoparticles) aggregated with the polymer sheet, and Figure 10.7d indicates the composite of [(polymer)-(Au nanoparticles)]. The surface morphology of [(MeO-PEG-NH)-(L-GluA)]-PCL based polymer nanoparticles has been optimized and presented in Figure 10.8. The pore size and shape appeared to be irregular due to molecular imprinting and interatomic interactions. Sometimes, intermolecular attractions also lead to form spherical nanoparticles and the whole mechanism can be termed as self-assembly between active functional groups present in the polymers. There are highly inconsistent pores with smooth and rough surfaces because of the solvent effect and sonication methods. A non-contact mode AFM operation has been carried out

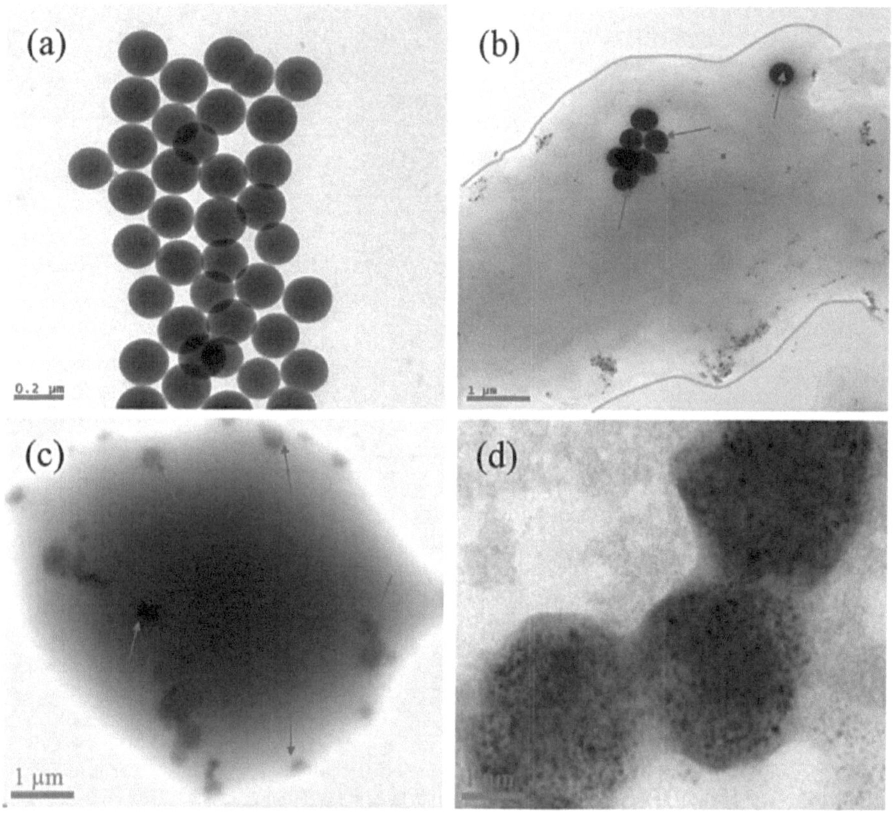

FIGURE 10.7 Transmission electron microscopic images of silica and polymeric nanoparticles. Image (a) elucidates the spherical-shaped silica nanoparticle with a smooth surface, image (b) shows the silica nanoparticle embedded inside the polymer matrix, and images (c and d) show the gold nanoparticle embedded within the polymer sheet and illustrate the aggregates of polymer-gold nanoparticles (Amgoth et al., 2016).

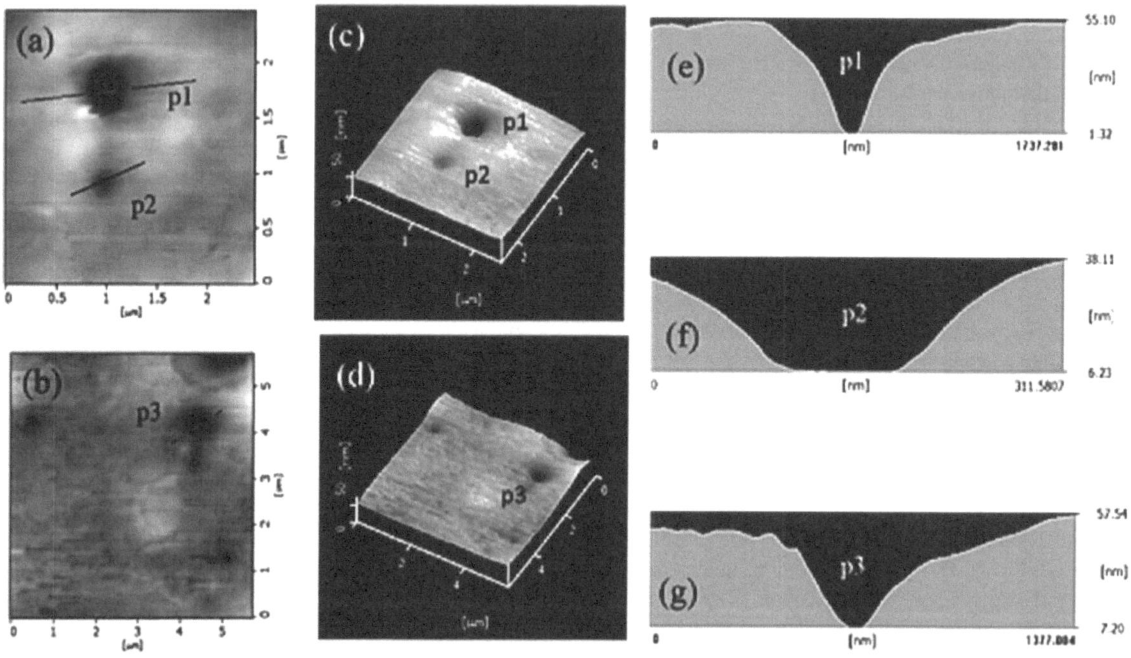

FIGURE 10.8 Atomic force microscopic images acquired on the surface of the porous polymer nanoparticles. Images (a and b) elucidate the physical appearance of the phase(s) of the surface of the particles, images (c and d) correspond to the 3-dimensional morphology of the surface of the porous polymer nanoparticles, and images (e, f, g) illustrate the pore profile and surface properties for pores p1, p2, and p3, marked in the phase and 3-dimensional images (Preetz et al., 2010; Tian and Hammond, 2006; Amgoth et al., 2016).

to acquire the topography of the nanoparticles. However, tip or probe size and magnification followed by the resolution play a crucial role in acquiring the topography of the nanoparticles and polymer capsules.

However, wet chemical methods and sample preparation procedures also play a vital role in exhibiting a considerable change in the size, shape, and surface morphology of polymer nanoparticles. Furthermore, surface properties can be studied using the profilometer and line scanning mode operation of AFM and TEM instruments. Figure 10.8 illustrates the surface properties (smooth and rough) followed by the pore profiles for particular pores labeled as pore1 (p1), pore2 (p2), and pore3 (p3) and their corresponding pore features. Figure 10.8e–g shows pore width and depth as well as the surface of the nanoparticles. The uniform pores, consistent particle size, and shapes can be optimized by optimizing several parameters such as sonication, solvent, pH, and concentration of the system. However, appeared pores and their size distribution can be modified based on our requirements. Amphiphilic (combination of hydrophilic and hydrophobic) comb-like dendrimers were developed from the block copolymers (BCPs) synthesized from the [(MeO-PEG-NH)-(L-GluA)]-PCL. The hydrophobic PCL block and a hydrophilic [(MeO-PEG-NH)-(L-GluA)] block were modified with the help of dry THF (tetrahydrofuran) and dispersed in IPA (isopropanol) for microscopic characterizations and morphological analysis. They exhibited comb-like dendrimer nanochannels with high-persistence length of the rod-like helical backbone architecture (Perrier et al., 2011). In dry THF solvent, the active functional groups were self-assembled into spherical micelles with the hydrophobic comb-like dendritic exterior shell. The [(MeO-PEG-NH)-(L-GluA)]-PCL-based dendritic morphology enhances the stability of the dendrimers, resulting in low critical solution temperature (LCST) and low critical micelle concentration (LCMC) of approximately ~10^{-8} mM. The size of these dendrimers was pH- and temperature-dependent due to the presence of carboxylic acid (COOH), ester (R-COO-R), and hydroxyl (–OH) groups at the surfaces of the dendrimers and at the end of the branches. Furthermore, the nanoscale clustering and complexation of the periphery of nanoparticles and capsules has been performed through the modification of active functional groups as potential substrates for subsequent functionalization and surface modifications. However, the attachment or functionalization of a polymer compound with functional groups or ligands can also lead to the development of porous or dendritic nanoparticles to improve the surface features based on our requirements (Perrier et al., 2011; Amgoth et al., 2016). There are several methods for the design and development of polymer nanoparticles with considerable surface area and surface modification feasibilities. The factors affecting the surface modification in polymer nanoparticles are shown in Figure 10.9. The post-insertion method is used for the preparation of dendrimer nanoparticles and lipid nanocapsules (LNCs) with a chemical reactive surface.

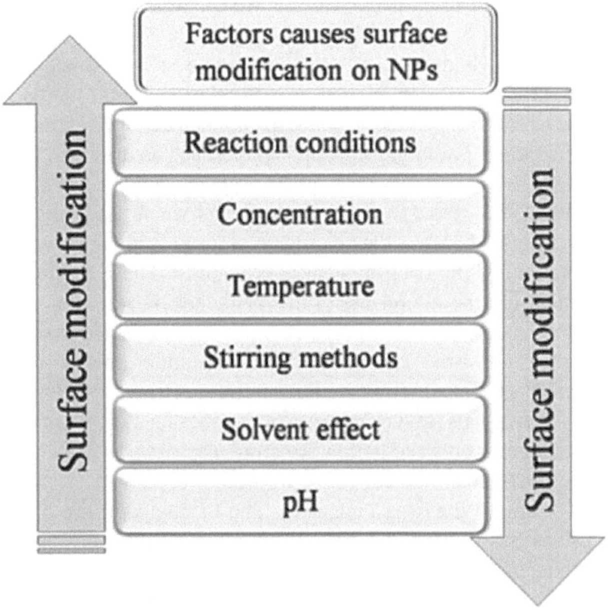

FIGURE 10.9 The schematic shows the factors causing surface modification in polymer nanoparticles.

This method was designed for the grafting of ligands to the polymer compounds, and later it was extended also to the development of other particles. The addition of surfactants such as IGEPAL, etc., to the Si precursor (TEOS; tetraethyl orthosilicates), can change the surface and size of the SiO_2 nanoparticles. The dispersion solvents, temperature conditions, concentration of the nanoparticles, and other environmental factors also affect the morphology and surface of the nanoparticles. The concentration of the nanoparticles is expressed in percentage (%) of the number of surfactants used for the formulation methods. The post-insertion method for the synthesis and preparation of nanoparticles was performed at RT under continuous magnetic stirring. The final product was used for the solvent evaporation through the rotary evaporation, and a well-dried product was used for characterizations such as XRD, DSC, TGA, BET, etc. However, the particle size was measured by the dynamic light scattering (DLS) method. However, these dendritic and smooth surface-area nanoparticles were obtained through time correlation functions analyzed by the auto-correlation method for a better understanding of their properties, such as surface area, porosity, and hyper-branching nature. During the preparation of polymer nanoparticles or dendrimers, one can focus on the experimental conditions such as the ratio between both reactants, the quantity of solvent, and external stimuli factors such as the stirring methods, sonication, magnetic fields, electric field applications, pH, temperature, viscosity, density, and molecular weight of reactants. The rate of reaction can be tuned or controlled based on requirements. For example, reaction completes faster at room temperature and basic medium with pH ~9. Functional groups such as hydroxyl, ester, acid, and amide show effects on the surface of nanoparticles at their basic pH levels. Thus silica nanoparticles with a size of 50–100 nm are developed with 'sol-gel' methods and the size can be increased or even reduced based on the modified chemical synthesis methods followed by the core principle from the Stöbers 'sol-gel' synthesis methods. There are other methods to reduce the size of nanoparticles. These include a media milling process followed by ball milling and rigorous stirring methods. Furthermore, during the media milling or ball milling processes, interatomic/molecular collisions or bombardments take place and this leads to a reduction in the size of nanoparticles. Sometimes, dipole–dipole interactions, electrostatic interactions, or van der Waals attractions also cause self-assembly between polymer molecules for the formation of nanoparticles and capsules, and those can be easily disintegrated or ruptured with ultra-sonication and other agitation methods. However, porous polymeric nanocapsules and nanoparticles have a potential role in biological applications such as drug and gene delivery (Mora-Huertas et al., 2010). Polymers, i.e., PCL, PLGA, PNIPAM, PS, and PEG are advantageous when designing the nanocapsules because they can form network structures, which helps loading the drug molecules efficiently. These properties aid in sustaining the release of the medicines. Furthermore, the mass of the synthesized polymer has been confirmed through the matrix-assisted laser desorption ionization/time of flight (MALDI-TOF) and high-resolution mass spectrometry (HRMS) characterizations. In the synthesis of polymer nanocapsules with defined mesopores and hollow structure (cavities/or voids), core-shell (host-guest) nature is also important to absorb, encapsulate, or entrap the drug molecules. The biocompatibility of the capsules has been characterized through standard MTT assay. Further, the drug loading and release behaviors of the nanocapsules have been studied according to the method we developed (Gunti et al., 2017; Tian and Hammond, 2006). The rigid morphology or close compacted structure of nanoparticles or nanocapsules with covalent bonding cannot allow us to modify their surface even at rigorous conditions. Loosely bound functional groups allow to modify the surface of the nanoparticles based on our applications. The homogeneous mixture of ingredients present in the polymer nanoparticles leads to the formation of a close-compacted and smooth surface on the particles. However, the development of a vinyl ether-functionalized poly phosphoester as a template for multiple post-polymerization conjugation core-shell-type degradable polymeric nanoparticles was appended in the various publications, and these help to design and develop polymer nanoparticles with surface-modifiable properties (Lim et al., 2014). Also, the reactions involved in the synthesis followed by the polymerization and the polymerization methods at the interface of two immiscible liquids aid in the formation of nanoparticles with featured surfaces. This needs to be reviewed carefully before we commit to perform a reaction. The confirmation and confinement of two reactants followed by the formation of nanoparticles with systemic interface have the advantage to tune their surface properties. Phase transitions and thermal stability can help to improve the reaction kinetics, higher yields, and selectivity of nanoparticles. However, the presence of a liquid interface can accelerate the reaction through the phase-transfer catalyst and is employed to draw a reaction for the development of nanoparticles based on phase transfer phenomena. Furthermore, the use of immiscible systems in emulsions offers an easy phase separation and the formulation of desired nanoparticles. However, a brief overview on low molecular weight and low-density polymer complexes shows the significant proximity of the interface in emulsions, and this strategy can be used for the efficient production of nano and microparticles for various applications (Piradashvili et al., 2016). Interestingly, the morphology of polymer nanoparticles depends on the self-organization between functional groups of polymers and supra-molecular regions and stereoselectivity (Babu et al., 2010; Mann et al., 2016; Gao et al., 2016). Several mechanisms were proposed for the formation of nanoscale pores within the nanoparticles because of the removal of some of the BCP fragments from the network of PCL. The size of BCP-based polymer nanoparticles can be reduced by removing blocks from the parent blocks. As PCL is hydrophobic, BCP spheroids may be dissolved out from the PCL network structure and create the pores of the equivalent size of BPC spheroids. Moreover, as per results appended in Figure 10.6, temperature change affects the pore size and morphology as well (Mann et al., 2016; Gao et al., 2016). Therefore, the lower temperature method is favorable for the formation of the capsules having a larger pore size. The pore size distribution has been calculated by the freehand polygon counting method considering 50 pores from each particle. As Figure 10.9 corroborates, there are several external stimuli-responsive factors such as pH, heat, ionic strength, and electric and magnetic field which affects the size, shape, and surface properties of the polymer nanoparticles.

REFERENCES

Amgoth, C. 2016. Synthesis and characterization of polymeric nanoparticles and capsules as payload for anticancer drugs and nanomedicines. *Materials Today: Proceedings* 3(10):3833–3837. doi: 10.1016/j.matpr.2016.11.036

Amgoth, C., Dharmapuri, G., Kalle, A. M., Paik, P. 2016. Nanoporous capsules of block co-polymers of [(MeO-PEG-NH)-b-(L-GluA)]-PCL for the controlled release of anticancer drugs for therapeutic applications. *Nanotechnology* 27(12):125101. doi: 10.1088/0957-4484/27/12/125101

Amgoth, C., Joshi, S. 2017. Thermosensitive block copolymer [(PNIPAM)-b-(glycine)] thin film as protective layer for drug loaded mesoporous silica nanoparticles. *Materials Research Express* 4(10):105306. doi: 10.1088/2053-1591/aa91eb

Argyo, C., Weiss, V., Brauchle, C., Bein, T. 2014. Multifunctional mesoporous silica nanoparticles as a universal platform for drug delivery. *Chemistry of Materials* 26(1):435–451. doi: 10.1021/cm402592t

Ashjari, M., Khoee, S., Mahdavian, A. R. 2012. Controlling the morphology and surface property of magnetic/cisplatin-loaded nanocapsules via W/O/W double emulsion method. *Colloids and Surfaces. Part A: Physicochemical and Engineering Aspects* 408:87–96. doi: 10.1016/j.colsurfa.2012.05.035

Babu, S. S., Mohwald, H., Nakanishi, T. 2010. Recent progress in morphology control of supramolecular fullerene assemblies and its applications. *Chemical Society Reviews* 39(11):4021–4035. doi: 10.1039/c000680g

Baudry, D. B., Brachais, L., Cretu, A., Gattin, R., Loupy, A., Stuerga, D. 2003. Synthesis of polycaprolactone by microwave irradiation – An interesting route to synthesize this polymer via green chemistry. *Environmental Chemistry Letters* 1(1):19–23. doi: 10.1007/s10311-002-0005-4

Bergbreiter, D. E., Case, B. L., Liu, Y. S., Caraway, J. W. 1998. Poly (*N*-isopropylacrylamide) soluble polymer supports in catalysis and synthesis. *Macromolecules* 31(18):6053–6062. doi: 10.1021/ma980836a

Bertram, J. P., Jay, S. M., Hynes, S. R., Robinson, R., Criscione, J. M., Lavik, E. B. 2009. Functionalized poly (lactic-co-glycolic acid) enhances drug delivery and provides chemical moieties for surface engineering while preserving biocompatibility. *Acta Biomaterialia* 5(8):2860–2871. doi: 10.1016/j.actbio.2009.04.012

Bradburne, C. E., Delehanty, J. B., Gemmill, K. B., Mei, B. C., Mattoussi, H., Susumu, K., Blanco-Canosa, J. B., Dawson, P. E., Medint, I. L. 2013. Cytotoxicity of quantum dots used for in vitro cellular labeling: Role of QD surface ligand, delivery modality, cell type, and direct comparison to organic fluorophores. *Bioconjugate Chemistry* 24(9):1570–1583. doi: 10.1021/bc4001917

Branco, M. C., Schneider, J. P. 2009. Self-assembling materials for therapeutic delivery. *Acta Biomaterialia* 5(3):817–831. doi: 10.1016/j.actbio.2008.09.018

Chang, K. Y., Lee, Y. D. 2009. Ring-opening polymerization of e-caprolactone initiated by the antitumor agent doxifluridine. *Acta Biomaterialia* 5(4):1075–1081. doi: 10.1016/j.actbio.2008.11.010

Chaubey, N., Sahoo, A. K., Chattopadhyay, A., Ghosh, S. S. 2014. Silver nanoparticle loaded PLGA composite nanoparticles for improving therapeutic efficacy of recombinant IFNγ by targeting the cell surface. *Biomaterials Science* 2(8):1080–1089. doi: 10.1039/c3bm60251f.V

Chen, J., Liu, M., Gong, H., Huang, Y., Chen, C. 2011. Synthesis and self-assembly of thermoresponsive PEG-b-PNIPAM-b-PCL ABC triblock copolymer through the combination of atom transfer radical polymerization, ring-opening polymerization, and click chemistry. *Journal of Physical Chemistry. Part B* 115(50):14947–14955. doi: 10.1021/jp208494w

Delcea, M., Möhwald, H., Skirtach, A. G. 2011. Stimuli-responsive LbL capsules and nanoshells for drug delivery. *Advanced Drug Delivery Reviews* 63(9):730–747. doi: 10.1016/j.addr.2011.03.010

Fadel, T. R., Sharp, F. A., Vudattu, N., Ragheb, R., Garyu, J., Kim, D., Hong, E., Li, N., Haller, G. L., Pfefferle, L. D., Justesen, S., Harold, K. C., Fahmy, T. M. 2014. A carbon nanotube-polymer composite for T-cell therapy. *Nature NANO Technology* 9(8):639–647. doi: 10.1038/NNANO.2014.154

Gao, W., Shi, Y., Zuo, L., Fan, W., Liu, T. 2016. Rough-surfaced molybdenum carbide nanobeads grown on graphene-coated carbon nanofibers membrane as free-standing hydrogen evolution reaction electrocatalyst. *Materials Today Chemistry* 1–2:32–39. doi: 10.1016/j.mtchem.2016.10.003

Gunti, R., Dharmapuri, G., Doddapaneni, S., Amgoth, C. 2017. Composite BME-Au nanoparticles: Chemopreventive effect on skin carcinoma and inhibition on leukemia blood cancer cells. *Advanced Materials Letters* 8:1057–1064. doi: 10.5185/amlett.2017.1733

Guo, J., Yang, W., Wang, C., He, J., Chen, J. 2006. Poly (*N*-isopropylacrylamide)-coated luminescent/magnetic silica microspheres: Preparation, characterization, and biomedical applications. *Chemistry of Materials* 18(23):5554–5562. doi: 10.1021/cm060976w

Gustafson, T. P., Lim, Y. H., Flores, J. A., Heo, J. S., Zhang, F., Zhang, S., Samarajeewa, S., Raymond, J. E., Wooley, K. L. 2014. Holistic assessment of covalently labeled core–shell polymeric nanoparticles with fluorescent contrast agents for theranostic applications. *Langmuir* 30(2):631–641. doi: 10.1021/la403943w

Kassi, E., Constantinou, M. S., Patrickios, C. S. 2013. Group transfer polymerization of biobased monomers. *European Polymer Journal* 49(4):761–767. doi: 10.1016/j.eurpolymj.2012.11.012

Lai, C.-Y., Trewyn, B. G., Jeftinija, D. M., Jeftinija, K., Xu, S., Jeftinija, S., Lin, V. S.-Y. 2003. A mesoporous silica nanosphere-based carrier system with chemically removable CdS nanoparticle caps for stimuli-responsive controlled release of neurotransmitters and drug molecules. *Journal of the American Chemical Society* 125(15):4451–4459. doi: 10.1021/ja028650l

Lai, W., Zhuang, J., Que, X., Fu, L., Tang, D. 2014. Mesoporous nanogold–MnO2–poly-(o-phenylenediamine) hollow microspheres as nanotags and peroxidase mimics for sensing biomolecules. *Biomaterials Science*. doi: 10.1039/c3bm60284b

Li, J., Cong, H., Li, L., Zheng, S. 2014. Thermoresponsive improvement of poly (*N*-isopropylacrylamide) hydrogels via formation of poly (sodium p-styrene sulfonate) nanophases. *ACS Applied Materials and Interfaces* 16(7):13677–11368. doi: 10.1021/am503148v

Li, L., Chen, S., Xu, L., Bai, Y., Nie, Z., Liu, H., Qi, L. 2014. Template-free synthesis of uniform mesoporous SnO_2 nanospheres for efficient phosphopeptide enrichment. *Journal of Materials Chemistry B* 2(9):1121–1124. doi: 10.1039/c3tb21617a

Li, R., Li, L., Han, Y., Gai, S., He, F., Yang, P. 2014. Core-shell structured Gd2O3:Ln@*m*SiO2 hollow nanospheres: Synthesis, photoluminescence and drug release properties. *Journal of Materials Chemistry B*. doi: 10.1039/C3TB21718C

Li, X., Wang, Z., Li, Y., Bian, K., Yin, T., Gao D. 2018. Self-assembly of bacitracin-gold nanoparticles and their toxicity analysis. *Materials Science and Engineering C* 82:310–316. doi: 10.1016/j.msec.2017.07.053

Lim, Y. H., Heo, G. S., Rezenom, Y. H., Pollack, S., Raymond, J. E., Elsabahy, M., Wooley, K. L. 2014. Development of a vinyl ether-functionalized polyphosphoester as a template for multiple postpolymerization conjugation chemistries and study of core degradable polymeric nanoparticles. *Macromolecules* 47(14):4634–4644. doi: 10.1021/ma402480a

Liu, W., Wen, S., Shen, M., Shi, X. 2014. Doxorubicin-loaded poly (lactic-co-glycolic acid) hollow microcapsules for targeted drug delivery to cancer cells. *New Journal of Chemistry* 38(8):3917–3924. doi: 10.1039/c4nj00672k

Magnusson, J. P., Saeed, A. O., Trillo, F. F. A., Alexander, C. 2011. Synthetic polymers for biopharmaceutical delivery. *Polymer Chemistry* 2(1):48–59. doi: 10.1039/c0py00210k

Mann, S. K., Dufour, A., Glass, J. J., De Rose, R., Kent, S. J., Such, G. K., Johnston, A. P. R. 2016. Tuning the properties of pH responsive nanoparticles to control cellular interactions in vitro and ex vivo. *Polymer Chemistry* 7(38):6015–6024. doi: 10.1039/c6py01332e

Mona, M. A., Mottaleb, A., Neumann, D., Lamprecht, A. 2011. Lipid nanocapsules for dermal application: A comparative study of lipid-based versus polymer-based nanocarriers. *European Journal of Pharmaceutics and Biopharmaceutics* 79(1):36–42. doi: 10.1016/j.ejpb.2011.04.009

Mora-Huertas, C. E., Fessi, H., Elaissari, A. 2010. Polymer-based nanocapsules for drug delivery. *International Journal of Pharmaceutics* 385(1–2):113–142. doi: 10.1016/j.ijpharm.2009.10.018

Nuruzatulifah, A. M., Nizam, A. A., Ain, N. M. N. 2016. Synthesis and characterization of polystyrene nanoparticles with covalently attached fluorescent dye. *Materials Today: Proceedings* 3S:S112–S119. doi: 10.1016/j.matpr.2016.01.015

Ortegaa, M. J. S., Garcíac, J. M. P., Goycoolea, F. M., Vinuesa, J. L. O. 2011. Chitosan nanocapsules: Effect of chitosan molecular weight and acetylation degree on electrokinetic behaviour and colloidal stability. *Colloids and Surfaces, Part B: Biointerfaces* 82(2):571–580. doi: 10.1016/j.colsurfb.2010.10.019

Park, H. S., Gong, M. S., Knowles, J. C. 2012. Synthesis and biocompatibility properties of polyester containing various diacid based on isosorbide. *Journal of Biomaterials Applications* 27(1):99. doi: 10.1177/0885328212447245

Perrier, T., Fouchet, F., Bastiat, G., Saulnier, P., Benoît, J. P. 2011. OPA quantification of amino groups at the surface of lipidic nanocapsules (LNCs) for ligand coupling improvement. *International Journal of Pharmaceutics* 419(1–2):266–270. doi: 10.1016/j.ijpharm.2011.07.028

Piradashvili, K., Alexandrino, E. M., Wurm, F. R., Landfester, K. 2016. Reactions and polymerizations at the liquid–liquid interface. *Chemical Reviews* 116(4):2141–2169. doi: 10.1021/acs.chemrev.5b00567

Preetz, C., Hauser, A., Hause, G., Kramer, A., Mader, A. 2010. Application of atomic force microscopy and ultrasonic resonator technology on nanoscale: Distinction of nanoemulsions from nanocapsules. *European Journal of Pharmaceutical Sciences* 39(1–3):141–151. doi: 10.1016/j.ejps.2009.11.009

Puskas, J. E., Seo, K. S., Sen, M. Y. 2011. Green polymer chemistry: Precision synthesis of novel multifunctional poly (ethylene glycol)s using enzymatic catalysis. *European Polymer Journal* 47(4):524–534. doi: 10.1016/j.eurpolymj.2010.10.015

Quadir, M. A., Morton, S. W., Deng, Z. J., Shopsowitz, K. E., Murphy, R. P., Epps, III, T. H., Hammond, P. T. 2014. PEG–polypeptide block copolymers as pH-responsive endosome-solubilizing drug nanocarriers. *Molecular Pharmaceutics* 11(7):2420–2430. doi: 10.1021/mp500162w

Salmaso, S., Caliceti, P. 2013. Self-assembling nanocomposites for protein delivery: Supramolecular interactions of soluble polymers with protein drugs. *International Journal of Pharmaceutics* 440(1):111–123. doi: 10.1016/j.ijpharm.2011.12.029

Shen, C., Shen, B., Liu, X., Yuan, H. 2018. Nanosuspensions based gel as delivery system of nitrofurazone for enhanced dermal bioavailability. *Journal of Drug Delivery Science and Technology* 43:1–11. doi: 10.1016/j.jddst.2017.09.012

Song, J. H., Kretzschmar, H. 2009. Assembled surface-anisotropic colloids as a template for a multistage catalytic membrane reactor. *ACS Applied Materials and Interfaces* 1(8):1747–1754. doi: 10.1021/am900286k

Sun, T., Zhang, Y. S., Pang, B., Hyun, D. C., Yang, M., Xia, Y. 2014. Engineered nanoparticles for drug delivery in cancer therapy. *Angewandte Chemie International Edition* 53(46):12320–12364. doi: 10.1002/anie.201403036

Teixeira, M., Alonso, M. J., Pinto, M. M. M., Barbosa, C. M. 2005. Development and characterization of PLGA nanospheres and nanocapsules containing xanthone and 3-methoxyxanthone. *European Journal of Pharmaceutics and Biopharmaceutics* 59(3):491–500. doi: 10.1016/j.ejpb.2004.09.002

Tian, L., Hammond, P. T. 2006. Comb-dendritic block copolymers as tree-shaped macromolecular amphiphiles for nanoparticle self-assembly. *Chemistry of Materials* 18(17):3976–3984. doi: 10.1021/cm060232i

Venturini, C. G., Jager, E., Oliveira, C. P., Bernardi, A., Ana, M. O., Battastini, Guterres, G. S., Pohlmann, A. R. 2011. Formulation of lipid core nanocapsules. *Colloids and Surfaces. Part A: Physicochemical and Engineering Aspects* 375(1–3):200–208. doi: 10.1016/j.colsurfa.2010.12.011

Wang, H., Garakani, T. M., Krappitz, T., Rijn, P. V., Boker, A. 2013. Morphology control and surface functionalization of protein-SiO_2 hybrid capsules. *Journal of Materials Chemistry B* 1(46):6427–6433. doi: 10.1039/C3TB21013H

Wang, K., Hu, Q., Zhu, W., Zhao, M., Ping, Y., Tang, G. 2015. Structure-invertible nanoparticles for triggered co-delivery of nucleic acids and hydrophobic drugs for combination cancer therapy. *Advanced Functional Materials* 25(22):3380–3392. doi: 10.1002/adfm.201403921

Wang, X., Hu, Y., Song, L., Yang, H., Xing, W., Lu, H. 2011. In situ polymerization of graphene nanosheets and polyurethane with enhanced mechanical and thermal properties. *Journal of Materials Chemistry* 21(12):4222. doi: 10.1039/C0JM03710A

Wang, X., Salick, M. R., Wang, X., Cordie, T., Han, W., Peng, Y., Li, Q., Turng, L. S. 2013. Poly (ε-caprolactone) nanofibers with a self-induced nanohybrid Shish-Kebab structure mimicking collagen fibrils. *Biomacromolecules* 14(10):3557–3569. doi: 10.1021/bm400928b

Wang, Y., Nepal, D., Geckeler, K. E. 2005. Hollow porous carbon nanospheres with large surface area and stability, assembled from oxidized fullerenes. *Journal of Materials Chemistry* 15(10):1049–1054. doi: 10.1039/b413988g

Wu, C., Baldursdottir, S., Yang, M., Mu, H. 2018. Lipid and PLGA hybrid microparticles as carriers for protein delivery. *Journal of Drug Delivery Science and Technology* 43:65–72. doi: 10.1016/j.jddst.2017.09.006

Yin, Z. C., Wang, Y. L., Wang, K. 2018. A pH-responsive composite hydrogel beads based on agar and alginate for oral drug delivery. *Journal of Drug Delivery Science and Technology* 43:12–18. doi: 10.1016/j.jddst.2017.09.009

Yuan, L., Wang, Y., Pan, M., Rempel, G. L., Pan, Q. 2013. Synthesis of poly (methyl methacrylate) nanoparticles via differential microemulsion polymerization. *European Polymer Journal* 49(1):41–48. doi: 10.1016/j.eurpolymj.2012.10.005

11

Cellulose Fibers and Nanocrystals: Preparation, Characterization, and Surface Modification

Djalal Trache, Ahmed Fouzi Tarchoun, Mehdi Derradji, Oussama Mehelli, M. Hazwan Hussin, and Wissam Bessa

CONTENTS

11.1 Introduction

In today's age, it is undeniable and well accepted that the innovation and development of new materials can improve the quality of life. Several industrial and government sectors as well as other consumers are increasingly demanding more sustainable and environmentally friendly materials and more effective procedures. To meet this growing trend, it is necessary to use sustainable,renewable, biodegradable, and non-petroleum-based resources that present low environmental and animal/human health and safety risks (Rajinipriya et al., 2018; Kim, Shim, et al., 2015; Isikgor and Becer, 2015; Kargarzadeh et al., 2017; Pandey et al., 2015; Oksman and Bismarck, 2014). Among a number of natural sources, cellulose is a virtually inexhaustive biopolymer and holds an important place in abundant organic feedstocks (Trache, 2017; Vazquez et al., 2015). This ubiquitous and abundant resource can be found in a wide range of living species encompassing wood, agricultural biomass, marine animals, and some bacteria. The composition and the structure of cellulose depend closely on the source.

The exploration of cellulose has received much attention owing to its outstanding features such as renewability, biodegradability, flexibility, functionality, and high mechanical performance. Researchers have been manipulating the hierarchical structure design that spans from nanoscopic to microscopic scales (Trache et al., 2017; Moon, Schueneman, and Simonsen, 2016; Moon et al., 2011). However, futuristic cellulose-based materials and their widespread uses cannot simply be achieved using traditional cellulose. Therefore, researchers and industries focus on the preparation of nanocellulose in significant amounts from different sources and employing various strategies (Trache, 2018; Shaghaleh, Xu, and Wang, 2018; Rajinipriya et al., 2018; Phanthong et al., 2018; Nascimento et al., 2018; Klemm et al., 2018). With their nanoscale diameters, cellulose nanofibers displayed prominent characteristics such as high strength, excellent stiffness, high surface area, low density, low coefficient of thermal expansion, dimensional stability, elongation morphology, and ease of bio-conjugation. These properties essentially depend on the source and the isolation processes. The opportunity of producing cellulose nanofibers with various features is considered a fairly exciting topic, which can promote the exploration of unexplored biomass (Phanthong et al., 2018; Chen et al., 2018; Wang et al., 2017; Vasconcelos et al., 2017; Reiniati, Hrymak, and Margaritis, 2017; Lamaming et al., 2017; Kargarzadeh et al., 2017). On the other hand, the presence of several −OH side chemical groups on the surface allows nanocellulose to be functionalized, which may comprise converting it to thiol, aldehyde, amine, and carboxylic acid groups (Dufresne, 2013a, b). Further functionalizations such as grafting larger macromolecules (e.g., proteins or polymers) and smaller molecules (e.g., metal oxide nanoparticles) can also be applied.

The advantages of the 3-dimensional hierarchical nanostructure of nanocellulose and its physicochemical features at nanometric scale open novel prospects in numerous applications such as barrier films, pharmaceuticals, biomedical implants, flexible displays, electroactive polymers, batteries, supercapacitors, fibers and textiles, and templates for electronic components, drug delivery, antimicrobial films, and many others (Thakur, 2015b; Klemm et al., 2018; Klemm et al., 2011). Typically, there exist two types of cellulose nanofibers: cellulose nanocrystal (CNC) and cellulose nanofibrils (CNFs) (Trache, 2018; Kargarzadeh, Mariano, et al., 2018; Azeredo, Rosa, and Mattoso, 2017). CNFs are produced by chemical, mechanical, or combined treatments. They are composed of amorphous and crystalline domains. The important differences between these cellulose nanofibers are their crystallinity and dimension. However, CNC is usually obtained by using acid hydrolysis, although, currently, various other processes have been established. These treatments can remove the disordered (amorphous) parts of purified cellulose fibers leaving behind ordered regions. The produced CNCs consist of rod-like, cylindrical, and elongated particles having dimensions of 4–70 nm (width) and between 100 nm and several micrometers (length).

More recently, various research works have been conducted in different universities, as well as in government and industry laboratories focusing on optimizing the production procedures to decrease production costs and enhance yields and quality. Besides, a number of reviews, books, and patents have been issued in the last few years covering several facets associated to CNC, encompassing preparation procedures, characterization, surface modification, self-assembly of suspensions, processing, and applications (Rajinipriya et al., 2018; Phanthong et al., 2018; Klemm et al., 2018; Kargarzadeh, Mariano, et al., 2018; Kargarzadeh, Huang, et al., 2018; Chen et al., 2018; Wang et al., 2017; Trache et al., 2017; Kargarzadeh et al., 2017; Jawaid, Boufi, and HPS, 2017; Golmohammadi et al., 2017; Du et al., 2017; De France, Hoare, and Cranston, 2017; Singla et al., 2016; Moon, Schueneman, and Simonsen, 2016; Hoeng, Denneulin, and Bras, 2016; Abitbol et al., 2016; Vazquez et al., 2015; Thakur, 2015b; Kim, Shim, et al., 2015; Charreau, L Foresti, and Vázquez, 2013).

This chapter reviews advanced research activities on main cellulose feedstocks and the foremost extraction methods. The extraction processes and properties of cellulose nanocrystals are provided. In addition, the key properties and the surface modification of CNCs are discussed.

11.2 Cellulose Fibers: Structure and Chemistry

Cellulose, a fascinating structural component and sustainable raw material, is the ubiquitous and most abundant biopolymer existing on Earth (Trache et al., 2017; Trache, 2017; Trache, Khimeche, et al., 2016). To meet the growing demand for biocompatible and environmentally friendly materials, this inexhaustible renewable biopolymer is regularly regenerated by nature in relative short periods of time. The worldwide production of this biomacromolecule is estimated to be between 10^{10} and 10^{11} t each year. However, only a small amount of 6×10^9 t is explored by several industry sectors such as textile, paper, chemical, and material industries (Lavoine et al., 2012). It can be found in different kind of living species ranging from wood, annual plats, sea animals, algae, fungi, and some bacteria (Trache, Hussin, et al., 2016). Table 11.1 displays various cellulose sources. Cellulose is a prominent raw material for the fabrication of various chemicals, and is considered as a strong candidate to substitute petroleum-based polymers owing to its outstanding features (Shaghaleh, Xu, and Wang, 2018). This biopolymer, as a white powder, was discovered and isolated by Anselm Payen (Siqueira, Bras, and Dufresne, 2010; Habibi, Lucia, and Rojas, 2010). A few years later, Herman Staudinger established the chemical structure of cellulose (Borges et al., 2015). Recently, numerous books (Kargarzadeh et al., 2017; Thakur, 2015b,a; Pandey et al., 2015; Wertz, Mercier, and Bédué, 2010; Postek et al., 2013) and review articles (Trache et al., 2017; Trache, Hussin, et al., 2016; Eyley and Thielemans, 2014; Moon et al., 2011; Klemm et al., 2011; Habibi, Lucia, and Rojas, 2010; Klemm et al., 2005; Klemm et al., 2018; Phanthong et al., 2018) have focused on summarizing the state of the art on this fascinating biomacromolecule. Therefore, only some significant information will be given in this review. From the production point of view, cellulose can be broadly obtained with a specific dimension by enzymatic, chemical, and/or mechanical treatments of cellulosic feedstocks such as wood, cotton, and annual plants (Trache, Hussin, et al., 2016). In plant cell walls, cellulose is biosynthesized from the mutual action of biopolymerization spinning and crystallization, where the processes are orchestrated by specific enzymatic terminal

TABLE 11.1

Various Sources for the Production of Cellulose Fibers

Source group	Sources
Hardwood	Eucalyptus, elm, birch
Softwood	Pine, spruce, cedar
Annual plants/ Agricultural residues	Oil palm, hemp, jute, sisal, alfa, kenaf, begasse, corn, sunflower, bamboo, canola, wheat, rice, pineapple leaf and core, peanut shells, potato peel, garlic straw residues
Animal	Tunicates
Bacteria	*Alcaligenes*, *Achromobacter*, and *Gluconacetobacter*
Algae	Green, gray, red, yellow-green, brown algae

complexes (TCs) situated in the plasma membrane (Klemm et al., 2018). During the process, TCs go through the membrane in ways that match with the orientation of the microtubules. The latter influence the arrangement of cellulose deposition (Somerville, 2006). The obtained cellulose chains are aggregated and joined in a continuous manner to generate long threadlike bundles of microfibrils. Depending on their origin, the microfibril diameters range from 2 to 20 nm for lengths that can reach several tens of microns (Klemm et al., 2018). The synthesis method is not completely known, but the microfibril structure infers that their biosynthesis and assembly encompass the coordination of 36 active sites (Vincent, 2002). In contrast, this biopolymer can be fabricated directly by a bottom-up procedure, where it is biosynthesized from glucose using specific bacteria (Trache et al., 2017; Wei et al., 2014). This bacterial cellulose is produced as a form-stable water-containing nanofiber network with cross-sectional dimensions ranging from 20 to 100 nm (Vasconcelos et al., 2017). In this bacterium, the enzyme cellulose synthase is available on the cytoplasmic membrane, and the obtained cellulose is produced extracellularly. Bacterial cellulose is naturally fabricated in a pure form without requiring rigorous processing to eliminate contaminants (Lin and Dufresne, 2014). However, lignocellulosics contain cellulose, hemicelluloses, lignins, and minor amounts of extractives and trace components.

It is worth noting that the amounts and features of cellulose are thoroughly influenced by the natural source and the isolation procedure (Trache, Hussin, et al., 2016). These biopolymers are composed of repeating β (1,4)-bound D-glucopyranosyl units (anhydroglucose unit, AGU) in the $^{4}C_{1}$-chain configuration, in which every monomer unit is corkscrewed at 180° compared to its neighbors (Figure 11.1) (Rojas, 2016). The monomers are assembled by condensation such that glycosidic oxygen bridges associate the AGUs. Three hydroxyl groups are encountered in each anhydroglucose unit: a primary group at C6 and two secondary groups at C2 and C3 (Rojas, 2016). The chemical groups impart cellulose some of specific features such as biodegradability, chirality, and hydrophilicity (Klemm et al., 2005). These hydroxyl groups allow chemical functionalization by esterification, carboxymethylation, etherification, hydroxypropylation, and cyanoethylation of cellulose to generate various derivatives via the reactions of some, or all, of the OH chemical groups (Huang et al., 2016; Fang et al., 2016; Chen, Zhu, et al., 2016; Boujemaoui et al., 2015; Klemm et al., 1998).

Naturally, cellulosic chains present a degree of polymerization of approximately 10000 AGUs in wood-derived cellulose and about 15000 units in native cellulose cotton (George and Sabapathi, 2015). Owing to the variety of OH groups on the AGU along the skeleton, there are extensive hydrogen linkages, as schematized in Figure 11.2, among individual cellulose chains (intra- and inter-molecular bonds) (Shaghaleh, Xu, and Wang, 2018). Consequently, the crystallization of numerous cellulosic chains provides insoluble microfibrils and two structural domains, i.e., ordered (crystalline) and paracrystalline (amorphous) domains (Trache, 2018; Trache et al., 2017; Trache, Hussin, et al., 2016). The interactions existing between the OH chemical groups and oxygen atoms of neighboring ring molecules are responsible for some characteristics such as hierarchical organization, multiscale microfibrillated structure, and highly cohesive nature (Lavoine et al., 2012; Brinchi et al., 2013).

Commonly, cellulose presents four polymorphs *vis.* cellulose I, II, III, and IV (Figure 11.3) (Lavoine et al., 2012; Trache et al., 2017; Trache, Hussin, et al., 2016). Cellulose I, native cellulose, is the conventional form found in nature. In 1984, Atalla and Vander Hart attested that this cellulose I can be categorized in two allomorphs (Rajinipriya et al., 2018), I_{α} and I_{β}.

Cellulose is insoluble in most known solvents owing to the chains' rigidity and the presence of polar/non-polar chemical groups. Furthermore, cellulose does not melt, which limits its direct use for common polymers employing the available industrial processes (Oksman and Bismarck, 2014). In addition, its crystallinity renders it recalcitrant to acid/base hydrolysis, thus making its treatment challenging. For these reasons, appropriate combinations of different treatments should be used or developed to either functionalize the fibers or produce nanofibers.

11.3 Cellulose Sources

Cellulose can be produced from various cellulosic sources including cotton, lignocellulosic biomass, marine animals, fungi, and bacteria (Kargarzadeh et al., 2017; Trache, 2017; Singla et al., 2016; Moon, Schueneman, and Simonsen, 2016; Gupta et al., 2016; Thakur and Voicu, 2016; Dufresne, 2013b; Agbor et al., 2011). Table 11.1 summarizes the most used raw materials for its preparation. Nowadays, cotton and wood are clearly the most important industrial sources of cellulose, being the main raw materials utilized for the manufacturing of biofuels, bioethanol, and biochemicals. However, competition among several industries such as pulp and papers, furniture sectors, and building materials, in addition to the use of cotton for the textile industry and the combustion of wood for energy, have increased the demand of such sources (Trache, 2017). This challenge, however, pushed the scientific community to explore other sources such as agricultural wastes, and forest and industrial residues (Rajinipriya

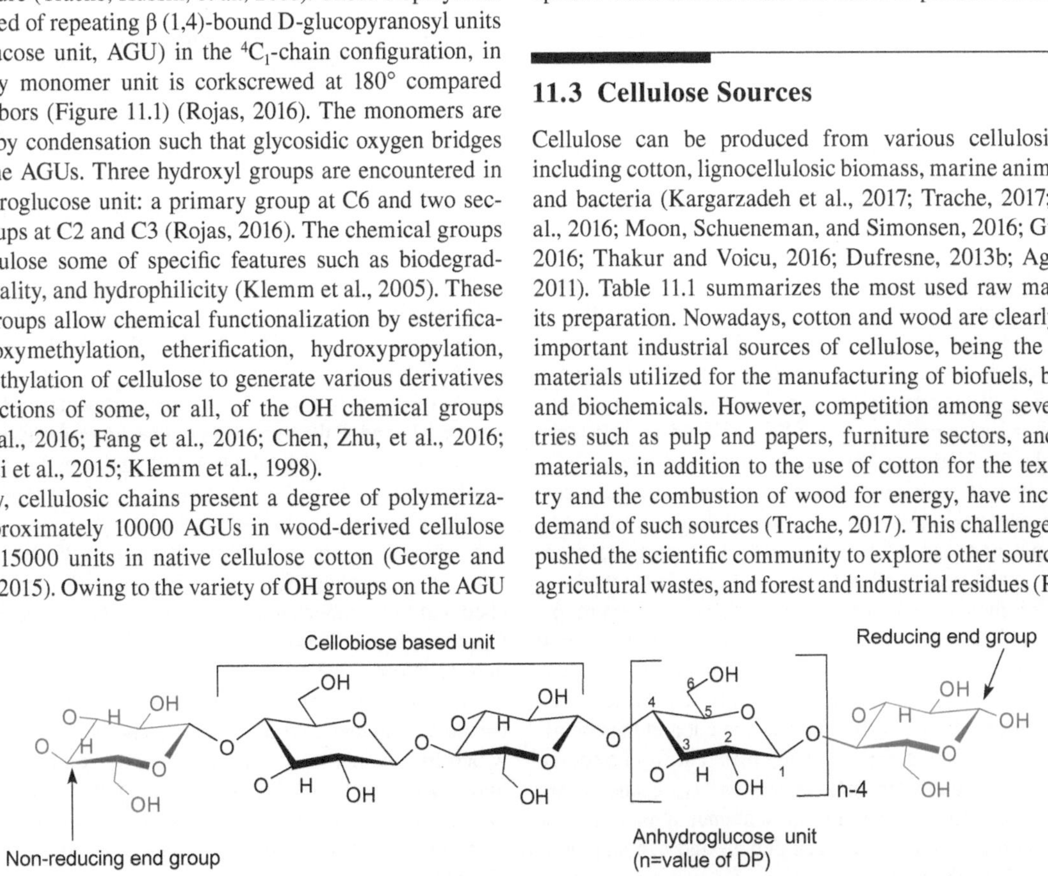

FIGURE 11.1 Molecular structure of cellulose showing the number of carbon atoms, the reducing end with a hemiacetal, and the non-reducing end with a free hydroxyl at C4. Reprinted with permission (Trache, Hussin, et al., 2016), Copyright © Elsevier Limited.

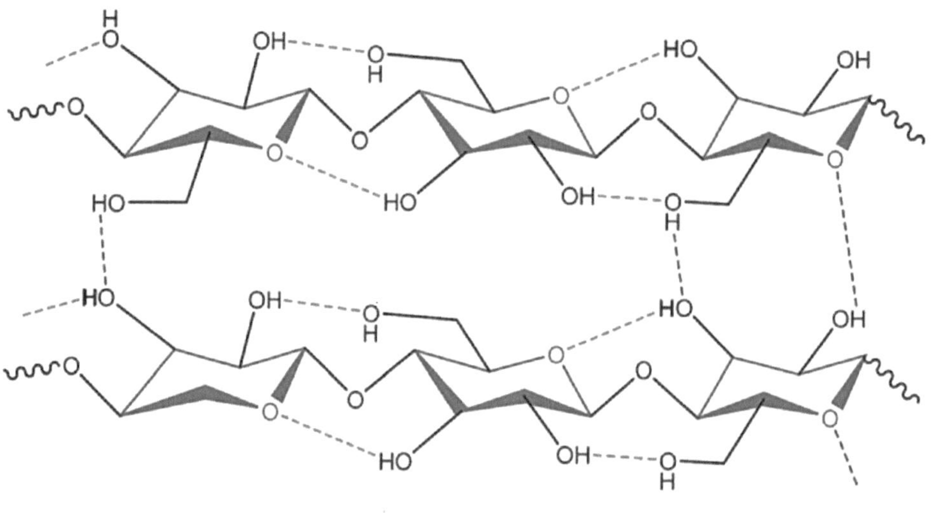

FIGURE 11.2 Intra- and inter-molecular hydrogen bonds in the molecular structure of cellulose. Reprinted with permission (Shaghaleh, Xu, and Wang, 2018), Copyright © The Royal Society of Chemistry.

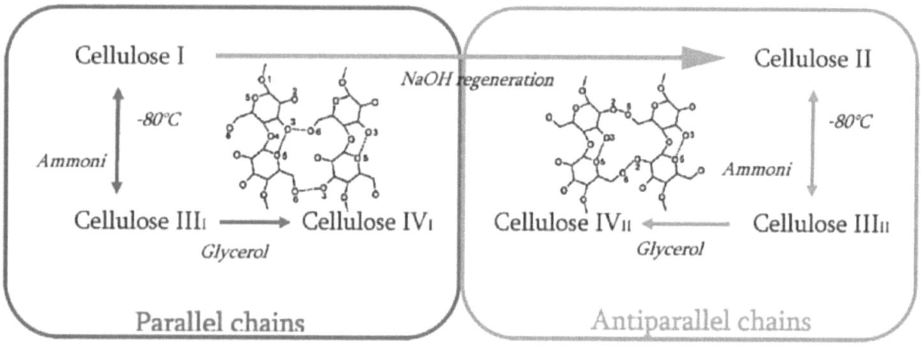

FIGURE 11.3 Polymorphs of cellulose and the main steps to produce them. Reprinted with permission from (Lavoine et al., 2012), Copyright © Elsevier Limited.

et al., 2018; Phanthong et al., 2018). Figure 11.4 depicts the hierarchical representation of the key sources from which cellulose can be isolated. Specifically, these natural sources can replace petroleum-based materials owing to their exceptional environmentally friendly features. Moreover, forest residues, and agricultural and industrial wastes, might be employed as feedstocks or fuel for the fabrication of high-value-added products.

It is obvious that pulps produced from woody sources are currently the most significant feedstocks for cellulose. Nevertheless, annual plants and agricultural residues have attracted more attention in recent years owing to cost and environmental concerns. These lignocellulosics can be categorized into five groups: (1) seed or fruit, (2) bast or stem, (3) grass, (4) leaf, and (5) straw fibers (Trache et al., 2017). The cell wall structure of lignocellulosic biomass is constituted by three kinds of polymers, i.e., cellulose, hemicellulose, and lignin. The content and the chemical composition of such components are different from a source to another due to the difference in species, type, and origin (Thakur, 2015a,b). Lignin, the third most abundant natural polymer, plays the role of binder between and around hemicellulose and cellulose. This abundant cross-linked aromatic polymer provides strength and protection to the plant by forming a matrix surrounding cellulose and hemicellulose (Ragauskas et al., 2014; Kirk, 2018). Lignin is basically constituted by three phenylpropane monomers, i.e., p-coumaryl, coniferyl, and sinapyl alcohols (Isikgor and Becer, 2015). On the other hand, hemicelluloses are hetero-polysaccharides constituted by different branched chains of monomers such as hexoses and pentoses (Delidovich, Leonhard, and Palkovits, 2014). The amorphous structure of hemicelluloses is easily hydrolyzed and can be dissolved in alkali solutions. In addition to the cross-links with lignin, hemicelluloses adhere to cellulose fibrils through hydrogen bonds and van der Waals interactions (Phanthong et al., 2018). Broadly, cellulose can be isolated from biomass sources using the top-down chemical and/or mechanical treatments. An efficient procedure to remove lignin, hemicelluloses, and other extractives provides pure cellulose.

Other living organisms including, animals, algae, and bacteria can be used to fabricate cellulose (Trache et al., 2017). Tunicates, which live mainly in oceans, are the only animal source of cellulose (Zhao et al., 2015; Cao et al., 2017). It is reported that tunicate cellulose exhibits various chemical functions in a number of tunicate classes and species, generating different structures. Normally, tunicate cellulose is constituted of pure cellulose Iβ (Trache et al., 2017). Another important cellulose source is algae (green, red, gray, yellow-green) (Moon et al., 2011). Algae,

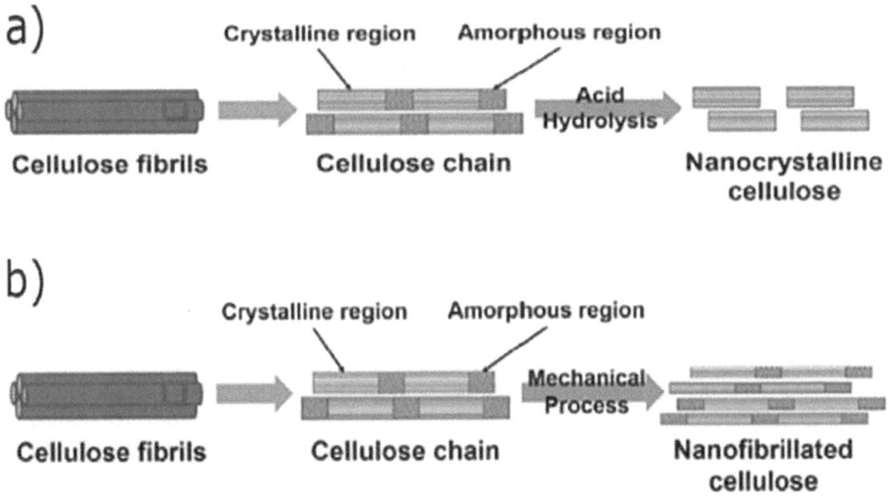

FIGURE 11.4 Schematic of nanocelluloses a) cellulose nanocrystals, b) nanofibrillated cellulose. Reprinted with permission from (Phanthong et al., 2018), Copyright © Elsevier Limited.

composed of cellulose and agar, were revealed as being a practical source for various applications owing to their high carbohydrate content. The production of red algae has increased from 5.3 million tons in 2006 to 10.8 million in 2011 (Kim, Wi, et al., 2015). The crystallinity of the cellulose derived from algae is high and can exceed 95 % (George and Sabapathi, 2015).

Some bacteria have also been reported to synthesize cellulose. The main strain is called Acetobacter xylinus (Mohammadkazemi, Azin, and Ashori, 2015; Reiniati, Hrymak, and Margaritis, 2017; Picheth et al., 2017). The synthesis process avoids the chemical treatments used for the elimination of hemicellulose and lignin in plant-derived celluloses. Other bacteria have been employed to produce cellulose as well.

11.4 Cellulose Isolation Methods

Cellulose can be isolated using different pretreatment and purification procedures, depending on its source. A suitable combination of treatments generates pure cellulose. The latter can be subsequently used to produce cellulose nanocrystals.

11.4.1 Cellulose from Lignocellulosic Materials

Lignocellulose is a bio-composite that contains cellulosic and non-cellulosic components such as hemicelluloses, lignin, extractives, and other components. To recover pure cellulose and remove non-cellulosic components, the pretreatment and purification of biomass are crucial steps. It is well known that lignin is the hardest chemical ingredient to remove from biomass and is considered as one of the main factors to optimize the pretreatment process and obtain pure cellulose. Currently, diverse techniques are available to isolate cellulose from lignocellulosics, i.e., physical, chemical, physicochemical, biological, and combined processes (Pandey et al., 2015; Abdul Khalil et al., 2014; Agbor et al., 2011; Hassan, Williams, and Jaiswal, 2018). Chemical, physical, and physicochemical pretreatments are the most investigated techniques among all reported pretreatment methods. Some common chemical or physicochemical pretreatment methods include acid, alkali, organosolv, ammonia fiber, hydrothermal, ozonolysis, carbon dioxide explosion, ionic liquid, and wet oxidation (Behera et al., 2014; Jönsson and Martín, 2016; Rabemanolontsoa and Saka, 2016). However, before the application of these methods, it is necessary to remove dirt/impurities, wax, pectin, and water-soluble extractives. This step is usually performed by boiling the lignocellulosic material in a toluene/ethanol mixture after a water-washing process.

Chemical pulping is mainly applied for the treatment of lignocellulosic materials, and the major technologies used are kraft, sulfite, soda, organosolv, and ionic liquid (Galkin and Samec, 2016; Jönsson and Martín, 2016). In the kraft process, lignin and parts of hemicellulose are degraded into black liquor using NaOH and Na_2S. The sulfite process, however, uses an aqueous mixture of bisulfite and sulfite to isolate hemicellulose and lignin from cellulose. The soda process is typically utilized for the pulping of non-wood plants. Usually, the raw material is digested at 413–443 K with an aqueous solution of NaOH; hydroquinone can be employed as additive to decrease the degradation of carbohydrates as well. On the other hand, the organoslov process, based on organic solvents such as methanol, ethanol, acetic acid, and peroxyformic acid, or their mixtures with water at a temperature ranging from 453 to 473 K, can be used for the delignification process. The utilization of ionic liquids (ILs) is an alternative approach. ILs disrupt the non-covalent interactions between the different components of lignocellulosics and lead to a better separation without causing important degradation.

Physical/mechanical techniques require high levels of energy. Several approaches have been developed to extract cellulose fibrils from different lignocellulosics. Some of the most important processes are high-intense ultrasonication, cryocrushing, comminution, micro-fluidization, and high-pressure homogenization (Trache et al., 2017).

The biological process is mainly related to the action of fungi that are able to produce enzymes to attack hemicelluloses, lignin, and polyphenols existing in the lignocellulosic material (Rabemanolontsoa and Saka, 2016; Behera et al., 2014). This process has received much attention owing to its advantages over other methods, such as high yield of desired product, low energy consumption, and absence of toxic compounds. Nevertheless, this process exhibits techno-economic challenges because it is very slow and necessitates of specific checks of the growth conditions and of a huge space to be performed. Thus, further studies are required to efficiently optimize this ecofriendly method.

11.4.2 Cellulose from Animals, Algae, and Bacteria

Tunicates are marine invertebrate sea animals. They are the only animal species that generate cellulose in the outer tissues, known as tunic, from which a purified fraction named tunicin can be produced (Zhao et al., 2015). Tunicate cellulose can be combined in the form of microfibrils constituted by pure cellulose Iβ allomorph, similar to lignocellulosic biomass cellulose (Cao et al., 2017; Sacui et al., 2014). Usually, the prehydrolysis-kraft process was employed for the isolation of cellulose (Koo et al., 2002). Typically, tunicate tunic can be recovered by removing the interior organs of the animal, freeze-drying them and milling them in powder. The lipids, ash, and some sugars other than glucose can be eliminated using a prehydrolysis procedure. Subsequently, the purified residue will be subjected to the kraft-cooking step in order to remove further proteins and some residual sugars. This is followed by the bleaching process to eliminate the residual proteins and some chromophoric components, and finally by purification.

Algae are potential sources of pure cellulose (Chen, Lee, et al., 2016; Bettaieb et al., 2015). Commonly, this kind of biomass needs to be washed to eliminate eventual contamination. It is then dried and ground into powder. The latter will be subjected to dewaxing and bleaching to recover pure cellulose.

Pure and highly crystalline cellulose can be produced from bacteria (Vasconcelos et al., 2017; Reiniati, Hrymak, and Margaritis, 2017; Picheth et al., 2017; Campano et al., 2016; Mohammadkazemi, Azin, and Ashori, 2015; Keshk, 2014; Huang, Zhu, et al., 2014; Dugan, Gough, and Eichhorn, 2013; Charreau, L Foresti, and Vázquez, 2013; Hirai et al., 2008). Acetobacter xylinum can generate cellulose microfibrils in the form of thick, clear, flat pellicles that float on the surface of the growth medium. These cellulosic pellicles are composed of some impurities that can be eliminated by hydrolysis using alkaline solution. Various processes have been developed to synthesize bacterial cellulose.

11.5 Overview of Cellulose Nanofibers

Recently, cellulose nanomaterials (CNMs) have attracted a lot of attention and are extensively investigated owing to their interesting characteristics (Chen et al., 2018; Nascimento et al., 2018; Du et al., 2017; Wang et al., 2017; De France, Hoare, and Cranston, 2017; Azeredo, Rosa, and Mattoso, 2017; Kargarzadeh, Huang, et al., 2018; Golmohammadi et al., 2017; Phanthong et al., 2018). CNMs can be categorized in a number of subclasses depending on their production methodology, shape, size, and function (Jawaid, Boufi, and HPS, 2017; Kargarzadeh et al., 2017). Broadly, CNMs can be divided into nanostructure materials (cellulose microcrystals and cellulose microfibrils) and cellulose nanofibers (Trache et al., 2017).

The expression of cellulose nanofibers designates a crystallite or cellulose fibril comprising of at least one dimension in the nanometric scale. The development and emergence of cellulose nanofibers has received much attention during these last few years from both industrial and academic communities owing to its potential for a broad spectrum of uses and several outstanding characteristics such as ecofriendliness, renewability, light weight, and low coefficient of thermal expansion (Charreau, L. Foresti, and Vázquez, 2013; Miller, 2016; Trache, 2018; Phanthong et al., 2018; Nascimento et al., 2018; Klemm et al., 2018; Kargarzadeh, Huang, et al., 2018; Chen et al., 2018; Reiniati, Hrymak, and Margaritis, 2017; Picheth et al., 2017; Kargarzadeh et al., 2017). The features of the particles and tiny fibers play a prominent role for the establishment of the new bio economy. This significant interest is obvious from the number of scientific research papers on the field, which has risen considerably in the last two decades. Broadly, many topics related to cellulose nanofibers have been treated, including their production form different sources employing various isolations methods, modification, and characterization, as well as their use in widespread applications. Currently, extensive research work is being conducted over the world in order to optimize the preparation procedures with lower costs, consistency, improved yields, and better quality.

Cellulose nanofibers can be split up into two types (Kargarzadeh, Mariano, et al., 2018; Trache et al., 2017). The first one concerned cellulose nanofibrils with different terminologies, comprising of nanoscale-fibrillated cellulose, nanofibrillar cellulose, fibril aggregates, nanofibrilated cellulose, nanofibrils, cellulosic fibrillar fines, nanofibers and occasionally microfibrils or microfibrillated cellulose, while the second type involved nanocrystalline cellulose with numerous nomenclatures, encompassing cellulose nanowhiskers, cellulose nanocrystals, cellulose whiskers, nanorods and cellulose crystallites. Additional classes of cellulose nanofibers materials such as bacterial cellulose, amorphous nancellulose and cellulose nanoyarn can be found as well. The terminology, which will be maintained in this chapter, is in accordance with the Technical Association of the Pulp and Paper Industry (TAPPI) standard recommendations (TAPPI, 2017).

The outstanding properties of cellulose nanofibers such as the nanometer scale, high specific surface area, thermal stability, non-toxic, prospect mechanical features, easy processing, high aspect ratio and stiffness have built up new opportunities for developing of new type of cellulose nanofibers-based systems. In addition, the various structure of cellulose nanofibers and their surface chemistry properties, which can be developed from several cellulosic sources through various production approaches, have allowed the development of different kind of materials or devices (Abitbol et al., 2016; Chen et al., 2018). These latter have been used in biosensors and bio-imaging, biomimetic materials, pharmaceutical binders, liquid crystals, reinforcing of polymer composites, templates, substrate for printing electronics, support for catalysts and immobilization of enzymes, optically transparent materials, low-calorie food additives, carrier vehicles for systems controlled release, emulsion stabilizers, rheological modifiers, aerogels, energy storage, etc.

Cellulose nanofibers is an attractive raw material for various reactions owing to its simple structure as well as its reactive surface containing hydroxyl groups. This reactivity can be tailored by the addition of specific functional groups (Eyley and Thielemans, 2014; Nascimento et al., 2018). Several modifications can be performed through derivatization reactions such as acetylation, amidation, carbamation, cationization, silanization, oxidation, Diels-Alder, huisgen cycloaddition, grafting from, grafting onto, oxidation and so forth.

The following sections of this chapter will only focus on one type of cellulose nanofibers, i.e., cellulose nanocrystals, where the preparation procedures, characterization methods, and properties, as well as the surface modification of these nanomaterials, will be treated.

11.6 Cellulose Nanocrystals: Preparation Methods

Cellulose nanocrystals are cellulose nanofibers with high strength, which are often isolated from cellulose fibrils. They are rod or needle like shape with diameter of 4-70 nm and length from 100-6000 nm based on the source and the isolation process (Trache, 2018). Nanocrystalline cellulose contains 100% of cellulose with high crystallinity around 54-88% (Phanthong et al., 2018). These nanocrystals can be extracted from a widespread of raw materials that are primarily subjected to diverse pretreatment procedures for the complete/partial removing of lignin, hemicelluloses and extractives (Chen et al., 2018; Trache et al., 2017; Kargarzadeh et al., 2017; Jawaid, Boufi, and HPS, 2017; Agbor et al., 2011). Cellulose comprises crystalline domains and disordered regions in different ratios. An appropriate combination treatment can be used to remove the amorphous regions and recuperate cellulose nanocrystals (Ng et al., 2015; Trache et al., 2017; Dufresne, 2013b; Postek et al., 2013). The disordered or paracrystalline regions act as structural defects and are selectively hydrolyzed, whereas the ordered parts, which present a good resistance to the process, remain intact. The obtained nanocrystals are short and exhibit short aspect ratios and a larger relative crystallinity with respect to that of cellulose nanofibrils.

The fabrication of cellulose nanocrystals has received tremendous attention in recent decades. Over the last two decades, an increasing pool of research works has emerged describing different preparation methodologies of nanocrystalline cellulose. A number of processes have been established and other continue to be developed to produce cellulose nanocrystals in an economic/sustainable manner with preferred features. Many approaches have been described to isolate cellulose nanocrystals, namely, chemical acid hydrolysis (Kontturi et al., 2016; Du et al., 2016; Chen, Zhu, et al., 2016; Thakur, 2015b; Jonoobi et al., 2015; Liu et al., 2014; Tang et al., 2011), mechanical refining (Pandey et al., 2015; Amin et al., 2015), enzymatic hydrolysis (Anderson et al., 2014; Xu et al., 2013; Chen et al., 2012), oxidation method (Vazquez et al., 2015; Sun et al., 2015; Visanko et al., 2014), ionic liquid treatment (Lazko et al., 2016; Tan, Hamid, and Lai, 2015; Mao et al., 2015), subcritical water hydrolysis (Novo et al., 2016; Novo et al., 2015), plasma-based technology, and combined processes (Chowdhury and Hamid, 2016; Tang et al., 2014; Lu et al., 2013). The benefits and shortcomings of each process have recently been reported by Trache et al. (Trache et al., 2017). However, some post-treatments, such as centrifugation, solvent elimination, washing, sonication, neutralization, purification, filtration, fractionation, surface modifications, and stabilization need to be followed to finally recover the nanocrystalline cellulose product (Trache et al., 2017; Klemm et al., 2011).

The acid hydrolysis process remains the most employed method for the isolation of cellulose nanocrystals (Nascimento et al., 2018; Chen et al., 2018). It commonly requires acids such as sulfuric, phosphoric, hydrochloric, and hydrobromic liquid acids, solid and gaseous acids, organic acids, or a mixture of inorganic and organic acids (solid/liquid/gaseous/organic/inorganic acids).

The most common mechanism of acidic hydrolysis is based on the production of hydronium ions in water, which act as catalysts (Figure 11.5) (Mao et al., 2017; Trache, Hussin, et al., 2016). The latter rapidly protonate the glycosidic oxygen (pathway I). Protonation might also occur to the pyranic oxygen, generating the breakage of the anhydroglucose ring. This step allows the regeneration of the catalyst proton. A partial protonation from both oxygen atoms occurred during the conformation restriction of the cellulosic chain. The formation of carbonation via a unimolecular step is one of the most important steps during the hydrolysis process. The mechanism takes place through pathway I or II. This process leads to the separation of the crystalline regions and the degradation of the paracrystalline (amorphous) domains. The disordered parts in the polymer chains are easily available to the hydrolysis process owing to kinetic force and steric hindrance. The parameters involved for the production of cellulose nanocrystals can change their geometry and crystallinity index, as well as their aspect ratio.

Calvert was the first author who reported the hydrolysis of cellulose in 1855 (Mao et al., 2017). The main acid hydrolysis with H_2SO_4, known as the Scholler process, was reported in the 1920s. The utilization of hydrochloric acid for cellulose hydrolysis is called Bergius process (Trache, Hussin, et al., 2016). In 1951, Rånby produced stable colloidal suspensions of cellulose using H_2SO_4 (Nascimento et al., 2018). This acid reacts with the surface OH chemical groups of cellulose generating charged surface sulfate esters that allow the dispersion of the individual cellulose nanocrystal bundles in aqueous medium (Chen et al., 2018). Nevertheless, the existence of such groups decreases the thermal stability. Since the early investigations, experimental conditions such as liquid to solid ratio, hydrolysis process, temperature, time, pressure, and source have been tailored in the attempt to optimize the product quality and yields. The obtained nanocrystals may exhibit different shapes, depending on their biological source and the isolation process (Klemm et al., 2011).

Cellulose nanocrystals present several advantageous features such as low density, better mechanical properties, high surface area, elongation morphology, and ease of bio-conjugation. These advantages have recently motivated substantial research programs on the preparation of cellulose nanocrystals, and initiatives for the commercial exploitation of these nanofibers have arisen (Nascimento et al., 2018; Trache et al., 2017). Four commercial producers currently fabricate cellulose nanocrystals at capacities beyond pilot plant scale: CelluForce (Canada, 1000 kg/day), American process (USA, 500 kg/day), Melodea/Holmen (Sweden, 100 kg/day), and Alberta Innovates (Canada, 10 kg/day). Further research facilities are currently producing CNC as well, and various new laboratories and pilot-scale operations have appeared, such as Forest Products Lab (USA, 10 kg/

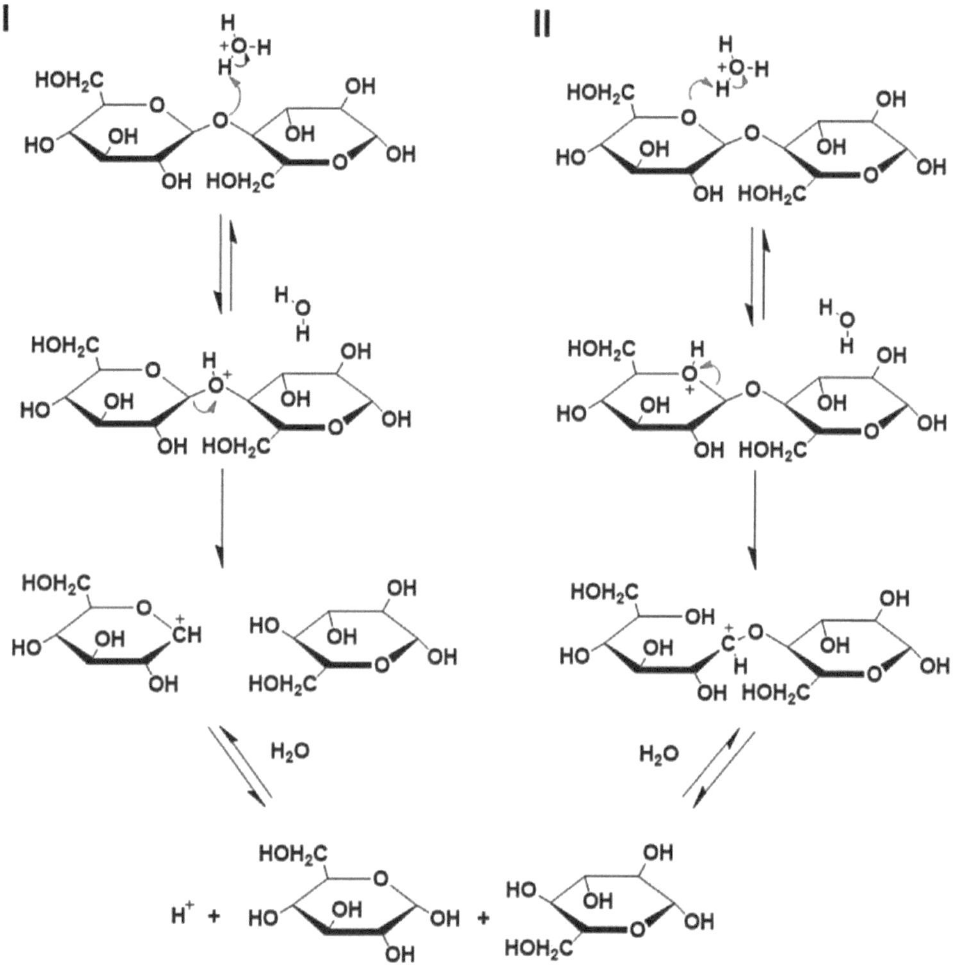

FIGURE 11.5 Mechanism of the acid catalyzed hydrolysis of ß-1,4glucan. Pathway I is initiated by the protonation of the glycosidic oxygen, while pathway II by the protonation of the ring oxygen. Pathway I is dominant over pathway II. Reprinted with permission from (Mao et al., 2017), Copyright © American Chemical Society.

day), India Council for Agricultural Research (India, 10 kg/day), Blue Goose Biorefineries (Canada, 10 kg/day) and FPInnovation (Canada, 3 kg/day). However, the employment of alternative sources of cellulose to prepare large-scale cellulose nanocrystals remains timid. In this sense, the price of these nanofibers is expected to decrease with the utilization of cheaper sources of pulp and with the improvement of the isolation process.

11.7 Characterization and Properties of Cellulose Nanocrystals

The appropriate characterization of cellulose nanocrystal–based materials, which allows an in-depth understanding of their properties, requires the joined employment of various analytical tools. The characterization of cellulose nanocrystals has been widely described by scientists and experts in the literature starting from the preparation of samples and the establishment of the experimental conditions to the discussion of the results (Huang, Chang, et al., 2014; Foster et al., 2018). The main analytical techniques used for such purposes involve spectroscopic, microscopic, and thermal methods. These kinds of procedures are primarily utilized (i) to point out the size and shape of CNCs, (ii) to study their physicochemical properties, and (iii) to assess the surface of modified cellulose nanocrystals. It was reported that CNCs displayed various morphologies depending on the source of isolation as well as on the procedure of extraction (Vasconcelos et al., 2017; Neto et al., 2016; Lemke et al., 2012; Trache et al., 2017; Thakur, 2015b). Scanning electron microscopy (SEM), transmission electron microscopy (TEM), and atomic force microscopy (AFM) are well-established tools to investigate the size, shape, and morphology of such nanofibers. Spectroscopic techniques such as Fourier transform infrared spectroscopy (FTIR), Raman, X-ray diffraction (XRD), and nuclear magnetic resonance (NMR) have been extensively used for the determination of the crystallinity of CNCs (Karimi and Taherzadeh, 2016; Foston, Hubbell, and Ragauskas, 2011). Good results have been obtained with these techniques. Thermal properties are considered as crucial features to understand the behavior of CNCs, as well as their suitable uses in several applications. They properties can be easily accessed using thermogravimetric analysis (TGA) and differential scanning calorimetry (DSC). These analytical techniques have been revealed to be effective for such goals. Other properties, such as mechanical, rheological, and optical, among others, can also be determined using different analytical tools (Huang, Chang, et al., 2014; Foster et al., 2018). The following part will be limited to some important methods, i.e., spectroscopic techniques (FTIR, XRD), microscopic methods (SEM, AFM, and TEM), and thermal analysis tool (TGA).

11.7.1 Fourier Transform Infrared Spectroscopy (FTIR)

Infrared (IR) spectroscopy is a well-known analytical method that has served scientists for over 50 years. It is a high-throughput analysis that can offer both quantitative and qualitative information about the chemical features of CNCs extracted from natural sources (García-García et al., 2018; Naduparambath et al., 2018). The absorption of IR light caused by the excitation from the ground vibrational energy level to a higher one provides valuable information on molecular structure, interactions, and assemblies (Siesler et al., 2008; Tasumi, 2014). The main advantages of such a technique concern the wide diffusion, limited costs, and high selectivity and sensibility with the relatively fast and easy acquisition time of the spectra. However, a possible drawback is that it requires some manipulation precautions of the resulting spectrum. The presence of CO_2 and humidity can lead to the absorption of electromagnetic radiation, which usually necessitate to acquire a background spectrum just prior to the characterization and subtract it from the sample spectrum. Sometimes, the baseline should be adjusted by moving the spectrum to zero for the domains where no absorbance is found. Also, to compare the FTIR results properly, the normalization procedure can also be applied. Usually, the IR wavelength range is split up into three domains, i.e., near-IR (12800–4000 cm^{-1}), mid-IR (4000–400 cm^{-1}), and finally far-IR (400–10 cm^{-1}).

FTIR has been extensively utilized for determining CNC composition and structure. The mid-IR is often preferred for functional group analysis and has the benefit of having high throughput. Characteristic IR vibrational modes are listed in Table 11.2 and an exemplary FTIR spectrum of CNCs derived from rice husks is depicted in Figure 11.6. The typical functional groups detected in CNCs consist mostly of hydroxyl (O–H), alkyl (C–H), β-glycosidic linkage of glucose rings (C–O–C), etc. (Trache, Khimeche, et al., 2016; Trache et al., 2013; Islam et al., 2017). An efficient isolation process can be evaluated by the absence of the characteristic bonds of lignin and hemicellulose in the FTIR spectra.

TABLE 11.2
The Most Important FTIR Spectral Peak Assignments for CNC

Band position (cm^{-1})	Band assignment	Vibration type
800–900	C–H	Rocking vibration
1000–1100	C–O–C	Stretching vibration of pyranose ring
1100–1300	C–O	Stretching vibration of cellulose
~1300	$C–H_2$	Rocking vibration
2900–3000	C–H	Symmetrical stretching vibration
3300–3400	O–H	Stretching vibration

In addition, FTIR can also be used to measure the crystallinity index of CNCs by determining the relative peak heights of areas (Park et al., 2010; Fan, Dai, and Huang, 2012; Karimi and Taherzadeh, 2016). This method provides satisfactory results compared to those obtained by other methods such as XRD and NMR (Ju et al., 2015). The OH groups in CNCs are the key factors for the intra- and inter-molecular hydrogen bonds and generate ordered and disordered parts. In FTIR spectra, the characteristic band situated between 1420 and 1430 cm^{-1} (A_{1430}) is attributed to a s symmetric Ch_2 bending vibration and named crystalline band, whereas the absorption band found at around 893–898 cm^{-1} (A_{898}) is attributed to C-O-C stretching at the β-glycosidic linkage, known as amorphous band. Consequently, the crystallinity index corresponds to the ratio of A_{1430} on A_{898}.

11.7.2 X-Ray Diffraction Analysis

XRD is a powerful nondestructive tool to analyze crystalline or semi-crystalline materials. It offers appropriate information on structure, phases, crystal orientation, and other structural parameters such as crystallinity, crystal defect, and average crystallite size (Bunaciu, Udriştioiu, and Aboul-Enein, 2015). The XRD diffraction pattern is the fingerprint of periodic atomic arrangements in a given material. The powder XRD, widely used to determine the crystallinity degree, can be carried out on homogenized and finely ground samples. This method presents the benefit of not necessitating any pretreatment of the specimen. Moreover, unlike in destructive analysis methods (e.g., thermal methods), the sample does not go through any physicochemical modification during the spectrum recording and consequently it can be recovered and used for other analyses.

XRD provides data directly associated to both crystalline and disordered regions of CNC. Strong signals are attributed the crystalline regions, whereas the broader and weak signals of the diffraction patters are assigned to the amorphous domains. CNC crystallinity is crucial to determine the thermal and mechanical features of CNC-based products.

Typically, three methods can be applied to compute the crystallinity index, based on the XRD outputs. The most famous method is the one developed by Segal et al., in which the crystallinity is simply computed by dividing the height of (I_{002}) peak (at around 22°) and the intensity of diffraction of the amorphous part at (I_{am}) 18°, as given below (Equation 11.1). Despite the extensive use of such approach to compute the crystallinity of cellulose, it presents some disadvantages (French and Cintrón, 2013; Ju et al., 2015) which need to be taken into account during any measurement.

$$CrI(\%) = \frac{(I_{002} - I_{am})}{I_{002}} \times 100 \tag{11.1}$$

The second method, called the deconvolution (curve fitting) technique, employs the concept that five crystalline planes of cellulose, corresponding to (101), (10 –1), (021), (002), and (404) are scattered on the amorphous domains (Karimi and Taherzadeh, 2016; Sathitsuksanoh and Renneckar, 2018). Thus, the obtained data can be deconvoluted. Gaussian, Lorentzian, and Voigt are the common functions utilized for such purpose. The main challenge of this method is the choice of the proper peaks.

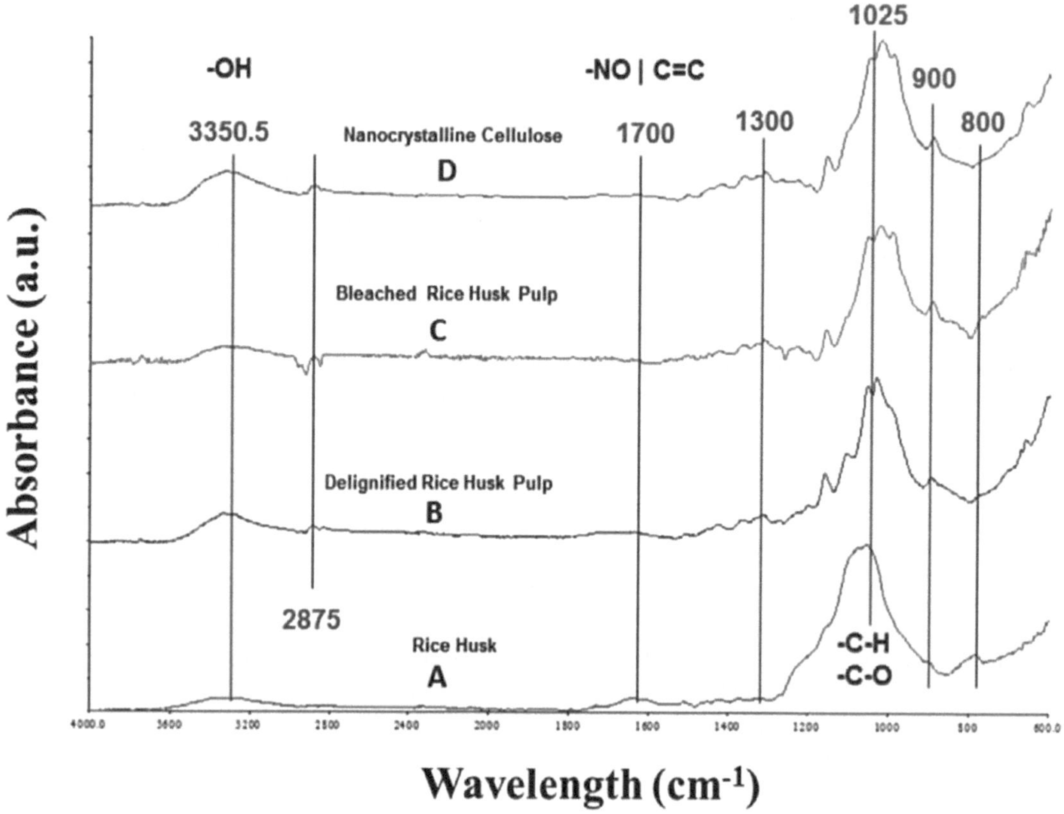

FIGURE 11.6 FT-IR spectra of RH (A), D-RHP (B), B-RHP (C), and NCC (D) samples. Reprinted with permission from (Islam et al., 2017), Copyright © Elsevier Limited.

The last method, commonly known as Ruland–Vonk, or method of subtraction of amorphous domains, allows the determination of the crystallinity index by dividing the area of the crystalline region above the amorphous profile, as a standard, by the area of the total domain (Ciolacu, Ciolacu, and Popa, 2011). Usually the amorphous profile is determined by employing a pattern obtained from a disordered compound (Ju et al., 2015). Numerous studies have reported the crystallinity index of CNCs. Table 11.3 displays some examples of the crystallinity of CNCs derived from various sources.

11.7.3 Scanning Electron Microscopy (SEM)

SEM is a versatile tool employed in several industrial and research labs. This technique can perform the analysis of morphology and microstructure, as well as surface characterization. It is a surface imaging method that takes advantage of the secondary electrons and the backscattered electrons emitted from the sample after its interaction with the focused beam of accelerated electrons generated by the microscope to build 2-dimensional images. When it is equipped with energy-dispersive spectroscopy (EDX), SEM allows the determination of the elemental composition with 1–3 % accuracy (Karimi and Taherzadeh, 2016). Metallic samples provide high contrast as a high interaction potential of the higher density atoms with the electron beam. Consequently, they are easily detected. However, several nanomaterials, due to their organic nature, are essentially invisible to SEM because they are made up of low electron density and non-conductive atoms, and do not reflect an electron beam sufficiently. Therefore, to mitigate such challenge, sputter coating with a thin layer conductive material (200–300 Å) such as gold or platinum is commonly utilized to produce a conductive layer on the sample. This approach decreases thermal damage, prevents surface charging, and enhances the secondary electron signal necessitated in the electronic microscope (Crucho and Barros, 2017; Michler, 2008).

SEM images are usually explored to access the structural and dimensional appearance of CNCs (Kargarzadeh et al., 2017; Huang, Chang, et al., 2014; Dufresne, 2013b). Several authors have demonstrated that the preparation procedure influences the surface morphology in terms of surface smoothness and the size of cellulose nanofibers (Nascimento et al., 2014; Lamaming et al., 2015). Also, the removal of the cementing materials, such hemicellulose, lignin, wax, and pectin, can be clearly detected where an increase of the specific surface manifests (Nascimento et al., 2014; Johar, Ahmad, and Dufresne, 2012; Cherian et al., 2011). The CNC samples often present a shortening in fiber and rough surface. Sometimes, blender mats of CNC clusters can be obtained due to the resolution of the SEM (Foster et al., 2018). Alternatively, transmission electron microscopy can overcome such drawbacks.

11.7.4 Transmission Electron Microscopy (TEM)

TEM is one of the most powerful tools for characterizing nanomaterials. It can deliver direct images and chemical information

TABLE 11.3
Some Properties of CNC Particles Reported by Several Authors

Crystallinity index (%)	Length (nm)	Diameter (nm)	T_{onset} of decomposition (°C)	Reference
69.4	100–500	20–45	240	(Liu et al., 2017)
85.9	130 ± 30.23	9 ± 1.96	185.78	(Ilyas, Sapuan, and Ishak, 2018)
88	210–480	15–50	220	(Oun and Rhim, 2015)
91.4	132 ± 55	8 ± 3	237	(Reid, Villalobos, and Cranston, 2016)
89.4	245 ± 24	23 ± 10	260	(Cheng et al., 2017)
67	187 ± 42	12 ± 1	/	(Marett, Aning, and Foster, 2017)
72	400–600	4–8	200	(Yao et al., 2017)
70.62	307	8	200	(Coelho et al., 2018)
90	139.7–423.4	20.9–46.8	/	(Shaheen and Emam, 2018)
95.8	75–80	15–20	200	(Tan, Hamid, and Lai, 2015)

at a special resolution down to the level of atomic dimensions (<1 nm) (Michler, 2008). In a classical TEM analysis, an accelerated electron beam is transmitted through a very thin sample, and the interaction of the electrons with the material generates elastic and inelastic scattering. The most widely employed technique to measure the size of nanomaterials, TEM can provide bright- or dark-field images, leading to precise information of particle size even at nanoscale. The very small operative wavelength of the electron is the key factor for the large magnification of TEM, from 50× to 10^6× (Mitić et al., 2017; Lin et al., 2014). Adequate specimen preparation allows to obtain better images. Compared to SEM, TEM samples require further manipulation such as diluting in a suspension and adding supplementary surfactant dispersants, as well as incorporating a heavy metal stain. This latter approach is considered as the main manner to mitigate contrast issues. Usually, stains can be incorporated just after the deposition of the specimen onto an appropriate substrate. Positive and negative stains can be applied. Negative stains are the more common way owing to its simplicity. The usually employed negative stains encompass ammonium molybdate, phosphotungstic acid, uranyl acetate, and other heavy metal solutions, including vanadium-based products (Kargarzadeh et al., 2017; Foster et al., 2018).

Different cellulose nanocrystals produced from different sources have been already analyzed by TEM (Figure 11.7), in which ribbon, needle, or rod-like shapes have been obtained with widths of 4–70 nm and length between 75 nm and several micrometers (Table 11.3), respectively (Trache, 2018; Trache et al., 2017). These various shapes are caused by the hydrolysis treatment, which splits up the bonds that link the microfibrils to amorphous regions. Consequently, the disordered domain, which depicts as structural defeats, affects the transverse cleavage of microfibrils into nanocrystals.

11.7.5 Atomic Force Microscopy (AFM)

AFM is a versatile tool to study the topography, dispersion, aggregation, and physicochemical features of cellulosic materials at nanometer resolution. This technique does not necessitate oxide-free, electrically conductive surfaces for analysis, such as scanning tunneling microscopy (STM). Different scanning modes can be utilized in AFM, such as contact, noncontact, and tapping modes. An AFM tool contains a micro-machined cantilever with a sharp tip at one end. The interactions between the tip and the sample surface, either by electrostatic or van der Waals repulsion, as well as attraction, lead to the deflection of the cantilever over the surface of the samples to generate a 3-dimensional

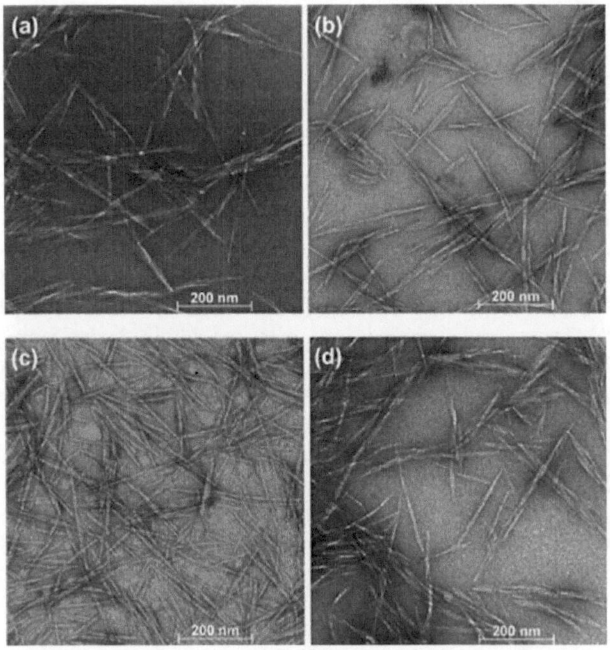

FIGURE 11.7 Transmission electron microscopy (TEM) images of cellulose nanocrystals obtained from elephant grass samples: (a) pretreated with concentrated alkaline peroxide and hydrolyzed for 40 min (NCCP40); (b) pretreated under the same conditions, but hydrolyzed for 60 min (NCCP60); (c) pretreated with diluted alkaline peroxide and hydrolyzed for 60 min (NCDP60), and (d) pretreated with acid-alkali and hydrolyzed for 60 min (NCAA60). Reprinted with permission from (Nascimento and Rezende, 2018), Copyright © Elsevier Limited.

image with a vertical resolution of 0.5 nm (Morita et al., 2015; Dufrêne et al., 2017; Zhang, Zheng, and Cosgrove, 2016).

AFM presents some advantages over SEM and TEM, as no specific preparation is required, such as staining, dehydration, or metal coating, which means that the native state of CNCs is usually maintained (Karimi and Taherzadeh, 2016). This method offers a more accurate and precise analysis of the detailed structure and the thickness the individual whiskers with respect to TEM (Neto et al., 2013). In addition, AFM permits the discrimination of individual crystallites from the agglomerated structures (Silvério et al., 2013). Table 11.3 presents length and width of CNCs prepared from some natural sources.

11.7.6 Thermogravimetric Analysis (TGA)

TGA, one of the oldest thermal analysis methods, is a characterization technique that has been widely employed in the investigation of material science. It allows several applications, such as the determination of the decomposition patters, degradation mechanisms, and stability, as well as reaction kinetics and the determination of organic/inorganic contents, volatile components, etc. It is a suitable method for monitoring the physicochemical properties of samples subjected to heating programs (static or dynamic) in a chosen atmosphere (inert or reactive) by following the mass change (gain or loss) (Cai et al., 2018). A thermogravimetric analyzer consists of a furnace containing a sample holder supported by a precise microbalance where a sample pan can be placed. The mass of the specimen is continuously measured during the experiment. Based on the variation of the sample mass with respect to temperature/time, the differential thermogravimetry (DTG) curve can be plotted. The employment of TGA presents several advantages, such as simplicity, minimal amount of sample, high accuracy, and precise control (Bach and Chen, 2017).

TGA has been widely employed to investigate the thermal behavior of CNCs (Naduparambath et al., 2018; Coelho et al., 2018; García-García et al., 2018; Reddy and Rhim, 2014). Usually, CNC samples demonstrate two stages of mass loss within the temperature range of 20–700 °C. The first stage ascribed to the evaporation of water occurred at around 60–140 °C. The second stage, however, occurred within a wide range: 200–450 °C, which can be assigned to complex processes such as dehydration, decarboxylation, depolymerization, and decomposition of glycosyl units followed by the formation of a charred residue (Trache et al., 2014; Lédé, 2012; Leszczyńska et al., 2018). Table 11.3 displays some decomposition temperatures of CNC derived from various natural sources. Typically, these nanomaterials present high thermal stability owing to the high crystallinity of CNCs, because these nanoparticles are compacted in a very orderly way, which prevents the melting of CNC particles (Kalia et al., 2011).

11.8 Surface Modification of Cellulose Nanocrystals

The functionalization of cellulose nanocrystals is currently of interest due to their improved properties once produced. Various modification routes have been proposed lately in order to obtain the functionalization of cellulose nanocrystals (Habibi, 2014; Eyley and Thielemans, 2014). Some of the improved properties of the modified cellulose nanocrystals are their hydrophilic or hydrophobic features. It is understood that the hydrophilic or hydrophobic features of modified cellulose nanocrystals may provide different interfacial reaction, including different energy of adsorption, for example in a water–oil emulsion system. Generally, the modification of cellulose nanocrystals can be done through physical adsorption of surfactant or reaction with polymers (Hu, Patten, et al., 2015). The existence of active sites in cellulose makes the modification process possible under certain conditions (Kargarzadeh, Mariano, et al., 2018). The modification of cellulose nanocrystals can be categorized in three different groups: (i) surface native functional groups, (ii) adsorption of functional groups, and (iii) covalently attached molecules (Salajková, Berglund, and Zhou, 2012; Moon et al., 2011; Fortunati et al., 2012). In this chapter, a few modification methods of cellulose nanocrystals will be discussed in detail.

11.8.1 Covalent Modification

The mechanical properties of cellulose nanocrystals are very dependent on their crystalline structure. Therefore, they should avoid the softening or the liquid retention of the cellulose nanocrystals so as to retain the crystalline structure intact. In order to maintain the best properties of cellulose nanocrystals, the modification normally deals with a grafting process in which the cellulose nanocrystals will be grafted with different chemical compounds, such as polymers. Some examples of grafting techniques include esterification, etherification, silylation, oxidation and radical reaction (Kargarzadeh, Mariano, et al., 2018). Apart from that, various studies have shown the surface functionalization of nanocrystalline cellulose via grafting with acrylamide (Yang and Ye, 2012), tetra alkyl ammonium (Trifol et al., 2016), rosin (de Castro et al., 2016) and ascorbic acid (Filpponen and Argyropoulos, 2010).

11.8.1.1 Esterification

In general, the esterification of cellulose nanocrystals can be depicted as the formation of new ester bonds during the reaction of carboxylic acid with alcohol, as shown in Figure 11.8. Acetylation is one of a few esterification techniques in which acetyl-$COCH_3$ are introduced on the surface of the cellulose

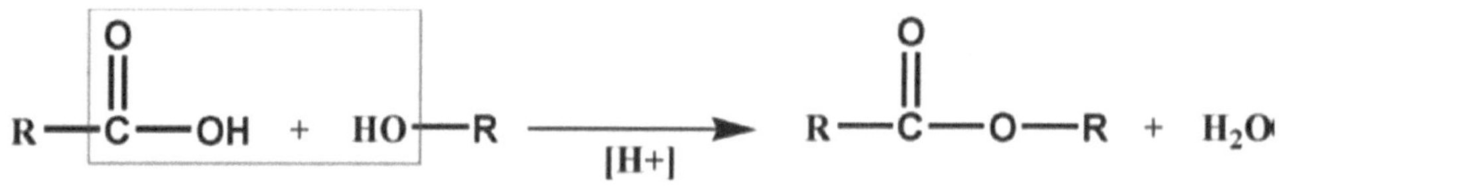

FIGURE 11.8 General scheme of esterification reaction.

nanocrystals. In this case, cellulose will be converted to cellulose acetate (CA) by reacting it with acetic anhydride under acidic conditions (Jonoobi et al., 2010; Lin et al., 2011). Various types of catalysts have been studied by many researchers for the acylation of cellulose nanocrystals, such as trimethylamine (de Menezes et al., 2014), imidazole (Pires et al., 2015), and lipase (Božič et al., 2015). One of the most common esterification reactions, which involves cellulose nanocrystals, takes place during the isolation of CNCs using sulfuric acid, and is sometimes referred to as sulfonation reaction. The side reaction of sulfonation takes place during the hydrolysis of cellulose nanocrystals and forms organosulfates (Kargarzadeh, Mariano, et al., 2018). Theoretically, the sulfuric acid will attack the C6 position of cellulose and later result in the removal of water (Figure 11.9).

The esterification reaction is particularly dependent on the reaction conditions (Bondeson, Mathew, and Oksman, 2006; Teodoro et al., 2011). In addition, other parameters, such as the biomass source and treatment routes, can also be considered.

11.8.1.2 Silylation

Similar to the esterification process, silylation is the reaction in which the silyl group (R_3Si–) reacts either with alcohols, amines, or carboxylic acids and later forms the covalent bond. In the past, Gousse et al. (Gousse et al., 2004) studied a series of alkyldimethyl chlorosilanes used as stabilizers for CNCs in organic solvent, as shown in Figure 11.10. Due to the newly developed hydrophobic characteristics from the long chain structures, the CNC particles were observed to be soluble in some hydrophobic organic solvents. Nevertheless, silylation results in the particle chain on the surface being soluble in the reaction medium, hence it diminishes the cellulose microfibrillar features (Gousse et al., 2004).

Recent findings have reported that silylation could improve the later application of cellulose nanocrystals, for example by increasing mechanical features (Pei, Zhou, and Berglund, 2010), attaining antimicrobial activity (Saini et al., 2016), and enhancing the wettability of aerogels in non-polar solvents (Aulin et al., 2010).

11.8.1.3 Etherification

Etherification is a reaction that involves the formation of new ester bond (R-O-R) on the surface of cellulose comprising OH chemical groups (Kargarzadeh, Mariano, et al., 2018). In practice, the etherification process plays a major role in the production of carboxymethyl cellulose. For cellulose nanocrystals, the dissolution of cellulose should be avoided since etherification is presumed to be performed on the surface of cellulose only. Hence, no specific position of the hydroxyl groups is preferred for this type of reaction. In 2015, Bae and Kim conducted studies on the reaction of alkyl bromide with nanocellulose (Figure 11.11) (Bae and Kim, 2015). In the studies, they reported that the long hydrophobic chains were grafted on the surface of nanocellulose by reacting the hydroxyl groups of cellulose with a sequential removal of bromide ions from alkyl bromide. A similar technique was adopted by Hassan et al. (Hassan et al., 2012) for the modification of grated terpyridine, which produced a highly fluorescent nanocellulose material. The etherification of cellulose nanocrystals can be carried out by copolymerization with acrylic derivatives as well (Anirudhan and Rejeena, 2014) and epoxy (Dufresne and Belgacem, 2013). Furthermore, cationic CNCs can be prepared by reacting the hydroxyl groups on CNCs with epoxy in the presence of 2,3-epoxypropyl-trimethyl ammonium chloride molecules (Hasani et al., 2008; Ho et al., 2011). These resulted positively charge CNCs, with addition features such as gelling and high viscosity.

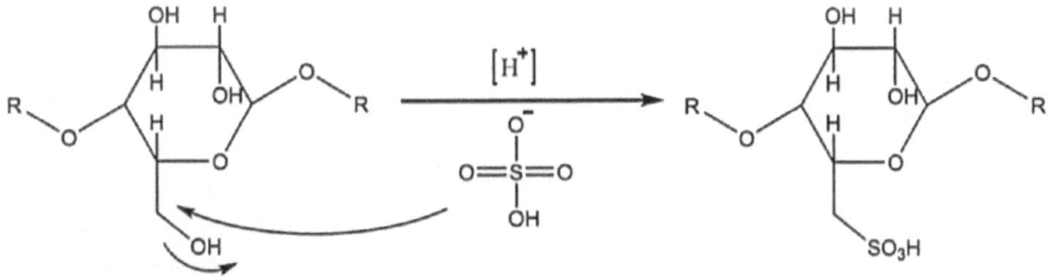

FIGURE 11.9 Scheme for the cellulose sulfonation reaction during acid hydrolysis.

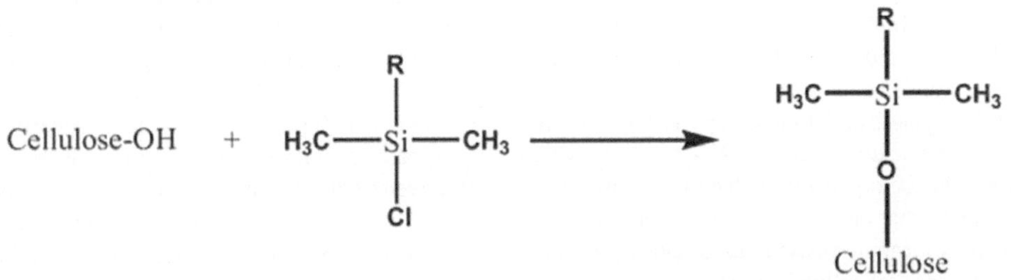

FIGURE 11.10 Scheme for the reaction of cellulose with silane derivatives. Here, R represents various alkyldimethylchlorosilanes such as C_3H_7, C_4H_9, C_8H_{17}, and $C_{12}H_{25}$.

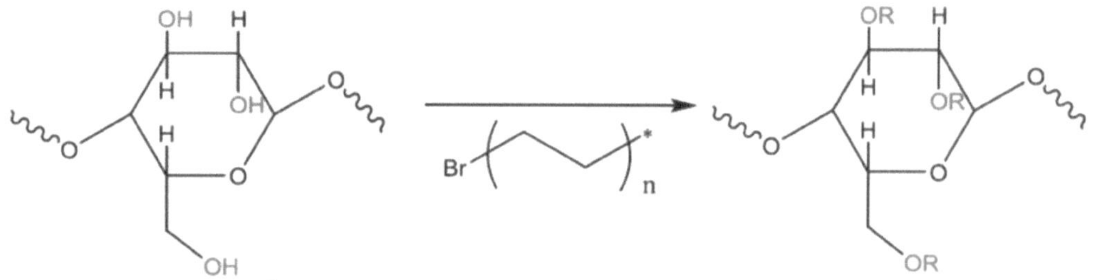

FIGURE 11.11 Example of an alkyl bromide reaction on a cellulose surface.

11.8.2 Non-Covalent Modification

The modification on the surface of cellulose nanocrystals can also be applied via adsorption of molecules such as surfactants, oligomers, or copolymers. The electrostatic and van der Waals or hydrogen bond interaction between these additives with the surface of nanocellulose will impart the hydrophobic features of the particles (Habibi, 2014). It was believed that surfactants were used to stabilize the CNC particles in non-polar solvent (Heux, Chauve, and Bonini, 2000). In addition, surfactants can change the surface properties of CNCs and enhance their dispersion and stability (Kargarzadeh, Mariano, et al., 2018). The influence of surfactants in CNC systems, especially on their dispersion, has been studied elsewhere, for example in water (Hu, Ballinger, et al., 2015) and organic solvents (Abitbol, Marway, and Cranston, 2014; Chi and Catchmark, 2017). The application of charged surfactants on the adsorption of CNCs for iridescent properties has also been recently investigated, with the conclusion that cationic surfactants are more practical to use since the interaction of positively charged surfactants with the surface of CNCs is more favorable than that of anionic surfactants (Bardet, Belgacem, and Bras, 2015). Furthermore, Ben Azouz et al. (Ben Azouz et al., 2011) have studied the applicability of uncharged surfactant interaction with CNCs. It was revealed that the viscosity of uncharged polyethylene oxide (PEO) was reduced by the addition of a certain quantity of CNCs into the matrix system. This shows that some parts of the PEO polymer chains interacted with the CNC molecules.

11.8.3 Mercerization

Mercerization is not an actual surface modification method. It is a step of the reorganization of cellulose polymorphs used to achieve the desired structure. It was first developed by John Mercer, where the cellulose type I polymorph is changed to cellulose type II by treating it with 17 % of NaOH solution (Dinand et al., 2002). According to Okano and Sarko (Okano and Sarko, 1984), the transformation of the cellulose type I (parallel chain) crystalline phase to cellulose type II (anti-parallel chain) surpasses the intermediate state of Na-cellulose, which consists of swollen cellulose organization due to fiber hydration. Additionally, the transformation process is said to be dependent on the presence of an amorphous phase (Kargarzadeh, Mariano, et al., 2018). The anti-parallel cellulose type II possesses lower energy levels and is more stable (more amorphous phase) compared than cellulose type I (more crystalline phase). The Na-cellulose intermediate immediately transforms to cellulose type II with the removal of NaOH. In fact, the conversion of cellulose type I to type II is irreversible and strongly dependent on the amount of disordered region.

Other than that, previous reports have stated that mercerization and ball milling techniques can produce CNCs having type II polymorph (Yue et al., 2012; Nge, Lee, and Endo, 2013). Type II CNC nanoparticles will give different morphologies (spherical or irregular nanoparticles) depending on the regeneration and hydrolysis of mercerized type I CNCs.

Conclusion

The research on cellulose and the preparation of cellulose nanocrystals have considerably increased over the last few years. The advantages of CNCs over other nanomaterials are notable, in that they are renewable and ecofriendly, and their features can be changed and tuned owing to the large amount of hydroxyl groups on the surface. In this review, the state of the art of numerous aspects of cellulose nanocrystals and their significance were discussed. Various extraction procedures of cellulose nanocrystals were described, together with preparation and modification methods. Several characteristics of this renewable nanomaterial, such as chemical composition, morphology, crystallinity, and thermal properties, were also reported. The different isolation procedures of cellulose nanocrystals as well as their characterization have been optimized for several years, and better methodologies that possibly focus on a more important scale are being established. Although several commercial-scale applications have recently been started, the approaches nowadays are too costly and require quite a long time. Thus, there is an opportunity to address the critical need to develop new platforms which will lead to another generation of cellulose-nanocrystal–based products. This can be achieved by employing powerful concepts of nanoscience along with a full understanding of the production, properties, and behavior of this type of cellulose nanofibers. Also, in order to ensure higher effectiveness in the extensive utilization of cellulose nanocrystals, additional efforts are needed to enlarge the application of these nanofibers in several fields. We expect that cellulose nanocrystals will provide scientists and technologists with further fascinating options in the upcoming years.

Acknowledgments

The authors wish to thank their parental institutes for providing the necessary facilities to accomplish this work.

REFERENCES

Abdul Khalil, H., Y. Davoudpour, M. N. Islam, et al. 2014. "Production and modification of nanofibrillated cellulose using various mechanical processes: A review." *Carbohydrate Polymers* 99:649–665.

Abitbol, T., H. Marway, and E. D. Cranston. 2014. "Surface modification of cellulose nanocrystals with cetyltrimethylammonium bromide." *Nordic Pulp and Paper Research Journal* 29(1):46–57.

Abitbol, T., A. Rivkin, Y. Cao, et al. 2016. "Nanocellulose, a tiny fiber with huge applications." *Current Opinion in Biotechnology* 39:76–88.

Agbor, V. B., N. Cicek, R. Sparling, A. Berlin, and D. B. Levin. 2011. "Biomass pretreatment: Fundamentals toward application." *Biotechnology Advances* 29(6):675–685.

Amin, K. N. M., P. K. Annamalai, I. C. Morrow, and D. Martin. 2015. "Production of cellulose nanocrystals via a scalable mechanical method." *RSC Advances* 5(70):57133–57140.

Anderson, S. R., D. Esposito, W. Gillette, J. Zhu, U. Baxa, and S. E. Mcneil. 2014. "Enzymatic preparation of nanocrystalline and microcrystalline cellulose." *Tappi Journal* 13(5):35–41.

Anirudhan, T. S., and S. R. Rejeena. 2014. "Poly (acrylic acid-co-ac rylamide-co-2-acrylamido-2-methyl-1-propanesulfonic acid) -grafted nanocellulose/poly (vinyl alcohol) composite for the in vitro gastrointestinal release of amoxicillin." *Journal of Applied Polymer Science* 131(17):40699.

Aulin, C., J. Netrval, L. Wågberg, and T. Lindström. 2010. "Aerogels from nanofibrillated cellulose with tunable oleophobicity." *Soft Matter* 6(14):3298–3305.

Azeredo, H. M., M. F. Rosa, and L. H. C. Mattoso. 2017. "Nanocellulose in bio-based food packaging applications." *Industrial Crops and Products* 97:664–671.

Bach, Q.-V., and W.-H. Chen. 2017. "Pyrolysis characteristics and kinetics of microalgae via thermogravimetric analysis (TGA): A state-of-the-art review." *Bioresource Technology* 246:88–100.

Bae, J. H., and S. H. Kim. 2015. "Alkylation of mixed micro- and nanocellulose to improve dispersion in polylactide." *Polymer International* 64(6):821–827.

Bardet, R., N. Belgacem, and J. Bras. 2015. "Flexibility and color monitoring of cellulose nanocrystal iridescent solid films using anionic or neutral polymers." *ACS Applied Materials and Interfaces* 7(7):4010–4018.

Behera, S., R. Arora, N. Nandhagopal, and S. Kumar. 2014. "Importance of chemical pretreatment for bioconversion of lignocellulosic biomass." *Renewable and Sustainable Energy Reviews* 36:91–106.

Ben Azouz, K., E. C. Ramires, W. Van den Fonteyne, N. El Kissi, and A. Dufresne. 2011. "Simple method for the melt extrusion of a cellulose nanocrystal reinforced hydrophobic polymer." *ACS Macro Letters* 1(1):236–240.

Bettaieb, F., R. Khiari, A. Dufresne, M. F. Mhenni, and M. N. Belgacem. 2015. "Mechanical and thermal properties of *Posidonia oceanica* cellulose nanocrystal reinforced polymer." *Carbohydrate Polymers* 123:99–104.

Bondeson, D., A. Mathew, and K. Oksman. 2006. "Optimization of the isolation of nanocrystals from microcrystalline cellulose by acid hydrolysis." *Cellulose* 13(2):171.

Borges, J., J. Canejo, S. Fernandes, P. Brogueira, and M. Godinho. 2015. "Cellulose-based liquid crystalline composite systems." In: *Nanocellulose Polymer Nanocomposites: Fundamentals and Applications*, edited by V. K. Thakur, 215–235. Wiley-Scrivener.

Boujemaoui, A., S. Mongkhontreerat, E. Malmström, and A. Carlmark. 2015. "Preparation and characterization of functionalized cellulose nanocrystals." *Carbohydrate Polymers* 115:457–464.

Božič, M., V. Vivod, S. Kavčič, M. Leitgeb, and V. Kokol. 2015. "New findings about the lipase acetylation of nanofibrillated cellulose using acetic anhydride as acyl donor." *Carbohydrate Polymers* 125:340–351.

Brinchi, L., F. Cotana, E. Fortunati, and J. Kenny. 2013. "Production of nanocrystalline cellulose from lignocellulosic biomass: Technology and applications." *Carbohydrate Polymers* 94(1):154–169.

Bunaciu, A. A., E. G. UdriŞTioiu, and H. Y. Aboul-Enein. 2015. "X-ray diffraction: Instrumentation and applications." *Critical Reviews in Analytical Chemistry* 45(4):289–299.

Cai, J., D. Xu, Z. Dong, et al. 2018. "Processing thermogravimetric analysis data for isoconversional kinetic analysis of lignocellulosic biomass pyrolysis: Case study of corn stalk." *Renewable and Sustainable Energy Reviews* 82:2705–2715.

Campano, C., A. Balea, A. Blanco, and C. Negro. 2016. "Enhancement of the fermentation process and properties of bacterial cellulose: A review." *Cellulose* 23(1):57–91.

Cao, L., D. Yuan, C. Xu, and Y. Chen. 2017. "Biobased, self-healable, high strength rubber with tunicate cellulose nanocrystals." *Nanoscale* 9(40):15696–15706.

Charreau, H., M. L. Foresti, and A. Vázquez. 2013. "Nanocellulose patents trends: A comprehensive review on patents on cellulose nanocrystals, microfibrillated and bacterial cellulose." *Recent Patents on Nanotechnology* 7(1):56–80.

Chen, L., J. Zhu, C. Baez, P. Kitin, and T. Elder. 2016. "Highly thermal-stable and functional cellulose nanocrystals and nanofibrils produced using fully recyclable organic acids." *Green Chemistry* 18(13):3835–3843.

Chen, W., H. Yu, S.-Y. Lee, T. Wei, J. Li, and Z. Fan. 2018. "Nanocellulose: A promising nanomaterial for advanced electrochemical energy storage." *Chemical Society Reviews* 47(8):2837–2872.

Chen, X., X. Deng, W. Shen, and L. Jiang. 2012. "Controlled enzymolysis preparation of nanocrystalline cellulose from pretreated cotton fibers." *BioResources* 7(3):4237–4248.

Chen, Y. W., H. V. Lee, J. C. Juan, and S.-M. Phang. 2016. "Production of new cellulose nanomaterial from red algae marine biomass *Gelidium elegans*." *Carbohydrate Polymers* 151:1210–1219.

Cheng, M., Z. Qin, Y. Chen, S. Hu, Z. Ren, and M. Zhu. 2017. "Efficient extraction of cellulose nanocrystals through hydrochloric acid hydrolysis catalyzed by inorganic chlorides under hydrothermal conditions." *ACS Sustainable Chemistry and Engineering* 5(6):4656–4664.

Cherian, B. M., A. L. Leão, S. F. de Souza, et al. 2011. "Cellulose nanocomposites with nanofibres isolated from pineapple leaf fibers for medical applications." *Carbohydrate Polymers* 86(4):1790–1798.

Chi, K., and J. M. Catchmark. 2017. "Enhanced dispersion and interface compatibilization of crystalline nanocellulose in polylactide by surfactant adsorption." *Cellulose* 24(11): 4845–4860.

Chowdhury, Z. Z., and S. B. A. Hamid. 2016. "Preparation and characterization of nanocrystalline cellulose using ultrasonication combined with a microwave-assisted pretreatment process." *BioResources* 11(2):3397–3415.

Ciolacu, D., F. Ciolacu, and V. I. Popa. 2011. "Amorphous cellulose—Structure and characterization." *Cellulose Chemistry and Technology* 45(1):13–21.

Coelho, C. C., M. Michelin, M. A. Cerqueira, et al. 2018. "Cellulose nanocrystals from grape pomace: Production, properties and cytotoxicity assessment." *Carbohydrate Polymers* 192:327–336.

Crucho, C. I., and M. T. Barros. 2017. "Polymeric nanoparticles: A study on the preparation variables and characterization methods." *Materials Science and Engineering: Part C* 80:771–784.

de Castro, D. O., J. Bras, A. Gandini, and N. Belgacem. 2016. "Surface grafting of cellulose nanocrystals with natural antimicrobial rosin mixture using a green process." *Carbohydrate Polymers* 137:1–8.

De France, K. J., T. Hoare, and E. D. Cranston. 2017. "Review of hydrogels and aerogels containing nanocellulose." *Chemistry of Materials* 29(11):4609–4631.

de Menezes, A. J., E. Longo, F. L. Leite, and A. Dufresne. 2014. "Characterization of cellulose nanocrystals grafted with organic acid chloride of different sizes." *Journal of Renewable Materials* 2(4):306–313.

Delidovich, I., K. Leonhard, and R. Palkovits. 2014. "Cellulose and hemicellulose valorisation: An integrated challenge of catalysis and reaction engineering." *Energy and Environmental Science* 7(9):2803–2830.

Dinand, E., M. Vignon, H. Chanzy, and L. Heux. 2002. "Mercerization of primary wall cellulose and its implication for the conversion of cellulose I→ cellulose II." *Cellulose* 9(1):7–18.

Du, H., C. Liu, X. Mu, et al. 2016. "Preparation and characterization of thermally stable cellulose nanocrystals via a sustainable approach of FeCl3-catalyzed formic acid hydrolysis." *Cellulose* 23(4):2389–2407.

Du, X., Z. Zhang, W. Liu, and Y. Deng. 2017. "Nanocellulose-based conductive materials and their emerging applications in energy devices-A review." *Nano Energy* 35:299–320.

Dufrêne, Y. F., T. Ando, R. Garcia, et al. 2017. "Imaging modes of atomic force microscopy for application in molecular and cell biology." *Nature Nanotechnology* 12(4):295.

Dufresne, A. 2013a. "Nanocellulose: A new ageless bionanomaterial." *Materials Today* 16(6):220–227.

Dufresne, A. 2013b. *Nanocellulose: From Nature to High Performance Tailored Materials*. Walter de Gruyter.

Dufresne, A., and M. N. Belgacem. 2013. "Cellulose-reinforced composites: From micro-to nanoscale." *Polímeros* 23(3):277–286.

Dugan, J. M., J. E. Gough, and S. J. Eichhorn. 2013. "Bacterial cellulose scaffolds and cellulose nanowhiskers for tissue engineering." *Nanomedicine* 8(2):287–298.

Eyley, S., and W. Thielemans. 2014. "Surface modification of cellulose nanocrystals." *Nanoscale* 6(14):7764–7779.

Fan, M., D. Dai, and B. Huang. 2012. "Fourier transform infrared spectroscopy for natural fibres." In: *Fourier Transform-Materials Analysis*, edited by S. Mohammed Salih. InTech.

Fang, W., S. Arola, J.-M. Malho, E. Kontturi, M. B. Linder, and P. i. Laaksonen. 2016. "Noncovalent dispersion and functionalization of cellulose nanocrystals with proteins and polysaccharides." *Biomacromolecules* 17(4):1458–1465.

Filpponen, I., and D. S. Argyropoulos. 2010. "Regular linking of cellulose nanocrystals via click chemistry: Synthesis and formation of cellulose nanoplatelet gels." *Biomacromolecules* 11(4):1060–1066.

Fortunati, E., M. Peltzer, I. Armentano, L. Torre, A. Jiménez, and J. Kenny. 2012. "Effects of modified cellulose nanocrystals on the barrier and migration properties of PLA nano-biocomposites." *Carbohydrate Polymers* 90(2):948–956.

Foster, E. J., R. J. Moon, U. P. Agarwal, et al. 2018. "Current characterization methods for cellulose nanomaterials." *Chemical Society Reviews* 47(8):2609–2679.

Foston, M. B., C. A. Hubbell, and A. J. Ragauskas. 2011. "Cellulose isolation methodology for NMR analysis of cellulose ultrastructure." *Materials* 4(11):1985–2002.

French, A. D., and M. S. Cintrón. 2013. "Cellulose polymorphy, crystallite size, and the Segal crystallinity index." *Cellulose* 20(1):583–588.

Galkin, M. V., and J. S. Samec. 2016. "Lignin valorization through catalytic lignocellulose fractionation: A fundamental platform for the future biorefinery." *ChemSusChem* 9(13):1544–1558.

García-García, D., R. Balart, J. Lopez-Martinez, M. Ek, and R. Moriana. 2018. "Optimizing the yield and physico-chemical properties of pine cone cellulose nanocrystals by different hydrolysis time." *Cellulose* 25(5):2925–2938.

George, J., and S. Sabapathi. 2015. "Cellulose nanocrystals: Synthesis, functional properties, and applications." *Nanotechnology, Science and Applications* 8:45.

Golmohammadi, H., E. Morales-Narvaez, T. Naghdi, and A. Merkoci. 2017. "Nanocellulose in sensing and biosensing." *Chemistry of Materials* 29(13):5426–5446.

Gousse, C., H. Chanzy, M. Cerrada, and E. Fleury. 2004. "Surface silylation of cellulose microfibrils: Preparation and rheological properties." *Polymer* 45(5):1569–1575.

Gupta, V., P. Carrott, R. Singh, M. Chaudhary, and S. Kushwaha. 2016. "Cellulose: A review as natural, modified and activated carbon adsorbent." *Bioresource Technology* 216:1066–1076.

Habibi, Y. 2014. "Key advances in the chemical modification of nanocelluloses." *Chemical Society Reviews* 43(5):1519–1542.

Habibi, Y., L. A. Lucia, and O. J. Rojas. 2010. "Cellulose nanocrystals: Chemistry, self-assembly, and applications." *Chemical Reviews* 110(6):3479–3500.

Hasani, M., E. D. Cranston, G. Westman, and D. G. Gray. 2008. "Cationic surface functionalization of cellulose nanocrystals." *Soft Matter* 4(11):2238–2244.

Hassan, M. L., C. M. Moorefield, H. S. Elbatal, G. R. Newkome, D. A. Modarelli, and N. C. Romano. 2012. "Fluorescent cellulose nanocrystals via supramolecular assembly of terpyridine-modified cellulose nanocrystals and terpyridine-modified perylene." *Materials Science and Engineering: Part B* 177(4):350–358.

Hassan, S. S., G. A. Williams, and A. K. Jaiswal. 2018. "Emerging technologies for the pretreatment of lignocellulosic biomass." *Bioresource Technology* 262:310–318.

Heux, L., G. Chauve, and C. Bonini. 2000. "Nonflocculating and chiral-nematic self-ordering of cellulose microcrystals suspensions in nonpolar solvents." *Langmuir* 16(21):8210–8212.

Hirai, A., O. Inui, F. Horii, and M. Tsuji. 2008. "Phase separation behavior in aqueous suspensions of bacterial cellulose nanocrystals prepared by sulfuric acid treatment." *Langmuir* 25(1):497–502.

Ho, T., T. Zimmermann, R. Hauert, and W. Caseri. 2011. "Preparation and characterization of cationic nanofibrillated cellulose from etherification and high-shear disintegration processes." *Cellulose* 18(6):1391–1406.

Hoeng, F., A. Denneulin, and J. Bras. 2016. "Use of nanocellulose in printed electronics: A review." *Nanoscale* 8(27):13131–13154.

Hu, Z., S. Ballinger, R. Pelton, and E. D. Cranston. 2015. "Surfactant-enhanced cellulose nanocrystal pickering emulsions." *Journal of Colloid and Interface Science* 439:139–148.

Hu, Z., T. Patten, R. Pelton, and E. D. Cranston. 2015. "Synergistic stabilization of emulsions and emulsion gels with water-soluble polymers and cellulose nanocrystals." *ACS Sustainable Chemistry and Engineering* 3(5):1023–1031.

Huang, J., P. R. Chang, N. Lin, and A. Dufresne. 2014. *Polysaccharide-Based Nanocrystals: Chemistry and Applications*. John Wiley & Sons.

Huang, P., Y. Zhao, S. Kuga, M. Wu, and Y. Huang. 2016. "A versatile method for producing functionalized cellulose nanofibers and their application." *Nanoscale* 8(6):3753–3759.

Huang, Y., C. Zhu, J. Yang, Y. Nie, C. Chen, and D. Sun. 2014. "Recent advances in bacterial cellulose." *Cellulose* 21(1):1–30.

Ilyas, R., S. Sapuan, and M. Ishak. 2018. "Isolation and characterization of nanocrystalline cellulose from sugar palm fibres (*Arenga pinnata*)." *Carbohydrate Polymers* 181:1038–1051.

Isikgor, F. H., and C. R. Becer. 2015. "Lignocellulosic biomass: A sustainable platform for the production of bio-based chemicals and polymers." *Polymer Chemistry* 6(25):4497–4559.

Islam, M. S., N. Kao, S. N. Bhattacharya, R. Gupta, and P. K. Bhattacharjee. 2017. "Effect of low pressure alkaline delignification process on the production of nanocrystalline cellulose from rice husk." *Journal of the Taiwan Institute of Chemical Engineers* 80:820–834.

Jawaid, M., S. Boufi, and H. P. S. Abdul Khalil 2017. *Cellulose-Reinforced Nanofibre Composites: Production, Properties and Applications*. Woodhead Publishing.

Johar, N., I. Ahmad, and A. Dufresne. 2012. "Extraction, preparation and characterization of cellulose fibres and nanocrystals from rice husk." *Industrial Crops and Products* 37(1):93–99.

Jonoobi, M., J. Harun, A. P. Mathew, M. Z. B. Hussein, and K. Oksman. 2010. "Preparation of cellulose nanofibers with hydrophobic surface characteristics." *Cellulose* 17(2):299–307.

Jonoobi, M., R. Oladi, Y. Davoudpour, et al. 2015. "Different preparation methods and properties of nanostructured cellulose from various natural resources and residues: A review." *Cellulose* 22(2):935–969.

Jönsson, L. J., and C. Martín. 2016. "Pretreatment of lignocellulose: Formation of inhibitory by-products and strategies for minimizing their effects." *Bioresource Technology* 199:103–112.

Ju, X., M. Bowden, E. E. Brown, and X. Zhang. 2015. "An improved X-ray diffraction method for cellulose crystallinity measurement." *Carbohydrate Polymers* 123:476–481.

Kalia, S., A. Dufresne, B. M. Cherian, et al. 2011. "Cellulose-based bio-and nanocomposites: A review." *International Journal of Polymer Science* 2011:1–35.

Kargarzadeh, H., I. Ahmad, S. Thomas, and A. Dufresne. 2017. *Handbook of Nanocellulose and Cellulose Nanocomposites*. John Wiley & Sons.

Kargarzadeh, H., J. Huang, N. Lin, et al. 2018. "Recent developments in nanocellulose-based biodegradable polymers, thermoplastic polymers, and porous nanocomposites." *Progress in Polymer Science* 87:197–227.

Kargarzadeh, H., M. Mariano, D. Gopakumar, et al. 2018. "Advances in cellulose nanomaterials." *Cellulose* 25(4):2151–2189.

Karimi, K., and M. J. Taherzadeh. 2016. "A critical review of analytical methods in pretreatment of lignocelluloses: Composition, imaging, and crystallinity." *Bioresource Technology* 200:1008–1018.

Keshk, S. M. 2014. "Bacterial cellulose production and its industrial applications." *Journal of Bioprocessing and Biotechniques* 4(2):1000150.

Kim, H. M., S. G. Wi, S. Jung, Y. Song, and H.-J. Bae. 2015. "Efficient approach for bioethanol production from red seaweed *Gelidium amansii*." *Bioresource Technology* 175:128–134.

Kim, J.-H., B. S. Shim, H. S. Kim, et al. 2015. "Review of nanocellulose for sustainable future materials." *International Journal of Precision Engineering and Manufacturing-Green Technology* 2(2):197–213.

Kirk, T. K. 2018. *Lignin Biodegradation: Microbiology, Chemistry, and Potential Applications: Volume I*. CRC Press.

Klemm, D., E. D. Cranston, D. Fischer, et al. 2018. "Nanocellulose as a natural source for groundbreaking applications in materials science: Today's state." *Materials Today*. doi: 10.1016/j.mattod.2018.02.001.

Klemm, D., B. Heublein, H. P. Fink, and A. Bohn. 2005. "Cellulose: Fascinating biopolymer and sustainable raw material." *Angewandte Chemie International Edition* 44(22):3358–3393.

Klemm, D., F. Kramer, S. Moritz, et al. 2011. "Nanocelluloses: A new family of nature-based materials." *Angewandte Chemie International Edition* 50(24):5438–5466.

Klemm, D., B. Philipp, T. Heinze, U. Heinze, and W. Wagenknecht. 1998. "Comprehensive cellulose chemistry, Vol. 1." In: *Functionalization of Cellulose: Functionalization of Cellulose*. Weinheim: Wiley-VCH.

Kontturi, E., A. Meriluoto, P. A. Penttilä, et al. 2016. "Degradation and crystallization of cellulose in hydrogen chloride vapor for high-yield isolation of cellulose nanocrystals." *Angewandte Chemie International Edition* 55(46):14455–14458.

Koo, Y. S., Y. S. Wang, S. H. You, and H. D. Kim. 2002. "Preparation and properties of chemical cellulose from ascidian tunic and their regenerated cellulose fibers." *Journal of Applied Polymer Science* 85(8):1634–1643.

Lamaming, J., R. Hashim, C. P. Leh, and O. Sulaiman. 2017. "Properties of cellulose nanocrystals from oil palm trunk isolated by total chlorine free method." *Carbohydrate Polymers* 156:409–416.

Lamaming, J., R. Hashim, O. Sulaiman, C. P. Leh, T. Sugimoto, and N. A. Nordin. 2015. "Cellulose nanocrystals isolated from oil palm trunk." *Carbohydrate Polymers* 127:202–208.

Lavoine, N., I. Desloges, A. Dufresne, and J. Bras. 2012. "Microfibrillated cellulose–its barrier properties and applications in cellulosic materials: A review." *Carbohydrate Polymers* 90(2):735–764.

Lazko, J., T. Sénéchal, A. Bouchut, et al. 2016. "Acid-free extraction of cellulose type I nanocrystals using Brønsted acid-type ionic liquids." *Nanocomposites* 2(2):65–75.

Lédé, J. 2012. "Cellulose pyrolysis kinetics: An historical review on the existence and role of intermediate active cellulose." *Journal of Analytical and Applied Pyrolysis* 94:17–32.

Lemke, C. H., R. Y. Dong, C. A. Michal, and W. Y. Hamad. 2012. "New insights into nano-crystalline cellulose structure and morphology based on solid-state NMR." *Cellulose* 19(5):1619–1629.

Leszczyńska, A., P. Radzik, K. Haraźna, and K. Pielichowski. 2018. "Thermal stability of cellulose nanocrystals prepared by succinic anhydride assisted hydrolysis." *Thermochimica Acta* 663:145–156.

Lin, N., and A. Dufresne. 2014. "Nanocellulose in biomedicine: Current status and future prospect." *European Polymer Journal* 59:302–325.

Lin, N., J. Huang, P. R. Chang, J. Feng, and J. Yu. 2011. "Surface acetylation of cellulose nanocrystal and its reinforcing function in poly (lactic acid)." *Carbohydrate Polymers* 83(4):1834–1842.

Lin, P.-C., S. Lin, P. C. Wang, and R. Sridhar. 2014. "Techniques for physicochemical characterization of nanomaterials." *Biotechnology Advances* 32(4):711–726.

Liu, Y., H. Wang, G. Yu, Q. Yu, B. Li, and X. Mu. 2014. "A novel approach for the preparation of nanocrystalline cellulose by using phosphotungstic acid." *Carbohydrate Polymers* 110:415–422.

Liu, Z., X. Li, W. Xie, and H. Deng. 2017. "Extraction, isolation and characterization of nanocrystalline cellulose from industrial kelp (*Laminaria japonica*) waste." *Carbohydrate Polymers* 173:353–359.

Lu, Z., L. Fan, H. Zheng, Q. Lu, Y. Liao, and B. Huang. 2013. "Preparation, characterization and optimization of nanocellulose whiskers by simultaneously ultrasonic wave and microwave assisted." *Bioresource Technology* 146:82–88.

Mao, J., H. Abushammala, N. Brown, and M. Laborie. 2017. "Comparative assessment of methods for producing cellulose I nanocrystals from cellulosic sources." In: *Nanocelluloses: Their Preparation, Properties, and Applications.* ACS Symposium Series.

Mao, J., B. Heck, G. Reiter, and M.-P. Laborie. 2015. "Cellulose nanocrystals' production in near theoretical yields by 1-butyl-3-methylimidazolium hydrogen sulfate ([Bmim] HSO 4)–mediated hydrolysis." *Carbohydrate Polymers* 117:443–451.

Marett, J., A. Aning, and E. J. Foster. 2017. "The isolation of cellulose nanocrystals from pistachio shells via acid hydrolysis." *Industrial Crops and Products* 109:869–874.

Michler, G. H. 2008. *Electron Microscopy of Polymers.* Springer Science & Business Media.

Miller, J. 2016. "Cellulose nanomaterials production-state of the industry." Accessed 11-11-2016, http://www.tappinano.org/media/1114/cellulose-nanomaterials-production-state-of-the-industry-dec-2015.pdf.

Mitić, Ž., A. Stolić, S. Stojanović, et al. 2017. "Instrumental methods and techniques for structural and physicochemical characterization of biomaterials and bone tissue: A review." *Materials Science and Engineering: Part C* 79:930–949.

Mohammadkazemi, F., M. Azin, and A. Ashori. 2015. "Production of bacterial cellulose using different carbon sources and culture media." *Carbohydrate Polymers* 117:518–523.

Moon, R. J., A. Martini, J. Nairn, J. Simonsen, and J. Youngblood. 2011. "Cellulose nanomaterials review: Structure, properties and nanocomposites." *Chemical Society Reviews* 40(7):3941–3994.

Moon, R. J., G. T. Schueneman, and J. Simonsen. 2016. "Overview of cellulose nanomaterials, their capabilities and applications." *JOM* 68(9):2383–2394.

Morita, S., F. J. Giessibl, E. Meyer, and R. Wiesendanger. 2015. *Noncontact Atomic Force Microscopy*, Vol. 3. Springer.

Naduparambath, S., T. Jinitha, V. Shaniba, M. Sreejith, A. K. Balan, and E. Purushothaman. 2018. "Isolation and characterisation of cellulose nanocrystals from sago seed shells." *Carbohydrate Polymers* 180:13–20.

Nascimento, D. M., J. S. Almeida, A. F. Dias, et al. 2014. "A novel green approach for the preparation of cellulose nanowhiskers from white coir." *Carbohydrate Polymers* 110:456–463.

Nascimento, D. M., Y. L. Nunes, M. C. Figueirêdo, et al. 2018. "Nanocellulose nanocomposite hydrogels: Technological and environmental issues." *Green Chemistry* 20(11):2428–2448.

Nascimento, S. A., and C. A. Rezende. 2018. "Combined approaches to obtain cellulose nanocrystals, nanofibrils and fermentable sugars from elephant grass." *Carbohydrate Polymers* 180:38–45.

Neto, W. P. F., J.-L. Putaux, M. Mariano, et al. 2016. "Comprehensive morphological and structural investigation of cellulose I and II nanocrystals prepared by sulphuric acid hydrolysis." *RSC Advances* 6(79):76017–76027.

Neto, W. P. F., H. A. Silvério, N. O. Dantas, and D. Pasquini. 2013. "Extraction and characterization of cellulose nanocrystals from agro-industrial residue–soy hulls." *Industrial Crops and Products* 42:480–488.

Ng, H.-M., L. T. Sin, T.-T. Tee, et al. 2015. "Extraction of cellulose nanocrystals from plant sources for application as reinforcing agent in polymers." *Composites Part B: Engineering* 75:176–200.

Nge, T. T., S.-H. Lee, and T. Endo. 2013. "Preparation of nanoscale cellulose materials with different morphologies by mechanical treatments and their characterization." *Cellulose* 20(4):1841–1852.

Novo, L. P., J. Bras, A. García, N. Belgacem, and A. A. Curvelo. 2015. "Subcritical water: A method for green production of cellulose nanocrystals." *ACS Sustainable Chemistry and Engineering* 3(11):2839–2846.

Novo, L. P., J. Bras, A. García, N. Belgacem, and A. A. da Silva Curvelo. 2016. "A study of the production of cellulose nanocrystals through subcritical water hydrolysis." *Industrial Crops and Products* 93:88–95.

Okano, T., and A. Sarko. 1984. "Mercerization of cellulose. I. X-ray diffraction evidence for intermediate structures." *Journal of Applied Polymer Science* 29(12):4175–4182.

Oksman, K., and A. Bismarck. 2014. *Handbook of Green Materials: Processing Technologies, Properties and Applications (in 4 Volumes)*, Vol. 5. World Scientific.

Oun, A. A., and J.-W. Rhim. 2015. "Effect of post-treatments and concentration of cotton linter cellulose nanocrystals on the properties of agar-based nanocomposite films." *Carbohydrate Polymers* 134:20–29.

Pandey, J., H. Takagi, A. Nakagaito, and H. Kim. 2015. *Handbook of Polymer Nanocomposites. Processing, Performance and Application.* Springer.

Park, S., J. O. Baker, M. E. Himmel, P. A. Parilla, and D. K. Johnson. 2010. "Cellulose crystallinity index: Measurement techniques and their impact on interpreting cellulase performance." *Biotechnology for Biofuels* 3(1):10.

Pei, A., Q. Zhou, and L. A. Berglund. 2010. "Functionalized cellulose nanocrystals as biobased nucleation agents in poly (l-lactide)(PLLA)–Crystallization and mechanical property effects." *Composites Science and Technology* 70(5):815–821.

Phanthong, P., P. Reubroycharoen, X. Hao, G. Xu, A. Abudula, and G. Guan. 2018. "Nanocellulose: Extraction and application." *Carbon Resources Conversion* 1(1):32–43.

Picheth, G. F., C. L. Pirich, M. R. Sierakowski, et al. 2017. "Bacterial cellulose in biomedical applications: A review." *International Journal of Biological Macromolecules* 104(A):97–106.

Pires, P. A., N. I. Malek, T. C. Teixeira, T. A. Bioni, H. Nawaz, and O. A. El Seoud. 2015. "Imidazole-catalyzed esterification of cellulose in ionic liquid/molecular solvents: A multi-technique approach to probe effects of the medium." *Industrial Crops and Products* 77:180–189.

Postek, M. T., R. J. Moon, A. W. Rudie, and M. A. Bilodeau. 2013. *Production and applications of cellulose.* Peachtree Corners: TAPPI Press.

Rabemanolontsoa, H., and S. Saka. 2016. "Various pretreatments of lignocellulosics." *Bioresource Technology* 199:83–91.

Ragauskas, A. J., G. T. Beckham, M. J. Biddy, et al. 2014. "Lignin valorization: Improving lignin processing in the biorefinery." *Science* 344(6185):1246843.

Rajinipriya, M., M. Nagalakshmaiah, M. Robert, and S. Elkoun. 2018. "Importance of agricultural and industrial waste in the field of nanocellulose and recent industrial developments of wood based nanocellulose: A review." *ACS Sustainable Chemistry and Engineering* 6(3):2807–2828.

Reddy, J. P., and J.-W. Rhim. 2014. "Isolation and characterization of cellulose nanocrystals from garlic skin." *Materials Letters* 129:20–23.

Reid, M. S., M. Villalobos, and E. D. Cranston. 2016. "Benchmarking cellulose nanocrystals: From the laboratory to industrial production." *Langmuir* 33(7):1583–1598.

Reiniati, I., A. N. Hrymak, and A. Margaritis. 2017. "Recent developments in the production and applications of bacterial cellulose fibers and nanocrystals." *Critical Reviews in Biotechnology* 37(4):510–524.

Rojas, O. J. 2016. *Cellulose Chemistry and Properties: Fibers, Nanocelluloses and Advanced Materials*, Vol. 271. Springer.

Sacui, I. A., R. C. Nieuwendaal, D. J. Burnett, et al. 2014. "Comparison of the properties of cellulose nanocrystals and cellulose nanofibrils isolated from bacteria, tunicate, and wood processed using acid, enzymatic, mechanical, and oxidative methods." *ACS Applied Materials and Interfaces* 6(9):6127–6138.

Saini, S., M. N. Belgacem, M.-C. B. Salon, and J. Bras. 2016. "Non leaching biomimetic antimicrobial surfaces via surface functionalisation of cellulose nanofibers with aminosilane." *Cellulose* 23(1):795–810.

Salajková, M., L. A. Berglund, and Q. Zhou. 2012. "Hydrophobic cellulose nanocrystals modified with quaternary ammonium salts." *Journal of Materials Chemistry* 22(37):19798–19805.

Sathitsuksanoh, N., and C. Renneckar. 2018. "Characterization methods and techniques." In: *Introduction to Renewable Biomaterials: First Principles and Concepts*, edited by A. S. Ayoub and L. A. Lucia. John Wiley & Sons.

Shaghaleh, H., X. Xu, and S. Wang. 2018. "Current progress in production of biopolymeric materials based on cellulose, cellulose nanofibers, and cellulose derivatives." *RSC Advances* 8(2):825–842.

Shaheen, T. I., and H. E. Emam. 2018. "Sono-chemical synthesis of cellulose nanocrystals from wood sawdust using Acid hydrolysis." *International Journal of Biological Macromolecules* 107(B):1599–1606.

Siesler, H. W., Y. Ozaki, S. Kawata, and H. M. Heise. 2008. *Near-Infrared Spectroscopy: Principles, Instruments, Applications.* John Wiley & Sons.

Silvério, H. A., W. P. F. Neto, N. O. Dantas, and D. Pasquini. 2013. "Extraction and characterization of cellulose nanocrystals from corncob for application as reinforcing agent in nanocomposites." *Industrial Crops and Products* 44:427–436.

Singla, R., A. Guliani, A. Kumari, and S. K. Yadav. 2016. "Nanocellulose and nanocomposites." In: *Nanoscale Materials in Targeted Drug Delivery, Theragnosis and Tissue Regeneration*, 103–125. Springer.

Siqueira, G., J. Bras, and A. Dufresne. 2010. "Cellulosic bionanocomposites: A review of preparation, properties and applications." *Polymers* 2(4):728–765.

Somerville, C. 2006. "Cellulose synthesis in higher plants." *Annual Review of Cell and Developmental Biology* 22:53–78.

Sun, B., Q. Hou, Z. Liu, and Y. Ni. 2015. "Sodium periodate oxidation of cellulose nanocrystal and its application as a paper wet strength additive." *Cellulose* 22(2):1135–1146.

Tan, X. Y., S. B. A. Hamid, and C. W. Lai. 2015. "Preparation of high crystallinity cellulose nanocrystals (CNCs) by ionic liquid solvolysis." *Biomass and Bioenergy* 81:584–591.

Tang, L.-r., B. Huang, W. Ou, X.-r. Chen, and Y.-d. Chen. 2011. "Manufacture of cellulose nanocrystals by cation exchange resin-catalyzed hydrolysis of cellulose." *Bioresource Technology* 102(23):10973–10977.

Tang, Y., S. Yang, N. Zhang, and J. Zhang. 2014. "Preparation and characterization of nanocrystalline cellulose via low-intensity ultrasonic-assisted sulfuric acid hydrolysis." *Cellulose* 21(1):335–346.

TAPPI. 2017. *Standard Terms and Their Definition for Cellulose Nanomaterial.*

Tasumi, M. 2014. *Introduction to Experimental Infrared Spectroscopy: Fundamentals and Practical Methods.* John Wiley & Sons.

Teodoro, K. B., E. d. M. Teixeira, A. C. Corrêa, A. d. Campos, J. M. Marconcini, and L. H. Mattoso. 2011. "Whiskers de fibra de sisal obtidos sob diferentes condições de hidrólise ácida: Efeito do tempo e da temperatura de extração." *Polímeros* 21(4):280–285.

Thakur, V. K. 2015a. *Lignocellulosic Polymer Composites: Processing, Characterization, and Properties.* John Wiley & Sons.

Thakur, V. K. 2015b. *Nanocellulose Polymer Nanocomposites: Fundamentals and Applications.* John Wiley & Sons.

Thakur, V. K., and S. I. Voicu. 2016. "Recent advances in cellulose and chitosan based membranes for water purification: A concise review." *Carbohydrate Polymers* 146:148–165.

Trache, D. 2017. "Microcrystalline cellulose and related polymer composites: Synthesis, characterization and properties." In: *Handbook of Composites from Renewable Materials*, edited by V. K. Thakur, M. Kumari Thakur, and M. R. Kessler, 61–92. Scrivener Publishing LLC.

Trache, D. 2018. "Nanocellulose as a promising sustainable material for biomedical applications." *AIMS Materials Science* 5(2):201–205. doi: 10.3934/matersci.2018.2.201.

Trache, D., A. Donnot, K. Khimeche, R. Benelmir, and N. Brosse. 2014. "Physico-chemical properties and thermal stability of microcrystalline cellulose isolated from Alfa fibres." *Carbohydrate Polymers* 104:223–230.

Trache, D., M. H. Hussin, C. T. H. Chuin, et al. 2016. "Microcrystalline cellulose: Isolation, characterization and bio-composites application – A review." *International Journal of Biological Macromolecules* 93(Pt A):789–804. doi: 10.1016/j.ijbiomac.2016.09.056.

Trache, D., M. H. Hussin, M. M. Haafiz, and V. K. Thakur. 2017. "Recent progress in cellulose nanocrystals: Sources and production." *Nanoscale* 9(5):1763–1786.

Trache, D., K. Khimeche, A. Donnot, and R. Benelmir. 2013. "FTIR spectroscopy and X-ray powder diffraction characterization of microcrystalline cellulose obtained from alfa fibers." *MATEC Web of Conferences* 3:1023–1023.

Trache, D., K. Khimeche, A. Mezroua, and M. Benziane. 2016. "Physicochemical properties of microcrystalline nitrocellulose from Alfa grass fibres and its thermal stability." *Journal of Thermal Analysis and Calorimetry* 124(3):1485–1496.

Trache, D., Tarchoun, A. F., Derradji, M., Hamidon, T. S., Masruchin, N., Brosse, N., & Hussin, M. H. 2020. "Nanocellulose: from fundamentals to advanced applications." *Frontiers in Chemistry*, 8:392.

Trifol, J., D. Plackett, C. Sillard, et al. 2016. "A comparison of partially acetylated nanocellulose, nanocrystalline cellulose, and nanoclay as fillers for high-performance polylactide nanocomposites." *Journal of Applied Polymer Science* 133(14):43257.

Vasconcelos, N. F., J. P. A. Feitosa, F. M. P. da Gama, et al. 2017. "Bacterial cellulose nanocrystals produced under different hydrolysis conditions: Properties and morphological features." *Carbohydrate Polymers* 155:425–431.

Vazquez, A., M. L. Foresti, J. I. Moran, and V. P. Cyras. 2015. "Extraction and production of cellulose nanofibers." In: *Handbook of Polymer Nanocomposites. Processing, Performance and Application*, 81–118. Springer.

Vincent, J. F. 2002. "Survival of the cheapest." *Materials Today* 5(12):28–41.

Visanko, M., H. Liimatainen, J. A. Sirviö, J. P. Heiskanen, J. Niinimäki, and O. Hormi. 2014. "Amphiphilic cellulose nanocrystals from acid-free oxidative treatment: Physicochemical characteristics and use as an oil–water stabilizer." *Biomacromolecules* 15(7):2769–2775.

Wang, X., C. Yao, F. Wang, and Z. Li. 2017. "Cellulose-based nanomaterials for energy applications." *Small* 13(42):1702240.

Wei, H., K. Rodriguez, S. Renneckar, and P. J. Vikesland. 2014. "Environmental science and engineering applications of nanocellulose-based nanocomposites." *Environmental Science: Nano* 1(4):302–316.

Wertz, J.-L., J. P. Mercier, and O. Bédué. 2010. *Cellulose Science and Technology*. Switzerland: CRC Press.

Xu, Y., J. Salmi, E. Kloser, et al. 2013. "Feasibility of nanocrystalline cellulose production by endoglucanase treatment of natural bast fibers." *Industrial Crops and Products* 51:381–384.

Yang, J., and D. Y. Ye. 2012. "Liquid crystal of nanocellulose whiskers' grafted with acrylamide." *Chinese Chemical Letters* 23(3):367–370.

Yao, J., H. Huang, L. Mao, Z. Li, H. Zhu, and Y. Liu. 2017. "Structural and optical properties of cellulose nanocrystals isolated from the fruit shell of *Camellia oleifera* Abel." *Fibers and Polymers* 18(11):2118–2124.

Yue, Y., C. Zhou, A. D. French, et al. 2012. "Comparative properties of cellulose nano-crystals from native and mercerized cotton fibers." *Cellulose* 19(4):1173–1187.

Zhang, T., Y. Zheng, and D. J. Cosgrove. 2016. "Spatial organization of cellulose microfibrils and matrix polysaccharides in primary plant cell walls as imaged by multichannel atomic force microscopy." *The Plant Journal* 85(2):179–192.

Zhao, Y., Y. Zhang, M. E. Lindström, and J. Li. 2015. "Tunicate cellulose nanocrystals: Preparation, neat films and nanocomposite films with glucomannans." *Carbohydrate Polymers* 117:286–296.

12

Protein and Peptide Nanoparticles: Preparation and Surface Modification

K. Vinay, S. Neha, and S. S. Maitra

CONTENTS

12.1 Introduction

A diverse variety of active molecules is available with an important role in the cure and prevention of diseases, but due to biological and chemical instability and limited bioavailability, these molecules are unable to get therapeutic value (Chen et al., 2011; Esmaili et al., 2011). The past few decades have seen protein or peptide nanoparticles as promising candidates. Active molecules can be used by various methods including immobilization, adsorption, and encapsulation (Orive et al., 2004). As compared to similar counterparts, nanoparticles provide several benefits (Panyam and Labhasetwar, 2003; Reis et al., 2006). Nanoparticles can control drug release and increase drug efficacy. As carriers, they have advantages, including: (1) they can be administrated through nasal, oral, intra-ocular, and parental ways, (2) they have higher drug loading capabilities with preserved drug activity, and (3) they can be used by binding specific molecules on the surface of nanoparticles (Mohanraj and Chen, 2006). In recent years, protein nanoparticles have become very popular due to their lower toxicity and their biodegradability (Jahanshahi and Babaei, 2008).

Due to their amphipathic nature, proteins are considered as suitable candidates (Marty et al., 1978). Nanoparticles can be synthesized from proteins that dissolve in water, such as bovine serum albumin, and proteins that are not soluble in water, such as gliadin. (Weber et al., 2000; Ezpeleta et al., 1996). Nanoparticles synthesized from proteins have advantages such as biodegradability, the ability to modify surface features, and ease of manipulation (Weber et al., 2000; Weber, Kreuter, and Langer, 2000). Figure 12.1 illustrates the different types of proteins that can be used for the preparation of protein nanoparticles. The low toxicity of animal proteins is an advantage over synthetic polymers (Leo et al., 1997). On the other hand, the disadvantage of using animal proteins is the greater risk of contamination (Pathak and Thassu, 2016). In comparison to animal proteins, plant proteins are less expensive and less toxic (Ezpeleta et al., 1996). It is known that protein nanoparticles find a variety of uses in therapeutics. The shape and size of the nanoparticles determine the type of tissue to which they can be applied therapeutically (Zhang, Chan, and Leong, 2013; Sharma, 2009). For example, gelatin nanoparticles can be used to carry and release drugs to the brain. Therefore, it

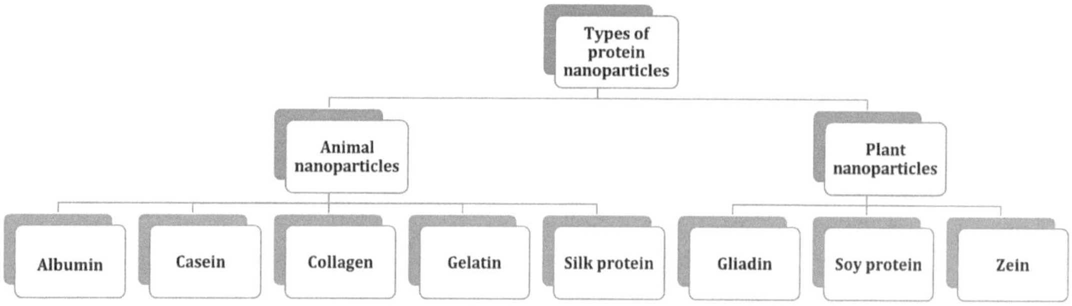

FIGURE 12.1 Types of proteins used to prepare protein nanoparticles.

finds use in brain tissue therapy. (Zhang, Chan, and Leong, 2013; Sharma, 2009; Cupaioli et al., 2014; Peluffo et al., 2015).

When nanoparticles are designed for drug delivery systems, various parameters need to be controlled. These parameters include the size of the nanoparticle, the release of the active component, and the surface properties. These parameters are required for accurate and specific drug action (Jahanshahi, Zhang, and Lyddiatt, 2005; Soppimath et al., 2001). Drug delivery through nanoparticles offers certain advantages (Coester, Nayyar, and Samuel, 2006). The first advantage is that particle morphology, charge, and size can be controlled (Verma and Garg, 2001). Secondly, the molecules are released slowly (Moghimi, 2006). Thirdly, the solubility and stability of the drugs are improved (Langer, 1998). Nanoparticles can be engineered for prolonged drug circulation and enhanced cellular uptake (Gref et al., 1994). Nanoparticles selection depends on various parameters. These include (a) nanoparticle size, (b) solubility and stability, (c) release in the delivery systems, (d) charge on the nanoparticle, (e) biocompatibility, and (f) toxicity (Kreuter et al., 1995).

12.2 Parameters for the Preparation of Protein Nanoparticles

The preparation and characteristics of nanoparticles are affected by the physico-chemical properties of the protein. Figure 12.2 illustrates the factors affecting the preparation of protein nanoparticles.

12.2.1 Protein Composition

Protein composition is an important factor in nanoparticle preparation. The characteristics of nanoparticles are affected by variations in the molecular weight of the proteins (Pathak and Thassu, 2016). In plant proteins, various pigments are present in which the molecular weight fractions vary. Therefore, plant proteins are first purified before the preparation of nanoparticles (Pathak and Thassu, 2016).

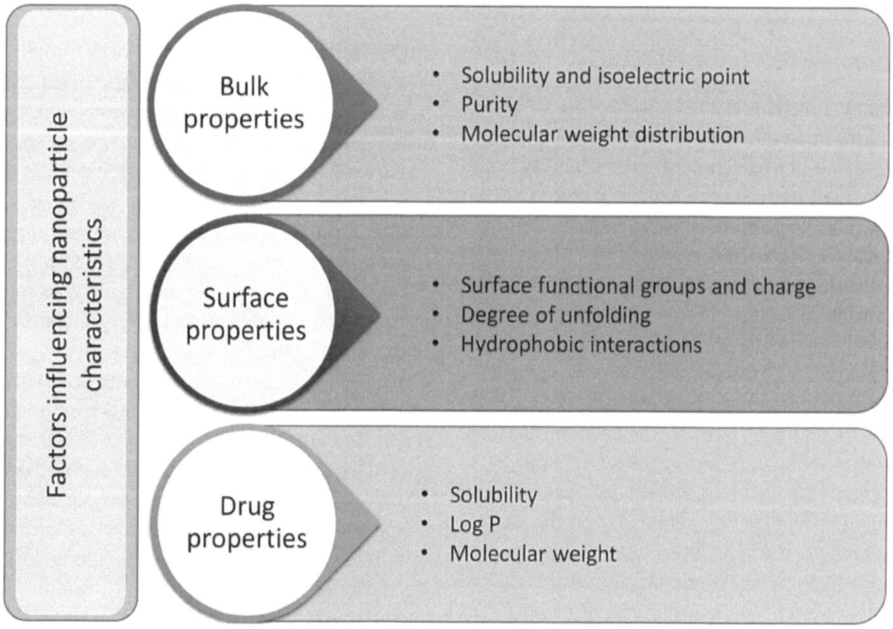

FIGURE 12.2 Parameters affecting the formation of nanoparticles.

12.2.2 Protein Solubility

Proteins have varying solubilities in both non-aqueous and aqueous solvents. This solubility depends on the polarity of the solvents (Wang and Uludag, 2008). The solubility of the proteins is an important factor in determining the role and specificities of the nanoparticles. Therefore, the solubility of the protein determines the characteristics and method of nanoparticle preparation.

12.2.3 Surface Properties

Proteins have various functional groups. Therefore, these functional groups can be modified according to the required group specificities. The modifications result in alterations of parameters such as bio-distribution, drug loading, stability, and biocompatibility. Changes in the functional group of the nanoparticles result in the exposure of the groups (Pathak and Thassu, 2016).

12.2.4 Properties of Drugs

Generally, drug loading can be performed either by interaction with proteins by covalent or non-covalent interactions, or by encapsulation inside the nanoparticles. In the case of gelatin nanoparticles, hydrophilic drugs reported higher encapsulation efficiency (Vandervoort and Ludwig, 2004).

12.3 Methods of Preparation

Protein nanoparticles can be prepared using various methods. The selection of proteins depends on several parameters. These include amino acid composition, as well as the physical and chemical properties of the protein to be selected. An increase in the unfolding of protein and hydrophobic interactions are the factors that need to be considered while preparing protein nanoparticles. Based on crosslinking, composition, and concentration, proteins undergo conformational changes (Pathak and Thassu, 2016). Most of the methods of protein nanoparticle preparation involve the precipitation of the macromolecules. Precipitation in the proteins can be induced by varying temperature, salinity, and pH (Miladi et al., 2014).

12.3.1 Desolvation

Desolvation is a process where agents such as alcohols are added to the aqueous protein solution. In the desolvation process, a desolvating agent is added to the therapeutic protein. To stabilize the formed nanoaggregate in the process, glutaraldehyde, a crosslinking agent, is used, which results in the stabilization of the nanoparticles (Lee et al., 2014; Estrada and Champion, 2015). Anionic polymer can be used to improve chemical crosslinking (Zhang et al., 2008). Desolvation results in conformational changes in the protein, which assumes a coil conformation (Elzoghby, Elgohary, and Kamel, 2015). The advantage of crosslinking the proteins is that they protect the coacervates formed and make the nanoparticles denser. Crosslinking of the amino groups of the protein makes the nanoparticles denser and protects the coacervates. Figure 12.3 illustrates protein nanoparticles preparation by the desolvation process. The examples where the nanoparticles are produced using the desolvation process include whey, albumin, silk, gliadin, and gelatin. Some processes require a second desolvation process. This helps to achieve smaller nanoparticles and uniform particle size. There are disadvantages in the extensive use of the desolvation technique. In the process, toxic crosslinkers are used, which results in the denaturing of the protein structure (Elzoghby, Elgohary, and Kamel, 2015). The activities of the proteins can be measured by enzymatic assays using ELISA (Kundu et al., 2010; Zhang et al., 2008; Lee et al., 2014).

In gelatin nanoparticle preparation, the heterogeneity in the molecular weight of gelatin results in a wide size range. Therefore, a second solvation step is required to produce smaller nanoparticles (Coester et al., 2000). In the first desolvation step, high-molecular-weight gelatin is precipitated, and low-weight gelatin is removed. In the second step, the high-molecular-weight gelatin is desolvated and redissolved. A single-step desolvation

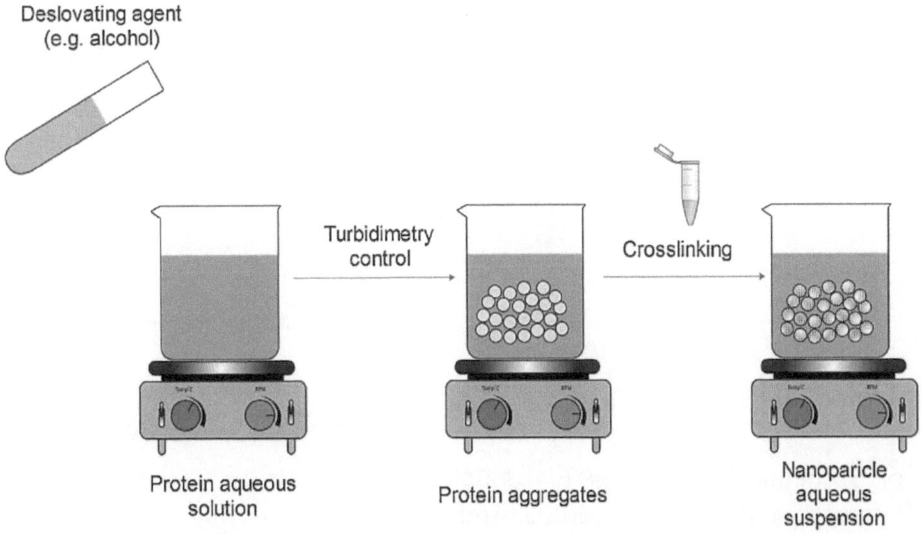

FIGURE 12.3 Protein nanoparticle preparation by desolvation. (Reprinted with permission from [Tarhini, Greige-Gerges, and Elaissari, 2017)].

can also be used to prepare protein nanoparticles. In the process, the gelatin molecules are sufficiently uncharged to prevent aggregation (Ofokansi et al., 2010). In a study on human serum albumin, the desolvation process was characterized and optimized for the preparation of nanoparticles (Weber, Kreuter, and Langer, 2000). Glutaraldehyde was used in different amounts to stabilize the nanoparticles after desolvation. The study demonstrated that the size of the particles does not depend on the amount of crosslinker, but on the amount of desolvating agent. The number of amino groups on the surface of human serum albumin was reduced when the volume of glutaraldehyde was increased. Gülseren et al. studied the incorporation of zinc in whey protein nanoparticles (Gülseren, Fang, and Corredig, 2012). The study demonstrated that the size of the protein nanoparticles increased with the ratio of ethanol to water and with the amount of $ZnCl_2$. The prepared nanoparticles were stable at 22 °C for 30 days, and showed high efficiency towards incorporation.

Arroyo-Maya et al. evaluated the stability and structure of assembled α-Lactalbumin nanoparticles which were prepared by two processes, viz. desolvation and crosslinking (Arroyo-Maya et al., 2014). The study used techniques such as transmission electron microscopy and spectroscopy to understand the shape of the nanoparticles. Recently, albumin nanoparticles were prepared by a rapid and simple method (Jahanban-Esfahlan, Dastmalchi, and Davaran, 2016). The study focused on an improved desolvation process for drug delivery systems for 100 nm particle size in controlled conditions. The method produced nanoparticles of 100 nm in size and polydispersity below 0.2.

12.3.2 Crosslinking

Crosslinking methods are used to increase the stability of protein and achieve sustained drug delivery. There are several types of crosslinkers which are usually employed in the preparation of protein-based nanoparticles, such as thermal, chemical, enzymatic, and ionic crosslinkers (Elzoghby, Elgohary, and Kamel, 2015). Figure 12.4 illustrates the ionic and chemical crosslinking processes of protein nanoparticles. Langer et al. used 8 % glutaraldehyde aqueous solution for crosslinked nanoparticles. The studies were conducted to analyze the crosslinking effect on the isoelectric point and size of nanoparticles by using a different concentration of glutaraldehyde (Langer et al., 2003). The integrity of the protein nanoparticles is maintained by crosslinking, which is an essential parameter. It has been reported that crosslinking with conventional crosslinkers is toxic and causes harm to the environment. Conventional crosslinkers, such as glutaraldehyde, are usually toxic and environmentally unfriendly. Moreover, the washing step is time-consuming, and requires a dialysis process (Varca, Queiroz, and Lugão, 2016). In a study conducted by Achilli et al., protein nanoparticles were prepared by dynamic aggregation and ionizing-induced radiation (Achilli et al., 2015). Two independent steps were used to prepare the nanoparticles. These include ethanol-based bovine serum albumin aggregation and gamma radiation-based crosslinking. The study conducted by Varca et. al. reported the development of nanoparticles based on doses of irradiation observing the presence and absence of solvents. Bovine serum albumin in phosphate buffer was irradiated with gamma cells in a phosphate buffer. The study evaluated the exposition of bovine serum albumin to 2.5–10 kGy. It was reported that bovine serum albumin nanoparticles can be prepared by adjusting the particle size (Varca, Queiroz, and Lugão, 2016). In immunotherapy, chitosan was crosslinked with scorpion proteins to study the structural properties (Soares et al., 2017). In the study, the scorpion venom was entrapped in chitosan nanoparticles. In a different study, fluorescent nonvaccines were prepared using ovalbumin proteins (Dong et al., 2018). The nonvaccine,

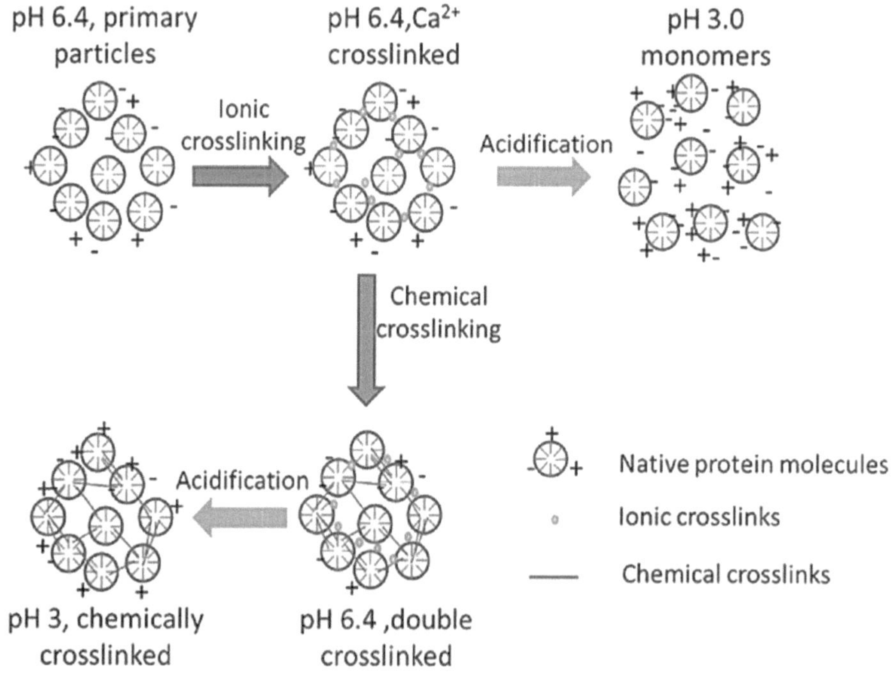

FIGURE 12.4 Ionic and chemical crosslinking processes of nanoparticles. (Printed with permission from [Xu, Teng, and Wang, 2016)].

along with additional carriers, was able to carry antigens and to monitor the in vivo delivery of antigens.

12.3.3 Coacervation

Coacervation reduces the protein solubility, leading to phase separation (Mohanty et al., 2005). The process results in desired nanoparticles formation forming phase separation (Elzoghby, Elgohary, and Kamel, 2015). Figure 12.5 illustrates the preparation of protein nanoparticles using coacervation. Coacervation can be employed to prepare nanoparticles from gelatin, legumin, and silk. Various factors such as polymer ratio, polymer concentration, weight, pH, temperature, and molecular conformation affect the morphology of the nanoparticles (Gummel et al., 2008; Priftis et al., 2013; Hedayati, Jahanshahi, and Attar, 2012; Hosseini et al., 2013; Aryee and Nickerson, 2014; Wang, Cheng, and Xu, 2008). Studies have demonstrated that the dissolution of polymers can be improved by increasing the ionic strength and it also affects the binding affinity of polyelectrolytes (Chollakup et al., 2010; Weinbreck et al., 2004). In a study, soluble nanoparticles were fabricated using a complex coacervation technique by mixing high and low molecular weight poly-D-lysine with bovine serum albumin. The study demonstrated that salt concentration and polyelectrolytes mass influenced the size of the particles. The formation of the complex coacervation is controlled by two major steps, which include (a) the entanglement of polyelectrolytes and kinetics of diffusion, and (b) conformation changes leading to the thermodynamic reorganization of the previously created aggregates and to disentanglement, which causes instability in the coacervates (Maldonado, Sadeghi, and Kokini, 2017).

12.3.4 Emulsification

An emulsion can be prepared either by single emulsions (oil-in-water) or double emulsions (water-in-oil-in-water). For the preparation of emulsions, homogenization and ultrasonication are required, then the solvent is evaporated (Sailaja, Amareshwar, and Chakravarty, 2011). Figure 12.6 illustrates the preparation of protein nanoparticles using single and double emulsion methods. During the preparation of gelatin nanoparticles using coacervation, alcohol and salt addition results in the preparation of desired nanoparticles. Upon the addition of aqueous gelatin solution to the sodium sulfate gelatin, nanoparticles in a size range from 600 nm to 1000 nm are produced. The addition of natural salt or alcohol promotes coacervation, which results in the desired nanoparticles. Gelatin nanoparticles in the size range of 600–1000 nm can be prepared by slow addition of sodium sulfate to the aqueous gelatin solution containing surfactant, and isopropanol is added to precipitate sodium sulfate (Lu et al., 2004). In the process of gelatin crosslinking, nanoparticles were performed by glutaraldehyde. In another study, thermal crosslinking of denatured whey protein was used to stabilize the Pickering emulsion at 80 °C for 15 min. The results obtained from the study may provide applicability of the whey protein isolate nanoparticles in environment-friendly and food-grade applications (Wu et al., 2015).

In a study, a double emulsion nanoparticle technique was used to encapsulate bovine serum albumin (Martinez et al., 2017). The outcomes from the study demonstrated that the nanoparticle size distribution is affected by the emulsion viscosity, which is demonstrated by a decrease in the polydispersity index on decreasing the concentration of poly(ε-caprolactone). Another study performed by Zhu et al. focused on the development of ion, pH, and temperature-dependent emulsion gels (Zhu et al., 2018). In the study, gel strength was controlled by altering pH, temperature, and ionic concentration values. The effects of ions, pH, and temperature on the droplet size distribution, rheological properties, microstructures, and ζ-potential of these systems were characterized. The addition of the whey protein isolate had a positive effect on gel strength, with a 2 % whey protein isolate as the strongest gel. In a different study, on chitosan, stabilized nanoparticles were prepared by encapsulating Pickering emulsions (Dammak and José do Amaral Sobral, 2018). The hesperidin-loaded emulsion stability was evaluated by investigating various factors such as oil-to-water ratio, lecithin concentration, and chitosan nanoparticles. The long shelf life of the emulsions was demonstrated by the study. Ye et al. reported the characterization of hydrophobic phytoglycogen nanoparticles, which were stabilized by oil-in-water Pickering emulsion (Ye et al., 2018). Octenyl succinate phytoglycogen nanoparticles were used as a stabilizer to prepare the emulsions. Also, the rheological properties and microstructure of emulsions with an oil-in-water ratio of 1 to 0.1 were investigated. The study demonstrated the possibility of producing Pickering emulsion to be used in food industries employing nanoparticles made from phytoglycogen. Richter et al. demonstrated the physical characteristics, potential use, and concept of formation of Pickering emulsions obtained by a novel method (Richter et al., 2018). In this process, emulsification and nanoprecipitation were

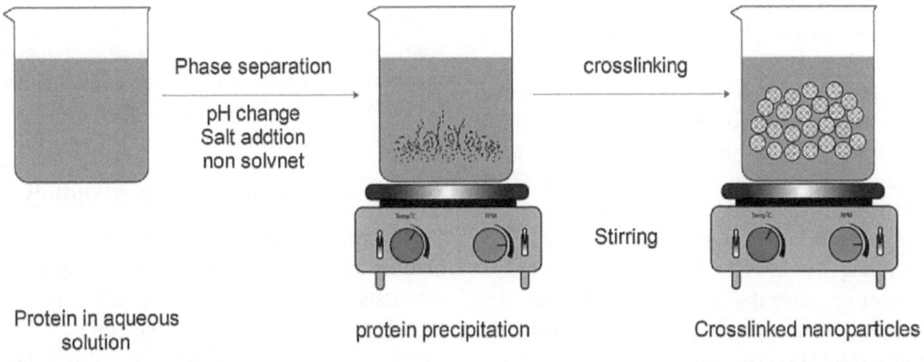

FIGURE 12.5 Protein nanoparticle preparation by coacervation. (Reprinted with permission from [Tarhini, Greige-Gerges, and Elaissari, 2017)].

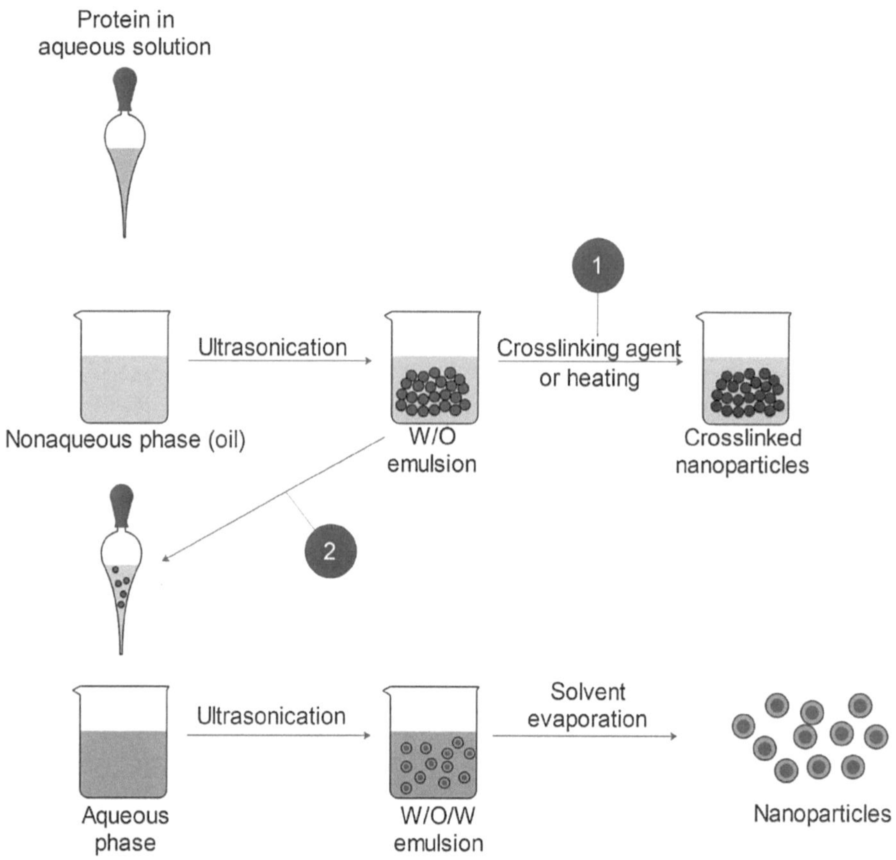

FIGURE 12.6 Protein nanoparticle preparation by emulsion. (1) Single emulsion and (2) Double emulsion. (Printed with permission from [Tarhini, Greige-Gerges, and Elaissari, 2017)].

combined. In the study, it was observed that oil droplets from the Pickering emulsion were unable to separate out nanoparticles after centrifugation. In a study conducted by Richter et al., soy peptide nanoparticles were prepared from soy peptide using ultrasound. The study also investigated the potential of produced nanoparticles as stabilizers for emulsions (Richter et al., 2018). The obtained results demonstrated that nanoparticles based on peptide can be utilized for preparing stable oil-to-water emulsion systems as bifunctional and effective emulsifiers.

The emulsification solvent evaporation technique is used to prepare gelatin nanoparticles. These nanoparticles, in the range of 100–400 nm, were prepared based on a single emulsion. In the method, the aqueous phase was mixed with the oil phase (Bajpai and Choubey, 2006a, b; Cascone et al., 2002) or paraffin oil (Choubey and Bajpai, 2010) then crosslinking with glutaraldehyde was performed (Choubey and Bajpai, 2010). In another approach, bioactive insulin-loaded gelatin nanoparticles were prepared by the water-to-water emulsion technique (Zhao et al., 2012). The reverse-phase microemulsion method includes the addition of aqueous gelatin solution to surfactant solution, followed by crosslinking through the addition of glutaraldehyde, and evaporation to remove the n-hexane for the recovery of gelatin nanoparticles (Gupta et al., 2004). The advantage of a microemulsion system is that the size of the nanoparticles can be controlled (Hou, Kim, and Shah, 1988).

12.3.5 Nanoprecipitation

This method needs two solvents which are miscible to each other. Nanoprecipitation occurs upon the addition of protein to non-solvents. Figure 12.7 illustrates the preparation of protein-based nanoparticles using nanoprecipitation. The chemical and physical properties of the synthesized nanoparticles are affected by factors such as injection rate, agitation speed, etc. The size of the nanoparticles is affected by the addition of surfactant (Rao and Geckeler, 2011). Precipitation involves the addition of water to ethanol slowly, followed by the addition of gelatin crosslinker (Lee et al., 2012; Quintanar-Guerrero et al., 1998). Nanoprecipitation is a rapid, easy to perform, and straightforward technique. It does not involve aqueous oily interfaces; therefore, sonication is not required (Khan and Schneider, 2013; Bilati, Allémann, and Doelker, 2005).

12.3.6 Nanoparticles Auto Assembly

The auto assembly of nanoparticles can also be used for the production of gelatin nanoparticles. Figure 12.8 illustrates the preparation of peptide nanoparticles from insoluble soy peptide aggregates. The production of gelatin nanoparticles through self-assembly includes: (i) Chemical modification: Kim and Byun first proposed self-assembled nanoparticles in which gelatin is

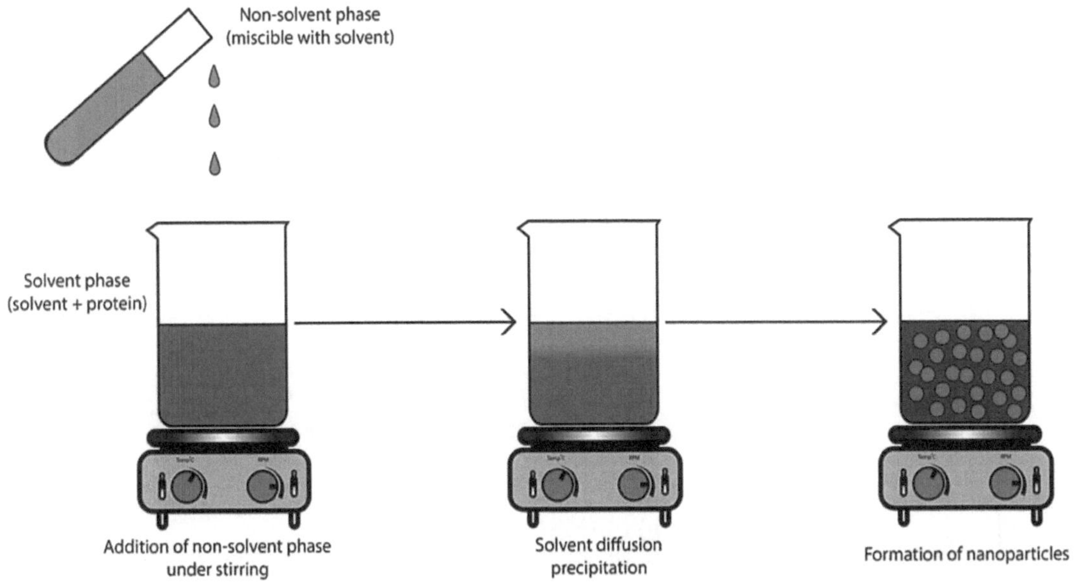

FIGURE 12.7 Preparation of protein-based nanoparticles using nanoprecipitation. (Printed with permission from [Tarhini et al., 2018]).

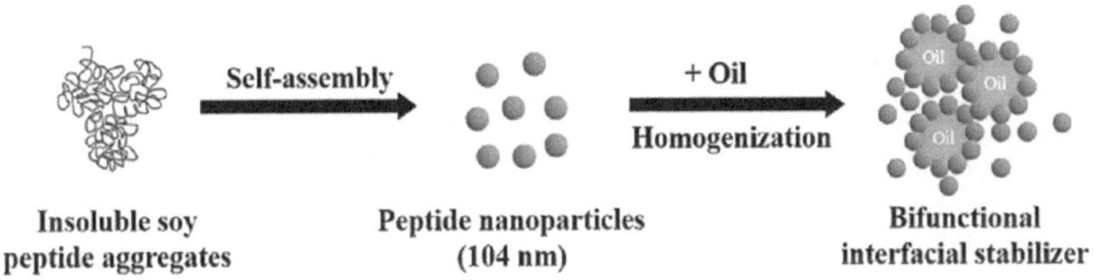

FIGURE 12.8 Preparation of peptide nanoparticles from insoluble soy peptide aggregates. (Modified with permission from [Zhang et al., 2018]).

coupled to deoxycholic acid and polyethene glycol (Kim and Byun, 1999). Hydrophilic gelatin or recombinant human gelatin can be chemically modified by utilizing hydrophobic groups (Won et al., 2011; Li, Liu, and Chen, 2011). (ii) Simple mixing: This method involves a direct mixing of drug solutions and gelatin in which the major force is hydrogen bonding (Chen et al., 2010; Li and Gu, 2011).

12.3.7 Coating Layer by Layer

This technique involves coating the gelatin nanoparticles using oppositely charged polyelectrolytes. Figure 12.9 illustrates the preparation of double-chambered nanoparticles using the layer-by-layer technique. These polyelectrolytes can be polyallylamine, polystyrene, or polyglutamic acid (Shutava et al., 2009; Ai, Jones, and Lvov, 2003). Nano-sized aggregates can be prepared by creating protein micelles above the critical solution temperature and threshold concentration. The obtained nano-sized scale crosslinked micelles are stable upon dilution and resistant to variations in experimental conditions, e.g., ionic strength, pH, solvents, etc., and shear forces (Batrakova et al., 2006).

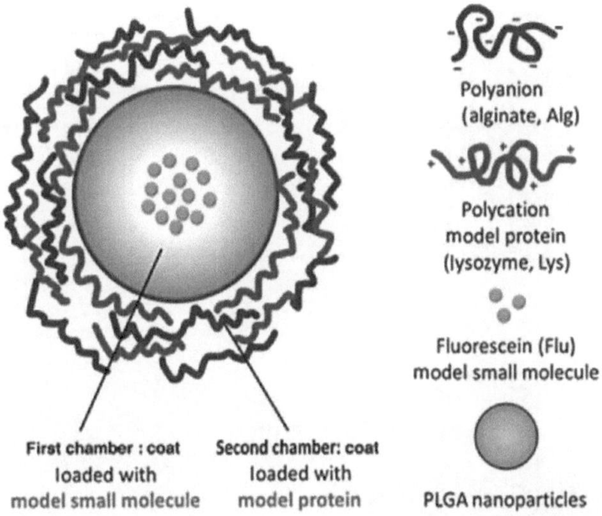

FIGURE 12.9 Preparation of double-chambered nanoparticles using the layer-by-layer technique. (Printed with permission from [Sakr, Jordan, and Borchard, 2015)].

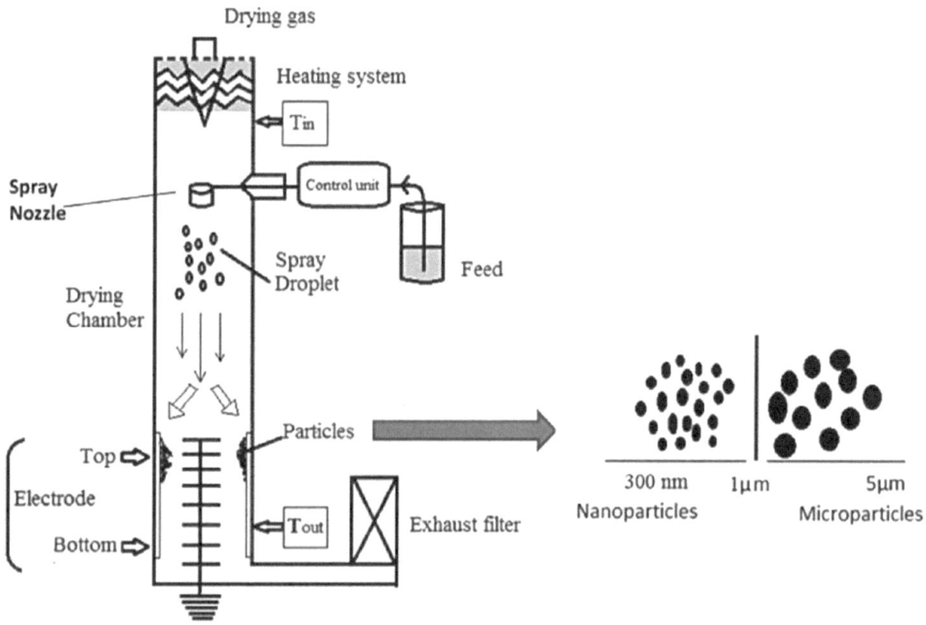

FIGURE 12.10 Nanospray-drying technique overview. (Reprinted with Creative Commons Attribution [CC BY] license from [Haggag and Faheem, 2015]).

12.3.8 Spray Drying

This technique is used for molecules that are heat sensitive. It prevents the protein from degradation. Nanoparticle characteristics such as particle size, bulk density, and flow properties can be easily controlled. Spray drying can be performed in four steps: spray air contact, dried-end product separation, conversion of the suspension in spray, and drying (Lee et al., 2011; Vauthier and Bouchemal, 2009). This technique can be used to prepare bovine serum albumin nanoparticles. The technique was also used for drug delivery applications (Lee et al., 2011). Figure 12.10 describes the design of a typical nanospray drier. The spray drier can produce small droplets due to vibration. In the study conducted by Lee et al. for bovine serum albumin production, the statistical experimental design method was used to study the effects of surfactant concentration, bovine serum albumin solution concentration, and spray mesh size.

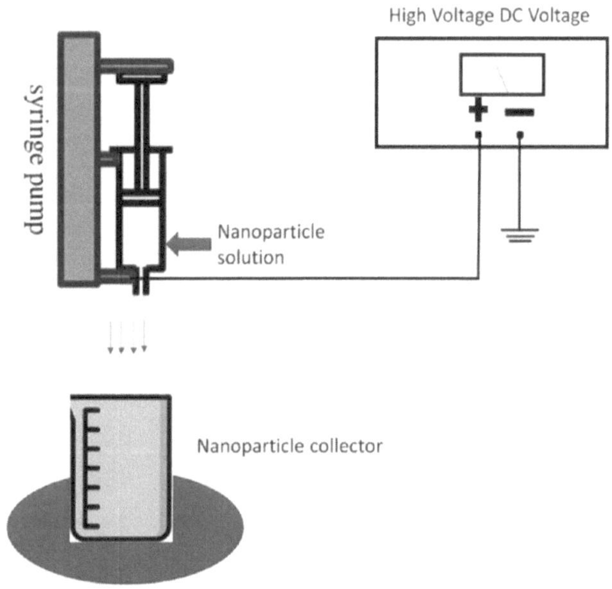

FIGURE 12.11 Illustration of the electrospraying technique.

12.3.9 Electrospray

This technique utilizes a small nozzle to pull a protein by applying an electric field. Figure 12.11 illustrates the electrospray technique. In this method, the sprayed jet moves to the collector to evaporate the solvent. In order to completely evaporate the solvent, the distance must be long enough; if it is not, protein particles will fuse together (Davidov-Pardo, Joye, and McClements, 2015). Compared to other conventional methods, the electrospray technique is more advantageous. By using this technique, the nanoparticles generated do not require any surfactant or template, and moreover, electrospraying leads to self-dispersion (Elzoghby, Elgohary, and Kamel, 2015). This method has been used to produce insulin nanoparticles in active form (Gomez et al., 1998). It was also used for the proliferation enhancement of antigen-specific T cells by the synthesis of peptide nanoparticles (Furtmann et al., 2017). The results from the study demonstrated the applicability of the technique to produce monodisperse nanoparticles.

12.3.10 Salting Out

This technique involves water-miscible solvent separation from the aqueous solution. In the process, salting-out agents, such as a salt of calcium or magnesium, are used. After that, dilution with aqueous solution forms nanospheres. Ultimately, cross filtration is used to remove the salting-out of the agent. (Nagavarma et al., 2012). Figure 12.12 illustrates the salting-out technique for nanoparticle

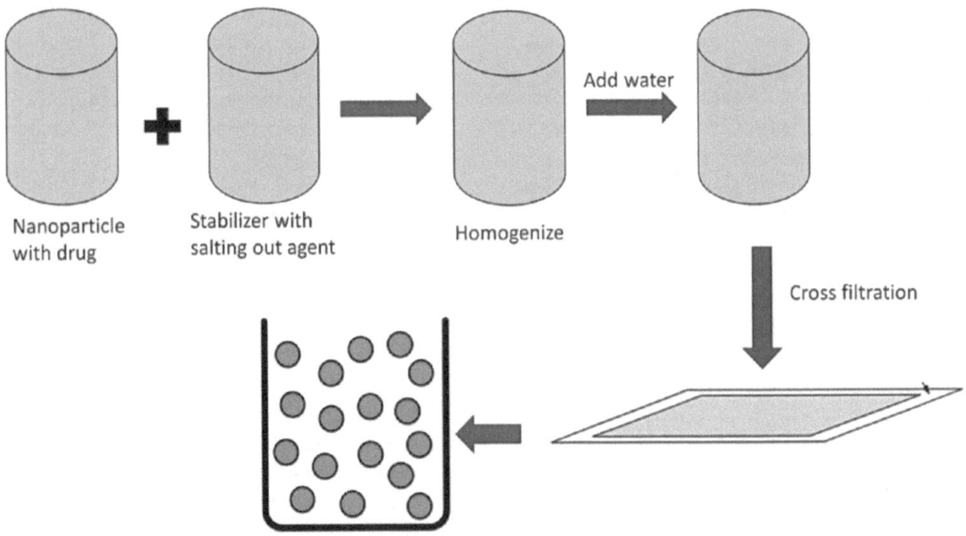

FIGURE 12.12 Salting out: a technique for nanoparticle preparation.

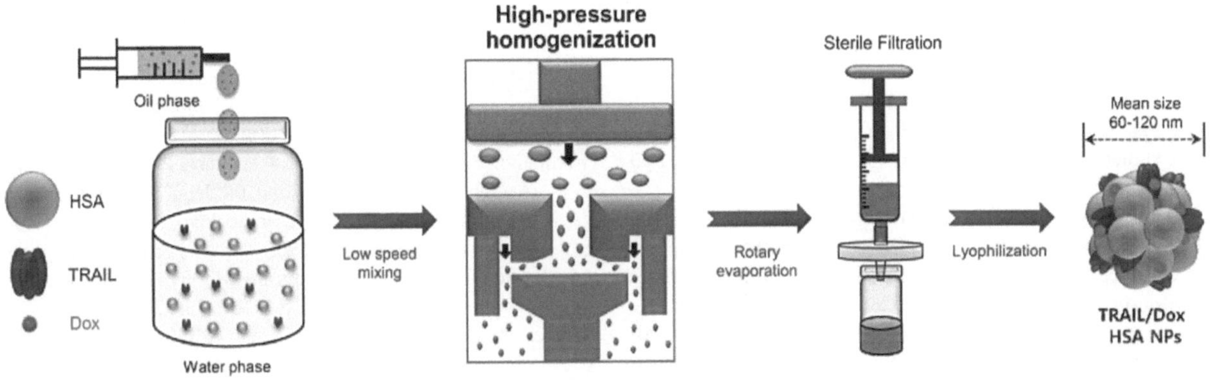

FIGURE 12.13 Preparation of nanoparticles using the NAB™ technology. (Printed with permission from [Thao le et al., 2016]).

preparation. This technique has advantages, e.g., it reduces the stress on protein carriers. The drawbacks of this technique are associated with its restriction to the costly washing step of nanoparticles as well as to lipophilic drugs (Nagavarma et al., 2012).

12.3.11 Albumin-Bound Nanoparticle Preparation

This technique involves mixing selected drugs with albumin and subsequently passing the solution through a high-pressure jet. In the process, nanoparticles in a size range of 100–200 nm are formed (Desai, 2007). Figure 12.13 illustrates a schematic diagram presentation of nanoparticle preparation using the NAB™ technology. This technology includes disulfide bounding via homogenization, which is advantageous to the traditional methods that utilize solvents (Elzoghby, Elgohary, and Kamel, 2015).

Conclusion

Although the methods described in this chapter for protein nanoparticle preparation have already been used to prepare nanoparticles, the studies reported to date suggest that more research is required to make these methods efficient for use in various applications. Also, these methods require optimization to perform efficiently.

REFERENCES

Achilli, E, G Casajus, M Siri, C Flores, S Kadłubowski, Sdel V Alonso, and Mariano Grasselli. 2015. "Preparation of protein nanoparticle by dynamic aggregation and ionizing-induced crosslinking." *Colloids and Surfaces A: Physicochemical and Engineering Aspects* 486:161–171.

Ai, Hua, Steven A Jones, and Yuri M Lvov. 2003. "Biomedical applications of electrostatic layer-by-layer nano-assembly of polymers, enzymes, and nanoparticles." *Cell Biochemistry and Biophysics* 39(1):23.

Arroyo-Maya, Izlia J, Humberto Hernández-Sánchez, Esmeralda Jiménez-Cruz, Menandro Camarillo-Cadena, and Andrés Hernández-Arana. 2014. "α-Lactalbumin nanoparticles prepared by desolvation and crosslinking: Structure and stability of the assembled protein." *Biophysical Chemistry* 193:27–34.

Aryee, Felix NA, and Michael T Nickerson. 2014. "Effect of pH, biopolymer mixing ratio and salts on the formation and stability of electrostatic complexes formed within mixtures of lentil protein isolate and anionic polysaccharides (κ-carrageenan and gellan gum)." *International Journal of Food Science and Technology* 49(1):65–71.

Bajpai, AK, and Jyoti Choubey. 2006a. "Design of gelatin nanoparticles as swelling controlled delivery system for chloroquine phosphate." *Journal of Materials Science: Materials in Medicine* 17(4):345–358.

Batrakova, Elena, Tatiana Bronich, Joseph Vetro, and Alexander V Kabanov. 2006. *Polymer Micelles as Drug Carriers.*

Bilati, Ugo, Eric Allémann, and Eric Doelker. 2005. "Development of a nanoprecipitation method intended for the entrapment of hydrophilic drugs into nanoparticles." *European Journal of Pharmaceutical Sciences* 24(1):67–75.

Cascone, Maria Grazia, Luigi Lazzeri, Claudia Carmignani, and Zhouhai Zhu. 2002. "Gelatin nanoparticles produced by a simple W/O emulsion as delivery system for methotrexate." *Journal of Materials Science: Materials in Medicine* 13(5):523–526.

Chen, Meiwan, Zhangfeng Zhong, Wen Tan, Shengpeng Wang, and Yitao Wang. 2011. "Recent advances in nanoparticle formulation of oleanolic acid." *Chinese Medicine* 6(1):20.

Chen, Yu-Chi, Yu Shu-Huei, Guo-Jane Tsai, Deh-Wei Tang, Fwu-Long Mi, and Yu-Ping Peng. 2010. "Novel technology for the preparation of self-assembled catechin/gelatin nanoparticles and their characterization." *Journal of Agricultural and Food Chemistry* 58(11):6728–6734.

Chollakup, Rungsima, Wirasak Smitthipong, Claus D Eisenbach, and Matthew Tirrell. 2010. "Phase behavior and coacervation of aqueous poly (acrylic acid)– poly (allylamine) solutions." *Macromolecules* 43(5):2518–2528.

Choubey, Jyoti, and AK Bajpai. 2010. "Investigation on magnetically controlled delivery of doxorubicin from superparamagnetic nanocarriers of gelatin crosslinked with genipin." *Journal of Materials Science: Materials in Medicine* 21(5):1573–1586.

Coester, CJ, K Langer, H Von Briesen, and J Kreuter. 2000. "Gelatin nanoparticles by two step desolvation a new preparation method, surface modifications and cell uptake." *Journal of Microencapsulation* 17(2):187–193.

Coester, Conrad, Paras Nayyar, and John Samuel. 2006. "In vitro uptake of gelatin nanoparticles by murine dendritic cells and their intracellular localisation." *European Journal of Pharmaceutics and Biopharmaceutics* 62(3):306–314.

Cupaioli, Francesca A, Fabio A Zucca, Diana Boraschi, and Luigi Zecca. 2014. "Engineered nanoparticles. How brain friendly is this new guest?." *Progress in Neurobiology* 119:20–38.

Dammak, Ilyes, and Paulo José do Amaral Sobral. 2018. "Formulation optimization of lecithin-enhanced pickering emulsions stabilized by chitosan nanoparticles for hesperidin encapsulation." *Journal of Food Engineering* 229:2–11. doi: 10.1016/j.jfoodeng.2017.11.001.

Davidov-Pardo, Gabriel, Iris J Joye, and David Julian McClements. 2015. "Chapter nine-food-grade protein-based nanoparticles and microparticles for bioactive delivery. Fabrication, characterization, and utilization." *Advances in Protein Chemistry and Structural Biology* 98:293–325.

Desai, N. 2007. *Nanoparticle Albumin Bound (Nab) Technology: Targeting Tumors Through the Endothelial gp60 Receptor and SPARC.* Vol. 3.

Dong, Xia, Zhiting Sun, Jie Liang, Hai Wang, Dunwan Zhu, Xigang Leng, Chun Wang, Deling Kong, and Feng Lv. 2018. "A visible fluorescent nanovaccine based on functional genipin crosslinked ovalbumin protein nanoparticles." *Nanomedicine: Nanotechnology, Biology and Medicine* 14(4):1087–1098.

Elzoghby, Ahmed O, Mayada M Elgohary, and Nayra M Kamel. 2015. "Chapter six – Implications of protein- and peptide-based nanoparticles as potential vehicles for anticancer drugs." In: *Advances in Protein Chemistry and Structural Biology*, edited by Rossen Donev, 169–221. Academic Press.

Esmaili, Mansoore, S Mahmood Ghaffari, Zeinab Moosavi-Movahedi, Malihe Sadat Atri, Ahmad Sharifizadeh, Mohammad Farhadi, Reza Yousefi, Jean-Marc Chobert, Thomas Haertlé, and Ali Akbar Moosavi-Movahedi. 2011. "Beta casein-micelle as a nano vehicle for solubility enhancement of curcumin; food industry application." *LWT-Food Science and Technology* 44(10):2166–2172.

Estrada, LPH, and JA Champion. 2015. "Protein nanoparticles for therapeutic protein delivery." *Biomaterials Science* 3(6):787–799.

Ezpeleta, Isabel, Juan M Irache, Serge Stainmesse, Christiane Chabenat, Jacques Gueguen, Yves Popineau, and Anne-Marie Orecchioni. 1996. "Gliadin nanoparticles for the controlled release of all-trans-retinoic acid." *International Journal of Pharmaceutics* 131(2):191–200.

Furtmann, B, J Tang, S Kramer, T Eickner, F Luderer, G Fricker, A Gomez, B Heemskerk, and PS Jahn. 2017. "Electrospray synthesis of poly(lactide-co-glycolide) nanoparticles encapsulating peptides to enhance proliferation of antigen-specific CD8(+) T cells." *Journal of Pharmaceutical Sciences* 106(11):3316–3327. doi: 10.1016/j.xphs.2017.06.013.

Gomez, A, D Bingham, L de Juan, and K Tang. 1998. "Production of protein nanoparticles by electrospray drying." *Journal of Aerosol Science* 29(5):561–574. doi: 10.1016/S0021-8502(97)10031-3.

Gref, Ruxandra, Yoshiharu Minamitake, Maria Teresa Peracchia, Vladimir Trubetskoy, Vladimir Torchilin, and Robert Langer. 1994. "Biodegradable long-circulating polymeric nanospheres." *Science* 263(5153):1600–1603.

Gülseren, İbrahim, Yuan Fang, and Milena Corredig. 2012. "Zinc incorporation capacity of whey protein nanoparticles prepared with desolvation with ethanol." *Food Chemistry* 135(2):770–774.

Gummel, Jérémie, François Boué, Daniel Clemens, and Fabrice Cousin. 2008. "Finite size and inner structure controlled by electrostatic screening in globular complexes of proteins and polyelectrolytes." *Soft Matter* 4(8):1653–1664.

Gupta, Ajay Kumar, Mona Gupta, Stephen J Yarwood, and Adam SG Curtis. 2004. "Effect of cellular uptake of gelatin nanoparticles on adhesion, morphology and cytoskeleton organisation of human fibroblasts." *Journal of Controlled Release* 95(2):197–207.

Haggag, YA, and AM Faheem. 2015. "Evaluation of nano spray drying as a method for drying and formulation of therapeutic peptides and proteins." *Frontiers in Pharmacology* 6:140. doi: 10.3389/fphar.2015.00140.

Hedayati, Rouhollah, Mohsen Jahanshahi, and Hussain Attar. 2012. "Fabrication and characterization of albumin-acacia nanoparticles based on complex coacervation as potent nanocarrier." *Journal of Chemical Technology and Biotechnology* 87(10):1401–1408.

Hosseini, Seyed Mohammad Hashem, Zahra Emam-Djomeh, Seyed Hadi Razavi, Ali Akbar Moosavi-Movahedi, Ali Akbar Saboury, Mohammad Amin Mohammadifar, Asgar Farahnaky, Maliheh Sadat Atri, and Paul Van der Meeren. 2013. "Complex coacervation of β-lactoglobulin–κ-Carrageenan aqueous mixtures as affected by polysaccharide sonication." *Food Chemistry* 141(1):215–222.

Hou, MJ, M Kim, and DO Shah. 1988. "A light scattering study on the droplet size and interdroplet interaction in microemulsions of AOT-oil-water system." *Journal of Colloid and Interface Science* 123(2):398–412.

Jahanban-Esfahlan, A, S Dastmalchi, and S Davaran. 2016. "A simple improved desolvation method for the rapid preparation of albumin nanoparticles." *International Journal of Biological Macromolecules* 91:703–709. doi: 10.1016/j.ijbiomac.2016.05.032.

Jahanshahi, M, Z Zhang, and A Lyddiatt. 2005. "Subtractive chromatography for purification and recovery of nano-bioproducts." *IEE Proceedings-Nanobiotechnology*.

Jahanshahi, Mohsen, and Zahra Babaei. 2008. "Protein nanoparticle: A unique system as drug delivery vehicles." *African Journal of Biotechnology* 7(25).

Khan, Saeed Ahmad, and Marc Schneider. 2013. "Improvement of nanoprecipitation technique for preparation of gelatin nanoparticles and potential macromolecular drug loading." *Macromolecular Bioscience* 13(4):455–463.

Kim, Kyung Jin, and Youngro Byun. 1999. "Preparation and characterizations of self-assembled pegylated gelatin nanoparticles." *Biotechnology and Bioprocess Engineering* 4(3):210–214.

Kreuter, Jörg, Renad N Alyautdin, Dimitri A Kharkevich, and Alexei A Ivanov. 1995. "Passage of peptides through the blood-brain barrier with colloidal polymer particles (nanoparticles)." *Brain Research* 674(1):171–174.

Kundu, Joydip, Yong-Il Chung, Young Ha Kim, Giyoong Tae, and SC Kundu. 2010. "Silk fibroin nanoparticles for cellular uptake and control release." *International Journal of Pharmaceutics* 388(1–2):242–250.

Langer, K, S Balthasar, V Vogel, N Dinauer, H Von Briesen, and D Schubert. 2003. "Optimization of the preparation process for human serum albumin (HSA) nanoparticles." *International Journal of Pharmaceutics* 257(1):169–180.

Langer, Robert. 1998. "Drug delivery and targeting." *Nature* 392(6679):5–10.

Lee, Eun Ju, Saeed Ahmad Khan, Joong Kon Park, and Kwang-Hee Lim. 2012. "Studies on the characteristics of drug-loaded gelatin nanoparticles prepared by nanoprecipitation." *Bioprocess and Biosystems Engineering* 35(1–2):297–307.

Lee, Hong Jai, Hee Ho Park, Kim Jeong Ah, Ju Hyun Park, Jina Ryu, Jeongseon Choi, Jongmin Lee, Won Jong Rhee, and Tai Hyun Park. 2014. "Enzyme delivery using the 30Kc19 protein and human serum albumin nanoparticles." *Biomaterials* 35(5):1696–1704.

Lee, Sie Huey, Desmond Heng, Wai Kiong Ng, Hak-Kim Chan, and Reginald BH Tan. 2011. "Nano spray drying: A novel method for preparing protein nanoparticles for protein therapy." *International Journal of Pharmaceutics* 403(1):192–200.

Leo, Eliana, Maria Angela Vandelli, Riccardo Cameroni, and Flavio Forni. 1997. "Doxorubicin-loaded gelatin nanoparticles stabilized by glutaraldehyde: Involvement of the drug in the cross-linking process." *International Journal of Pharmaceutics* 155(1):75–82.

Li, Wei-Ming, Dean-Mo Liu, and San-Yuan Chen. 2011. "Amphiphilically-modified gelatin nanoparticles: Self-assembly behavior, controlled biodegradability, and rapid cellular uptake for intracellular drug delivery." *Journal of Materials Chemistry* 21(33):12381–12388.

Li, Zheng, and Liwei Gu. 2011. "Effects of mass ratio, pH, temperature, and reaction time on fabrication of partially purified pomegranate ellagitannin– gelatin nanoparticles." *Journal of Agricultural and Food Chemistry* 59(8):4225–4231.

Lu, Ze, Teng-Kuang Yeh, Max Tsai, Jessie L-S Au, and M Guill Wientjes. 2004. "Paclitaxel-loaded gelatin nanoparticles for intravesical bladder cancer therapy." *Clinical Cancer Research* 10(22):7677–7684.

Maldonado, L, R Sadeghi, and J Kokini. 2017. "Nanoparticulation of bovine serum albumin and poly-d-lysine through complex coacervation and encapsulation of curcumin." *Colloids and Surfaces, Part B: Biointerfaces* 159:759–769.

Martinez, NY, PF Andrade, N Duran, and S Cavalitto. 2017. "Development of double emulsion nanoparticles for the encapsulation of bovine serum albumin." *Colloids and Surfaces, Part B: Biointerfaces* 158:190–196. doi: 10.1016/j.colsurfb.2017.06.033.

Marty, JJ, RC Oppenheim, and P Speiser. 1978. "Nanoparticles – a new colloidal drug delivery system." *Pharmaceutica Acta Helvetiae* 53(1):17–23.

Miladi, K, D Ibraheem, M Iqbal, S Sfar, H Fessi, and A Elaissari. 2014. "Particles from preformed polymers as carriers for drug delivery." *Excli Journal* 13:28.

Moghimi, SM. 2006. "Recent developments in polymeric nanoparticle engineering and their applications in experimental and clinical oncology. Anti-Cancer Agents in Medicinal Chemistry 6:553–61." *Current Pharmaceutical Design* 12(36):4729–4749.

Mohanraj, VJ, and Y Chen. 2006. "Nanoparticles – a review." *Tropical Journal of Pharmaceutical Research* 5(1):561–573.

Mohanty, Biswaranjan, VK Aswal, J Kohlbrecher, and HB Bohidar. 2005. "Synthesis of gelatin nanoparticles via simple coacervation." *Journal of Surface Science and Technology* 21(3/4):149.

Nagavarma, BVN, Hemant KS Yadav, A Ayaz, LS Vasudha, and HG Shivakumar. 2012. "Different techniques for preparation of polymeric nanoparticles – A review." *Asian Journal of Pharmaceutical and Clinical Research* 5(3):16–23.

Ofokansi, Kenneth, Gerhard Winter, Gert Fricker, and Conrad Coester. 2010. "Matrix-loaded biodegradable gelatin nanoparticles as new approach to improve drug loading and delivery." *European Journal of Pharmaceutics and Biopharmaceutics* 76(1):1–9.

Orive, Gorka, Alicia R Gascon, Rosa Ma Hernández, Alfonso Domínguez-Gil, and José Luis Pedraz. 2004. "Techniques: New approaches to the delivery of biopharmaceuticals." *Trends in Pharmacological Sciences* 25(7):382–387.

Panyam, Jayanth, and Vinod Labhasetwar. 2003. "Biodegradable nanoparticles for drug and gene delivery to cells and tissue." *Advanced Drug Delivery Reviews* 55(3):329–347.

Pathak, Yashwant, and Deepak Thassu. 2016. *Drug Delivery Nanoparticles Formulation and Characterization*. Vol. 191. CRC Press.

Peluffo, Hugo, Ugutz Unzueta, María Luciana Negro-Demontel, Zhikun Xu, Esther Váquez, Neus Ferrer-Miralles, and Antonio Villaverde. 2015. "BBB-targeting, protein-based nanomedicines for drug and nucleic acid delivery to the CNS." *Biotechnology Advances* 33(2):277–287.

Priftis, Dimitrios, Katie Megley, Nicolas Laugel, and Matthew Tirrell. 2013. "Complex coacervation of poly (ethylene-imine)/polypeptide aqueous solutions: Thermodynamic and rheological characterization." *Journal of Colloid and Interface Science* 398:39–50.

Quintanar-Guerrero, David, Eric Allemann, Hatem Fessi, and Eric Doelker. 1998. "Preparation techniques and mechanisms of formation of biodegradable nanoparticles from preformed polymers." *Drug Development and Industrial Pharmacy* 24(12):1113–1128.

Rao, J Prasad, and Kurt E Geckeler. 2011. "Polymer nanoparticles: Preparation techniques and size-control parameters." *Progress in Polymer Science* 36(7):887–913.

Reis, Catarina Pinto, Ronald J Neufeld, António J Ribeiro, and Francisco Veiga. 2006. "Nanoencapsulation I. Methods for preparation of drug-loaded polymeric nanoparticles." *Nanomedicine: Nanotechnology, Biology and Medicine* 2(1):8–21.

Richter, AR, JPA Feitosa, HCB Paula, FM Goycoolea, and RCM de Paula. 2018. "Pickering emulsion stabilized by cashew gum-poly-l-lactide copolymer nanoparticles: Synthesis, characterization and amphotericin B encapsulation." *Colloids and Surfaces, Part B: Biointerfaces* 164:201–209. doi: 10.1016/j.colsurfb.2018.01.023.

Sailaja, A, P Amareshwar, and P Chakravarty. 2011. "Different techniques used for the preparation of nanoparticles using natural polymers and their application." *International Journal of Pharmacy and Pharmaceutical Sciences* 3 Suppl 2:45–50.

Sakr, Omar S, Olivier Jordan, and Gerrit Borchard. 2015. "Novel layer-by-layer deposition technique for the preparation of double-chambered nanoparticle formulations." *Journal of Pharmaceutical Sciences* 104(8):2637–2640. doi: 10.1002/jps.24507.

Sharma, Hari S. 2009. *Nanoneuroscience and Nanoneuropharmacology*. Vol. 180. Elsevier.

Shutava, Tatsiana G, Shantanu S Balkundi, Pranitha Vangala, Joshua J Steffan, Rebecca L Bigelow, James A Cardelli, D Patrick O'Neal, and Yuri M Lvov. 2009. "Layer-by-layer-coated gelatin nanoparticles as a vehicle for delivery of natural polyphenols." *ACS Nano* 3(7):1877–1885.

Soares, Karla S, Alice R Rocha, Oliveira, Alessandra Daniele-Silva, Fiamma Glaucia-Silva, Ana Luiza P Caroni, Matheus F Fernandes-Pedrosa, and Arnóbio A da Silva-Júnior. 2017. "Self-assembled scorpion venom proteins cross-linked chitosan nanoparticles for use in the immunotherapy." *Journal of Molecular Liquids* 241:540–548.

Soppimath, Kumaresh S, Tejraj M Aminabhavi, Anandrao R Kulkarni, and Walter E Rudzinski. 2001. "Biodegradable polymeric nanoparticles as drug delivery devices." *Journal of Controlled Release* 70(1–2):1–20.

Tarhini, M, I Benlyamani, S Hamdani, G Agusti, H Fessi, H Greige-Gerges, A Bentaher, and A Elaissari. 2018. "Protein-based nanoparticle preparation via nanoprecipitation method." *Materials (Basel)* 11(3). doi: 10.3390/ma11030394.

Tarhini, Mohamad, Hélène Greige-Gerges, and Abdelhamid Elaissari. 2017. "Protein-based nanoparticles: From preparation to encapsulation of active molecules." *International Journal of Pharmaceutics* 522(1):172–197. doi: 10.1016/j.ijpharm.2017.01.067.

Thao le, Q, HJ Byeon, C Lee, S Lee, ES Lee, YW Choi, HG Choi, ES Park, KC Lee, and YS Youn. 2016. "Doxorubicin-bound albumin nanoparticles containing a TRAIL protein for targeted treatment of colon cancer." *Pharmaceutical Research* 33(3):615–626. doi: 10.1007/s11095-015-1814-z.

Vandervoort, J, and A Ludwig. 2004. "Preparation and evaluation of drug-loaded gelatin nanoparticles for topical ophthalmic use." *European Journal of Pharmaceutics and Biopharmaceutics* 57(2):251–261.

Varca, Gustavo HC, Rodrigo G Queiroz, and Ademar B Lugão. 2016. "Irradiation as an alternative route for protein crosslinking: Cosolvent free BSA nanoparticles." *Radiation Physics and Chemistry* 124:111–115.

Vauthier, Christine, and Kawthar Bouchemal. 2009. "Methods for the preparation and manufacture of polymeric nanoparticles." *Pharmaceutical Research* 26(5):1025–1058.

Verma, Rajan K, and Sanjay Garg. 2001. "Drug delivery technologies and future directions." *Pharmaceutical Technology* 25(2):1–14.

Wang, Guilin, and Hasan Uludag. 2008. "Recent developments in nanoparticle-based drug delivery and targeting systems with emphasis on protein-based nanoparticles." *Expert Opinion on Drug Delivery* 5(5):499–515.

Wang, Jin Cheng, Si Hao Chen, and Zi Cheng Xu. 2008. "Synthesis and properties research on the nanocapsulated capsaicin by simple coacervation method." *Journal of Dispersion Science and Technology* 29(5):687–695.

Weber, C, C Coester, J Kreuter, and K Langer. 2000. "Desolvation process and surface characterisation of protein nanoparticles." *International Journal of Pharmaceutics* 194(1):91–102.

Weber, C, J Kreuter, and K Langer. 2000. "Desolvation process and surface characteristics of HSA-nanoparticles." *International Journal of Pharmaceutics* 196(2):197–200.

Weinbreck, Fanny, Roland HW Wientjes, Hans Nieuwenhuijse, Gerard W Robijn, and Cornelus G de Kruif. 2004. "Rheological properties of whey protein/gum arabic coacervates." *Journal of Rheology* 48(6):1215–1228.

Won, Young-Wook, Sun-Mi Yoon, Chung Hee Sonn, Kyung-Mi Lee, and Yong-Hee Kim. 2011. "Nano self-assembly of recombinant human gelatin conjugated with α-tocopheryl succinate for Hsp90 inhibitor, 17-AAG, delivery." *ACS Nano* 5(5):3839–3848.

Wu, Jiande, Mengxuan Shi, Wei Li, Luhai Zhao, Ze Wang, Xinzhong Yan, Willem Norde, and Yuan Li. 2015. "Pickering emulsions stabilized by whey protein nanoparticles prepared by thermal cross-linking." *Colloids and Surfaces, Part B: Biointerfaces* 127:96–104. doi: 10.1016/j.colsurfb.2015.01.029.

Xu, Ruoyang, Zi Teng, and Qin Wang. 2016. "Development of tyrosinase-aided crosslinking procedure for stabilizing protein nanoparticles." *Food Hydrocolloids* 60:324–334. doi: 10.1016/j.foodhyd.2016.04.009.

Ye, Fan, Ming Miao, Steve W Cui, Bo Jiang, Zhengyu Jin, and Xingfeng Li. 2018. "Characterisations of oil-in-water Pickering emulsion stabilized hydrophobic phytoglycogen nanoparticles." *Food Hydrocolloids* 76:78–87. doi: 10.1016/j.foodhyd.2017.05.003.

Zhang, Sufeng, Guilin Wang, Xiaoyue Lin, Maria Chatzinikolaidou, Herbert P Jennissen, Marcus Laub, and Hasan Uludağ. 2008. "Polyethylenimine-coated albumin nanoparticles for BMP-2 delivery." *Biotechnology Progress* 24(4):945–956.

Zhang, Ying, Hon Fai Chan, and Kam W Leong. 2013. "Advanced materials and processing for drug delivery: The past and the future." *Advanced Drug Delivery Reviews* 65(1):104–120.

Zhang, Yuanhong, Feibai Zhou, Mouming Zhao, Lianzhu Lin, Zhengxiang Ning, and Baoguo Sun. 2018. "Soy peptide nanoparticles by ultrasound-induced self-assembly of large peptide aggregates and their role on emulsion stability." *Food Hydrocolloids* 74:62–71. doi: 10.1016/j.foodhyd.2017.07.021.

Zhao, Ying-Zheng, Xing Li, Cui-Tao Lu, Yan-Yan Xu, Hai-Feng Liu, Dan-Dan Dai, Lu Zhang, Chang-Zheng Sun, Wei Yang, and Xiao-Kun Li. 2012. "Experiment on the feasibility of using modified gelatin nanoparticles as insulin pulmonary administration system for diabetes therapy." *Acta Diabetologica* 49(4):315–325.

Zhu, Yuqing, Xing Chen, David Julian McClements, Liqiang Zou, and Wei Liu. 2018. "pH-, ion- and temperature-dependent emulsion gels: Fabricated by addition of whey protein to gliadin-nanoparticle coated lipid droplets." *Food Hydrocolloids* 77:870–878. doi: 10.1016/j.foodhyd.2017.11.032.

13

Recent Advances in Glycolipid Biosurfactants at a Glance: Biosynthesis, Fractionation, Purification, and Distinctive Applications

Rohini Kanwar and S.K. Mehta

CONTENTS

13.1 Introduction

The diversification of nanotechnology into the biomedical realm has enormously widened the scope of targeting a diverse array of drugs and bioactives in comparison to conventional drug delivery systems. The advancement in the development of nanocarriers such as micelles, vesicles, niosomes, liposomes, microemulsions, nanoemulsions, polymeric nanoparticles, lipid nanoparticles, etc. has also escalated the demand to quash the commercialization conundrum [1]. In colloidal assemblies, the surfactant generally plays a crucial role in the stabilization of the fabricated structures; e.g., in nanoemulsions, the surfactant assists in reducing the interfacial tension and facilitates its formation [2]. Moreover, during storage, the aggregation of the droplets is prevented as repulsive interactions are facilitated between them [3, 4]. Consequently, it has become a prerequisite to exploit environment friendly surfactants, commonly known as "biosurfactants", to stabilize the nanostructures, and to commercially benefit of the fabricated products [5–7].

Biosurfactants (BS) have recently been brought to the forefront as propitious molecules owing to their novel structures, biological nature, versatility, and distinct properties. They are impending to many therapeutic and biomedical (antimicrobial) applications [8]. BS encompass a diverse group of amphiphilic molecules with distinguishing chemical structures. They are produced by numerous aerobically growing microorganisms such as bacteria, yeast, and fungi, and especially by bacteria growing on water-immiscible substrates or in an aqueous medium from a carbon source feedstock (e.g., hydrocarbons, carbohydrates, oils, fats, or mixtures thereof). These molecules, being formed as secondary metabolites, facilitate the transport of nutrients, interfere in microbe-host interactions and quorum-sensing mechanisms, or act as biocide agents, thereby performing a critical role in the survival of their producing microorganisms [9]. Advantages such as microbial origin, mild production conditions, lower eco-toxicity, higher biodegradability, versatile biological activity, lower critical micellar concentration, higher surface activity, ecological acceptability, specificity at high temperatures, pH, saline conditions, high selectivity, high foaming capacity, excellent ability to form molecular assembly and liquid crystals, etc. offers an edge to microbial BS in comparison to chemically prepared conventional surfactants [10–12].

BS are mainly categorized according to their nature and chemical composition. Rosenberg and Ron [13] elaborated BS into two broad categories, namely: a) low molecular weight molecules which assist in reducing the surface and interfacial tension (include glycolipids, lipopeptides, and fatty acids/phospholipids); b) high molecular weight molecules that adhere tightly to the surfaces, thus acting as better stabilizing agents for emulsions (comprise of polymeric and surfactant particles) [14–16].

In this chapter, we have elaborated one important class of BS, i.e,. glycolipid BS, owing to their significant advances. This article was formulated with a focus on their microbial origin, structural characteristics, and isolation from microorganisms, as well as on the huge advancement in their application sphere.

Microbial glycolipid BS are the most promising BS amongst all, owing to their excellent surface active properties (because of low molecular weight), high productivity from renewable sources, versatile biochemical properties, low critical micellar concentration, and high application potential [17]. They are composed of a carbohydrate head (one or two sugar residues, mainly glucose and galactose, in either α or β configuration) which is attached to a lipid backbone/tail. They constitute the backbone of biological structures (such as cell membranes) and different metabolic functions. They were found to be involved in the uptake of low polarity hydrocarbons by microorganisms. They are capable of improving the physical properties of nanocarriers and facilitating site-specific targeting through carbohydrate–cell protein interactions [18]. The pH-driven self-assembly behaviour of microbial glycolipids such as micelles, vesicles, fibres, and bilayers make them valuable alternatives for synthetic-type surfactants [19–21].

Depending upon their microbial origin and structural characteristics, these BS are categorized mainly into four groups:

1. Rhamnolipids (RLs)
2. Sophorolipids (SLs)
3. Trehalose lipids (TLs)
4. Mannosylerythritol lipids (MELs)

The structure of glycolipid BS basically comprises of a hydrophilic head (mono- or disaccharide unit) linked to a hydrophobic tail (fatty acid chain–derived tail). The amphipathic structure present in glycolipids reduces the surface and interfacial tension between solids, liquids, and gases, thereby leading to the formation of micelles and emulsions in the liquid systems [22]. Certain external variables have been found to affect the production of glycolipids in terms of both quantity and type, e.g., culture conditions such as carbon source, temperature, pH, stirring, nitrogen, availability of oxygen, concentration of salts, etc. [15, 23].

Various microorganisms have been used for the production of glycolipids in the past few decades, e.g., Alcanivorax borkumensis, Arthrobacter sp., Candida antarctica T34, Candida apicola IMET 43747, Corynebacterium sp., Pseudomonas aeruginosa GL-1, Pseudomonas aeruginosa UW-1, Pseudozyma fusiformata, Pseudomonas putida, R. erythropolis, Rhodococcus ruber, Streptococcus mitis, Tsukamurella sp., VKM Y-2821, etc. [24]. But a specific class of organisms has been employed for the synthesis of a particular type of glycolipids (as listed in Table 13.1). In recent times, glycolipid BS have grabbed considerable attention due to their distinctive applications such as bioremediation, biofuel production, oil recovery from soil, hard-surface cleansing, and drug delivery [25, 26], out of which a few renowned applications have been listed in Table 13.1.

13.2 Biosynthesis and Physiochemical Aspects of Glycolipid BS

Glycolipid BS synthesis can be carried out in three ways (Figure 13.1):

I) The lipid fraction as well as the sugar fraction can be synthesized from the substrate.
II) The lipid fraction can be derived from the carbon source (hydrophobic source), whereas the sugar fraction can be synthesized *de novo.*
III) The glycolipid fraction can be derived from the carbon source, while the lipid fraction can be synthesized *de novo* [22].

TABLE 13.1

Different Glycolipid BS Production by Different Types of Microorganisms with Renowned Applications

Glycolipid BS	Producing Microorganisms	Applications (References)
Rhamnolipid	Pseudomonas aeruginosa, Burkholderia sp., Serratia rubidea.	Bioremediation and enhanced oil recovery, pest control, therapeutics, pharmaceuticals, cosmetics, stimulation of the biodegradation and uptake of poorly soluble substrates [16, 27, 28].
Sophorolipids	Candida bombicola, Candida apicola, Rhodotorula bogoriensis, Wickerhamiella domericqiae, Candida bogoriensis, Candida batistae, Candida lipolytica.	Potential use in skin moisturizer, antiviral activity against HSV and influenza virus, sperm-immobilizing activities, liver cancer treatment [29].
Trehalose lipids	Rhodococcus erythropolis, Rhodococcus opacus, Arthrobacter paraffineus, Rhodococcus ruber, Mycobacterium spp., Nocardia sp., Corynebacterium spp., Micrococcus luteus.	Hydrocarbon bioremediation, improved oil recovery, cleaning of oil storage tanks, growth inhibition and induction of differentiation in human promyelocytic leukemia cell lines on the basis of hydrophobic moiety structure [28], electronic printing, micro-electronics, viral research, magnetic recording [30].
Mannosylerythritol lipids	Pseudozyma sp. like Pseudozyma antartica, P. aphidis, P. crassa, P. graminicola, P. hubeiensis, P. parantarctica, P. rugulosa, P. shanxiensis, P. siamensis, P. tsukubaensis; Candida antartica, Schizonella melanogramma, Ustialgo sp. like Ustialgo cynodontis, Ustialgo scitaminea.	Gene transfection, pharmaceutics [28], growth arrest and apoptosis of tumor cells [31].

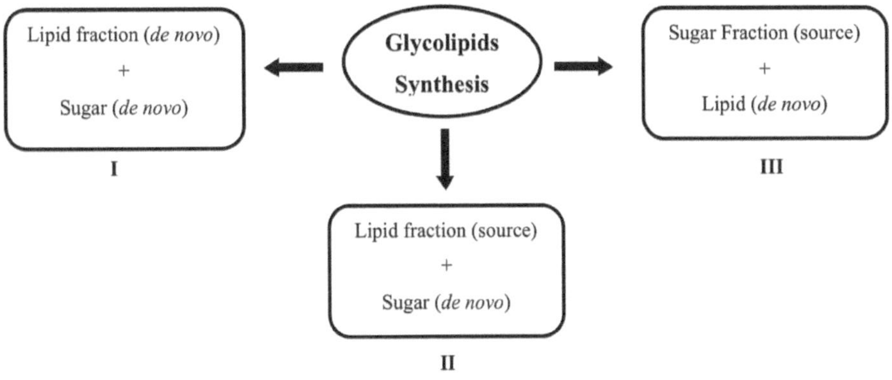

FIGURE 13.1 Synthetic pathway for glycolipid BS.

The synthesis of glycolipids has been found to be dependent on the polarity of the carbon source, which actually affects the specificity of the synthesis routes in each organism [28]. Further ahead, biosynthesis of different types of glycolipids with their physiochemical aspects has been outlined in detail [28].

13.2.1 *Rhamnolipids (RLs)*

RLs comprise of one or two L- rhamnose units linked through a β-glycoside bond to up to three hydroxyl fatty acid groups (containing 8 to 14 carbon atoms chain length). They are described as a mixture of four congeners: α-L-rhamnopyranosyl-α-L-rhamno pyranosyl-β-hydroxydecanoate (Rha-Rha-C_{10}) and α-L-rhamno pyranosyl-α-L-rhamnopyranosyl-β-hydroxydecanoyl-β-hydro xydecanoate (Rha-Rha-C_{10}-C_{10}) with their mono-RL congeners Rha-C_{10}-C_{10} and Rha-C_{10}. With the development of sensitive analytical techniques, a wide range of RL congeners and homologues, i.e., more than 60 structures, have already been identified [32]. RLs are capable of exhibiting excellent surface active properties including emulsifying, foaming, dispersing, and penetrating actions. Despite being an anionic surfactant, they show a lower CMC of 10^{-4} to 10^{-5} M. Due to their significant antiviral activity against herpes simplex virus type I and II, high antimicrobial activity against bacteria and phytopathogenic fungi, and immunological activity, this class of glycolipids has become of the utmost importance [33].

13.2.1.1 Biosynthesis of RLs

Generally, the biosynthesis of RLs is carried out in three steps:

1. Lipid fraction synthesis
2. Sugar synthesis
3. Dimerization by enzymes

But at the genetic level, the biosynthesis of RLs in Pseudomonas aeruginosa (in which rhl was found responsible for producing RL) can be described in three consecutive stages (Figure 13.2).

RhlA (codified by the rhlA gene) synthesizes fatty acid dimers (HAAs), which have been thereafter used in combination with dTDP-L-rhamnose as precursors to produce mono-RL using RhlB rhamnosyltransferase. In the last step, RhlC uses mono-RL, together with dTDP-L-rhamnosein, as substrates in the production of di-RLs [34]. Furthermore, it has been seen that a bicistronic operon rhlAB codifies the first two enzymes, whereas rhlC is located in a different part of the genome fabricating a mix of mono- and di –RL [35]. From the above synthesis, both mono-RL and di-RL have been obtained.

13.2.2 *Sophorolipids (SLs)*

A SL constitutes a dimeric carbohydrate such as a sophorose unit (a glucose disaccharide with β-1, 2 bond) linked through a β-bond to a long-chain hydroxylated fatty acid (16 to 18 carbon atoms chain length). It is possible to perform SL structure alterations. For instance, the sophorose unit can be acetylated at 6' – or 6" – position, which can reduce the hydrophilicity of SLs and simultaneously enhance their cytokine and antiviral stimulating effects, whereas the hydroxyl fatty acid can be attached to the sophorose unit at the terminal/subterminal position, and the carboxylic end of the fatty acid can either be left free or esterified at the 4" position (lactone form). Depending on the culture conditions, SLs can be non-ionic or anionic in nature. They exhibit a 40–100 mg/L critical micellar concentration with the ability of lowering the aqueous surface tension down to 40 mN/m.

13.2.2.1 Biosynthesis of SLs

Candida bombicola and Candida apicola have been targeted (or grown) onto hydrophobic substrates such as alkanes, as they possess the necessary enzymes required for the terminal oxidation of alkanes, thereby generating fatty acids which are further metabolized through β oxidation. After obtaining fatty acids, a number of steps has been laid down to obtain SL [36].

SLs can be categorized into two types: acidic SLs and lactone SLs (Figure 13.3). Acidic SLs exhibit better foaming properties, while lactone SLs are more efficient in reducing the surface tension and act as better antimicrobial agents.

13.2.3 *Trehalose Lipids (TLs)*

Trehalose is a glucose disaccharide with α, α-1, 1-glycoside bond. In its structure, trehalose disaccharide is bound to the carbon 6 and 6' of the hydroxylated fatty acids in α- and β- configurations (chains may vary from 20 to 90 carbon atoms). These lipids evidently showcase antitumor, antiviral, and anti microbial activities, particularly the trehalose 6,6' dimycolate.

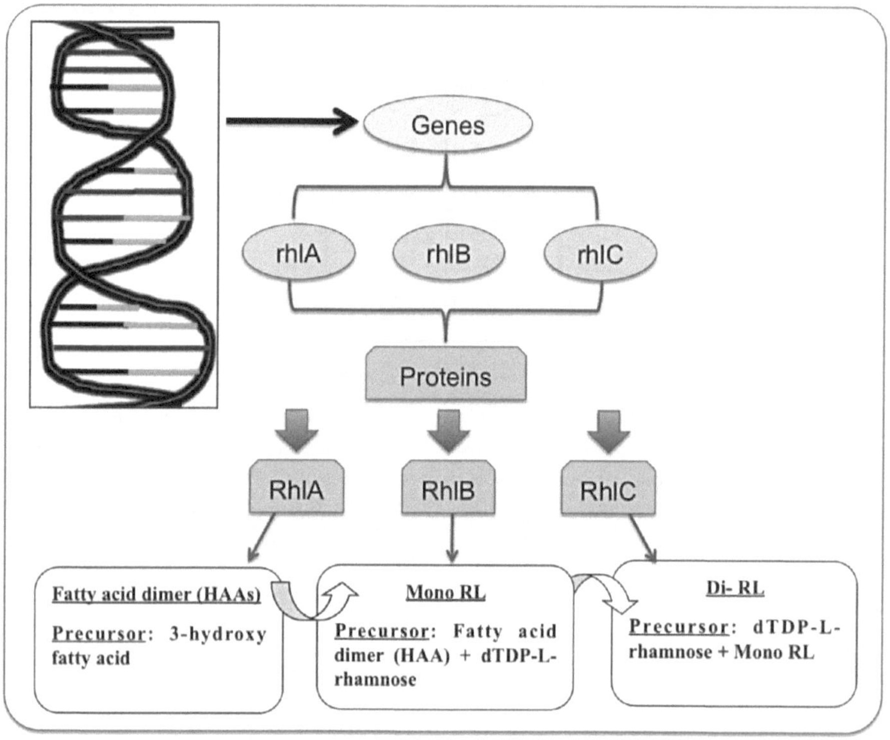

FIGURE 13.2 Biosynthetic pathway for sophorolipid (SL) synthesis.

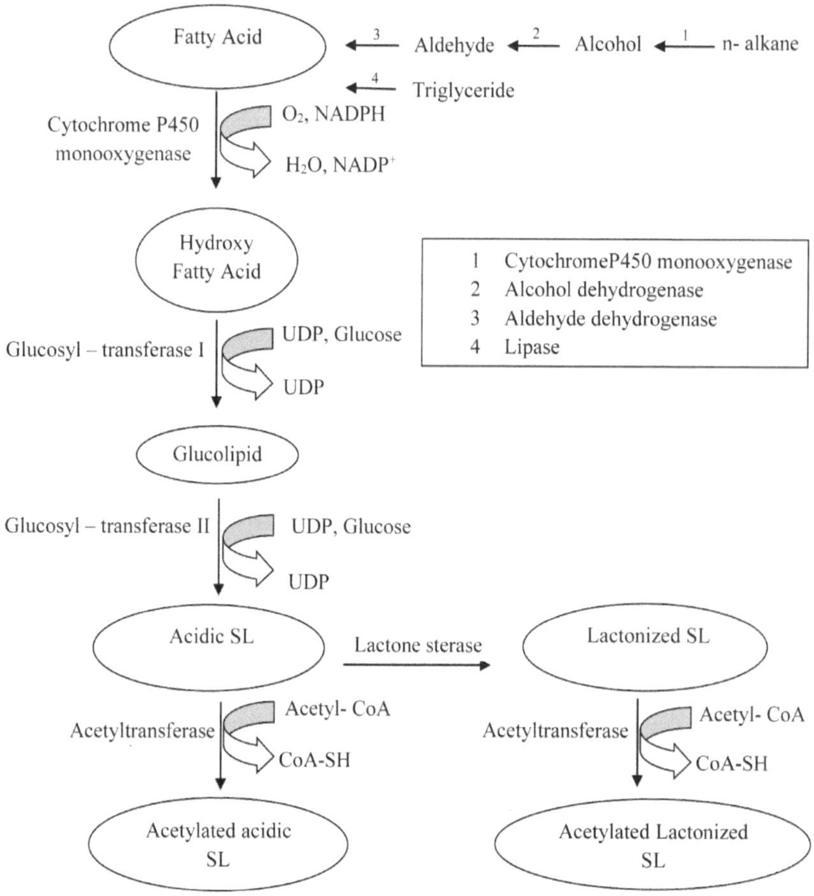

FIGURE 13.3 Biosynthetic pathway for sophorolipid (SL) synthesis.

13.2.3.1 Biosynthesis of TLs

In the Rhodococcus erythropolis organism, TL synthesis depends on the carbon source. Initially, the synthesis of lipids begins with the condensation of fatty acids through a Claisen-type reaction which leads to the formation of mycolates or, we can say, to activation as thiol esters, as shown in Equation (13.1):

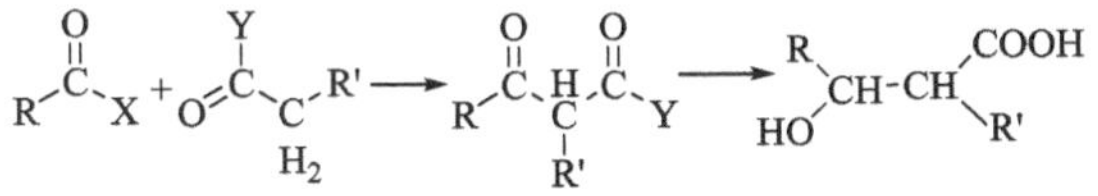
(13.1)

where X and Y are the activating groups, depicting the biosynthetic pathway for the formation of mycolates.

The β-oxidation of hydrophobic carbon sources produces the long-chain fatty acid, which thereby makes up the largest part of the glycolipid molecule.

The sugar residue trehalose-6-phosphate has been synthesized by catalysing trehalose-6-phosphate synthetase, which links two units of D-Glycopyranosyl in C1 and C1'. UDP- glucose and glucose-6-phosphate probably act as intermediate precursors, as shown in Equation (13.2).

$$\text{UDP-Glucose} + \text{Glucose-6-Phosphate} \rightarrow \text{Trehalose-6-Phosphate} + \text{UDP} \quad (13.2)$$

Finally, Trehalose-6-phosphate leads to the synthesis of trehalose mycolates, i.e., trehalose-mono and dicorynomycolate based on n-alkane induction studies with Rhodococcus erythropolis [37].

A biosynthetic diagram (Figure 13.4) shows an independent synthesis of the corynomycolic acid moiety and the trehalose moiety, which subsequently esterified to produce trehalose mycolates.

13.2.4 Mannosylerythritol Lipids (MELs)

MELs comprise of 4-*O*-*β*-D-mannopyranosyl-*meso*-erythritol as the hydrophilic group and a fatty acid/acetyl group as the hydrophobic moiety. The number and position of acetyl groups on mannose or erythritol and the fatty acid chain length may impart structural changes to MELs. These lipid structures firmly depend on their microbial origin and non-ionic nature. They also display significant antitumor and differentiation-inducing activities against human leukemia cells.

13.2.4.1 Biosynthesis of MELs

Organisms such as Candida antarctica and basidiomycetes of the Pseudozyma genus are employed for the biosynthesis of MEL glycolipids. One of the most important steps in the synthesis of MEL glycolipids is the conversion of long-chain fatty

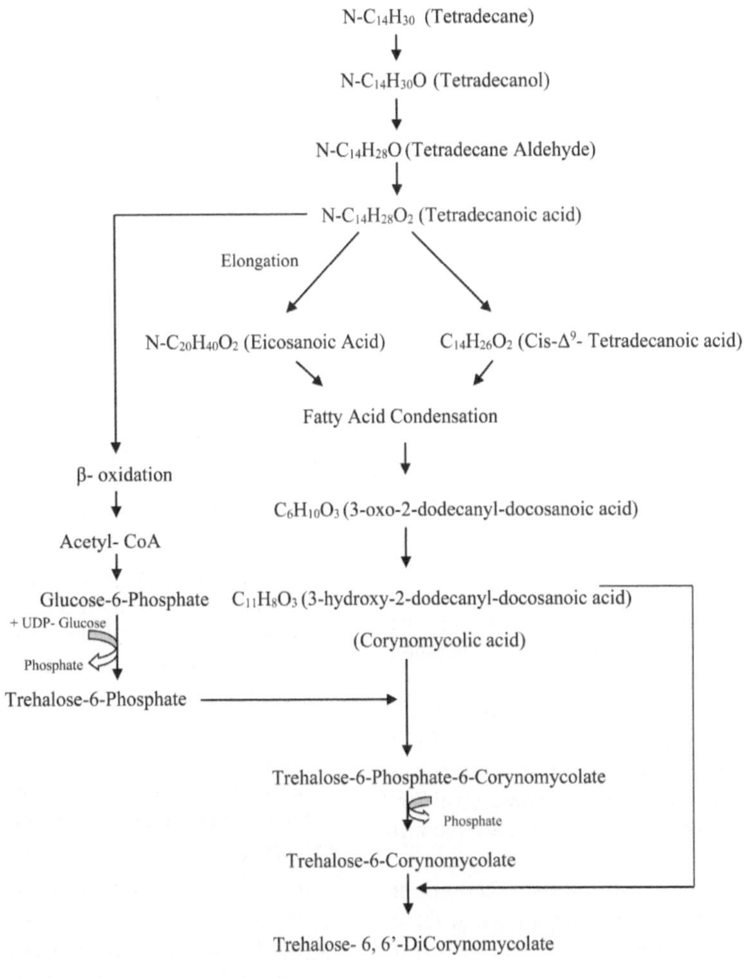

FIGURE 13.4 Biosynthetic pathway for trehalose lipid (TL) synthesis.

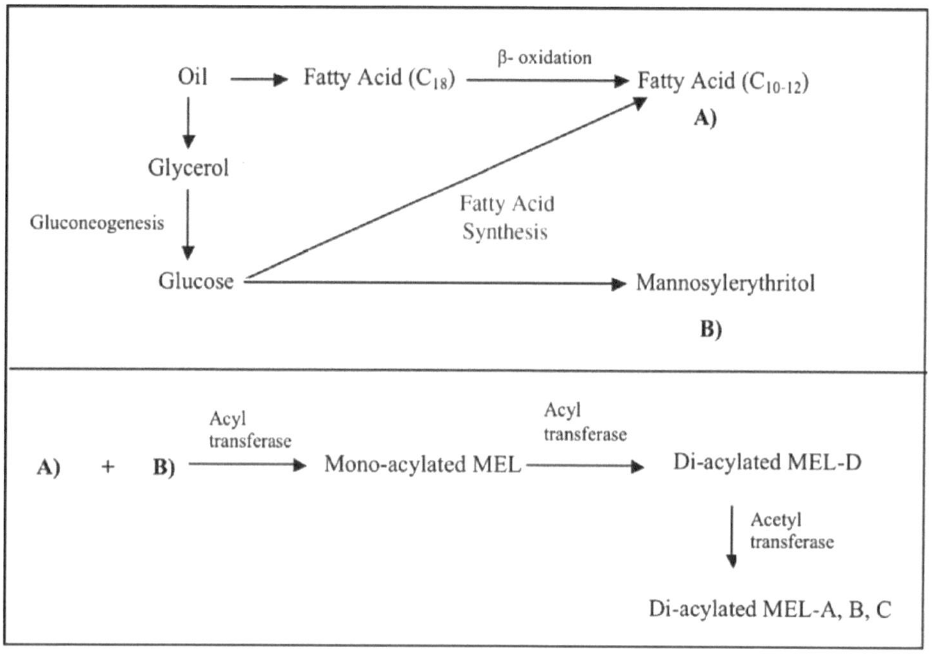

FIGURE 13.5 Biosynthetic pathway for mannosylerythritol lipid (MEL) synthesis.

acids into short-chain fatty acids (Figure 13.5) by β-oxidation [38]. This chain shortening is quite different from the three pathways generally recognized in the microorganisms, since it involves *de novo* synthesis, chain elongation, and intact incorporation [32].

13.3 Microbial Glycolipid Fractionation and Purification

Despite the potential physiochemical and biological properties of glycolipids BS, there lies a problem in their commercialization due to their high production cost. To overcome this limitation, research is being conducted to improve the fermentation and downstream processes in order to obtain glycolipid recovery with high yield, better purity, and efficient production costs [28]. Thus, extraction and purification methods play a key role in deciding the ultimate route for obtaining the glycolipids either in pure form or as selected mixtures, and to understand their specific biological role.

Different downstream processes can be employed for the extraction of glycolipid BS from microorganisms. A detailed methodology has been described in three steps.

In the first step (I), preliminary separation of the microorganisms from the fermentation media is required. Here, bacteria or yeast cells containing fermentation media have been taken into consideration to discuss the extraction methods. There are different routes available to reach a crude lipid extract (i.e, end product of step I).

These are classically removed by centrifugation (convenient with large volumes); however, being heavier than water, these glycolipids can even be simply decanted after heating or pH adjustment (if necessary).

In the case of specific classes of glycolipids,

SLs: Can be extracted for instance by dissolving the glycolipid in ethanol and later by centrifugation to remove cells [39].

RLs: Can be extracted after acidification or addition of salt-like aluminium sulfate into the fermentation media followed by precipitation [40, 41].

TLs: Can be isolated by performing diafiltration followed by isopropanol precipitation [42], whereas acid precipitation can be utilized for succinoyl TL [43].

In the second step (II), the resulted crude extract, being contaminated with the lipids or carbohydrate compounds, is further isolated using various techniques such as solvent extraction, adsorption, membrane filtration [44], foam fractionation [45], and selective crystallisation [46]. The details of the preferable route followed by each glycolipid are given ahead.

SLs: To isolate them (after step I), solvent extraction can be performed using ethyl acetate [47]. An additional extraction is carried out with apolar solvents (such as pentane, n-hexane, etc.) to remove the residual lipid substrates [48].

RLs: Their isolation can be carried out using ethyl acetate or a mixture of $CHCl_3/CH_3OH$ (2:1, v/v) [41, 49]. Even adsorption methods on resin or activated carbon can be employed for their isolation [50, 51].

TLs: Here, a mixture of $CHCl_3/CH_3OH$ (2 : 1, v/v), methyl *t*-butyl ether, or a mixture of ethyl acetate/CH_3OH (8 : 1, v/v) solvents can be employed for their isolation [30, 52].

MELs: For their recovery, amberlite XAD followed by solvent extraction and heating can be employed [53, 54].

In addition, if a preliminary acidification step is performed, then the extraction yield can be improved, thereby reducing its solubility in water.

In the third step (III), the concentrated glycolipid mixture thus obtained is employed to various chromatographic techniques for further separation of individual glycolipids. Various techniques in this regard are described below:

Reversed phase or silica gel column chromatography can be employed for the purification of SLs, RLs [55–57], and MELs. By this technique, MELs can be purified by the gradient elution of a mixture of $CHCl_3/CH_3OCH_3$ [54, 58], whereas, for TL purification with $CHCl_3/CH_3OH$ (2 : 1, v/v), the solvent system can be utilized [59].

Silica-gel TLC plates with a solvent mixture of $CHCl_3/CH_3OH$ (9 : 1, v/v) and subsequently a mixture of $CHCl_3/CH_3OH/H_2O/CH_3COOH$ (65 : 15 : 2 : 0.1 v/v) can be utilized for RL purification [60].

The ion exchange separation technique based on the ionization properties of glycolipids seems interesting. An RL negative charge at high pH values can also be purified by this method [41, 61].

The solid support-free liquid–liquid separation technique is based on the separation of solutes depending on their partition coefficients between two immiscible solvent systems [62]. For glycolipid purification, this technique appears to be promising and efficient due to its ability to purify complex mixtures of homologous compounds.

13.4 Microbial Glycolipid Distinctive Applications

The inherent inimitable properties of glycolipid BS such as surface activity (i.e., emulsifying, dispersing, solubilising, foaming, penetrating, wetting, etc.) and self-assembly not only advanced the research efforts in the direction of overcoming various limitations associated with this class, but also widened the scope of their utilization in various other fields, for instance in soil remediation, cell cycle regulation, cold storage, immunoglobulin sensing, gene delivery, skin care, and many more [23, 32]. This chapter elaborates the present state of the art of the therapeutic applications of glycolipid BS.

The tendency of glycolipid BS to affect the adhesion of microorganisms, disrupting cell membranes, changing the physical membrane structure, modifying the protein conformation, etc., has inspired varied studies to account for their plausible therapeutic applications. A few major applications are discussed below:

- **The potential of glycolipids BS as antitumor agents**

  The glycolipid BS have become of paramount interest due to their ability to act as antitumor agents, which control a variety of mammalian cell functions thereby interfering with cancer progression processes [63]. For example, RLs showcase antiproliferative effects against human breast cancer cells and inhibit the phagocytic response of macrophages at low concentrations [22]. Even SL BS inhibits growth and induces apoptosis against liver cancer H7402 cells. Further, succinoyl TL inhibits growth and differentiation against promyelocytic leukemia HL60 cells. Likewise, MEL BS triggers both apoptotic and differentiation mechanisms. Even MEL BS inhibits the growth and differentiation activities in several cancer lines such as myelogenous leukemia K562.

- **The potential of glycolipid BS as health care and therapeutic agents**

  Owing to their inherent versatile biochemical functions, environmentally friendly features, excellent moisturizing properties, and skin compatibility, they have successfully carved a niche for themselves in the personal-care market where products such as toothpastes, contact lens solutions, deodorants, antacids, antidandruff shampoos, etc. utilize large amounts of synthetic surfactants. For example, Kao Co. Ltd. use SLs as humectants in cosmetic make-up brands, i.e., Sofina [64]. The mycolates are appropriate for applications in pastes, creams, films, and sticks.

  Nowadays, great efforts are being made to develop pharmaceutically acceptable excipients-based formulations for drug delivery purposes. Exploiting natural surfactants in the formulations serves to create greener alternatives over their synthetic counterparts. For example, Müller *et al.* [5] demonstrated the efficacy of RL BS in formulating non-toxic drug-loaded nanoparticles for dermal route of administration. Cheow and Hadinoto [6] examined the effects of RL BS on the release of drug encapsulated in lipid-polymer-coated hybrid nanoparticles. Nguyen *et al.* [29] fabricated a biocompatible microemulsion comprising of lecithin in combination with RL and SL BS. Bai *et al.* [7] exploited RL BS to formulate effective nanoemulsions and claimed that these BS can efficienlty replace synthetic surfactants in certain commercial applications. Ahmad *et al.* [4] proved that glycolipid-stabilized nanoemulsions can be potential candidates for drug delivery purposes. Bharali *et al.* [65] reported the composite of colloidal silver nanoparticles/RL, which shows prominent antibacterial and chemotactic activity in comparison to all of its individual precursor components. Kanwar and co-workers [66] for the first time fabricated SL-based solid lipid nanoparticles using poloxamer as the surfactant by employing the solvent-injection method. Furthermore, they prepared SL-based nanostructured lipid carriers [67] and achieved high encapsulation and *in-vitro* release for rifampicin. Apart from these known applications, many more efforts are being laid down to completely understand these assemblies.

Conclusion

An insight into the surfactant world depicts a new field of biosurfactants of microbial origin, which have been developed with the objective of environmental sustainability. Glycolipids, in spite of their inherent surface activity (due to their chemical structure), have also been found to show biological activities – antifungal, antibacterial, anti-carcinogenic, immune-modulating, etc. Thus, glycolipid BS have become a feasible alternative to be utilized in bioremediation, biofuel production, hard-surface cleansing, and biomedical areas. Glycolipid-based nanosystems have gained significant attention owing to their biodegradability and, most importantly, pH-driven self-assembly behavior. The only problem that needs to be tackled is their production and extraction cost, which researchers have tried to solve to a great extent by utilizing various downstream processes. This field seems to provide great scope for future researchers as the inherent potential associated with the glycolipids BS class is yet to be explored completely.

Acknowledgments

Rohini Kanwar gratefully acknowledges CSIR, New Delhi. S. K. Mehta is thankful to DST PURSE II.

REFERENCES

1. Mishra, B. B. T. S., Patel, B. B., and Tiwari, S. 2010. Colloidal nanocarriers: A review on formulation technology, types and applications toward targeted drug delivery. *Nanomedicine* 6(1):9–24.
2. Kralova, I., and Sjoblom, J. 2009. Surfactants used in food industry: A review. *J Dispers Sci Technol* 30(9):1363–83.
3. Jafari, S. M., Assadpoor, E., He, Y., and Bhandari, B. 2008. Re-coalescence of emulsion droplets during high-energy emulsification. *Food Hydrocoll* 22(7):1191–202.
4. Ahmad, N., Ramsch, R., Llinàs, M., Solans, C., Hashim, R., and Tajuddin, H. A. 2014. Influence of nonionic branched-chain alkyl glycosides on a model nano-emulsion for drug delivery systems. *Colloids Surf B Biointerfaces* 115:267–74.
5. Müller, F., Hönzke, S., Luthardt, W. O., et al. 2017. Rhamnolipids form drug-loaded nanoparticles for dermal drug delivery. *Eur J Pharm Biopharm* 116:31–7.
6. Cheow, W. S., and Hadinoto, K. 2012. Lipid–polymer hybrid nanoparticles with rhamnolipid-triggered release capabilities as anti-biofilm drug delivery vehicles. *Particuology* 10(3):327–33.
7. Bai, Long, and McClements, D. J. 2016. Formation and stabilization of nanoemulsions using biosurfactants: Rhamnolipids. *J Colloid Interface Sci* 479:71–9.
8. Makkar, R. S., and Cameotra, S. S. 2002. An update on the use of unconventional substrates for biosurfactant production and their new applications. *Appl Microbiol Biotechnol* 58(4):428–34.
9. Gudiya, E. J., Rangarajan, V., Sen, R., and Rodrigues, L. R. 2013. Potential therapeutic applications of biosurfactants. *Trends Pharmacol Sci* 34(12):667–75.
10. Edwards, K. R., Lepo, J. E., and Lewis, M. A. 2003. Toxicity comparison of biosurfactants and synthetic surfactants used in oil spill remediation to two estuarine species. *Mar Pollut Bull* 46(10):1309–16.
11. Martins, P. C., and Martins, V. G. 2018. Biosurfactant production from industrial wastes with potential remove of insoluble paint. *Int Biodeterior Biodegrad* 127:10–6.
12. Unás, J. H., de Alexandria Santos, D., Azevedo, E. B., and Nitschke, M. 2018. *Brevibacterium luteolum* biosurfactant: Production and structural characterization. *Biocatal Agric Biotechnol* 13:160–7.
13. Rosenberg, E., and Ron, E. Z. 1999. High- and low-molecular-mass microbial surfactants. *Appl Microbiol Biotechnol* 52(2):154–62.
14. Healy, M. G., Devine, C. M., and Murphy, R. 1996. Microbial production of biosurfactants. *Resour Conserv Recycl* 18(1–4):41–57.
15. Nitschke, M.,and Coasta, S. G. V. A. O. 2007. Biosurfactants in food industry. *Trends Food Sci Technol* 18(5):252–59.
16. Varjani, S. J., and Upasani, V. N. 2017. Critical review on biosurfactant analysis, purification and characterization using rhamnolipid as a model biosurfactant. *Bioresour Technol* 232:389–97.
17. Cameotra, S. S., and Makkar, R. S. 2004. Recent applications of biosurfactants as biological and immunological molecules. *Curr Opin Microbiol* 7(3):262–6.
18. Benvegnu, T., Lemiègre, L., Ballet, C., Portier, Y., and Plusquellec, D. 2014. Glycolipid-based nanosystems for the delivery of drugs, genes and vaccine adjuvant applications. *Carbohydr Chem* 40:341–77.
19. Baccile, N., Cuvier, A. S., Prévost, S., et al. 2016. Self-assembly mechanism of pH-responsive glycolipids: Micelles, fibers, vesicles, and bilayers. *Langmuir* 32(42):10881–94.
20. Baccile, N., Selmane, M., Griel, P. L., et al. 2016. pH-driven self-assembly of acidic microbial glycolipids. *Langmuir* 32(25):6343–59.
21. Kaname, Y., Hiroyuki, M., Shoko, K.,Toshimi, S., and Seiji, I. 2007. Formation of self-assembled glycolipid nanotubes with bilayer sheets. *J Nanosci Nanotechnol* 7(3):960–4.
22. Cortés-sánchez, A. D. J., Hernández-sánchez, H., and Jaramillo-flores, M. E. 2013. Biological activity of glycolipids produced by microorganisms: New trends and possible therapeutic alternatives. *Microbiol Res* 168(1):22–32.
23. Kitamoto, D., Morita, T., Fukuoka, T., Konishi, M., and Imura, T. 2009. Self-assembling properties of glycolipid biosurfactants and their potential applications. *Curr Opin Colloid Interface Sci* 14(5):315–28.
24. Geys, R., Soetaert, W., and Van Bogaert, I. 2014. Biotechnological opportunities in biosurfactant production. *Curr Opin Biotechnol* 30:66–72.
25. Bognolo, G. 1999. Biosurfactants as emulsifying agents for hydrocarbons. *Colloids Surf A Physicochem Eng Asp* 152(1–2):41–52.
26. Ron, E. Z., and Rosenberg, E. 2002. Biosurfactants and oil bioremediation. *Environ Biotechnol* 13(3):249–52.
27. Sekhon, K. K., and Rahman, P. K. S. M. 2014. Rhamnolipid biosurfactants – Past, present and future scenario of global market. *Front Microbiol* 5:1–7.
28. Nguyen, T. T. L., Edelen, A., Neighbors, B., and Sabatini, D. A. 2010. Biocompatible lecithin-based microemulsions with rhamnolipid and sophorolipid biosurfactants: Formulation and potential applications. *J Colloid Interface Sci* 348(2):498–504.
29. Hubert, J., Plé, K., Hamzaoui, M., et al. 2012. New perspectives for microbial glycolipid fractionation and purification processes. *C R Chim* 15(1):18–28.
30. Marques, A. M., Pinazo, A., Farfan, M., et al. 2009. The physicochemical properties and chemical composition of trehalose lipids produced by *Rhodococcus erythropolis* 51T7. *Chem Phys Lipids* 158(2):110–7.

31. Rangarajan, V., Sen, R., Gudina, E. J., and Rodrigues, L. R. 2013. Potential therapeutic applications of biosurfactants. *Trends Pharmacol Sci* 34(12):667–75.
32. Kitamoto, D., Isoda, H., and Nakahara, T. 2002. Functions and potential applications of glycolipid biosurfactants – From energy-saving materials to gene delivery carriers. *J Biosci Bioeng* 94(3):187–201.
33. Abdel-Mawgoud, A. M., Lépine, F., and Déziel, E. 2010. Rhamnolipids: Diversity of structures, microbial origins and roles. *Appl Microbiol Biotechnol* 86(5):1323–36.
34. Soberón-Chávez, G., and Maier, R. M. 2011. Biosurfactants: A general overview. *Microbiol Monogr* 20:1–11.
35. Ron, E. Z., and Rosenberg, E. 2001. Minireview: Natural roles of biosurfactants. *Environ Microbiol* 3(4):229–36.
36. Van Bogaert, I. N. A., Saerens, K., and Vandamme, E. J. 2007. Microbial production and application of sophorolipids. *Appl Microbiol Biotechnol* 76(1):23–34.
37. Lang, S., and Philp, J. C. 1998. Surface-active lipids in rhodococci. *Antonie Leeuwenhoek* 74(1–3):59–70.
38. Morita, T., Konishi, M., Fukuoka, T., Imura, T., and Kitamoto, D. 2006. Analysis of expressed sequence tags from the anamorphic basidiomycetous yeast, *Pseudozyma antarctica*, which produces glycolipid biosurfactants, mannosylerythritol lipids. *Yeast* 23(9):661–71.
39. Fleurackers, S. J. J., Van Bogaert, I. N. A., and Develter, D. 2010. On the production and identification of medium-chained sophorolipids. *Eur J Lipid Sci Technol* 112(6):655–62.
40. Deziel, E., Lepine, F., Dennie, D., Boismenu, D., Mamer, O. A., and Villemur, R. 1999. Liquid chromatography/mass spectrometry analysis of mixtures of rhamnolipids produced by *Pseudomonas aeruginosa* strain 57RP grown on mannitol or naphthalene. *Biochim Biophys Acta* 1440(2–3):244–52.
41. Schenk, T., Schuphan, I., and Schmidt, B. 1995. High-performance liquid chromatographic determination of the rhamnolipids produced by *Pseudomonas aeruginosa*. *J Chromatogr A* 693(1):7–13.
42. Bryant, F. O. 1990. Improved method for isolation of biosurfactant glycolipids from *Rhodococcus* sp. strain H13A. *Appl Environ Microbiol* 56(5):1494–6.
43. Uchida, Y., Tsuchiya, R., Chino, M., Hirano, J., and Tabuchi, T. 1989. Extracellular accumulation of mono-succinoyl and di-succinoyl trehalose lipids by a strain of *Rhodococcus erythropolis* grown on n-alkanes. *Agric Biol Chem* 53:757–63.
44. Häussler, S., Nimtz, M., Domke, T., Wray, V., and Steinmetz, I. 1998. Purification and characterization of a cytotoxic exolipid of *Burkholderia pseudomallei*. *Infect Immun* 66(4):1588–93.
45. Helvaci, S. S., Peker, S., and Ozdemir, G. 2004. Effect of electrolytes on the surface behavior of rhamnolipids R1 and R2. *Colloids Surf B Biointerfaces* 35(3–4):225–33.
46. Hu, Y. M., and Ju, L. K. 2001. Purification of lactonic sophorolipids by crystallization. *J Biotechnol* 87(3):263–72.
47. Ashby, R. D., and Solaiman, D. 2010. The influence of increasing media methanol concentration on sophorolipid biosynthesis from glycerol-based feed stocks. *Biotechnol Lett* 32(10):1429–37.
48. Cavalero, D. A., and Cooper, D. G. 2003. The effect of medium composition on the structure and physical state of sophorolipids produced by *Candida bombicola* ATCC 22214. *J Biotechnol* 103(1):31–41.
49. Trummler, K., Effenberger, F., and Syldatk, C. 2003. An integrated microbial/enzymatic process for production of rhamnolipids and L-(+)-rhamnose from rapeseed oil with *Pseudomonas* sp DSM 2874. *Eur J Lipid Sci Technol* 105(10):563–71.
50. Haba, E., Pinazo, A., Jauregui, O., Espuny, M. J., Infante, M. R., and Manresa, A. 2003. Physicochemical characterization and antimicrobial properties of rhamnolipids produced by *Pseudomonas aeruginosa* 47T2 NCBIM 40044. *Biotechnol Bioeng* 81(3):316–22.
51. Abalos, A., Pinazo, A., Infante, M. R., Casals, M., Garcia, F., and Manresa, A. 2001. Physicochemical and antimicrobial properties of new rhamnolipids produced by *Pseudomonas aeruginosa* AT10 from soybean oil refinery wastes. *Langmuir* 17(5):1367–71.
52. Kuyukina, M. S., Ivshina, I. B., Philp, J. C., Christofi, N., Dunbar, S. A., and Ritchkova, M. I. 2001. Recovery of *Rhodococcus biosurfactants* using methyl tertiary-butyl ether extraction. *J Microbiol Methods* 46(2):149–56.
53. Rau, U., Hammen, S., Heckmann, R., Wray, V., and Lang, S. 2001. Sophorolipids: A source for novel compounds. *Ind Crops Prod* 13(2):85–92.
54. Rau, U., Nguyen, L. A., Roeper, H., Koch, H., and Lang, S. 2005. Downstream processing of mannosylerythritol lipids produced by *Pseudozyma aphidis*. *Eur J Lipid Sci Technol* 107(6):373–80.
55. Lin, S. C. 1996. Biosurfactants: Recent advances. *J Chem Technol Biotechnol* 66(2):109–20.
56. Davila, A. M., Marchal, R., Monin, N., and Vandecasteele, J. P. 1993. Identification and determination of individual sophorolipids in fermentation products by gradient elution high-performance liquid chromatography with evaporative light-scattering detection. *J Chromatogr* 648(1):139–49.
57. Sim, L., Ward, O. P., and Li, Z. Y. 1997. Production and characterisation of a biosurfactant isolated from *Pseudomonas aeruginosa* UW-1. *J Ind Microbiol Biotechnol* 19(4): 232–38.
58. Fukuoka, T., Morita, T., Konishi, M., Imura, T., and Kitamoto, D. 2007. Characterization of new glycolipid biosurfactants, tri-acylated mannosylerythritol lipids, produced by Pseudozyma yeasts. *Biotechnol Lett* 29(7):1111–8.
59. Smyth, T. J. P., Perfumo, A., Marchant, R., and Banat, I. M. 2010. Isolation and analysis of low molecular weight microbial glycolipids. In: K. N. Timmis (Ed.), *Handbook of Hydrocarbon and Lipid Microbiology* (pp. 3705–23). Springer-Verlag, Berlin Heidelberg.
60. Monteiro, S. A., Sassaki, G. L., de Souza, L. M., et al. 2007. Molecular and structural characterization of the biosurfactant produced by Pseudomonas aeruginosa DAUPE 614. *Chem Phys Lipids* 147(1):1–13.
61. Heyd, M., Kohnert, A., Tan, T. H., et al. 2008. Development and trends of biosurfactant analysis and purification using rhamnolipids as an example. *Anal Bioanal Chem* 391(5):579–90.
62. Berthod, A. 2002. Counter current chromatography, the support-free liquid stationary phase. In: D. Barcelo (Ed.), *Comprehensive Analytical Chemistry*, Vol. XXXVIII, Elsevier, Amsterdam.
63. Inès, M., and Dhouha, G. 2015. Glycolipid biosurfactants: Potential related biomedical and biotechnological applications. *Carbohydr Res* 416:59–69.

64. Yamane, T. 1987. Enzyme technology for the lipid industry. An engineering overview. *J Am Oil Chem Soc* 64(12):1657–62.
65. Bharali, P., Saikia, J. P., Paul, S., and Konwar, B. K. 2013. Colloidal silver nanoparticles/rhamnolipid (SNPRL) composite as novel chemotactic antibacterial agent. *Int J Biol Macromol* 61:238–42.
66. Kanwar, R., Gradzielski, M., and Mehta, S. K. 2018. Biomimetic solid lipid nanoparticles of sophorolipids designed for antileprosy drugs. *J Phys Chem B* 122(26):6837–45.
67. Kanwar, R., Gradzielski, M., Prevost, S., Appavou, M. S., and Mehta, S. K. 2019. Experimental validation of biocompatible nanostructured lipid carriers of sophorolipid: Optimization, characterization and in-vitro evaluation. *Colloids Surf B Biointerfaces.* (doi: 10.1016/j.colsurfb.2019.06.036).

14

Insight into Covalent/Non-Covalent Functionalization of Silica Nanoparticles for Neurotherapeutic and Neurodiagnostic Agents

Anup K. Srivastava, Babita Kaundal, Garima Khanna, Subhasree Roy Choudhury, and Surajit Karmakar

CONTENTS

14.1 Introduction

Silicon (Si) is the second most abundant element on the earth crust exist as silicates. Silicon oxides is explored later and undergoes significant innovations in the context of their applications.[1] These silicon compounds can be tuned for flexible diameter, desirable surface functionality, mechanical strength, compatibility, and stability by modulating the synthesis route.[2] The crystalline structural arrangement categorizes silicon compounds into crystalline and amorphous forms. Solid crystalline α-quartz, β-quartz, cristobalite, and tridymite display defined lattice arrangement, whereas amorphous silica is devoid of any lattice arrangements.[3] The last decade evidences the enormous progresses made with the novel route for the synthesis of silica nanospheres (SNPs) and mesoporous silica nanoparticles (MSNs).[4] For the first time, in 2001, the MSNs of MCM-41 type silica materials were used for drug delivery application following a vast innovation in synthesis turned its use in the progressive areas of nanotechnology and nanomedicine.[5, 6] Multi-functional silica nanoparticles fabricated using bottom-up synthesis chemistry, endows a flexible physicochemical characteristics which makes them suitable for smart and superior drug delivery vehicles and bioimaging agents. MSNs with porous structures can condense a higher content of bioactive agents and precisely deliver them at the disease site with minimum nonspecific side effects. MSNs with large-diameter (>100 nm) pore size are categorized as microporous, the ones with pore-size diameter in the range of 2 to 100 nm as mesoporous, whereas, when the size is <2 nm, they are classified as nanoporous nanomaterials.[7] These porous MSNs of desirable size can carry a massive load of the cargo with advantages of controlled and long-term release due to steric hindrance caused by condensation inside the pores.[8] Beyond the specific surface area and pore volume, the surface of the MSNs is crucial and enabling the material for biotargeting or stimuli-responsiveness. Critical covalent and noncovalent chemical interactions provide the specific functionality to the material surface which determines the biological fate in drug delivery applications.[9] This functionalization makes the material disease-specific, and the inbuilt stealth characteristics enabling these materials to resident for longer time in the bloodstream. In addition to the surface functionality other properties like shape, surface architecture and diameter of the nanoparticles determine the enhanced permeability and retention (EPR) at disease site.[10] Particles with a diameter under 200 nm in size can easily internalize in cancer cells, whereas the particles with a diameter under 15 nm can only penetrate the blood–brain barrier (BBB) and localized in the brain.[11] These particles in sub-10 nm diameter can easily permeate the kidneys and have very low systemic resident time.(Figure 14.1).

Neurotherapeutics and neurodiagnostics are two highly progressive and challenging areas in clinical health care. The increasing incidence of neurological diseases including Alzheimer's, Parkinson's, Huntington, stroke, traumatic brain injury (TBI), and brain cancer are socially and economically impacting people's livelihoods, and giving rise to higher morbidity and mortality.[12–15] Despite the scientific advancements, some therapeutic and diagnostic platforms are still facing challenges due to invasiveness or undesirable efficacy. The BBB is a major obstacle in the retention of systematically administered therapeutics; however, major advancements in the fabrication of smart or targeted therapeutic options means they can either penetrate

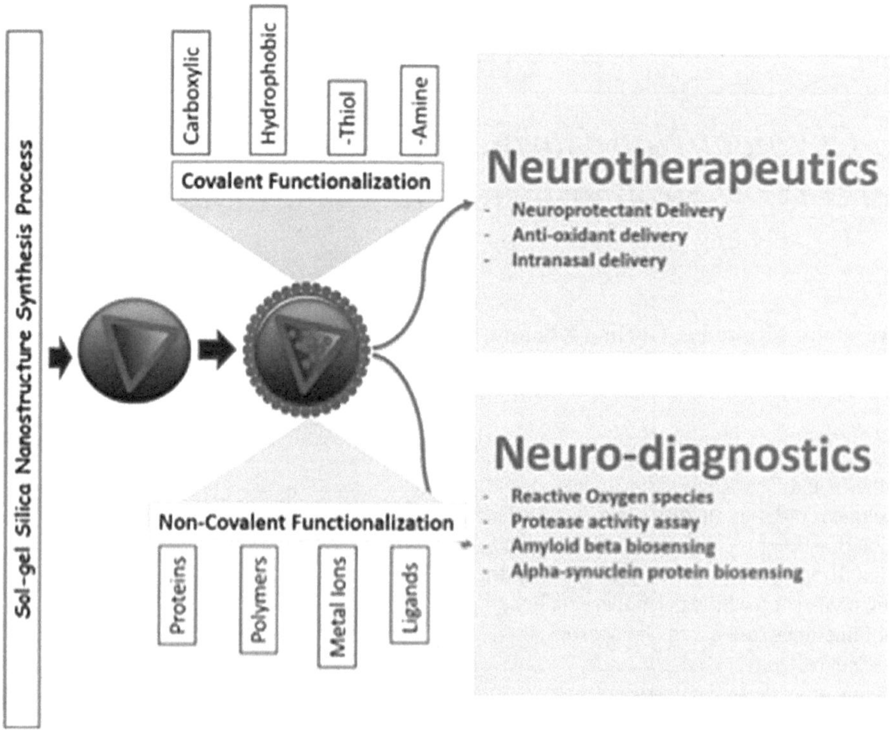

FIGURE 14.1 The general representation of the steps involve the sol-gel–based synthesis of the silica nanostructures. The covalent and non-covalent functionalization generates a particular functionality on the surface of silica nanostructures. The specific biomedical application of the fabricated silica nanostructures in neurotherapeutics and neurodiagnostic applications.

or bypass this morphological barrier. The BBB comprises a multicellular interface of endothelial cells, pericytes, astrocytes, and neuronal cells which separate the peripheral blood capillary side form brain tissue. Functionally, the BBB allows the translocation of highly selective molecules into the brain at physiologically relevant concentration and rejects around 98 % of the neuroactive compounds due to the distinct morphology and catalytic characteristics of endothelial cells, forming a tight junction to minimize the endocytic activity.[16] In the last two decades, nanoengineered delivery vehicles serving as BBB shuttles have been developed. They have enormous potential to circumvent BBB without affecting the integrity of the barrier.[17] The surface functionalization of drug delivery nanomaterials with chimeric proteins, cationized albumin, IgG, apolipoprotein A and E,[18] receptor-associated protein (RAP),[19] transferrin (Tf), leptin, and lactoferrin[20, 21] displayed superior BBB penetration potential.[22] Irrespective of this procedure, MSNs possess the ability to directly penetrate the BBB without any of BBB translocation signal. This characteristic of MSNs enables them to have enormous potential in neuroprotectant deliveries, bioimaging and diagnostics applications in brain disease. Thus, these silica-based materials inherit a great potential of post synthesis functionalization enables it for superior multimodal therapeutic and diagnostic applications. (Table 14.1)

The present review summarizes the current progress in collective fabrication, including synthesis route, post-synthesis processing, and covalent and non-covalent chemistry-based surface functionalization of the silica nanostructures. A special efforts has been made on elaborating the fine chemistry of the surface functionalization which enables the material for superior neurotherapeutic and neurodiagnostic applications in brain injuries associated with Alzheimer's and Parkinson's disease as well as brain stroke. The particular surface functionalization accommodates multifunction aspects in the single materials are discussed in the drug delivery and diagnostics part. The silica based nanomaterials are emerging as a robust candidate and owe a great potential for innovation further in area of neurotherapeutic and neurodiagnostic applications. (Figure 14.2).

14.2 Common Synthesis and Characterization

The spherical shape of the silica nanostructures are always favoured for easy surface functionalization and target specificity in biomedical applications. Two well-established procedures, namely Stöber synthesis and reverse microemulsion method, are being followed for the preparation of silica nanoparticles. The isotropic and thermodynamically controlled water-in-oil (w/o) microemulsion composed of a single-phase ionic/nonionic surfactant, oil, and water as homogeneous mixture serves as nanoreactor for oriented particle growth in the synthesis.[23] The negatively charged silica precursor interacts with the foreign molecules via electrostatic interactions, helping to build the desirable characteristics in the prepared material. This method is preferred to obtain highly uniform and monodispersed SNPs, as it overcome the associated limitations such as extracellular leaching of the loaded molecules and the toxicity pertain by residual surfactants present in the prepared silica nanomaterials.

TABLE 14.1

Selected Silica Nanomaterials, Surface Functionalization, and Respective Neurotherapeutic and Neurodiagnostic Applications

Nanomaterial	Functionalization	Application	Reference
Mesoporous silica nanomaterials (MSNs)	Functionalized analogs, AP-T, aminopropyl groups, and MP-T, mercaptopropyl groups	Human neuroblastoma	68
Silica nanoparticles (SNPs)	MicroRNA-34a using anti-disialoganglioside GD2	Neuroblastoma tumor growth by targeted delivery	69
MSNs	Folic acid (MSM-FOL)	Antineoplastic therapy	4
Silica shells	Fluorescein isothiocyanate (FITC), cationic polyelectrolyte, polyethyleneimine (PEI)	Fluorescent magnetic bio probes	70
SNPs	--	Alzheimer's disease	71
Silica nanopills	--	Neuroblastoma	72
MSNs	Gold nanorods	Cervical cancer cells, fibroblast cells, human umbilical vein endothelial cells, and neuroblastoma cells	73
Silica-based hybrid materials	Polyethylene glycol (PEG)	Neuroblastoma SH-SY5Y, glioma U25	43
MSNs	Curcumin and chrysin	Nose-to-brain olfactory drug delivery	47
SNPs	PEI	Murine neuroblastoma	74
Silica hybrid materials	Bioactive chlorogenic acid	Neuroblastoma	42
SNPs	--	Neuroblastoma	75
MSNs	PAMAM dendrimer	Neuroblastoma	41
MSNs	Polydopamine and graphene oxide layers	Drug delivery	51
MSNs	PEI	Gene delivery Neuro-2 A cells	37

On the other hand, the Stober method is preferred to obtain a highly monodispersed and uniform solid and core-shell silica nanospheres. The silicates (siloxanes), as a precursor of the reaction, undergo kinetically controlled hydrolysis, and the condensation reaction catalyzes by ammonium hydroxide in a variable water and ethanol solvent mixture.[24] The individual

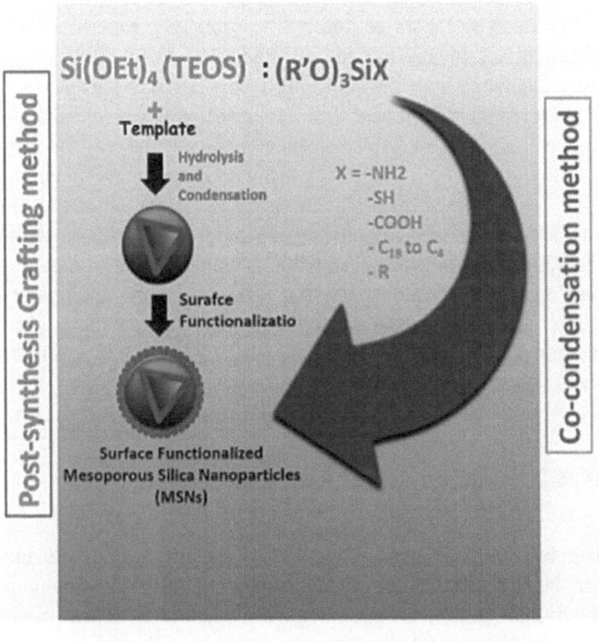

FIGURE 14.2 The outline of the covalent functionalization of the silica nanostructures by following the post-synthesis surface functionalization method and alternative co-condensation of the hydrolyzed TEOS and terminal functionality containing the silane group in a particular ratio.

components and their ratio play a critical role in the determination of particle diameter and dispersity. This method is used to obtain desirable molecules on the surface or as the core of the particle. The covalent surface functionalization of silica nanomaterials are achieved by co-condensing the desirable function group-containing siloxane with TEOS or by substantially adding this substituted siloxane into the reaction mixture. The small molecule or fluorophore covalently attached with 3-aminopropyltriethoxysilane (APTES) and (3-Mercaptopropyl) trimethoxysilane (MPTMS) are substituted siloxane preferably used in the synthesis process.[25] The magnetic silica nanoparticles are obtained by using magnetite (Fe_3O_4) nanoparticle as a seed triggers the hydrolysis and fast condensation of the silica precursor. These magnetic silica nanoparticles appear as core-shell structure incorporates multiple functionalities on the surface for both targeting and enhanced biocompatibility with stable magnetic effect. The primary focus of the silica research centers on the innovation in the synthesis methodology with desirable surface functionality. In order to achieve this, soft and hard templating methods are applied to get hollow and porous silica nanoparticles. Since the first synthesis demonstrated the use of cationic surfactant, cetyltrimethylammonium bromide (CTAB) templated the condensation of the hydrolyzed siloxane or organosiloxane precursor in the ethanol/water mixture or in buffer solutions. Use of non-ionic surfactant such as pluronic-F123 and a mixture of CTAB and pluronic-F123 are as soft template for preparing a highly biocompatible silica nanomaterials, which has significantly lower toxicity. Highly stable silica nanostructures are obtained by hard templating of carbon nanomaterials following calcination at high-temperature. Notably, all the templating methods produce hollow and porous silica nanostructures categorized as microporous, mesoporous, and nanoporous based on the mean pore size distribution pattern. Porous silica material with uniform hexagonal/cylindrical pore arrangement (1.5–10 nm), higher pore volume (>0.6 cm^2g^{-1}), and exceptional physical/chemisorption capacity due to larger specific surface area (~700–1500 m^2g^{-1}) was termed mobil composition of material number 41 (MCM-41). The efforts made to further increase the pore size distribution up to 30 nm categorized this as Santa Barbara amorphous type material (SBA-15). One common characteristic of the above-mentioned porous materials is their three-dimensional hexagonal pore structures. In addition, the cubic pore structure pattern was obtained in MCM-48 and MCM-50.

The prepared silica nanostructures are characterized for morphological and structural features, which includes measurement of mean hydrodynamic diameter (dh), polydispersity index (PDI) and surface zeta potential, which estimate mean diameter, uniformity in size and surface charge of the nanoparticles. Scanning electron microscopy (SEM), transmission electron microscopy (TEM), and atomic force microscopy (AFM) are use to image the morphology, shape, size, and surfaces of the nanomaterials. High-resolution TEM provides the crystalline parameter, pore diameter, and distribution pattern. X-ray diffraction (XRD) study is performed to estimate the crystalline feature of the silica nanomaterials. The more crucial information of surface functionality and the interaction including covalent, non-covalent, and ionic holding these functional groups on the surface are easily investigated by Fourier-transformed infrared spectroscopy (FT-IR), X-ray photoelectron spectroscopy (XPS), and nuclear magnetic resonance (NMR) measurements.

14.3 Covalent Attachment of Functionalities to Silica Surface

14.3.1 Functionalization by Co-Condensation Method

The one-pot condensation method is one of the facile routes for obtaining readily functionalized silica nanoparticles. In this procedure, TEOS precursor (TEOS, TMOS) was mixed with desirable terminal functional group or organic chain (R'O)3SiR containing organosilane in a optimized ratio and condensed in the presence of the structure directing cationic surfactant. The obtained solid silica or MSNs display considerable purity and require a robust analysis to confirm the incorporation and distribution of the desirable functionality on the material. One of the greatest benefits of the condensation method is the uniform functionalization of the material, including the inner wall and outer surface as well as the pores of MSNs. The functional assay demonstrates that at precise ratio and condensation conditions, the functionality does not affect the pore size distribution and specific surface area of the prepared materials. It is worth considering the consequences of the improper ratio, or huge terminal trialkoxysilane groups leading to a more amorphous-like random arrangement of the crystal lattice. It severely affects and changes the hexagonal/cubic pore structure to unsymmetrical structures. In general, the molar ratio of the co-precursor should be 10–40% of TEOS/TMOS, but a slight increase in the molar concentration of the co-precursor proceeds with fast hydrolysis and self-condensation instead of crosslinking with the hydrolyzed TEOS. The major disadvantage of this method, even when the ratio of precursor and co-precursors are precisely optimized, is that the post-synthesis processing leaches out the templated surfactant. Template removal is favourably performed by dialysis or ultrasonication with the solvent extraction method. Instead, the calcination step in the original process can end with undesirable or distorted materials. Several procedures have been demonstrated to attach the amine functionality on the silica surface, resulting in the significant basicity of the materials. The facile reaction process inducing dihydroxylation of the silanol group in the presence of ammonia at elevated temperature drastically undergoes a displacement reaction, replacing the hydroxyl group with amine functionality. However, the significant enhancement in the overall nitrogen content and increased basicity confirm the functionalization process on mesoporous silica materials (SBA-15 and FSM-16). These reactions produce a new class of silica materials characterized as silica oxynitrides (MSON).[26] Recently, ~50 nm diameter silica nanospheres are prepared by following the reverse microemulsion method were amine functionalized using 3-aminopropyltriethoxysilane (APTES) and N1-(3-trimethoxysilylpropyl) diethylenetriamine (DETAS) in particular ratio. The NMR and fluorescamine interaction assay confirm the correlation between the percentage surface coverage of amine functionality and biological response in the context of macrophages (RAW264.7) cell death and lung inflammation. The result revealed that only a displacement of 10 % overall silanol group (–Si-OH) with amine (–Si-NH2) is sufficient to completely overcome the associated toxicity on in-vitro and in-vivo models. The relevance and the interactome of the functionalized silica material in biological systems critically determine their compatibility and efficacy.[27] Following identical procedures, the superhydrophobic and hydrophobic silica nanoparticles were prepared by considering variable carbon-containing silanes. The 1:4 ratio of the hexadecyltrimethoxysilane (C_{16}) and TMOS was used as a precursor in the sol-gel method to produce a superhydrophobic materials. In addition, the variable carbon-containing silanes were incubated with TMOS to produce a range of particles with a hydrophobicity gradient.[28–31] The conjugation of phospholipids on hydrophobic MSNs surface enables it for high intensity focused ultrasound guided imaging and therapeutic applications.[32] The thiol functionalization on the silica nanostructures enables the material for further non-covalent interaction with a plethora of biomolecules, and this is implemented in catalytic, bio-separation, diagnostics, and drug delivery applications. Therefore, co-condensation of TEOS and 3-mercaptopropyl trimethoxysilane (MPTMS) produces highly dispersed and nano-sized silica nanoparticles.[5, 33]

14.3.2 Functionalization by the Post-Synthesis Grafting Method

The desirable surface functionality on the pristine silica surface can be achieved by covalently attaching the siloxane-tagged functional group in post-synthesis reactions. Extra mono/multiple layers of homo- or hetero-functional groups including amine, sulfhydryl, carboxyl, alkyl, and methyl acrylate are generally attached. In the post-synthesis functionalization process, the organosilane ($[R'O]_3SiR$) is reacted with the free silanol group present on the surface of silica nanoparticles. The prime benefit of the procedure is that the desirable functionality attached at the place of R with a silane chain can be covalently grafted on the silica nanostructures. The organosilane concentration and the pore diameter of MSNs are critical parameters and improper consideration gives an insufficient surface coverage of functionalized MSNs. Narrowing and complete blocking of the pores in MSNs is one of the major limitation creates major insufficiency in drug delivery applications. These post-coating methodologies provide reactive sites for bioconjugation, and in some cases enhance the colloidal stability of the functionalized nanoparticles. The surface functional groups enable the material for biorecognition and further conjugation with covalent (carbodiimide, disulfide, and succinimidyl ester hydrolysis coupling) or non-covalent (polyethylene glycol and polyethyleneimine) modification.

14.4 Non-Covalent Chemistry for Silica Nanoparticle Functionalization

Nanomedicines is at nanoscale dimension mimic the chemistry naturally persist in biomolecular structures. Nature-inspired biomolecules including protein, carbohydrates, nucleic acids, and others maintain rich chemical functionality and operate based on non-covalent interactions such as van der Waals, steric, hydrogen bonding, hydrophobic, and ionic interactions. In general, chemical reactions inspired from nature is the basis of the bioactive molecules loading, polymer coating, and attachment of the targeting ligands on the nanoparticles surface.[34]

Recently, a new strategy of pH and redox-responsive, hydrophilic/hydrophobic drugs loaded mesoporous silica nanoparticles based drug delivery system are demonstrated which displayed a good colloidal stability and controlled release of loaded drugs.[35] The surface of the drug-loaded MSNs are non-covalently functionalized with a self-assembled copolymer containing pyridine sulfide (PDS), 2-(diisopropylamino) ethyl methacrylate (DPA), and polyethylene glycol (PEG). Substantial bio-responsive changes in the surface charge of fabricated PEG-PDS-MSNs triggers both the targeting and the drug release. Immediately after administration the DPA moiety maintains the negative surface charge of the vehicle at physiological pH. The event at which the acidic pH is encountered, these DPA gets protonated, turns into positive charge and undergoes cellular internalization via endocytic pathways. The accumulated cationic polymer in the early endosome undergoes swelling and further lowers the compartment pH to trigger early endosomal release. The swelling of the cationic polymer promotes the cytoplasmic release of the drugs and confers maximum bioavailability to the bioactive compounds. The sequential and spatiotemporal release of the loaded drug omits several associated limitations such as colloidal stability, nonspecific targeting, and unwanted leakage of the drug. The strategy can potentially targets the multi-drug-resistant disease site with co-loading and simultaneous delivery of the hydrophilic doxorubicin hydrochloride and hydrophobic camptothecin.[28, 36]

14.5 Application of Functionalized Silica Nanoparticles

14.5.1 Precise Neuro-Delivery

The decorated MSNs are demonstrated as ideal nanocarrier system, overcoming many challenges of targeted drug delivery applications. MSNs is major part of the construct play multiple functions, including the protection of the desirable gene while passing through vessels and extracellular spaces, easily undergo endocytosis, triggers early endosomal release in the cells, and directs the gene towards nuclear pore complex. The MSNs based nonviral delivery system is preferred over viral particle-mediated delivery due to higher transfection efficiency, lower immunogenicity and advantages of comparatively fewer preparatory steps. However, conventional delivery systems comprise a mixture of cationic polymers such as polyethyleneimine, polyamidoamine, polyallylamine, or mixture of them in a particular ratio. MSNs are a superior delivery nanovesicle for drugs, nucleic acids, peptides, and proteins with the advantages of safety, biodegradability, and desirable control on the release utilizing flexible chemistry. The superior transfection efficiency of the MSN-based gene delivery system is achieved by functionalizing the MSNs with cationic polymers. The positive charges on MSNs favour a higher concentration of nucleic acid, including DNA, SiRNA, and plasmids inside the porous architecture. For instance, phosphonate-modified MSN (PMSNs) condensed with GFP pDNA and grafted with PEI are fabricated (165–215 nm). The ethidium bromide (EtBr)-based dye exclusion assay confirms the pDNA concentration in the nanoparticles. Co-delivery of a lysosomotropic agent Chloroquine and cationic MSNs enhances transfection efficiency which is significantly higher compared to bare MSNs in neuronal (Neuro 2A) cells.[37] The non-covalent chemistry-based grafting of the PMSN-PEI significantly enhances the transfection efficiency in neuronal cells and proved to be a superior gene delivery vehicle. However, the effect of uncontrolled and prolonged exposure of the silica nanostructures is highly dependent on the dimension and targeting method. The optimal exposure of the silica nanoparticles during brain development at postnatal day P1 to P7 displayed significant inflammation and suppressed the precursor proliferation in the hippocampal dentate gyrus (DG). This adverse effect of the silica nanostructures compromised the social interaction behaviour, inducing an anxious behaviour in adulthood without any visible effect on other motor functions.[38] The organically modified silica (ORMOSIL) nanoparticles can penetrate the living brain and accumulate in neuronal cell bodies and axonal terminals. The assessment in the Drosophila model doesn't display the genotoxicity or effect on neuron death or any interference with the neuronal physiological processes.[39] The nanoporous silica nanoparticles (NPSNPs) of <100 nm, loaded with brain-derived neurotrophic factor (BNDF), display long-term release (up to 80 days), triggering the resilience and survival of spiral ganglion in-vitro.[40]

In another approach, PAMAM dendrimer grafter MSNs are developed following hydrolysis of the TEOS under acid (HCl)-catalyzed reaction. The preparation method is different, with earlier sol-gel and ammonia-catalyzed reaction being carried out at higher pH. The resulting organic-inorganic hybrid nanostructures display satisfactory drug loading and release capacity and lower neuronal cell toxicity.[41] In another approach, the organic-inorganic silica nanostructures constituting phenolic antioxidant chlorogenic acid in the matrices trigger a superior antioxidative defense in the neuronal cells.[42] Similarly, the entrapped quercetin in MSNs with enormous biocompatibility triggers an anti-oxidation effect in PC-12 and SH-SY5Y cells.[43] The curcumin and RhoG-DsRed plasmid non-covalently tagged to TAT peptides are loaded in MSNs shows synergistic effects due to control release of curcumin pertain ROS quenching thus protective effect and collectively reduces the neuronal cell death. On the other hand, it enhances the gene expression associated with lamellipodia and filopodia formation for axonal outgrowth. The collective dual targeting strategy of the neurodegeneration is preferred over monotherapy, which is insufficient to either produce a positive effect or generate tolerance with time.[44]

The response of neuronal cells towards the differently functionalized silica microspheres is demonstrated. The microspheres with $-NH_3$, $-NH_2$, $-COOH$, and $-OH$ groups on the surface are exposed to the primary cortical neuron (PCN) monolayer culture and are also directly injected into the mice brain via intracranial injection. Indeed, the direct injection of the 1, 1.5 and 2 μm diameter microspheres in the striatum doesn't show well distribution pattern and the internalization into the brain macrophages. In addition to the diameter, surface functionality and charges severely affect the cell membrane association as well as cellular internalization. The amino-functionalized (zeta potential; +40 mV) microspheres are preferably internalized in the lesser period of the 24 hr, compared to $-COOH$ and $-OH$ (zeta potential; −70 mV) functionalized microspheres which undergo a lesser degree of association with the cell membrane, and a lesser degree of internalization even within 48 hr. The study revealed that both the size and the surface functionality critically define the preference toward phagocytosis by microglia and neuronal cells, and also determine the overall distribution in the brain.[45] The systemic entry of silica nanostructures is exposed to the barrier (BBB), which is a confined morphological barrier between the peripheral circulation and the central nervous system (CNS). The systemic exposure of SiNPs adversely damage the BBB integrity by disrupting tight junction and associated cytoskeleton, enhances ROS, induce the secretion of vascular endothelial growth factor (VEGF), triggers aquaporin-4 expression on endothelial and astrocytes cells which is leading to paracellular opening of the BBB. Hence, this mild disruption of the BBB allows the silica-based nanomedicine to circumvent BBB and established as the mechanism for targeted brain delivery. [46] The BBB is highly selective for the molecular transportation from peripheral system to the brain, thus it is a major obstacle in nanoparticles delivery to the brain. In order to avoid the drug delivery route of BBB a "nose-to-brain" drug delivery approach is demonstrated by nasal deposition of the therapeutics. The practice started with the first case of intranasal delivery of insulin and neuroprotective compounds to treat Alzheimer's disease in 1997. The first success attracted more research and advanced the studies in this area because this method anatomically bypasses the BBB and delivers the drug molecules directly into the CNS via an olfactory pathway. This method was further integrated with nanotechnology when curcumin- and chrysin-loaded MSNs were shown to have a therapeutic effect after intra-nasal (IN) route administration.[47]

The dopamine enriched silica reservoir is implanted inside mid brain of hemiparkinsonian rats shows slow release of dopamine and progressive improvement in the rotational motion behaviour without any cardinal motor symptoms. The reservoir potentially protected the PD progression as visible in the reversal of behavioural studies, also confirmed in the histopathological assessments.[48] In one of the such experiment the MSN is coated with gold nanoparticles (5 nm) covalently conjugated with α-SYN via C-terminal end of cysteine. The final raspberry-like appearance of the nanostructures covering 75 % of the surface area of MSNs demonstrated a controlled release of doxorubicin in the response to the divalent cation, including calcium ions. The specific and pathologically relevant metal ions interact with surface-coated alpha-synuclein and trigger the cargo release. This platform is utilizing a protein of neurological origin to construct a ligand responsive and highly dynamic drug delivery system.[49]

14.5.2 Biosensing and Neurodiagnostics

The covalent and non-covalent functionalization of the silica nanostructures dramatically changes the interaction and interfacial behavior. On systemic exposure, SNPs encountered serums proteins and formed a continuous multiple layer on the nanostructure surface known as protein nanoparticle corona. Diameter, shape, and surface charge are the physicochemical characteristics determining the surface coverage and content of the protein adsorbed on the surface. These non-covalent interactions dynamically determine the colloidal stability and fate of the SNPs in biological systems. This surface adsorption behavior changes the conformation of the intrinsically disordered proteins and leads to spontaneous aggregation on the SNPs surface. This interfacial behaviour of SNPs and proteins allows functionalized SNPs to probe the pathological transformation in the tau proteins. Tau is an unstructured microtubule-binding protein that provides stability and is responsible for memory consolidation in the brain. The circular dichroism-based measurement of ellipticity in the native tau proteins and tau-SNPs indicated conformational structuring of the random coiled region on the surface of the NPs. The partial folding of the tau protein on the surface of the SNPs marked enhanced hydrophobic regions assessed by enhanced ANS dye fluorescence. The similar concentration of the SNPs causing alterations in the tau conformation induces neuronal toxicity, reactive oxygen species (ROS) generation, apoptotic pathway activation, and neuronal cell death. The biophysical and biological relevance of the SNPs are crucially linked to the interfacial interaction of the biological moieties.[50]

The FITC-labelled APTES conjugate-based covalent functionalization of the MSNs displayed highly stable and fluorescent silica materials. The well-known anticancer drug cisplatin (CPT) mechanistically induces DNA damage in cancerous cells with marked nephrotoxicity as a side effect. CPT, loaded in the fluorescent MSNs, which are further grafted with layers of polydopamine, and graphene oxide dramatically enhance the particle stability, displaying a sustained release of CPT with superior efficacy in neuroblastoma cells. The PDA/GO layers on the MSNs are stabilized by electrostatic interactions among the charged moieties. The carbodiimide-based covalent coupling of EGFR antibody leads to MSNs-CPT@A-F@PDA@EGFRab and MSNs-CPT@A-F@PDA@GO@EGFRab, which can be targeted to the desirable disease site without inducing any significant non-specific toxicity.[51]

In recent years, silica nanostructures are functionalized with multiple tracing agents have significantly advances it to be use for multimodal photosensitizer-based therapy and biosensing applications. Recently, the high-molecular (10 kDa) and low-molecular (3 kDa) weight biotinylated dextran amine (BDA) was covalently conjugated on the thiolated core-shell silica surface. The specific preference of high molecular weight BDA towards axonal terminals and low molecular weight in the neuronal cell bodies makes it a suitable ligand for tracing. In more detail, the fabrication of the SPION@SiO_2(FITC)-BDA was performed via substantial processes of (1) superparamagnetic iron oxide (SPION) core as MRI contrasting agent; (2) Coating of TEOS-FITC and second TEOS-SH layer provides fluorescence characteristics and specificity to the material; (3) covalent binding of the BDA on the surface. The substantial fabrication of the silica on the SPION significantly reduced the associated toxicity without compromising the MRI contrast. The developed core-shell and multimodal axon tracing agent demonstrated the non-invasive, real-time retrograde, and anterograde pathways in neuronal cells. The highly specific, sensitive, and stable system displayed several advantages over the conventional invasive microscopic observations of the neuronal tissues. The development significantly improved with real-time histochemical observations of the axonal pathways.[52]

Amyloid beta (Aβ) as key protein implicated in the most prevalent neurodegenerative Alzheimer's disease (AD).[53] The fragment of Aβ is generated after proteolytic cleavage of the amyloid precursor protein (APP) by β-secretase and γ-secretase.[54] The selectivity of these proteases determines whether Aβ undergoes amyloidogenic or non-amyloidogenic pathway.[55] Thus, the level of these proteases is considered as a therapeutic and diagnostic target in AD.[56] The monitoring of the proteases activity is critical for the assessment of the molecular biomarker of the AD and also serves as a model for screening neuroprotectant or efficacy of anti-AD therapeutics. The poly (methacrylic acid) (PMAA) brushes-like nanostructures were fabricated on an amino-functionalized silica nanoparticle following the surface-initiated atom transfer radical polymerization (ATRP) method. These brushes were covalently linked with fluorescence-tagged cleavable peptides and abbreviated as "SiNPs-g-PMAA-Peptide". Encountering specific proteases, these short peptides undergo enzymatic cleavage, and the release of FITC tags in a solvent was quantitatively estimated. This rationally designed functionalized silica nanostructure-based protease detection system functions in multiple biological fluids with the limit of detection (LOD) of 1.1 pmol. Therefore, silica nanoparticle as a base material was fabricated to develop a highly sensitive peptide-tagged polymer brush-based protease assay method.[57] The Aβ-functionalized silica nanoparticles (Aβ-SiNaPs) were shown to detect a concentration as small as 16 fmol using the surface-based fluorescence intensity distribution analysis (sFIDA) method. This highly robust method sensitively detected Aβ oligomers in the real AD patient samples.[58–60]

The metal ions are known to play a significant role in the pathophysiology of AD and trigger the aggregation of Aβ, generating reactive oxygen species (ROS) and hydrogen peroxide (H_2O_2). Therefore, gold nanoparticle-capped MSN-based H_2O_2 responsive metal chelator CQ delivery significantly reduces the Cu^{2+}-based Aβ aggregations, ameliorating the cellular stress, membrane integrity, microtubular network, and apoptotic pathways activation. MSN, as the base of nanoformulation, displayed consistent BBB penetration irrespective of the gold capping on the surface.[61, 62] Another study with catechin-loaded PEGylated MSNs regulated the generation of ROS and RNS as potent neuroprotectants.[63] The dual monoclonal anti-tau functionalized silica-based hybrid magnetic nanoparticles and polyclonal anti-tau immobilized gold nanoparticle-based sandwich assay displayed a trivial, selective, and ultrasensitive tool to detect the tau protein. The core-shell magnetic silica nanoparticles surface was grafted with poly (2-hydroxyethyl methacrylate) via surface-mediated reversible addition fragmentation chain transfer (RAFT) polymerization, further covalently linked to monoclonal anti-tau antibody, specifically interacts with tau and further collected using magnets. The bio-separated but intact tau protein interacts and forms a sandwich with SERS-active gold nanoparticles. The complete assay comprises two steps, first, recognition and bio-separation; secondly, SERS-based detection. The linear correlation between tau concentration and SERS signal delivered a highly sensitive detection of a minimum of 25 fmol concentration of tau in biofluids.[64]

The graphene oxide quantum dots (GOQDs)-embedded silica molecularly imprinted polymer (SMIP) are developed as a sensitive fluorescent chemical nano-sensor for the detection of entacapone. Entacapone, a selective and reversible inhibitor of the catechol-o-methyl transferase and prescribed as anti-PD therapy, is detected using a developed sensor with a limit of detection of 0.31 umol.[65] A smart gated nanodevice triggering the cargo delivery stimulated by acetylcholinesterase activity was demonstrated for PD regulation. Janus Au-mesoporous silica nanoparticles work as a dual face complex, the Au-face accompanies the acetylcholinesterase, whereas the supramolecular beta cyclodextrin-benzimidazole inclusions interact on the MSNs face. The Au-face containing enzyme degrade the acetylcholine via hydrolysis, resulting in a slight reduction in local pH, which induces the dissociation of supramolecular structures on the MSN face to release the cargos. The proof of concept of the developed nanodevice was demonstrated for PD therapy.[66]

The alpha-synuclein-coated silica nanoparticles (α-SYN-SiNPs)-based standard provides an automated quantitation of oligomeric conformation in the body fluids via the sFIDA method. On the basis of statistical consideration, this method can detect femtomolar α-SYN concentration in the real samples.[67]

Conclusion

The current progress and recent innovations in the area of silica nanostructures fabrication are anticipated to achieve further advancements due to a controlled and facile synthesis process, as well as the ease with which the desirable size, shape, and surface functionality are achieved. In the present systemic review, we have elaborated the conventional and current novel synthesis routes of silica nanoparticles. The distinct sol-gel procedures include soft templating following hydrothermal treatment and calcination, solvent etching methods to get the plethora of the solid silica nanospheres, mesoporous silica nanospheres, and organosilica nanomaterials. We detailed the co-condensation and post-synthesis grafting-based covalent functionalization methodology fabricating the material with an amine, carboxy, thiol, hydrophobic, and organic functionality on the surface. Furthermore, attachment of targeted ligands on the silica nanostructures are heavily rely on non-covalent interactions thereby plays a pivotal role in biomedical applications. Among the plethora of demonstrated applications, we have shown the correlation between surface functionality and biomedical fate in the context of neurotherapeutic delivery and neurodiagnostic modalities.

REFERENCES

1. Liang, Y., et al., Mechanisms of silicon-mediated alleviation of abiotic stresses in higher plants: A review. *Environmental Pollution*, 2007. **147**(2): p. 422–428.
2. Bharti, C., et al., Mesoporous silica nanoparticles in target drug delivery system: A review. *Int J Pharm Investig*, 2015. **5**(3): p. 124–33.
3. Smith, D.K., Opal, cristobalite, and tridymite: Noncrystallinity versus crystallinity, nomenclature of the silica minerals and bibliography. *Powder Diffraction*, 2013. **13**(1): p. 2–19.
4. Ceresa, C., et al., Functionalized mesoporous silica nanoparticles: a possible strategy to target cancer cells reducing peripheral nervous system uptake. *Curr Med Chem*, 2013. **20**(20): p. 2589–600.
5. Zhang, Q., et al., Functionalized Mesoporous Silica Nanoparticles with Mucoadhesive and Sustained Drug Release Properties for Potential Bladder Cancer Therapy. *Langmuir*, 2014. **30**(21): p. 6151–6161.
6. Vallet-Regi, M., et al., A New Property of MCM-41: Drug Delivery System. *Chemistry of Materials*, 2001. **13**(2): p. 308–311.
7. Alothman, Z.A., A Review: Fundamental Aspects of Silicate Mesoporous Materials. *Materials*, 2012. **5**(12): p. 2874–2902.
8. Li, J., et al., Effects of pore size on in vitro and in vivo anticancer efficacies of mesoporous silica nanoparticles. *RSC Advances*, 2018. **8**(43): p. 24633–24640.
9. Matsumoto, A., et al., Surface Functionalization and Stabilization of Mesoporous Silica Spheres by Silanization and Their Adsorption Characteristics. *Langmuir*, 2002. **18**(10): p. 4014–4019.
10. Chen, F., et al., In vivo tumor targeting and image-guided drug delivery with antibody-conjugated, radiolabeled mesoporous silica nanoparticles. *ACS Nano*, 2013. **7**(10): p. 9027–39.
11. Betzer, O., et al., The effect of nanoparticle size on the ability to cross the blood–brain barrier: an in vivo study. *Nanomedicine*, 2017. **12**(13): p. 1533–1546.
12. Sahni, J.K., et al., Neurotherapeutic applications of nanoparticles in Alzheimer's disease. *Journal of Controlled Release*, 2011. **152**(2): p. 208–231.
13. Hall, E.D., R.A. Vaishnav, and A.G. Mustafa, Antioxidant Therapies for Traumatic Brain Injury. *Neurotherapeutics*, 2010. **7**(1): p. 51–61.
14. Dickey, A.S. and A.R. La Spada, Therapy development in Huntington disease: From current strategies to emerging opportunities. *American Journal of medical Genetics. Part A*, 2018. **176**(4): p. 842–861.
15. AlDakheel, A., L.V. Kalia, and A.E. Lang, Pathogenesis-targeted, disease-modifying therapies in Parkinson disease. *Neurotherapeutics: the Journal of the American Society for Experimental NeuroTherapeutics*, 2014. **11**(1): p. 6–23.
16. Bell, R.D. and M.D. Ehlers, Breaching the blood-brain barrier for drug delivery. *Neuron*, 2014. **81**(1): p. 1–3.
17. Dong, X., Current Strategies for Brain Drug Delivery. *Theranostics*, 2018. **8**(6): p. 1481–1493.
18. Kreuter, J., et al., Covalent attachment of apolipoprotein A-I and apolipoprotein B-100 to albumin nanoparticles enables drug transport into the brain. *Journal of Controlled Release*, 2007. **118**(1): p. 54–58.
19. Pan, W., et al., Efficient transfer of receptor-associated protein (RAP) across the blood-brain barrier. *Journal of Cell Science*, 2004. **117**(21): p. 5071–5078.
20. Banks, W.A., Leptin transport across the blood-brain barrier: implications for the cause and treatment of obesity. *Curr Pharm Des*, 2001. **7**(2): p. 125–33.
21. Huang, F.-Y.J., et al., In vitro and in vivo evaluation of lactoferrin-conjugated liposomes as a novel carrier to improve the brain delivery. *International Journal of Molecular Sciences*, 2013. **14**(2): p. 2862–2874.
22. Pardridge, W.M., Receptor-Mediated Peptide Transport through the Blood-Brain Barrier*. *Endocrine Reviews*, 1986. **7**(3): p. 314–330.
23. Sun, B., G. Zhou, and H. Zhang, Synthesis, functionalization, and applications of morphology-controllable silica-based nanostructures: A review. *Progress in Solid State Chemistry*, 2016. **44**(1): p. 1–19.
24. Han, Y., et al., Unraveling the Growth Mechanism of Silica Particles in the Stöber Method: In Situ Seeded Growth Model. *Langmuir*, 2017. **33**(23): p. 5879–5890.
25. Wang, X., et al., Synthesis, Properties, and Applications of Hollow Micro-/Nanostructures. *Chemical Reviews*, 2016. **116**(18): p. 10983–11060.
26. Bendjeriou-Sedjerari, A., et al., A well-defined mesoporous amine silica surface via a selective treatment of SBA-15 with ammonia. *Chem Commun (Camb)*, 2012. **48**(25): p. 3067–3069.
27. Hsiao, I.L., et al., Biocompatibility of Amine-Functionalized Silica Nanoparticles: The Role of Surface Coverage. *Small*, 2019. **15**(10): p. e1805400.
28. Palanikumar, L., et al., Noncovalent Surface Locking of Mesoporous Silica Nanoparticles for Exceptionally High Hydrophobic Drug Loading and Enhanced Colloidal Stability. *Biomacromolecules*, 2015. **16**(9): p. 2701–2714.
29. Baek, S., et al., Effect of Hydrophobic Silica Nanoparticles on the Kinetics of Methane Hydrate Formation in Water-in-Oil Emulsions. *Energy & Fuels*, 2019. **33**(1): p. 523–530.
30. Giasuddin, A.B.M., et al., One-Step Hydrophobic Silica Nanoparticle Synthesis at the Air/Water Interface. *ACS Sustainable Chemistry & Engineering*, 2019. **7**(6): p. 6204–6212.
31. Sriramulu, D., et al., Synthesis and Characterization of Superhydrophobic, Self-cleaning NIR-reflective Silica Nanoparticles. *Scientific Reports*, 2016. **6**: p. 35993.
32. Blum, N.T., et al., Temperature-Responsive Hydrophobic Silica Nanoparticle Ultrasound Contrast Agents Directed by Phospholipid Phase Behavior. *ACS Applied Materials & Interfaces*, 2019. **11**(17): p. 15233–15240.

33. Maria Claesson, E. and A.P. Philipse, Thiol-functionalized silica colloids, grains, and membranes for irreversible adsorption of metal(oxide) nanoparticles. *Colloids and Surfaces A: Physicochemical and Engineering Aspects*, 2007. **297**(1): p. 46–54.
34. Doane, T. and C. Burda, Nanoparticle mediated non-covalent drug delivery. *Advanced drug delivery reviews*, 2013. **65**(5): p. 607–621.
35. Palanikumar, L., et al., Noncovalent Polymer-Gatekeeper in Mesoporous Silica Nanoparticles as a Targeted Drug Delivery Platform. *Advanced Functional Materials*, 2015. **25**(6): p. 957–965.
36. Palanikumar, L., et al., Spatiotemporally and Sequentially-Controlled Drug Release from Polymer Gatekeeper–Hollow Silica Nanoparticles. *Scientific Reports*, 2017. **7**: p. 46540.
37. Zarei, H., et al., Enhanced gene delivery by polyethyleneimine coated mesoporous silica nanoparticles. *Pharm Dev Technol*, 2019. **24**(1): p. 127–132.
38. Fu, J., et al., Silica nanoparticle exposure during the neonatal period impairs hippocampal precursor proliferation and social behavior later in life. *Int J Nanomedicine*, 2018. **13**: p. 3593–3608.
39. Barandeh, F., et al., Organically Modified Silica Nanoparticles Are Biocompatible and Can Be Targeted to Neurons In Vivo. *PLOS ONE*, 2012. **7**(1): p. e29424.
40. Schmidt, N., et al., Long-term delivery of brain-derived neurotrophic factor (BDNF) from nanoporous silica nanoparticles improves the survival of spiral ganglion neurons in vitro. *PLOS ONE*, 2018. **13**(3): p. e0194778.
41. Yesil-Celiktas, O., et al., Synthesis of silica-PAMAM dendrimer nanoparticles as promising carriers in Neuro blastoma cells. *Anal Biochem*, 2017. **519**: p. 1–7.
42. Catauro, M. and S. Pacifico, Synthesis of Bioactive Chlorogenic Acid-Silica Hybrid Materials via the Sol-Gel Route and Evaluation of Their Biocompatibility. *Materials (Basel)*, 2017. **10**(7).
43. Catauro, M., et al., Entrapping quercetin in silica/polyethylene glycol hybrid materials: Chemical characterization and biocompatibility. *Mater Sci Eng C Mater Biol Appl*, 2016. **68**: p. 205–212.
44. Cheng, C.S., et al., Codelivery of Plasmid and Curcumin with Mesoporous Silica Nanoparticles for Promoting Neurite Outgrowth. *ACS Appl Mater Interfaces*, 2019. **11**(17): p. 15322–15331.
45. Wallace, V.J., et al., Neurons Internalize Functionalized Micron-Sized Silicon Dioxide Microspheres. *Cell Mol Neurobiol*, 2017. **37**(8): p. 1487–1499.
46. Xing, Y., et al., Mesoporous polydopamine nanoparticles with co-delivery function for overcoming multidrug resistance via synergistic chemo-photothermal therapy. *Nanoscale*, 2017. **9**(25): p. 8781–8790.
47. Lungare, S., K. Hallam, and R.K. Badhan, Phytochemical-loaded mesoporous silica nanoparticles for nose-to-brain olfactory drug delivery. *Int J Pharm*, 2016. **513**(1–2): p. 280–293.
48. Lopez, T., et al., Treatment of Parkinson's disease: nanostructured sol-gel silica-dopamine reservoirs for controlled drug release in the central nervous system. *Int J Nanomedicine*, 2010. **6**: p. 19–31.
49. Lee, D., et al., Ca2+-dependent intracellular drug delivery system developed with "raspberry-type" particles-on-a-particle comprising mesoporous silica core and alpha-synuclein-coated gold nanoparticles. *ACS Nano*, 2014. **8**(9): p. 8887–95.
50. Roshanfekrnahzomi, Z., et al., Silica nanoparticles induce conformational changes of tau protein and oxidative stress and apoptosis in neuroblastoma cell line. *Int J Biol Macromol*, 2019. **124**: p. 1312–1320.
51. Tran, A.V., et al., Targeted and controlled drug delivery by multifunctional mesoporous silica nanoparticles with internal fluorescent conjugates and external polydopamine and graphene oxide layers. *Acta Biomater*, 2018. **74**: p. 397–413.
52. Du, Y., et al., Development of multifunctional nanoparticles towards applications in non-invasive magnetic resonance imaging and axonal tracing. *J Biol Inorg Chem*, 2017. **22**(8): p. 1305–1316.
53. Masters, C.L. and D.J. Selkoe, Biochemistry of amyloid β-protein and amyloid deposits in Alzheimer disease. *Cold Spring Harbor perspectives in medicine*, 2012. **2**(6): p. a006262–a006262.
54. Haass, C., et al., Targeting of cell-surface β-amyloid precursor protein to lysosomes: alternative processing into amyloid-bearing fragments. *Nature*, 1992. **357**(6378): p. 500–503.
55. Citron, M., β-secretase as a target for the treatment of Alzheimer's disease. *Journal of Neuroscience Research*, 2002. **70**(3): p. 373–379.
56. Jung-Eun, C. and K. Jin Ryoun, Recent Approaches Targeting Beta-Amyloid for Therapeutic Intervention of Alzheimer's disease. *Recent Patents on CNS Drug Discovery (Discontinued)*, 2011. **6**(3): p. 222–233.
57. Wu, Y., et al., Versatile Functionalization of Poly(methacrylic acid) Brushes with Series of Proteolytically Cleavable Peptides for Highly Sensitive Protease Assay. *ACS Appl Mater Interfaces*, 2017. **9**(1): p. 127–135.
58. Kravchenko, K., et al., Analysis of anticoagulants for blood-based quantitation of amyloid beta oligomers in the sFIDA assay. *Biol Chem*, 2017. **398**(4): p. 465–475.
59. Herrmann, Y., et al., sFIDA automation yields sub-femtomolar limit of detection for Abeta aggregates in body fluids. *Clin Biochem*, 2017. **50**(4–5): p. 244–247.
60. Hulsemann, M., et al., Biofunctionalized Silica Nanoparticles: Standards in Amyloid-beta Oligomer-Based Diagnosis of Alzheimer's Disease. *J Alzheimers Dis*, 2016. **54**(1): p. 79–88.
61. Yang, L., et al., Gold nanoparticle-capped mesoporous silica-based H2O2-responsive controlled release system for Alzheimer's disease treatment. *Acta Biomater*, 2016. **46**: p. 177–190.
62. Geng, J., et al., Mesoporous silica nanoparticle-based H2O2 responsive controlled-release system used for Alzheimer's disease treatment. *Adv Healthc Mater*, 2012. **1**(3): p. 332–6.
63. Halevas, E., C.M. Nday, and A. Salifoglou, Hybrid catechin silica nanoparticle influence on Cu(II) toxicity and morphological lesions in primary neuronal cells. *J Inorg Biochem*, 2016. **163**: p. 240–249.
64. Zengin, A., U. Tamer, and T. Caykara, A SERS-based sandwich assay for ultrasensitive and selective detection of Alzheimer's tau protein. *Biomacromolecules*, 2013. **14**(9): p. 3001–9.
65. Ahmadi, H., F. Faridbod, and M. Mehrzad-Samarin, Entacapone detection by a GOQDs-molecularly imprinted silica fluorescent chemical nanosensor. *Anal Bioanal Chem*, 2019. **411**(5): p. 1075–1084.
66. Llopis-Lorente, A., et al., Enzyme-Controlled Nanodevice for Acetylcholine-Triggered Cargo Delivery Based on Janus Au-Mesoporous Silica Nanoparticles. *Chemistry*, 2017. **23**(18): p. 4276–4281.

67. Herrmann, Y., et al., Nanoparticle standards for immuno-based quantitation of alpha-synuclein oligomers in diagnostics of Parkinson's disease and other synucleinopathies. *Clin Chim Acta*, 2017. **466**: p. 152–159.
68. Di Pasqua, A.J., et al., Cytotoxicity of mesoporous silica nanomaterials. *J Inorg Biochem*, 2008. **102**(7): p. 1416–23.
69. Tivnan, A., et al., Inhibition of neuroblastoma tumor growth by targeted delivery of microRNA-34a using anti-disialoganglioside GD2 coated nanoparticles. *PLoS One*, 2012. **7**(5): p. e38129.
70. Pinheiro, P.C., et al., Fluorescent Magnetic Bioprobes by Surface Modification of Magnetite Nanoparticles. *Materials (Basel)*, 2013. **6**(8): p. 3213–3225.
71. Yang, X., et al., Uptake of silica nanoparticles: neurotoxicity and Alzheimer-like pathology in human SK-N-SH and mouse neuro2a neuroblastoma cells. *Toxicol Lett*, 2014. **229**(1): p. 240–9.
72. Alba, M., et al., Silica Nanopills for Targeted Anticancer Drug Delivery. *Small*, 2015. **11**(36): p. 4626–31.
73. Das, M., D.K. Yi, and S.S. An, Analyses of protein corona on bare and silica-coated gold nanorods against four mammalian cells. *Int J Nanomedicine*, 2015. **10**: p. 1521–45.
74. Babaei, M., et al., Promising gene delivery system based on polyethylenimine-modified silica nanoparticles. *Cancer Gene Ther*, 2017. **24**(4): p. 156–164.
75. Yang, Y., et al., Silica nanoparticles induced intrinsic apoptosis in neuroblastoma SH-SY5Y cells via CytC/Apaf-1 pathway. *Environ Toxicol Pharmacol*, 2017. **52**: p. 161–169.

15

Fabrication and Functionalization of Ionic Liquids

Neha Jindal and Kulvinder Singh

CONTENTS

15.1 Introduction

Recently, ionic liquids have gained excellent attention in the field of novel material investigation as innovative fluids. As the name suggests, these materials are ionic in nature and may or may not be liquid depending on their melting point. Generally, these compounds are composed of ions that melt below 100 °C. Ionic liquids are usually molten salts or molten oxides. A typical ionic liquid that possesses stability in both air and water is composed of nonmetallic cations, i.e., N-alkylpyridinium N and N'-dialkylimidazolium along with different anions such as thrifluroactate, tetrachloroaluminate, triflate, chloride, bromide, iodide, nitrate, hexaflurophosphate, tetrafluoroborate, etc. Due to large variation in size of heavier cations and small anions, the packing of the crystal lattice is not uniform; it is, instead, random, and is also present in various inorganic salts. For this reason, these salts appear to be liquid at room temperature (Wei and Ivaska, 2008). These ionic liquids are also called "designer solvents" because of their capability to reveal different physiochemical properties, i.e., density, hydrophobicity, viscosity, polarity, solvation, melting point, etc. reliant on their cation and anion along with the alkyl functionalization of cation (Plechkova and Seddon, 2008; Plechkova et al., 2010; Sun et al., 2009). It was in 1914 when Paul Walden first synthesized an ionic liquid, i.e., ethyl ammonium nitrate, by neutralizing ethyl amine with concentrated nitric acid with a melting point of 12 °C. Who knew that this class of salt would become a major contributor to the scientific community after almost one century. This ionic liquid possesses properties almost similar to those of water, i.e., it is colorless and odorless with exceptionally high viscosity. In terms of chemical properties, its ionic conductance is almost similar to that of cations and anions (Wang et al., 2017). After almost forty years, the second generation of ionic liquids was synthesized by mixing alkyl pyridinium chlorides with aluminum trichloride, but these ionic liquids possessed a major drawback concerning their stability in moisture as well as weaker control over tailoring the acidity and basicity (Hurley and WIer, 1951a; Fedorov and Kornyshev, 2014). A further step was made in the 70s, using chloroaluminate as anion and trialkyl ammonium as cation and leading to the first ionic liquid with a melting point below room temperature, as reported by Osteryoung and his coworkers (Chum et al., 1975; Robinson and Osteryoung, 1979). C.L. Hussey, one of the pioneers in ionic liquids, published lots of articles in various reputable journals on the electrochemical response of ionic liquids that contain aluminum ion as cation and its uses as a solvent in the electrochemical studies of various electroactive species (Hussey et al., 1979; Carpio et al., 1979; Wilkes et al., 1982; Laher and Hussey, 1982; Scheffler et al., 1983; Laher and Hussey, 1983a; Laher and Hussey, 1983b; Scheffler and Hussey, 1984; Hussey and O/ye, 1984; Sanders et al., 1986a; Hussey et al., 1985; Hitchcock et al., 1986; Appleby et al., 1986; Sanders et al., 1986b; Sun et al., 1987; Hussey, 1988). Ionic liquids are sensitive to air and moisture; therefore, the experiments on these liquids were carried out under inert atmosphere until the 90s, when, for the first time, an air and moisture stable ionic liquid was synthesized using reactants, i.e., 1-ethyl-3-methylimidazolium cation with tetrafluoroborate anion (Wei and Ivaska, 2008; Wilkes and Zaworotko, 1992). After this report, various ionic liquids were developed by different research groups. These are not limited to chloroaluminate salt, and are air and/or water sable (Marsh et al., 2004; Wang et al., 2017). Today, a number of reports have been published on ionic liquids in numerous reputable journals, and are increasing exponentially day by day, exceeding the annual growth of other scientific areas (Mei et al., 2019). This reveals that more and more scientific communities are engaged in the research on ionic liquids with successful outcomes. Ionic liquids, with their wide liquid range along with their capabilities of tailoring shape, size functionality, and low melting point, offer a wide control over different opportunities. This cannot be obtained through conventional solvent systems. Tailoring the nature, composition, and modification of ions generates a noble class of ionic liquids. These have unique or desirable physicochemical properties, which enable them to carry out a specific applications; therefore, they are called "functionalized ionic liquids" or "task-specific ionic liquids" (Plechkova et al., 2010; Wang et al., 2017; Plechkova and Seddon, 2008; Fedorov and Kornyshev, 2014; McFarlane et al., 2005; Wilkes, 2002). The possible combinations for the functionalization of ionic liquids are very high (Kaur and Singh, 2017). These well-defined physiochemical properties make them suitable for different task-specific applications in which conventional solvents fail to generate the desired results, such as high electrical properties, electrochemical potential windows, high thermal stability, low nucleophilicity, etc. (Zhou et al., 2014). In this context, the emerging multidisciplinary fields for the study of ionic liquids include environmental science, material engineering, chemical engineering, electrochemistry, organic chemistry, etc. Therefore, the present chapter focuses on the fabrication, classification, physiochemical properties, functionalization, and applications of ionic liquids.

15.2 Classification of Ionic Liquids

Ionic liquids are broadly divided into three generations, i.e., first, second, and third generation.

15.2.1 First Generation

First-generation ionic liquids came into existence in 1941, when the first ionic liquid was isolated, and continued until 1990. This was the time when ionic liquids started gaining popularity in the research community. This generation of ionic liquids was dedicated to dialkylimidazolium and alkylpyridium cations stabilized with chloroaluminate and other metal halides (Du et al., 2012). The ionic liquids of this generation faced critical hurdles such as poor stability, high moisture sensitivity, high toxicity, oxygen sensitivity, etc. Therefore, these ionic liquids can be handled in an inert atmosphere (Wells and Coombe, 2006; Endres and Zein El Abedin, 2006). All these drawbacks limit their utility in practical applications and constrain their further progress.

15.2.2 Second Generation

A new generation of ionic liquids came out after 10 years of research, i.e., second-generation ionic liquids. Moisture- and air-sensitive anions have been replaced by halide ions such as chloride, bromide, iodide, and other anions such as tetrafluoroborate, benzoate, and hexaflurophosphate anion, which provide stability both in water and in air (Kaur and Singh, 2017). In addition, this class of ionic liquids provides a better viscosity, excellent solubility in different solvents, and low melting point. However, toxicity is the major hurdle which remains in second-generation ionic liquids. Still, not every ionic liquid is toxic in nature, as toxicity majorly depends on the constituents of the ionic liquids. Therefore, one can plan the fabrication of particular ionic liquid considering the toxic effect of the constituents for a particular application. This generation has attracted the attention of the research community to a greater extent, and plays a critical role in different fields of science for novel applications.

15.2.3 Third Generation

Abbott et al. first introduced a new class of ionic liquids, i.e., third-generation ionic liquids, also called deep eutectic solvents. This generation of ionic liquids utilized more hydrophobic anions including alkylphosphates, alkylsulfates, organic acids, amino acids, sugars, etc., along with choline as cation, as compared to the other two generations. Deep eutectic solvents are synthesized by the hydrogen bond donor and hydrogen bond acceptor which are linked with each other via interaction by hydrogen bonding. In this context, multicomponent mixtures are also available in the literature. Instead of using ionic species, these deep eutectic solvents can be formed by neutral molecules such as sugar. Sucrose, glucose, and fructose have been used to synthesize the deep eutectic solvents in which no cation or anions are present. Deep eutectic solvents synthesized from natural molecules are also called natural deep eutectic solvents. These ionic liquids are comparatively easier and simpler in terms of synthesis protocol as compared to other ionic liquids. The typical reaction protocol undergoes the heating of nontoxic mixtures in a proper ratio to form a uniform liquid with 100 % utilization ratio. Deep eutectic solvents show non flammability, biodegradability, water stability, a wide liquid range, low toxicity, and low vapor pressure.

15.3 Properties of Ionic Liquids

15.3.1 Melting Point

The most imperative physical property of ionic liquids that can be correlated with the composition and structure of ionic liquids is the melting point of the ionic liquids. The melting point mainly depends on the nature of the cation and anion counterparts. Usually, larger ions tend to decrease the melting points, e.g., with the increase in the size of anions, the melting point decreases due to weaker columbic interactions between the crystal lattices of ionic liquids. Symmetrically substituted cations contribute to a higher melting point as in the case of 1,3-dialkylimidazolium in contrast to asymmetrical ones. In addition, with the increase in alkyl substitution (up to 8–10 carbon), the melting point decreases, then increases with further substitution. Ngo et al. also reported that the replacement of heavier and asymmetrical cations leads to a decrease in the melting point (Ngo et al., 2000). Ionic liquids carrying a smaller number or no hydrogen bonding usually have lower melting points; however, ionic liquids having substantial hydrogen-bonded anions (such as acetate ion) have melting points comparable to those ionic liquids that have highly delocalized anions which are unable to exhibit H-bonding (Bistrifluoromethylsulfonyl imide anion). These interactions not only govern the melting point but also direct the dissolution of other substances in ionic liquids.

15.3.2 Viscosity of Ionic Liquids

Viscosity is of enormous importance if ionic liquids are to be explored as solvents during a chemical reaction or for any other application. Typically, their viscosities are higher than those of other molecular solvents. They have been categorized as Newtonian fluids as the viscosity of most of the liquids remains constant with an increase in the shear rate (Huddleston et al., 2001; Dzyuba and Bartsch, 2002; Carda–Broch et al., 2003). At room temperature, the viscosity of ionic liquids usually varies from 10 cP to 500 cP. The viscosity of ionic liquids is highly dependent on temperature. For instance, in 1-butyl-3-methylimidazolium hexafluorophosphate, viscosity increases by 27 % on varying the temperature between 298 and 293 K (Visser et al., 2000). Ionic liquids with high viscosities are detrimental to the progress of the reaction. The diffusion rate of the redox reaction is diminished in liquids with high viscosity. On addition of solvents, the viscosity of ionic liquid changes dramatically. Consequently, the ones having higher viscosity can be made suitable for use by the addition of solvents having lower viscosity, e.g., the viscosity of 1-octyl-3-methylimidazolium iodide decreases exponentially upon addition of dichloromethane. Structural variations also affect the viscosity. In protic ionic liquids, the structures of anions affect viscosity more than the structures of cations (Greaves et al., 2006). In substituted imidazolium ionic liquids, aromatic rings stacking leads to higher viscosity (Greaves and Drummond, 2007; Nazari et al., 2013). When escalating the size of cations by enhancing the ring number in lactam-based ionic liquids, the viscosity increases due to an increase in cation–anion interaction.

15.3.3 Density of Ionic Liquids

The most important physical property that can be easily determined in the case of ionic liquids is density. The literature reveals that the densities of ionic liquids are usually higher than that of water. They are less receptive to variations in temperature. Upon raising the temperature from 298 to 303 K, the density is lowered only by 0.3 % (50.0 : 50.0 mol %) 1-Ethyl-3-methylimidazolium chloride/ aluminum trichloride (Baker et al., 2001). For symmetric and asymmetric series of ionic liquids, the density decreases with an increase in the chain length of alkyl, which is attributed to the increasing fraction of methylene groups ($-CH_2$). In different non-haloaluminate ionic liquids, the density of ionic liquids is increased by an increase in the mass of anion for same set of cations. The density is also influenced by the nature of the organic cation.

15.3.4 Diffusion and Conductivity

These are among the critical parameters of ionic liquids that should be known before exploring their various applications. The diffusion coefficients of ionic liquids have been generally calculated by electrochemical or NMR methods. It has been established that the diffusion of an ion is strongly interrelated with the constituent counter ion (Noda et al., 2001). Apparently, as the name suggests, ionic liquids are expected to have a very high conductive nature as these liquids are composed of ions. However, this is not the exact scenario. Ionic liquids have considerably lower conductivity than electrolytes in the molten state or solution phase. Their conductivities are usually similar to those of non-aqueous solvents or electrolytes (~10 mS cm^{-1}). The decrease in conductivity can be attributed to the reduced ion mobility, which is due to the large size of ions as well as to the decrease in the overall charge carried by the charge carriers via ion pairing and/or ion aggregation. On increasing the size of cations, conductivity decreases due to their lower mobility. Thus, the diffusion coefficients are dependent on the ionic pairs and aggregates present.

15.3.5 Solubility and Solvation in Ionic Liquids

The most important and prominent property of ionic liquids is their utilization as organic solvents for solvation and reaction medium, which helps in replacing the toxic volatile organic solvents. It is due to this property that ionic solvents are well known as green solvents and promote green chemistry. These greener solvents carry similar characteristics to those in dipolar, aprotic, and short-chain alcohols used as solvents, particularly dimethyl sulfoxide, dimethyl formamide, methanol, ethanol, etc. (Sheldon, 2001; Welton, 1999). The polarities of the green solvents vary in the range of solvents having water and chlorine as components. By varying constituent ions, the solvation properties of ionic liquids can be manipulated according to the requirements, e.g. when a change of anion from chloride ion to hexafluorophosphate ion takes place, the ionic liquid miscible with water becomes completely immiscible. Lipophilicity can be customized by the degree of cation substitution. On increasing the alkyl substitution, the lipophilicity increases, which results in higher solubility of both hydrocarbons as well as non-polar organic moieties.

15.3.6 Thermal Stability

The stability of ionic liquids, particularly thermal stability, is the most important factor when the ionic liquids are tested for various applications. The onset of thermal decomposition temperature provides the characteristics of the thermal stability of ionic liquids and is explored with the aid of thermal gravimetric analysis instruments. In addition to this, Fourier transform infrared spectroscopy, mass spectrometry, nuclear magnetic resonance, density functional theory, and even X-ray photoelectron spectroscopy experiments are commonly used (Longo et al., 2016; Liu et al., 2015; Xue et al., 2014). The nucleophilicity of anions affects the thermal stability of ionic liquids. In the case of imidazolium ionic liquids, the trend of variations in thermal stability for different anions follows the order: acetate < chloride < bromide < iodide < nitrate <hydrogen tetraoxosulfate < tosylate anion < tetra fluoroborate < bistriflimide anion < triflate anion < hexafluorophosphate < perchlorate, while the thermal stability of ionic liquids having the same anion follows the order: imidazolium > pyridine > pyrrolidinium > phosphonium.

15.4 Synthesis and Functionalization of Ionic Liquids

In 1914, the first attempt was made to synthesize room-temperature ionic liquids by the reaction of ethylammonium nitrate, which is the neutralization reaction of ethyl amine using concentrated nitric acid with the synthesized ionic liquid melting at 12 °C (Walden, 2019). Ionic liquids did not gain any interest until the development of binary ionic liquids derived from mixing 1,3-dialkylimidazolium chloride or N alkylpyridinium and aluminum chloride (Chum et al., 1975; Wilkes, 2002). These ionic liquids have been classified into two categories, i.e., simple salts made up of a single cation and anion, while the binary ionic liquids where equilibrium exists between cations and anions, e.g. ethylammonium nitrate, are simple ionic liquids. On the other hand, the ionic liquids synthesized from 1,3-dialkylimidazolium chlorides and aluminum chloride are the binary ionic liquids that carry various ionic species with different melting points and other properties such as viscosity, stability, etc., depending on the mole fractions of cations and anions used. (Figure 15.1).

The required cation can be fabricated either by the protonation reaction of amine by the desired acid, or by the reaction of amine with haloalkane via quaternization reactions, followed by heating. The anion transfer reaction can be performed by the reaction of halide salts with Lewis acid to form ionic liquids. Aluminum chloride–based ionic liquids are highly explored and extensively used (Wilkes et al., 1982; Robinson and Osteryoung, 1979; Hurley and WIer, 1951b). These ionic liquids involve a simple mixing of Lewis acid and metal halide salt, resulting in the formation of one or more ionic liquids depending on the ratio of two constituents, as shown in the reactions given below.

$$[\text{emim}]^+ \text{Cl}^- + \text{AlCl}_3 \leftrightarrow [\text{emim}]^+ [\text{AlCl}_4]^- \quad (15.1)$$

$$[\text{emim}]^+ [\text{AlCl}_4]^- + \text{AlCl}_3 \leftrightarrow [\text{emim}]^+ [\text{Al}_2\text{Cl}_7]^- \quad (15.2)$$

$$[\text{emim}]^+ [\text{Al}_2\text{Cl}_7]^- + \text{AlCl}_3 \leftrightarrow [\text{emim}]^+ [\text{Al}_3\text{Cl}_{10}]^- \quad (15.3)$$

When the concentration of $AlCl_3$ exceeds [emim]Cl, the ionic liquid is acidic in nature, while the opposite case results in the formation of basic ionic liquids. However, when the concertation of both reactants is the same, it leads to the development of a neutral charge over the ionic liquids. In addition of $AlCl_3$, other Lewis acid were also used for the formation of ionic liquids, including indium chloride, cuprous chloride, boron trichloride, etc. (Böhm and Herrmann, 2000; Williams et al., 1987; Chauvin et al., 1995; "Catalyst Comprising Indium Salt and Organic Ionic Liquid and Process for Friedel-Crafts Reactions" 2002).

Anion metathesis is the technique of choice for the synthesis of ionic liquids which are both air and water stable, generally derived from 1,3-dialkylimidazolium cation. The typical synthetic route involves the reaction of metal halide with salts of potassium/sodium/silver of nitrates, nitrites, tetrafluoroborates, sulfates, acetates, or the appropriate anions of free acids. This leads to the formation of a wide variety of ionic liquids (Hasan et al., 1999; Larsen et al., 2000; MacFarlane et al., 2002; Bonhôte et al., 1996; Lancaster et al., 2001; Holbrey and Seddon, 1999; Wilkes and Zaworotko, 1992; Cammarata et al., 2001; Fuller et al., 1994; Huddleston et al., 1998). From the above discussion, it is clear that a wide variety of ionic liquids can be formulated by the simple combination of a number of anions and cations, which is roughly estimated to be 10^{18}. Different novel methodologies, as well as nonconventional techniques, have been developed so far,

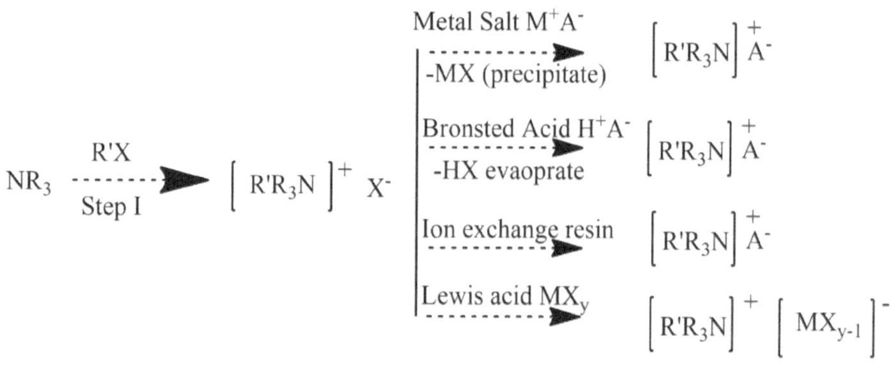

FIGURE 15.1 Different routes for the fabrication of ionic liquids.

i.e., microwave, ultrasound irradiation, which can be used alone or in combination with other methodologies for the improvement in the desired properties, reducing reaction time, improving cost and energy, improving yield, etc. (Lévêque et al., 2002; Deetlefs and Seddon, 2003; Lévêque et al., 2006). In addition, the recent improvement in the one-pot reaction protocol, which is solventless, efficient, less time consuming, etc., improves the drawback of ionic liquids. In particular, it cuts down the cost of ionic liquids, thus encouraging the research community to explore more in the various fields of science (Varma and Namboodiri, 2001a; Varma and Namboodiri, 2001b; Vu et al., 2007).

15.4.1 Synthesis of Task-Specific Ionic liquids or Functionalized Ionic Liquids

Functionalized or task-specific ionic liquids can be synthesized using conventional methodology via the substitution of halide ion from organic salt by phosphine and imidazole, resulting in the formation of the desired functional moiety in the vicinity of ionic liquids. This substitution reaction is carried forward by anion exchange. (Figure 15.2).

This functionalization can be helpful in performing various applications of ionic liquids such as catalysis, synthesis, organic reactions, supercapacitors, conductive materials, nanoparticles synthesis, etc. This methodology is appropriate for the fabrication of ionic liquids that are stable towards bases, but due to the higher basicity of the imidazolium ring, the elimination reaction of hydrogen halide, or Hoffmann elimination reaction, takes place in certain cases. Particularly, the functional groups have been functionalized directly to the imidazolium ring via the quaternization route, i.e. fluorous, chain, allyl, alkyne, thiol carboxyl, hydroxyl, etc. (Feng et al., 2007; Fei et al., 2004; Zhao et al., 2005; Fraga-Dubreuil and Bazureau, 2001; Merrigan et al., 2000). In this context, a new method has been developed by Wasserscheid et al. for the fabrication of task-specific ionic liquids using the Michael reaction, in which the acid from the anion protonates the nucleophile, followed by incorporation into the ionic liquids, as shown in Figure 15.3 (Wasserscheid et al., 2003).

The functionalization of the -OH group on ionic liquids can be achieved by two protocols, which are given below.

Before Holbrey et al., a one pot, facile, high yielding and a simple method for hydroxyl functionalization could not been achieved fully. In this reaction procedure, the nucleophile, i.e., imidazole, is allowed to react with the epoxide molecule, resulting in the ring opening without further polymerization (Holbrey et al., 2003) as shown in Figure 15.4.

In the second methodology, developed by Bao et al., four reactions, i.e., glyoxal, formaldehyde, ammonia, and amino acids have been used as precursors, leading to the formation of optically active hydroxyl functionalized task-specific ionic liquids (Bao et al., 2002). The detailed reaction is shown in Figure 15.5.

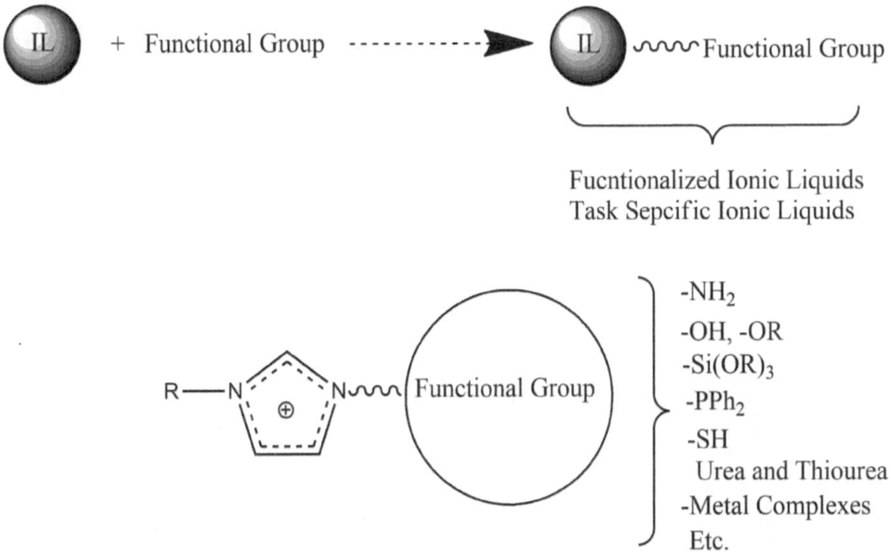

FIGURE 15.2 Synthesis of task-specific or functionalized ionic liquids.

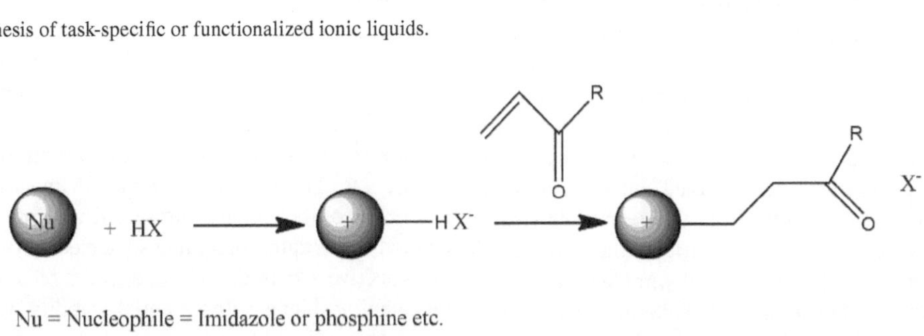

FIGURE 15.3 Reaction of keto-functionalized ionic liquids.

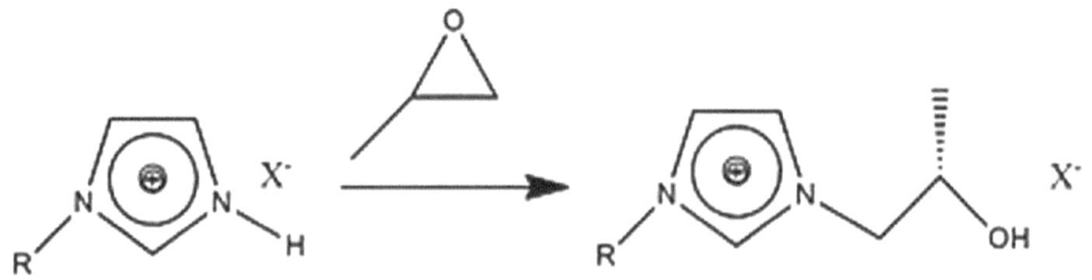

FIGURE 15.4 Synthesis of hydroxyl-functionalized chiral ionic liquids.

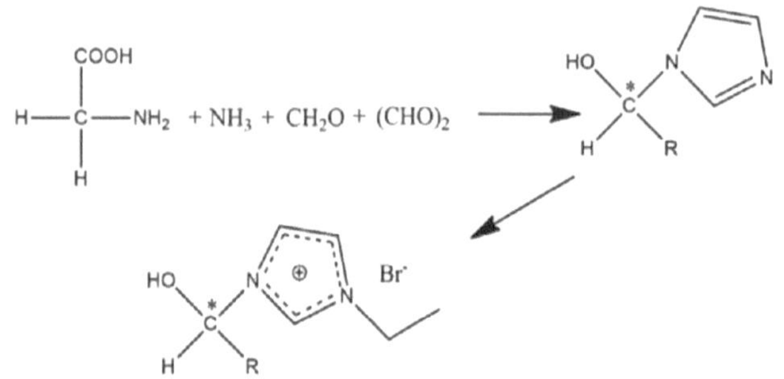

FIGURE 15.5 Reaction scheme for chiral ionic liquids.

15.5 Applications of Ionic Liquids

15.5.1 Application in Electrochemistry

The most exciting specialty of ionic liquids is that their physicochemical properties can be tuned by the combination of different cations with particular anions (Wasserscheid and Welton, 2007). In addition, the viscosities of the hydrophobic ionic liquids are strongly dependent on the quantity of water present (Widegren et al., 2005). There is a lot of literature on the electrochemical properties of ionic liquids which reveals the wide potential window of ~5.5 V as compared to almost all aqueous electrolytes (Suarez et al., 1997; Seddon et al., 2000; Koch et al., 1996; Fuller et al., 1997; Quinn et al., 2002; Galiński et al., 2006; Chaudhary et al., 2013). Their unmatched properties, such as thermal and electrochemical stability and excellent ionic conductivity, along with non-flammability, makes them suitable candidates for electrochemical studies that include supercapacitors (Stenger-Smith et al., 2002; Sato et al., 2004; Liu et al., 2006), fuel cells (Noda et al., 2003), actuators (Zhou et al., 2003), batteries (Seki et al., 2006; Fung and Zhu, 2002), electrochemical sensors (Sun and Armstrong, 2010; Buzzeo et al., 2004a; Baker et al., 2005; Buzzeo et al., 2004b; Shiddiky and Torriero, 2011), etc. The exceptional high conductance of ionic liquids makes them very special in the field of electrochemistry. In particular, the conductance of ionic liquids ranges from 0.1 to 18 mS cm^{-1} (Galiński et al., 2006). In this context, the conductivities of various ionic liquids based on dialkylimidazolium cations have been studied and are present in the literature (Galiński et al., 2006). Both the nature and structure of anions and cations significantly affects the conductivity of the ionic liquids, e.g., [1-ethyl-3-methylimidazolium][trifl uoroacetate] showed the conductivity of 9.6 mS cm^{-1}, while [1-et hyl-3-methylimidazolium][heptafluorobutanoate] reveals a conductivity of 2.7 mS cm^{-1}. In addition, when cations of different imidazolium form ionic liquids with same anion shows the conductivities ranging from 0.98 to 8.8 mS cm^{-1}. Ionic liquids are electroactive in nature, showing oxidation and/or reduction on the application of applied potential. Murray and his team developed a number of ionic liquids that are electroactive in nature by functionalizing ferrocene to the imidazolium counterpart, and explained their electron transport properties (Balasubramanian et al., 2006; Wang et al., 2008). The major problem in using ionic liquids in electrochemistry is the non-availability of reference potential, so the comparison of ionic liquids with the conventional solvent is a little difficult. In this context, Wedd and his team used cobaltocene as reference and calculated the potential of 1-Butyl-3-methylimidazolium hexafluorophosphate. Its value is comparable with acetonitrile (Hultgren et al., 2002). The voltammetric studies of ferrocene-derived ionic liquids were also conducted by drop casting or modification of working electrodes and showed that all the processes are diffusion-controlled and chemically reversible (Hultgren et al., 2002).

15.5.2 Ionic Liquids as Ion-Sensitive Electrodes

Ion-sensitive electrodes play an important role in electroanalytical techniques, which are based on the modification of the membrane in the potentiometer. The potential of the solution is measured when the membrane modified electrode is dipped in the solution along with a reference electrode and connected with potentiostat (Mamińska et al., 2006). The microelectrode, i.e., silver/silver chloride, is placed in a 2–5 cm glass cylinder open on both sides. Both ends of the cylinder are sealed with a PVC membrane that contain ionic liquids providing solid electrolytes internally. These ion-sensitive electrodes provide good reliability and excellent repeatability along with potential stability, which shows robust achievements in the development of novel ion sensitive electrodes. The ionic liquid-based ion sensitive electrodes were thoroughly examined, and their thermodynamics have been explained by Kakuichi and Yoshimatsu (Kakiuchi and Yoshimatsu, 2006). Several reports are available in the literature which reveal the use of ionic liquids as ion-sensitive electrodes (Coll et al., 2005; Shvedene et al., 2006; Ortuño et al., 2014; Wardak and Lenik, 2013). Ionic liquids have been found to be excellent molecules for the fabrication of ion-selective electrode membranes due to their ionic and plasticizing nature. In this context, fullerene-modified carbon nanotubes and ionic liquids (1-Butyl-3-methyimidazolium tetrafluoroborate) have been successfully applied as ion-sensitive electrodes for the detection of diazepam in real samples such as tablets, urine, and serum with linearity ranges of 0.3–700 µM and a detection limit of 87 nM. Miao et al. fabricated 1-methyl-2-butylthioimidazolium bis (trifluoromethanesulphonyl) imide-based ion sensitive electrodes for the detection of mercury (II), and the detection limit turns out to be 4.1×10^{11} mol L^{-1} with excellent selectivity (Miao et al., 2018).

15.5.3 Voltammetric Sensors

Along with ion-sensitive electrodes, ionic liquids have also been widely used as voltammetric sensors in electroanalytical methods with the aid of cyclic voltammetry. In this context, minute concentrations of chloride ions were detected using different methods of voltammetry such as cathodic stripping, square wave, and linear sweep voltammetry (Villagrána et al., 2004). Wang et al. have synthesized a novel ionic liquid, i.e., 1-[3-(N-pyrrolyl) propyl]-3-butylimidazolium bromide, and functionalized it on the surface of a graphene oxide sheet. The prepared composite has been used as an electrochemical mediator for the electrochemical sensing of bisphenol (Y. Wang et al., 2018). Composites of ionic liquids with nanoparticles have also gained a lot of interest for the development of electrochemical sensors, e.g., cobalt disulfide decoded with ionic liquid (1-(3-aminopropyl)– 3-methylimidazolium bromide) for the electrochemical sensing of hydrazine with a detection limit of 0.39 µM. The recovery rate for hydrazine sensing is quite high in lake water samples (Luan et al., 2018). The effective utilization of ionic liquids in electrochemical sensors is due to their high communication of electron transfer as well as to the presence of functional groups that specifically target a particular analyte for electrochemical sensing (Beytur et al., 2018). Features like = high electron communication, high stability towards a wide potential range, water stability, etc. enhance the utilization of ionic liquids in the electrochemical sensing of various toxic chemicals/ions (Chaiyo et al., 2018; Mert et al., 2018; Alavi-Tabari et al., 2018; Zhu et al., 2018; Huang et al., 2018; X. Wang et al., 2018).

15.5.4 Ionic Liquids in Supercapacitors

Ionic liquids also play an important role in supercapacitor applications, particularly in the electrical double-layer type, due to their wide working potential window as compared to conventional electrolytes, and by providing charged species at the electrolyte/electrode interface rather than diffusion. Due to this, ionic liquids have appeared as promising candidates for their use as electrolyte in supercapacitor application, but unfortunately there are only a handful of articles on the fundamental studies of electrolyte/electrode interface in supercapacitors (Vatamanu et al., 2010; Céline Merlet et al., 2013; Fedorov and Kornyshev, 2014; Forse et al., 2015; Vatamanu et al., 2011; Cline Merlet et al., 2011; Fedorov and Kornyshev, 2008). Ionic liquids were not only explored as electrolytes but also utilized as electrolytic capacitors (Song et al., 2006). In addition, only a few research groups are engaged in the pure electrolytic application of supercapacitors. The price of ionic liquids may be one reason, but it has been dropping dramatically in recent years due to their growing demand in various laboratories and industries. The price of ionic liquids is not due to the reagent used in the synthesis but to their purification steps. In addition, due to the high charge in ionic liquid moieties, the possibility of ionic contamination is also very high (Eftekhari, 2017). Therefore, the current research on ionic liquids is limited to the common ionic liquids that are indeed suitable choice for supercapacitor electrolytes (Eftekhari, 2017). Commonly utilized cations for ionic liquids for supercapacitor application are given as tetraethylammonium, 1-butyl-1-methylpyrrolidinium, N-propyl-N-methylpyrrolidinium, 1-butyl-3-methylimidazolium, 1-ethyl-3-methylimidazolium, etc., while the counter anions are bis(trifluoromethylsulfonyl)imide, bis(fluorosulfonyl)imide, hexafluorophosphate, tetrafluoroborate, bromide, chloride, etc. (Song et al., 2006; Eftekhari, 2017).

15.5.5 Application of Ionic Liquids in Industry

Ionic liquids find numerous applications in industrial chemistry. Some of them are highlighted in this chapter.

15.5.5.1 BASIL Process

The most successful example of the usage of ionic liquids in industries is BASIL™, i.e., the biphasic acid scavenging utilizing ionic liquid) process. In 2002, it was first announced publicly on the BASF site, Ludwigshafen, Germany. The BASIL™ process is utilized for the synthesis of alkoxyphenylphosphines. The synthesis protocol used triethylamine as an acid scavenger that has been formed during the reaction, but the reaction mixture is difficult to work up as the byproduct, triethylammonium chloride forms a dense solution with insoluble paste. This can be eliminated by replacing triethylamine with 1-methylimidazole as ionic liquids. This process requires much smaller reactors as compared to the previous one, with an increase in yield from 50 to 98 %, and the catalyst is recycled, which was immediately recognized through an ECN innovation award in 2004.

15.5.5.2 Replacing Phosgene

A German company, BASF, utilized ionic liquids in different directions. It has been observed that hydrogen chloride may act as chlorinating agent against arenes (Rogers and Seddon, 2005; Reichardt, 1979). BASF revealed that hydrogen chloride in ionic liquids substitutes phosgene, e.g., the reaction of butan-1,4-diol with phosgene gives 1,4-dichlorobutane. Through the same reaction, when explored with hydrogen chloride in place of phosgene, four products have been formed with a minor concentration of 1,4-dichlorobutane, while the major concentrations are of tetrahydrofuran and 1-chlorobutan-4-ol.

15.5.6 Ionic Liquids in Environmental Application

Environmental challenges in the recent years have shown quite drastic changes in the world. Almost every part of the world is struggling with various problems in the environment, with the contamination of the water systems being one of the major challenges. Presently, various research groups are actively engaged in the water-treatment area, using a number of techniques. These treatments include electrochemical methods such as flotation coagulation–flocculation, filtration, adsorption, ion exchange, chemical precipitation, etc. (Reichardt, 1979). These techniques carry various drawbacks that include maintenance, process limitations, cost, interference, etc. Ionic liquids, on the other hand, due to the charge they carry, play an important role in the removal of toxic heavy metal ions from industrial discarded water, and are more effective than conventionally used adsorption materials/techniques. This can be attained by the desired functionalized-task specific ionic liquids (Rajendran, 2010;Visser et al., 2002). The utilization of task-specific ionic liquids as a potential candidate for the extraction of heavy metal ions has produced extraordinary results (Rajendran, 2010; Visser et al., 2002). In this context, Visser et al. (2002) have designed and developed a number of ionic liquids to extract mercury and cadmium from contaminated water. These hydrophobic ionic liquids are highly heavy-metal-ions adsorbent as compared to conventional adsorbents that bind with the metal ion quite effectively and remove it from the aqueous medium when treating waste water (Fuerhacker et al., 2012). In addition, sewage sludge carries nitrogen and phosphorous, which can be efficiently applied as valuable fertilizers. Therefore, many countries have imposed severe legal restrictions regarding the metal ions when sludge has been used for other applications. For this reason, the removal of heavy metal ions from sludge is necessary before it can be utilized for fertilization or irrigation applications. In terms of ionic liquids, a lot of research has gone into the application of ionic liquids for the removal of toxic ions in aqueous media in standard conditions, and only a few reports are devoted to the direct application of ionic liquids for the removal of toxic metal ions from real waste-water samples (Mert et al., 2018; Rajendran, 2010). There is a vast area to be explored on the removal of toxic chemicals from sludge for the utilization of water in various activities.

Conclusion

The present chapter focuses on the introduction, properties, synthesis, functionalization, and applications of ionic liquids. Although ionic liquids are known as being an excellent replacement for conventional solvents, they also finds a major place in wide applications such as environmental remediation, supercapacitors, batteries, sensing, catalysts, etc. Due to so much diversity in the applications of ionic liquids, the focus is progressively shifting from academics to industrialists, as their utilization has led to numerous sustainable developments. Flexibility in changing the physiological and chemical parameters by modifying the ionic liquids opens a window for the organic chemist to design a ionic liquid as per the demand of the applications. The conclusion is that there are a lot more possibilities to explore in various fields of science by functionalizing ionic liquids.

REFERENCES

Alavi-Tabari, Seyed A.R., Mohammad A. Khalilzadeh, and Hassan Karimi-Maleh. 2018. "Simultaneous Determination of Doxorubicin and Dasatinib as Two Breast Anticancer Drugs Uses an Amplified Sensor with Ionic Liquid and ZnO Nanoparticle." *Journal of Electroanalytical Chemistry* 811(February). Elsevier: 84–88. doi:10.1016/J.JELECHEM.2018.01.034.

Appleby, Denise, Charles L. Hussey, Kenneth R. Seddon, and Janet E. Turp. 1986. "Room-Temperature Ionic Liquids as Solvents for Electronic Absorption Spectroscopy of Halide Complexes." *Nature* 323(6089). Nature Publishing Group: 614–616. doi:10.1038/323614a0.

Baker, Gary A., Sheila N. Baker, Siddharth Pandey, and Frank V. Bright. 2005. "An Analytical View of Ionic Liquids." *The Analyst* 130(6). The Royal Society of Chemistry: 800. doi:10.1039/b500865b.

Baker, Sheila N., Gary A. Baker, Maureen A. Kane, and Frank V. Bright. 2001. "The Cybotactic Region Surrounding Fluorescent Probes Dissolved in 1-Butyl-3-Methylimidazolium Hexafluorophosphate: Effects of Temperature and Added Carbon Dioxide." American Chemical Society. doi:10.1021/JP0103528.

Balasubramanian, Ramjee, Wei Wang, and Royce W. Murray. 2006. "Redox Ionic Liquid Phases: Ferrocenated Imidazoliums." *Journal of the American Chemical Society*. doi:10.1021/JA0625327.

Bao, Weiliang, Zhiming Wang, and Yuxia Li. 2002. "Synthesis of Chiral Ionic Liquids from Natural Amino Acids." American Chemical Society. doi:10.1021/JO020503I.

Beytur, Murat, Faruk Kardaş, Onur Akyıldırım, Abdullah Özkan, Bahar Bankoğlu, Haydar Yüksek, Mehmet Lütfi Yola, and Necip Atar. 2018. "A Highly Selective and Sensitive Voltammetric Sensor with Molecularly Imprinted Polymer Based Silver@Gold Nanoparticles/Ionic Liquid Modified Glassy Carbon Electrode for Determination of Ceftizoxime." *Journal of Molecular Liquids* 251(February). Elsevier: 212–217. doi:10.1016/J.MOLLIQ.2017.12.060.

Böhm, Volker P. W., and Wolfgang A. Herrmann. 2000. "Nonaqueous Ionic Liquids: Superior Reaction Media for the Catalytic Heck-Vinylation of Chloroarenes." *Chemistry – A European Journal* 6(6). John Wiley & Sons, Ltd: 1017–1025. doi:10.1002/(SICI)1521-3765(20000317)6:6<1017::AID-CHEM1017>3.0.CO;2-8.

Bonhôte, Pierre, Ana-Paula Dias, Nicholas Papageorgiou, Kuppuswamy Kalyanasundaram, and Michael Grätzel. 1996. "Hydrophobic, Highly Conductive Ambient-Temperature Molten Salts." *Inorganic Chemistry* 35(5). American Chemical Society: 1168–1178. doi:10.1021/ic951325x.

Buzzeo, Marisa C., Russell G. Evans, and Richard G. Compton. 2004a. "Non-Haloaluminate Room-Temperature Ionic Liquids in Electrochemistry—A Review." *ChemPhysChem* 5(8). John Wiley & Sons, Ltd: 1106–1120. doi:10.1002/cphc.200301017.

Buzzeo, Marisa C., Christopher Hardacre, and Richard G. Compton. 2004b. "Use of Room Temperature Ionic Liquids in Gas Sensor Design." American Chemical Society. doi:10.1021/AC040042W.

Cammarata, L., S. G. Kazarian, P. A. Salter, and T. Welton. 2001. "Molecular States of Water in Room Temperature Ionic Liquids. Electronic Supplementary Information Available."

See http://www.rsc.org/suppdata/cp/b1/b106900d/. *Physical Chemistry Chemical Physics* 3(23). The Royal Society of Chemistry: 5192–5200. doi:10.1039/b106900d.

Carda–Broch, S., A. Berthod, and D. W. Armstrong. 2003. "Solvent Properties of the 1-Butyl-3-Methylimidazolium Hexafluorophosphate Ionic Liquid." *Analytical and Bioanalytical Chemistry* 375(2). Springer-Verlag: 191–199. doi:10.1007/s00216-002-1684-1.

Carpio, Ronald A., Lowell A. King, Richard E. Lindstrom, John C. Nardi, and Charles L. Hussey. 1979. "Density, Electric Conductivity, and Viscosity of Several N-Alkylpyridinium Halides and Their Mixtures with Aluminum Chloride." *Journal of the Electrochemical Society* 126(10). The Electrochemical Society: 1644. doi:10.1149/1.2128768.

"Catalyst Comprising Indium Salt and Organic Ionic Liquid and Process for Friedel-Crafts Reactions." 2002, October. https://patents.google.com/patent/US7928031.

Chaiyo, Sudkate, Eda Mehmeti, Weena Siangproh. Thai Long Hoang, Hai Phong Nguyen, Orawon Chailapakul, and Kurt Kalcher. 2018. "Non-Enzymatic Electrochemical Detection of Glucose with a Disposable Paper-Based Sensor Using a Cobalt Phthalocyanine–Ionic Liquid–Graphene Composite." *Biosensors and Bioelectronics* 102(April). Elsevier: 113–120. doi:10.1016/J.BIOS.2017.11.015.

Chaudhary, Ganga Ram, Shafila Bansal, Priya Saharan, Pratibha Bansal, and S. K. Mehta. 2013. "Applications of Surface Modified Ionic Liquid/Nanomaterial Composite in Electrochemical Sensors and Biosensors." *BioNanoScience* 3(3). Springer US: 241–253. doi:10.1007/s12668-013-0094-5.

Chauvin, Yves, Sandra Einloft, and Helene Olivier. 1995. "Catalytic Dimerization of Propene by Nickel-Phosphine Complexes in 1-Butyl-3-Methylimidazolium Chloride/AlEtxCl3-x (x = 0, 1) Ionic Liquids." *Industrial and Engineering Chemistry Research* 34(4): 1149–1155. doi:10.1021/ie00043a017.

Chum, Helena L., V. R. Koch, L. L. Miller, and R. A. Osteryoung. 1975. "Electrochemical Scrutiny of Organometallic Iron Complexes and Hexamethylbenzene in a Room Temperature Molten Salt." *Journal of the American Chemical Society* 97(11). American Chemical Society: 3264–3265. doi:10.1021/ja00844a081.

Coll, Carmen, Roberto H. Labrador, Ramón Martínez Mañez, Juan Soto, Félix Sancenón, María-Jesús Seguí, and Enrique Sanchez. 2005. "Ionic Liquids Promote Selective Responses towards the Highly Hydrophilic Anion Sulfate in PVC Membrane Ion-Selective Electrodes." *Chemical Communications* 24(June). The Royal Society of Chemistry: 3033. doi:10.1039/b503154k.

Deetlefs, Maggel, and Kenneth R. Seddon. 2003. "Improved Preparations of Ionic Liquids Using Microwave Irradiation." This work was presented at the Green Solvents for Catalysis meeting held in Bruchsal, Germany, 13–16th October 2002. *Green Chemistry* 5(2). The Royal Society of Chemistry: 181–186. doi:10.1039/b300071k.

Du, Jiang, Jian Qi, Dan Wang, and Zhiyong Tang. 2012. "Facile Synthesis of Au@TiO2 Core–Shell Hollow Spheres for Dye-Sensitized Solar Cells with Remarkably Improved Efficiency." *Energy and Environmental Science* 5(5). The Royal Society of Chemistry: 6914. doi:10.1039/c2ee21264a.

Dzyuba, Sergei V., and Richard A. Bartsch. 2002. "Influence of Structural Variations in 1-Alkyl(Aralkyl)-3-Methylimidazolium Hexafluorophosphates and Bis(Trifluoromethylsulfonyl)Imides on Physical Properties of the Ionic Liquids." *ChemPhysChem* 3(2). John Wiley & Sons, Ltd: 161–166. doi:10.1002/1439-7641(20020215)3:2<161::AID-CPHC161>3.0.CO;2-3.

Eftekhari, Ali. 2017. "Supercapacitors Utilising Ionic Liquids." *Energy Storage Materials* 9(October). Elsevier: 47–69. doi:10.1016/J.ENSM.2017.06.009.

Endres, Frank, and Sherif Zein El Abedin. 2006. "Air and Water Stable Ionic Liquids in Physical Chemistry." *Physical Chemistry Chemical Physics* 8(18). The Royal Society of Chemistry: 2101. doi:10.1039/b600519p.

Fedorov, Maxim V., and Alexei A. Kornyshev. 2008. "Towards Understanding the Structure and Capacitance of Electrical Double Layer in Ionic Liquids." *Electrochimica Acta* 53(23). Pergamon: 6835–6840. doi:10.1016/J.ELECTACTA.2008.02.065.

Fedorov, Maxim V., and Alexei A. Kornyshev. 2014. "Ionic Liquids at Electrified Interfaces." *Chemical Reviews* 114(5). American Chemical Society: 2978–3036. doi:10.1021/cr400374x.

Fei, Zhaofu, Dongbin Zhao, Rosario Scopelliti, and Paul J. Dyson. 2004. "Organometallic Complexes Derived from Alkyne-Functionalized Imidazolium Salts." American Chemical Society. doi:10.1021/OM034248J.

Feng, Guo Ren, Jia Jian Peng, Hua Yu Qiu, Jian Xiong Jiang, Lan Tao, and Guo Qiao Lai. 2007. "Synthesis of Novel Greener Functionalized Ionic Liquids Containing Appended Hydroxyl." *Synthetic Communications* 37(16). Taylor & Francis Group: 2671–2675. doi:10.1080/00397910701465230.

Forse, Alexander C., John M. Griffin, Céline Merlet, Paul M. Bayley, Hao Wang, Patrice Simon, and Clare P. Grey. 2015. "NMR Study of Ion Dynamics and Charge Storage in Ionic Liquid Supercapacitors." *Journal of the American Chemical Society* 137(22). American Chemical Society: 7231–7242. doi:10.1021/jacs.5b03958.

Fraga-Dubreuil, Joan, and Jean Pierre Bazureau. 2001. "Grafted Ionic Liquid-Phase-Supported Synthesis of Small Organic Molecules." *Tetrahedron Letters* 42(35). Pergamon: 6097–6100. doi:10.1016/S0040-4039(01)01190-X.

Fuerhacker, Maria, Tadele Measho Haile, Daniel Kogelnig, Anja Stojanovic, and Bernhard Keppler. 2012. "Application of Ionic Liquids for the Removal of Heavy Metals from Wastewater and Activated Sludge." *Water Science and Technology* 65(10). IWA Publishing: 1765–1773. doi:10.2166/wst.2012.907.

Fuller, Joan, Richard T. Carlin, Hugh C. De Long, and Dustin Haworth. 1994. "Structure of 1-Ethyl-3-Methylimidazolium Hexafluorophosphate: Model for Room Temperature Molten Salts." *Journal of the Chemical Society, Chemical Communications* 3(January). The Royal Society of Chemistry: 299. doi:10.1039/c39940000299.

Fuller, Joan, Richard T. Carlin, and Robert A. Osteryoung. 1997. "The Room Temperature Ionic Liquid 1-Ethyl-3-Methylimidazolium Tetrafluoroborate: Electrochemical Couples and Physical Properties." *Journal of the Electrochemical Society* 144(11). The Electrochemical Society: 3881. doi:10.1149/1.1838106.

Fung, Y. S., and D. R. Zhu. 2002. "Electrodeposited Tin Coating as Negative Electrode Material for Lithium-Ion Battery in Room Temperature Molten Salt." *Journal of the Electrochemical Society* 149(3). The Electrochemical Society: A319. doi:10.1149/1.1448501.

Galiński, Maciej, Andrzej Lewandowski, and Izabela Stępniak. 2006. "Ionic Liquids as Electrolytes." *Electrochimica Acta* 51(26): 5567–5580. doi:10.1016/j.electacta.2006.03.016.

Greaves, Tamar L., and Calum J. Drummond. 2007. "Protic Ionic Liquids: Properties and Applications." American Chemical Society. doi:10.1021/CR068040U.

Greaves, Tamar L., Asoka Weerawardena, Celesta Fong, Irena Krodkiewska, and Calum J. Drummond. 2006. "Protic Ionic Liquids: Solvents with Tunable Phase Behavior and Physicochemical Properties." American Chemical Society. doi:10.1021/JP0634048.

Hasan, Masihul, Ivan V. Kozhevnikov, M. Rafiq, H. Siddiqui, Alexander Steiner, and Neil Winterton. 1999. "Gold Compounds as Ionic Liquids. Synthesis, Structures, and Thermal Properties of N,N'-Dialkylimidazolium Tetrachloroaurate Salts." American Chemical Society. doi:10.1021/IC990657P.

Hitchcock, Peter B., Thamer J. Mohammed, Kenneth R. Seddon, Jalal A. Zora, Charles L. Hussey, and E. Haynes Ward. 1986. "1-Methyl-3-Ethylimidazolium Hexachlorouranate(IV) and 1-Methyl-3-Ethylimidazolium Tetrachlorodioxo-Uranate(VI): Synthesis, Structure, and Electrochemistry in a Room Temperature Ionic Liquid." *Inorganica Chimica Acta* 113(2). Elsevier: L25–L26. doi:10.1016/S0020-1693(00)82244-6.

Holbrey, John D., and Kenneth R. Seddon. 1999. "The Phase Behaviour of 1-Alkyl-3-Methylimidazolium Tetrafluoroborates; Ionic Liquids and Ionic Liquid Crystals." *Journal of the Chemical Society, Dalton Transactions* 0(13). Royal Society of Chemistry: 2133–2140. doi:10.1039/a902818h.

Holbrey, John D., Megan B. Turner, W. Matthew Reichert, and Robin D. Rogers. 2003. "New Ionic Liquids Containing an Appended Hydroxyl Functionality from the Atom-Efficient, One-Pot Reaction of 1-Methylimidazole and Acid with Propylene Oxide." *Green Chemistry* 5(6). The Royal Society of Chemistry: 731. doi:10.1039/b311717k.

Huang, Qing, Wei Li, Tian Wu, Xiaoling Ma, Kai Jiang, and Xianbo Jin. 2018. "Monoethanolamine-Enabled Electrochemical Detection of H2S in a Hydroxyl-Functionalized Ionic Liquid." *Electrochemistry Communications* 88(March). Elsevier: 93–96. doi:10.1016/J.ELECOM.2017.12.024.

Huddleston, Jonathan G., Ann E. Visser, W. Matthew Reichert, Heather D. Willauer, Grant A. Broker, and Robin D. Rogers. 2001. "Characterization and Comparison of Hydrophilic and Hydrophobic Room Temperature Ionic Liquids Incorporating the Imidazolium Cation." *Green Chemistry* 3(4). The Royal Society of Chemistry: 156–164. doi:10.1039/b103275p.

Huddleston, Jonathan G., Heather D. Willauer, Richard P. Swatloski, Ann E. Visser, and Robin D. Rogers. 1998. "Room Temperature Ionic Liquids as Novel Media for 'Clean' Liquid–Liquid Extraction." *Chemical Communications* 0(16). Royal Society of Chemistry: 1765–1766. doi:10.1039/A803999B.

Hultgren, Victoria M., Andrew W. A. Mariotti, Alan M. Bond, and Anthony G. Wedd. 2002. "Reference Potential Calibration and Voltammetry at Macrodisk Electrodes of Metallocene Derivatives in the Ionic Liquid [Bmim][PF6]." American Chemical Society. doi:10.1021/AC015729K.

Hurley, Frank H., and Thomas P. WIer. 1951a. "The Electrodeposition of Aluminum from Nonaqueous Solutions at Room Temperature." *Journal of the Electrochemical Society* 98(5). The Electrochemical Society: 207. doi:10.1149/1.2778133.

Hurley, Frank H., and Thomas P. WIer. 1951b. "Electrodeposition of Metals from Fused Quaternary Ammonium Salts." *Journal of the Electrochemical Society* 98(5). The Electrochemical Society: 203. doi:10.1149/1.2778132.

Hussey, C. L. 1988. "Room Temperature Haloaluminate Ionic Liquids. Novel Solvents for Transition Metal Solution Chemistry." *Pure and Applied Chemistry* 60(12). De Gruyter: 1763–1772. doi:10.1351/pac198860121763.

Hussey, C. L., L. A. King, and R. A. Carpio. 1979. "The Electrochemistry of Copper in a Room Temperature Acidic Chloroaluminate Melt." *Journal of the Electrochemical Society* 126(6). The Electrochemical Society: 1029. doi:10.1149/1.2129169.

Hussey, Charles L., and Harald A. ye. 1984. "Transport Numbers in Molten Acidic Aluminum Chloride-1-Methyl-3-Ethylimidazolium Chloride." *Journal of the Electrochemical Society* 131(7). The Electrochemical Society: 1621. doi:10.1149/1.2115920.

Hussey, Charles L., John R. Sanders, and Harald A. O/ye. 1985. "Transport Numbers in the Basic Aluminum Chloride-1-Methyl-3-Ethylimidazolium Chloride Ionic Liquid." *Journal of the Electrochemical Society* 132(9). The Electrochemical Society: 2156. doi:10.1149/1.2114308.

Kakiuchi, Takashi, and Takahiro Yoshimatsu. 2006. "A New Salt Bridge Based on the Hydrophobic Room-Temperature Molten Salt." *Bulletin of the Chemical Society of Japan* 79(7). The Chemical Society of Japan 公益社団法人 日本化学会: 1017–1024. doi:10.1246/bcsj.79.1017.

Kaur, Navneet, and Vasundhara Singh. 2017. "Current Status and Future Challenges in Ionic Liquids, Functionalized Ionic Liquids and Deep Eutectic Solvent-Mediated Synthesis of Nanostructured TiO_2: A Review." *New Journal of Chemistry* 41(8). The Royal Society of Chemistry: 2844–2868. doi:10.1039/C6NJ04073J.

Koch, V. R., L. A. Dominey, C. Nanjundiah, and M. J. Ondrechen. 1996. "The Intrinsic Anodic Stability of Several Anions Comprising Solvent-Free Ionic Liquids." *Journal of the Electrochemical Society* 143(3). The Electrochemical Society: 798. doi:10.1149/1.1836540.

Laher, T. M., and C. L. Hussey. 1982. "Electrochemical Studies of Chloro Complex Formation in Low-Temperature Chloroaluminate Melts. 1. Iron(II), Iron(III), and Nickel(II)." *Inorganic Chemistry* 21(11). American Chemical Society: 4079–4083. doi:10.1021/ic00141a040.

Laher, T. M., and C. L. Hussey. 1983a. "Electrochemical Studies of Chloro Complex Formation in Low Temperature Chloroaluminate Melts. 2. Silver(I)." *Inorganic Chemistry* 22(9). American Chemical Society: 1279–1283. doi:10.1021/ic00151a004.

Laher, T. M., and C. L. Hussey. 1983b. "Copper(I) and Copper(II) Chloro Complexes in the Basic Aluminum Chloride-1-Methyl-3-Ethylimidazolium Chloride Ionic Liquid." *Inorganic Chemistry* 22(22). American Chemical Society: 3247–3251. doi:10.1021/ic00164a016.

Lancaster, N. Llewellyn, Thomas Welton, and G. Brent Young. 2001. "A Study of Halide Nucleophilicity in Ionic Liquids." *Journal of the Chemical Society, Perkin Transactions 2* 12(November). The Royal Society of Chemistry: 2267–2270. doi:10.1039/b107381h.

Larsen, Anna S., John D. Holbrey, Fook S. Tham, and Christopher A. Reed. 2000. "Designing Ionic Liquids: Imidazolium Melts with Inert Carborane Anions." *Journal of the American Chemical Society*. doi:10.1021/JA0007511.

Lévêque, Jean-Marc, Simon Desset, Joel Suptil, Claude Fachinger, Micheline Draye, Werner Bonrath, and Giancarlo Cravotto.

2006. "A General Ultrasound-Assisted Access to Room-Temperature Ionic Liquids." *Ultrasonics Sonochemistry* 13(2). Elsevier: 189–193. doi:10.1016/J.ULTSONCH.2005.09.001.

Lévêque, Jean-Marc, Jean-Louis Luche, Christian Pétrier, Rudy Roux, and Werner Bonrath. 2002. "An Improved Preparation of Ionic Liquids by Ultrasound." *Green Chemistry* 4(4). The Royal Society of Chemistry: 357–360. doi:10.1039/B203530H.

Liu, Hongtao, Ping He, Zhiying Li, Yang Liu, and Jinghong Li. 2006. "A Novel Nickel-Based Mixed Rare-Earth Oxide/Activated Carbon Supercapacitor Using Room Temperature Ionic Liquid Electrolyte." *Electrochimica Acta* 51(10). Pergamon: 1925–1931. doi:10.1016/J.ELECTACTA.2005.06.034.

Liu, Shuangyue, Yu Chen, Yang Shi, Haitao Sun, Zhengyu Zhou, and Tiancheng Mu. 2015. "Investigations on the Thermal Stability and Decomposition Mechanism of an Amine-Functionalized Ionic Liquid by TGA, NMR, TG-MS Experiments and DFT Calculations." *Journal of Molecular Liquids* 206(June). Elsevier: 95–102. doi:10.1016/J.MOLLIQ.2015.02.022.

Longo, Luiz S., Emily F. Smith, and Peter Licence. 2016. "Study of the Stability of 1-Alkyl-3-Methylimidazolium Hexafluoroantimonate(V) Based Ionic Liquids Using X-Ray Photoelectron Spectroscopy." *ACS Sustainable Chemistry and Engineering* 4(11). American Chemical Society: 5953–5962. doi:10.1021/acssuschemeng.6b00919.

Luan, Feng, Shuang Zhang, Dandan Chen, Kun Zheng, and Xuming Zhuang. 2018. "CoS2-Decorated Ionic Liquid-Functionalized Graphene as a Novel Hydrazine Electrochemical Sensor." *Talanta* 182(May). Elsevier: 529–535. doi:10.1016/J.TALANTA.2018.02.031.

MacFarlane, D. R., S. A. Forsyth, J. Golding, and G. B. Deacon. 2002. "Ionic Liquids Based on Imidazolium, Ammonium and Pyrrolidinium Salts of the Dicyanamide Anion." *Green Chemistry* 4(5). The Royal Society of Chemistry: 444–448. doi:10.1039/b205641k.

Mamińska, Renata, Artur Dybko, and Wojciech Wróblewski. 2006. "All-Solid-State Miniaturised Planar Reference Electrodes Based on Ionic Liquids." *Sensors and Actuators. Part B: Chemical* 115(1). Elsevier: 552–557. doi:10.1016/J.SNB.2005.10.018.

Marsh, K. N., J. A. Boxall, and R. Lichtenthaler. 2004. "Room Temperature Ionic Liquids and Their Mixtures—A Review." *Fluid Phase Equilibria* 219(1). Elsevier: 93–98. doi:10.1016/J.FLUID.2004.02.003.

McFarlane, J., W. B. Ridenour, H. Luo, R. D. Hunt, D. W. DePaoli, and R. X. Ren. 2005. "Room Temperature Ionic Liquids for Separating Organics from Produced Water." *Separation Science and Technology* 40(6). Taylor & Francis Group: 1245–1265. doi:10.1081/SS-200052807.

Mei, Meng, Xiaojia Huang, and Lei Chen. 2019. "Recent Development and Applications of Poly (Ionic Liquid)s in Microextraction Techniques." *TrAC Trends in Analytical Chemistry* 112(March). Elsevier: 123–134. doi:10.1016/J.TRAC.2019.01.003.

Merlet, Céline, Benjamin Rotenberg, Paul A. Madden, and Mathieu Salanne. 2013. "Computer Simulations of Ionic Liquids at Electrochemical Interfaces." *Physical Chemistry Chemical Physics* 15(38). The Royal Society of Chemistry: 15781. doi:10.1039/c3cp52088a.

Merlet, Céline, Mathieu Salanne, Benjamin Rotenberg, and Paul A. Madden. 2011. "Imidazolium Ionic Liquid Interfaces with Vapor and Graphite: Interfacial Tension and Capacitance from Coarse-Grained Molecular Simulations." *The Journal of Physical Chemistry C* 115(33). American Chemical Society: 16613–16618. doi:10.1021/jp205461g.

Merrigan, Travis L., Eleanor D. Bates, Scott C. Dorman, and James H. Davis Jr. 2000. "New Fluorous Ionic Liquids Function as Surfactants in Conventional Room-Temperature Ionic Liquids." *Chemical Communications* 20(January). The Royal Society of Chemistry: 2051–2052. doi:10.1039/b005418f.

Mert, Samet, Bahar Bankoğlu, Abdullah Özkan, Necip Atar, and Mehmet Lütfi Yola. 2018. "Electrochemical Sensing of Ractopamine by Carbon Nitride Nanotubes/Ionic Liquid Nanohybrid in Presence of Other β-Agonists." *Journal of Molecular Liquids* 254(March). Elsevier: 8–11. doi:10.1016/J.MOLLIQ.2018.01.066.

Miao, Juan, Xin Wang, Yunchang Fan, Jing Li, Lina Zhang, Guitao Hu, Can He, and Can Jin. 2018. "Determination of Total Mercury in Seafood by Ion-Selective Electrodes Based on a Thiol Functionalized Ionic Liquid." *Journal of Food and Drug Analysis* 26(2). Elsevier: 670–677. doi:10.1016/J.JFDA.2017.08.004.

Nazari, Shidokht, Stanley Cameron, Michel B. Johnson, and Khashayar Ghandi. 2013. "Physicochemical Properties of Imidazo-Pyridine Protic Ionic Liquids." *Journal of Materials Chemistry A* 1(38). The Royal Society of Chemistry: 11570. doi:10.1039/c3ta12022h.

Ngo, Helen L., Karen LeCompte, Liesl Hargens, and Alan B. McEwen. 2000. "Thermal Properties of Imidazolium Ionic Liquids." *Thermochimica Acta* 357–358(August). Elsevier: 97–102. doi:10.1016/S0040-6031(00)00373-7.

Noda, Akihiro, Md. Abu, Bin Hasan Susan, Kenji Kudo, Shigenori Mitsushima, Kikuko Hayamizu, and Masayoshi Watanabe. 2003. "Brønsted Acid–Base Ionic Liquids as Proton-Conducting Nonaqueous Electrolytes." American Chemical Society. doi:10.1021/JP022347P.

Noda, Akihiro, Kikuko Hayamizu, and Masayoshi Watanabe. 2001. "Pulsed-Gradient Spin–Echo 1H and 19F NMR Ionic Diffusion Coefficient, Viscosity, and Ionic Conductivity of Non-Chloroaluminate Room-Temperature Ionic Liquids." American Chemical Society. doi:10.1021/JP004132Q.

Ortuño, Joaquín A., Francisca Tomás-Alonso, and Aurora M. Rubio. 2014. "Ion-Selective Electrodes Based on Ionic Liquids." *Ionic Liquids in Separation Technology* (January). Elsevier, 275–299. doi:10.1016/B978-0-444-63257-9.00009-2.

Plechkova, Natalia V., Robin D. Rogers, and Kenneth R. Seddon, eds. 2010. *Ionic Liquids: From Knowledge to Application*, Vol. 1030. ACS Symposium Series. Washington, DC: American Chemical Society. doi:10.1021/bk-2009-1030.

Plechkova, Natalia V., and Kenneth R. Seddon. 2008. "Applications of Ionic Liquids in the Chemical Industry." *Chemical Society Reviews* 37(1). The Royal Society of Chemistry: 123–150. doi:10.1039/B006677J.

Quinn, Bernadette M., Zhifeng Ding, Roger Moulton, and Allen J. Bard. 2002. "Novel Electrochemical Studies of Ionic Liquids." American Chemical Society. doi:10.1021/LA011458X.

Rajendran, A. 2010. "Applicability of an Ionic Liquid in the Removal of Chromium from Tannery Effluents: A Green Chemical Approach." *African Journal of Pure and Applied Chemistry* 4. http://www.academicjournals.org/ajpac.

Reichardt, C. 1979. *Solvent Effects in Organic Chemistry.* Verlag Chemie. https://patents.google.com/patent/CA2473839C/en.

Robinson, J., and R. A. Osteryoung. 1979. "An Electrochemical and Spectroscopic Study of Some Aromatic Hydrocarbons in the Room Temperature Molten Salt System Aluminum Chloride-n-Butylpyridinium Chloride." *Journal of the American Chemical Society* 101(2). American Chemical Society: 323–327. doi:10.1021/ja00496a008.

Rogers, Robin D., and Kenneth R. Seddon, eds. 2005. *Ionic Liquids IIIB: Fundamentals, Progress, Challenges, and Opportunities*, Vol. 902. ACS Symposium Series. Washington, DC: American Chemical Society. doi:10.1021/bk-2005-0902.

Sanders, John R., Edmund H. Ward, and Charles L. Hussey. 1986a. "Aluminium Bromide-1-Methyl-3-Ethylimidazolium Bromide Ionic Liquids: Densities, Viscosities, Electrical Conductivities, and Phase Transitions." *ECS Proceedings Volumes* 1986–1(January). The Electrochemical Society: 307–316. doi:10.1149/198601.0307PV.

Sanders, John R., Edmund H. Ward, and Charles L. Hussey. 1986b. "Erratum: 'Aluminum Bromide-1-Methyl-3-Ethylimidazolium Bromide Ionic Liquids. I. Densities, Viscosities, Electrical Conductivities, and Phase Transitions." [J. Electrochem. Soc., 133: 325 (1986)]. *Journal of the Electrochemical Society* 133(7). The Electrochemical Society: 1526. doi:10.1149/1.2152158.

Sato, Takaya, Gen Masuda, and Kentaro Takagi. 2004. "Electrochemical Properties of Novel Ionic Liquids for Electric Double Layer Capacitor Applications." *Electrochimica Acta* 49(21). Pergamon: 3603–3611. doi:10.1016/J.ELECTACTA.2004.03.030.

Scheffler, Towner B., and Charles L. Hussey. 1984. "Electrochemical Study of Tungsten Chloro Complex Chemistry in the Basic Aluminum Chloride-1-Methyl-3-Ethylimidazolium Chloride Ionic Liquid." *Inorganic Chemistry* 23(13). American Chemical Society: 1926–1932. doi:10.1021/ic00181a027.

Scheffler, Towner B., Charles L. Hussey, Kenneth R. Seddon, Christopher M. Kear, and Phillip D. Armitage. 1983. "Molybdenum Chloro Complexes in Room-Temperature Chloroaluminate Ionic Liquids: Stabilization of Hexachloromolybdate(2-) and Hexachloromolybdate(3-)." *Inorganic Chemistry* 22(15). American Chemical Society: 2099–2100. doi:10.1021/ic00157a001.

Seddon, Kenneth R., Annegret Stark, and María-José Torres. 2000. "Influence of Chloride, Water, and Organic Solvents on the Physical Properties of Ionic Liquids." *Pure and Applied Chemistry* 72(12). De Gruyter: 2275–2287. doi:10.1351/pac200072122275.

Seki, Shiro, Yo Kobayashi, Hajime Miyashiro, Yasutaka Ohno, Akira Usami, Yuichi Mita, Masayoshi Watanabe, and Nobuyuki Terada. 2006. "Highly Reversible Lithium Metal Secondary Battery Using a Room Temperature Ionic Liquid/Lithium Salt Mixture and a Surface-Coated Cathode Active Material." *Chemical Communications* 5(January). The Royal Society of Chemistry: 544–545. doi:10.1039/B514681J.

Sheldon, Roger. 2001. "Catalytic Reactions in Ionic Liquids." *Chemical Communications* 23(November). The Royal Society of Chemistry: 2399–2407. doi:10.1039/b107270f.

Shiddiky, Muhammad J. A., and Angel A. J. Torriero. 2011. "Application of Ionic Liquids in Electrochemical Sensing Systems." *Biosensors and Bioelectronics* 26(5). Elsevier: 1775–1787. doi:10.1016/J.BIOS.2010.08.064.

Shvedene, Natalia V., Denis V. Chernyshov, Maria G. Khrenova, Andrey A. Formanovsky, Vladimir E. Baulin, and Igor V. Pletnev. 2006. "Ionic Liquids Plasticize and Bring Ion-Sensing Ability to Polymer Membranes of Selective Electrodes." *Electroanalysis* 18(13–14). John Wiley & Sons, Ltd: 1416–1421. doi:10.1002/elan.200603537.

Song, Ye, Xufei Zhu, Xinlong Wang, and Mingjie Wang. 2006. "Characteristics of Ionic Liquid-Based Electrolytes for Chip Type Aluminum Electrolytic Capacitors." *Journal of Power Sources* 157(1). Elsevier: 610–615. doi:10.1016/J.JPOWSOUR.2005.07.085.

Stenger-Smith, John D., Cynthia K. Webber, Nicole Anderson, Andrew P. Chafin, Kyukwan Zong, and John R. Reynolds. 2002. "Poly(3,4-Alkylenedioxythiophene)-Based Supercapacitors Using Ionic Liquids as Supporting Electrolytes." *Journal of the Electrochemical Society* 149(8). The Electrochemical Society: A973. doi:10.1149/1.1485773.

Suarez, Paulo A. Z., Vânia M. Selbach, Jeane E. L. Dullius, Sandra Einloft, Clarisse M. S. Piatnicki, Denise S. Azambuja, Roberto F. de Souza, and Jairton Dupont. 1997. "Enlarged Electrochemical Window in Dialkyl-Imidazolium Cation Based Room-Temperature Air and Water-Stable Molten Salts." *Electrochimica Acta* 42(16). Pergamon: 2533–2535. doi:10.1016/S0013-4686(96)00444-6.

Sun, Hui, Dongju Zhang, Chengbu Liu, and Changqiao Zhang. 2009. "Geometrical and Electronic Structures of the Dication and Ion Pair in the Geminal Dicationic Ionic Liquid 1,3-Bis[3-Methylimidazolium-Yl]Propane Bromide." *Journal of Molecular Structure: THEOCHEM* 900(1–3). Elsevier: 37–43. doi:10.1016/J.THEOCHEM.2008.12.024.

Sun, I. Wen, Edmund H. Ward, and Charles L. Hussey. 1987. "Reactions of Phosgene with Oxide-Containing Species in a Room-Temperature Chloroaluminate Ionic Liquid." *Inorganic Chemistry* 26(26). American Chemical Society: 4309–4311. doi:10.1021/ic00273a007.

Sun, Ping, and Daniel W. Armstrong. 2010. "Ionic Liquids in Analytical Chemistry." *Analytica Chimica Acta* 661(1). Elsevier: 1–16. doi:10.1016/J.ACA.2009.12.007.

Varma, Rajender S., and Vasudevan V. Namboodiri. 2001a. "An Expeditious Solvent-Free Route to Ionic Liquids Using Microwaves." *Chemical Communications* 7(January). The Royal Society of Chemistry: 643–644. doi:10.1039/b101375k.

Varma, Rajender S., and Vasudevan V. Namboodiri. 2001b. "Solvent-Free Preparation of Ionic Liquids Using a Household Microwave Oven." *Pure and Applied Chemistry* 73(8). De Gruyter: 1309–1313. doi:10.1351/pac200173081309.

Vatamanu, Jenel, Oleg Borodin, and Grant D. Smith. 2010. "Molecular Insights into the Potential and Temperature Dependences of the Differential Capacitance of a Room-Temperature Ionic Liquid at Graphite Electrodes." *Journal of the American Chemical Society* 132(42). American Chemical Society: 14825–14833. doi:10.1021/ja104273r.

Vatamanu, Jenel, Oleg Borodin, and Grant D. Smith. 2011. "Molecular Simulations of the Electric Double Layer Structure, Differential Capacitance, and Charging Kinetics for *N* -Methyl- *N* -Propylpyrrolidinium Bis(Fluorosulfonyl) Imide at Graphite Electrodes." *Journal of Physical Chemistry. Part B* 115(12). American Chemical Society: 3073–3084. doi:10.1021/jp2001207.

Villagrána, Constanza, Craig E. Banks, Christopher Hardacre, and Richard G. Compton. 2004. "Electroanalytical Determination of Trace Chloride in Room-Temperature Ionic Liquids." American Chemical Society. doi:10.1021/AC030375D.

Visser, Ann E., Richard P. Swatloski, W. Matthew Reichert, Scott T. Griffin, and Robin D. Rogers. 2000. "Traditional Extractants in Nontraditional Solvents: Groups 1 and 2 Extraction by Crown Ethers in Room-Temperature Ionic Liquids." American Chemical Society. doi:10.1021/IE000426M.

Visser, Ann E., Richard P. Swatloski, W. Matthew Reichert, Rebecca Mayton, Sean Sheff, Andrzej Wierzbicki, Jr., James H. Davis, and Robin D. Rogers. 2002. "Task-Specific Ionic Liquids Incorporating Novel Cations for the Coordination and Extraction of Hg2+ and CD2+: Synthesis, Characterization, and Extraction Studies." American Chemical Society. doi:10.1021/ES0158004.

Vu, Peter D., Andrew J. Boydston, and Christopher W. Bielawski. 2007. "Ionic Liquids via Efficient, Solvent-Free Anion Metathesis." *Green Chemistry* 9(11). The Royal Society of Chemistry: 1158. doi:10.1039/b705745h.

Walden, P. 2019. "Ueber die Molekulargrösse und Elektrische Leitfähigkeit Einiger Geschmolzenen Salze." *Bulletin de l'Académie Impériale Des Sciences de St.-Pétersbourg. VI Série* 8(6) (1914): 405–422. Accessed June 18. http://www.mathnet.ru/php/archive.phtml?wshow=paper&jrnid=im&paperid=6491&option_lang=eng.

Wang, Binshen, Li Qin, Tiancheng Mu, Zhimin Xue, and Guohua Gao. 2017. "Are Ionic Liquids Chemically Stable?" *Chemical Reviews* 117(10). American Chemical Society: 7113–7131. doi:10.1021/acs.chemrev.6b00594.

Wang, Wei, Ramjee Balasubramanian, and Royce W. Murray. 2008. "Electron Transport and Counterion Relaxation Dynamics in Neat Ferrocenated Imidazolium Ionic Liquids." *The Journal of Physical Chemistry C* 112(46). American Chemical Society: 18207–18216. doi:10.1021/jp806132j.

Wang, Xing, Yanying Wang, Xiaoxue Ye, Tsunghsueh Wu, Hongping Deng, Peng Wu, and Chunya Li. 2018. "Sensing Platform for Neuron Specific Enolase Based on Molecularly Imprinted Polymerized Ionic Liquids in Between Gold Nanoarrays." *Biosensors and Bioelectronics* 99(January). Elsevier: 34–39. doi:10.1016/J.BIOS.2017.07.037.

Wang, Yanying, Chunya Li, Tsunghsueh Wu, and Xiaoxue Ye. 2018. "Polymerized Ionic Liquid Functionalized Graphene Oxide Nanosheets as a Sensitive Platform for Bisphenol A Sensing." *Carbon* 129(April). Pergamon: 21–28. doi:10.1016/J.CARBON.2017.11.090.

Wardak, Cecylia, and Joanna Lenik. 2013. "Application of Ionic Liquid to the Construction of Cu(II) Ion-Selective Electrode with Solid Contact." *Sensors and Actuators. Part B: Chemical* 189(December). Elsevier: 52–59. doi:10.1016/J.SNB.2012.12.065.

Wasserscheid, Peter, Birgit Drießen-Hölscher, Roy van Hal, H. Christian Steffens, and Jörg Zimmermann. 2003. "New, Functionalised Ionic Liquids from Michael-Type Reactions—A Chance for Combinatorial Ionic Liquid Development." *Chemical Communications* 16(July). The Royal Society of Chemistry: 2038–2039. doi:10.1039/B306084E.

Wasserscheid, Peter, and Thomas Welton, eds. 2007. *Ionic Liquids in Synthesis*. Wiley. doi:10.1002/9783527621194.

Wei, Di, and Ari Ivaska. 2008. "Applications of Ionic Liquids in Electrochemical Sensors." *Analytica Chimica Acta* 607(2). Elsevier: 126–135. doi:10.1016/J.ACA.2007.12.011.

Wells, Andrew S., and Coombe, Vyvyan T. 2006. "On the Freshwater Ecotoxicity and Biodegradation Properties of Some Common Ionic Liquids." American Chemical Society. doi:10.1021/OP060048I.

Welton, Thomas. 1999. "Room-Temperature Ionic Liquids. Solvents for Synthesis and Catalysis." American Chemical Society. doi:10.1021/CR980032T.

Widegren, Jason A., Arno Laesecke, and Joseph W. Magee. 2005. "The Effect of Dissolved Water on the Viscosities of Hydrophobic Room-Temperature Ionic Liquids." *Chemical Communications* 12(March). The Royal Society of Chemistry: 1610. doi:10.1039/b417348a.

Wilkes, John S. 2002. "A Short History of Ionic Liquids—From Molten Salts to Neoteric Solvents." *Green Chemistry* 4(2). The Royal Society of Chemistry: 73–80. doi:10.1039/b110838g.

Wilkes, John S., Joseph A. Levisky, Robert A. Wilson, and Charles L. Hussey. 1982. "Dialkylimidazolium Chloroaluminate Melts: A New Class of Room-Temperature Ionic Liquids for Electrochemistry, Spectroscopy and Synthesis." *Inorganic Chemistry* 21(3). American Chemical Society: 1263–1264. doi:10.1021/ic00133a078.

Wilkes, John S., and Michael J. Zaworotko. 1992. "Air and Water Stable 1-Ethyl-3-Methylimidazolium Based Ionic Liquids." *Journal of the Chemical Society, Chemical Communications* 13(January). The Royal Society of Chemistry: 965. doi:10.1039/c39920000965.

Williams, Stephen D., J. P. Schoebrechts, J. C. Selkirk, and G. Mamantov. 1987. "A New Room Temperature Molten Salt Solvent System: Organic Cation Tetrachloroborates." *Journal of the American Chemical Society* 109(7): 2218–2219. doi:10.1021/ja00241a069.

Xue, Zhimin, Yuwei Zhang, Xiao-qin Zhou, Yuanyuan Cao, and Tiancheng Mu. 2014. "Thermal Stabilities and Decomposition Mechanism of Amino- and Hydroxyl-Functionalized Ionic Liquids." *Thermochimica Acta* 578(February). Elsevier: 59–67. doi:10.1016/J.TCA.2013.12.005.

Zhao, Dongbin, Zhaofu Fei, T. Geldbach, R. Scopelliti, G. Laurenczy, and P. Dyson. 2005. "Allyl-Functionalised Ionic Liquids: Synthesis, Characterisation, and Reactivity." *Helvetica Chimica Acta* 88(3). John Wiley & Sons, Ltd: 665–675. doi:10.1002/hlca.200590046.

Zhou, Dezhi, Geoffrey M. Spinks, Gordon G. Wallace, Churat Tiyapiboonchaiya, Douglas R. MacFarlane, Maria Forsyth, and Jiazeng Sun. 2003. "Solid State Actuators Based on Polypyrrole and Polymer-In-Ionic Liquid Electrolytes." *Electrochimica Acta* 48(14–16). Pergamon: 2355–2359. doi:10.1016/S0013-4686(03)00225-1.

Zhou, Qing, Xingmei Lu, Suojiang Zhang, and Liangliang Guo. 2014. "Physicochemical Properties of Ionic Liquids." In *Ionic Liquids Further UnCOILed*, 275–307. Hoboken, NJ, USA: John Wiley & Sons, Inc. doi:10.1002/9781118839706.ch11.

Zhu, Xudong, Yanbo Zeng, Zulei Zhang, Yiwen Yang, Yunyun Zhai, Hailong Wang, Lingyu Liu, Jian Hu, and Lei Li. 2018. "A New Composite of Graphene and Molecularly Imprinted Polymer Based on Ionic Liquids as Functional Monomer and Cross-Linker for Electrochemical Sensing 6-Benzylaminopurine." *Biosensors and Bioelectronics* 108(June). Elsevier: 38–45. doi:10.1016/J.BIOS.2018.02.032.

16

Fabrication and Functionalization of Other Inorganic Nanoparticles and Nanocomposites

Kiranmai Mandava and Uma Rajeswari B.

CONTENTS

16.1 Iron (Fe) and Iron Oxide Nanoparticles (IONPs)

Iron has four unpaired electrons, whereas iron oxide has two unpaired electrons. Due to the presence of four unpaired electrons, iron behaves as a magnet and iron oxide behaves as a paramagnetic material because it has only two unpaired electrons. The particle size of super magnetic iron nanoparticles ranges from 50 to 180 nm (Preeti *et al.*, 2017). At present, the application of iron oxide nanoparticles (IONPs) in biomedical, medical, and other fields appears to be increasing fast. IONPs, with their super paramagnetic properties, have been used in separation, tracking, and cell labelling, they have also been employed as therapeutic agents in cancer therapy and as diagnostic agents (Heo *et al.*, 2014).

16.1.1 Fabrication of Magnetic Nanoparticles for Biomedical Applications

Chemically magnetic nanoparticles can be synthesized using solution techniques or from aerosol/vapor phases. Nanocomposites consisting of magnetic nanoparticles are dispersed in submicron-sized organic or inorganic matrixes that usually have a spherical shape. Magnetic nanoparticles are produced by solution techniques such as precipitation, co-precipitation, microemulsions, polyol techniques, thermolysis of precursors, hydrothermal reactions, and sol-gel

and flow injection synthesis. Other solution techniques have been developed for use in *in vivo* applications, such as the synthesis of magnetite/maghemite particles prepared by oxidation of apoferitin with trimethylamino-N-oxide, which was loaded with various amounts of Fe (II) ions, of special interest in the use of dendrimers as templating hosts for the production of magnetic nanoparticles (Tartaj *et al.*, 2003). Along with sonochemical-assisted synthesis, electrochemical methods have also been employed for the production of maghemite nanoparticles. (Sophie *et al.*, 2008). However, the most common method for the production of magnetic nanoparticles is the chemical co-precipitation technique using iron salts.

Apart from these, aerosol techniques such as laser pyrolysis and spray are significant because these technologies are continuous chemical processes, allowing for high-rate and well-defined nanoparticle production. Their high rate of yield can anticipate a promising future for the synthesis of magnetic nanoparticles useful in both *in vivo* and *in vitro* applications. Various techniques have been discussed; however, the synthesis of super magnetic nanoparticles is a complex process due to their colloidal nature, complex purification process, and difficulty in designing experimental conditions leading to a monodisperse population of magnetic grains of uniform size (Carvalho *et al.*, 2013). More recently, one can use composites that consist of super paramagnetic nanocrystals dispersed in submicron diamagnetic matrixes having long sedimentation times in the absence of a magnetic field. Functionality and biocompatibility are the advantages of using diamagnetic matrixes. Other methods, such as the deposition and encapsulation of magnetic nanoparticles with polymeric and inorganic matrixes, were also used for production. Magnetic resonance imaging, magnetic cell separation, or magneto relaxometry are the biomedical applications which utilize the magnetic properties of the nanoparticles in magnetic fields, but the function is size-dependent. Currently used common methods for the fractionation of magnetic fields are centrifugation (Sjogren *et al.*, 1997) and size-exclusion chromatography (Nunes *et al.*, 1987). These methods separate the particles based on their non-magnetic properties, such as density or size (Massart, 1981).

16.1.2 Functionalization of Magnetic Nanoparticles

Magnetic nanoparticles have many unique properties such as super paramagnetic high coercivity, low Curie temperature, high magnetic susceptibility, small size, and low toxicity. Magnetic IONPs possess high surface energies due to a large surface to volume ratio. Consequently, they tend to aggregate to minimize the surface energies. Moreover, naked IONPs have high chemical reactivity and they are susceptible to atmospheric oxidation, especially magnetite, resulting in a loss of magnetism and dispersibility. Therefore, strategies comprise grafting by coating with organic molecules (including small organic molecules or surfactants) (Gupta and Gupta, 2005), polymers (Gupta *et al.*, 2004), and biomolecules (Tiefenauer *et al.*, 1993), or coating with an inorganic layer such as silica (Ashtari *et al.*, 2005), metal or nonmetal elementary substances (Mandal *et al.*, 2005), metal oxide, or metal sulfide (Natile and Glisenti *et al.*, 2002) to improve the stability of nanoparticles which can also be used for further functionalization to bind to ligands (Meiwu *et al.*, 2008). One of the experimentally easiest and most versatile ways to functionalize nanoparticles is the avidin-biotin coupling strategy. Because many ligands are commercially available in a biotinylated form, this method is often used for research purposes. In this method, nanoparticles would be functionalized in a layer by layer pattern. The main disadvantages are agglomeration during layer by layer assembly and subsequent purification (Amstad *et al.*, 2009). Proteins are one of the most promising materials of functionalization among several potential biocompatible and biodegradable substances serving as a protective layer on super paramagnetic iron oxide nanoparticles (SPIONs) (Koneracka *et al.*, 1999). They can provide biocompatibility, high solubility, and hydrophilicity. Several materials have been introduced as coating agents to improve the colloidal stability of the ferrofluid, namely bovine serum albumin (BSA), starch, and poly (ethylene glycol) (PEG). BSA was one among several types of protein molecules used for the immobilization of the surface of nanoparticles. Special functional groups are required on the surface of the nanoparticles to covalently attach the BSA molecules (Mikhaylova, 2004). However, the hydrothermal approach was used to produce magnetite (Fe_3O_4) nanoparticles with relatively uniform size and size distribution under specific conditions of elevated temperature and pressure (Ge *et al.*, 2009). Pramanik and coworkers developed an approach to form 3-aminopropyltriethoxysilane (APTS)-coated IONPs by adding APTS into the mixture solution of iron (II) and iron (III) ions for controlling the co-precipitate formation of IONPs. In this method, IONPs can be further silanized by APTS by taking advantage of the rich hydroxy groups on the IONP surfaces, which can lead to the synthesis of IONPs with surface amino groups (Wu *et al.*, 2008).

Functionalization of nanoparticles with surfactant molecules, such as the self-assembly of surfactants on flat substrates, can be used to address the challenge of nanomaterials which need functionalized nanoparticles with molecular engineered interfacial interactions. When these surfactant molecules have the appropriate substitute at their ω-position, interparticle interactions may provide the driving force for the self-assembly of three-dimensional structures (Ulman *et al.*, 1995). Self-assembled monolayers of different adsorbents on solid surfaces have become one of the central themes in modern materials science and molecular level engineering. They can be helpful in producing efficient and economically functionalized nanoparticles. Recently, this technology has been used to synthesize platinum as well as gold, palladium, and iron nanoparticles (Ulman, 1996). However, it is clear that such nanoparticles will not find a wide range of use due to their relatively high price. But they can be utilized as sensor or bioassay applications.

Due to monodispersity in size, shape, and composition, high-moment super paramagnetic nanoparticles (SPMNPs) prepared in organic phase are ideal for biomedical applications. However, these SPMNPs are normally stabilized by surfactants such as oleic acid/oleylamine, which are hydrophobic in nature. So, such hydrophilic nanoparticles need to be functionalized via surfactant exchange or surfactant addition for biological applications (Shin *et al.*, 2009).

In comparison to the above methods, the bifunctional ligand exchange strategy provides better colloidal stability of the magnetic nanoparticles in physiological conditions due to strong interactions between the bidentate or multidentate functional groups with the iron oxide surface (Xu *et al.*, 2004). As well as in situ coating becomes well known as it requires no surfactant exchange process to disperse the NP in aqueous media owing to presence of multidentate ligands that lead to the development of biocompatible and thermodynamically stable dispersions of NP in aqueous media. Magnetic nanoparticles have been made in the presence of a variety of carboxylates (i.e, citric acid), or other multi-dentate ligands to obtain thermodynamically stable and biocompatible dispersions of nanoparticles in aqueous media (Laurent *et al.*, 2008). Super paramagnetic iron oxide nanoparticles (SPIONPs), functionalized with certain targeting moieties, can be used for the imaging of molecular markers associated with disease development. However, molecular markers in small lesions usually present at a low level, which makes it very challenging for the SPIONPs to generate detectable contrast in MRI. Therefore, synthesizing SPIONPs with substantial signal enhancement to improve sensitivity has been the troubleshooting issue in SPIONP-based disease detection and diagnosis. Various types of SPIONPs with different cores and surface coating were synthesized, and few are currently under clinical trials, which determines a signal enhancement (Gillis *et al.*, 2002; Brooks *et al.*, 2001). The functionalization of magnetic microspheres with specific ion-exchange separation groups such as dimethyl aminoethyl, carboxymethyl, sulphopropyl, and a few hydrophobic groups, can be performed in a similar fashion as a typical functionalized agarose/cellulose microsphere. These synthetic strategies are based on the activation of polysaccharide in alkali followed by the addition of a chloride-carbon bond containing compounds such as chloroacetic acid or 2-chloroethyldiethylamine (Boeden *et al.*, 1991). The hydrophobic group can be attached via n-alkylamines/carbonyl schiff base coupling (Perka *et al.*, 1976).

Magnetic nanoparticles are a major class of nanoscale materials with the potential ability to revolutionize current clinical diagnostic and therapeutic techniques (Mishra *et al.*, 2009; Such *et al.*, 2009; Kral *et al.*, 2009). Moreover, other applications of magnetic nanoparticles are being widely studied, including magnetically assisted gene therapy, magnetically enhanced transfection, magnetically induced hyperthermia, and magnetic-force-based tissue engineering (Corchero *et al.*, 2009). Recently, iron oxide nanoparticles functionalized with dendrimers were developed for the magnetic resonance imaging of tumors (Shi *et al.*, 2008). Other applications include molecular imaging, which is one of the most reliable applications of targeted IONPs. The modified cellular enzyme-linked immune sorbent assay (ELISA), is named cellular magnetic linked immune sorbent assay (C-ELISA) and has been developed as an application of MRI for *in vitro* clinical diagnosis (Burtea *et al.*,2006). Another important application is the functionalization for *in vitro* protein or cell separation, drug delivery, and treatment of hyperthermia (Safarik *et al.*, 2004; Xu *et al.*, 2004). SPIONs with biocompatible polymers having a controlled size can also be used as targeted delivery vehicles (Wahajuddin et al., 2012). Feridex is a type of SPION which is the representative contrast agent in MRI (Bulte *et al.*, 2004). In targeted drug delivery systems, IONPs can be used as immobilizing agents for bioactive substances (Laurent *et al.*, 2008).

16.1.3 Nanocomposites of MNPs

Various biological methods, such as proteins transferring, anti-carcinoembryonic antigen monoclonal antibody rch24, herceptin, and chlorotoxin, have been conjugated onto IONP surfaces. Unfortunately, these ligands tend to display immunogenicity, and the biological macromolecules used are very expensive and not available for many types of cancer, which thereby limits their applications. This is largely a result of difficulties related to the *in vivo* stability and macrophase uptake of many folic acid, modified magnetic nanoparticles, which are the most widely used ligand for targeting cancer. To overcome the problems in the magnetic resonance imaging of a tumor, dendrimers are recently being used for the surface modification of IONPs in imaging applications. Dendrimers, especially poly(amidoamine) dentdimers, have shown the ability to conjugate targeting ligands, imaging agents, and drug molecules for targeted cancer therapeutics (Shi *et al.*, 2008). The hydrophilic polymers that can stabilize the particles in aqueous media and also provide functional groups for conjugation of biological molecules are attractive materials because of their ready availability. Thus, there is a need to enlarge the portfolio of IONP-stabilizing and functionalizing polymers, especially with vinyl polymers which are relatively easy to generate under laboratory conditions. Commercially available polymers do not possess the reactive groups necessary for more sophisticated bioconjugations with high selectivity and reactivity. To incorporate functional groups with higher reactivity and selectivity to polymer-coated IONPs, additional modification steps are inevitable. In this context, living radical polymerization techniques such as reversible addition-fragmentation chain transfer polymerization (RAFT) offer an excellent platform for the fabrication of polymer coatings on IONPs via grafting approaches by using heterotelechelic polymer (Cyrille B *et al.*, 2009). As well as aqueous dispersed super magnetic IONP at physiological salt concentrations lead to development of iron oxide cores which can be coated with polymers to form dispersant type of polymer coating. Commercially available super paramagnetic IONP intended for magnetic labeling, cell separation purposes, and as magnetic resonance contrast agents are typically coated with sugars such as dextran or synthetic polymers such as silicone (Jung CW *et al.*, 1995). Nanoparticles have been used as fillers both in polymeric nanocomposites – to improve mechanical, electrical, and optical properties – and in metallic nanocomposites – to control electro deposition. Polymer nanocomposites reinforced with inorganic nanoparticles have attracted a lot of interest due to their lightness, homogeneity, cost-effective-process availability, and tunable physical properties. Vinyl-ester resin polymers with IONPs possesses high resistance to moisture and chemicals and good mechanical properties. Furthermore, the functional groups of the polymers surrounding the nanoparticles make these nanocomposites good candidates for various applications in the biomedical area (Zhanhu G *et al.*, 2008).

16.2 Copper Nanoparticles

Among the different metal particles, copper nanoparticles have attracted considerable interest because of their unique physical, chemical, optical, catalytic, mechanical, and electrical properties; moreover, their low-cost preparation has been of great interest recently. Metallic copper is a rather noble material among the elements of group II. Copper is the most abundant material in the Earth's crust, but at the same time the least stable in its metallic form. The price to properties ratio of this element is excellent and makes it very suitable, in a similar way to silver in large scale applications. Copper nanoparticles possess a wide range of applications in the fields of metallurgy, catalysis, and nano and optoelectronics (Dhas *et al.*, 1998; Vitulli *et al.*, 2002; Zhou, 2004; Quaranta *et al.*, 2004).

16.2.1 Problems in the Fabrication of Copper Nanoparticles

Stability and reactivity make the integration of copper nanoparticles in new-generation nano devices difficult. The most common methods employed for the preparation and stabilization of copper nanoparticles are the chemical, physical, and biological methods. To avoid stability problems, the reduction methods are usually preferred in inert-atmosphere (argon, nitrogen) inorganic solvents, micro emulsion systems, and in the presence of surfactants (Pileni, 1998; Kapoor *et al.*, 2003; 2004; Song *et al.*, 2004; Wu *et al.*, 2006).

16.2.2 Methods of Fabrication of Copper Nanoparticles

16.2.2.1 Chemical Methods (Lisiecki, 1993)

These include chemical reduction, electrochemical techniques, photochemical reduction, thermal decomposition, and dissociation with ionic fluids. Among chemical methods, reduction is the most frequently applied method for the preparation of stable colloidal dispersions in organic solvents (Wu *et al.*, 2006). A simple and relatively cost-effective method for obtaining copper nanoparticles is the dissociation of micro sized copper particles with ionic fluids (Kim *et al.*, 2009).

16.2.2.2 Physical Methods

Physical methods include laser ablation, gas evaporation, and exploding wire (Zhou *et al.*, 2008)

16.2.2.3 Physicochemical Method

Apart from the ones descripted above, an interesting method is the formation of copper nanoparticles by arc discharge between two copper cylindrical electrodes submerged into a liquid phase. Here a physical process coexists with a chemical one, the reaction of plasma with the solution (Xie *et al.*, 2004). As well as dual plasma, the single-step process is another interesting and recent method using arc discharge for obtaining surface-stable copper nanoparticles (Tavares and Coulombes, 2011).

16.2.2.4 Biological Methods

Living organisms have huge potential for the fabrication of nanoparticles for wide applications (Ghorbani *et al.*, 2011). Moreover, only a few of the reported synthetic methods have the potential of producing stable copper nanoparticles due to the higher oxidation rate (Aslam *et al.*, 2002). Because of the limitations of physical and chemical methods, green nanotechnology has attracted significant attention because it enhances the quality of the environment, reduces pollution, and conserves natural and non-renewable resources. However, many reaction conditions and toxic chemicals may not be suitable for biological and biochemical application (Rajender, 2011). Therefore, various green stabilizing agents such as polyphenols, enzymes, citric acid, vitamins (B, C, D, K), biodegradable polymers, and silica are able to stabilize and functionalize metallic nanoparticles without showing any deleterious effect on the environment and biosystems.

Among the above-mentioned methods, the chemical method is the most applicable to synthesize copper nanoparticles. Toxicity is a drawback of chemical methods; in addition, many of these methods consume high energy, although, there is a need to synthesize copper nanoparticles by chemical methods rarely. In contrast, biological methods are alternative methods performed in environmental conditions and without consuming energy. But the time they require to synthesize copper nanoparticles is higher in comparison with chemical methods. Physical methods are usually fast and do not involve any toxic chemicals. The process is faster than other methods, but the high cost of the equipment is the general disadvantage (Ghorbani, 2014).

16.2.3 Coating of Copper Nanoparticles with Protective Agents

Nanoparticles of copper produced by all of the reported methods need protective agents such as amphiphilic molecules or polymers, which, forming a layer on the surface of the reduced species, protect the particle from both oxidation and self-aggregation. The coating is the finest approach to overcome the reactivity and stability of copper nanoparticles.

An interesting alternative way of protecting copper particles is by covering them with another, more inert metal. The concept of core-shell nanoparticles comprising a core or inner material and a shell or outer material is very widely used. Many possible combinations of organic and inorganic materials can be employed. In this type of core-shell, metal is covered with suitable capping organic discrete or polymeric molecules (Severance *et al.*, 2009; Benavente, *et al.*, 2013). Apart from the methods we mentioned, other chemical methods such as self-assembly in sol-gel (Yesh chenko *et al.*, 2007), selective chemical etching, etc. (Meziani *et al.*, 2009), have been extensively used to fabricate metal nano-structures. In the sol-gel method, copper nanoparticles were grown in silica matrix by a sol-gel-prepared porous matrix impregnated with copper nitrate followed by post-annealing.

To date, carbon and silica seem to be the best polymers for capping nanoparticles. Researchers have developed several techniques such as modified Kratschmer–Huffman arc evaporation geometries (Seraphens, 1993), hydrocarbon decomposition (Klug, 2002), arc discharge in deionized water (Ang *et al.*, 2004a), hydrolysis of tetraethoxysilicate (Klug *et al.*, 2003), the plasma polymerization method (He *et al.*, 2002), and plasma torch synthesis (Turgut *et al.*, 1999) to coat nanoparticles. As a long-term plan, a new method was developed to coat nanoparticles online. The term online coating implies that nanoparticles fabricated by the flow-levitation method directly enter a chamber in which carbon and hydrogen (CH) plasma produced by a hollow-cathode glow discharge are sprayed onto the surfaces of the nanoparticles simultaneously (Haile *et al.*, 2004). Copper nanoparticles were prepared by reducing copper acetate with sodium borohydride in the presence of alkyne ligands, which can effectively cap and protect the nanoparticles. These nanoparticles exhibited apparent electrocatalytic activity in oxygen reduction in alkaline media, and were markedly better than those reported earlier with poly- or single-crystalline copper electrodes (Keliu *et al.*, 2014). The antimicrobial effects of metal nanoparticles have been attracting the researchers' attention from both medical and technological standpoints. Copper nanoparticles have synergized the antimicrobial activity toward a broad spectrum of microorganisms, including pathogenic bacteria (Gunawan *et al.*, 2011). Due to their antibacterial potency, copper nanoparticles have been widely used for biomedical tools, hospital equipment, storage equipment, household materials, food processing, and antifouling paints (Mann *et al.*, 2002). Apart from the above applications, copper nanoparticles can also be used as super strong materials (Wang, *et al.*, 1999), heat transfer systems (Mandal *et al.*, 2006), sensors (Kang *et al.*, 2007), and pastes. Conductive inks containing copper nanoparticles can be used as a substituent for very expensive noble metals employed in printed electronics, and used to prepare transmissive thin films for displays of electronic items (mobile, laptops, LED etc) (Athanassiou *et al.*, 2007).

16.2.4 Functionalization

However, pure metal nanoparticles are difficult to prepare and have poor stability for electrolysis due to their susceptibility to atmosphereic and solution oxidization (Aoun *et al.*, 2004). Recently, nickel and copper nanoparticles have been used for sequential deposition on multi-walled carbon nanotube–modified electrodes using a simple electrochemical method as an effective non-enzymatic sensor, or the catalytic oxidation of glucose (Kuo-chiang *et al.*, 2013). Moreover, recently, dimethylglyoxime-functionalized copper nanoparticles were fabricated by a microwave method to construct a new glucose sensor (Xu *et al.*, 2006). When copper nanoparticles were functionalized with organic/inorganic coatings or well defined capping molecules on an electrode, they could decrease the surface ohmic drop (Khulbe *et al.*, 1980) so as to prevent the dissolution of the copper oxide and increase the stability of the electrode (Luo, 1994). Recently, it has reported that the functionalization of copper nanoparticles on plasma-activated high-density polyethylene via dithiol inter layer shows magnetic properties (Reznickova *et al.*, 2016).

Nowadays, researchers have focused on the most common type of catalysts comprising metal oxide nanoparticles deposited on inorganic or organic porous supports leading to the development of stabilized highly dispersed catalytic active phases. They have prepared catalysts with different contents of copper deposited on a functional two-dimensional hexagonal SBA-15-like mesostructure containing the original polyether groups of the surfactant covalently bound to the silica framework, and a dispersing agent for the copper precursors to be subsequently impregnated/infiltrated into the support porosity, respectively by using incipient wetness impregnation followed by mild drying and the solvent-free melt infiltration method (Rudolf *et al.*, 2005). Since numerous oxide nanoparticles are manufactured industrially by flame synthesis, carbon-protected copper nanoparticles have been developed by the flame spray-based process (Athanassious *et al.*, 2006). Moreover to prolong the life span of magnetic nanoparticles and to avoid undesired effects such as aggregation in aqueous solution and organic solvents, researchers have recently developed biocompatible stabilizing agents such as biodegradable polymers and enzymes for the surface functionalization of magnetic nanoparticles (Jurate, 2011).

Recently, amine-functionalized nanoparticles were prepared by the surface functionalization of copper-engineered nanomaterials using dimethylaminopyridine by meliorum technologies. The nanomaterials were used to assess the impact of functionalization on physiological performance in a euryhaline fish, *Fundulus heteroelitus*, and to characterize the effects of salinity on engineered nanomaterial toxicity (Black *et al.*, 2017). Recently, a simple and eco-friendly method for copper nanoparticles capped with L-cystein in aqueous solution has been developed. L-cystein was used as a functionalization agent because of its active groups, such as ammonia and carboxylate ion, which make it a promising candidate for capping and functionalization, especially in biologically applications (Nikhil Kumar, 2016). To prevent nanosized copper organosol from aggregation, polyvinyl pyrrolidine–protected organosol under refluxing condition (Curtis *et al.*, 1988) and alkane thiolate–protected copper clusters (Chen *et al.*, 2001) were prepared. Self-assembled monolayers of alkane thiolate on copper particles using thiols of different chain lengths in alcohol (Sung *et al.*, 2000) have been developed. Furthermore, a new method has been reported in which cystein-functionalized copper organosols were developed. This was found to be an effective catalyst for the synthesis of octyl phenyl ether (Ang *et al.*, 2004; Panigrahi *et al.*, 2006). The functionalization of nanoparticles with selective biomolecules for eco-friendly antibacterial applications will improve the stability, aqueous dispersion, and control of eco-toxicological effects which require clarification prior to their industrial applications (Veerapandian *et al.*, 2012).

16.2.5 Nanocomposites

The high chemical reactivity of fabricated nanoparticles is found to be the main reason for aggregation, which reduces the specific surface area and the conductivity interfacial free energy, thereby diminishing the overall reactivity (Pomogailo and Keselman, 2005; He *et al.*, 2007). In recent years, nanoparticles and polymer composites have become significant owing to their nanosize and large surface area. In order for the copper nanoparticles to be protected from oxidation, they have been encapsulated in carbon/silica polyethylene and pure carbon as nanocomposite (Nasibulin *et al.*, 2005). Recent reports have shown that copper nanoparticles/polymer composites exhibit promising antifungal and antibacterial properties. A copper nanoparticle functionalized with an acyclic group, which stabilizes it, is another approach by which one can overcome the stability problem. It has been reported that copper nanoparticles functionalized with acrylic groups are less susceptible to aggregation within a matrix. The acrylic functionality of the nanoparticles can be copolymerized with other acrylics (Kelechi *et al.*, 2008).

16.2.6 Copper Oxide Nanoparticles (CONPs)

CONPs have been attracting a lot of attention because of the importance of copper in modern technologies and of its ready availability. They exhibit improved activity in catalytic, optical, electrical, and mechanical properties compared with ordinary copper oxide due to their peculiar chemical and physical properties, such as the superiority of the quantum size effect, surface effect, macroscopic quantum tunneling effect, and volume effect (Xi *et al.*, 2010; He *et al.*, 2007; Motogoshi *et al.*, 2010).

16.2.6.1 Fabrication of CONPs

Well-defined CONPs can be fabricated by different ways and parameters like solvents, surfactants, methods, and temperature, and precursors are used to control the shape and size of the desired nanoparticles (Suleiman *et al.*, 2013). CONPs have been prepared through synthetic methods such as sonochemical synthesis with the support of ultrasounds, with a reaction time up to 30 min and calcination at 600–700 °C (Wongpisutpaisan *et al.*, 2011), thermal decomposition of the oxalate precursor (Ghane *et al.*, 2010), copper acetyl acetonate (Nasibulin *et al.*, 2001), simple and economical wet chemical method (Ethiraj and Kang, 2012), aqueous precipitation method (Lanje *et al.*, 2010), hydrothermal microwave procedure after thermal treatment at 393 k for 1 h (Volanti *et al.*, 2008), modified sol-gel technique using sodium dodecyl sulphate as a surfactant (Nithya *et al.*, 2014), chemical reduction method (Karthik and Kannappan, 2013), calcinations at a high temperature from 300 to 400 °C (Srivastava *et al.*, 2013), mechanochemical process (Ayask *et al.*, 2015), reverse micelle technique (Rani *et al.*, 2014), simple sol-gel method (Radhakrishnan and Beena, 2014), simple precipitation technique (Ahamed *et al.*, 2014), pulsed laser ablation technique (Gondal *et al.*, 2012), quick precipitation method (Wua *et al.*, 2010), electrochemical method (Waichal *et al.*, 2012), spray pyrolysis (Etefagh *et al.*, 2013), solid–liquid arc discharge process (Yao *et al.*, 2015), and alcothermal method (Tetsuya Kida *et al.*, 2007), among others.

To achieve strong antibacterial features, a novel composition of CONP : iron nanoparticles with sizes ranging from 10 to 60 nm was synthesized successfully by the sol-gel method at 400 °C (Hoseinia *et al.*, 2015). So far, different methods have been used to fabricate CONPs. The green synthesis of CONPs usesplants and microorganisms such as *Gloriosa superba* leaves (Raja Naikaa *et al.*, 2015), brown algae (*Bifurcaria bifurcate*) extract (Abboud *et al.*, 2014), flower extracts of *Cassia alata* L. (Jayalakshmi and Yogamoorthi, 2014), *Centella asiatica* leaf extract (Devi *et al.*, 2014), fig leaves (*Ficus carica*) (Gultekin *et al.*, 2016), *Arbutus anedo* leaf extract (Yanping *et al.*, 2018), *Leucaena leucophala* L. leaves extract (Yogesh *et al.*, 2017), *Hibiscus rosa-sinensis* flower extract (Rajendran *et al.*, 2018), *Albizia lebbeck* leaf extract (Jayakumarai *et al.*, 2016), and extracts of *E.coli* (Ghorbani *et al.*, 2015).

16.2.6.2 CONP Nanocomposites

Due to the industrial revolution, in recent decades, different materials have been used as adsorbents for the treatment of dye-containing waste water. Metal-based nanoparticles, due to their large surface area, can be applied for the removal of toxic materials. Besides, surface functionalization enhances adsorption capacity to a large extent. Recently, copper oxide-zinc oxide nanoparticles composite was synthesized and functionalized using 3-aminopropyltrimethoxysilane by employing the genetic programming model to be used as adsorbent for the removal of unwanted and toxic materials from waste water (Mahmoodi *et al.*, 2015). A new nanocomposite of white graphene-copper oxide functionalized using hexagonal boron nitride was reported to improve the stability of nanocompostes (Paulose *et al.*, 2017). CONPs have attracted a lot of attention because of their wide range of applications, such as high-temperature super conductors (Yip and Sauls, 1992), sensors (Kim *et al.*, 2008; Umar *et al.*, 2009), catalysis (Yang *et al.*, 2010), optical (Yu *et al.*, 2004), electrical (Podhajecky *et al.*, 1985), giant magnetic resistance material (Musa *et al.*, 1998; Zheng *et al.*, 2000), gas sensors (Ishihara *et al.*, 1998; Tamaki *et al.*, 1998), preparation of organic-inorganic nanostructure composites, and solar energy transformation (Kumar *et al.*, 2001). The functionalization of nanoparticles has a significant effect on the characteristic features of copper oxide filled vinyl-ester resin composites (Zhanhu *et al.*, 2007). Polymeric nanocomposites embedded with inorganic nanoparticles have attracted a lot of interest due to their high homogeneity, flexible processability, and tunable physical properties (Nadagouda *et al.*, 2009). The functional groups of the polymer surrounding the nanoparticles enable these nanocomposites to be better candidates for various applications such as site-specific molecule targeting in biomedical areas (Nadagouda and Varma, 2008). Copper oxide nanoparticles functionalized with a bifunctional coupling agent such as methacryloxypropyl trimethoxysilanes (MPs) were used to fabricate vinyl-ester resin polymeric nanocomposites (Polshetliwar and Varma, 2010).

16.3 Palladium Nanoparticles

Nowadays, palladium nanoparticles are beingstudied because of their efficient heterogenous and homogenous catalytic nature and ferromagnetic properties (Horinouchi *et al.*, 2006). In addition to catalysis, the propensity of Pd to adsorb hydrogen has also led to their use in hydrogen storage Tobiska *et al.*, 2001), sensing applications (Pintar *et al.*, 2001), and water denitrification process.

16.3.1 Fabrication of Palladium Nanoparticles

16.3.1.1 Polyol Method

Nowadays, the polyol method has been used to produce palladium nanoparticles of controlled size and shape using silver nitrate. A low content of silver nitrate should be used to have well-controlled sizes and shapes. The salts of palladium chloride and sodium palladium chlorate were used as good precursors. In addition to silver nitrate, ferric chloride and sodium iodide were possibly used as efficient agents for the size and shape control. Ethylene glycol was used as both a reductant and a solvent (Nguyen *et al.*, 2010).

16.3.1.2 Microemulsion Technique

In comparison with other methods of synthesis, the w/o (water in oil) microemulsion technique is feasible and has an isotropic liquid medium with nanosized water droplets dispersed in a continuous oil phase and stabilized by surfactant molecules in a water by oil interface. They not only act as microreactors, but also inhibit the excess aggregation of particles by the adsorption of surfactants on the particle surface when the particle size approaches to that of the water pool. As a result, the particles are very fine in size and monodispersed (Cheng *et al.*, 2001).

16.3.1.3 Other Methods of Fabrication

Other synthetic approaches, such as chemical reduction of palladium chloride by sodium tetra borate and by arc discharge, magnetic stirring at 80 °C of palladium (II) acetate ($Pd(OAc)_2$), reduction in super critical carbon dioxide of $Pd(OAc)_2$, thermally induced reduction of palladium complex Pd(fod)2 where fod is 2,2-dimethyl-6,6,7,7,8,8,8-heptafluoro-3,5-octanedionate, and sonochemical reduction of palladium chloride have been using for a long time in the presence of various surfactant molecules to control the formation of undesirable agglomerates of palladium nanoparticles (Abderrafik et al., 2006). Researchers have developed palladium nanoparticles of monodisperse sizes and well-defined morphologies. These methods employ numerous solvents, reducing agents, and stabilizers. The stabilizers were used as capping agents (block copolymers, dendrimers, polymers, etc.), to protect the nanoparticles from agglomeration. Taking precautions to work under an inert atmosphere (argon, nitrogen) is quite common in these synthetic methods to prevent the surface oxidation of nanoparticles. This procedure uses reducing palladium (II) in five different solvents, which can work both as protective molecules and as reducing agents. Physical reducing agents such as photoreduction and ultrasonic radiation, and additional chemicals such as sodium citrate and sodium borohydrate, have been reported (Roceo *et al.*, 2011).

Other methods employed for the synthesis of palladium nanoparticles, such as microwave-assisted synthesis, sol-gel technique, and chemical reduction, exhibit a lack of precise control over nucleation, mixing, and growth, which subsequently shows an effect on the particle size, size distribution, and reproducibility. Microreactor technology is a part of the process intensification used in chemical synthesis. It improves the chemical and physical properties of the nanoparticles (Sharada *et al.*, 2016). The synthesis of the palladium nanoparticles, achieved by combining the methods of electro spinning and reduction of metal salts in a hydrazine solution, yields large surface area fibers with well-separated nanosized palladium nanoparticles (Mustafa, 2004).

A method to form metal nanoparticles, particularly an electrochemical process for synthesizing palladium nanoparticles, was developed utilizing an aqueous solution containing a polysaccharide and a metal salt for the electrochemical deposition of a polysaccharide film possessing metal nanoparticles on an electrically conductive substrate. In this method, an electrical potential was applied between an electrically conductive substrate and the aqueous solution to form the nanoparticles (Halad *et al.*, 2014). A new technique, laser ablation, was reported for the synthesis of palladium nanoparticles in distilled water without any surfactant and ethyl acetate in presence or absence of tetraoctyl ammonium bromide (TC_8ABr) respectively. This can be used to stabilize palladium nanoparticles ablated in liquid (Hwang *et al.*, 2000). Recently, a new method was developed to form palladium nanoparticles with well-controlled size and shape. In this method, 3 nm gold is used as a seed colloid to synthesize palladium nanoparticles in solution and on a surface (Chen *et al.*, 2010a).

16.3.2 Biological Methods

Conventionally palladium nanoparticles are synthesized via various physical or chemical methods with the support of toxic and hazardous reducing and stabilizing agents. To overcome these problems, biocompatible and environmentally benign biological techniques are superseding the traditional physical and wet chemical methods. The plant-mediated synthesis of palladium nanoparticles has gained a lot of attention recently compared to the use of microorganisms and marine organisms (Konishi *et al.*, 2007) as green reductants owing to their ready availability, better cost effectiveness, rapid process, and the ability to use in large scale biosynthesis. Plant-based catalysis using banana peel extract, leaf extract of *Perilla frutescens*, *Pulicaria glutinosa*, *Origanum vulgare*, *Euphorbia granulate*, *Ocimum sanctum*, soya bean leaf extract, *Diopyros kaki* leaf, *Cinnamon camphora*, *Cinnamon zeylanicum* bark, *Curcuma longa* tuber (Petla *et al.*, 2014), and coffee and tea extracts (Nadagouda and Varma, 2008) to synthesize palladium nanoparticles was reported. Unfortunately, the majority of these reported plant-based catalysts often suffer from limitations such as high-reaction temperatures (>120 °C) and high catalyst loadings (up to 12 mol %), and are used only for a mono catalytic system. To overcome these problems, a new biological source, namely *Garcinia pedunculata Roxb.* aqueous leaf extract was used to fabricate palladium nanoparticles which bestow multifunctional roles; viz, bioreduction palladium salts aid in multiple catalytic reactions (Hazarika *et al.*, 2017).

16.3.3 Functionalization

The functionalization of the palladium nanoparticles is important for efficient catalysis. Various efficient palladium catalyst precursors have been developed in recent years for effective cross-coupling in organic reactions.

16.3.3.1 Functionalization Iminophosphines

Phosphines especially electron rich iminophosphines are bidentate ligands used for the functionalization of palladium nanoparticles with both hard and soft donor atoms, and are expected to exhibit hemilabile behavior when coordinated to palladium. A palladium–phosphine complex supported on silica-coated magnetic nanoparticles was reported as a highly active and recyclable catalyst for the Suziki–Miyaura cross-coupling reaction Shylesh *et al.*, 2010). A new catalyst was synthesized by gathering the advantages of heterogenous catalysis, magnetic separation, and enhanced catalytic activity of palladium in the presence of phosphine ligands (Natália *et al.*, 2010).

16.3.3.2 Functionalization by Thiol

In addition to the method described above, the preparation of monolayered protected clusters had a significant impact in chemistry, physics, and material sciences. The preparation of thiol-protected PNPs capped by 11-mercaptoundecanoic acid, 9-mercapto-1-nonanol, or 1-dodecanethiol has been reported to achieve high densities of either carboxylic or hydroxy functionalities at their surface. This method is more advantageous than the reported one-phase (Cargnello *et al.*, 2011) and two-phase procedures for the synthesis of thiol-protected palladium nanoparticles. Recently, a new stabilizer was used to protect fabricated palladium nanoparticles.

Thesynthesisofopticallyactive2,2'-bis(diphenylphosphino)-1,1'-bisnaphthyl-protected palladium nanoparticles with smaller cores, narrower size distribution, and difference in particle size between monophosphine Ph_3P and biphosphine-stabilized palladium nanoparticles was reported. This was used for the asymmetric hydrosilization of styrene which had not hydrosilized by biphosphine ligand (Tamura and Fujihara, 2003).

16.3.4 Nanocomposites

Palladium-catalyzed cross-coupling reactions have been of strategic importance in organic chemistry. Since their discovery in the 1970s due to the lack of recyclability and potential contamination from the residual metals in the reaction product, a significant effort in which a new composite material of palladium nanocatalyst fixed to solid graphene support has developed by using the microwave irradiation method. These nanocomposites are recyclable. The fabrication of a palladium-based membrane composite in which palladium nanoparticles are packed in the interstices of an intermediate support layer via vacuum-assisted electro-less plating, in which metal is deposited at the activated layer of palladium nanoparticles, was reported. These composites are used for the continuous separation process coupled with the catalytic process (Siamaki *et al.*, 2011). Apart from the nanocomposites mentioned above, new palladium nanocomposites such as palladium nanoparticles encapsulated in amine-functionalized mesoporous metal-organic frameworks have been emerging as a very promising functional material for the heterogenous catalysis of aryl chlorides by direct anionic exchange. As well as synthesizing palladium nanocomposites using amino-functionalized mesoporous metal organic frameworks (Huang *et al.*, 2012), a new method was demonstrated to fabricate small palladium nanoparticles (3.0 ± 0.4 nm, 3 wt % Pd) in a redox active metal organic framework in the absence of extra reducing and capping agents which can enhance the hydrogen absorption capacity of metal organic frameworks. These metal organic frameworks were prepared by vapor deposition of $[(n^5 - C_5H_5)\ Pd\ (n^3 - C_3H_5)]$ followed by reduction with hydrogen gas (Kolmakov *et al.*, 2005). Nanocomposites of PNPs within the titanium-based metal organic frameworks MIL-125[38] and $NH_{2\_}$ MIL-125, also reported (Martis *et al.*, 2013). Palladium nanoparticles can be fabricated in porous coordination polymers as nanocomposites, which may enhance the hydrogen storage capacity of porous coordination polymers and catalyze organic reactions (Young *et al.*, 2008). A study was undertaken to synthesize in situ palladium nanoparticle-filled carbon nanotubes using an arc-discharge solution process in which metallic particles were encapsulated inside the nanotubes to increase the hydrogen storage potential of carbon nanotubes (Debasis *et al.*, 2004). A new palladium-small arc Go-CNT nanocomposite was fabricated by depositing palladium nanoparticles onto a reduced graphene oxide and carbon nanotube support by using the hydrothermal method. Thus far, this material is the most effective nanohybrid catalyst for nanoparticle reduction, and this method is environmentally friendly, low cost, and easy to scale-up (Tai *et al.*, 2013).

16.4 Magnesium Nanoparticles

Magnesium has attracted strong interest because of its high-performance applications among hydrogen storage materials due to its light weight, which increases the storage capacity.

16.4.1 Major Fabrication Methods of Magnesium Nanoparticles

Traditionally, magnesium nanoparticles can be fabricated mainly by four synthetic methods: cytochemistry (Sergeev, 2003), gas-phase synthesis (Kooi *et al.*, 2006), deposition from ethereal solutions (Conner *et al.*, 1956), and high-temperature molten salts such as magnesium chloride (Kisza *et al.*, 1995). New methods have also been used to fabricate magnesium nanoparticles. Sonoelectrochemistry is a unique method to prepare nanoparticles of metals having a large negative reduction potential such as magnesium and aluminium. Ultrasound waves mechanically reduce the magnesium to nanosize (Haas and Gedanken, 2008). The hydrogen storage capacity of magnesium decreases due to the atmospheric oxidation. To avoid this problem, magnesium nanoparticles produced by the gas evaporation method under the gas pressures of helium have an average diameter of 3.6–8 nm (Ogaway *et al.*, 2008). Among the various methods of fabrication, the mechanical milling process has been found to be an efficient way of producing magnesium nanoparticles which use magnesia and graphite powders as raw materials (Azonano, 2013). Laser ablation is another useful technique without the use of any surface-reactive reagents or dispersants. It can be used to control the compositions, production rate, particle size, and morphology of the laser ablated product (Tran *et al.*, 2018). Magnesium nanoparticles with a high capacity to store hydrogen were discovered using a gas-phase technique (Krishna *et al.*, 2010). The hydrogen plasma metal reaction has been known as a good method to prepare magnesium nanoparticles of the desired size. In this method, acetylene (alkyne) was introduced into the arc plasma to regulate the growth of magnesium nanoparticles (Zhang *et al.*, 2011). Recently, functionalized magnetic magnesium ferrite ($MgFe_2O_4$) nanoparticles were synthesized by the sol-gel auto combustion method. This is to be used as supporting material for the immobilization of enzymes over other materials because of lower-mass transfer resistance (Sharma *et al.*, 2017).

16.4.2 Magnesium Nanocomposites

Due to the energetic efficiency of the hydro/dehydrogenation reactions of magnesium, these require a high temperature of more than 350 °C and high pressure of hydrogen gas due to the low surface activity of magnesium. To improve the surface activity the nanoparticles with a metalorganic framework was reported which can absorb and desorb the hydrogen gas at room temperature (Ogawa *et al.*, 2012). Nanocomposites of magnesium nanoparticles into a metal-organic framework can improve the H_2 adsorption performance of the material (Lim *et al.*, 2012). As well as the magnesium nanocomposites of the metal-organic framework, the encapsulation of magnesium nanoparticles in a polymer matrix allowed the achievement of a speedy uptake of hydrogen with high capacity (Makridis *et al.*, 2013). Magnesium nanoparticles have been attracting the attention of researchers due not only to their hydrogen storage capacity but also because of their many biomedical applications including drug delivery, magnetic separation, magnetic resonance, and hyperthermia imaging contrast media (Pankhurst *et al.*, 2003). Due to the high reactivity of magnesium nanoparticles at the lowest size, which reduces hydrogen storage capacity, the preparation of magnesium ultrafine nanoparticles with minimum impurities and high efficiency is still challenging.

16.5 Calcium Nanoparticles

Inorganic nanomaterials such as calcium phosphate, tricalcium phosphate, calcium carbonate, hydroxyl apatite (Itokazu *et al.*,1998; Paul and Sharma, 2003), colloidal gold, silicon, iron oxide, and layered double hydroxide are being applied in modern pharmaceutical and medication areas (Ginebra *et al.*, 2006; Li *et al.*, 2010) due to the controlled release of drugs from the delivery system (Fadeel and Garcia-Bennet, 2010; Moreno *et al.*, 2012).

16.5.1 Fabrication of Calcium Nanoparticles

16.5.1.1 Calcium Dihydrogen Phosphate Nanoparticles

Calcium dihydrogen phosphates have been used as tooth substitutes and artificial bones. They are also good candidates for applications in catalysis and ion exchangers due to their substitution and unique surface structures. Moreover, microemulsions, particularly w/o emulsions, have received a lot of attention as reaction media for the synthesis of calcium dihydrogen phosphate nanoparticles (Lai *et al.*, 2005). Various other processes employed to produce calcium phosphate nanopowders include chemical co-precipitation (Mavis *et al.*, 2010), sol-gel process (Weng and Bapsta, 1999; Russell *et al.*, 1996), spray pyrolysis (Suchanek and Yoshimura, 1998), hydrothermal synthesis (Suchanek and Yoshimura, 1998), emulsion processing (Koumoulidis *et al.*, 2003), mechanochemical method (Wang *et al.*, 1997), and auto combustion method (Chin *et al.*, 2002).

16.5.1.1.1 Disadvantages of the Traditional Methods of Fabrication

Traditional fabrication methods have been suffering from drawbacks such as lack of control of synthesis parameters, low specific surface area of powders, poor reproducibility, etc. Alternative preparative methods such as plasma spraying (Dong *et al.*, 2003), pulsed laser deposition, and flame pyrolysis have been established (Madler *et al.*, 2002).

16.5.1.2 Calcium Carbonate Nanoparticles

Calcium carbonate is one of the most common and abundant inorganic biominerals and has a long history of utilization in numerous fields. Calcium carbonate nanoparticles have been employed in different fabrication areas. Owing to their low toxicity, bioavailability, and slow biodegradability, these nanoparticles have been used for controlled drug delivery, protein encapsulation, and biosensing applications (Peng *et al.*, 2010). So far, a few methods, including the reversed microemulsion method (Li and Mann, 2002), double emulsion technique (Gupta *et al.*, 2000), decomposition of cockle shells (Wang *et al.*, 2006), oil in water microemulsion method using high-pressure homogenization (Qian *et al.*, 2011), two-membrane dialysis system (Hu *et al.*, 2004), chemical precipitation method (Ueno *et al.*, 2005), flame synthesis (Huber *et al.*, 2005), and biomineralization and reactive precipitation (Casanova and Higuita, 2011) using a high-pressure jet homogenizer have been reported in the synthesis of calcium carbonate nanoparticles. Among these, the emulsion techniques and chemical precipitation have been generally used for the preparation of drug-loaded calcium carbonate nanoparticles (Shafiu *et al.*, 2013).

16.5.1.3 Pharmaceutical Applications

Owing to its large surface to volume ratio, biocompatible characteristics, and high hydrophilicity (Solmaz *et al.*, 2015; Dizaj and Barzegar-Jalali *et al.*, 2015), nanosized calcium carbonate shows greater advantages and novel characteristics such as the tendency to aggregate and a much larger specific surface area.

These properties constitute the following applications:

Enzyme (α-amylase, lipase) (Shan *et al.*, 2007; Peng *et al.*, 2001) immobilization by physical adsorption.

The α-amylase was immobilized on glutaraldehyde-activated silanized calcium carbonate nanoparticles by using the covalent binding method (Demir *et al.*, 2012).

Photophysical enhancements of dyes without the use of toxic heavy metals:

Because of the development of nanoparticle-based multifunctional biocompatible and biodegradable fluorescent technologies for controlled dye and drug delivery, recently a new nanoencapsulation system based on calcium phosphate was developed to retain the photophysical enhancements of dyes without the use of toxic heavy metals (Petrov *et al.*, 2005). Spherical calcium nanoparticles were synthesized and cyanine dye (Cy_3) was encapsulated via ethanol into the microemulsion and subsequent precipitation of the calcium nanoparticles around the dye. The particles were adhered to 20 μm porous silica spheres in a bare HPLC column (Morgan *et al.*, 2008).

Oral delivery of insulin:

To overcome the problems of oral delivery of insulin, many particulate delivery systems using polymers have been evaluated. But none of these seem to be suitable for use as a commercial oral insulin formulation. Calcium phosphate proved to be highly compatible with insulin (Hari *et al.*, 2009); therefore calcium phosphate–based nanoparticles were developed and modified by conjugating it with polyethylene glycol to make it hydrophilic, thereby increasing its insulin entrapment efficiency (Paul and Sahrma, 2007). In this method, polyethylene glycol was conjugated with the calcium nanoparticles after activating the phosphate group by carbodiimide chemistry.

In a drug delivery system, calcium phosphate–based hybrid nanoparticles are promising candidates because of their biocompatibility. Hybrid nanoparticles composed of calcium phosphate were prepared using organic additives and/or templates such as surfactants (Ramachandran *et al.*, 2008), polyelectrolytes (Ramachandran *et al.*, 2008; Welzel *et al.*, 2004a), block copolymers (Welzel *et al.*, 2004b), and reverse micelle droplets (Kakizawa *et al.*, 2004), cross-linked polymer micelles (Roy *et al.*, 2003), and liposomes (Schmidt and Ostafin, 2002). However, there has been limited success; therefore, a novel method – nanogel-templated mineralization – was used to prepare nanohybrid materials of calcium phosphate (Sugawara *et al.*, 2006).

Magnetic nanoparticles are of great interest owing to their potential biomedical applications (Wang *et al.*, 2006; Kim *et al.*, 2006), but there is hardly any report on tissue repair via magnetic nanoparticles. There is a new strategy to enhance bone tissue regeneration via the integration of super magnetic nanoparticles into calcium phosphate ceramics (Wu *et al.*, 2010).

16.5.2 Nanocomposites

Human bones are a natural nanocomposite consisting of nanosized bone apatite crystals and collagen-rich organic matrix. Therefore, synthetic biometric composite materials consisting of bioceramic such as hydroxyl apatite and polymers such as polyethylene have been developed as bone substitutes because they combine the ductility and processability of polymer matrixes with the osteoconductivity of bioceramic particles (Wang, 2003). Nowadays, a composite approach with calcium phosphate and poly (hydroxybutyrate-co-hydroxyvelarate) (PHBV) for the fabrication of biodegradable porous scaffolds for bone tissue engineering has been developed (Bin *et al.*, 2010). Apart from the above mentioned nanocomposite, a new combination of three-dimensional nanocomposite scaffolds based on calcium phosphate (Ca-P)/poly(hydroxybutyrate-co-hydroxyvalerate) (PHBV) and carbonated hydroxyapatite (CHAp)/poly(l-lactic acid) (PLLA) nanocomposite microspheres was fabricated for bone tissue engineering via the selective laser sintering method (Duan *et al.*, 2010).

In addition to their use in the area of biomedicine, calcium phosphate–based composite nanoparticles have also shown to be attractive candidates for bioimaging and therapeutic delivery applications. Composite nanoparticles utilizing lipids and polymers in combination with calcium phosphate cores or shells have promised to image with organic dyes and lanthanides and to deliver oligonucleotides and a variety of drug molecules (Chen *et al.*, 2010b; Li *et al.*, 2010). As well as the above nanocomposites, a new composite of porous calcium phosphate has been combined with reinforcing whiskers or glass particles to yield composites with calcium and phosphate ion release and good mechanical properties for caries inhibition (Hockin *et al.*, 2011).

16.6 Iridium and Iridium Oxide Nanoparticles

Platinum-group metals are especially utilized as catalysts for many important industrial reactions. In particular, iridium has been of great interest to electrochemists because of the electrochemical behavior of the iridium single crystal (Motoo *et al.*,

1984). Iridium nanoparticles can be fabricated by chemical vapor deposition (Gutsch *et al.*, 2005) and microemulsion techniques (Yasser *et al.*, 2015). The formation of aggregates due to the high fabrication temperature in chemical vapor deposition can be overcome by the microemulsion technique. In this technique, agglomeration and particle size can be controlled by the saturation of the reactive surfaces with functionalized organic molecules. Nevertheless, using such compounds might influence the electrocatalytic behavior of the metal nanoparticles by blocking active sites. A novel method under electrochemical conditions was developed to overcome the problems associated with traditional methods. In this method, iridium nanospheres were deposited electrically onto glassy carbon substrates from aqueous solution under electrochemical conditions (García and Koper, 2011). Basically, iridium nanoparticles were prepared by thermal hydrolysis of ionized iridium (III) precursor hexachloridium ($(Ir\ Cl_6)^{2-}$ in basic medium (Zhao *et al.*, 2011; Ioroi *et al.*, 2000). It has revealed that when the metal nanoparticles are grown on metal oxide substrates, their catalytic activity can be significantly improved (Diao *et al.*, 2010; Diao *et al.*, 2007; Comotti *et al.*, 2006) compared to that of pure metal or metal oxide catalysts. To develop a highly active noble metal catalyst, it is desirable to prepare metals and their oxide nanoparticles with precise control of their composition, size, and morphology. Photochemical synthesis is a convenient way to control the composition, size, and morphology of nanoparticles (Xue *et al.*, 2008; Nakajima *et al.*, 2014).

Another method of photo powdered hydrolysis was reported to prepare IrO_x nanoparticles and Ir/IrO_x nanocomposite. A partial reduction of IrO_x to iridium on the surface of IrO_x nanoparticles is useful to prepare this nanocomposite (Di *et al.*, 2015). Iridium nanoparticles can be fabricated in one phase in surfactant-free conditions using tetrahydrofuron as a solvent (Yee *et al.*, 1999). A method was reported to synthesize stable iridium nanoparticles by employing sodium S-dodecylthiosulfate as a ligand precursor using sodium borohydride reduction (Diego *et al.*, 2015). Recently, research has focused on the immobilization of nanomaterials. A kind of nitrogen-doped carbon nanomaterials has been explored with glucose and dicyandiamide as carbon and nitrogen sources to immobilize iridium catalyst for alcohol condensation in water (Di *et al.*, 2016).

16.7 Titanium Dioxide Nanoparticles (Titania, TONPs)

16.7.1 Fabrication of TONPs

Nano-structured titania materials are of interest due to their technical applications in photoelectric conversion and photocatalysis (Aguado *et al.*, 2002; Nakashima *et al.*, 2002; Zakeeruddin *et al.*, 2002). A variety of methods have been developed for fabricating titania nanoparticles, including the sol-gel process, hydrolysis technique, and gas condensation (Lin *et al.*, 1997; Xianyu *et al.*, 2001; Xie *et al.*, 2002; Xiaoli *et al.*, 2003). Self-assembled monolayer is the most feasible and widely used technique for producing materials in the nanosize scale range. It has the advantage of fabricating in a layer by layer fashion (Ulman, 1996; Burnside *et al.*, 1998; Chen *et al.*, 2001).

A simple and versatile hydrothermal synthetic rout to in situ functionalized TONPs was developed using titanium(IV) isopropoxide as precursor and selected silane coupling agents such as 3-aminopropyltriethoxysilate, 3-(2-aminoethylamino)propyldimethoxymethylsilane and n-decyltriethoxysilane(Dalod *et al.*, 2017). A new method of optimized ultrasonic-assisted dispersion of unfunctionalized TONPs into epoxy resin is reported. A novel top-down approach was developed to fabricate single crystal titania nanostructures, and they were functionalized with biological and organic molecules such as biotin, DNA, streptavidin, and biocompatible polyethylene glycol (Ha *et al.*, 2016).

16.7.2 Functionalization of TONPs

Titania-polymer composites are functionalized by surface modification with hydroxyl functional groups to improve the adhesion between nanofiber and material to be coated (Sundarrajan and Ramakrishna, 2010). Recently, titania has attracted considerable attention due to its enhancement of the antifouling propensity and permselectivity of thin film nanocomposite polyamide membranes and owing to its low production cost, high chemical and thermal stability, and, most importantly, to its photocatalytic activity upon ultra violet irradiation (Khataee and Mansoori, 2011; Lee and Park, 2013; Ong *et al.*, 2016; Rajaeian *et al.*, 2015).

16.7.3 Nanocomposites

Nanocomposite membranes of titania were fabricated by the electro spinning of polymer solution and electro spraying nanoparticle suspensions simultaneously. Here, the polymer solution was prepared by mixing cellulose and polyethylene terephthalate in appropriate solvents, and the nanoparticle suspension was prepared by mixing TONPs dissolved in methanol and surface modifier. TONPs are the most widely used nanoparticles to prepare thin film nanocomposite membranes.

Another approach to incorporate TONPs into the polyamide matrix using oleic acid in heptanes as a surface modifier is via biphasic solvothermal reaction. A trimesoyl chloride-heptane solution is added to titania suspension in heptanes to fabricate a thin film nanocomposite (TFN) polyamide (PA) membrane (Khorshidi *et al.*, 2018). The membranes used in desalination equipment have been prepared by super hydrophilic TFN membranes. Due to the poor chemical stability of conventional polyamide membranes, different membranes prepared with carboxylated titania and hydroxylated titania are being used as a nanofiller to improve the performance of the resultant TFN membranes in view of vapor permeance and selectivity (Arthanareeswaran and Thanikaivelan, 2010; Hamid *et al.*, 2011; Zhao *et al.*, 2012). A new thermally stable and optically active poly(ester-imide) was synthesized via direct polyesterification of N,N'-(pyromellitoyl)-bis-(L-tyrosinedimethylester) and N-trimellitylimido-L-methionine using tosyl chloride/pyridine/N,N'-dimethylformamide system as a condensing agent. After PEI/titanium bionanocomposites were prepared using the modified nanosized TiO_2 via sonochemical reaction, which can accelerate hydrolysis and improve the dispersion of the nanoparticles in the polymer matrix (Mallakpour *et al.*, 2011).

16.8 Tin Oxide Nanoparticles

Tin oxide is considered as a wide-band-gap, n-type semiconductor. It has excellent optical, electronic, and chemical properties, which enable it to be used for several applications such as electrode luminescence and gas sensing, as well as being employed as anode material for lithium ion batteries and in photocatalysis (Gnanam and Rajendran, 2010).

16.8.1 Fabricating Methods of Tin Oxide Nanoparticles

It is well known that the gas-sensing characteristics of tin oxide can be dramatically altered by the morphological and microstructural features of the sensing elements. There has been little research into using spherical-like tin oxide nanoparticles as gas sensors (Chen *et al.*, 2006). Various methods have been reported regarding the fabrication of tin oxide nanoparticles. Most studies have used a sol-gel process involving different precursors such as tin alkoxide and various Sn salts (Cabot *et al.*, 2001; Bose *et al.*, 2002; Rumyantseva *et al.*, 2005). Sonochemical methods are also used to form nano crystalline tin oxide with a particle size as low as 3 nm. A solution approach was employed using tin chloride as starting material heated in ethylene glycol via a reflux process (Jiang *et al.*, 2005). Recently, a hydrothermal method was reported to prepare mesoporous tin oxide nanocrystals by performing hydrolysis of tin chlorate pentahydrate in deionized water under refluxing conditions to yield tin oxide precipitates, followed by a hydrothermal treatment to yield tin oxide nanoparticles (Fujihara *et al.*, 2007). A new hydrothemal method for the preparation of the spherical tin oxide nanoparticles was performed using alcohol and water for gas-sensing applications (Hui and Yeh, 2007). In addition to the above methods, solvothermal (Han *et al.*, 2009), microwave (Krishnakumar *et al.*, 2008), and co-precipitation techniques (Yu *et al.*, 2006) were also reported for the fabrication of tin oxide nanoparticles. Research focused on tin oxide nanoparticles using co-precipitation, and synthesizing nanoparticles have been used to fabricate a working electrode (acetylcholinesterase, tin oxide, and carboxylated multi-walled carbon nanotubes/copper) for pesticide detection in water samples (Dhull, 2018).

Tin oxide, an n-type semiconductor, has been extensively used as gas sensor. Therefore, a sensor was developed using tin oxide nanoparticles fabricated by a chloride solution combustion synthesis method to detect volatile organic compound indoor pollutants. Nevertheless, the major drawback of this gas sensor is that it is unable to distinguish a specified compound when they are exposed to a mixture of reducing gases (Norouz-Oliaee *et al.*, 2010). To overcome this disadvantage, the tin oxide sensor was doped with samarium oxide (Sm_2O_3), a rare earth metal oxide, using the chloride solution combustion synthesis method (Habibzadeh *et al.*, 2010; Ahmadnia-Feyzabada *et al.*, 2013). A new method was also reported to improve the gas-sensing properties by functionalizing networked tin oxide nanowires with silver and palladium nanoparticles using X-ray radiolysis (Choi *et al.*, 2012).

Various methods have been developed for the fabrication of tin oxide nanoparticles, but the formation of SnO_2 nanorods via aerosol-assisted chemical vapor deposition without catalyst seeds and their in situ functionalization with gold nanoparticles have been developed recently for the fabrication of chemo-resistive gas sensors (Vallejos *et al.*, 2016). Moreover, the gas-sensing properties of functional metal oxide layers depend on a multitude of parameters, including valance concentration, layer morphology and thickness, size and shape of the nano/microstructure, and porosity. The use of tin oxide inks (colloidal suspensions of tin oxide nanoparticles) in combination with inkjet printing was reported to develop a layer by layer deposition technique which can control the thickness and composition of gas-sensitive layers (Gao *et al.*, 2017). A new method was developed in which porous tin oxide nanotubes were synthesized alternatively via a reactive-template method using manganese oxide nanorods as the sacrificial template. Synthesized tin oxide nanotubes were more often used to fabricate a chemical gas-sensing device of superior performance when compared to bulk tin oxide materials (Zhang *et al.*, 2013).

To improve the photocatalytic properties of tin oxide nanoparticles and carbon quantum dots, a nanocomposite was fabricated by surface functionalization of tin oxide nanoparticles through composite formation with carbon quantum dots to exploit the photocatalytic properties of both for the degradation of malachite green dye (Javed *et al.*, 2016).

16.9 Selenium Nanoparticles (SeNPs)

16.9.1 Uniqueness of Selenium and Its Nanoparticles

- Selenium is an essential micronutrient and has been gaining importance in medicine for the effective management of diseases.
- Selenium in a nano form that has gained importance as a possible supplement in the treatment of diseases due to its unique properties.
- Nanoselenium was observed to have better absorption, biocompatibility, bioefficacy, and lower toxicity compared to various organic and inorganic forms of selenium.
- Novel selenium nanoparticles have been gaining attention as potential drug carriers due to their excellent biological activities.
- Nano forms of selenium can be used for various applications due to their advantages over the bulk form due to their low toxicity, better reactivity, and low dosage.
- Among semiconductors, selenium is the most important one in view of its applications as rectifier and in solar cells, xenografts, and photographic exposure meters (Chen *et al.*, 2003).

16.9.2 Fabrication of SeNPs

Various methods have been reported regarding the preparation of SeNPs. Among them, physical, vapor-phase diffusion, vapor-deposition, and a few wet chemical methods are the most significant so far (Gates *et al.*, 2000; Mayers *et al.*, 2003). A new method of synthesizing SeNPs through the reduction of the aqueous selenious acid solution by sodium borohydride was reported (Nath *et al.*, 2004).

The ever-increasing demand for portable electronics and the low-energy limitations of lithium-ion batteries have led to the search for new cathode materials. Selenium is a promising cathode material due to chemical properties similar to those of sulfur, but its conductivity is approximately 20 orders of magnitude greater than that of sulfur (II). Although Se possesses several advantages, the Se cathode suffers from the shuttle effect to a certain degree. Therefore, to improve the electrochemical performance of selenium cathode-based composite, SeNPs are encapsulated in a reduced graphene oxide with a selenium content as high as 80% (w/w) to improve the capacity and energy density of the lithium-selenium battery (Peng *et al.*, 2015).

16.9.3 Biological Synthesis of SeNPs

Inspired by the potential applications of SeNPs, researchers have carried out the fabrication of SeNPs in an eco-friendly way. For instance, the fabrication was obtained by using the bacteria *Zooglea ramigera* (Srivastava *et al.*, 2013) and *Pseudomonas alcaliphila* (Srivastava *et al.*, 2011). SeNPs with various sizes and shapes can be synthesized and purified economically from redox systems.

16.9.4 Applications of Functionalized SeNPs in Cancer Therapy

However, pure SeNPs will aggregate completely in aqueous medium, especially under physiological conditions, resulting in poor anticancer activity (Vekariya *et al.*, 2012). Therefore, a simplistic, organic solvent-free and pH-aided strategy for fabricating curcumin-functionalized SeNPs was developed to overcome their shortfalls in cancer chemoprevention (Yu *et al.*, 2016).

In cancer treatment, nanoparticles functionalized with polysaccharides could interact specifically with biological targets on the surface of tumor cells. SeNPs capped with a monosaccharide sialic acid showed distinguished cell internalization and selectivity between tumor cells and normal cells. Due to the stability problem of sialic acid, SeNPs functionalized with spirulina polysaccharide (*Spirulina platensis*) have been reported to show significantly enhanced cell-penetrating and apoptosis-inducing abilities of selenium nanoparticles in A375 human melanoma cells (Yang *et al.*, 2012). A new method was reported to wrap the SeNPs by using glucan AF1 from fruiting bodies of *Auricularia auricular-judae* to improve stability (Zhaohua *et al.*, 2017).

In addition to above techniques, recently, SeNPs functionalized with oridonin, the main pharmacological active substance of *Rabdosia rubescens*, were prepared to improve the anticancer activity of oridonin against different cancer cells (Jiang *et al.*, 2016). Previously, it was reported that the functionalization of SeNPs improves biological activity. A water-soluble polysaccharide-protein complex of *Polyporus rhinoceros* for the surface functionalization of SeNPs has shown a significant effect in the reduction of tumor growth (Wu *et al.*, 2013).

16.9.5 Other Applications of Functionalized SeNPs

The control of toxic heavy metal pollution is a growing demand to remove heavy metals from waste water prior to its release into the water ecosystem. Copper is one of the most toxic metals, and it leads to headache, nausea, respiratory problems, abdominal pain, liver and kidney failure, and finally gastrointestinal bleeding. However, some nanomaterials in their own forms as adsorbents suffer from aggregation to remove copper from the water. Therefore, a technology of nanomaterials immobilization has developed to improve the dispersion of nanoparticles. Cotton is one of the most important fibers used as a supporting material. Recently, a novel supported nonadsorbent (nano-selenium-supported cotton) has been developed by loading SeNPs with sucrose on adsorbent cotton to adsorb copper from waste water (Huang *et al.*, 2015). Selenium/ruthenium nanoparticles via capping with of L-cystein were prepared to treat Alzheimer's disease. Recently, aminoacids were also used for the surface functionalization of SeNPs for better stabilization in *in vivo* applications (Yang *et al.*, 2014). Moreover, mushroom polysaccharide protein complexes, ATP, glutathione, folic acid, pectin mixed alginate/pectin, ovalbumin, β-lactobulin, sodium alginate, and leaf extract from lemon plants were also used for the functionalization of SeNPs (Chaudhary *et al.*, 2014).

16.10 Zirconium Oxide (ZrO_2) Nanoparticles

16.10.1 Fabrication

ZrO_2 is a wide-band-gap transition metal oxide which can be widely used in the synthesis of structural ceramic devices, catalysts, gas sensors, and optoelectronic devices (Grover *et al.*, 2007; Tahir *et al.*, 2007; Rozoa *et al.*, 2008; Liang *et al.*, 2009). So far, a variety of chemical and physical techniques have been developed, including hydrothermal (Seok *et al.*, 2008; Espinoza-González *et al.*, 2011; Machmudah *et al.*, 2014), sol-gel (Ehrhart *et al.*, 2006; Heshmatpour *et al.*, 2011), precipitation (Beena *et al.*, 2006), chemical vapor deposition (Torres-Huerta *et al.*, 2009; Keskinen *et al.*, 2004), hydrolysis (Tai *et al.*, 2004), and spray pyrolysis (Nimmo *et al.*, 2002) techniques. The microwave plasma technique (Dittmar *et al.*, 2008) and sputtering (Rozoa *et al.*, 2008) have been developed to synthesize ZrO_2 nanoparticles. Due to limiting factors such as high reaction temperature, longer reaction time, toxic reagents, byproducts, and high cost of production, a modified thermal treatment method has been used. In this method, a solution containing metal precursor and capping mediator is directly submitted to calcination, thus eliminating the drying process and reducing the preparation time and energy consumption (Keiteb *et al.*, 2016).

Among physical and chemical methods, electric arc discharge in liquid (distilled water) is an attractive method because of the simplicity of its apparatus building, high throughput, and cost-effective procedure (Ashkarran *et al.*, 2011). Moreover, a new method, co-precipitation, was developed to fabricate ZrO_2 nanoparticles with higher surface area (85 m^2/g) in comparison to other methods (Alaei *et al.*, 2014).

A facile and environmentally friendly synthetic sonochemical preparation of nanosized zirconium (IV) complex in the presence of methanol and monoethylene glycol (MEG) as a solvent and its use as a new precursor for the preparation of pure metal ZrO_2 nanoparticles were reported (Ranjbar *et al.*, 2016). Zirconium dioxide nanomaterials, including nanoparticles and their derived films, were synthesized using a new surfactant self-assembling sol-gel route, which involves the complexation of the zirconium alkoxide precursor with acetyl acetone as a chelating agent and tween 20 as a structural directing agent. Our results have demonstrated that the self-assembling sol-gel method synthesizes ZrO2 films with improved structural properties, which are considered responsible for the enhancement of catechol adsorption on the surface of ZrO_2 and fast electron transfer. This leads to a very interesting electrochemical behavior of catechol detection, including high selectivity and sensitivity, excellent repeatability and stability, and good reversibility and linearity over a wide range of catechol concentration (Chen *et al.*, 2011).

16.10.2 Functionalization

16.10.2.1 Stability Aspect of ZrO_2 Nanoparticles

Pure ZrO_2 has three well known polymorphs: monoclinic, tetragonal, and cubic. The monoclinic form is thermodynamically stable at room temperature, but it transforms reversibly to a tetragonal structure above 117 °C. High temperature polymorphs (cubic and tetragonal) have to be stabilized at lower temperatures because of their diverse applications.

- Yttria is a good stabilizer and yttria-stabilized ZrO_2 plays an important role among the doped alloys of ZrO_2, owing to its exceptional properties. Consequently, yttria-stabilized ZrO_2 nanoparticles were prepared by the method of co-precipitation (Maridurai *et al.*, 2016).
- Zirconia-based ceramics have gained considerable interest for several applications (e.g., solid electrolytes in fuel cells and oxygen sensors, thermal barrier coatings, and biomaterials for dental and orthopedic applications) due to their high mechanical strength, improved fracture toughness, and easy affordability. Generally, one of the main disadvantages of ceramic materials is their brittleness, i.e., low fracture toughness. Electrophoretic deposition has recently gained interest in the ceramics processing due to its wide versatility, allowing the fabrication of laminated and graded ceramics, fiber-reinforced ceramic matrix composites, coatings, and nanostructures. Researchers have been working at producing several ceramic structures such as multilayered specimens and 3-dimensional-shape structures using electrophoretic deposition. This technique was applied to the dental materials field in order to obtain dental crowns (Frank *et al.*, 2014).
- Moreover, a cost-effective, simple, environmentally benign, and scalable approach was developed for the preparation of isostructural ZrO_2 nanoparticles. In this regard, chitosan, an aminopolysaccaride polymer (a derivative of the polysaccharide biopolymer chitin), found in nature in the exoskeletons of shrimps and lobsters, has been found to assume favorable properties, including being biodegradable, nontoxic, inexpensive, and readily acquired from marine wastes (Al Sagheer *et al.*, 2009). Moreover, hydroxyl, amine, and carbonyl functionalities render chitosan a material with potential affinity toward metal ion immobilization (Ngah *et al.*, 2005; Guibal, 2005).
- In a similar vein, polyvinyl alcohol (PVA), a synthetic polymer that has been used as the organic phase in the preparation of various organic/inorganic hybrid films (Pirzada *et al.*, 2012), is analogously capable of immobilizing metal ionic species due to its hydroxylation and hydrophilicity. In this method, both chitosan and PVA were implemented in a simple, eco-friendly, and cost-effective sol-gel process to selectively produce zirconium nanoparticles assuming pure or mixed crystalline phases. The method applied involves the preparation of chitosan-ZrOx (polymer of zirconium oxide) and polyvinyl alcohol (PVA)ZrOx hybrid films, and the subsequent calcination of the films to recover quantitatively the ZrOx content in the form of zirconia nanoparticles in specific crystalline modification(s) (Bumajdad *et al.*, 2018).

16.10.3 Nanocomposites

The dispersion of nanometer-sized inorganic particles into the organic polymer matrix has a significant effect on the properties of nanocomposites. Several methods have been developed to synthesize ZrO_2/polymernanocomposite particles in the past decades.

- ZrO_2/polymer nanocomposite was synthesized by the inductively coupled polymerization method (He *et al.*, 2004).
- ZrO_2/polymethyl methacrylate nanocomposite particles are prepared by in situ polymerization of methyl methacrylate (MMA) in the presence of ZrO_2 nanoparticles, which are modified beforehand by the silane coupling agent γ-methacryloxy propyl trimethoxy silane (Fangqiang *et al.*, 2013).
- Novel nanocomposites were synthesized in a combination of grapheme oxide with Fe_3O_4-ZrO_2 and used to degrade organic species in polluted water (Nagi M El-Shafai *et al.*, 2018).
- Nanocomposites developed by using an ZrO2 solution to synthesize electroless plated nickel-phosphorous-vanadium-zirconium oxide (Ni-P-V-ZrO_2) nanocomposite coated with improved mechanical properties such as hardness, good wear, and corrosion resistance were reported (Yang *et al.*, 2011).
- Nowadays, ZrO_2 nanocomposites are used to develop glucose biosensors using the self-assembly method. Here, the biosensor was constructed by immobilizing Gox in cetyl trimethyl ammonium bromide/polyethylene glycol (PEG)/ZrO_2 nanocomposite film (Ahmad *et al.*, 2013).
- Nanocomposite coatings have been developed to improve the mechanical properties of nanocomposites. Electrodeposition is an important method to coat nanocomposites because of its easy operation, low cost, and temperature (Low *et al.*, 2006).
- Ni-ZrO_2 nanocomposite coating with excellent properties could be prepared by the addition of ZrO_2 in the composite electrodeposition (Wang *et al.*, 2005; Hou *et al.*, 2006).
- A new method was reported to produce Ni-ZrO_2 nanocomposite coatings by ultrasonic electrodeposition (Yu-Jun Xue *et al.*, 2015). These coatings can exhibit excellent corrosion resistance (Tian *et al.*, 2013).

- Owing to the similar properties of ZrO_2 and ceria (CeO_2) in terms of physical, chemical, and thermal stability (Ellawala *et al.*, 2015), ZrO_2-CeO_2 nanocomposites can be used as a substitute in many ceria application fields.
- A method was developed to prepare inorganic CeO_2-coated ZrO_2 composites or ZrO_2-core cera shell composites by the chemical precipitation method in situ (Hou *et al.*, 2016).
- Natural dyes have been used to sensitize TiO2 nanocrystalline solar cells, but they still need pigment purification and co-adsorption of other compounds. To overcome these obstacles, nanocrystalline ZrO2-TiO2 films sensitized with a bioorganic dye, chlorophyll extracted from the green leaves of *Chromolaena odorata*, were synthesized by the precipitation method (Pai and Nair, 2015).

16.11 Zinc (Zn) and Zinc Oxide (ZnO) Nanoparticles (Zinc Oxide Nanoparticles)

Zinc oxide is a versatile functional compound that has a diverse group of growth morphologies, such as nanocombs, nanorings, nanohelixes/nanosprings, nanobelts, nanowires, and nanocages. Zinc oxide nanoparticles have received considerable attention due to their antimicrobial, high catalytic, UV blocking, photochemical, and biosensor activities (Meruvu *et al.*, 2011; Sharma *et al.*, 2010).

16.11.1 Fabrication Methods

Zinc and zinc oxide nanoparticles can be fabricated by various methods. Among these solid-vapor processes is the facile thermal evaporation technique in which condensed or powdered source material(s) is/are vaporized at elevated temperatures, and then the resultant vapor phase condenses under certain conditions of temperature, pressure, atmosphere, substrate, etc., to form the desired products (Dai *et al.*, 2003).

Among the synthesis methods available to synthesize zinc oxide nanoparticles, the sol-gel method is relatively simple and works at a lower temperature when compared to other methods such as chemical vapor deposition, hydrothermal, and plasma, which require high temperatures in the process of synthesis. The process involves heating the mixture to stoichiometric reagents to make the ions move so that the desired solid product is obtained (Alias *et al.*, 2010).

Economically, the established methods used to fabricate zinc oxide nanoparticles such as sol-gel, evaporative decomposition of the solution, hydrothermal, wet chemical synthesis, and template-assisted growth (Pal *et al.*, 2005) cannot be employed to produce pure zinc oxide nanoparticles with the desired shape, crystallinity, and large active surface area. In order to use these nanoparticles in agriculture and in the environment, a simple, cost-effective fabrication process to synthesize zinc oxide nanoparticles, which can provide a high yield with reasonable purities, is required (Jie *et al.*, 2004). Therefore, a simple, cost-effective, and functional method such as the solution-combustion method has been employed to fabricate zinc oxide nanoparticles, with a mixture of ethanol and ethylene glycol (v/v, 60/40) as the solvent and zinc acetate as the zinc source (Ali *et al.*, 2013).

Ball milling is a simple, effective, and inexpensive technique for the synthesis of nanomaterials. It avoids organic solvents, high energy consumption, and gaseous emissions, as well as a long processing time (Bolokang *et al.*, 2015; Chen *et al.*, 1999). Nowadays, nanotechnology is an important and emerging tool for the development of eco-friendly and reliable methodology for the fabrication of nanoscale materials by using biological sources (Gilaki, 2010). The green synthesis of zinc oxide nanoparticles has increasingly become a matter of interest since traditional methods are expensive and explosive (Mason *et al.*, 2012). Various plant extracts such as *M. koenigui* leaf extract (Elumalai *et al.*, 2015), *Physalis alkekengi* L. (Qu *et al.*, 2011b), *Sedum alfredi Hance* (Qu *et al.*, 2011), leaf extract of *Parthenium hysterophorous* (Rajiv *et al.*, 2013), *Hibiscus rosa-sinensis* (Sharmila Devi *et al.*, 2014), *Caltropis gigantean*, and *Green tea (Camellia sinensis)* (Vidya *et al.*, 2013), were used for the biosynthesis of zinc oxide nanoparticles (Senthilkumar and Sivakumar, 2014).

16.11.2 Biological Synthesis

A new method on one-pot synthesis has been developed to synthesize zinc oxide nanoparticles using the chemical and green methods. The aqueous leaf extract of *Corriandrum sativum* as an ecofriendly agent was used along with zinc acetate and sodium hydroxide as a surrogate for chemical method. (Gnanasangeetha *et al.*, 2013). Even though various methods have been developed for the fabrication of zinc oxide nanoparticles, the challenge remains to produce monodisperse zinc oxide nanoparticles.

16.11.3 Functionalization

To overcome this issue, a method was developed to produce zinc oxide nanoparticles by employing a biotemplate, the cage-shaped protein apoferritin, in which the cavities worked as chemical reaction vessels with uniform size (Yoko *et al.*, 2012). Free-base porphyrin and zinc-porphyrin are promising alternatives for photochemical and photoelectrochemical applications due to their ultrafast electron injection, slow charge recombination kinetics, and high absorption coefficients. Therefore, the carboxylic group containing base porphyrins would be used to functionalize zinc oxide nanoparticles for sensitizing the photoelectrochemical efficiency of zinc oxide nanoparticles to be used as biosensors (Tu *et al.*, 2011). Recently, efforts have made in the dispersion and surface functionalization of three metal oxide nanoparticles such as titania, zinc oxide, and ceria for photocatalytic functions, sunscreen applications, and ultraviolet blocking. The formation and surface functionalization of zinc oxide nanoparticles can often be achieved at near room temperature using zinc nitrate and tris(hydroxymethyl)aminomethane, which is a polydentate ligand that can be used as a functionalizing agent (Faure *et al.*, 2013).

In present days, three-dimensional nanostructures with complex morphology and high dimensionality have received great research interest (Zhong *et al.*, 2006). These structures with a large void space are more advantageous for various applications (Lou *et al.*, 2008). Due to their exceptional catalytic activity, noble metals such as gold, platinum, and palladium are often loaded onto semiconductors to improve sensor performances.

TABLE 16.1

Various Fabrication Methods of Inorganic Nanoparticles

Fabrication method	Nanoparticles	Reference
Precipitation	MNP; CO; ZrO_2	Tartaj *et al.*, 2003; Lanje *et al.*, 2010; Ahamed *et al.*, 2014; Wua *et al.*, 2010; Seok Lee *et al.*, 2008
Co-precipitation	MNP; Ca; Sn	Tartaj *et al.*, 2003; Mavis *et al.*, 2000; Yu *et al.*, 2006
Microemulsion	MNP	Tartaj *et al.*, 2003
Polyol	MNP; PNP	Tartaj *et al.*, 2003; Nguyen *et al.*, 2010
Thermolysis/ Thermal decomposition	MNP; Cu; CO	Tartaj *et al.*, 2003; Lisiecki *et al.*,1993; Ghane *et al.*, 2010; Nasibulin *et al.*, 2001
Hydrothermal	MNP; Ca; Ir; Sn; ZrO_2; Zinc oxide	Tartaj *et al.*, 2003; Suchanek *et al.*, 1998; Zhao *et al.*, 2011; Cabot *et al.*, 2001; Seok Lee *et al.*, 2008; Pal *et al.*, 2005
Sol-gel	MNP; Cu; CO; Ca; Pt; ZrO_2; Zinc oxide	Tartaj *et al.*, 2003; Benavente *et al.*, 2013; Radhakrishnan and Beena, 2014; Volanti *et al.*, 2008; Weng and Bapsta, 1999; Russell *et al.*, 1996; Lin *et al.*, 1997; Beena *et al.*, 2006; Chen *et al.*, 2011; Alias *et al.*, 2010
Flow-injection	MNP	Tartaj *et al.*, 2003
Oxidation	MNP	Tartaj *et al.*, 2003
Chemical reduction	Cu; CO; PNP	Lisiecki *et al.*, 1993; Karthik and Kannappan, 2013; Roceo *et al.*, 2011
Photochemical reduction	Cu; PNP; Ir	Lisiecki *et al.*,1993; Roceo *et al.*, 2011; Xue *et al.*, 2008; Nakajima *et al.*, 2014
Electrochemical	MNP; Cu; CO; PNP	Sophie *et al.*, 2008; Lisiecki and Pileni, 1993; Waichal *et al.*, 2012; Halad *et al.*, 2014
Solid arc discharge	Cu; CO; PNP ZrO_2	Aslam *et al.*, 2002; Yao *et al.*, 2005; Rehspringer *et al.*, 2006; Ali *et al.*, 2011
Pulsed laser ablation	Cu; CO; PNP; Mg	Ghorbani, 2014; Gondal *et al.*, 2013; Hwang *et al.*, 2000; Tran *et al.*, 2008
Gas evaporation	CuNP	Ghorbani, 2014
Exploding wire	CuNP	Ghorbani, 2014
Sonochemical	MNP; CO; PNP; Sn; ZrO_2	Sophie *et al.*, 2008; Wongpisutpaisan *et al.*, 2011; Roceo *et al.*, 2011; Jiang *et al.*, 2005; Ranjbar *et al.*, 2016
Spray pyrolysis	CO; Ca; ZrO_2	Etefagh *et al.*, 2013; Suchanek *et al.*, 1998; Nimmo *et al.*, 2002
Selective chemical etching	CuNP	Yeshchenko *et al.*, 2007
Wet chemical method	CONP	Ethiraj and Kang, 2012
Calcination	CONP	Srivastava *et al.*, 2013
Mechanochemical	CONP CaNP	Ayask *et al.*, 2015 Wang *et al.*, 1997
Reverse micelle technique	CONP	Rani *et al.*, 2014
Alcothermal method	CONP	Tetsuya Kida *et al.*, 2017
Microwave assisted	PNP SnNP	Sharada *et al.*, 2016 Krishnakumar *et al.*, 2008
Electro spinning combined with reduction	PNP	Mustafa *et al.*, 2004
Microemulsion technique	PNP; Ca; Ir	Cheng *et al.*, 2001; Koumoulidis *et al.*, 2003; Kim *et al.*, 2000; Qian *et al.*, 2005 Yasser *et al.*, 2005
Cytochemistry	PNP	Sergeev, 2003
Gas-phase synthesis/Chemical vapor deposition	PNP; Ir; Sn; ZrO_2	Kooi *et al.*, 2006; Krishna et al., 2010; Gutsch *et al.*, 2005; Vallejos *et al.*, 2016; Torres-Huerta *et al.*, 2009
Deposition	PNP	Conner *et al.*,1957
Sonoelectrochemistry	Mg	Haas and Gedanken, 2008
Mechanical milling	Mg; Zinc oxide	Azonano, 2013; Bolokang *et al.*, 2015
Sol-gel auto combustion method	Mg	Abhishek Sharma *et al.*, 2017
Auto combustion method	Calcium	Chin *et al.*, 2002
Plasma spraying	Calcium	Dong *et al.*, 2003
Flame pyrolysis	Calcium	Madler *et al.*, 2002
Reversed microemulsion	Calcium; ZrO_2	Li, 2002; Tai *et al.*, 2004
Double emulsion technique	Calcium	Gupta *et al.*, 2000
Two membrane dialysis	Calcium	Hu *et al.*, 2004
Flame synthesis	Calcium	Huber *et al.*, 2004
Gas condensation	PtNP	Xianyu *et al.*, 2001
Solvothermal	Sn	Han *et al.*, 2001
Combustion synthesis	Sn	Ahmadnia-Feyzabada *et al.*, 2013

(Continued)

TABLE 16.1 (CONTINUED)

Various Fabrication Methods of Inorganic Nanoparticles

Fabrication method	Nanoparticles	Reference
Radiolysis	Sn	Choi and Kim, 2012
Layer deposition technique	Sn	Gao *et al.*, 2017
Reactive template method	Sn	Zhang *et al.*, 2013
Physical vapor deposition method	Sn	Qian *et al.*, 2010
Vapor diffusion method	Sn	Qian *et al.*, 2010
Microwave plasma technique	ZrO_2	Dittmar *et al.*, 2008
Sputtering	ZrO_2	Rozoa *et al.*, 2008
Thermal evaporation	Zinc oxide	Dai *et al.*, 2003
Solution combustion method	Zinc oxide	Ali *et al.*, 2013

Therefore, functionalizing three-dimensional hierarchically porous nanomaterials with gold nanoparticles can help to find novel properties (Liu *et al.*, 2011) (Table 16.1).

REFERENCES

Abboud, Y., Saffaj, T., Chagraoui, A., et al. 2014. Biosynthesis, characterization and antimicrobial activity of copper oxide nanoparticles produced using brown alga extract (*Bifurcaria bifurcata*). *Applied Nanoscience* 4(5): 571–576.

Abderrafik, N., Jean, L.R., Khatmai, D. 2006. Synthesis of palladium nanoparticles by sonochemical reduction of palladium (II) nitrate in aqueous solution. *The Journal of Physical chemistry* B 110(1): 383–389.

Aguado, J., van Grieken, R., Lopez-Munoz, M.J., Marugan, J. 2002. Removal of cyanides in wastewater by supported TiO_2-based photo- catalysts. *Catalysis Today* 75(1–4): 95.

Ahamed, M., Alhadlaq, H.A., Khan, M.A., Karuppiah, P., Al-Dhabi, N.A. 2014. Synthesis, characterization, and antimicrobial activity of copper oxide nanoparticles. *Journal of Nanomaterials* 2014: 1–4.

Ahmad, N.M., Abdullah, J., Ramli, N.I., et al. 2013. Characterization of ZrO2/PEG composite film as immobilization matrix for glucose oxidase. *International Journal of Materials and Metallurgical Engineering* 7(8): 611–614.

Ahmadnia-Feyzabada, Sadegh, Mortazavia, Yadollah, Khodadadia, Abbas Ali, Hemmati, S. 2013. Sm2O3 doped-SnO2 nanoparticles, very selective and sensitive to volatile organic compounds. *Sensors and Actuators. Part B* 181: 910–918.

Ali R. Siamaki, Abd El Rahman S. Khder, Victor Abdelsayed. et.al. 2011. Microwave-assisted synthesis of palladium nanoparticles supported on graphene:A highly active and recyclable catalyst for carbon–carbon cross-coupling reactions. *Journal of Catalysis* 279: 1–11.

Al Sagheer, F.A., Al-Sughayer, M.A., Muslim, S., Elsabee, M.Z. 2009. Extraction and characterization of chitin and chitosan from marine sources in Arabian Gulf. *Carbohydrate Polymers* 77(2): 410–419.

Alaei, Mahshad, Rashidi, Ali Morad, Bakhtiari, Iida. 2014. Preparation of high surface area ZrO_2 nanoparticles. *Iranian Journal of Chemistry and Chemical Engineering* 33: 2.

Ali, Mohammad Amdad, Idris, Mahmudur Rahman, Emran Quayum, Md. 2013. Fabrication of ZnO nanoparticles by solution combustion method for the photocatalytic degradation of organic dye. *Journal of Nanostructure in Chemistry* 3(1): 36.

Alias, S.S., Ismail, A.B., Mohamad, A.A. 2010. Effect of pH on ZnO nanoparticle properties synthesized by sol-gel centrifugation. *Journal of Alloys and Compounds* 499(2): 231–237.

Amstad, E., Gillich, T., Bilecka, I., Textoe, M., Reimhult, E. 2009. Ultrastable iron oxide nanoparticle colloidal suspensions using dispersants with catechol-derived anchor groups. *Nanoletters* 9(12): 4042–4048.

Ang, K.H., Alexandrou, I., Mathur, N.D., Amaratunga, G.A.J., Haq, S. 2004a. The effect of carbon encapsulation on the magnetic properties of Ni nanoparticles produced by arc discharge in de-ionized water. *Nanotechnology* 15: 5.

Ang, T.P., Wee, T.S.A., Chin, W.S. 2004b. Three dimensional self-assembled monolayer (3D SAM) of alkanethiols on copper nanoclusters. *The Journal of Physical Chemistry. Part B* 108(30): 11001.

Aoun, S.B., Dursun, Z., Koga, T., Bang, G.S., Sotomura, T., Taniguchi, I. 2004. Effect of metal ad-layer on Au (III) electrodes on electrocatalytic oxidation of glucose in an alkaline solution. *Journal of Electroanalytical Chemistry* 567(2): 175.

Arthanareeswaran, G., Thanikaivelan, P. 2010. Fabrication of cellulose acetate-zirconia hybrid membranes for ultrafiltration applications: Performance, structure and fouling analysis. *Separation Purification Technology* 74(2): 230–235.

Ashkarran, Ali Akbar, Aghigh, Seyed Mahyad, Ahmadi Afshar, Seyedeh Arezoo, Kavianipour, Mona, Ghoranneviss, Mahmoud. 2011. Synthesis and characterization of ZrO2 nanoparticles by an arc discharge method in water. *Synthesis and Reactivity in Inorganic, Metal-Organic, and Nano-Metal Chemistry* 41: 425–428.

Ashtari, P., He, X., Wang, K., Gong, P. 2005. An efficient method for recovery of target ssDNA based on amino-modified silica-coated magnetic nanoparticles. *Talanta* 67(3): 548–554.

Aslam, M., Gopakumar, G., Shoba, T.L., et al. 2002. Formation of Cu and Cu_2O nanoparticles by variation of the surface ligand: Preparation, structure and insulating- to- metallic transition. *Journal of Colloid and Interface Science* 255(1): 79–90.

Athanassiou, E.K., Grass, R.N., Stark, W.J. 2007. Large scale production of carbon coated copper nanoparticles for sensor applications. *Nanotechnology* 17(6): 1668–1673.

Ayask, H.K., Khaki, J.V., Sabzevar, M.H. 2015. Facile synthesis of copper oxide nanoparticles using copper hydroxide by mechanochemical process. *Journal of Ultrafine Grained and Nanostructured Materials* 48: 37–44.

AzoNano. 2013. Magnesium (mg) nanoparticles - Properties, applications. US Research Nanomaterials, Inc. https://www.azonano.com/Suppliers.aspx?SupplierID=2494 (accessed June 20 2013).

Beena, T., Kalpesh, S., Basha, Sk., Raksh, V.J. 2006. Synthesis of nanocrystalline zirconia using sol-gel and precipitation techniques. *Industrial and Engineering Chemistry Research* 45(25): 8643–8650.

Benavente, E., Lozano-Zarto, H., Gonzales, G. 2013. Fabrication of copper nanoparticles: Advances in synthesis, morphology control, and chemical stability. *Recent Patents on Nanotechnology* 7(2): 108–132.

Bera, D., Kuiry, S.C., McCutchen, M., et al. 2004. In-situ synthesis of palladium nanoparticles-filled carbon nanotubes using arc-discharge in solution. *Chemical Physics Letters* 386(4–6): 364–368.

Bin, D., Min, W., Wen, Y. Z. et.al. 2010. Three-dimensional nanocomposite scaffolds fabricated via selective laser sintering for bone tissue engineering. *Acta Biomaterialia* 6: 4495–4505.

Black, Merryl N., Henry, Elenor F., Olivia, A., et al. 2017. Environmentally relevant concentrations of amine-functionalized copper nanoparticles exhibit different mechanisms of bioactivity in *Fundulus heteroclitus* in fresh and brackish water. *Nanotoxicology* 11(8): 1–16.

Boeden, H.F., Pommerrning, K., Becker, M., et al. 1991. Bead cellulose derivatives as supports for immobilization and chromatographic purification of proteins. *Journal of Chromatography* 552(1–2): 389–414.

Bolokang, A.S., Cummings, F.R., Dhonge, B.P., et al. 2015. Characteristics of the mechanical milling on the room temperature ferromagnetism and sensing properties of TiO2 nanoparticles. *Applied Surface Science* 331: 362–372.

Bose, A.C., Kalpana, D., Thangadurai, P., Ramasamy, S. 2002. Synthesis and characterization of nanocrystalline SnO2 and fabrication of lithium cell using nano-SnO2. *Journal of Power Sources* 107(1): 138.

Brooks, R.A., Moiny, F., Gillis, P. 2001. On T2-shortening by weakly magnetized particles: The chemical exchange model. *Magnetic Resonance in Medicine* 45(6): 1014–1020.

Bulte J.W., Kraitchman D.L. Iron oxide MR contrast agents for molecular and cellular imaging. *NMR Biomed.* 2004; 17: 484–499.

Bumajdad, Ali, Nazeer, Ahmed Abdel, Sagheer, Fakhreia Al., Shamsun Nahar, Mohamed, Zaki, I. 2018. Controlled synthesis of ZrO2 nanoparticles with tailored size, morphology and crystal phases via organic/inorganic hybrid films. *Scientific Reports* 8: 3703.

Burnside, S.D., Shklover, V., Barbe, C., et al. 1998. Self-organization of TiO2 nanoparticles in thin films. *Chemical Materials* 10(9): 2419.

Burtea, C., Laurent, S., Vander Elst, L., Muller, R.N. 2006. Specific E-selectin targeting with a superparamagnetic with MRI contrast agent. *Contrast Media and Molecular Imaging* 1(1): 15–22.

Cabot, A., Dieguez, A., Romano-Rodriguez, A., Morate, R., Barsan, N. 2001. Hydrothermal synthesis of SnO2 nanoparticles and their gas-sensing of alcohol. *Sensors and Actuators. Part B* 79(98).

Cargnello, Matteo, Wieder, Noah L., Canton, Patrizia, et al. 2011. A versatile approach to the synthesis of functionalized thiol-protected palladium nanoparticles. *Chemistry of Materials* 23(17): 3961–3969.

Carvalho, M.D., Henriques, F., Ferreira, L.P., et al. 2013. Ironoxide nanoparticles: The influence of synthesis method and size on composition and magnetic properties. *Journal of Solid State Chemistry* 20: 144–152.

Casanova, H., Higuita, L.P. 2011. Synthesis of calcium carbonate nanoparticles by reactive precipitation using a high pressure jet homogenizer. *Chemical Engineering Journal* 175: 569–578.

Chaudhary, S., Umar, A., Mehta, S.K. 2014. Surface functionalized selenium nanoparticles for biomedical applications. *Journal of Biomedical Nanotechnology* 10(10): 3004–3042.

Chen, Hongjun, Wei, Gang, Ispas, Adriana, Hickey, Stephen, G. 2010a. Synthesis of palladium nanoparticles and their applications for surface-enhanced Raman scattering and electrocatalysis. *Journal of Physical Chemistry C* 114(50): 21976–21981.

Chen, S., Sommers, J.M. 2001. Alkanethiolate-protected copper nanoparticles: Spectroscopy, electrochemistry, and solid-state morphological evolution. *Journal of Physical Chemistry. Part B* 105(37): 8816–8820.

Chen, Y., Gerald, J.F., Williams, J.S., Willis, P. 1999. Mechanochemical synthesis of boron nitride nanotubes. *Journal of Metastable and Nanocrystalline Materials* 2–6: 173–178.

Chen, Y., Liu, W., Ye, C., Yu, L., Qi, S. 2001. Preparation and characterization of self-assembled alkanephosphate monolayers on glass substrate coated with nano-TiO2 thin film. *Materials Research Bulletin* 36(15): 2605.

Chen, Y., Lunsford, S.K., Song, Y., et al. 2011. Synthesis, characterization and electrochemical properties of mesoporous zirconia nanomaterials prepared by self-assembling sol–gel method with Tween 20 as a template. *Chemical Engineering Journal* 170(2–3): 518–524.

Chen, Y., Sun, Q., Li, H. 2003. Synthesis of novel selenium tubular structures. *Chemical Letters* 32(5): 448–449.

Chen, Y.C., Ostafin, A., Mizukami, H. 2010b. Synthesis and characterization of pH sensitive carboxy SNARF-1 nanoreactors. *Nanotechnology* 21(21): 215503.

Chen, Y.J., Nie, L., Xue, X.Y., et al. 2006. High capacity and excellent cycling stability of single-walled carbon nanotube/SnO2SnO2 core-shell structures as Li-insertion materials. *Applied Physical Letters* 88. http://www.ncbi.nlm.nih.gov/pubmed/083105.

Cheng, C.W., Ting, C.H., Dong, H.C. 2001. Synthesis of palladium nanoparticles in water-in-oil microemulsions. *Colloids and Surfaces A Physicochemical and Engineering Aspects* 189(1-3): 145–154.

Cheon, Young Eun, Suh, Myunghyun Paik. 2008. Multifunctional fourfold interpenetrating diamondoid network: Gas separation and fabrication of palladium nanoparticles. *Chemistry-A European Journal* 14(13): 3961–3967.

Chiu, Hui-Chi, Yeh, Chen-Sheng. 2007. Hydrothermal synthesis of SnO_2 nanoparticles and their gas-sensing of alcohol. *Journal of Physical Chemistry C* 111(20): 7256–7259.

Choi, Sun-Woo, Kim, Sang Sub. 2012. Enhancement of gas sensing properties by functionalization of networked SnO2 nanowires with metal nanoparticles selenium nanoparticles (SeNP). *IMCS 2012 – The 14th International Meeting on Chemical Sensors*, 722–724.

Chun, Y., Ying-kui, G., Mi-lin, Z. 2010. Thermal decomposition and mechanical properties of hydroxylapatite ceramic. *Transactions of Nonferrous Metals Society of China* 20(2): 254–258.

Comotti, M., Li, W.C., Spliethoff, B., Schüth, F. 2006. Support effect in high activity gold catalysts for CO oxidation. *Journal of American Chemical Society* 128(3): 917–924.

Conner, J.H., Reid, W.E., Wood, G.B. 1957. Electrodeposition of metals from organic Solutions V. Electrodeposition of magnesium and magnesium alloys. *Journal of the Electrochemical Society* 104(1): 38–41.

Corchero, J., Villaverde, A. 2009. Biomedical applications of distally controlled magnetic nanoparticles. *Trends in Biotechnology* 27(8): 468–476.

Curtis, A.C., Duff, D.G., Edwards, P.P., et al. 1988. A morphology – Selective copper organosol. *Angewandte Chemie International Edition* 27(11): 1530.

Cyrille, B., Bulmus, A., Pavis. 2009. The stabilization and bio-functionalization of iron oxide nanoparticles using heterotelechelic polymers. *Journal of Material Chemistry* 99: 111–123.

Dai, Z.R., Pan, Z.W., Wang, Z.L. 2003. Novel nanostructures of functional oxides synthesized by thermal evaporation. *Advanced Functional Materials* 13(1): 9–24.

Dalod, A.R., Henriksen, L., Grande, T., Einarsrud, M.A. 2017. Functionalized TiO2 nanoparticles by single-step hydrothermal synthesis: The role of the silane coupling agents. *Beilstein Journal of Nanotechnology* 8: 304–312.

Debasis, B., Suresh, C., Kuiry. et al. 2004. In-situ synthesis of palladium nanoparticles-filled carbon nanotubes using arc-discharge in solution. *Chemical Physics Letters* 386: 364–368.

Demir, M.M., Gulgun, M.A., Menceloglu, Y.Z., et al. 2004. Pd nanoparticles by electrospinning from poly (acrylonitrile-co-acrylic acid)-pdcl$_2$ solutions. Relation between preparation conditions, particle size and catalytic activity. *Macromolecules* 37(5): 1787–1792.

Demir, Serap, Go, Sevda Burcu, Vezir Kahraman, Memet. 2012. *a*-Amylase immobilization on functionalized nano CaCO3 by covalent attachment. *Starch/Staerke* 64(1): 3–9.

Devi, H.S., Singh, T.D. 2014. Synthesis of copper oxide nanoparticles by a novel method and its application in the degradation of methyl orange. *Advances in Electrical and Electronic Engineering* 4: 83–88.

Dhas, N.A., Raj, C.P., Gedanken, A. 1998. Synthesis, characterization, and properties of metallic copper nanoparticles. *Chemistry of Materials* 10(5): 1446–1452.

Dhull, Vikas. 2018. Fabrication of AChE/SnO2-cMWCNTs/Cu nanocomposite-based sensor electrode for detection of methyl parathion in water. *International Journal of Analytical Chemistry* 2018: 1–7.

Diao, P., Zhang, D., Guo, M., Zhang, Q. 2007. Electrocatalytic oxidation of CO on supported gold nanoparticles and submicroparticles: Support and size effects in electrochemical systems. *Journal of Catalysis* 250(2): 247–253.

Diao, P., Zhang, D.F., Wang, J.Y., Zhang, Q. 2010. Electrocatalytic activity of supported gold nanoparticles toward CO oxidation: The perimeter effect of gold–support interface. *Electrochemical Communications* 12(11): 1622–1625.

Diego, J. Gavial, et al. 2015. Mechanistic insights into the formation of dodecanethiolate-stabilized magnetic iridium nanoparticles. Thiosulphate vs thiol ligands. *Journal of Physical Chemistry.*

Dittmar, A., Hoang, D.L., Martin, A. 2008. TPR and XPS characterization of chromia-lanthana-zirconia catalyst prepared by impregnation and microwave plasma enhanced chemical vapour deposition methods. *Thermochimica Acta* 470(1–2): 40–46.

Dizaj, Solmaz Maleki, Barzegar-Jalali, Mohammad. 2015. Calcium carbonate nanoparticles as cancer drug delivery system. *Expert Opinion on Drug Delivery* 12(10): 1–12.

Di Xu, Peng Diao, Tao Jin et.al. 2015. Iridium Oxide Nanoparticles and Iridium/Iridium Oxide Nanocomposites: Photochemical Fabrication and Application in Catalytic Reduction of 4-Nitrophenol. *ACS Applied Material Interfaces* 7(30): 16738–16749.

Dong, Z.L., Khor, K.A., Quek, C.H., White, T.J., Cheang, P. 2003. TEM and STEM analysis on heat-treated and in vitro plasma-sprayed hydroxyapatite/Ti-6Al-4V composite coatings. *Biomaterials* 24(1): 97–105.

Duan, B., Wang, M., Zhou, W.Y., Cheung, W.L. 2010. Three-dimensional nanocomposite scaffolds fabricated via selective laser sintering for bone tissue engineering. *Acta Biomaterialia* 6(12): 4495–4505.

Ehrhart, G., Capoen, B., Robbe, O., Boy, P., Turrell, S., Bouazaoui, M. 2006. Structural and optical properties of n-propoxide sol–gel derived ZrO2 thin films. *Thin Solid Films* 496(2): 227–233.

El-Shafai, N.M., El-Khouly, M.E., El-Kemary, M., Ramadan, M.S., Masoud, M.S. 2018. Graphene oxide–metal oxide nanocomposites: Fabrication, characterization and removal of cationic rhodamine B dye. *RSC Advances* 8(24): 13323–13332.

Ellawala, K.C. Pradeep, Habu, T., Tooriyama, H., et al. 2015. Ultra-simple synthetic approach to the fabrication of CeO_2–ZrO2 mixed nanoparticles into homogeneous, domain, and core–shell structures in mesoporous spherical morphologies using supercritical alcohols. *The Journal of Supercritical Fluids* 97: 217–223.

Elumalai, K., Velmurugan, S., Ravi, S., Kathiravan, V., Ashokkumar, S. 2015. Biofabrication of ZnO nanoparticles using leaf extract of curry leaf (*Murraya Koenigii*) and its antimicrobial activities. *Materials Science in Semiconductor Processing* 34: 365–372.

Espinoza-Gonzalez, R.A., Diaz-Droguett, D.E., Avila, J.I., Gonzalez-Fuentes, C.A., Fuenzalida, V.M. 2011. Hydrothermal growth of zirconia nanobars on zirconium oxide. *Materials Letters* 65(14): 2121–2123.

Etefagh, R., Azhir, E., Shahtahmasebi, N. 2013. Synthesis of CuO nanoparticles and fabrication of nanostructural layer biosensors for detecting *Aspergillus niger* fungi. *Scientia Iranica* 20: 1055–1058.

Ethiraj, A.S., Kang, D.J. 2012. Synthesis and characterization of CuO nanowires by a simple wet chemical method. *Nanoscale Research Letters* 7(1): 70.

Fadeel, B., Garcia-Bennett, A.E. 2010. Better safe than sorry: Understanding the toxicological properties of inorganic nanoparticles manufactured for biomedical applications. *Advanced Drug Delivery Reviews* 62(3): 362–374.

Fangqiang, F.A.N., Zhengbin, X.I.A., Qingying, L.I., Zhong, L.I., Huanqin, C.H.E.N. 2013. ZrO2/PMMA nanocomposites: Preparation and its dispersion in polymer matrix. *Chinese Journal of Chemical Engineering* 21(2): 113–120.

Faure, Bertrand, Salazar-Alvarez, German, Ahniyaz, Anwar, et al. 2013. Dispersion and surface functionalization of oxide nanoparticles for transparent photocatalytic and UV-protecting coatings and sunscreens. *Science and Technology of Advanced Materials* 14(2): 023001.

Frank, Stefan, Mochales, Carolina, Heimann, Martin, et al. 2014. Electrophoretic deposition of zirconia nanoparticles. *NanoScience and Technology* 1(1): 1–5.

Fujihara, S., Maeda, T., Ohgi, H., et al. 2004. Hydrothermal routes to prepare nanocrystalline mesoporous SnO2 having high thermal stability. *Langmuir* 20(15): 6476.

Gao, H., Lyu, X., Wöllenstein, J., Palzer, S. 2017. Layer by layer deposition of colloidal SnO2 nano particles. *Proceedings* 1(4): 318.

García, G., Koper, M.T.M. 2011. Carbon monoxide oxidation on Pt single crystal electrodes: Understanding the catalysis for low temperature fuel cells. *ChemPhysChem* 12(11): 2064–2072.

Gates, B., Yin, Y., Xia, Y.A. 2000. A solution-phase approach to the synthesis of uniform nanowires of crystalline selenium with lateral dimensions in the range of 10–30 nm. *Journal of the American Chemical Society* 122(50): 12582.

Ge, S., Shi, X., Sun, K., et al. 2009. A facile hydrothermal synthesis of iron oxide nanoparticles with tunable magnetic properties. *Journal of Physical Chemistry* 113(31): 13593–13599.

Ghane, M., Sadeghi, B., Jafari, A.R., Pakenjhad, A.R. 2010. Synthesis and characterization of a Bi-Oxide nanoparticle ZnO/CuO by thermal decomposition of oxalate precursor method. *International Journal of NANO Dimensions* 1: 33–40.

Ghorbani, H.R., Safekordi, A.A., Attar, H., et al. 2011. Biological and non-biological methods for silver nanoparticles synthesis. *Chemical and Biochemical Engineering Quarterly* 25(3): 317–326.

Ghorbani, Hamid Reza. 2014. Biological and non-biological methods for fabrication of copper nanoparticles. *Chemical Engineering Communications*: 1463–1467.

Ghorbani, Hamid Reza, Fazeli, Iman, Fallahi, Ali Asghar. 2015. Biosynthesis of copper oxide nanoparticles using extract of *E.coli*. *Oriental Journal of Chemistry* 31(1): 515–517.

Gilaki, M. 2010. Retracted: Biosynthesis of silver nanoparticles using plant extracts. *Journal of Biological Sciences* 10(5): 465–467.

Gillis, P., Moiny, F., Brooks, R.A. 2002. On T-2-shortening by strongly magnetized spheres: A partial refocusing model. *Magnetic Resonance in Medicine* 47(2): 257–263.

Ginebra, M., Traykova, T., Planell, J. 2006. Calcium phosphate cements as bone drug delivery systems: A review. *Journal of Controlled Release* 113(2): 102–110.

Gnanam, S., Rajendran, V. 2010. Synthesis of tin oxide nano particles by sol-gel process: Effect of solvents on the optical properties. *Journal of Sol-Gel Science and Technology* 53(3): 555–559.

Gnanasangeetha, D., Saralathambavani, D. 2013. One pot synthesis of zinc oxide nanoparticles via chemical and green method. *Research Journal of Material Sciences* 1(7): 1–8.

Gondal, M.A., Drmosh, Q., Saleh, T.A. 2012. Synthesis of nickel oxide nanoparticles using pulsed laser ablation in liquids and their optical characterization. *Applied Surface Science* 258(18): 6982–6986.

Gondal, M.A., Qahtan, T.F., Dastageer, M.A., Saleh, T.A., Maganda, Y.W. 2013. Synthesis and characterization of copper oxides nanoparticles via pulsed laser ablation in liquid. *High Capacity Optical Networks and Emerging/ Enabling Technologie*: 146–150.

Grover, V., Shukla, R., Tyagi, A.K. 2007. Facile synthesis of ZrO2 powders: Control of morphology. *Scripta Materialia* 57(8): 699–702.

Guibal, E. 2005. Heterogeneous catalysis on chitosan-based materials: A review. *Progress in Polymer Science* 30(1): 71–109.

Gultekin, Demet Demirci, Gungor, Azize Alayli, Onem, Hicran, Babagil, Aynur, Nadaroğlu, Hayrunnisa. 2016. Synthesis of copper nanoparticles using a different method: Determination of their antioxidant and antimicrobial activity. *Journal of Turkish Chemical Society, Section A: Chemistry* 3(3): 623–636.

Gunawan, C., Teoh, W.Y., Marquis, C.P., Amal, R. 2011. Cytotoxic origin of copper (II) oxide nanoparticles: Comparative studies with micron-sized particles, leachate and metal salts. *ACS Nanomaterials* 5(9): 7214–7225.

Guo, Z., Liang, X., Pereira, T., Scaffaro, R., Hahn, H.T. 2007. CuO nanoparticle filled vinyl-ester resin nanocomposites: Fabrication, characterization and property analysis. *Composites Science and Technology* 67(10): 2036–2044.

Gupta, A.K., Gupta, M. 2005. Synthesis and surface engineering of iron oxide nanoparticles for biomedical applications. *Biomaterials* 26(18): 3995–4021.

Gupta, A.K., Wells, S. 2004. Surface-modified superparamagnetic nanoparticles for drug delivery: Preparation, characterization, and cytotoxicity studies. *Nanobioscience* 3: 66–73.

Gupta, R. 2004. *Synthesis of Precipitated Calcium Carbonate Nanoparticles Using Modified Emulsion Membranes*. Georgia Institute of Technology.

Gupta, R. 2000. Synthesis of precipitated calcium carbonate nanoparticles using modified emulsion membranes. *Gorgia Institute of Technology*. Ph D thesis.

Gutsch, A., Mühlenweg, H., Krämer, M. 2005. Tailor-made nanoparticles via gas-phase synthesis. *Small* 1(1): 30–46.

Ha, Seungkyu, Janissen, Richard, Ye, Yera., et al. 2016. Tunable top-down fabrication and functional surface coating of single-crystal titanium dioxide nanostructures and nanoparticles. *Nanoscale* 1–3: 9.

Haas, Iris and Gedanken, Aharon. 2008. Synthesis of metallic magnesium nanoparticles by sonoelectrochemistry. *Chemical Communications*: 1795–1797.

Habibzadeh, S., Khodadadi, A.A., Mortazavi, Y. 2010. CO and ethanol dual selective sensor of Sm2O3-doped SnO2 nanoparticles synthesized by microwave-induced combustion. *Sensors and Actuators. Part B* 144(1): 131–138.

Haile, L., Jiangshan, L., Tang, Y., et al. 2004. Production of copper nanoparticles by the flow-levitation method. *Nanotechnology* 15(12): 1866–1869.

Halad, P., Prasanth Kumar, J., Michael, A. 2014. *Synthesis of Palladium Nanoparticles*. US 9,932,685 B2.

Han, Z., Guo, N., Li, F. et al. 2001. Solvothermal preparation and morphological evaluation of stannous oxide powders. *Materials Letters* 48: 99–103.

Hamid, N.A.A., Ismail, A.F., Matsuura, T., et al. 2011. Morphological and separation performance study of polysulfone/titanium dioxide (PSF/TiO2) ultrafiltration membranes for humic acid removal. *Desalination* 273(1): 85–92.

Han, Z., Guo, N., Li, F., et al. 2001. Solvothermal preparation and morphological evaluation of stannous oxide powders. *Materials Letters* 48(2): 99–103.

Hari S., Muddana, Thomas, T. et.al. 2009. Photophysics of Cy3-Encapsulated Calcium Phosphate *Nanoparticles*. Nanoletters. 9(4).

Hazarika, Munmi, Borah, Debajit, Bora, Popymita, Silva, Ana R., Das, Pankaj. 2017. Biogenic synthesis of palladium nanoparticles and their applications as catalyst and antimicrobial agent. *PLOS ONE*.

He, F., Zhao, D., Liu, J., Roberts, C.B. 2007. Stabilization of Fe-Pd nanoparticles with sodium carboxy methyl cellulose for enhanced transport and dechlorination of trichloroethylene in soil and ground water. *Industrial and Engineering Chemical Research* 46(1): 29–34.

He, P., Lian, J., Wang, L.M., Van Ooji, W.J., Shi, D. 2002. *Materials Research Society Symposia Proceedings* 703. California.

He, W., Guo, Z.G., Pu, Y.K., Yan, L.T., Si, W.J. 2004. Polymer coating on the surface of zirconia nanoparticles by inductively coupled plasma polymerization. *Applied Physics Letters* 85(6): 896–889.

He, Y. 2007. A novel solid-stabilized emulsion approach to CuO nanostructures microspheres. *Material Research Bulletin* 42(1): 190–195.

Heo, Dong Nyoung, Min, Kung Hyun, Choi, Gi Hyun, et al. 2014. Scale-up production of theranostic nanoparticles. In: *Cancer Theranostics*, eds., Xiaoyuan Chen, Stephen Wong, 457–470. Academic Press.

Heshmatpour, F., Aghakhanpour, R.B. 2011. Synthesis and characterization of nanocrystalline zirconia powder by simple sol-gel method with glucose and fructose as organic additives. *Powder Technology* 205(1–3): 193–200.

Hockin, H.K., Xua, J.L., Moreaua, L.S., Chow, L.C. 2011. Nanocomposite containing amorphous calcium phosphate nanoparticles for caries inhibition. *Dental Materials* 27(8): 762–769.

Horinouchi, S., Yamanoi, Y., Yonezawa, T., Mouri, T., Nishihara, H. 2006. Hydrogen storage capacity of isocyanide-stabilized palladium nanoparticles. *Langmuir* 22(4): 1880–1884.

Hoseinia, B.A., Shahtahmasebia, B., Rezaee-Roknabadia, M., et al. 2011. *Fabrication of Cuo:Fe Nanoparticles by Sol-Gel Method and Study of Structural and Antibacterial Properties.* Paper presented at 2nd Conferences on Application of Nanotechnology in Sciences. Engineering and Medicine: Mashhad-Iran.

Hou, F.Y., Wang, W., Guo, H.T. 2006. Effect of the dispersibility of ZrO2 nanoparticles in Ni–ZrO2 electroplated nanocomposite coatings on the mechanical properties of nanocomposite coatings. *Applied Surface Science* 252(10): 3812–3817.

Hou, Xilu, Qiu, Kehui, Zhang, Min, et al. 2016. Preparation and characterization of zirconia-core ceria-shell composites and its formation mechanism. *2nd International Conference on Machinery, Materials Engineering, Chemical Engineering and Biotechnology.* Atlantis Press.

Hu, Z., Deng, Y., Sun, Q. 2004. Synthesis of precipitated calcium carbonate nanoparticles using a two-membrane system. *Colloid Journal* 66(6): 745–750.

Huang, L., Tong, X., Li, Y., Teng, J., Bai, Y. 2015. Preparation of a novel supported selenium nanoparticles adsorbent and its application for copper removal from aqueous solution. *Journal of Chemical and Engineering Data* 60(1): 151–160.

Huang, Yuanbiao, Liu, Songjuan, Lin, Zujin, et al. 2012. Facile synthesis of palladium nanoparticles encapsulated in amine-functionalized mesoporous metal–organic frameworks and catalytic for dehalogenation of aryl chlorides. *Journal of Catalysis* 292: 111–117.

Huber, M., Stark, W.J., Loher, S., et al. 2005. Flame synthesis of calcium carbonate nanoparticles. *Chemical Communications*: 648–650.

Hwang, C.B., Fu, Y.S., Lu., Y.L., et al. 2000. Synthesis, characterization, and highly efficient catalytic reactivity of suspended palladium nanoparticles. *Journal of Catalysis* 195(2): 336–341.

Ioroi, T., Kitazawa, N., Yasuda, K., Yamamoto, Y., Takenaka, H. 2000. Iridium oxide/platinum electrocatalysts for unitized regenerative polymer electrolyte fuel cells. *Journal of Electrochemical Society* 147(6): 2018–2022.

Ishihara, T., Higuchi, M., Takagi, T., et al. 1998. Preparation of CuO thin films on porous $BaTiO_3$ by self-assembled multibilayer film formation and application as a CO_2 sensor. *Journal of Materials Chemistry* 8(9): 2037–2042.

Ishihara, T., Kometani, K., Hashida, M., Takita, Y. 1999. Application of mixed oxide capacitor to the selective carbon dioxide sensor. *Journal of Electrochemical Society* 138: 1173–1176.

Itokazu, M., Sugiyama, T., Ohno, T., Wada, E., Katagiri, Y. 1998. Development of porous apatite ceramic for local delivery of chemotherapeutic agents. *Journal of Biomedical Material Research* 39(4): 536–538.

Javed, Rida, Nawaz, Faisal, Sohail, Muhammad, Ahmad, Iqbal. 2016. Fabrication of SnO2/CQDs composite for photocatalytic degradation of malachite green dye. *Journal of Contemporary Research in Chemistry* 1(1): 42–50.

Jayakumarai, G., Gokulpriya, C., Sudhapriya, R., Sharmila, G., Muthukumaran, C. 2016. Phytofabrication and characterization of monodisperse copper oxide nanoparticles using *Albizia lebbeck* leaf extract. *Applied Nanoscience* 5(8): 1017–1021.

Jayalakshmi, Yogamoorthi, A. 2014. Green synthesis of copper oxide nanoparticles using aqueous extract of flowers of *Cassia alata* and particles characterisation. *International Journal of Nanomaterials and Biostructures* 4: 66–71.

Jiang, L., Sun, G., Zhou, Z., et al. 2005. Size-controllable synthesis of monodispersed SnO2 nanoparticles and application in electrocatalysts. *Journal of Physical Chemistry. Part B* 109(18): 8774–8778.

Jiang, P. I., Jiye C.A.I., MO(MO), Hua, J.I.N. et.al. 2016. Ordonin functionalized selenium nanoparticles and method of preparation thereof. US 2016/0257694 A1.

Jie, J., Wang, G., Wang, Q., et al. 2004. Synthesis and characterization of aligned ZnO nanorods on porous aluminium oxide template. *The Journal of Physical Chemistry. Part B* 108(32): 11976–11980.

Jung, C.W., Jacobs, P. 1995. Physical and chemical properties of superparamagnetic iron oxide nanoparticles MR contrast agents: Ferimoxide, ferumoxtran and ferumoxsil. *Magnetic Resonance Imaging* 13(5): 661–674.

Jurate, V., Rajender, S.V. 2011. Green synthesis of metal nanoparticles: Biodegradable polymers and enzymes in stabilization and surface functionalization. *Chemical Science* 2: 837.

Kakizawa, Y., Miyata, K., Furukawa, S., Kataoka, K. 2004. Size-controlled formation of a calcium phosphate-based organic–inorganic hybrid vector for gene delivery using poly(ethylene glycol)-block-poly(aspartic acid). *Advanced Materials* 16(18): 699–702.

Kang, X., Mai, Z., Zou, X., Cai, P., Mo, J. 2007. A sensitive non-enzymatic glucose sensor in alkaline media with a copper nanocluster/multiwall carbon nanotube-modified glassy carbon electrode. *Analytical Biochemistry* 363(1): 143–150.

Kapoor, S., Mukherjee, T. 2003. Photochemical formation of copper nanoparticles in PVP. *Chemical Physics Letters* 370(1–2): 83–87.

Karthik, A.D., Kannappan, G. 2013. Synthesis of copper precursor, copper and its oxide nanoparticles by green chemical reduction method and its antimicrobial activity. *Journal of Applied Pharmaceutical Sciences* 3: 016–021.

Keiteb, A.S., Saion, E., Zakaria, A., Soltani, N. 2016. Structural and optical properties of zirconia nanoparticles by thermal treatment synthesis. *Journal of Nanomaterials* 1–6.

Kelechi, C., Anyaogu, Andrei, V., et al. 2008. Synthesis, characterization and antifouling potential of functionalized copper nanoparticles. *Langmuir* 24(8): 4340–4344.

Keliu, Y.S., Shaowei, C., Chen, S. 2014. Electrocatalytic activities of alkyne-functionalized copper nanoparticles in oxygen reduction in alkaline media. *Journal of Power Sources* 268: 469–475.

Keskinen, H., Moravec, P., Smolik, J., et al. 2004. Preparation of ZrO2 fine particles by CVD process: Thermal decomposition of zirconium tert-butoxide vapor. *Journal of Materials Science* 39(15): 4923–4929.

Khataee, A., Mansoori, G.A. 2011. *Nanostructured Titanium Dioxide Materials*. World Scientific.

Khorshidi, B., Biswas, I., Ghosh, T., Thundat, T., Sadrzadeh, M. 2018. Robust fabrication of thin film polyamide-TiO2 nanocomposite membranes with enhanced thermal stability and anti-biofouling propensity. *Scientific Reports* 8(1): 784.

Khulbe, K.C., Mann, R.S., Manoogian, A. 1980. Behaviour of nickel-copper alloy in hydrogenation, orthohydrogen-parahydrogen conversion and H_2-D_2 exchange reaction. *Chemical Reviews* 80(5): 417–428.

Kim, J., Kang, S.W., Mun, S.H., Kang, Y.S. 2009. Facile synthesis of copper nanoparticles by ionic liquids and its application to facilitated olefin transport membranes. *Industrial and Engineering Chemistry Research* 48(15): 7437–7441.

Kim, J., Lee, J.E., Lee, J., et al. 2006. Magnetic fluorescent delivery vehicle using uniform mesoporous silica spheres embedded with monodisperse magnetic and semiconductor nanocrystals. *Journal of the American Chemical Society* 128(3): 688–689.

Kim, W., Zhang, Q.W., Saito, F. 2000. Mechanochemical synthesis of hydroxyapatite from $Ca(OH)_2$-P_2O_5 and CaO-$Ca(OH)_2$-P_2O_5 mixtures. *Journal of Material Science* 35(21): 5401–5405.

Kim, Y.S., Hwang, I.S., Kim, S.J., Lee, C.Y., Lee, J.H. 2008. CuO nanowire gas sensors for air quality control in automotive cabin. *Sensors and Actuators. Part B* 135(1): 298–303.

Kisza, A., Kaźmierczak, J., Børresen, B., Haarberg, G.M., Tunold, R. 1995. Kinetics and mechanism of the magnesium electrode reaction in molten magnesium chloride. *Journal of Applied Electrochemistry* 25(10): 940–946.

Klug, K.L. 2002. PhD. Dissertation, Northwestern University.

Klug, K.L., Dravid, V.P., Johnson, D.L. 2003. Silica-encapsulated magnetic nanoparticles formed by a combined arc evaporation/chemical vapor deposition technique. *Journal of Material Research* 18(4): 988.

Kolmakov, A., Klenov, D.O., Lilach, Y., et al. 2005. Enhanced gas sensing by individual SnO2 nanowires and nanobelts functionalized with Pd catalyst particles. *Nano Letters*.

Koneracka, M., Kopcansky, P., Antalik, M., et al. 1999. Immobilization of proteins and enzymes to fine magnetic particles. *Journal of Magnetism and Magnetic Materials* 210(1–3): 427–430.

Konishi, Y., Ohno, K., Saitoh, N., et al. 2007. Bioreductive deposition of platinum nanoparticles on the bacterium Shewanella algae. *Journal of Biotechnology* 128(3): 648–653.

Kooi, B.J., Palasantzas, G., De Hosson, J.T.M. 2006. Gas-phase synthesis of magnesium nanoparticles: A high-resolution transmission electron microscopy study. *Applied Physics Letters* 89(16): 161914.

Koumoulidis, G.C., Katsoulidis, A.P., Ladavos, A.K., et al. 2003. Preparation of hydroxyapatite via microemulsion route. *Journal of Colloid and Interface Science* 259(2): 254–260.

Kral, V., Sotola, J., Neuwirth, P., et al. 2006. Nanomedicine-current status and perspectives: A big potential or just a catch word? *Chemicke Listy* 100: 4–9.

Krishna, G., Kooi, B.J., Palasantzas, G., et al. 2010. Thermal stability of gas phase magnesium nanoparticles. *Journal of Applied Physics* 107(5): 7.

Krishnakumar, T., Pinna, N., Kumari, K.P., Perumal, K., Jayaprakash, R. 2008. Microwave-assisted synthesis and characterization of tin oxide nanoparticles. *Materials Letters* 62(19): 3437–3440.

Kumar, R.V., Elgamiel, R., Diamant, Y., Gedanken, A., Norwig, J. 2001. Sonochemical preparation and characterization of nanocrystalline copper oxide embedded in poly (vinyl alcohol) and its effect on crystal growth of copper oxide. *Langmuir* 17(5): 1406–1410.

Kuo-chiang, L., Yu-ching, L., Shen-Ming, C. 2013. A highly sensitive non-enzymatic glucose sensor based on multi-walled carbon nanotubes decorated with nickel and copper nanoparticles. *Electrochimica Acta* 961: 164–172.

Lai, Chen, Tang, ShaoQiu, Wang, YingJun, Wei, Kun. 2005. Formation of calcium phosphate nanoparticles in reverse microemulsions. *Materials Letters* 59(2–3): 210–214.

Lanje, A.S., Sharma, S.J., Pode, R.B., Ningthoujam, R.S. 2010. Synthesis and optical characterization of copper oxide nanoparticles. *Advances in Applied Science Research* 1: 36–40.

Laurent, S., Forge, D., Port, M., et al. 2008. Magnetic iron oxide nanoparticles: Synthesis, stabilization, vectorization, physicochemical characterizations, and biological applications. *Chemical Reviews* 108(6): 2064–2110.

Lee, S.Y., Park, S.J. 2013. TiO_2 photocatalyst for water treatment applications. *Journal of Industrial and Engineering Chemistry* 19(6): 1761–1769.

Li, J., Chen, Y.C., Tseng, Y.C., Mozumdar, S., Huang, L. 2010. Biodegradable calcium phosphate nanoparticle with lipid coating for systemic Si RNA delivery. *Journal of Controlled Release* 142(3): 416–428.

Li, M., Mann, S. 2002. Emergent nanostructures: water-induced mesoscale transformation of surfactant stabilized amorphous calcium carbonate nanoparticles in reverse microemulsions. *Advanced Functional Materials* 12(1112): 773–779.

Li, Y., Liu, D., Ai, H., et al. 2010. Biological evaluation of layered double hydroxides as efficient drug vehicles. *Nanotechnology* 21(10): 105101.

Liang, L., Xu, Y., Wu, D., Sun, Y.A. 2009. Simple sol–gel route to ZrO2 films with high optical performances. *Materials Chemistry and Physics* 114(1): 252–256.

Lim, D.W., Yoon, J.W., Ryu, K.Y., Suh, M.P. 2012. Mg nanocrystals embedded in a metal–organic framework: Hybrid hydrogen storage with synergistic effect on physic and chemisorptions. *Angewandte Chemie* 51(39): 9814–9817.

Lin, H., Keng, C., Tung, C. 1997. Gas-sensing properties of nanocrystalline TiO2. *NANO Structured Materials* 9(1–8): 747.

Lisiecki, I., Pileni, M.B. 1993. Synthesis of copper metallic cluster using reverse micelles as microreactors. *Journal of American Chemical Society* 115: 3887–3896.

Liu, D., Chen, X., Xu, G., et al. 2016. Iridium nanoparticles supported on hierarchical porous N-doped carbon: An efficient water-tolerant catalyst for bio-alcohol condensation in water.

Liu, Xianghong, Zhang, Jun, Wang, Liwei, et al. 2011. 3D hierarchically porous ZnO structures and their functionalization by Au nanoparticles for gas sensors. *Journal of Materials Chemistry* 21(2): 349–356.

Lou, X.W., Archer, L.A., Yang, Z.C. 2008. Hollow micro/nanostructures: Synthesis and applications. *Advanced Materials* 20(21): 3987–4019.

Low, C.T.J., Wills, R.G.A., Walsh, F.C. 2006. Electrodeposition of composite coatings containing nanoparticles in a metal deposit. *Surface and Coatings Technology* 201(1–2): 371–383.

Luo, P., Kuwana, T. 1994. Nickel-titanium alloy electrode as a sensitive and stable detector for carbohydrates. *Analytical Chemistry* 66(17): 2775–2782.

Machmudah, S., Widiyastuti, W., Prastuti, O.P., et al. 2014. Synthesis of ZrO2 nanoparticles by hydrothermal treatment. *AIP Conference Proceedings* 1586: 166–172.

Madler, L., Kammler, H.K., Mueller, R., Pratsinis, S.E. 2002. Controlled synthesis of nanostructured particles by flame spray pyrolysis. *Journal of Aerosol Science* 33(2): 369–389.

Mahmoodi, Niyaz Mohammad, Chamani, Hooman, Kariminia, Hamid-Reza. 2015. Functionalized copper oxide-zinc oxide nanocomposite: Synthesis and genetic programming model of dye adsorption. *Desalination and Water Treatment* 57(40): 18755–18769.

Makridis, S.S., Gkanas, E., Panagakos, G., et al. 2013. Polymer-stable magnesium nanocomposites prepared by laser ablation for efficient hydrogen storage. *International Journal of Hydrogen Storage* 38(26): 11530–11535.

Mallakpour, S., Zeraatpisheh, F., Sabzalian, M.R. 2011. Sonochemical-assisted fabrication of biologically active chiral poly (ester-imide)/TiO2 bio nanocomposites derived from L-methionine and L-tyrosine amino acids. *eXPRESS Polymer Letters* 5(9): 825–837.

Mandal, D., Bolander, M.E., Mukhopadhyay, D., Sarkar, G., Mukherjee, P. 2006. The use of microorganisms for the formation of metal nanoparticles and their application. *Applied Microbiology and Biotechnology* 69(5): 485–492.

Mandal, M., Kundu, S., Ghosh, S.K., et al. 2005. Magnetite nanoparticles with tunable gold or silver shell. *Journal of Colloid and Interface Sciences* 286(1): 187–194.

Mann, E.L., Nathan, A., James, Chisholm, S.W. 2002. Copper toxicity and cyanobacteria ecology in the Sargasso Sea. *Limnology and Oceanography* 47(4): 976–988.

Maria Chong, A.S., Zhao, X.S. 2004. Functionalized nanoporous silicas for the immobilization of penicillin acylase. *Applied Surface Science* 237(1–4): 398–404.

Maridurai, Thirupathy, Balaji, Dhanapal, Sagadevan, Suresh. 2016. Synthesis and characterization of yttrium stabilized zirconia nanoparticles. *Materials Research* 19(4): 812–816.

Martis, M., Mori, K., Fujiwara, K., Ahn, W., Yamashita, H. 2013. Amine-functionalized MIL125 with imbedded palladium nanoparticles as an efficient catalyst for dehydrogenation of formic acid at ambient temperature. *Journal of Physical Chemistry C* 117(44): 22805–22810.

Mason, C., Vivekanandan, S., Misra, M., Mohanty, M.K. 2012. Switchgrass (Panicum virgatum) extract mediated green synthesis of silver nanoparticles. *World Journal of Nanoscience and Engineering* 2(2): 47–52.

Massart, R. 1981. Preparation of aqueous magnetic liquids in alkaline and acidic media. *IEEE Transactions on Magnetics* 17(2): 1247–1248.

Mavis, B., Tas, A.C. 2000. Dip coating of calcium hydroxyapatite on Ti-6Al-4V substrates. *Journal of the American Ceramic Society* 83(4): 989.

Mayers, B., Jiang, X., Sunderland, D., Cattle, B., Xia, Y. 2003. Hollow nanostructures of platinum with controllable dimensions can be synthesized by templating against selenium nanowires and colloids. *Journal of American Chemical Society* 125(44): 13364.

Meiwu, Q., Chagzhong, J., Jiang, C. 2008. Magnetic iron oxide nanoparticles: Synthesis and surface functionalization strategies. *Nanoscale Research Letters* 3(11): 397–415.

Meruvu, S., Hugendubler, L., Muller, E. 2011. Regulation of adipocyte differentiation by the zinc finger protein. ZNF638. *The Journal of Biological Chemistry* 286(30): 26516–26523.

Meziani, M.J., Bunker, C.E., Lu, F., et al. 2009. Formation and properties of stabilized aluminum nanoparticles. *ACS Applied Materials and Interfaces* 1(3): 703–709.

Mikhaylova, M., Kim, D.K., Berry, C.C., et al. 2004. BSA immobilization on amine-functionalized supraparamagnetic iron oxide nanoparticles. *Chemical Materials* 16(12): 2344–2354.

Mishra, B., Patel, B.B., Tiwari, S. 2009. Colloidal nanocarriers: A review on formulation technology, types and applications toward targeted drug delivery. *Nanomedicine: Nanotechnology, Biology and Medicine* 6(1): 9–24.

Moreno-Vega, A.I., Gomez-Quintero, T., Nunez-Anita, R.E., Acosta-Torres, L., Castaño, V. 2012. Polymeric and ceramic nanoparticles in biomedical applications. *Journal of Nanotechnology* 2012: 1–10.

Morgan, T.T., Muddana, H.S., Altinoglu, E.I., et al. 2008. Encapsulation of organic molecules in calcium phosphate nanocomposite particles for intracellular imaging and drug delivery. *Nano Letters* 8(12): 4108–4115.

Motogoshi, R., Oku, T., Suzuki, A., et al. 2010. Fabrication and characterization of cuprous oxide. Fullerene solar cells. *Synthesis of Metals* 160: 1219–1222.

Motoo, S., Furuya, N. 1984. Hydrogen and oxygen adsorption on Ir (111), 100 and (110) planes. *Journal of Electroanalytical Chemistry and Interfacial Electrochemistry* 167: 309–315.

Muddana, H.S., Morgan, T.T., Adair, J.H. and Butler, P.J. 2009. Photophysics of Cy3-encapsulated calcium phosphate nanoparticles. *Nanoletters* 9(4): 1559–1566.

Musa, A.O., Akomolafe, T., Carter, M.J. 1998. Production of cuprous oxide, a solar cell material, by thermal oxidation and a study of its physical and electrical properties. *Solar Energy Materials and Solar Cells* 51(3–4): 305–316.

Mustafa, M. Demir, Mehmet, A. Gulgun et.al. 2004. Pd nanoparticles by electrospinning from poly (acrylonitrile-co-acrylic acid)-pdcl$_2$ solutions. Relation between preparation conditions, particle size and catalytic activity. *Macromolecules* 37: 1787–1792.

Nadagouda, M.N., Polshetliwar, V., Varma, R.S. 2009. Self-assembly of palladium nanoparticles: Synthesis of nanobelts, nanoplates and nanotrees using vitamin B_1, and their application in carbon–carbon coupling reactions. *Journal of Material Chemistry* 19(14): 2026–2031.

Nadagouda, M.N., Varma, R.S. 2008. Green synthesis of silver and palladium nanoparticles at room temperature using coffee and tea extract. *Green Chemistry* 10(8): 859–886.

Nakajima, T., Shinoda, K., Tsuchiya, T. 2014. UV-assisted nucleation and growth of oxide films from chemical solutions. *Chemical Society Reviews* 43(7): 2027–2041.

Nakashima, T., Ohko, Y., Tryk, D.A., Fujishima, A. 2002. Decomposition of endocrine-disrupting chemicals in water by use of TiO2 photocatalysts immobilized on polytetrafluoroethylene mesh sheets. *Journal of Photochemistry and Photobiology, Part A* 6051: 1.

Nasibulin, A.G., Ahonen, P.P., Richard, O., Kauppinen, E.I., Altman, I.S. 2001. Copper and copper oxide nanoparticle formation by chemical vapor nucleation from copper (II) acetylacetonate. *Journal of Nanoparticle Research* 3(5/6): 383–498.

Nasibulin, A.G., Shurygina, L.I., Kauppinen, E.I. 2005. Synthesis of nanoparticles using vapor-phase decomposition of copper (II) acetylacetonate. *Colloid Journal* 67: 1–20.

Natália, J.S., Costa, Pedro, K., et al. 2010. A single-step procedure for the preparation of palladium nanoparticles and a phosphine-functionalized support as catalyst for Suzuki cross coupling reactions. *Journal of Catalysis* 276(2): 382–389.

Nath, S., Ghosh, S.K., Panigahi, S., Thundat, T., Pal, T. 2004. Synthesis of selenium nanoparticle and its photocatalytic application for decolorization of methylene blue under UV irradiation. *Langmuir* 20(18): 7880–7883.

Natile, M.M., Glisenti, A. 2002. Surface reactivity of NiO: Interaction with methanol. *Chemistry of Materials* 14–12(12): 4895–4903.

Ngah, W.S.W., Ab Ghani, S., Kamari, A. 2005. Adsorption behavior of Fe(II) and Fe(III) ions in aqueous solution on chitosan and cross-linked chitosan beads. *Bioresource Technology* 96(4): 443–450.

Nguyen, V.L., Nguyen, D.C.H., Hirata, V., et al. 2010. Chemical synthesis and characterization of palladium nanoparticles. *Advances in Natural Sciences: Nanoscience and Nanotechnology* 1(3): 035012–035016.

Nikhil, J., Zhong, L.W., Tapan, K.S., et al. 2000. Seed-mediated growth method to prepare cubic copper nanoparticles. *Current Science* 79(9): 1367–1370.

Nikhil Kumar, L., Sheobachan, U. 2016. Facile and green synthesis of highly stable L-cysteine functionalized copper nanoparticles. *Applied Surface Sciences* 16: 31155–31152.

Nimmo, W., Hind, D., Ali, N.J., Hampartsoumian, E., Milne, S.J. 2002. The production of ultrafine zirconium oxide powders by spray pyrolysis. *Journal of Materials Science* 37(16): 3381–3387.

Nithya, K., Yuvasree, P., Neelakandeswari, N., et al. 2014. Preparation and characterization of copper oxide nanoparticles. *International Journal of Chemical Technology and Research* 6: 2220–2222.

Norouz-Oliaee, S., Khodadadi, A.A., Mortazavi, Y., Alipour, S. 2010. Highly selective Pt/SnO2 sensor to propane or methane in presence of CO and ethanol, using gold nanoparticles on Fe2O3 catalytic filter. *Sensors and Actuators. Part B* 147(2): 400–405.

Nunes, A.C., Yu, Z.C. 1987. Fractionation of a water-based ferrofluid. *Journal of Magnetism and Magnetic Materials* 65(2–3): 265–268.

Ogawa, Satoshi, Fujimoto, Taishi, Tsukada, Chie, et al. 2012. *Spectroscopic Study of Hydrogen Storage of Mg-Pd Nanocomposite Material*. Brno, Czech Republic: EU.

Ogaway, S., Niwa, H., Nomoto, T., Yagi, S. 2010. Fabrication and characterization of magnesium nanoparticle by gas evaporation method. *e - Journal of Surface Science and Nanotechnology* 8: 246–249.

Ong, C.S., Goh, P.S., Lau, W.J., Misdan, N., Ismail, A.F. 2016. Nanomaterials for biofouling and scaling mitigation of thin film composite membrane: A review. *Desalination* 393: 2–15.

Pai, Asha R., Nair, Bipin. 2015. Synthesis and characterization of a binary oxide ZrO2–TiO2 and its application in chlorophyll dye-sensitized solar cell with reduced graphene oxide as counter electrodes. *Bulletin of Material Science* 38(5): 1129–1133.

Pal, U., Santiago, P. 2005. Controlling the morphology of ZnO nanostructures in a low-temperature hydro thermal process. *The Journal of Physical Chemistry. Part B* 109(32): 15317–15321.

Panigrahi, Sudipa, Kundu, Subrata, Basu, Soumen, et al. 2006. Cysteine functionalized copper organosol: Synthesis, characterization and catalytic application. *Nanotechnology* 17(21): 5461–5468.

Pankhurst, Q.A., Connolly, J., Jones, S.K., Dobson, J. 2003. Application of magnetic nanoparticles in biomedicine. *Journal of Physics. Part D* 36(13): 67–181.

Paul, W., Sharma, C.P. 2003. Ceramic drug delivery: A perspective. *Journal of Biomaterials Applications* 17(4): 253–264.

Paul, W., Sharma, C.P. 2007. Tricalcium phosphate delayed release formulation: A proof of concept study. *Journal of Pharmaceutical Sciences* 97: 875–882.

Paulose, S., Raghavan, R., George, B.K. 2017. Functionalized white graphene – Copper oxide nanocomposite: Synthesis, characterization and application as catalyst for thermal decomposition of ammonium perchlorate. *Journal of Colloid and Interface Science* 494: 64–73.

Peng, C., Zhao, Q., Gao, C. 2010. Sustained delivery of doxorubicin by porous CaCO3 and chitosan/alginate multilayers-coated CaCO3 microparticles. *Colloids Surface Physicochemical Engineering Aspects* 353(2–3): 132–139.

Peng, L.F., Liu, X.X., Yang, G.Y. 2001. Solid-phase glycerolysis of palm oil for producing monoacylglycerol by immobilized lipase using CaCO3 powder as carrier. *China Surfactant Detergent and Cosmetics* 5: 13–16.

Peng, Xiang, Wang, Lei, Zhang, Xuming, et al. 2015. Reduced graphene oxide encapsulated selenium nanoparticles for high-power lithium-selenium battery cathode. *Journal of Power Sources* 288: 214–220.

Perka, J., Slamberg, J., Hradi, J. 1976. Chemical transformation of polymers, XIX ion exchange derivatives of bead cellulose. *Angewandte Makromolekulare Chemie* 53(1): 73–80.

Petla, Ramesh Kumar, Vivekanandhan, Singaravelu, Misra, Manjusri, et al. 2012. Soybean (*Glycine max*) leaf extract based green synthesis of palladium nanoparticles. *Journal of Biomaterials and Nanobiotechnology* 3: 14–19.

Petrov, A.I., Volodkin, D.V., Sukhorukov, G.B. 2005. Protein–calcium carbonate co-precipitation: A tool for protein encapsulation. *Biotechnology Progress* 21(3): 918–925.

Pileni, M.P. 1998. Optical properties of nanosized particles dispersed in colloidal solutions or arranged in 2D or 3D superlactices. *New Journal of Chemistry* 22(7): 693–702.

Ping, Z., Liu, T., Xu, H., et al. 2017. Construction of highly stable selenium nanoparticles embedded in hollow nanofibers of polysaccharide and their antitumor activities. *Nano Research* 10(11): 3775–3789.

Pintar, A., Batista, J., Levec, J. 2001. Catalytic denitrification: Direct and indirect removal of nitrates from potable water. *Catalysis Today* 66(2–4): 503–510.

Pirzada, T., Arvidson, S.A., Saquing, C.D., Shah, S.S., Khan, S.A. 2012. Hybrid silica–PVA nanofibers via sol-gel electrospinning. *Langmuir* 28(13): 5834–5844.

Podhajecky, P., Klapste, B., Novak, P., et al. 1985. The influence of preparation conditions on the electrochemical behavior of CuO in a Li/CuO cell. *Journal of Power Sources* 14(4): 269–275.

Polshetliwar, V., Varma, R.S. 2010. Green chemistry by nano-catalysis. *Green Chemistry* 12(5): 743–754.

Pomogailo, A.D and Keslelman, V. 2005. *Metallopolymer Nanocomposites*. Berlin Heidelberg, New York: Springer.

Preeti, K.M., Aruna, S., Lafuente, J.V., et al. 2017. Intravenous administration of functionalized magnetic iron oxide nanoparticles does not induce CNS injury in the rat: Influence of spinal cord trauma and cerebrolysin treatment. In nanomedicine in central nervous system injury and repair. *International Review of Neurobiology* 137: 47–63.

Qian, K., Shi, T., Tang, T., et al. 2011. Preparation and characterization of nanosized calcium carbonate as controlled release pesticide carrier for validamycin against *Rhizoctonia solani*. *Microchimica Acta* 173(1–2): 51–57.

Qu, J., Luo, C., Hou, J. 2011a. Synthesis of ZnO nanoparticles from Zn-hyper accumulator (*Sedum alfredii Hance*) plants. *Micro and Nano Letters* 6(3): 174–176.

Qu, J., Yuan, X., Wang, X., Shao, P. 2011b. Zinc accumulation *and* synthesis of ZnO nanoparticles using *Physalis alkekengi* L. *Environment and Pollution* 159(7): 1783–1788.

Quaranta, A., Ceccato, R., Menato, C.R., et al. 2004. Formation of copper nanocrystals in alkali-lime silica glass by means of different reducing agents. *Journal of Non-Crystalline Solids* 345(46): 671–675.

Radhakrishnan, A.A., Beena, B.B. 2014. Structural and optical absorption analysis of CuO nanoparticles. *Indian Journal of Advanced Chemical Sciences* 2: 158–161.

Raja Naikaa, H., Lingarajua, H., Manjunathb, K., et al. 2015. Green synthesis of CuO nanoparticles using *Gloriosa superba* L. extract and their antibacterial activity. *Journal of Taibah University for Science* 9(1): 7–12.

Rajaeian, B., Heitz, A., Tade, M.O., Liu, S. 2015. Improved separation and antifouling performance of PVA thin film nanocomposite membranes incorporated with carboxylated TiO2 nanoparticles. *Journal of Membrane Science* 485: 48–59.

Rajender, S.V. 2011. Cheminform abstract: Green synthesis of metal nanoparticles: Biodegradable polymers and enzymes in stabilization and surface functionalization. *Chemical Science* 2: 837–846.

Rajendran, A., Siva, E., Dhanraj, C., Senthilkumar, S. 2018. A green and facile approach for the synthesis copper oxide nanoparticles using *Hibiscus rosa-sinensis* flower extracts and its antibacterial activities. *Journal of Bioprocessing and Biotechnology* 8(3): 1–4.

Rajiv, P., Sivaraj, R., Rajendran, V. 2013. Biofabrication of zincoxide nanoparticles using leaf extract of *Parthenium hysterophorus* L. and its size-dependent antifungal activity against plant fungal pathogens. *Spectrochimica Acta, Part A – Molecular and Biomolecular Spectroscopy* 112: 384–387.

Ramachandran, Rukmani, Paul, Willi, Sharma, Chandra, P. 2008. Synthesis and characterization of pegylated calcium phosphate nanoparticles for oral insulin delivery. *Journal of Biomedical Materials Research, Part B.*

Rani, R., Kumar, H., Salar, R.K., Purewal, S.S. 2014. Antibacterial activity of copper oxide nanoparticles against gram-negative bacteria strain synthesized by reverse micelle technique. *International Journal of Pharmaceutical Research and Development* 6: 72–78.

Ranjbar, M., Yousefi, M., Lahooti, M., Heydar Mahmoudi Najafi, S., Malekzadeh, A. 2016. Synthesis of pure monoclinic zirconia nanoparticles using ultrasound cavitation technique. *Journal of Particle Science and Technology* 2: 69–77.

Rehspringer, J.-L., Djameleddine, K., Khatmi, D. 2006. Synthesis of palladium nanoparticles by sonochemical reduction of palladium (II) nitrate in aqueous solution. *The Journal of Physical Chemistry. Part B* 110(1): 383–389.

Reznickova, A., Orendac, M., Koiska, Z., et al. 2016. Copper nanoparticles functionalized PE: Preparation characterization and magnetic properties. *Applied Surface Science* 390: 728–734.

Roceo, R., Samantha, K.R.L., Ana, L.F.O., et al. 2011. Aerobic synthesis of palladium nanoparticles. *Reviews on Advanced Material Science* 27(1): 31–42.

Roy, I., Susmita, M., Amarnath, M et al. 2003. Calcium phosphate nanoparticles as novel non-viral vectors for targeted gene delivery. *The International Journal of Pharmacy* 250(1): 25–33.

Rozoa, C., Jaque, D., Fonseca, L.F., Garcia Sole, J. 2008. Luminescence of rare earth-doped Si–ZrO2 co-sputtered films. *Journal of Luminescence* 128(7): 1197–1204.

Rudolf, C., Mazilu, I., Chirieac, A., et al. 2005. Copper nanoparticles supported on polyether functionalized mesoporous silica. Synthesis and application as hydrogenation catalysts. *Environmental Engineering and Management Journal* 14(2): 399–408.

Rumyantseva, M.N., Gaskov, A.M., Rosman, N., Pagnier, T., Morante, J.R. 2005. Raman surface vibration modes in nanocrystalline SnO_2: Correlation with gas sensor performances. *Chemical Materials* 17(4): 893.

Russell, S.W., Luptak, K.A., Suchicital, C.T.A., Alford, T.L., Pizziconi, V.B. 1996. Chemical and structural evolution of Sol-gel-derived hydroxyapatite thin films under rapid thermal processing. *Journal of the American Ceramic Society* 79(4): 837.

Safarik, J., Safarikova, M. 2004. Magnetic techniques for the isolation and purification of proteins and peptides. *Biomagnetic Research and Technology* 2(1): 7.

Schmidt, H.T., Ostafin, A.E. 2002. Liposome directed growth of calcium phosphate nanoshells. *Advanced Materials* 14(7): 532.

Senthilkumar, S.R., Sivakumar, T. 2014. Green tea (*Camellia sinensis*) mediated synthesis of zinc oxide nanoparticles and studies on their antimicrobial activities. *International Journal of Pharmacy and Pharmaceutical Sciences* 6(6): 461–465.

Seok Lee, W., Woog Kim, S., Heon Koo, B., Sik Bae, D. 2008. Synthesis and microstructure of Y2O3-doped ZrO2–CeO2 composite nanoparticles by hydrothermal process. *Colloids and Surfaces. Part A* 313–314: 100–104.

Seraphens, Z.D. 1993. Selective encapsulation of the carbides of yttrium and titanium into carbon nanoclusters. *Applied Physics Letters* 63(15): 2073–2075.

Sergeev, G.B. 2003. Cryochemistry of metal nanoparticles. *Journal of Nanoparticle Research* 5(5–6): 529–537.

Severance, C.L., Mc Bride, R., Finley, M. 2009. *Nano Inks for Imparting EMI Shielding to Windows*. US20090084599 A1.

Shafiu Kamba, A., Ismail, M., Tengku Ibrahim, T.A., et al. 2013. A pH-sensitive, biobased calcium carbonate aragonite nanocrystal as a novel anticancer delivery system. *BioMed Research International* 0: 1–10.

Shan, D., Wang, S.X., Xue, H.G., Cosnier, S. 2007. Direct electrochemistry and electrocatalysis of hemoglobin entrapped in composite matrix based on chitosan and CaCO3 nanoparticles. *Electrochemical Communications* 9: 529–534.

Shan, D., Wang, Y.N., Xue, H.G., Cosnier, S. 2009. Sensitive and selective xanthin amperometric sensors based on calcium carbonate nanoparticles. *Sensors and Actuators. Part B* 136(2): 510–516.

Sharada, S., Prashant, L.S., Kumar, P., R., et al. 2016. Synthesis of palladium nanoparticles using continuous flow microreactor. *Colloids and Surfaces A: Physicochemical and Engineering Aspects* 498: 297–304.

Sharma, A., Kumar, A., Meena, K.R., et al. 2017. Fabrication and functionalization of magnesium nanoparticle for lipase immobilization in n-propyl gallate synthesis. *Journal of King Saud University: Science* 29(4): 536–546.

Sharma, D., Rajput, J., Kaith, B., Kaur, M., Sharma, S. 2010. Synthesis of ZnO nanoparticles and study of their antibacterial and antifungal properties. *Thin Solid Films* 519(3): 1224–1229.

Sharmila Devi, R., Gayathri, R. 2014. Green synthesis of zinc oxide nanoparticles by busing *Hibiscus rosa-sinensis*. *International Journal of Current and Engineering and Technology* 4(4): 2444–2446.

Shin, J., Anisur, R.M., Ko, M.K., et al. 2009. Hollow manganese oxide nanoparticles as multifunctional agents for magnetic resonance imaging and drug delivery. *Angewandte Makromolekulare Chemie* 48(2): 321–324.

Solmaz, M.D., Mohammad, B. J. 2015. Calcium carbonate nanoparticles as cancer drug delivery system. *Expert Opinion Drug Delivery* 12(10).

Shylesh, S., Wang, L., Thiel, W.R. 2010. Palladium (II)-phosphine complexes supported on magnetic nanoparticles: Filtration-free, recyclable catalysts for Suzuki–Miyaura cross-coupling reactions. *Advanced Synthesis and Catalysis* 352(2–3): 425–432.

Siamaki, A.R., Khder, A.E.R.S., Abdelsayed, V., El-Shall, M.S., Gupton, B.F. 2011. Microwave-assisted synthesis of palladium nanoparticles supported on graphene: A highly active and recyclable catalyst for carbon–carbon cross-coupling reactions. *Journal of Catalysis* 279(1): 1–11.

Sjogren, C.E., Johansson, C., Naevestad, A., et al. 1997. Crystal size and properties of superparamagnetic iron oxide (SPIO) particles. *Magnetic Resonance Imaging* 15(1): 55–67.

Song, X.Y., Sun, S.X., Zhang, W.M., Yin, Z. 2004. A method for the synthesis of spherical copper nanoparticles in the organic phase. *Journal of Colloid and Interface Sciences* 273(2): 463–469.

Sophie, L., Delphine, F., Marc, P., et al. 2008. Magnetic iron oxide nanoparticles: Synthesis, stabilization, vectorization, physicochemical characterizations, and biological applications. *Chemical Reviews* 108(6): 2064–2110.

Srivastava, N., et al. 2011. Biosynthesis of structural characterization of selenium nanoparticles mediated by *P. alcalipha*. *Colloids and Surface, Part B* 88(1): 196–201.

Srivastava, N., Mukhopadhyay, M. 2013. Biosynthesis of structural characterization of selenium nanoparticles mediated by *Zooglea ramigera*. *Powder Technology* 244: 26–29.

Srivastava, S., Kumar, M., Agrawal, A., Dwivedi, S.K. 2013. Synthesis and characterization of copper oxide nanoparticles. *IOSR Journal of Applied Physics* 5(4): 61–65.

Such, W.H., Suslick, K.S., Stucky, G.D., Suh, Y.H. 2009. Nanotechnology, nanotoxicology and neuroscience. *Progressive Neurobiology* 87(3): 133–170.

Suchanek, W., Yoshimura, M. 1998a. Preparation of fibrous, porous hydroxyapatite ceramics from hydroxyapatite whiskers. *Journal of the American Ceramic Society* 81(3): 765–767.

Suchanek, W., Yoshimura, M. 1998b. Processing and properties of hydroxyapatite-based biomaterials for use as hard tissue replacement implants. *Journal of Materials Research* 13(1): 94–117.

Sugawara, Ayae, Yamane, Setsuko, Akiyoshi, Kazunari. 2006. Nanogel-templated mineralization: Polymer-calcium phosphate hybrid nanomaterials. *Macromolecular Rapid Communications* 27(6): 441–446.

Suleiman, M., Mousa, M., Hussein, A., et al. 2013. Copper(II)-oxide nanostructures: Synthesis, characterizations and their applications-review. *Journal of Materials and Environmental Science* 4: 792–797.

Sun, Tai, Zhang, Zheye, Xiao, Junwu, et al. 2013. Facile and green synthesis of palladium nanoparticles-graphene-carbon nanotube material with high catalytic activity. *Scientific Reports* 3: 2527.

Sundarrajan, Subramanian, Ramakrishna, Seeram. 2010. Fabrication of functionalized nanofiber membranes containing nanoparticles. *Journal of Nanoscience and Nanotechnology* 10(2): 1139–1147.

Sung, M.M., Sung, K., Kim, C.G., Lee, S.S., Kim, Y. 2000. Self-assembled monolayers of alkanethiols on oxidized copper surfaces. *Journal of Physical Chemistry. Part B* 104(10): 2273–2277.

Tahir, M.N., Gorgishvili, L., Li, J., et al. 2007. Facile synthesis and characterization of monocrystalline cubic ZrO2 nanoparticles. *Solid State Sciences* 9(12): 1105–1109.

Tai, C.Y., Hsiao, B.Y., Chiu, H.Y. 2004. Preparation of spherical hydrous-zirconia nanoparticles by low temperature hydrolysis in a reverse microemulsion. *Colloids and Surfaces A: Physicochemical and Engineering Aspects* 237(1–3): 105–111.

Tai, S., Zheye, Z., Junwu, X. et.al. 2013. Facile and Green Synthesis of Palladium Nanoparticles-Graphene-Carbon Nanotube Material with High Catalytic Activity. *Scientific Reports* 3: 2527.

Tamaki, J., Shimanoe, K., Yamada, Y., et al. 1998. Dilute hydrogen sulfide sensing properties of CuO–SnO2 thin film prepared by low-pressure evaporation method. *Sensory Actuators B* 49: 121–125.

Tamura, Masaru, Fujihara, Hisashi. 2003. Chiral bisphosphine BINAP-stabilized gold and palladium nanoparticles with small size and their palladium nanoparticle-catalyzed asymmetric reaction. *Journal of the American Chemical Society* 125(51): 15742–15743.

Tartaj, Pedro, Puertomorales, Mariadel. 2003. The preparation of magnetic nanoparticles for applications in biomedicine. *Journal of Physics. Part D* 36(13): 182–197.

Tavares, J., Coulombes. 2011. Dual plasma synthesis and characterization of a stable copper-ethylene glycol nano fluid. *Powder Technology* 210(2): 132–142.

Tetsuya Kida, W., Oka, T., Nagano, M., Ishiwata, Y., Zheng, X. 2007. Synthesis and application of stable copper oxide nanoparticle suspensions for nanoparticulate film fabrication. *Journal of the American Ceramic Society* 90(1): 107–110.

Tian, Y.F., Li, X.H., Ao, Z.H., Xue, Y.J. 2013. Corrosion resistance of Ni–ZrO2 nanocomposite coating prepared by pulse electrodeposition with rotating cathode in an ultrasonic field. *Applied Mechanics and Materials* 278–280: 422–425.

Tiefenauer, L.X., Kuhne, G., Andres, R.Y. 1993. Antibody-magnetite nanoparticles: In vitro characterization of a potential tumor-specific contrast agent for magnetic resonance imaging. *Bioconjugate Chemistry* 4(5): 347–353.

Tobiska, P., Hugon, O., Trouillet, A., Gagnaire, H. 2001. An integrated optic hydrogen sensor based on SPR on palladium. *Sensory Actuators* B 74(1-3): 168–172.

Torres-Huerta, A.M., Dominguez-Crespo, M.A., Ramirez-Meneses, E., Vargas-Garcia, J.R. 2009. MOCVD of zirconium oxide thin films: Synthesis and characterization. *Applied Surface Science* 255(9): 4792–4795.

Tran, X., Phuoc, B.H., Howard, D.V., Soong, Y., Chyu, M.K. 2008. Synthesis of mg(OH), MgO, and Mg nanoparticles using laser ablation of magnesium I water and solvents. *Optics and Lasers in Engineering* 46(11): 829–834.

Tu, Wenwen, Lei, Jianping, Wang, Peng, Ju, Huangxian. 2011. Photoelectrochemistry of free-base-porphyrin-functionalized zinc oxide nanoparticles and their applications in biosensing. *Chemistry* 17(34): 9440–9447.

Turgut, Z., Nuhfer, N.T., Piehler, H.R., McHenry, M.E. 1999. Magnetic properties and microstructural observations of oxide coated FeCo nanocrystals before and after compaction. *Journal of Applied Physics* 85(8): 4406.

Ueno, Y., Futagawa, H., Takagi, Y., Ueno, A., Mizushima, Y. 2005. Drug-incorporating calcium carbonate nanoparticles for a new delivery system. *Journal of Controlled Release* 103(1): 93–98.

Ulman, A. 1995. *Supramolecular Assemblies: Vision and Strategy*, eds., H. Kuhn, A. Ulman. Boston: Academic Press.

Ulman, A. 1996. Formation and structure of self-assembled monolayers. *Chemical Reviews* 96(4): 1533.

Umar, A., Rahman, M.M., Al-Hajry, A., Hahn, Y.B. 2009. Enzymatic glucose biosensor based on flower-shaped copper oxide nanostructures composed of thin nanosheets. *Electrochemical Communications* 11(2): 278–281.

Vallejos, S., Selina, S., Annanouch, I., et al. 2016. Micromachined gas sensors based on Au-functionalized SnO2 nanorods directly integrated without catalyst seeds. *Procedia Engineering* 168: 1078–1081.

Veerapandian, M., Sadhasivam, S., Choi, J., Kyusik, Y. 2012. Glucosamine functionalized copper nanoparticles. Preparation, characterization and enhancement of anti-bacterial activity by ultraviolet irradiation. *Chemical Engineering Journal* 209: 558–567.

Vekariya, K.K., Kaur, J.K., Tikoo. 2012. ERα signaling imparts chemotherapeutic selectivity to selenium nanoparticles in breast cancer. *Nanomedicine: Nanotechnology, Biology, and Medicine* 8(7): 1125.

Vidya, C., Hiremath, S., Chandraprabha, M.N., et al. 2013. Green synthesis of ZnO nanoparticles by *Calotropis gigantea*. *International Journal of Current and Engineering and Technology* 1: 118–120.

Vitulli, G., Bernini, M., Bertozzi, S., et al. 2002. Nanoscale copper particles derived from solvated Cu atoms in the activation of molecular oxygen. *Chemistry of Materials* 14(3): 1183–1186.

Volanti, D.P., Keyson, D., Cavalcante, L.S., et al. 2008. Synthesis and characterization of CuO flower-nanostructure processing by a domestic hydrothermal microwave. *Journal of Alloys and Compounds* 459(1–2): 537–542.

Wahajuddin, Arora, S. 2012. Superparamagnetic iron oxide nanoparticles: Magnetic nanoplatforms as drug carriers. *International Journal of Nanomedicine* 7: 3445–3457.

Waichal, R.P., Karvir, G.D., Patil, K.R., et al. 2012. Synthesis of cuprous oxide nanoparticles by electrochemical method and evaluation of the corresponding nanoparticle film for humidity sensing. *1st International Symposium, Physics and Technology of Sensors (ISPTS)*, Pune, India.

Wang, C., He, C., Tong, Z., et al. 2006. Combination of adsorption by porous CaCO3microparticles and encapsulation by polyelectrolyte multilayer films for sustained drug delivery. *International Journal of Pharmaceutics* 308(1–2): 160–167.

Wang, C.-c., Chen, D.-H., Hung, T.I.-c. 2001. Synthesis of palladium nanoparticles in water-in-oil microemulsions. *Colloids and Surfaces A: Physicochemical and Engineering Aspects* 189: 145–154.

Wang, C.K., Chern Lin, J.H., Ju, C.P., Ong, H.C., Chang, R.P.H. 1997. Structural characterization of pulsed laser-deposited hydroxyapatite film on titanium substrate. *Biomaterials* 18(20): 1331–1338.

Wang, C.Y., Zhou, Y., Chen, B., Liu, H.J., Mo, X. 1999. Preparation of shell-core Cu_2O-Cu nanocomposite particles and Cu nanoparticles in a new microemulsion system. *Journal of Colloid and Interface Science* 220: 468–470.

Wang, L., Yang, Z.M., Gao, J.H., et al. 2006. A biocompatible method of decorporation: Bisphosphonate-modified magnetite nanoparticles to remove uranyl ions from blood. *Journal of the American Chemical Society* 128(41): 13358–13359.

Wang, M. 2003. Developing bioactive composite materials for tissue replacement. *Biomaterials* 24(13): 2133–2151.

Wang, W., Hou, F.Y., Wang, H., Guo, H.T. 2005. Fabrication and characterization of Ni–ZrO2 composite nano-coatings by pulse electrodeposition. *Scripta Materialia* 53(5): 613–618.

Welzel, T., Meyer-Zaika, W., Epple, M. 2004a. Continuous preparation of functionalised calcium phosphate nanoparticles with adjustable crystallinity. *Chemical Communications* 0(10): 1204–1205.

Welzel, T., Radtke, I., Meyer-Zaika, W., Heumann, R., Epple, M. 2004b. Transfection of cells with custom-made calcium phosphate nanoparticles coated with DNA. *Journal of Materials Chemistry* 0(14): 2213.

Weng, W., Bapsta, J.L. 1999. Preparation and characterization of hydroxyapatite coatings on Ti6A14V alloy by a Sol-gel method. *Journal of the American Ceramic Society* 82(1): 27–32.

Wongpisutpaisan, N., Charoonsuk, P., Vittayakorn, N., Pecharapa, W. 2011. Sonochemical synthesis and characterization of copper oxide nanoparticles. *Energy Procedia* 9: 404–409.

Wu, C., Mosher, B.P., Zeng, T. 2006. One-step green route to narrowly dispersed copper nanocrystals. *Journal of Nanoparticle Research* 8(6): 965–969.

Wu, H., Zhu, H., Li, X., et al. 2013. Induction of apoptosis and cell cycle arrest in A549 human lung adenocarcinoma cells by surface-capping selenium nanoparticles: An effect enhanced by polysaccharide-protein complexes from Polyporus rhinocerus. *Journal of Agricultural and Food Chemistry* 61(41): 9859.

Wu, W., He, Q., Jiang, C. 2008. Magnetic iron oxide nanoparticles: Synthesis and surface functionalization strategies. *Nanoscale Research Letters* 3(11): 397–415.

Wu, Yao, Jiang, Wen, Wen, Xiantao, et al. 2010. A novel calcium phosphate ceramic–magnetic nanoparticle composite as a potential bone substitute. *Biomedical Materials* 5: 015001.

Wua, R., Ma, Z., Gua, Z., Yang, Y. 2010. Preparation and characterization of CuO nanoparticles with different size and morphology. *International Conference on Mechanic Automation and Control Engineering (MACE)*, Wuhan.

Xi, Y., Hu, C., Gao, P., et al. 2010. Morphology and phase selective synthesis of Cu_xO (X=1,2) nanostructures and their catalytic degradation activity. *Material Science and Engineering. Part B* 166(1): 113–117.

Xiangyangshi, S.H.W., Scott, D.S., Song, G., et al. 2008. Dendrimer-functionalized shell-crosslinked iron oxide nanoparticles for *in vivo* magnetic resonance imaging of tumors. *Advanced Materials* 20(9): 1671–1678.

Xianyu, W.X., Park, M.K., Lee, W.I. 2001. Thickness effect in the photocatalytic activity of TiO2 thin films derived from sol-gel process. *Korean Journal of Chemical Engineering* 18(6): 903–907.

Xiaoli, Y., Huixiang, S., Dahui, W. 2003. Photoelectrocatalytic degradation of phenol using a TiO2/Ni thin-film electrode. *Korean Journal of Chemical Engineering* 20(4): 679–684.

Xie, H., Zhang, Q., Xi, T., Wang, J., Liu, Y. 2002. Thermal analysis on nanosized TiO2 prepared by hydrolysis. *Thermochimica Acta* 381(1,3): 45–48.

Xie, S.Y., Ma, Z.J., Wang, C.F., et al. 2004. Preparation and self-assembly of copper nanoparticles via discharge of copper rod electrodes in a surfactant solution: A combination of physical and chemical processes. *Journal of Solid State Chemistry* 177(10): 3743–3747.

Xu, C., Xu, K., Gu, H., et al. 2004. Dopamine as a robust anchor to immobilize functional molecules on the iron oxide shell of magnetic nanoparticles. *Journal of the American Chemical Society* 126(32): 9938–9939.

Xu, D., Diao, P., Jin, T., et al. 2015. Iridium oxide nanoparticles and iridium/iridium oxide nanocomposites: Photochemical fabrication and application in catalytic reduction of 4-nitrophenol. *ACS Applied Material and Interfaces* 7(30): 16738–16749.

Xu, Qin, Zhao, Yu, Xu, Jin Zhong, Zhu, Jun-Jie. 2006. Preparation of functionalized copper nanoparticles were prepared by a microwave heating method to be used to construct a new glucose sensor. *Sensors and Actuators B Chemical* 114: 379–386.

Xue, C., Métraux, G.S., Millstone, J.E., Mirkin, C.A. 2008. Mechanistic study of photomediated triangular silver nanoprism growth. *Journal of American Chemical Society* 130(26): 8337–8344.

Xue, Y.-J., Liu, Chun-Yang, Li, Ji-Shun, Ma, Wei. 2015. Fabrication and tribological behavior of Ni-Zro2 nanocomposite coatings prepared by electrodeposition in ultrasonic field. *Proceedings of the ASME 2015 International Design Engineering Technical Conferences & Computers and Information in Engineering Conference*, Boston, MA, USA.

Yang, Fang, Tang, Quanming, Zhong, Xueyun, et al. 2012. Surface decoration by *Spirulina* polysaccharide enhances the cellular uptake and anticancer efficacy of selenium nanoparticles. *International Journal of Nanomedicine* 7: 835–844.

Yang, L., Chen, Q., Liu, Y., et al. 2014. Se/Ru nanoparticles as inhibitors of metal-induced aggregation in Alzheimer's disease. *Journal of Material Chemistry B* 2: 1977–1987.

Yang, S., Wang, C., Chen, L., Chen, S. 2010. Facile dicyandiamide-mediated fabrication of well-defined CuO hollow microspheres and their catalytic application. *Materials Chemistry and Physics* 120(2–3): 296–301.

Yang, Yongjian, Chen, Weiwei, Zhou, Chungen, Xu, Huibin, Gao, Wei. 2011. Fabrication and characterization of electroless Ni–P–ZrO2 nano-composite coatings. *Applied Nanoscience* 1(1): 19–26.

Yanping, Yu, Fei, Zhiqiang, Cui, Jing, et al. 2018. Biosynthesis of copper oxide nanoparticles and their *in vitro* cytotoxicity towards nasopharynx cancer (KB Cells) cell lines. *International Journal of Pharmacology* 14(5): 609–614.

Yao, W.T., Yu, S.H., Zhou, Y., et al. 2005. Formation of uniform CuO nanorods by spontaneous aggregation: Selective synthesis of CuO, Cu2O, and Cu nanoparticles by a solid-liquid phase arc discharge process. *The Journal of Physical Chemistry. Part B* 109(29): 14011–14016.

Yasser S, Khaled A., Soliman, Sun, S., Jacob, T., Kibler, L.A. 2015. Electrochemical fabrication of well-defined spherical iridium nanoparticles and electrocatalytic activity towards carbon monoxide adlayer oxidation. *Electrocatalysis* 6(4): 365–372.

Ye, P., Xu, Z.K., Che, A.F., Wu, J., Seta, P. 2006. Chitosan tethered poly (acrylonitrile –co-maleic acid) hollow fiber membrane for lipase immobilization. *Biomaterials* 26(32): 6394–6403.

Yee, Chanel K., Jordan, Rainer, Ulman, Abraham, et al. 1999. Novel one-phase synthesis of thiol-functionalized gold, palladium, and iridium nanoparticles using superhydride. *Langmuir* 15(10): 3486–3491.

Yesh chenko, O.A., Dmitruk, I.M., Dmytrukb, A.M., Alexcenko, A.A. 2007. Influence of annealing conditions on size and optical properties of copper nanoparticles embedded in silica matrix. *Materials Science and Engineering: Part B* 137(1–3): 247–254.

Yip, S.K., Sauls, J.A. 1992. Nonlinear Meissner effect in CuO superconductors. *Physical Review Letters* 69(15): 2264–2267.

Yogesh, B.A., Gotan, H., et al. 2017. Biosynthesis of copper oxide nanoparticles using leaves extract of *Leucaena leucocephala* L and their promising upshot against the selected human pathogens. *International Journal of Molecular and Clinical Biology* 7(1): 776–786.

Yoko, S., Mitsuhro, O., Ichiro, Y. 2012. Fabrication of zinc oxide semiconductor nanoparticles in the apoferritin cavity. *Crystal Growth and Design* 12(8): 4130–4134.

Yu, D., Wang, D., Yu, W., et al. 2006. Synthesis of ITO nanowires and nanorodes with corundum structure by a coprecipitation method. *Materials Letters* 58: 84–87.

Yu, S., Wang, Y., Zhang, W., et al. 2016. pH-assisted surface functionalization of selenium nanoparticles with curcumin to achieve enhanced cancer chemopreventive activity. *The Royal Society of Chemistry*: 1–3.

Yu, T., Cheong, F.C., Sow, C.H. 2004. The manipulation and assembly of CuO nanorods with line optical tweezers. *Nanotechnology* 15(12): 1732–1736.

Zakeeruddin, S.M., Nazeeruddin, Md.K., Humphry-Baker, R., et al. 2002. Design, synthesis, and application of amphiphilic ruthenium polypyridyl photosensitizers in solar cells based on nanocrystalline TiO2 films. *Langmuir* 18(3): 952.

Zhang, Jun, Guo, Jing, Xu, Hongyan, Cao, Bingqiang. 2013. Reactive-template fabrication of porous SnO2 nanotubes and their remarkable gas-sensing performance. *ACS Applied Material and Interfaces* 5(16): 7893–7898.

Zhang, Xuanzhou, Yang, Rong, Yang, Junzhi, et al. 2011. Synthesis of magnesium nanoparticles with superior hydrogen storage properties by acetylene plasma metal reaction. *International Journal of Hydrogen Energy* 36(8): 4967–4975.

Zhanhu, G., Kinny, L., Yutong, L., et al. 2008. Fabrication and characterization of iron-oxide nanoparticles reinforced vinyl-ester resin nanocomposites. *Composites Science and Technology* 68(6): 1513–1520.

Zhaohua, P., Ting, L., Hui, Xu. et.al. 2017. Construction of highly stable selenium nanoparticles embedded in hollow nanofibers of polysaccharide and their antitumor activities. *Nano Research*. 10: 3775–3789.

Zhao, S., Wang, P., Wang, C., Sun, X., Zhang, L. 2012. Thermostable PPESK/TiO2 nanocomposite ultrafiltration membrane for high temperature condensed water treatment. *Desalination* 299: 35–43.

Zhao, Y.X., Hernandez-Pagan, E.A., Vargas-Barbosa, N.M., Dysart, J.L., Mallouk, T.E. 2011. A high yield synthesis of ligand-free iridium oxide nanoparticles with high electrocatalytic activity. *Journal of Physical Chemistry Letters* 2(5): 402–406.

Zheng, X.G., Xu, C.N., Tomokiyo, Y., Tanaka, E., Yamada, H. Soejima, Y. 2000. Observation of charge stripes in cupric oxide. *Physical Review Letters* 85(24): 5170–5173.

Zhong, L.S., Hu, J.S., Liang, H.P., Cao, A.M., et al. 2006. Self-assembled 3D flower like iron oxide nanostructures and their application in water treatment. *Advanced Materials* 18(18): 2426–2431.

Zhou, D.W. 2004. Heat transfer enhancement of copper nanofluid with acoustic cavitation. *International Journal of Heat and Mass Transfer* 47(14–16): 3109–3117.

Zhou, R., Zhou, F., Hao, X., et al. 2008. Influences of surfactant (PVA) concentration and pH on the preparation of copper nanoparticles by electron beam irradiation. *Radiation Physics and Chemistry* 77(2): 169–173.

17

Clay/Non-Ionic Surfactant Hybrid Nanocomposites

Giuseppe Cavallaro, Giuseppe Lazzara, Stefana Milioto, Filippo Parisi, and Luciana Sciascia

CONTENTS

17.1 Introduction

Clay minerals are attracting the interest of researchers in materials science because they are available at low cost and with a very large variety of surface chemistry/properties and mesoscopic structure/morphology (Abdullayev et al., 2012; Wei et al., 2014). The high swelling capacity and the ability to adsorb significant molecules of different nature make clays suitable for application in different fields as lubricants, catalysts, diluents, flavor correctors, emulsifiers, rheological agents, and drug delivery modifiers (Savic-Gajic et al., 2014; Ambrogi et al., 2012; Bromberg et al., 2011; Calabrese et al., 2013a; Carretero, 2002; Choy et al., 2007; Lazzara, Riela, and Fakhrullin, 2017).

Surfactant molecules assumed a central role in colloidal science due to their different technological applications, such as the control of drug release, the removal of contaminants from soil, the synthesis of nanoparticles, detergency, etc. (Turco Liveri et al., 2012; Turco Liveri et al., 2007; Gennara Cavallaro et al., 2006; Giuseppe Cavallaro et al., 2015a; Hassan, Ruso, and Piñeiro, 2011; Fatma et al., 2015).

In proper solvents, they are able to self-aggregate, thus forming molecular clusters called micelles. In aqueous solution, these microstructured systems are highlighted by the simultaneous coexistence of hydrophilic and hydrophobic domains having a local order and the fluidity typical of liquids. Another interesting aspect deals with the possibility to study, in reproducible and tunable systems, complex phenomena such as the molecular confinement in microscopic regions, the solubilization and the reactivity in micro-heterogeneous media, etc. (Turco Liveri et al., 2009; Sciascia, Hauser, and Turco Liveri, 2008; Rossi et al., 2008; Sciascia, Lombardo, and Turco Liveri, 2007; Sciascia et al., 2010).

The surfactants are adsorbed preferentially at the interfaces (e.g., liquid/air, solid/liquid, and liquid/liquid) lowering the superficial tension and, therefore, diminishing the spontaneous tendency of the systems to achieve the minimal superficial area. Consequently, they are widely used as dispersing agents, stabilizing emulsions, and suspensions, or to modulate the wetting of liquids.

Non-ionic surfactants in particular attract the interest of both scientific and industrial communities, due to their non-toxicity and biocompatibility, which makes them very suitable for (above all) pharmacological and environmental applications (Jahan, Balzer, and Mosto, 2008).

Most non-ionic surfactants are polyglycerol alkyl ethers, glucosyl dialkyl ethers, crown-ethers, ester-linked surfactants, polyoxyethylene alkyl ethers (generally referred to as Brij), sorbitan esters (Spans), and polysorbates (Tweens). Since the World Health Organization (WHO) attested their high biocompatibility and safety profile (Muzzalupo et al., 2009; Jeevana Jyothi and Sreelakshmi, 2011), polysorbates have been widely applied in the biotechnology industry. The polar head group of polysorbates is a polyoxyethylene sorbitan, while the hydrophobic tail is an ester. Some structures of different polysorbates are reported in Figure 17.1.

Recently, research has also focused on a new class of surfactants, i.e., the amphiphilic polymers. During the last few years, such macromolecules have been increasingly employed in a variety of applications (e.g., control of the rheological properties, stabilization of latex, drug release, etc.) because they are at the same time polymers and amphiphiles (Chiappisi et al., 2012; Nguyen-Kim et al., 2016; Dintcheva et al., 2016). They are self-assembling molecules leading to the formation of several structures and are active at the interface. Within this class of polymeric surfactants, we will address our attention to the tri-block copolymers based on poly(ethylene oxide) (PEO) and poly(propylene oxide) (PPO), commercially available under the trade name of pluronics or synperonics. Their chemical structure is $HO[CH_2CH_2O]_a[CH_2CHO(CH_3)]_b[CH_2CH_2O]_aH$ and they are denoted as PEO-PPO-PEO or $EO_aPO_bEO_a$ (with a and b being the number of ethylene oxide and the propylene oxide units, respectively). Reverse architecture is also available where the hydrophilic PEO block is linked to the two PPO blocks. The reverse copolymers are represented as PPO-PEO-PPO or $PO_bEO_aPO_b$ where a and b have the same meaning as above. Some pluronics discussed in this chapter are collected in Table 17.1, where their structure, molecular weight, and abbreviation are reported. For the sake of simplicity, hereafter they will be referred to through abbreviations.

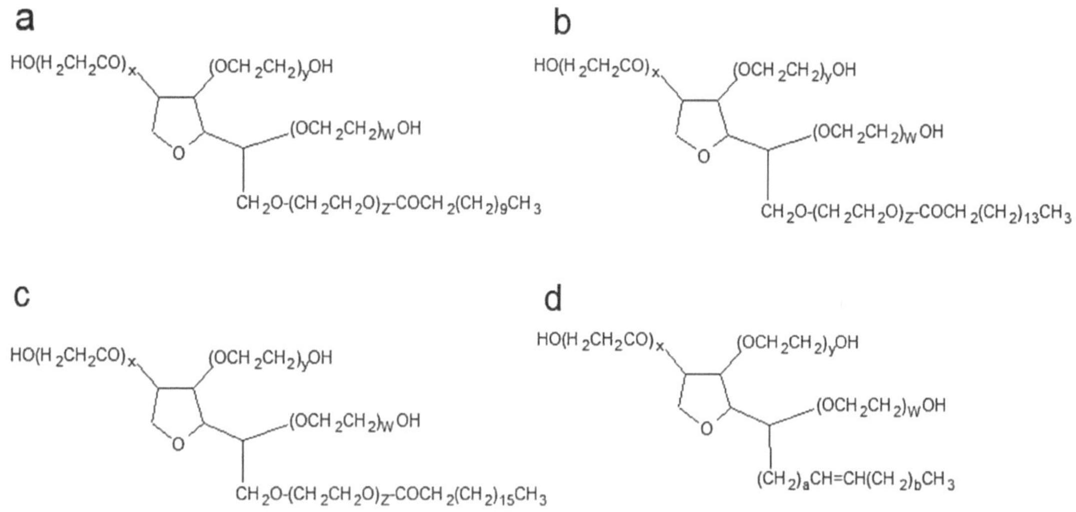

FIGURE 17.1 Chemical structures of (a) Tween 20; (b) Tween 40; (c) Tween 60; (d) Tween 80; a + b = 7 and x + y + w + z = 20.

TABLE 17.1
Structures and Abbreviations for Pluronics

Structure	Abbreviation	EO/PO (g/g)	Mw (g/mol)
$EO_{13}PO_{30}EO_{13}$	L64	0.66	2900
$EO_{76}PO_{29}EO_{76}$	F68	4.0	8400
$EO_{103}PO_{39}EO_{103}$	F88	4.0	11400
$EO_{132}PO_{50}EO_{132}$	F108	4.0	14600
$EO_{13}PO_{40}EO_{13}$	P85	0.50	4600
$EO_{16}PO_{39}EO_{16}$	P84	0.67	3700
$EO_{17}PO_{56}EO_{17}$	P103	0.46	4950
$EO_{20}PO_{68}EO_{20}$	P123	0.45	5800
$EO_{97}PO_{68}EO_{97}$	F127	2.16	12600
$EO_{1}PO_{17}EO_{1}$	L31	0.089	1100
$EO_{11}PO_{16}EO_{11}$	L35	1.0	1900
$PO_{8}EO_{23}PO_{8}$	10R5	1.1	1950
$PO_{14}EO_{24}PO_{14}$	17R4	0.65	2650
$PO_{19}EO_{33}PO_{19}$	25R4	0.66	3600
$EO_{11}PO_{28}EO_{11}$	PE6200	0.60	2600
$EO_{8}PO_{50}EO_{8}$	L92	0.24	3650
$EO_{6}PO_{39}EO_{6}$	L81	0.23	2750
$EO_{37}PO_{56}EO_{37}$	P105	1.0	6500
$EO_{67}PO_{39}EO_{67}$	F87	2.6	7700

The combination of clays and non-ionic surfactants is a promising strategy that opens up a very wide range of hybrid materials for research and industrial applications.

In the following paragraphs, a brief description of the structures, chemical physical properties, and relevant applications of clay minerals and non-ionic surfactants will be provided. Then the topic of the interaction between the two classes of materials will be discussed and some important applications of the hybrid nanocomposites will be presented.

17.2 Clay Minerals

Clay minerals are characterized by a layered structure consisting of two main structural units: the tetrahedral sheet, where a Si atom is coordinated with four oxygens, and the octahedral sheet, where Al and/or Mg cations are coordinated with six oxygens or hydroxyl groups.

Isomorphous substitutions generate a permanent negative charge on the basal planes of clay mineral crystals, which is balanced by the presence of counterions. In aqueous suspension, counterions may exchange with ions in the bulk solution. The total amount of cations adsorbed on the clay is called cation exchange capacity (CEC) (Luckham and Rossi, 1999) and is a measure of the capacity of clay minerals to adsorb charged molecules. Moreover, octahedral Al-OH/Mg-OH and tetrahedral Si-OH groups at the broken edges of the clay produce additional charges, which can be either positive or negative.

Clay minerals can be distinguished on the basis of the arrangement of the tetrahedral and octahedral sheets. The two main classes are briefly discussed in the following paragraphs:

1. Two-layer minerals (1 : 1 structure) consisting of one silicon tetrahedral layer and one aluminum octahedral layer

Within this structure, isomorphic substitutions occur rarely, and this gives these clays a low cation exchange capacity (<15–20 mEq/100 g) (López-Galindo, Viseras, and Cerezo, 2007). Because of the strong hydrogen bonds that develop between packets, the water cannot penetrate inside the layers. One of the main representative components of this class of phyllosilicates is kaolin. The interlayer space of this clay is about 7.2 Å thick and it is non-expandable due to strong hydrogen bonds, which inhibit the entry of water molecules or ions. The exchange capacity of kaolinite is therefore minimal (1–8 meq/100 g) (Tombácz and Szekeres, 2006). Only external surfaces are available for adsorption, and the effective surface area is very limited (10–30 m^2/g).

2. Three-layer minerals (1 : 2 structure) with a layer of aluminum octahedra intercalated between two layers of silicon tetrahedrons

This type of clay mineral can have a linear (smectites) or fibrous (sepiolite-palygorskite) three-dimensional structure.

In the smectite group, the interaction between the layers is due to common oxygens which bind the sheets. These are quite weak. Furthermore, the numerous isomorphous substitutions engender a high repulsive potential on the surface of the layers. This promotes the entry of water molecules, thus enlarging the c-spacing between the layers. Smectites are therefore characterized by the occurrence of expanding lattice, where all the layer surfaces are available for hydration and cation exchange.

Among the plethora of smectite structures, a great deal of interest (Wang, Du, and Luo, 2008; Sciascia et al., 2011; Ambrogi et al., 2012; Calabrese et al., 2013a) was recently focused on the montmorillonite mineral clay, $[(1/2Ca,Na)_{0.5-1}(Al,Mg,Fe)_{4-6}(Si,Al)_8O_{20}(OH)_4 \cdot nH_2O]$, due to its large specific surface area, high cations exchange capacity, easy availability, eco-friendliness, and non-toxicity.

The fibrous clays group is composed by palygorskite and sepiolite, which present a fibrous morphology originated from a 180° inversion occurring every six or four silicon tetrahedra, respectively. This arrangement produces a structure of chains aligned parallel to the axis with a T-O-T structure. The resulting tridimensional ordering creates open channels filled by zeolitic and crystallization water.

Although clays offer several advantages, the scientific community is always asking for new materials, with improved and modulable properties. This can be achieved by the opportune modification of mineral clay by intercalation with organic or inorganic compounds. In particular, the preparation of organoclays by the use of both polymeric and conventional surfactants is of great interest because it is possible to obtain new materials with improved mechanical and rheological properties, and able to control the adsorption and release of different classes of molecules for biomedical and technological application, especially in the field of wastewater decontamination and in the formulations of drug delivery systems (Lee and Tiwari, 2012; Rodrigues et al., 2013; Tămăşan, Radu, and Simon, 2013; Zampori et al., 2010; Calabrese et al., 2017; Giuseppe Cavallaro et al., 2014; Calabrese et al., 2013a; DeLisi et al., 2007; Loyens, Jannasch, and Maurer, 2005).

17.3 Surfactants in Water

Surfactants are amphiphilic molecules with a polar head group and a hydrophobic tail. In aqueous solution, at a given concentration (critical micellar concentration, CMC), they form aggregates called micelles. In the last few decades, water-surfactant binary systems were the object of a great number of detailed and systematic physicochemical studies which allowed to considerably enhance the knowledge in this field. A large number of experimental techniques (Dautzenberq, 1988) was used for this purpose.

The nature of the interactions that govern the formation of the micelles was evidenced through thermodynamic studies (De Lisi and Milioto, 1994; Zana, 1987). Volume, enthalpy, heat capacity, compressibility, etc., were simultaneously determined in order to unambiguously characterize the behaviour of the systems. For instance, volume and enthalpy are able to detect the micellization process while they are not sensitive to the structural transition of the aggregates (Zana, 1987). Heat capacity and compressibility are sensitive to both the micellization process and the changes in the micellar aggregate structure (Zana, 1987). Sophisticated apparatuses allowed to determine very accurate thermodynamic properties, well suited for modeling. On this basis, the standard free energy and the property change for the formation of micelles were determined. Structural insights were provided by scattering techniques. For instance, the small angle neutron scattering (SANS) methodology (Dautzenberq, 1988) provided insights on the size, shape, and degree of ionization in ionic surfactants.

As already mentioned, the scientific community is highly interested in the investigation of the behavior of non-ionic surfactants, which are frequently used in biomedical and environmental applications (Azeem, Anwer, and Talegaonkar, 2009; Som, Bhatia and Yasir, 2012).

Polysorbates, in particular, are largely employed due to their high hydrophilic–lipophilic balance (HLB) and low critical micelle concentration value, which result in very efficient surface activity at low concentrations. They tend to self-organize into different types of nanostructures (spherical or rod micelles, vesicles, oil-in-water emulsion), independently of surfactant concentration and solution conditions. The micellar aggregates of the polysorbates are characterized by three preferential solubilization sites, i.e., the polar head group, the palisade layer, and the micellar core. Polysorbates, like most non-ionic surfactants, are classified as GRAS (generally recognized as safe) substances and are mild to use (Azeem, Anwer, and Talegaonkar, 2009). Moreover, since they are not in the ionic state in the solution, they possess high stability and are less susceptible to the effects of strong electrolyte inorganic salts or acids and bases. They also have good compatibility with other types of surfactants and excellent solubility in both water and organic solvents.

17.4 Block Copolymers in Water

In the case of block copolymers, the situation is more complex because the interactions between the constituent blocks have to be taken into account; further on, it becomes more peculiar when these interactions involve a *good solvent* only for a given block, which, consequently, behaves like a *selective solvent*. This generates an amphiphilic behavior and, then, the tendency of the copolymer to self-organize in solvent media. The self-assembling process produces structures similar to those formed by the surfactants at low molecular weight such as micelles of various forms, mesophases of different geometries, etc. (Loh, 2002). The polar group commonly present in the block copolymers is poly(ethylene oxide) (PEO), which is highly soluble in water. The high sensitivity to temperature and solubility in water of the PEO allows to modulate the block copolymers' self-assembling ability by also varying the temperature. Among the copolymers containing the PEO block, there is the family of pluronics.

Most of the thermodynamic, spectroscopic, and structural studies of pluronics in water are available with a few copolymer compositions but in a very broad temperature range due to the high-temperature sensitivity of the copolymers. From these experiments, the temperature onset of the aggregates formation is determined and it is called critical micellar temperature (cmt) (Taboada et al., 2003). By studying a copolymer–water mixture as a function of the copolymer concentration, at a fixed temperature, the critical micellar concentration can be evaluated (De Lisi et al., 2006; Lazzara, Milioto, and Gradzielski, 2006).

The use of these copolymers in different fields increased significantly in recent years due to their commercial availability, low cost, and low toxicity (Somasundaran, 2015; Chiappisi et al., 2012; De Lisi et al., 2008; Dintcheva et al., 2016; Lazzara et al., 2009; Nguyen-Kim et al., 2016; Torchilin, 2005). They are flexible molecular frameworks where the PO and the EO segments of different masses coexist and their behavior can be easily and finely tuned for specific purposes.

Polymer therapeutics was revealed to be effective for the treatment of human diseases because amphiphilic block copolymers aggregates are able to solubilize water-insoluble drugs. Concerning the drug loading, $EO_aPO_bEO_a$ micelles are very effective with respect to several drugs (Torchilin, 2005). The importance of the block copolymers for drug delivery is also due to the improvement of the drug effectiveness and to the reduction of side effects. Oil-in-water microemulsions based on $EO_aPO_bEO_a$ emerged as good tools for the treatment of cardiotoxicity caused by bupivacaine (Varshney et al., 2004). Recently, (Batrakova et al., 2006) demonstrated that a nanocomposite formed by the block copolymer P85 and antineoplastic drug doxorubicin is able to prevent the development of multidrug resistance in the human breast carcinoma cell. In particular, the dose of drug tolerated by the cancer cells in the presence of the copolymer was 1000 times lower than in its absence. This is probably due to the effect of the bulk hydrophobicity and the chemical microstructure of the block copolymers of ethylene oxide and propylene oxide on the membrane disturbing ability (Demina et al., 2005).

(Gao, Fain, and Rapoport, 2004) reported a new modality of tumor chemotherapy, based on drug encapsulation in polymeric micelles and subsequent controlled drug release at a target site initiated by ultrasonic irradiation. Mice ovarian cancer was treated by using such a technique and micelles based on P105. Encouraging effects in the reduction of side effects were obtained.

17.5 Surfactant/Clay Mixtures

The studies on the interactions between clay and surfactant are usually not systematic in spite of the number of papers published on the subject. Structural and thermodynamic investigations (Loyens, Jannasch, and Maurer, 2005; De Lisi et al., 2008; De Lisi et al., 2005; Calabrese et al., 2016; Guégan, 2013; Bujdák, Hackett, and Giannelis, 2000; Amirianshoja et al., 2013) on clay/surfactant mixtures evidenced a rich pattern of behavior due to the self-assembling ability of both polymeric and conventional surfactants which confers an additional degree of complexity.

The adsorption of the traditionally used ionic surfactants was extensively investigated, while fewer studies focused on the non-ionic ones, although they recently proved to be more effective in the adsorption of different classes of compounds (Ghiaci, Kalbasi, and Abbaspour, 2007; Guegan et al., 2009; Deng, Dixon, and White, 2003;Shen, 2001). Moreover, non-ionic surfactant/clay hybrids answer the growing request for non-polluting, non-toxic, and biocompatible materials to be potentially used in pharmaceutical, biomedical, and environmental applications.

The uptake of non-ionic surfactants is mainly due to physical phenomena and it is very sensitive to changes in the experimental conditions, i.e., concentration, temperature, or structure of the adsorbent.

Figure 17.2 (Paria and Khilar, 2004) is a representation of the main orientation changes undergone by nonionic surfactants on increasing the concentration of adsorbate.

Five successive stages (I–V) can be recognized.

In the first stage (I), surfactants adsorb on clay as single monomeric molecules, through van der Waals interaction between the clay surface and the hydrophobic moieties of the surfactant. Since surfactant concentration is very low, adsorbate–adsorbate interactions are negligible.

In the second region (II), the gradual saturation of the clay surface is observed.

At this point, the adsorbate–adsorbate interactions become significant (stage III). Three different situations can occur, depending on the nature of both the adsorbent and the surfactant, namely:

A. Hydrophobic clay surface and short hydrophilic group of surfactants. In this case, polar heads are displaced from the surface by the alkyl chains of the neighbouring molecules.
B. Balance between hydrophobic and hydrophilic interactions. In this case, the surfactant remains flat on the clay surface.
C. Hydrophilic clay and long hydrophilic group of surfactants. In this case, alkyl chains are displaced from the surface.

When the surfactant concentration reaches a critical value, alkyl chains tend to aggregate.

If the adsorbent presents a hydrophobic surface, surfactants can either orient themselves vertically (stage IV(i)) or form micellar aggregates called hemimycelles (Luckham and Rossi, 1999) (stage IV (ii)), on the clay surface.

As in the case of hydrophilic surfaces, surfactants first orient themselves vertically, then, on increasing concentrations, two situations may occur: the adsorption of a second layer of molecules parallel to the first, promoted by the attraction between the alkyl chains (step V (i)), or the adsorption of micelles (step V (ii)) with the formation of so-called admicelles (Luckham and Rossi, 1999). It is worth highlighting that the formation of these structures changes the nature of the clay surface from hydrophilic to hydrophobic, thus allowing the adsorption of organic non-polar molecules.

Information about the adsorption mechanism can be obtained from the adsorption isotherms where the amount of surfactant adsorbed per unit mass of the substrate (C_s) vs. the equilibrium concentration (C_e) is reported.

Due to the aggregation behaviour, the adsorption isotherms of surfactant onto clay are characterized by different regions with different mechanisms, as reported in Figure 17.2b.

Obviously, the shape of the isotherm depends on the nature of both surfactant and clay (cases A, B, and C). Moreover, in most cases, experiments are performed at surfactant concentrations either below or above the CMC; therefore, not all isotherms

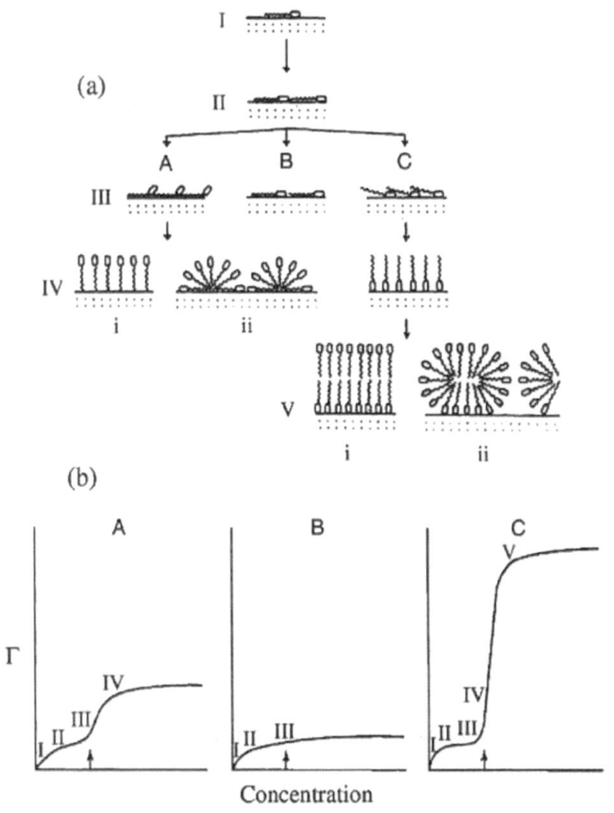

FIGURE 17.2 (a) Successive stages of adsorption of non-ionic surfactant, showing the orientation of molecules at the surface. (b) Adsorption isotherms corresponding to the adsorption sequences. Reproduced with permission from (Paria and Khilar, 2004).

reported in the literature present all the regions reported in Figure 17.2b.

The well-known Langmuir and Freundlich isotherms are classically applied (Calabrese et al., 2013b; Sánchez-Martín et al., 2008b; Praus and Turicová, 2007; Muherei and Junin, 2009) for the analysis of the sorption data of surfactant onto clays. However, these models are not always appropriate in the case of the adsorption of surfactants onto clays (Calabrese et al., 2016; Calabrese et al., 2017). In these cases, it is necessary to apply combinations of those isotherms, called dual-mode equations, given by the sum of two different contributions: a non-linear term, which can be either Langmuir-type or Freundlich-type, related to adsorption-like specific interactions, and a linear term, due to nonspecific interactions.

Further information about the adsorption process is obtained from kinetic studies. Although specific kinetic models were developed for the adsorption of non-ionic surfactants on a hydrophilic clay surface (Brinck, Jönsson, and Tiberg, 1998a, b; Paria and Khilar, 2004; Tiberg, 1996), in most applications, kinetic data are analyzed by means of the same model employed for adsorption processes, which are generally classified as adsorption diffusion models and adsorption reaction models (Qiu et al., 2009). Diffusion models are based on three consecutive steps: (i) external diffusion or film diffusion, (ii) internal or intra-particle diffusion, and (iii) adsorption and desorption of the adsorbate from the active sites of the clay. Reaction models are based on chemical reactions without considering the steps mentioned above. (Sciascia, Turco Liveri, and Merli, 2011) provided a detailed description of the adsorption kinetic models.

Since kinetic data are often affected by the presence of strong outliers, the analysis of the kinetic profiles requires the application of rigorous statistical criteria based on robust fitting methods (Merli, Sciascia, and Turco Liveri, 2010) aimed at achieving the correct modeling of the experimental data.

Beside equilibrium and kinetic studies, hybrid materials based on clays are usually characterized through X-ray diffraction (XRD) measurements, which represent one of the most effective techniques to obtain a full structural characterization of mineralogical phases (Merli and Sciascia, 2011; Sussich et al., 1998; Moneghini et al., 2008; Sussich, Princivalle, and Cesàro, 1999; Moneghini et al., 2000; Moneghini et al., 2010). Specifically, in the contest of organoclays characterization, XRD allows to investigate the basal spacing of clay materials and to propose the sites of interactions of the clay surface (Cui, Tarte, and Woo, 2008; Bertolino et al., 2016). It is interesting to highlight that, due to the formation of aggregates into the clay interlayer, the enlargement of the basal spacing is often strongly dependent on the surfactant concentration (Calabrese et al., 2016).

Additional informational useful for the characterization of surfactant/clay hybrid nanocomposites can also be obtained from thermogravimetric analysis (TG), which represents a very effective technique to investigate the microenvironment and arrangement of the surfactant onto the organoclays and to calculate the real surfactant loading (Giuseppe Cavallaro et al., 2018a; Giuseppe Cavallaro et al., 2018b; Bertolino et al., 2018; Giuseppe Cavallaro et al., 2016).

The functionalization of clay minerals with non-ionic surfactants represents a promising strategy to improve the performances of clay minerals in terms of both sorption capacity and stability of clay dispersions (Bertolino et al., 2016; Bertolino et al., 2017).

One of the main fields of application of these types of organoclays is the removal of organic pollutants from wastewaters (Calabrese et al., 2013b; Giuseppe Cavallaro et al., 2015b; Zhao et al., 2017). In this context, (Sciascia et al., 2019) reported the successful application of bio-organoclays, composed by a polyoxyethylene surfactant (Tween 20) and montmorillonite clay, to the treatment of olive mill wastewater samples. Due to the high content of phytotoxic organic compounds, these agricultural effluents constitute a serious and unsolved environmental problem. Hybrid composites were revealed as being more effective in decreasing the organic content in comparison with unmodified clay, thus highlighting that non-ionic surfactant functionalization is a promising route to obtain new tailor-made composites for environmental applications.

The performances of the Tween 20/montmorillonite composites in the field of controlled drug administration were also recently tested by encapsulating the drug cinnamic acid into the organoclays (Calabrese et al., 2017). The results obtained revealed that nanocomposites can be efficiently applied in the development of drug delivery systems, since they ensure a more prolonged drug release compared with free drugs, and allow for the complete release of the drug after oral administration.

Conclusion

From the arguments above, one may deduce that clay-non-ionic surfactant mixtures are good candidates for applications in which the enhancement of the hydrophobic solute and/or a strict control of the adsorption/desorption processes are required. Moreover, since these materials are biocompatible and have low environmental impact, they are particularly suitable for pharmaceutical, biomedical, and environmental applications. A full characterization of these systems, by means of thermodynamic, kinetic, and structural studies, is the first step towards a full comprehension of the mechanism underlying the formation of these composites and their employment in different application fields.

REFERENCES

Abdullayev, E., A. Joshi, W. Wei, Y. Zhao, and Y. Lvov. 2012. Enlargement of Halloysite Clay Nanotube Lumen by Selective Etching of Aluminum Oxide. *ACS Nano* 6(8): 7216–7226.

Ambrogi, V., L. Latterini, M. Nocchetti, C. Pagano, and M. Ricci. 2012. Montmorillonite as an Agent for Drug Photostability. *Journal of Materials Chemistry* 22(42): 22743–22749.

Amirianshoja, T., R. Junin, A. Kamal Idris, and O. Rahmani. 2013. A Comparative Study of Surfactant Adsorption by Clay Minerals. *Journal of Petroleum Science and Engineering* 101: 21–27.

Azeem, A., Md.K. Anwer, and S. Talegaonkar. 2009. Niosomes in Sustained and Targeted Drug Delivery: Some Recent Advances. *Journal of Drug Targeting* 17(9): 671–689.

Batrakova, E.V., D.L. Kelly, S. Li, Y. Li, Z. Yang, L. Xiao, D.Y. Alakhova, S. Sherman, V.Yu. Alakhov, and A.V. Kabanov. 2006. Alteration of Genomic Responses to Doxorubicin and Prevention of MDR in Breast Cancer Cells by a Polymer Excipient: Pluronic P85. *Molecular Pharmaceutics* 3(2): 113–123.

Bertolino, V., G. Cavallaro, G. Lazzara, M. Merli, S. Milioto, F. Parisi, and L. Sciascia. 2016. Effect of the Biopolymer Charge and the Nanoclay Morphology on Nanocomposite Materials. *Industrial and Engineering Chemistry Research* 55(27): 7373–7380.

Bertolino, V., G. Cavallaro, G. Lazzara, S. Milioto, and F. Parisi. 2017. Biopolymer-Targeted Adsorption onto Halloysite Nanotubes in Aqueous Media. *Langmuir* 33(13): 3317–3323.

Bertolino, V., G. Cavallaro, G. Lazzara, S. Milioto, and F. Parisi. 2018. Crystallinity of Block Copolymer Controlled by Cyclodextrin. *Journal of Thermal Analysis and Calorimetry* 132(1): 191–196.

Brinck, J., B. Jönsson, and F. Tiberg. 1998a. Kinetics of Nonionic Surfactant Adsorption and Desorption at the Silica–Water Interface: One Component. *Langmuir* 14(5): 1058–1071.

Brinck, J., B. Jönsson, and F. Tiberg. 1998b. Kinetics of Nonionic Surfactant Adsorption and Desorption at the Silica–Water Interface: Binary Systems. *Langmuir* 14(20): 5863–5876.

Bromberg, L., C.M. Straut, A. Centrone, E. Wilusz, and T.A. Hatton. 2011. Montmorillonite Functionalized with Pralidoxime as a Material for Chemical Protection Against Organophosphorous Compounds. *ACS Applied Materials and Interfaces* 3(5): 1479–1484.

Bujdák, J., E. Hackett, and E.P. Giannelis. 2000. Effect of Layer Charge on the Intercalation of Poly(Ethylene Oxide) in Layered Silicates: Implications on Nanocomposite Polymer Electrolytes. *Chemistry of Materials* 12(8): 2168–2174.

Calabrese, I., G. Cavallaro, G. Lazzara, M. Merli, L. Sciascia, and M.L. Turco Liveri. 2016. Preparation and Characterization of Bio-Organoclays Using Nonionic Surfactant. *Adsorption* 22(2): 105–116.

Calabrese, I., G. Cavallaro, C. Scialabba, M. Licciardi, M. Merli, L. Sciascia, and M.L. Turco Liveri. 2013a. Montmorillonite Nanodevices for the Colon Metronidazole Delivery. Special Section: Formulating Better Medicines for Children. *International Journal of Pharmaceutics* 457(1): 224–236.

Calabrese, I., G. Gelardi, M. Merli, G. Ritwo, L. Sciascia, and M.L. Turco Liveri. 2013b. New Tailor-Made Bio-Organoclays for the Remediation of Olive Mill Waste Water. *IOP Conference Series: Materials Science and Engineering* 47(1): 012040.

Calabrese, I., G. Gelardi, M. Merli, M.L. Turco Liveri, and L. Sciascia. 2017. Clay-Biosurfactant Materials as Functional Drug Delivery Systems: Slowing Down Effect in the In Vitro Release of Cinnamic Acid. *Applied Clay Science* 135: 567–574.

Carretero, M.I. 2002. Clay Minerals and Their Beneficial Effects upon Human Health. A Review. *Applied Clay Science* 21(3): 155–163.

Cavallaro, Gennara, G. Giammona, R. Lombardo, L. Sciascia, and M.L. Turco Liveri. 2006. Amphiphilic Derivatives of a Polyaspartamide: Their Aggregation and Solubilization Ability: Tensiometric and Spectrophotometric Studies. *Colloids and Surfaces. Part A: Physicochemical and Engineering Aspects* 289(1): 10–16.

Cavallaro, Giuseppe, G. Lazzara, S. Milioto, and F. Parisi. 2015a. Mixed Aggregates Based on Tetronic-Fluorinated Surfactants for Selective Oils Capture. *Colloids and Surfaces. Part A: Physicochemical and Engineering Aspects* 474: 85–91.

Cavallaro, Giuseppe, G. Lazzara, S. Milioto, and F. Parisi. 2015b. Hydrophobically Modified Halloysite Nanotubes as Reverse Micelles for Water-In-Oil Emulsion. *Langmuir* 31(27): 7472–7478.

Cavallaro, Giuseppe, G. Lazzara, S. Milioto, and F. Parisi. 2016. Steric Stabilization of Modified Nanoclays Triggered by Temperature. *Journal of Colloid and Interface Science* 461: 346–351.

Cavallaro, Giuseppe, G. Lazzara, S. Milioto, and F. Parisi. 2018b. Halloysite Nanotubes for Cleaning, Consolidation and Protection. 18(7–8).

Cavallaro, Giuseppe, G. Lazzara, S. Milioto, F. Parisi, V. Evtugyn, E. Rozhina, and R. Fakhrullin. 2018a. Nanohydrogel Formation Within the Halloysite Lumen for Triggered and Sustained Release. *ACS Applied Materials and Interfaces* 10(9): 8265–8273. Scopus.

Cavallaro, Giuseppe, G. Lazzara, S. Milioto, F. Parisi, and V. Sanzillo. 2014. Modified Halloysite Nanotubes: Nanoarchitectures for Enhancing the Capture of Oils from Vapor and Liquid Phases. *ACS Applied Materials and Interfaces* 6(1): 606–612.

Chiappisi, L., G. Lazzara, M. Gradzielski, and S. Milioto. 2012. Quantitative Description of Temperature Induced Self-Aggregation Thermograms Determined by Differential Scanning Calorimetry. *Langmuir* 28(51): 17609–17616.

Choy, J.-H., S.-J. Choi, J.-M. Oh, and T. Park. 2007. Clay Minerals and Layered Double Hydroxides for Novel Biological Applications. *Clays and Health* 36(1): 122–132.

Cui, L., N.H. Tarte, and S.I. Woo. 2008. Effects of Modified Clay on the Morphology and Properties of PMMA/Clay Nanocomposites Synthesized by In Situ Polymerization. *Macromolecules* 41(12): 4268–4274.

Dautzenberq, H. 1988. Surfactant Solutions. New Methods of Investigation. Hg. von Raoul Zana. ISBN 0-8247-7623-2. New York/Basel: Marcel Dekker. *Acta Polymerica* 39(8): 470–470.

De Lisi, R., M. Gradzielski, G. Lazzara, S. Milioto, N. Muratore, and S. Prévost. 2008. Aqueous Laponite Clay Dispersions in the Presence of Poly(Ethylene Oxide) or Poly(Propylene Oxide) Oligomers and Their Triblock Copolymers. *Journal of Physical Chemistry. Part B* 112(31): 9328–9336.

De Lisi, R., G. Lazzara, R. Lombardo, S. Milioto, N. Muratore, and M.L. Turco Liveri. 2005. Adsorption of Triblock Copolymers and Their Homopolymers at Laponite Clay/Solution Interface. Role Played by the Copolymer Nature. *Physical Chemistry Chemical Physics* 7(23): 3994–4001.

De Lisi, R., G. Lazzara, R. Lombardo, S. Milioto, N. Muratore, and M.L. Turco Liveri. 2006. Thermodynamic Behavior of Non-Ionic Tri-Block Copolymers in Water at Three Temperatures. *Journal of Solution Chemistry* 35(5): 659–678.

De Lisi, R., G. Lazzara, S. Milioto, and N. Muratore. 2007. Laponite Clay in Homopolymer and Tri-Block Copolymer Matrices: Thermal and Structural Investigations. *Journal of Thermal Analysis and Calorimetry* 87.

De Lisi, R., and S. Milioto. 1994. Thermodynamic Properties of Additive–Surfactant–Water Ternary Systems. *Chemical Society Reviews* 23(1): 67–73.

Demina, T., I. Grozdova, O. Krylova, A. Zhirnov, V. Istratov, H. Frey, H. Kautz, and N. Melik-Nubarov. 2005. Relationship Between the Structure of Amphiphilic Copolymers and Their Ability to Disturb Lipid Bilayers. *Biochemistry* 44(10): 4042–4054.

Deng, Y., J.B. Dixon, and G.N. White. 2003. Intercalation and Surface Modification of Smectite by Two Non-Ionic Surfactants. *Clays and Clay Minerals* 51(2): 150–161.

Dintcheva, N.T., G. Catalano, R. Arrigo, E. Morici, G. Cavallaro, G. Lazzara, and M. Bruno. 2016. Pluronic Nanoparticles as Anti-Oxidant Carriers for Polymers. *Polymer Degradation and Stability* 134: 194–201.

Fatma, N., M. Panda, W.H. Ansari, and Kabir-ud-Din. 2015. Solubility Enhancement of Anthracene and Pyrene in the Mixtures of a Cleavable Cationic Gemini Surfactant with Conventional Surfactants of Different Polarities. *Colloids and Surfaces. Part A: Physicochemical and Engineering Aspects* 467: 9–17.

Gao, Z., H.D. Fain, and N. Rapoport. 2004. Ultrasound-Enhanced Tumor Targeting of Polymeric Micellar Drug Carriers. *Molecular Pharmaceutics* 1(4): 317–330.

Ghiaci, M., R.J. Kalbasi, and A. Abbaspour. 2007. Adsorption Isotherms of Non-Ionic Surfactants on Na-Bentonite (Iran) and Evaluation of Thermodynamic Parameters. *Colloids and Surfaces. Part A: Physicochemical and Engineering Aspects* 297(1): 105–113.

Guégan, R. 2013. Self-Assembly of a Non-Ionic Surfactant onto a Clay Mineral for the Preparation of Hybrid Layered Materials. *Soft Matter* 9(45): 10913–10920.

Guegan, R., M. Gautier, J.-M. Beny, and F. Muller. 2009. Adsorption of a C10E3 Non-Ionic Surfactant on a Ca-Smectite. *Clays and Clay Minerals* 57(4): 502–509.

Hassan, N., J.M. Ruso, and Á. Piñeiro. 2011. Hydrogenated/Fluorinated Catanionic Surfactants as Potential Templates for Nanostructure Design. *Langmuir* 27(16): 9719–9728.

Jahan, K., S. Balzer, and P. Mosto. 2008. *Toxicity of Nonionic Surfactants* 110.

Jeevana, Jyothi, B., and K. Sreelakshmi. 2011. Design and Evaluation of Self-Nanoemulsifying Drug Delivery System of Flutamide. *Journal of Young Pharmacists* 3(1): 4–8.

Lazzara, G., S. Milioto, and M. Gradzielski. 2006. The Solubilisation Behaviour of Some Dichloroalkanes in Aqueous Solutions of PEO–PPO–PEO Triblock Copolymers: A Dynamic Light Scattering, Fluorescence Spectroscopy, and SANS Study. *Physical Chemistry Chemical Physics* 8(19): 2299–2312.

Lazzara, G., S. Milioto, M. Gradzielski, and S. Prevost. 2009. Small Angle Neutron Scattering, X-Ray Diffraction, Differential Scanning Calorimetry, and Thermogravimetry Studies to Characterize the Properties of Clay Nanocomposites. *The Journal of Physical Chemistry C* 113(28): 12213–12219.

Lazzara, G., S. Riela, and R.F. Fakhrullin. 2017. Clay-Based Drug-Delivery Systems: What Does the Future Hold? *Therapeutic Delivery* 8(8): 633–646.

Lee, S.M., and D. Tiwari. 2012. Organo and Inorgano-Organo-Modified Clays in the Remediation of Aqueous Solutions: An Overview. *Applied Clay Science* 59–60: 84–102.

Loh, W. 2002. Block Copolymer Micelles. In: A.T. Hubbard, ed., *Encyclopedia of Surface and Colloid Science*. Marcel Dekker.

López-Galindo, A., C. Viseras, and P. Cerezo. 2007. Compositional, Technical and Safety Specifications of Clays to Be Used as Pharmaceutical and Cosmetic Products. *Clays and Health* 36(1): 51–63.

Loyens, W., P. Jannasch, and F.H.J. Maurer. 2005. Poly(Ethylene Oxide)/Laponite Nanocomposites via Melt-Compounding: Effect of Clay Modification and Matrix Molar Mass. *Polymer* 46(3): 915–928.

Luckham, P.F., and S. Rossi. 1999. The Colloidal and Rheological Properties of Bentonite Suspensions. *Advances in Colloid and Interface Science* 82(1–3): 43–92.

Merli, M., and L. Sciascia. 2011. Iteratively Reweighted Least Squares in Crystal Structure Refinements. *Acta Crystallographica, Section A* 67(5): 456–468.

Merli, M., L. Sciascia, and M.L. Turco Liveri. 2010. Regression Diagnostics Applied in Kinetic Data Processing: Outlier Recognition and Robust Weighting Procedures. *International Journal of Chemical Kinetics* 42(10): 587–607.

Moneghini, M., B. Bellich, P. Baxa, and F. Princivalle. 2008. Microwave Generated Solid Dispersions Containing Ibuprofen. *International Journal of Pharmaceutics* 361(1–2): 125–130.

Moneghini, M., N. De Zordi, D. Solinas, S. MacChiavelli, and F. Princivalle. 2010. Characterization of Solid Dispersions of Itraconazole and Vitamin E TPGS Prepared by Microwave Technology. *Future Medicinal Chemistry* 2(2): 237–246. Scopus.

Moneghini, M., D. Voinovich, F. Princivalle, and L. Magarotto. 2000. Formulation and Evaluation of Vinylpyrrolidone/Vinylacetate Copolymer Microspheres with Carbamazepine. *Pharmaceutical Development and Technology* 5(3): 347–353.

Muherei, M., and R. Junin. 2009. Equilibrium Adsorption Isotherms of Anionic, Nonionic Surfactants and Their Mixtures to Shale and Sandstone. 3(2).

Muzzalupo, R., L. Tavano, C.O. Rossi, R. Cassano, S. Trombino, and N. Picci. 2009. Synthesis and Properties of Methacrylic-Functionalized Tween Monomer Networks. *Langmuir* 25(3): 1800–1806.

Nguyen-Kim, V., S. Prévost, K. Seidel, W. Maier, A.-K. Marguerre, G. Oetter, T. Tadros, and M. Gradzielski. 2016. Solubilization of Active Ingredients of Different Polarity in Pluronic® Micellar Solutions – Correlations Between Solubilizate Polarity and Solubilization Site. *Journal of Colloid and Interface Science* 477: 94–102.

Paria, S., and K.C. Khilar. 2004. A Review on Experimental Studies of Surfactant Adsorption at the Hydrophilic Solid–Water Interface. *Advances in Colloid and Interface Science* 110(3): 75–95.

Praus, P., and M. Turicová. 2007. A Physico-Chemical Study of the Cationic Surfactants Adsorption on Montmorillonite. *Journal of the Brazilian Chemical Society* 18(2): 378–383.

Qiu, H., L. Lv, B. Pan, Q. Zhang, W. Zhang, and Q. Zhang. 2009. Critical Review in Adsorption Kinetic Models. *Journal of Zhejiang University-Science. Part A* 10(5): 716–724.

Rodrigues, L.A.de S., A. Figueiras, F. Veiga, R.M. de Freitas, L.C.C. Nunes, E.C. da Silva Filho, and C.M. da Silva Leite. 2013. The Systems Containing Clays and Clay Minerals from Modified Drug Release: A Review. *Colloids and Surfaces, Part B: Biointerfaces* 103: 642–651.

Rossi, F., R. Lombardo, L. Sciascia, C. Sbriziolo, and M.L. Turco Liveri. 2008. Spatio-Temporal Perturbation of the Dynamics of the Ferroin Catalyzed Belousov–Zhabotinsky Reaction in a Batch Reactor Caused by Sodium Dodecyl Sulfate Micelles. *Journal of Physical Chemistry. Part B* 112(24): 7244–7250.

Sánchez-Martín, M.J., M.C. Dorado, C. del Hoyo, and M.S. Rodríguez-Cruz. 2008. Influence of Clay Mineral Structure and Surfactant Nature on the Adsorption Capacity of Surfactants by Clays. *Journal of Hazardous Materials* 150(1): 115–123.

Savic-Gajic, I., S. Stojiljkovic, I. Savic, and D. Gajic. 2014. Industrial Application of Clays and Clay Minerals. In: 379–402.

Sciascia, L., S. Casella, G. Cavallaro, G. Lazzara, S. Milioto, F. Princivalle, and F. Parisi. 2018. Olive Mill Wastewaters Decontamination Based on Organo-Nano-Clay Composites. *Ceramics International* (August 15). http://www.sciencedirect.com/science/article/pii/S027288421832217X.

Sciascia, L., M.J.B. Hauser, and M.L. Turco Liveri. 2008. Kinetic Evidence for the Incorporation of the [(Pentamethylcyclopentadienyl) (2,2′-Bipyridyl)(Aquo) Rhodium(III)] Complex into DPPC Vesicles. *Colloids and Surfaces. Part A: Physicochemical and Engineering Aspects* 322(1): 243–247.

Sciascia, L., R. Lombardo, and M.L. Turco Liveri. 2007. Nonlinear Response of a Batch BZ Oscillator to the Addition of the Anionic Surfactant Sodium Dodecyl Sulfate. 111(6).

Sciascia, L., F. Rossi, C. Sbriziolo, M.L. Turco Liveri, and R. Varsalona. 2010. Oscillatory Dynamics of the Belousov–Zhabotinsky System in the Presence of a Self-Assembling Nonionic Polymer. Role of the Reactants Concentration. *Physical Chemistry Chemical Physics* 12(37): 11674.

Sciascia, L., M.L. Turco Liveri, and M. Merli. 2011. Kinetic and Equilibrium Studies for the Adsorption of Acid Nucleic Bases onto K10 Montmorillonite. *Applied Clay Science* 53(4): 657–668.

Shen, Y.-H. 2001. Preparations of Organobentonite Using Nonionic Surfactants. *Chemosphere* 44(5): 989–995.

Som, I., K. Bhatia, and Mohd. Yasir. 2012. Status of Surfactants as Penetration Enhancers in Transdermal Drug Delivery. *Journal of Pharmacy and Bioallied Sciences* 4(1): 2–9.

Somasundaran, P., ed. 2015. *Encyclopedia of Surface and Colloid Science*, 3rd ed. CRC Press. https://www.taylorfrancis.com/books/9781466590618.

Sussich, F., F. Princivalle, and A. Cesàro. 1999. The Interplay of the Rate of Water Removal in the Dehydration of α,α-Trehalose. *Carbohydrate Research* 322(1–2): 113–119.

Sussich, F., R. Urbani, F. Princivalle, and A. Cesàro. 1998. Polymorphic Amorphous and Crystalline Forms of Trehalose. *Journal of the American Chemical Society* 120(31): 7893–7899.

Taboada, P., V. Mosquera, D. Attwood, Z. Yang, and C. Booth. 2003. Enthalpy of Micellisation of a Diblock Copoly(Oxyethylene/Oxypropylene) by Isothermal Titration Calorimetry. Comparison with the Van'T Hoff Value. *Physical Chemistry Chemical Physics* 5(12): 2625.

Tămăşan, M., T. Radu, and V. Simon. 2013. Spectroscopic Characterisation and In Vitro Behaviour of Kaolinite Polyvinyl Alcohol Nanocomposite. *Applied Clay Science* 72: 147–154.

Tiberg, F. 1996. Physical Characterization of Non-Ionic Surfactant Layers Adsorbed at Hydrophilic and Hydrophobic Solid Surfaces by Time-Resolved Ellipsometry. *Journal of the Chemical Society, Faraday Transactions* 92(4): 531–538.

Tombácz, E., and M. Szekeres. 2006. Surface Charge Heterogeneity of Kaolinite in Aqueous Suspension in Comparison with Montmorillonite. *Applied Clay Science* 34(1–4). Layer Charge of Clay Minerals Selected Papers from the Symposium on Current Knowledge on the Layer Charge of Clay Minerals Current Knowledge on the Layer Charge of Clay Minerals (October): 105–124.

Torchilin, V.P. 2005. Block Copolymer Micelles as a Solution for Drug Delivery Problems. *Expert Opinion on Therapeutic Patents* 15(1): 63–75.

Turco Liveri, M.L., M. Licciardi, L. Sciascia, G. Giammona, and G. Cavallaro. 2012. Peculiar Mechanism of Solubilization of a Sparingly Water Soluble Drug into Polymeric Micelles. Kinetic and Equilibrium Studies. *Journal of Physical Chemistry. Part B* 116(16): 5037–5046.

Turco Liveri, M.L., L. Sciascia, M. Allegra, L. Tesoriere, and M.A. Livrea. 2009. Partition of Indicaxanthin in Membrane Biomimetic Systems. A Kinetic and Modeling Approach. *Journal of Agricultural and Food Chemistry* 57(22): 10959–10963.

Turco Liveri, M.L., L. Sciascia, R. Lombardo, L. Tesoriere, E. Passante, and M.A. Livrea. 2007. Spectrophotometric Evidence for the Solubilization Site of Betalain Pigments in Membrane Biomimetic Systems. *Journal of Agricultural and Food Chemistry* 55(8): 2836–2840.

Varshney, M., T.E. Morey, D.O. Shah, J.A. Flint, B.M. Moudgil, C.N. Seubert, and D.M. Dennis. 2004. Pluronic Microemulsions as

Nanoreservoirs for Extraction of Bupivacaine from Normal Saline. *Journal of American Chemical Society* 126(16): 5108–5112.

Wang, X., Y. Du, and J. Luo. 2008. Biopolymer/Montmorillonite Nanocomposite: Preparation, Drug-Controlled Release Property and Cytotoxicity. 19(6).

Wei, W., R. Minullina, E. Abdullayev, R. Fakhrullin, D. Mills, and Y. Lvov. 2014. Enhanced Efficiency of Antiseptics with Sustained Release from Clay Nanotubes. *RSC Advances* 4(1): 488–494.

Zampori, L., P.G. Stampino, C. Cristiani, P. Cazzola, and G. Dotelli. 2010. Intercalation of Poly(Ethylene-Oxides) in Montmorillonite: Tailor-Made Nanocontainers for Drug Delivery Systems. *Applied Clay Science* 50(2): 266–270.

Zana, R. 1987. *Surfactant Solutions: New Methods of Investigation. Surfactant Science Series.* New York: M. Dekker. https://search.lib.virginia.edu/catalog/u707162.

Zhao, Q., H. Choo, A. Bhatt, S.E. Burns, and B. Bate. 2017. Review of the Fundamental Geochemical and Physical Behaviors of Organoclays in Barrier Applications. *Applied Clay Science* 142: 2–20.

18

Microorganism-Mediated Functionalization of Nanoparticles for Different Applications

Maheshkumar Prakash Patil and Gun-Do Kim

CONTENTS

18.1 Introduction

Nanotechnology is the most promising and recent area of research, and includes many academic disciplines such as physics, chemistry, biology, and engineering. The fastest growing interdisciplinary area of nanotechnology has become more attractive due to its applications in healthcare, medicines, environmental science, material science, solar energy, and so on. The term *nano* is adapted from the Greek word meaning "dwarf" and acts as a bridge between bulk materials and atomic or molecular structures (Patil and Kim, 2017). The concept of nanoparticles was first presented by Richard Feynman through his famous lecture entitled "There's plenty of room at the bottom" at the American Institute of Technology. Materials in nanoparticle form with more than one dimension and a size of about 100 nm have good activity compared to their bulk form. Nanoparticles can be synthesized by two different approaches: top-down and bottom-up (Fendler, 1998). In the top-down approach, bulk materials are broken down to nano-sized materials, while in the bottom-up approach atoms and molecules are assembled to molecular structures in nanometer range. The synthesis of nanoparticles carried out by different methods and the development of a green and eco-friendly nanoparticle synthesis process are still being researched. The synthesis of nanoparticles using biological entities has been performed for a long time. This mechanism is involved in the synthesis of nanoparticles, in phases such as activation, growth, and termination. The activation step includes the reduction of the metal ions, followed by the nucleation of the reduced metal atoms; the growth step includes the spontaneous coalescence of the small adjacent nanoparticle into larger size particles accompanied by thermodynamic stability; the termination step comprises the final shape of the nanoparticles (Shankar et al., 2016). Generally, nanoparticles are categorized on the bases of their morphology, dimensions, uniformity, composition, and agglomeration. The shape and size of nanoparticles play an important role in their functionality and toxicity on human beings and on the environment. Thin films (1 dimension) have useful in sensor devices and electronics; carbon nanotubes (2 dimension) are more stable and have high adsorption ability; and quantum dots (3 dimensions) are useful in bio-imaging. Nanoparticles can be categorized into metallic, oxide, and sulfide and, morphologically, into spherical, plate, and crystalline structures. This review presents an overview of the microorganism-mediated functionalization of different nanoparticles and their various applications on the basis of recent literature.

18.1.1 Nanoparticle Synthesis by Different Methods

The synthesis of nanoparticles is traditionally performed by chemical, physical, and biological methods, as shown in Figure 18.1. The biological and chemical methods of nanoparticles synthesis use the bottom-up approach (Pattekari et al., 2011; Patil et al., 2019). Physical and chemical methods utilize hazardous chemicals and high levels of pressure and energy. They are expensive and produce toxic byproducts. So, physical and chemical methods are disadvantageous for the synthesis of nanoparticles. In addition, the nanoparticles synthesized through these methods have limited application in the medical, healthcare, and cosmetic fields due to their toxicity for human beings and to their unstable nature (Iravani et al., 2014; Li et al., 2011). Modern methods, such as biological method, include the use of plants and microorganisms for the synthesis and functionalization of nanoparticles, as described in Figure 18.2. The nanoparticles synthesized through these methods are non-toxic to human beings and eco-friendly (Patil et al., 2018a; Patil et al., 2018b; Patil et al., 2018c). The use of plant extracts for nanoparticle synthesis is easier than that of microorganisms (Kumar et al., 2014; Kumar et al., 2017; Patil et al., 2016; Patil et al., 2017a; Patil et al., 2017b), but there is an issue regarding the phytochemical composition variations in plants depending on seasons. So, using microbes for the synthesis of nanoparticles is much more convenient than employing other methods.

18.1.2 Nanoparticles Synthesis Using Microorganisms

The development of new materials and methods of a reliable, eco-friendly, and non-toxic nature for medical and healthcare applications is still ongoing (Patil and Kim, 2017). Microorganisms such as bacteria, fungi, actinomycetes, algae, and viruses were used for the synthesis of metallic, oxides, sulfide, and other nanoparticles for different applications (Li et al., 2011; Patil and Kim, 2018; Ahmad et al., 2013). The synthesis of defined shapes, sizes, and enzyme- or substance-linked nanoparticles is practised by physical and chemical methods. These methods

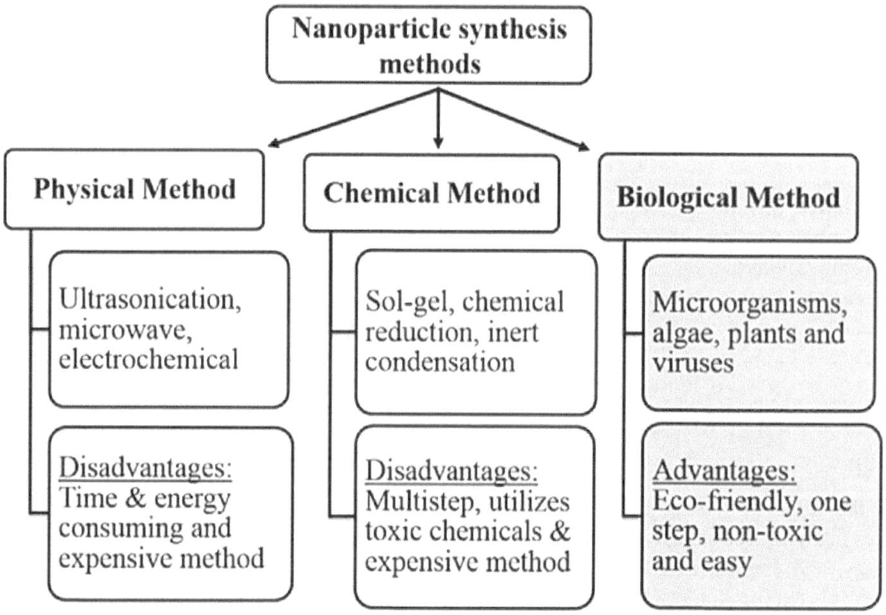

FIGURE 18.1 Schematic presentation of different methods for the synthesis of nanoparticles and their comparative advantages and disadvantages.

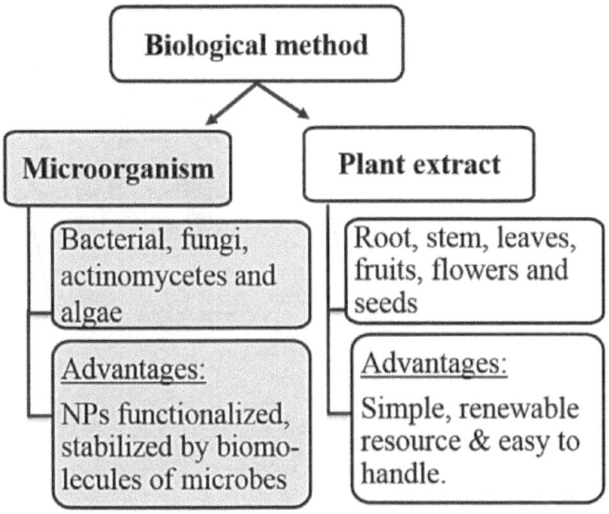

FIGURE 18.2 Flow chart of the biological method for nanoparticle synthesis.

produce toxic waste as a byproduct, which is not good to human beings or to the environment. This problem can be overcome by using microorganisms for nanoparticle synthesis. The functionalization and the nanoparticles synthesized in this process are identified as a biogenic, eco-friendly, and non-toxic (Hulkoti and Taranath, 2014).

18.2 Microbe-Mediated Synthesis of Nanoparticles

Microorganisms and inorganic materials have been in continuous communication since the beginning of life on Earth. The environment is rich in different forms of materials (complex, toxic, and non-toxic). Microbes have adapted by evolving, making chemical changes to survive. They are able to convert bulk-form material to nanoparticles. The main reason behind the survival of microorganisms in an inorganic-material-rich environment is the changes in the chemical forms of the materials from toxic to non-toxic form. In an environment with a high level of metals such as mainstream metal processing plants and naturally mineralized zones, microorganisms have developed a genetic and proteomic response to regulate homeostasis (Mergeay et al., 2003) and also harbor a metal-resistant cluster of genes which enable cell detoxification by efflux and reductive precipitation (Nies, 1999). In recent years, researchers have focused on the interaction between materials and microbial species. Their studies found that many microorganisms (eukaryotic and prokaryotic) are able to produce nanoparticles through either extracellular or intracellular processes (Patil and Kim, 2018). In this study, we focus on microorganisms such as bacteria, fungi, yeast, cyanobacteria, actinomycetes, algae, and fungus for the synthesis of nanoparticles and their applications in different fields. Microorganisms have been employed for the synthesis of nanoparticles such as silver (Ag), gold (Au), copper (Cu), zinc (Zn), iron (Fe), lead (Pb), titanium (Ti), aluminium (Al), cobalt (Co), cadmium (Cd) platinum (Pt), and many more because of their unique importance and properties (Hulkoti and Taranath, 2014; Ramkumar et al., 2017). The proposed microorganism-mediated synthesis of different nanoparticles, their characterization using different techniques, and their broad spectrum of application are presented in Figure 18.3.

18.2.1 Mechanism of Nanoparticle Formation

Several types of microorganisms (bacteria, fungi, and algae) have been reported for the biosynthesis of metallic and oxide forms of nanoparticles through two ways: intracellular and extracellular. The synthesis process is carried out by an enzymatic reaction or by means of biological agents. This phenomenon is known as biosynthesis. The biosynthesis of nanoparticles occurs inside and outside the cells of microorganisms called intracellular and extracellular, respectively. In general, in the intracellular synthesis of nanoparticles, ions are trapped into cell membranes or transported into the cytoplasm where cytoplasmic protein (enzyme) reduces ions (eg., Ag^+/Au^+) into nuclei (eg., Ag^0/Au^0). Later, these nuclei form a cluster into nanoparticles with different morphology. The intracellularly synthesized nanoparticles can be diffused out from the inside to the outside of cell cytoplasm. In the case of extracellular synthesis, a cell surface bounded and outside released proteins (enzymes) can be responsible for the reduction of ions into different morphological nanoparticles. The exact mechanism behind the formation of nanoparticles is not very clear just yet due to the involvement of different biochemical compositions of microbes and to the interaction of different microbes with different types of metals for the synthesis of nanoparticles. The ions transported into the cells are a known phenomenon. The proposed mechanisms of nanoparticle synthesis using microorganisms are demonstrated in Figure 18.4.

18.2.1.1 Intracellular Synthesis

The electrostatic attraction of positively charged ions towards negatively charged bacterial cells where transported intracellular and the enzymes involve in reduction process to synthesize nanoparticles. The mechanistic intracellular synthesis of AgNPs by bacteria was well explained by Klaus et al. (1999). The bacterium *Pseudomonas stutzeri* AG259 was isolated from

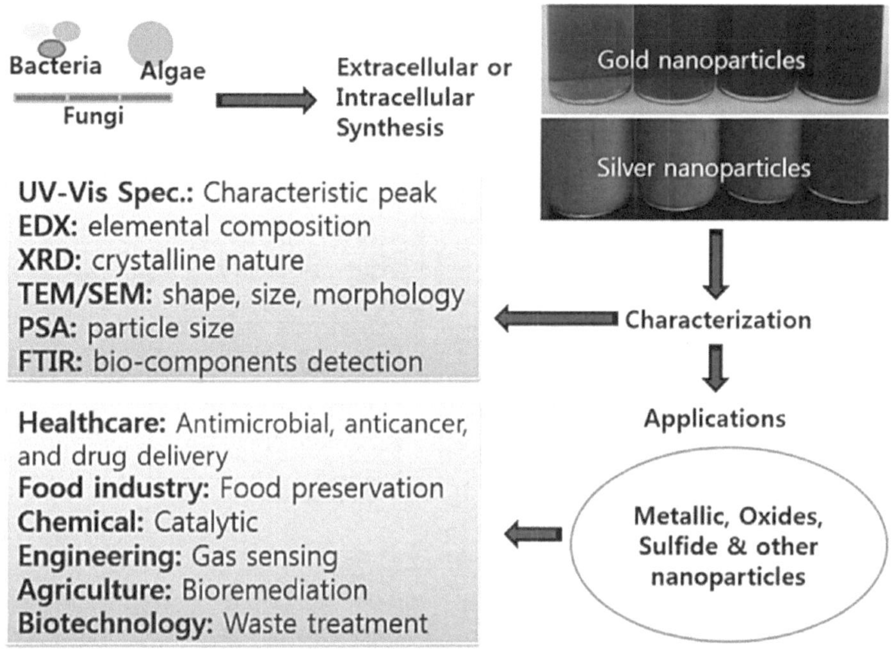

FIGURE 18.3 Schematic presentation of metallic nanoparticles synthesis using microorganisms; their characterization and applications in different fields.

a silver mine and placed in a silver nitrate solution, and the reduction of Ag^+ ions to AgNPs (hexagonal, triangular shapes) in a periplasmic space was observed. Similar observations for AuNPs were reported. In *Bacillus subtilis* 168, the intracellular accumulation and reduction of Au3+ were reported by Southam and Beveridge (Southam and Beveridge, 1996). Likewise, observations were made by Mukherjee et al. (2001b); the fungi *Verticillium sp.* Were able to synthesize AuNPs intra- and extracellularly. *Verticillium sp.* traps the gold ions in cytoplasm (intracellular), followed by enzymatic reduction and aggregation of metal atoms, and finally by the formation of nanoparticles and capping (Mukherjee et al., 2001b). In another study, the use of *Verticillium sp.* was reported for the synthesis of intracellular AgNPs, and it was concluded that the formation of AgNPs was caused by intracellular enzymes, not by peptides such as glutathione (Mukherjee et al., 2001a). Dameron et al. (1989) demonstrated the formation of intracellular sulfide nanoparticles in yeast; *Candida gabrata* and

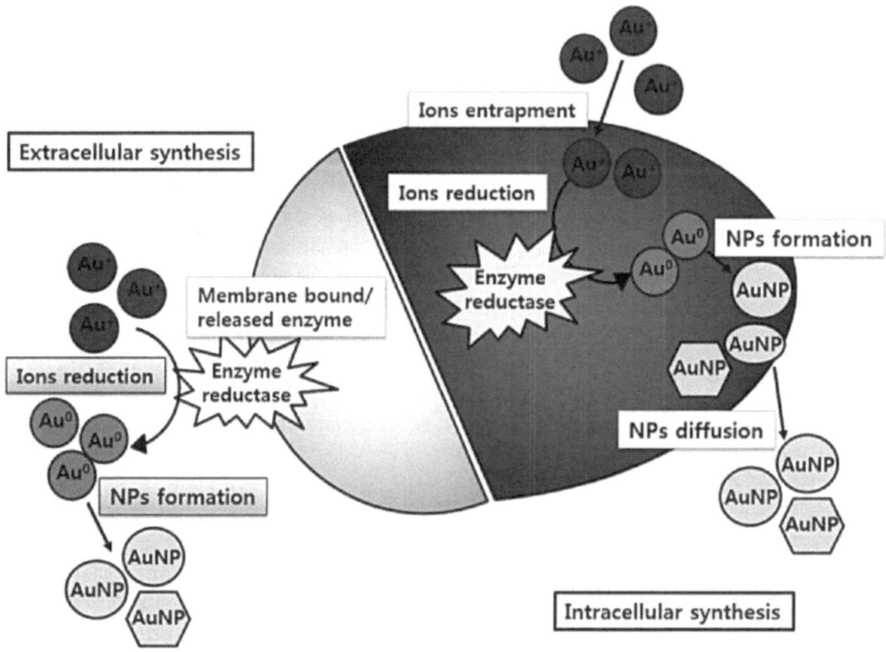

FIGURE 18.4 Demonstration of a proposed mechanism of nanoparticle synthesis using microorganisms.

TABLE 18.1

Synthesis of Biological Nanoparticles from Bacteria

Bacterial Species	Mechanism	NPs	Size (nm)	Shape	Reference
Rhodobacter sphaeroides	Ext	PbS	10.5 ± 0.5	Spherical	Bai and Zhang, 2009
Klebsiella pneumoniae	Ext	Au	10–15	Spherical	Prema et al., 2016
Klebsiella pneumoniae, Escherichia coli, Pseudomonas jessinii	Ext	Ag/Ag-Cl	ND	Polymorphic	Muller et al., 2016
Bacillus strain	Ext	Ag	42–92	Spherical	Das et al., 2014
Bacillus flexus	Ext	Ag	12, 65	Spherical, triangular	Priyadarshini et al., 2013
Lactobacillus kimchicus	Int	Au	5-30	Spherical	Markus et al., 2016
Lactobacillus sporogenes	Int	ZnO	145.70	Hexagonal	Mishra et al., 2013
Aeromonas hydrophila	Ext	ZnO	57.72	Spherical, oval	Jayaseelan et al., 2012
Citrobacter braakii	Ext	CdS	50–100	Spherical	Zhu et al., 2016
Escherichia coli	Int	CdS	5–10	Spherical	Yan et al., 2017
Bacillus amyloliquefaciens	Ext	TiO_2	22.11-97.28	Spherical	Khan and Fulekar, 2016
Aeromonas hydrophila	Ext	TiO_2	28–54	Spherical, uneven	Jayaseelan et al., 2013
Escherichia coli, Pseudomonas aeruginosa	Ext/Int	Fe	23 ± 1	Spherical	Crespo et al., 2017
Pyrobaculum islandicum	Ext	Fe, Mn	ND	Spherical	Kashefi and Lovley, 2000
Morganella morganii	Ext	Cu	ND	Spherical	Ramanathan et al., 2013

Note: Ext, extracellular; Int, intracellular; ND, not determined; NPs, nanoparticles.

Schizosaccharomyces pombe were able to synthesize cadmium sulfide (CdS). CdS coated with glutathione (a sulfide-rich peptide) was observed intracellularly.

18.2.1.2 Extracellular Synthesis

The extracellular synthesis of nanoparticles depends on the proteins (enzymes) bound on the cell surface or secreted extracellularly. Bacteria, fungi, and algae are reported for the extracellular synthesis of different nanoparticles (metallic, oxide, and sulfide) (Manivasagan et al., 2016). As it is accepted that the ions are converted into nanoparticles by microbial enzymes, there must be microbial enzymes able to generate/provide electrons for the reduction process. In general, the enzymes produced and secreted extracellularly are involved in the reduction process for the nanoparticle formation. The mechanism behind extracellular synthesis of AgNPs is due to enzyme nitrate reductase. The fungi Fusarium oxysporum biomass was added to an Ag^+ ions solution and found involvement of nitrate-dependent reductase and a shuttle quinone extracellular process in AgNPs formation. (Duran et al., 2005). Similarly, Gajbhiye et al. (2009) demonstrated the extracellular synthesis of AgNPs, a common fungus Alternaria alternate cell free extract added to $AgNO_3$ solution and formation of AgNPs due to cell-free extract contained enzymes were noticed. The extracellular synthesis of nanoparticles using bacteria is well explained by Shahverdi et al. (2007) enterobacteriaceae members (*Escherichia coli, Klebsiella pneumonia*, and *Enterobacter cloacae*) cell free supernatant is involved to reduce Ag^+ to AgNPs. Another bacterium, *Rhodopseudomonas capsulate*, is known to secrete cofactor NADH and NADH-dependent enzymes, which are involved in the extracellular synthesis of AuNPs by reducing gold ions (He et al., 2007). Another research study also supports the involvement of the extracellular proteins produced by *Stenotrophomonas* in the synthesis of AgNPs and AuNPs (Malhotra et al., 2013). The extracellular synthesis of TiO_2 nanoparticles by *Lactobacillus* sp. and *Saccharomyces cerevisiae* was reported. Microorganisms have a negative electro-kinetic potential which attracts positively charged ions, and membrane-bound oxidoreductase take part in the oxidation-reduction process for the formation of nanoparticles (Jha et al., 2009).

18.2.2 Bacteria in Synthesis of Nanoparticles

Several bacterial species reported for the synthesis of different nanoparticles are listed in Table 18.1. The main reason behind nanoparticles formation is the reduction of metal ions into nanoparticles. The chemical method for nanoparticles synthesis is more useful, but produces toxic byproducts, and the synthesized nanoparticles are toxic for human beings and the environment. Owing to these undesirable byproducts and toxicity, the biosynthesis of nanoparticles using bacteria is proposed as a better option. AuNP synthesis using *Bacillus subtilis* was first reported in 1980 (Beveridge and Murray, 1980). The biosynthesis of crystalline Ag-based nanoparticles (200 nm) was first reported by using *Pseudomonas stutzeri* AG259, a silver-resistant bacterium in which nanoparticles are deposited in vacuole-like granules in the periplasm, probably due to metal efflux and metal binding (Klaus et al., 1999). Later, Cu, Zn, Fe, Pb, Ti, Al, Co, Cd, and many more were biosynthesized using terrestrial isolate bacteria (Hulkoti and Taranath, 2014; Ahmed et al., 2017). At the present time, bacteria are being explored as potential bio-factories for the synthesis of different nanoparticles (Li et al., 2011; Manivasagan et al., 2016). Nair and Pradeep (Nair and Pradeep, 2002) studied the properties of *Lactobacillus* strains, which are commonly found in buttermilk and used for the biosynthesis of well-defined morphologies of Ag, Au, and Ag-Au alloy crystal. Malhotra and coworkers (2013) reported the use of a marine isolate *Stenotrophomonas* strain for the extracellular synthesis of AuNPs (10–50 nm) and AgNPs (40–60 nm). Fayaz et al. (2011) used thermophilic bacteria for the synthesis of Ag and Au nanoparticles: *Geobacillus stearothermophilus* cell-free extract was mixed with a metal salt solution and produced more polydisperse AgNPs compared

TABLE 18.2

Cyanobacteria and Actinomycete-Mediated Synthesis of Different Nanoparticles

	Species	Mechanism	NPs	Size (nm)	Shape	Reference
Cyanobacteria	*Lyngbya majuscule, Spirulina subsalsa*	Ext	Au	<20	Spherical	Chakraborty et al., 2009
	Plectonema boryanum	Ext/Int	Au	10–30	Octahedral	Lengke et al., 2006
	Oscillatoria willei	Ext	Ag	100–200	Spherical	MubarakAli et al., 2011
	Phormidium tenue	Ext	CdS	5	Spherical	MubarakAli et al., 2012
	Spirulina platensis	Ext	Ag, Au, Ag-Au	6–25	Spherical	Govindaraju et al., 2008
Actinomycetes	*Nocardiopsis* sp.	Ext	Ag	45	Spherical	Manivasagan et al., 2013
	Nocardiopsis sp.	Ext	Au	11	Spherical	Manivasagan et al., 2015
	Streptomyces albidoflavus	Ext/Int	Ag	14.5	Spherical	Shetty et al., 2012
	Rhodococcus sp.	Int	Au	5–25	Spherical	Ahmad et al., 2003
	Streptomyces sp.	Ext	Cu, Zn	100–150	ND	Usha et al., 2010

Note: Ext, extracellular; Int, intracellular; ND, not determined; NPs, nanoparticles.

with AuNPs. The synthesis of 10–15 nm AgNPs in extreme conditions (60 °C) on the cell wall of *Corynebacterium* sp. SH09 was also reported (Zhang et al., 2005). Further synthesis of an average 26 nm size of AgNPs with high concentration of silver nitrate by *Idiomarina* sp. PR58-8 (metal tolerant) was also reported (Seshadri et al., 2012). Singh et al. (2013) reported AgNPs synthesis using cell-free extract of *Acinetobacter calcoaceticus* and observed a variation in particle size depending on reaction parameters. The concentration of nanoparticles increases as the incubation time is extended. The concentration of salt solution also affects monodispersity and size, and increasing temperature causes the nanoparticle formation process to accelerate and also form smaller nanoparticles (Singh et al., 2013). *Escherichia coli* ABLE C is reported for intracellular synthesis of CdS nanocrystals. Parameters such as growth phase and strain type are important in the synthesis and to control the size and shape of nanoparticles (Sweeney et al., 2004).

18.2.3 Cyanobacteria and Actinomycetes in the Synthesis of Nanoparticles

Among the lower organisms, cyanobacteria are used for the bioremediation of precious and toxic metals. They are also involved in the transformation to a non-toxic form. *Lyngbya majuscula* is a cyanobacteria reported for nanoalloy synthesis (size: 5–25 nm). *Lyngbya* thallus was exposed to a silver and gold equimolar solution, and after 72 h intra- and extracellular synthesis of Au-Ag were recorded (Roychaudhury et al., 2016). Patel et al. (2015) used different cyanobacteria cell biomass and cell-free supernatant and observed active factors involvement in the formation of AgNPs to be extracellular polysaccharide and activation of which requires light. The different cyanobacteria and actinomycetes recently reported for nanoparticle synthesis are listed in Table 18.2.

Actinomycetes have been shown to be higher bacteria, but they were thought to be fungi for many years because they have filamentous forms; they are numerous and widely distributed in the soil and are next to bacteria in abundance. Actinomycetes have characteristics of both bacteria and fungi: they produce different kinds of proteins which contain amines and cysteine residues that bind with metal ions and are involved in nanoparticle synthesis. At present, actinomycetes are at the center of attention due to their ability to synthesize numerous secondary metabolites. The synthesis of mineral nanoparticles using actinomycetes *Thermomonospora* sp. in various conditions were reported by Rautaray et al. (2004). After that, many actinomycetes were used for the synthesis of different metal nanoparticles. The use of *Streptomyces hygroscopicus* BDUS 49 was reported for the extracellular synthesis of spherical AgNPs with a size of about 20–30 nm (Sadhasivam et al., 2010) and multidimensional AuNPs (Sadhasivam et al., 2012). *Streptomyces parvulus* SSNP11 was used for the protein-encapsulated spherical extracellular synthesis of AgNPs with a mean size of 2.1 nm (Prakasham et al., 2014), and *Streptomyces albidoflavus* CNP10 was employed for the intracellular and extracellular synthesis of AgNPs (Shetty et al., 2012). Actinomycetes mediated the reduction of silver ions by reducing nitrate to nitrite, and ammonium was also reported (Baker et al., 2013; Golinska et al., 2014). Among actinomycetes, *Streptomyces* was not the only bacterium reported for the synthesis of AgNPs. Other genera are also included, such as *Nocardiopsis* sp., *Actinomycetes* sp., and *Rhodococcus* sp. (Golinska et al., 2014). Other nanoparticles, such as Zn/Cu and manganese synthesis using *Streptomyces* sp., were also reported (Manimaran and Kannabiran, 2017).

18.2.4 Yeast and Fungi in the Synthesis of Nanoparticles

It has been known for a long time that, in nature, different nanoparticles are synthesized by microorganisms. For instance, the intracellular synthesis of magnetite is performed by magnetic bacteria (Sarikaya, 1999). Likewise, yeast, Candida glabrata, and Schizosaccharomyces pombe when challenged with cadmium, a intracellular nanocrystallites of CdS synthesized as a mechanism of detoxification. γ-glutamyl peptide are synthesized by yeast and cadmium bind to this peptide and forms a Cd-γglu peptide complex (Dameron et al., 1989). A silver-tolerant yeast strain, MKY3, was reported for the extracellular synthesis of AgNPs (2–5 nm), but it was observed that AgNPs were not synthesized when silver ions were exposed to the supernatant of yeast strain MKY3 (Kowshik et al., 2003). The yeasts reported in the synthesis of oxides and metal nanoparticles are listed in Table 18.3. A marine yeast, *Rhodosporidium diobovatum*, was used for the intracellular synthesis of lead sulfide (PbS) nanoparticles; the synthesized

TABLE 18.3

Yeast and Fungus-Mediated Synthesis of Different Nanoparticles

	Species	Mechanism	NPs	Size (nm)	Shape	Reference
Yeast	*Yarrowia lipolytica*	Int	Au	15	Hexagonal, triangular	Agnihotri et al., 2009
	Candida albicans	Ext	Ag	20–80	ND	Rahimi et al., 2016
	Candida utilis	Ext	Ag	20–80	Spherical	Waghmare et al., 2015
	Rhodosporidium diobovatum	Int	PbS	2–5	Spherical	Seshadri et al., 2011
	Candida albicans	Ext	ZnO	~ 20	Quasi-spherical	Shamsuzzaman et al., 2017
	Saccharomyces cerevisiae	Int	ZnS	30–40	Spherical	Mala and Rose, 2014
	Candida glabrata, Schizosaccharomyces pombe	Int	CdS	20 Å	Hexamer	Dameron et al., 1989
	Schizosaccharomyces pombe	Int	CdS	1–1.5	Hexagonal	Kowshik et al., 2002
	Saccharomyces cerevisiae	Ext	MnO_2	34	Spherical	Salunke et al., 2015
Fungi	*Fusarium solani*	Ext	Au	20–50	Spherical	Gopinath and Arumugam, 2014
	Rhizopus oryzae	Ext	Au	10	Spherical	Das et al., 2009
	Aspergillus foetidus	Ext	Au	10–40	Spherical	Roy et al., 2016
	Penicillium brevicompactum	Ext	Au	Various	Various	Mishra et al., 2011
	Penicillium chrysogenum	Int	Au	5–100	Spherical, triangular, rod-like	Sheikhloo and Salouti, 2011
	Phoma macrostoma	Int	Au	100–200	Spherical, triangular, rod-like	Sheikhloo and Salouti, 2012
	Aspergillus terreus	Ext	Ag	1–20	Spherical	Li et al., 2012
	Rhizopus stolonifer	Ext	Ag	9.47	Spherical	AbdelRahim et al., 2017
	Duddingtonia flagrans	Ext	Ag	11, 38	Quasi-spherical	Silva et al., 2017
	Neurospora crassa	Int	Ag and Au	11 and 32	Spherical	Castro-Longoria et al., 2011
	Aspergillus flavus	Int	Ag	2–22	Spherical	Vala et al., 2014
	Verticillium sp.	Int	Ag	25 ± 12	Spherical	Mukherjee et al., 2001a
	Cariolus versicolor	Ext	CdS	8–15	Spherical	Sanghi and Verma, 2009
	Aspergillus flavus	Ext	ZnS	18, 58.9	Spherical	Uddandarao and Raj MB, 2016
	Fusarium oxysporum	Ext/Int	Pt	10–100	Hexagonal, circle, pentagonal, square, rectangular	Riddin et al., 2006
	Colletotrichum sp.	Ext	S and Al	50 and 30	Spherical	Suryavanshi et al., 2017
	Fusarium oxysporum	Ext	CdS	5–20	Hexagonal	Ahmad et al., 2002

Note: Ext, extracellular; Int, intracellular; ND, not determined; NPs, nanoparticles.

PbS nanoparticles were crystalline. Their size was 2–5 nm and the elemental composition showed the presence of lead and sulfides in a 1 : 2 ratio. This indicates that the nanoparticles were covered by sulfur-rich peptides (Seshadri et al., 2011). The considerable differences in morphology, shape, size, location of particles, and dispersity of nanoparticles depends on the genera and type strains of yeasts. An intracellular synthesis of various morphologies of AuNPs by *Pichia jadinii* was also reported (Gericke and Pinches, 2006).

Fungi are eukaryotic, filamentous, and sporeforming organisms; they are a rich source of bioactive compounds and good candidates for the synthesis of different nanoparticles. Fungi can accumulate metals by physicochemical and biological mechanisms, including extracellular binding by metabolites and polymers, binding to specific polypeptides, and metabolism-dependent accumulation (Gade et al., 2010). Gericke and Pinches (2006) reported the intracellular synthesis of AuNPs by *Verticillium luteoalbum*, and also demonstrated that there are no changes of age on the morphology (shape and size) of the synthesized AuNPs but there will be a reduced number of AuNPs synthesis by old culture. The temperature, increasing up to 50 °C, causes a rapid increase in the AuNP synthesis rate (Gericke and Pinches, 2006). *Aspergillus niger* was incubated with silver nitrate solution; AgNPs were synthesized and aggregated on the surface of the cell wall within 72 h. The synthesized AgNPs were monodispersed with a size of 5–25 nm and stabilized by reducing the proteins or enzymes produced by fungi (Gade et al., 2008). Extracellular AgNP synthesis using *Penicillium brevicompactum* WA 2315 was reported and demonstrated that the NADH-dependent nitrate reductase enzyme was involved in the synthesis of AgNPs by reducing silver nitrate. Enzymes can act as carriers of electrons from NADH during the reduction process (Shaligram et al., 2009). In another study, *Fusarium oxysporum* f. sp. *lycopersici* was used for the synthesis of PtNPs. The synthesized PtNPs were 10–100 nm in size with square, pentagonal, rectangular, circular, and hexagonal shapes; bioaccumulation and synthesis were observed intracellularly on the cell wall or membrane, and also extracellularly in the surrounding medium (Riddin et al., 2006). *Cariolus versicolor* was also used for the extracellular synthesis of CdS, and it was confirmed that the thiol group of fungal proteins was mainly responsible for the production of stable CdSNPs (Sanghi and Verma, 2009). Other genera and species of fungi involved in the synthesis of different nanoparticles are listed in Table 18.3.

18.2.5 Algae in the Synthesis of Nanoparticles

There is another group of microorganisms belonging to the plant kingdom, namely algae, which is also used for the biosynthesis of metal nanoparticles. Table 18.4 includes a list of a few algae

TABLE 18.4
Algae-Mediated Synthesis of Different Nanoparticles

Species	Mechanism	NPs	Size (nm)	Shape	Reference
Turbinaria conoides, Sargassum tenerrimum	Ext	Au	27–35	Spherical	Ramakrishna et al., 2016
Pithophora oedogonia	Ext	Ag	25–44	Cubical, hexagonal	Sinha et al., 2015
Chlorella vulgaris	Ext	Ag	9.8 ± 5.7	Spherical	Ferreira et al., 2017
Chlorella vulgaris	Ext	Pd	5–20	Spherical	Arsiya et al., 2017
Padina gymnospora	Ext	Pt	5–50	Octahedral	Ramkumar et al., 2017
Sagassum muticum	Ext	ZnO	30–57	Hexagonal	Azizi et al., 2014

Note: Ext, extracellular; Int, intracellular; ND, not determined; NPs, nanoparticles.

reported for nanoparticle synthesis. Hosea et al. (1986) reported the accumulation of AuNPs (9–20 nm) in a dried-cell suspension of alga *Chlorella vulgaris*. The formation of phytochelatin-coated CdS nanocrystalites in a marine phytoplanktonic alga, *Phaeodactylum tricornutum*, has been reported in response to CdS; phytochelatin peptides are involved in cellular detoxification mechanisms and have the capacity to synthesize stable phytochelatin-metal complexes (Scarano and Morelli, 2003). The extracellular polysaccharides and intracellular polyphosphates of microalgae actively take part in metal sequestrations (Kaplan et al., 1987; Zhang and Majidi, 1994). In another research project, it was reported that *Spirulina platensis* is a potential candidate for the synthesis of AgNPs, AuNPs, and bimetallic Au core-AgNPs. The biomass of algae was exposed to an aqueous solution of silver and gold, and AgNPs (7–16 nm), AuNPs (6–10), and bimetallic (Ag: Au, 50 : 50 ratio; 17–25 nm) were synthesized extracellularly (Govindaraju et al., 2008). An eco-friendly, intracellular synthesis of AuNPs of 5–35 nm in size was reported by Senapati et al; AuNPs accumulated more on the cell wall than on the cell membrane of *Tetraselmis kochinensis*, possibly due to the enzymes responsible for nanoparticle synthesis being present on the cell wall and cytoplasmic membrane (Senapati et al., 2012). Mahdavi et al. (2013) reported the seaweed (*Sargassum muticum*) extract-mediated synthesis of magnetic iron oxide (Fe_3O_4) nanoparticles; the synthesized nanoparticles were cubical, and the average particle diameter was 18 ± 4 nm. Seaweed extracts, containing the hydroxyl, aldehyde, and sulphate groups, are involved in the bioreduction and stabilization of Fe_3O_4 nanoparticles. Likewise, different genera of algae (*Caulerpa, Chlamydomonas, Cystophora, Padina, Spyrogyra, Ulva, Phormidium, Bifurcaria*) were reported for the formation of nanoparticles of metal (Ag, Au), sulfides (CdS) and oxides (CuO) (Shankar et al., 2016).

18.3 Effects of Different Parameters in the Synthesis of Nanoparticles

Biosynthesis is a novel route in the synthesis of nanoparticles. It is carried out by employing a variety of different parameters including pH, temperature, reaction time, concentration of reducing agent, source of ions (metal salt), and type of microorganisms. It is well known that the biological activity of nanoparticles depends on their size and shape. The most important criteria in the selection of microorganisms for the synthesis of nanoparticles are their intrinsic properties, such as enzyme production, growth rate, and metabolic pathways. Muller and coworkers (2016) described the effects of the culture medium on the extracellular synthesis of AgNPs. The formation of bigger AgCl-NPs (about 100 nm) in media containing a high Cl– concentration to AgNO3 in comparison to low Cl– containing media by using *Klebsiella pneumoniae*, *Escherichia coli*, and *Pseudomonas jessinii*. Gurunathan and coworkers (2009) reported optimum reaction conditions and formation of smaller AgNPs using *E. coli*. To obtain optimum conditions, a different medium with a different concentration of $AgNO_3$, and varying temperature and pH was employed. Nitrate medium with a concentration of 5 mM $AgNO_3$ at 60 °C and pH value of 10.0 proved to be an optimum condition, and with these parameters only 30 min were required to obtain over 95 % conversion (formation of AgNPs from $AgNO_3$) using the culture supernatant of *E. coli*. In another research study, it was stated that the development of color during AuNP synthesis also depends on the pH value of the reaction (Seshadri et al., 2011). The formation of golden, pink, and purple colors at pH 2.0, 7.0, and 9.0 respectively, was observed during the AuNP synthesis using biomass of yeast (*Yarrowia lipolytica*) at 30 °C for 72 h. The reduction of gold is pH-dependent manner, formation of gold crystals in acidic pH and AuNPs (15 nm) at pH 7.0 and pH 9.0 were observed. Mishra et al. (2011) found that a pH of 5–8 favored the synthesis of AuNPs and the formation of larger AuNPs at increasing concentrations (beyond 2 mM) of gold salt using the cell-free supernatant of *Penicillium brevicompactum*. The authors also noticed that the reaction was faster with an increase in AuNP intensity for 3, 6, and 9 h, and the reaction was completed at 12 h. The formation of shapes and sizes also depends on the concentration of gold salt; AuNP sizes of 10–50 nm (spherical), 10–70 nm (spherical, triangular, and hexagonal), and 50–120 nm (spherical, triangular, and diamond-like) were observed at 1 mM, 2 mM, and 3 mM concentration of gold salt, respectively. Manivasagan et al. (2013) observed an increasing absorbance of AgNPs synthesized using *Nocardiopsis* sp. MBRC-1 for a reaction incubation time of 24, 48, 72, and 96 h. The formation of a light to brown color in the reaction solution from 24 to 96 h of incubation was reported. Gericke and Pinches (2006) noticed the effects of different parameters on the intracellular synthesis of AuNPs in *Verticillium luteoalbum*. pH is the most important parameter. A variation in the pH does not affect the synthesis, but it causes changes in the shape, size, and number of particles produced; at pH 3, the AuNPs were about 10 nm (spherical); at pH 5, they were larger (spherical, triangular, hexagonal, and rod-shaped); while, at pH 7 and 9, the AuNPs were bigger than the AuNPs synthesized at pH 5 and the shape was irregular and undefined. Similarly, temperature affects size and shape formation; at lower temperatures (25 °C), smaller AuNPs (about 10 nm, spherical) are formed, and at higher temperatures (35 and 50 °C) bigger AuNPs (about 100 nm, of irregular shape) are observed. Temperature accelerates the reaction rate, and at the beginning, smaller AuNPs are formed, but as time increases the shape and size of the nanoparticles changes (formation of larger AuNPs).

18.4 Characterization of Nanoparticles

After the synthesis of nanoparticles, determining the physicochemical properties is a very important step in order to evaluate the functional aspects along with the identification of the chemical composition, shape, size, surface area, and dispersity. Characterization can be performed by different analytical techniques, such as UV-visible spectroscopy (UV-vis), scanning electron microscopy (SEM), transmission electron microscopy (TEM), dynamic light scattering (DLS), energy-dispersive X-ray spectroscopy (EDXS), X-ray diffractometry (XRD), Fourier transform infrared spectroscopy (FT-IR), and zeta potential measurement.

18.4.1 UV-Visible Spectroscopy (UV-Vis)

For the identification of nanoparticles within the size ranges of 2–100 nm, 200–800 nm light wavelength is used. This method is reliable, rapid, sensitive, selective for different nanoparticles, and easy to handle. Moreover, it can be useful for the qualitative and quantitative analysis of nanoparticles. The UV-vis absorption peak is specific for different nanoparticles. A peak at 500–600 nm is attributed to the surface plasmon response (SPR) of the AuNPs (Link and El-Sayed, 1999; He et al., 2007), and one at 400–500 nm is attributed to the AuNPs (Gajbhiye et al., 2009; Malhotra et al., 2013; Sinha et al., 2015; Link et al., 1999). The UV-vis absorption peak changes with the changes in the sizes and shapes of nanoparticles (Link and El-Sayed, 1999; Link et al., 1999).

18.4.2 Scanning Electron Microscopy (SEM)

SEM is useful for surface imaging. This method is applied to learn more about nanoparticles using a beam of highly energetic electrons to probe objects on a very fine scale. SEM is capable of resolving the size, shape, distribution, and surface morphology of nanoparticles at micro- to nanoscales (Sanghi and Verma, 2009; Uddandarao and Raj, 2016; Sinha et al., 2015; Ramkumar et al., 2017). Along with SEM using a probe to get nanoparticles histogram is general practice to measure particles size. The internal structure cannot be resolved through SEM, but modern high-resolution SEM is able to define the morphology of nanoparticles below the level of 10 nm.

18.4.3 Transmission Electron Microscopy (TEM)

TEM is the preferred technique for the determination of the size, shape, distribution, and quantitative measurement of nanoparticles (Sanghi and Verma, 2009; Uddandarao and Raj, 2016; Manivasagan et al., 2013). Similarly to SEM, TEM is also useful to obtain histograms for the size distribution of nanoparticles by an additional probe (Uddandarao and Raj, 2016; Manivasagan et al., 2013). TEM has a capacity superior to that of SEM. It can resolve images of nanoparticles and provide better resolution. The limitations of TEM applications are the strict requirements of vacuum and the thin section of the sample (Manivasagan et al., 2013; Singh et al., 2013). In addition, the TEM sample preparation process is time consuming, and very important in order to obtain good quality images.

18.4.4 Dynamic Light Scattering (DLS)

DLS is a radiation scattering technique, useful to determine the size distribution of nanoparticles with a scale ranging from microns to nanometers in a suspension or solution. This technique depends on the interaction of light with particles, and the size of the nanoparticles obtained by this method is usually larger than the size obtained from TEM (Uddandarao and Raj, 2016; Ramkumar et al., 2017).

18.4.5 Energy-Dispersive X-Ray Spectroscopy (EDXS)

The chemical characterization or elemental analysis of nanoparticles is carried out by this technique. A SEM instrument with a probe (detector) is used to detect the elemental composition of the nanoparticles (Manivasagan et al., 2013). A detector converts X-ray energy into a voltage signal (keV). The absorption peak is specific for different metals, such as peak at 3 keV for silver (Mukherjee at al., 2001a; Manivasagan et al., 2013), and peak at 2.195 and 2.53 for gold (Das et al., 2009; Chakraborty et al., 2009).

18.4.6 X-Ray Diffraction (XRD)

This technique is applicable for the determination of the molecular and crystal structure of the synthesized nanoparticles (Zhu et al., 2016; Khan and Fulekar, 2016). The qualitative and quantitative determination of compounds and solution of chemical species, respectively, is also applicable. (Khan and Fulekar, 2016). XRD is also useful to measure the degree of crystallinity (Manivasagan et al., 2013). XRD can be useful to determine the structural features of organic and inorganic substances, biomolecules, superconductors, polymers, glasses, and so on. The analysis is based on the formation of diffraction patterns; each material has specific diffraction patterns and this patterns can be compared to a reference database in the JCPDS library (Joint Committee on Powder Diffraction Standards) (Manivasagan et al., 2013; Ferreira et al., 2017; Singh et al., 2013).

18.4.7 Fourier Transform Infra-Red Spectroscopy (FT-IR)

FT-IR is a popular technique that has been used for the detection of functional groups attached to the surface of nanoparticles and also to help to find out about the involvement of biological molecules in the synthesis of nanoparticles. (Sinha et al., 2015; Arsiya et al., 2017). It is useful to determine the involvement of functional groups in the reduction process of nanoparticle synthesis by comparing the substrate with synthesized nanoparticles (Manivasagan et al., 2013; Arsiya et al., 2017; Rautaray et al., 2004). This technique is reproducible, reliable, and accurate. It also has a favorable signal to noise ratio.

18.5 Applications of Microbe-Mediated Nanoparticles

Microbe-mediated nanoparticles (metals, oxides, and sulfides) possess a broad spectrum of biocidal activities towards a diverse range of human pathogens such as bacteria and fungi. Nanoparticles have been reported for many more applications in different fields such as medical, agricultural, environmental, and industrial, and all of them show excellent results. Antimicrobial, anticancer, diagnosis, biosensors, and catalysis are just a few examples of all the applications of nanoparticles discussed in this chapter.

18.5.1 Antibacterial Activity of Nanoparticles

The antimicrobial activity of nanoparticles is useful in healthcare, food packaging, textile industries, and many more environmental applications. The development of antibiotic-resistant bacteria has led to an excess of infectious diseases caused by these microbes. Microbes adopt immunity due to the consecutive exposure to antibiotics for many generations and develop a resistance to multidrugs and antibiotics by inactivation, alteration of target sites, and modifications in their metabolic pathways (Alekshun and Levy, 2007; Tang et al., 2014). Therefore, we need to look for substitute antibiotics with strong bacteriostatic and bactericidal activities.

Microbial synthesized nanoparticles have been reported for their significant antibacterial activity. Nanoparticles of smaller sizes (1–100 nm) have a higher surface-area-to-mass ratio which enhances the interaction with microorganisms and shows high efficiency (enhanced antibacterial activity) (Patil and Kim, 2017; Manivasagan et al., 2016). The exact mechanism behind the antimicrobial activity of AgNPs is still not clear. However, many theories have been proposed to explain the antimicrobial activity of AgNPs such as the positive charge of Ag^+ interacting with the negatively charged bacterial cell wall and then increasing the permeability of the cell by changing the structure of the cell. This may lead to the death of the cell (Dibrov et al., 2002; Sondi and Salopek-Sondi, 2004; Feng et al., 2000). Another theory is the formation of free radicals by AgNPs, and these radicals are responsible for the damage to the cell membrane. They make it porous and finally cause the death of the cells (Sondi and Salopek-Sondi, 2004; Feng et al., 2000; Kim et al., 2007). AgNPs interact with the thiol groups the inactivated enzymes and ultimately cause the death of the cells (Feng et al., 2000; Spadaro et al., 1974). AgNPs interact with the sulfur and phosphorous groups of DNA and create complications in DNA replication. This is an another theory behind the antimicrobial activity of AgNPs (Feng et al., 2000). Bacterial signal transduction depends on the phosphorylation of the protein substrate. AgNPs dephosphorylates the peptide substrate on tyrosine residue which cause inhibition of signal transduction and finally inhibit the growth of bacteria (Prabhu and Paulose, 2012). The schematic presentation of the antibacterial activities of AgNPs is illustrated in Figure 18.5.

Nanoparticles (metal, oxides, and sulfides) have been reported for bactericidal activities towards Gram-positive and Gram-negative pathogens (Hussain et al., 2016). According to Manivasagan et al. (2013), AgNPs (spherical, 45 ± 0.15 nm) synthesized using the culture supernatant of *Nocardiopsis* sp. MBRC-1 have a concentration-dependent antibacterial activity, with maximum growth inhibition activity (MIC; minimum inhibitory concentration) against *Bacillus subtilis* (7 µg/ml), followed by *Pseudomonas aeruginosa* (10 µg/ml), *Escherichia coli* (13 µg/ml), *Staphylococcus aureus* (14 µg/ml), *Enterococcus hirae* (16 µg/ml), and *Shigella flexneri* (18 µg/ml). Similarly, the antibacterial activity of AgNPs (spherical, 12 nm, and triangular, 65 nm) synthesized using the culture supernatant of *Bacillus flexus* S-27 against *Escherichia coli*, *Pseudomonas aeruginosa*, *Streptococcus pyogenes*, and *Bacillus subtilis*, were reported (Priyadarshini et al., 2013). AgNPs (spherical, 8–12 nm) synthesized using *Acinetobacter calcoaceticus* LRVP54 show antibacterial activity against Gram-positive (*Streptococcus mutans*, *Staphylococcus aureus*) and Gram-negative (*Enterobacter aerogenes*, *Pseudomonas aeruginosa*, *Escherichia coli*, *Shigella sonnie*, *Salmonella typhimurium*) pathogens. Gram-negative pathogens are more susceptible to AgNPs compared to Gram-positive ones; they found the highest synergy with vancomycin on *Enterobacter aerogenes* (Singh et al., 2013). In another research study, the antibacterial activity of AgNPs against different pathogens, such as *Escherichia coli*, *Staphylococcus aureus*, *Pseudomonas aeruginosa*, *Shigella flexneri*, *Bacillus subtilis*, *Vibrio cholera*, *Micrococcus luteus*, *Klebsiella pneumoniae*, *Enterococcus faecalis*, and *Salmonella typhimurium*, was reported (Rahimi et al., 2016; Sinha et al., 2015; Ferreira et al., 2017; Patel et al., 2015; Sadhasivam et al., 2010).

AuNPs have attracted significant interest over the last decade as antibacterial agents. Prema et al. (2016) reported the antibacterial activity of spherical AuNPs with a size of 10–15 nm against *Staphylococcus epidermidis*, *Escherichia coli*, *Staphylococcus aureus*, *Pseudomonas aeruginosa*, and *Bacillus cereus*; AuNPs show a higher zone of inhibition (25.60 mm) against *E. coli*. The high synergistic effect of synthesized AuNPs along with amoxicillin and streptomycin against *E. coli* followed by *S. aureus* was also observed. Extracellularly synthesized spherical and average size 11.57 ± 1.24 nm AuNPs using *Nocardiopsis* sp. have been

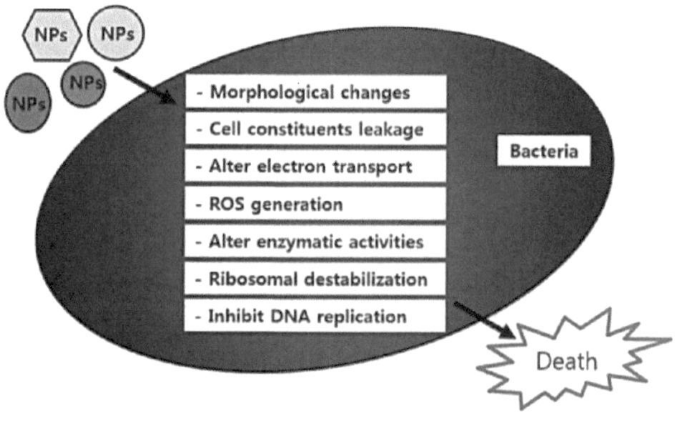

FIGURE 18.5 The mechanisms of the antibacterial activity of different metallic or oxide forms of nanoparticles are diagrammatically represented.

reported for their antibacterial activity against *Escherichia coli*, *Pseudomonas aeruginosa*, *Bacillus subtilis*, and *Staphylococcus aureus* (Manivasagan et al., 2015). In another research study, multidimensional AuNPs synthesized using *Streptomyces hygrocopicus* show excellent bactericidal activity against *Bacillus cereus*, *Staphylococcus aureus*, *Enterococcus faecalis*, *Staphylococcus epidermidis*, *Escherichia coli*, and *Salmonella typhimurium* with MIC values (μg/ml) of 64, 32, 256, 128, 32, and 32, respectively (Sadhasivam et al., 2012).

The nanoparticles of zinc oxide (ZnO) have been reported for their antibacterial potential. *Lactobacillus sporogens*-mediated ZnONPs were hexagonal with an average particle size of 145.7 nm, showing a growth inhibition of *Staphylococcus aureus* (Mishra et al., 2013). The nanoparticles synthesized using actinomycetes with bactericidal properties are used in the textile industry. Cu and Zn nanoparticles synthesized using *Streptomyces* sp. were coated onto cotton fabrics and tested against bacteria. CuNPs show 27 and 30 mm zones of bacteriostasis for *Escherichia coli* and *Staphylococcus aureus*, respectively. On the other hand, Zn nanoparticles show 27 and 30 mm zones of bacteriostasis for *Escherichia coli* and *Staphylococcus aureus*, respectively. These fabrics can be useful in hospitals to minimize or prevent infection (Usha et al., 2010). ZnO and titanium dioxide (TiO_2) nanoparticles synthesized using *Aeromonas hydrophila* show antibacterial activity against *Pseudomonas aeruginosa*, *Escherichia coli*, *Enterococcus faecalis*, *Streptococcus pyogenes*, and *Staphylococcus aureus* (Jayseelan et al., 2012; Jayseelan et al., 2013). Other nanoparticles, such as sulfur (SNPs) and aluminium oxide (AlNPs) synthesized using *Colletotrichum* sp. have also been reported. SNPs and AlNPs show MIC values (μg/ml) ranging from 250 to 700 and from 300 to 1000, respectively, against *Salmonella typhimurium*, *Chromobacterium violaceum*, and *Listeria monocytogenes* (Suryavanshi et al., 2017).

18.5.2 Antifungal Activities of Nanoparticles

Fungi are a serious concern for human health. They affect plants and animals and also damage the economy due to their mycotoxins (Sanchez-Hervas et al., 2008). Fungal infections are frequent in immunosuppressed patients, and overcoming such diseases is a long process, because there are very few commercial antifungal drugs (Kim et al., 2008). Therefore, there is a need to investigate and develop antifungal agents, which should be non-toxic, eco-friendly, and biocompatible. At this junction, different nanoparticles play an important role as anti-fungal agents against various disease-causing fungi. AgNPs show antifungal activity against different fungi, but to date the complete mechanism beyond antifungal activity remains unclear. The destruction of membrane integrity and the consequent inhibition of the budding process has been deemed responsible for the antifungal activity of AgNPs against *Candida albicans* (Kim et al., 2009). Fungus-mediated AgNPs were used as antifungals against *Phoma herbarum*, *Phoma glometra*, *Candida albicans*, *Trichoderma* sp., and *Fusarium semitectum*. Furthermore, the synergy of AgNPs with commercial antibiotics (fluconazole) increases the antifungal activity (Gajbhiye et al., 2009). The culture supernatant of *Nocardiopsis* sp. MBRC-1-mediated AgNPs (spherical, 45 ± 0.15 nm) has a concentration-dependent antifungal activity, with maximum growth inhibition activity (MIC) against *Candida albicans* (10 μg/ml), followed by *Aspergillus fumigates* (13 μg/ml), *Aspergillus niger* (16 μg/ml), and *Aspergillus brasiliensis* (18 μg/ml) (Manivasagan et al., 2013). Likewise, AuNPs synthesized using *Nocardiopsis* sp. MBRC-48 were used as antifungals against *Aspergillus niger*, *Aspergillus fumigates*, and *Aspergillus brasiliensis* (Manivasagan et al., 2015). Microbial-synthesized ZnO nanoparticles also exhibited good antifungal effects against *Aspergillus flavus*, *Aspergillus niger*, and *Candida albicans* with an MIC (μg/ml) value of 2.9 ± 0.01, 2.0 ± 0.04, and 0.9 ± 0.03, respectively (Jayseelan et al., 2012). *Bacillus licheniformis*-mediated CdSNPs, crystalline in nature and with sizes varying from 20 to 40 nm, were reported for their antifungal activity against food-borne pathogens including *Fusarium oxysporum*, *Aspergillus flavus*, and *Penicillium expansum* (Shivashankarappa and Sanjay, 2015). On the other hand, SNPs and AlNPs synthesized using *Colletotrichum* sp. also show antifungal activity against food-borne pathogens including *Aspergillus flavus* and *Fusarium oxysporum* (Suryavanshi et al., 2017). All these findings strongly suggest that different nanoparticles can be applicable in therapeutics as antifungal agents.

18.5.3 Anticancer Activity of Nanoparticles

In recent years, cancer has grown to be a major cause of death all over the world. Cancer is an uncontrolled cell proliferation with hysterical changes of enzymatic and biochemical parameters, which is the universal property of tumor cells. The over-expression of cellular growth can be arrested and regulated with systematic cell cycle mechanisms in cancerous cells by using microbe-functionalized nanoparticles as a novel controlling agent (Schrofel et al., 2014; Fariq et al., 2017). Recently, many research studies have reported that microbe-mediated nanoparticles have the potential to control cancer cell growth. The enhanced anticancer effect is due to secondary metabolites and other non-metal compositions in the synthesizing medium (Manivasagan et al., 2013; Manivasagan et al., 2015). The anticancer effect of metallic nanoparticles depends on the interaction between nanoparticles and cells, the cellular uptake, the uptake of nanoparticles by endocytosis, and the time, dose, size, and shape of nanoparticles (Soenen et al., 2012; Gong et al., 2015). Biosynthesized nanoparticles show their toxicity against cancer cells without affecting healthy cells; Pt nanoparticles are synthesized using cell-free extract of *Saccharomyces boulardii*. The synthesized nanoparticles are evaluated against MCF-7 (human breast cancer) and A431 (epidermoid carcinoma) cell lines (Borse et al., 2015). AgNPs synthesized using fungus were used for the dose-response elimination of human cervical cancer cells (HeLa). The morphological changes in cells due to the interactions with AgNPs are responsible for the death of the cells (Manivasagan et al., 2013). Similarly, AgNPs synthesized using *Cryptococcus laurentii* (BNM 0525) were found to be effective against MCF-7 and T47D breast cancer cells, while non-toxic to normal breast cells (MCF10-A) (Ortega et al., 2015). Besides, cancer cell lines observed higher endocytosis of AgNPs than normal cell lines. Manivasagan and coworkers (2015) reported that AuNPs synthesized using fungus have a good anticancer effect against HeLa cells. AuNPs treated cells show significant morphological changes and chromatin condensation, which is the characteristic of apoptosis. Roy et al. (2016) also reported the anticancer activity of AuNPs against A549 (lung carcinoma) cells. Furthermore, they observed that synthesized AuNPs have more inhibitory activity compared to gold ions. *Penicillium brevicompactum*-functionalized AuNPs reported for anti-tumorigenic activity on mouse mayo blast cancer (C_2C_{12}) cells. An increase in the concentration of AuNPs and interaction-time enhance cell death were also observed (Mishra et al., 2011). *Streptomyces bikiniensis*-mediated selenium nanorods reveal an ability to kill MCF-7 and Hep-G2 (liver cancer) human cancer cells. The possible mechanism suggested to explain their anticancer activity involved the mobilization of chromatin-bound copper followed by pro-oxidant action and consequently the death of MCF-7 and Hep-G2 cells (Ahmad et al., 2015). Mitochondria play an important role in controlling signaling pathways, but apoptosis regulators can damage their integrity (Kroemer et al., 1997; Green and Reed, 1998). The formation of ROS needs mitochondria which may activate an intrinsic caspase-dependent apoptotic pathway leading to cell death (Carlson et al., 2008). In the cytoplasm, the nanoparticles induce the ROS production, which has the ability to stimulate the apoptosis of exposed cells (Mittal and Pandey, 2014). In the end, it can reach the nucleus and damage the genomic DNA (Shukla et al., 2013). Therefore, through all these processes, the normal metabolism of the target cell can be disturbed, leading to cell death via apoptosis (DosSantos et al., 2014). The possible mechanisms of the anticancer activity of nanoparticles are diagrammatically represented in Figure 18.6.

18.5.4 Nanoparticles in Drug Delivery

Nanoparticles have been found to be convenient in targeted drug delivery, drug bio-availability, bioactivity, and stability. Delivering the drugs precisely and safely to their target sites at the right time to have a controlled release and achieve the maximum therapeutic effect is a key issue in the design and development of novel drug delivery systems. Drug-carrier nanoparticles must travel through blood–tissue barriers to reach the target cells and must enter in the target cells to contact the cytoplasmic target via specific transcytotic and endocytotic transport mechanisms across cellular barriers (Fedeel and Garcia-Bennett, 2010). Because of their very small size, nanoparticle drug carriers can bypass the skin, and due to their high surface-to-volume ratio, they show enhanced pharmacokinetics and biodistribution of drug (therapeutic agent), and thus reduce toxicity by their preferential accumulation at the target site (Paciotti et al., 2004). *Humicola* spp. (AAH-SCH-1) fungus-mediated monodispersed AuNPs conjugated with the anticancer drug doxorubicin have been reported for hepatic cancer treatment (Kumar et al., 2008). Similarly, *Humicola* sp.-mediated extracellular synthesis of gadolinium oxide nanoparticles and synthesized nanoparticles were loaded with the chemically modified anticancer drug Taxol to improve its potency against tumor cells (Khan et al., 2014). In another study, plant pathogenic fungus (*Helminthosporum solani*)-mediated AuNPs of different shapes (spherical, triangular, rod-like, pentagonal, star-shaped, and pyramidal) with size range from 2 to 70 nm were bioconjugated with doxorubicin. The bioconjugated doxorubicin was more readily taken into HEK293 cells with cytotoxicity comparable to doxorubicin (Kumar et al., 2008). Kundu et al. (2014) reported the extracellular synthesis of ZnONPs using bacteria; synthesized ZnONPs loaded with anthraquinone showed dose-dependent cytotoxicity against colon carcinoma cells (HT-29). It was suggested that they could be used as an anticancer drug carrier. Microbe-functionalized nanoparticles are supposed to be good candidates as drug-delivery carriers, but the determination of their biocompatibility and their toxicity to other cellular parts must be assured before execution.

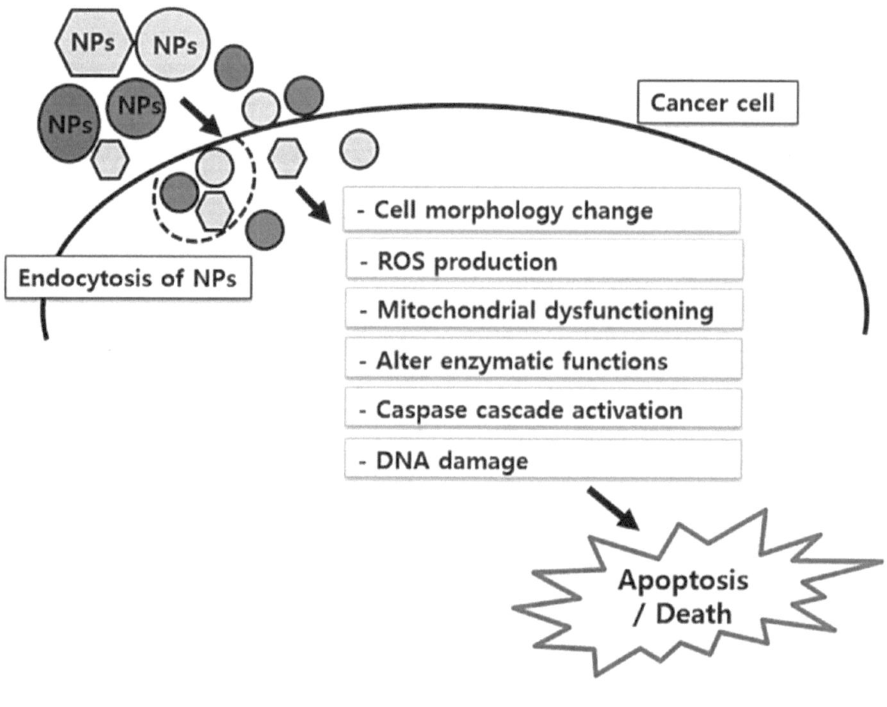

FIGURE 18.6 The mechanisms of the anticancer activity of different nanoparticles are diagrammatically represented.

18.5.5 Nanoparticles in the Food Industry

Food processing and packaging pass through many open processes, and during the handling of raw materials, manufacturing, and packaging, the chance of microbial contamination is high. To prevent such contamination during these processes, there is a need to develop biosensors to detect microbial contamination. Metal nanoparticles can be useful in the production of food packaging material, which can prevent microbial contamination (Cushen et al., 2012). *Trichoderma viride*-mediated AgNPs were incorporated into sodium alginate for the preparation of a thin film to coat carrot and pears. The thin film of AgNP-incorporated sodium alginate shows excellent antibacterial activity against *Staphylococcus aureus* and *Escherichia coli*, and also prevents a loss water and protein contents (Mohammed Fayaz et al., 2009). The uses of AgNPs in the food industry and in biomedical applications are increasing day by day. Yang et al. (2009) reported that the food storage material produced with AgNPs is responsible for the alteration of DNA replication and can control contamination. Metallic nanoparticles are being immobilized in food packaging materials for food preservation (Llorens et al., 2012).

18.5.6 Biosensors

The optical-electronic properties of metallic nanoparticles play an important role in biosensing applications. Extracellularly synthesized semiconductor monoclinic selenium (Se) nanoparticles using *Bacillus subtilis* were reported for H_2O_2 sensing application (Wang et al., 2010). Researchers suggested that SeNP-modified electrodes can be useful to detect levels of H_2O_2 in the clinical, pharmaceutical, and food industries, as well as in environmental applications. Ag-Au alloy nanoparticles are synthesized using yeast cells and applied for the biosensing of vanillin. Ag-Au alloy nanoparticle-modified glassy carbon electrodes were successfully applied to determine vanillin from vanilla beans and vanilla tea samples and suggested that this biosensor can be applicable in vanillin monitoring systems (Zheng et al., 2010). In another study, the quantum dots of CdSe/ZnS-conjugated ssDNA-fluorescent dye reported the detection of micrococcal nuclease activity in the culture medium of *S. aureus* by fluorescence microarray (Huang et al., 2008). The metallic nanoparticles or semiconductors quantum dots are applicable as chemical sensors and biosensors (Frasco and Chaniotakis, 2009).

18.5.7 Catalyst

Different nanoparticles have been intensively used to carry many reactions as catalysts or reductants to improve the reaction rate. The enhancement of the reaction rate by nanoparticles is due to their significant characteristics and large surface area. Environmental contamination through dyes and paint is extremely hazardous because of their non-biodegradable nature which causes crucial ecological problems, so there is a need to develop alternatives for the removal of such hazardous materials from the environment. Different metallic nanoparticles functionalized by microbes can be the solution to overcome this problem. For example, *Fusarium* sp. protein-coated AuNPs help reducing nitro-aromatic pollutants such as p-nitrophenol, o-nitrophenol, and o-nitroaniline in the presence of sodium borohydride (Guria et al., 2016). These AuNPs are recoverable, recyclable, and well stable without losing potential efficiency. Similarly, marine alga-functionalized AuNPs are able to enhance the degradation and reduction of 4-nitrophenol and p-nitroaniline in the presence of a reducing agent (sodium borohydride) (Ramakrishna et al., 2016). In another research, the catalytic activity of *Pycnoporus sanguineus* fungus-mediated AuNPs for the degradation of 4-nitroaniline was reported (Shi et al., 2015). *Ulva laculta* is a seaweed reported for the synthesis of AgNPs; synthesized AgNPs improve the photocatalytic degradation of synthetic organic dyes such as methylene orange (Kumar et al., 2013). The oxide form of titanium, titanium dioxide nanoparticles were synthesized using *Bacillus amyloliquefaciens* and helped in the degradation of reactive red 31 sulfonated dye (Khan and Fulekar, 2016). The ZnONPs synthesized using *Candida albicans* were used to enhance the rate of reaction in the synthesis of steroidal pyrazoline (Shamsuzzaman et al., 2017). ZnONPs worked as catalysts and improved the targeted molecule yield. All these reported studies suggest that different nanoparticles can be useful in increasing the rate of the reactions and preventing environmental contaminants being produced from chemical reactions and chemicals.

18.5.8 Agricultural Applications

Nano-biotechnology is a new and fast growing technology that has considerable importance also for agriculture. A large number of fungal pathogens and viral diseases affect crop yields and cause economic damage. Metallic nanoparticles might play an important role to overcome from these problems. For this aspect, Elbeshehy and group members synthesized AgNPs (77–92 nm) using *Bacillus* spp. The synthesized AgNPs had an antiviral effect against the bean yellow mosaic virus (Elbeshehy et al., 2015). In another study, the nanoemulsion of different oxide forms of nanoparticles such as aluminum oxide, titanium dioxide, and zinc oxide helped suppress different plant diseases such as weevil and grasserie in *Bombyx mori*. These diseases are caused by stored product pest *Sitophilus oryzae* and viruses such as baculovirus (*BmNPV*) (Goswami et al., 2010). Fungal diseases are responsible for the spoilage of crops, foods, and stored products. The antifungal activity of different nanoparticles can be applied to prevent the spoilage of agricultural products.

18.5.9 Bioremediation Applications

Microbe and microbe-mediated metallic nanoparticles are useful for cleaning up contaminated soil and ground water. The bioremediation of the radioactive waste caused by nuclear power plants and nuclear weapons production plants is possible by using nanoparticles. In another case, AgNPs were impregnated into cotton fabrics for biomedical applications, and during the process unbound AgNPs washed out with the wastewater from the plant and were discharged in the environment. In that aspect, *Chromobacterium violaceum* was reported for the bioremediation process of AgNPs. AgNPs were removed through the biosorption by bacteria from the effluent (Duran et al., 2010). Likewise, *Chromobacterium violaceum* was also reported for its ability to form water-soluble metal cyanide from metal-containing solids (Faramarzi et al., 2004). In another aspect, reducing the environmental contamination of Ag and Cu from computer circuits boards can be possible using the cyanogenic bioleaching bacterium *Bacillus megaterium* (Arshadi and Mousavi, 2015). New biologically inspired technologies have recently focused on the removal of precious metals and simultaneously on the prevention of ground water contamination. Pesticides and halogenated substances leek from the storage tanks of plants and are released from landfills. To remediate this, it is possible to use bio-palladium (Bio-Pd) nanoparticles (microbial produced) (Hennebel et al., 2012). Similarly, for water effluents from pharmaceutical companies contaminating ground water with a diverse range of chemicals, Bio-Pd and biogenic manganese oxides (BioMnOx), synthesized using *Pseudomonas putida* and *Shewanella oneidensis*, respectively, can be used. Bio-Pd and BioMnOx were reported for the removal of different micropollutants from pharmaceutical effluents (Forrez et al., 2011). All these studies indicate that microbial functionalized metallic nanoparticles are good candidates for the bioremediation process.

18.6 Challenges and Limitations

Despite the benefits of the microbe-mediated synthesis of metallic nanoparticles, there are still many limitations and challenges to be overcome before it can be applied practically. The most important challenge is the control of the shape, size, and particle size of different nanoparticles, because part of the mechanism behind the synthesis of nanoparticles remains unclear. Another limitation is that microbes need specific growth conditions; a fluctuation in growth parameters might change the enzymatic or morphological properties of microbes, which may interfere in the synthesis of nanoparticles. Different microbes show different properties of synthesized nanoparticles because of the different enzymatic properties of microorganisms. So we need to establish the exact mechanism behind nanoparticle synthesis. The stability of synthesized nanoparticles is the major challenge, as during storage nanoparticles can get aggregate or change their properties. In addition, it is also important to assess the toxicity of different metallic nanoparticles as well as their impact on the environment. Precious metals have importance in human consumption as well as in nature.

Conclusion

Microorganisms have the potential to synthesize different metallic nanoparticles for different applications. This chapter focuses mainly on the recent advancements in the research into the synthesis of different nanoparticles using microorganisms such as bacteria, cyanobacteria, actinomycetes, yeast, fungi, and algae, and on their significant applications in the fields of biomedicine, chemistry, and agriculture, etc. The traditional methods for the synthesis of nanoparticles are harmful to living things and to the environment. In order to prevent these dangers, the use of microorganisms for nanoparticles synthesis is the best alternative. Nanotechnology is a new field of science that leads to the development and synthesis of different nano-sized materials using different microbial groups, which also possess many biological and chemical properties such as antimicrobial, anticancer, and catalytic activities. They are suitable for some advanced biotechnological applications such as drug delivery, bioremediation, etc.

However, a lot of work is needed to produce nanoparticles in large scale and to control the shape and size of the nanoparticles synthesized through microbes. The synthesis of nanoparticles using microbes is eco-friendly, non-toxic, simple, and cost effective. Thanks to the recent advancements and ongoing efforts in the improvement of controlled synthesis, we hope that the implementation of these biosynthesis approaches on a large scale and their commercial application in biomedical, agricultural, and new drug discoveries will take place in the coming years.

Acknowledgments

This research was supported by Basic Science Research Program through the National Research Foundation of Korea (NRF) funded by the Ministry of Education (2018R1D1A1B07043388).

REFERENCES

AbdelRahim, K., Mahmoud, S. Y., Ali, A. M., Almaary, K. S., Mustafa, A. E. M. A., and Husseiny, S. M. 2017. Extracellular biosynthesis of silver nanoparticles using *Rhizopus stolonifer*. *Saudi J Biol Sci* 24(1):208–216. doi:10.1016/j.sjbs.2016.02.025.

Agnihotri, M., Joshi, S., Kumar, A. R., Zinjarde, S., and Kulkarni, S. 2009. Biosynthesis of gold nanoparticles by the tropical marine yeast *Yarrowia lipolytica* NCIM 3589. *Mater Lett* 63(15):1231–1234. doi:10.1016/j.matlet.2009.02.042.

Ahmad, A., Mukherjee, P., Mandal, D., Senapati, S., Khan, M. I., Kumar, R., and Sastry, M. 2002. Enzyme mediated extracellular synthesis of CdS nanoparticles by the fungus, *Fusarium oxysporum*. *J Am Chem Soc* 124(41):12108–12109. doi:10.1021/ja027296o.

Ahmad, A., Senapati, S., Khan, M. I., Kumar, R., Ramani, R., Srinivas, V., and Sastry, M. 2003. Intracellular synthesis of gold nanoparticles by a novel alkalotolerant actinomycetes, *Rhodococcus* species. *Nanotechnology* 14(7):824–828.

Ahmad, A., Senapati, S., Khan, M. I., Kumar, R., and Sastry, M. 2013. Extracellular biosynthesis of monodisperse gold nanoparticles by a novel extremophilic actinomycete, *Thermomonospora* sp. *Langmuir* 19(8):3550–3553. doi:10.4172/2157-7439.1000156.

Ahmad, M. S., Yasser, M. M., Sholkamy, E. N., Ali, A. M., and Mehanni, M. M. 2015. Anticancer activity of biostabilized selenium nanorods synthesized by *Streptomyces bikiniensis* strain Ess_amA-1. *Int J Nanomed* 10:3389–3401. doi:10.2147/IJN.S82707.

Ahmed, S., Annu, Chaudhry, S. A., and Ikramm, S. 2017. A review on biogenic synthesis of ZnO nanoparticles using plant and microbes: A prospect towards green chemistry. *J Photochem Photobiol B* 166:272–284. doi:10.1016/j.photobiol.2016.12.011.

Alekshun, M. N., and Levy, S. B. 2007. Molecular mechanisms of antibacterial multidrug resistance. *Cell* 128(6):1037–1050. doi:10.1016/j.cell.2007.03.004.

Arshadi, M., and Mousavi, S. M. 2015. Enhancement of simultaneous gold and copper extraction from computer printed circuits boards using *Bacillus megaterium*. *Bioresour Technol* 175:315–324. doi:10.1016/j.biortech.2014.10.083.

Arsiya, F., Sayadi, M. H., and Sobhani, S. 2017. Green synthesis of palladium nanoparticles using *Chlorella vulgaris*. *Mater Lett* 186:113–115. doi:10.1016/j.matlet.2016.09.101.

Azizi, S., Ahmad, M. B., Namvar, F., and Mohamad, R. 2014. Green biosynthesis of zinc oxide nanoparticles using brown marine macroalga *Sargassum muticum* aqueous extract. *Mater Lett* 116:275–277. doi:10.1016/j.matlet.2013.11.038.

Bai, H.-J., and Zhang, Z.-M. 2009. Microbial synthesis of semiconductor lead sulfide nanoparticles using immobilized *Rhodobacter sphaeroides*. *Mater Lett* 63(9–10):764–766. doi:10.1016/j.matlet.2008.12.050.

Baker, S., Harini, B. P., Rakshith, D., and Satish, S. 2013. Marine microbes: Invisible nanofactories. *J Pharm Res* 6(3):383–388.

Beveridge, T., and Murray, R. 1980. Sites of metal deposition in the cell wall of *Bacillus subtilis*. *J Bacteriol* 141(2):876–887.

Borse, V., Kaler, A., and Banerjee, U. C. 2015. Microbial synthesis of platinum nanoparticles and evaluation of their anticancer activity. *Int J Emerg Trends Electr Electron* 11(2):26–31.

Carlson, C., Hussain, S. M., Schrand, A. M., Braydich-Stolle, L. K., Hess, K. L., Jones, R. L., and Schlager, J. J. 2008. Unique cellular interaction of silver nanoparticles: Size-dependent generation of reactive oxygen species. *J Phys Chem B* 112(43):13608–13619. doi:10.1021/jp712087m.

Castro-Longoria, E., Vilcis-Nestor, A. R., and Avalos-Borja, M. 2011. Biosynthesis of silver, gold and bimetallic nanoparticles using the filamentous fungus *Neurospora crassa*. *Colloids Surf B Biointerfaces* 83(1):42–48. doi:10.1016/j.colsurfb.2010.10.035.

Chakraborty, N., Banerjee, A., Lahiri, S., Panda, A., Ghosh, A. N., and Pal, R. 2009. Biorecovery of gold using cyanobacteria and an eukaryotic alga with special reference to nanogold formation – A novel phenomenon. *J Appl Phycol* 21(1):145–152. doi:10.1007/s10811-008-9343-3.

Crespo, K. A., Baronetti, J. L., Quinteros, M. A., Paez, P. L., and Paraje, M. G. 2017. Intra- and extracellular biosynthesis and characterization of iron nanoparticles from prokaryotic microorganisms with anticoagulant activity. *Pharm Res* 34(3):591–598. doi:10.1007/s11095-016-2084-0.

Cushen, M., Kerry, J., Morris, M., Cruz-Romero, M., and Cummins, E. 2012. Nanotechnologies in the food industry–Recent developments, risks and regulation. *Trends Food Sci Technol* 24(1):30–46. doi:10.1016/j.tifs.2011.10.006.

Dameron, C. T., Reese, R. N., Mehra, R. K., Kortan, A. R., Carroll, P. J., Steigerwald, M. L., Brus, L. E., and Winge, D. R. 1989. Biosynthesis of cadmium sulphide quantum semiconductor crystallites. *Nature* 338(6216):596–597. doi:10.1038/338596a0.

Das, S. K., Das, A. R., and Guha, A. K. 2009. Gold nanoparticles: Microbial synthesis and application in water hygiene management. *Langmuir* 25(14):8192–8199.

Das, V. L., Thomas, R., Varghese, R. T., Soniya, E. V., Mathew, J., and Radhakrishnan, E. K. 2014. Extracellular synthesis of silver nanoparticles by the *Bacillus* strain CS 11 isolated from industrial area. *3 Biotech* 4(2):121–126.

Dibrov, P., Dzioba, J., Gosink, K. K., and Häse, C. C. 2002. Chemiosmotic mechanism of antimicrobial activity of Ag+ in *Vibrio cholerae*. *Antimicrob Agents Chemother* 46(8):2668–2670. doi:10.1128/AAC.46.8.2668-2670.2002.

Dos Santos, C. A., Seckler, M. M., Ingle, A. P., Gupta, I., Galdiero, S., Galdiero, M., Gade, A., and Rai, M. 2014. Silver nanoparticles: Therapeutical uses, toxicity, and safety issues. *J Pharm Sci* 103(7):1931–1944.

Duran, N., Marcato, P. D., Alves, O. L., Da Silva, J. P. S., De Souza, G. I. H., Rodrigues, F. A., and Esposito, E. 2010. Eco-system protection by effluent bioremediation: Silver nanoparticles impregnation in a textile fabrics process. *J Nanopart Res* 12(1):285–292. doi:10.1007/s11051-009-9606-1.

Duran, N., Marcato, P. D., Alves, O. L., De Souze, G. I., and Esposito, E. 2005. Mechanistic aspects of biosynthesis of silver nanoparticles by several *Fusarium oxysporum* strains. *J Nanobiotechnology* 3:8. doi:10.1186/1477-3155-3-8.

Elbeshehy, E. K. F., Elazzazy, A. M., and Aggelis, G. 2015. Silver nanoparticles synthesis mediated by new isolates of *Bacillus* spp., nanoparticles characterization and their activity against bean yellow mosaic virus and human pathogens. *Front Microbiol* 6:453. doi:10.3389/fmicb.2015.00453.

Fadeel, B., and Garcia-Bennett, A. E. 2010. Better safe than sorry: Understanding the toxicological properties of inorganic nanoparticles manufactured for biomedical applications. *Adv Drug Deliv Rev* 62(3):362–374. doi:10.1016/j.addr.2009.11.008.

Faramarzi, M. A., Stagars, M., Pensini, E., Krebs, W., and Brandl, H. 2004. Metal solubilization from metal-containing solid materials by cyanogenic *Chromobacterium violaceum*. *J Biotechnol* 113(1–3):321–326. doi:10.1016/j.biotech.2004.03.031.

Fariq, A., Khan, T., and Yasmin, A. 2017. Microbial synthesis of nanoparticles and their potential applications in biomedicine. *J Appl Biomed* 15(4):241–248. doi:10.1016/j.jab.2017.03.004.

Fayaz, A. M., Girilal, M., Rahman, M., Venkatesan, R., and Kalaichelvan, P. T. 2011. Biosynthesi of silver and gold nanoparticles using thermophilic *Geobacillus stearothermophilus*. *Process Biochem* 46:1958–1962. doi:10.1016/j.procbio.2011.07.003.

Fendler, J. H. 1998. *Nanoparticles and Nanostructure Films: Preparation, Characterization and Applications*. New York, Wiley. doi:10.1002/9783527612079.ch18.

Feng, Q. L., Wu, J., Chen, G. Q., Cui, F. Z., Kim, T. N., and Kim, J. O. 2000. A mechanistic study of the antibacterial effect of silver ions on *Escherichia coli* and *Staphylococcus aureus*. *J Biomed Mater Res* 52(4):662–668.

Ferreira, V. D. S., ConzFerreira, M. E., Lima, L. M. T. R., Feases, S., Souza, W. D., and Sant'Anna, C. 2017. Green production of microalgae-based silver chloride nanoparticles with antimicrobial activity against pathogenic bacteria. *Enzyme Microb Technol* 97:114–121. doi:10.1016/j.enzmictec.2016.10.018.

Forrez, I., Carballa, M., Fink, G., Wick, A., Hennebel, T., Vanhaecke, L., Ternes, T., Boon, N., and Verstraete, W. 2011. Biogenic metals for the oxidative and reductive removal of pharmaceuticals, biocides and iodinated contrast media in a polishing membrane bioreactor. *Water Res* 45(4):1763–1773. doi:10.1016/j.watres.2010.11.031.

Frasco, M. F., and Chaniotakis, N. 2009. Semiconductor quantum dots in chemical sensors and biosensors. *Sensors* 9(9):7266–7286. doi:10.3390/s90907266.

Gade, A., Ingle, A., Whiteley, C., and Rai, M. 2010. Mycogenic metal nanoparticles: Progress and applications. *Biotechnol Lett* 32(5):593–600.

Gade, A. K., Bonde, P., Ingle, A. P., Marcato, P. D., Duran, N., and Rai, M. K. 2008. Exploitation of *Aspergillus niger* for synthesis of silver nanoparticles. *J Biobased Mater Bioenergy* 2(3):243–247. doi:10.1166/jbmb.2008.401.

Gajbhiye, M., Kesharwani, J., Ingle, A., Gade, A., and Rai, M. 2009. Fungus-mediated synthesis of silver nanoparticles and their activity against pathogenic fungi in combination with fluconazole. *Nanomedicine* 5(4):382–386. doi:10.1016/j.nano.2009.06.005.

Gericke, M., and Pinches, A. 2006. Biological synthesis of metal nanoparticles. *Hydrometallurgy* 83(1–4):132–140. doi:10.1016/j.hydromet.2006.03.019.

Golinska, P., Wypij, M., Ingle, A. P., Gupta, I., Dahm, H., and Rai, M. 2014. Biogenic synthesis of metal nanoparticles from actinomycetes: Biomedical applications and cytotoxicity. *Appl Microbiol Biotechnol* 98(19):8083–8097. doi:10.1007/s00253-014-5953-7.

Gong, N., Chen, S., Jin, S., Zhang, J., Wang, P. C., and Liang, X.-J. 2015. Effects of the physicochemical properties of gold nanostructures on cellular internalization. *Regen Biomater* 2(4):273–280. doi:10.1093/rb/rbv024.

Gopinath, K., and Arumugam, A. 2014. Extracellular mycosynthesis of nanoparticles using *Fusarium solani*. *Appl Nanosci* 4(6):657–662. doi:10.1007/s13204-013-0247-4.

Goswami, A., Roy, I., Sengupta, S., and Debnath, N. 2010. Novel applications of solid and liquid formulations of nanoparticles against insect pests and pathogens. *Thin Solid Films* 519(3):1252–1257. doi:10.1016/j.tsf.2010.08.079.

Govindaraju, K., Basha, S. K., Kumar, V. G., and Singaravelu, G. 2008. Silver, gold and bimetallic nanoparticles production using single-cell protein (*Spirulina platensis*) geitler. *J Mater Sci* 43(15):5115–5122. doi:10.1007/s10853-008-2745-4.

Green, D. R., and Reed, J. C. 1998. Mitochondria and apoptosis. *Science-AAAS-Weekly Paper Edition* 281(5381):1309–1311.

Guria, M. K., Majumdar, M., and Bhattacharyya, M. 2016. Green synthesis of protein capped nano-gold particle: An excellent recyclable nano-catalyst for the reduction of nitro-aromatic pollutants at higher concentration. *J Mol Liq* 222:549–557. doi:10.1016/j.molliq.2016.07.087.

Gurunathan, S., Kalishwaralal, K., Vaidyanathan, R., Venkataraman, D., Pandian, S. R., Muniyandi, J., Hariharan, N., and Eom, S. H. 2009. Biosynthesis, purification and characterization of silver nanoparticles using *Escherichia coli*. *Colloids Surf B Biointerfaces* 74(1):328–335. doi:10.1016/j.colsurfb.2009.07.048.

He, S., Guo, Z., Zhang, Y., Zhang, S., Wang, J., and Gu, N. 2007. Biosynthesis of gold nanoparticles using the bacteria *Rhodopseudomonas capsulate*. *Mater Lett* 61(18):3984–3987. doi:10.1016/j.matlet.2007.01.018.

Hennebel, T., Corte, S. D., Verstraete, W., and Boon, N. 2012. Microbial production and environmental applications of Pd nanoparticles for treatment of halogenated compounds. *Curr Opin Biotechnol* 23(4):555–561. doi:10.1016/j.copbio.2012.01.007.

Hosea, M., Greene, B., McPherson, R., Henzl, M., Alexander, M. D., and Darnall, D. W. 1986. Accumulation of elemental gold on the alga *Chlorella vulgaris*. *Inorg Chim Acta* 123(3):161–165. doi:10.1016/S0020-1693(00)86339-2.

Huang, S., Xiao, Q., He, Z. K., Liu, Y., Tinnefeld, P., Su, X. R., and Peng, X. N. 2008. A high sensitive and specific QDs FRET Bioprobe for MNase. *Chem Commun (Camb)* 45(45):5990–5992. doi:10.1039/b815061c.

Hulkoti, N. I., and Taranath, T. C. 2014. Biosynthesis of nanoparticles using microbes – A review. *Colloids Surf B Biointerfaces* 121:474–483. doi:10.1016/j.colsurfb.2014.05.027.

Hussain, I., Singh, N. B., Singh, A., Singh, H., and Singh, S. C. 2016. Green synthesis of nanoparticles and its potential applications. *Biotechnol Lett* 38(4):545–560. doi:10.1007/s10529-015-2026-7.

Iravani, S., Korbekandi, H., Mirmohammadi, S. V., and Zolfaqhari, B. 2014. Synthesis of silver nanoparticles: Chemical, physical and biological methods. *Res Pharm Sci* 9(6):385–406.

Jayaseelan, C., Rahuman, A. A., Kirthi, A. V., Marimuthu, S., Santhoshkumar, T., Bagavan, A., Gaurav, K., Karthik, L., and Rao, K. V. B. 2012. Novel microbial route to synthesize ZnO nanoparticles using *Aeromonas hydrophila* and their activity against pathogenic bacteria and fungi. *Spectrochim Acta A* 90:78–84. doi:10.1016/j.saa.2012.01.006.

Jayaseelan, C., Rahuman, A. A., Roopam, S. M., Kirthi, A. V., Venkatesan, J., Kim, S.-K., Iyappan, M., and Siva, C. 2013. Biological approach to synthesize TiO_2 nanoparticles using *Aeromonas hydrophila* and its antibacterial activity. *Spectrochim Acta A* 107:82–89. doi:10.1016/j.saa.2012.12.083.

Jha, A. K., Prasad, K., and Kulkarni, A. R. 2009. Synthesis of TiO_2 nanoparticles using microorganisms. *Colloids Surf B Biointerfaces* 71(2):226–229. doi:10.1016/j.colsurfb.2009.02.007.

Kaplan, D., Christiaen, D., and Arad, S. M. 1987. Chelating properties of extracellular polysaccharides from *Chlorella* spp. *Appl Environ Microbiol* 53(12):2953–2956.

Kashefi, K., and Lovely, D. R. 2000. Reduction of Fe(III), Mn(IV), and toxic metals at 100°C by *Pyrobaculum islandicum*. *Appl Environ Microbiol* 66(3):1050–1056. doi:10.1128/AEM.66.3.1050-1056.2000.

Khan, R., and Fulekar, M. H. 2016. Biosynthesis of titanium dioxide nanoparticles using *Bacillus amyloliquefaciens* culture and enhancement of its photocatalytic activity for the degradation of a sulfonated textile dye reactive Red 31. *J Colloid Interface Sci* 475:184–191. doi:10.1016/j.jcis.2016.05.001.

Khan, S. A., Gambhir, S., and Ahmad, A. 2014. Extracellular biosynthesis of gadolinium oxide (Gd2O3) nanoparticles, their biodistribution and bioconjugation with the chemically modified anticancer drug Taxol. *Beilstein J Nanotechnol* 5:249–257. doi:10.3762/bjnano.5.27.

Kim, J. S., Kuk, E., Yu, K. N., Kim, J. H., Park, S. J., Lee, H. J., Kim, S. H., Park, Y. K., Park, Y. H., Hwang, C. Y., and Kim, Y. K. 2007. Antimicrobial effects of silver nanoparticles. *Nanomedicine* 3(1):95–101. doi:10.1016/j.nano.2006.12.001.

Kim, K.-J., Sung, W. S., Moon, S.-K., Choi, J.-S., Kim, J. G., and Lee, D. G. 2008. Antifungal effect of silver nanoparticles on dermatophytes. *J Microbiol Biotechnol* 18(8):1482–1884.

Kim, K.-J., Sung, W. S., Suh, B. K., Moon, S.-K., Choi, J.-S., Kim, J. G., and Lee, D. G. 2009. Antifungal activity and mode of action of silver nano-particles on *Candida albicans*. *Biometals* 22(2):235–242. doi:10.1007/s10534-008-9159-2.

Klaus, T., Joerger, R., Olsson, E., and Granqvist, C. G. 1999. Silver-based crystalline nanoparticles, microbially fabricated. *Proc Natl Acad Sci USA* 96(24):13611–13614.

Kowshik, M., Ashtaputre, S., Kharrazi, S., Vogel, W., Urban, J., Kulkarni, S. K., and Paknikar, K. M. 2003. Extracellular synthesis of silver nanoparticles by a silver-tolerant yeast strain MKY3. *Nanotechnology* 14(1):95–100. doi: S0957-4484(03)39921-0.

Kowshik, M., Deshmukh, N., Vogel, W., Urban, J., Kulkarni, S. K., and Paknikar, K. M. 2002. Microbial synthesis of semiconductor, their characterization, and their use in the fabrication of an ideal diode. *Biotechnol Bioeng* 78(5):583–588. doi:10.1002/bit.10233.

Kroemer, G., Zamzami, N., and Susin, S. A. 1997. Mitochondrial control of apoptosis. *Immunol Today* 18(1):44–51.

Kundu, D., Hazra, C., Chatterjee, A., Chaudhari, A., and Mishra, S. 2014. Extracellular biosynthesis of zinc oxide nanoparticles using Rhodococcus pyridinivorans NT2: Multifunctional textile finishing, biosafety evaluation and in vitro drug delivery in colon carcinoma. *J Photochem Photobiol B* 140:194–204. doi:10.1016/j.jphotobiol.2014.08.001.

Kumar, P., Govindaraju, M., Senthamilselvi, S., and Premkumar, K. 2013. Photocatalytic degradation of methyl orange dye using silver (Ag) nanoparticles synthesized from *Ulva lactuca*. *Colloids Surf B Biointerfaces* 103:658–661. doi:10.1016/j.colsurfb.2012.11.022.

Kumar, S. A., Peter, Y. A., and Nadeau, J. L. 2008. Facile biosynthesis, separation and conjugation of gold nanoparticles to doxorubicin. *Nanotechnology* 19(49):495101. doi:10.1088/0957-4484/19/49/495101.

Kumar, V., Jain, A., Wadhawan, S., and Mehta, S. K. 2017. Synthesis of biosurfactant-coated magnesium oxide nanoparticles for methylene blue removal and selective Pb2+ sensing. *IET Nanobiotechnol* 12(3):241–253.

Kumar, V., Kumar, A., Kumar, D., and Yadav, S. K. 2014. Biosurfactant stabilized anticancer biomolecule-loaded poly (d, l-lactide) nanoparticles. *Colloids Surf B Biointerfaces* 117:505–511.

Lengke, M. F., Fleet, M. E., and Southam, G. 2006. Morphology of gold nanoparticles synthesized by filamentous cyanobacteria from gold(I)-thiosulfate and gold(III)-chloride complexes. *Langmuir* 22(6):2780–2787. doi:10.1021/la052652c.

Li, G., He, D., Qian, Y., Guan, B., Gao, S., Cui, Y., Yakoyama, K., and Wang, L. 2012. Fungus-mediated green synthesis of silver nanoparticles using *Aspergillus terreus*. *Int J Mol Sci* 13(1):466–476. doi:10.3390/ijms13010466.

Li, X., Xu, H., Chen, Z.-S., and Chen, G. 2011. Biosynthesis of nanoparticles by microorganisms and their applications. *J Nanomater.* doi:10.1155/2011/270974.

Link, S., and El-Sayed, M. A. 1999. Size and temperature dependence of the plasmon absorption of colloidal gold nanoparticles. *J Phys Chem B* 103(21):4212–4217. doi:10.1021/jp984796o.

Link, S., Wang, Z. L., and El-Sayed, M. A. 1999. Alloy formation of gold–silver nanoparticles and the dependence of the plasmon absorption on their composition. *J Phys Chem B* 103(18):3529–3533. doi:10.1021/jp990387w.

Llorens, A., Lloret, E., Picouet, P. A., Trbojevich, R., and Fernandez, A. 2012. Metallic-based micro and nanocomposites in food contact materials and active food packaging. *Trends Food Sci Technol* 24(1):19–29. doi:10.1016/j.tifs.2011.10.001.

Mahdavi, M., Namvar, F., Ahmad, M. B., and Mohamad, R. 2013. Green biosynthesis and characterization of magnetic iron oxide (Fe3O4) nanoparticles using seaweed (*Sargassum muticum*) aqueous extract. *Molecules* 18(5):5954–5964. doi:10.3390/molecules18055954.

Mala, J. G. S., and Rose, C. 2014. Facile production of ZnS quantum dot nanoparticles by *Saccharomyces cerevisiae* MTCC 2918. *J Biotechnol* 170:73–78. doi:10.1016/j.jbiotec.2013.11.017.

Malhotra, A., Dolma, K., Kaur, N., Rathore, Y. S., Ashish, Mayilraj, S., and Chaudhury, A. R. 2013. Biosynthesis of gold and silver nanoparticles using a novel marine strain of *Stenotrophomonas*. *Bioresour Technol* 142:727–731. doi:10.1016/j.biortech.2013.05.109.

Manimaran, M., and Kannabiran, K. 2017. Actinomycetes-mediated biogenic synthesis of metal and metal oxide nanoparticles: Progress and challenges. *Lett Appl Microbiol* 64(6):401–408. doi:10.1111/lam.12730.

Manivasagan, P., Alam, M. S., Kang, K.-H., Kwak, M., and Kim, S.-K. 2015. Extracellular synthesis of gold nanoparticles by *Nocardiopsis* sp. and evaluation of its antimicrobial, antioxidant and cytotoxic activities. *Bioprocess Biosyst Eng* 38(6):1167–1177. doi:10.1007/s00449-015-1358-y.

Manivasagan, P., Nam, S. Y., and Oh, J. 2016. Marine microorganisms as potential biofactories for synthesis of metallic nanoparticles. *Crit Rev Microbiol* 42(6):1007–1019. doi:10.3109/1040841X.2015.1137860.

Manivasagan, P., Venkatesan, J., Senthilkumar, K., Sivakumar, K., and Kim, S.-K. 2013. Biosynthesis, antimicrobial and cytotoxic effect of silver nanoparticles using a novel *Nocardiopsis* sp. MBRC-1. *Bio Res Int* 2013:41–49. doi:10.1155/2013/287638.

Margeay, M., Monchy, S., Vallaeys, T., Auquier, V., Benotmane, A., Bertin, P., Taghavi, S., Dunn, J., Lelie, D. V. D., and Wattiez, R. 2003. *Ralstonia metallidurans*, a bacterium specifically adapted to toxic metals: Towards a catalogue of metal-responsive genes. *FEMS Microbiol Rev* 27(2–3):385–410.

Markus, J., Mathiyalagan, R., Kim, Y.-J., Abbai, R., Singh, P., Ahn, S., Perez, Z., Hurh, E. J., and Yang, D. C. 2016. Intracellular synthesis of gold nanoparticles with antioxidant activity by probiotic *Lactobacillus kimchicus* DCY51^T isolated from Korean kimchi. *Enzyme Microb Technol* 95:85–93. doi:10.1016/j.enzmictec.2016.08.018.

Mishra, A., Tripathy, S. K., Wahab, R., Jeong, S.-H., Hwang, I., Yang, Y.-B., Kim, Y.-S., Shin, H.-S., and Yun, H.-S. 2011. Microbial synthesis of gold nanoparticles using the fungus *Penicillium brevicompactum* and their cytotoxic effects against mouse mayo blast cancer C_2C_{12} cells. *Appl Microbiol Biotechnol* 92(3):617–630. doi:10.1007/s00253-011-3556-0.

Mishra, M., Paliwal, J. S., Singh, S. K., Selvarajan, E., Subathradevi, C., and Mohanasrinivasan, V. 2013. Studies on the inhibitory activity of biologically synthesized and characterized zinc oxide nanoparticles using *Lactobacillus sporogens* against *Staphylococcus aureus*. *J Pure Appl Microbiol* 7:1263–1268.

Mittal, S., and Pandey, A. K. 2014. Cerium oxide nanoparticles induced toxicity in human lung cells: Role of ROS mediated DNA damage and apoptosis. *BioMed Res Int.* doi:10.1155/2014/891934.

Mohammed Fayaz, A., Balaji, K., Girilal, M., Kalaichelvan, P. T., and Venkatesan, R. 2009. Mycobased synthesis of silver nanoparticles and their incorporation into sodium alginate films for vegetable and fruit preservation. *J Agric Food Chem* 57(14):6246–6252. doi:10.1021/jf900337h.

MubarakAli, D., Gopinath, V., Rameshbabu, N., and Thanjuddin, N. 2012. Synthesis and characterization of CdS nanoparticles using C-phycoerythrin from the marine cyanobacteria. *Mater Lett* 74:8–11. doi:10.1016/j.matlet.2012.01.026.

MubarakAli, D., Sasikala, M., Gunasekaran, M., and Thanjuddin, N. 2011. Biosynthesis and characterization of silver nanoparticles using marine cyanobacterium, *Oscillatoria willei*. *Dig J Nanomater and Biostruct* 6(2):385–390.

Mukherjee, P., Ahmad, A., Mandal, D., Senapati, S., Sainkar, S. R., Khan, M. I., Parischa, R., Ajayakumar, P. V., Alam, M., Kumar, R., and Sastry, M. 2001a. Fungus-mediated synthesis of silver nanoparticles and their immobilization in the mycelial matrix: A novel biological approach to nanoparticle synthesis. *Nano Lett* 1(10):515–519. doi:10.1021/nl0155274.

Mukherjee, P., Ahmad, A., Mandal, D., Senapati, S., Sainkar, S. R., Khan, M. I., Ramani, R., Parischa, R., Ajayakumar, P. V., Alam, M., Sastry, M., and Kumar, R. 2001b. Bioreduction of $AuCl_4^-$ ions by the fungus, *Verticillium* sp. and surface trapping of the gold nanoparticles formed. *Angew Chem Int Ed* 40(19):3585–3588.

Muller, A., Behsnilian, D., Walz, E., Graf, V., Hogekamp, L., and Greiner, R. 2016. Effect of culture medium on the extracellular synthesis of silver nanoparticles using *Klebsiella pneumoniae*, *Escherichia coli*, and *Pseudomonas jessinii*. *Biocatal Agric Biotechnol* 6:107–115. doi:10.1016/j.bcab.2016.02.012.

Nair, B., and Pradeep, T. 2002. Coalescence of nanoclusters and formation of submicron crystallites assisted by *Lactobacillus* strains. *Cryst Growth Des* 2(4):293–298. doi:10.1021/cg0255164.

Nies, D. H. 1999. Microbial heavy-metal resistance. *Appl Microbiol Biotechnol* 51(6):730–750.

Ortega, F. G., Fernandez-Baldo, M. A., Fernandez, J. G., Serrano, M. J., Sanz, M. I., Diaz-Mochon, J. J., Lorente, J. A., and Raba, J. 2015. Study of antitumor activity in breast cell lines using silver nanoparticles produced by yeast. *Int J Nanomed* 10:2021–2031. doi:10.2147/IJN.S75835.

Paciotti, G. F., Myer, L., Weinreich, D., Goia, D., Pavel, N., McLaughlin, R. E., and Tamarkin, L. 2004. Colloidal gold: A novel nanoparticle vector for tumor directed drug delivery. *Drug Deliv* 11(3):169–183. doi:10.1080/10717540490433895.

Patel, V., Berthold, D., Puranik, P., and Gantar, M. 2015. Screening of cyanobacteria and microalgae for their ability to synthesize silver nanoparticles with antibacterial activity. *Biotechnol Rep* 5:112–119. doi:10.1016/j.btre.2014.12.001.

Patil, M. P., and Kim, G.-D. 2017. Eco-friendly approach for nanoparticles synthesis and mechanism behind antibacterial activity of silver and anticancer activity of gold nanoparticles. *Appl Microbiol Biotechnol* 101(1):79–92. doi:10.1007/s00253-016-8012-8.

Patil, M. P., Bayaraa, E., Subedi, P., Piad, L. L. A., Tarte, N. H., and Kim, G.-D. 2019. Biogenic synthesis, characterization of gold nanoparticles using *Lonicera japonica* and their anticancer activity on HeLa cells. *J Drug Deliv Sci Technol* 51:83–90. doi:10.1016/j.jddst.2019.02.021.

Patil, M. P., Jin, X., Simeon, N. C., Palma, J., Kim, D., Ngabire, D., Kim, N.-H., Tarte, N. H., and Kim, G.-D. 2018a. Anticancer activity of *Sasa borealis* leaf extract-mediated gold nanoparticles. *Artif Cells Nanomed Biotechnol* 46(1):82–88. doi:10.1080/21691401.2017.1293675.

Patil, M. P., and Kim, G.-D. 2018. Marine microorganisms for synthesis of metallic nanoparticles and their biomedical applications. *Colloids Surf B* 172:487–495. doi:10.1016/j.colsurfb.2018.09.007.

Patil, M. P., Ngabire, D., Thi, P. H. H., Kim, M.-D., and Kim, G.-D. 2017a. Eco-friendly synthesis of gold nanoparticles and evaluation of their cytotoxic activity on cancer cells. *J Clust Sci* 28(1):119–132. doi:10.1007/s10876-016-1051-6.

Patil, M. P., Palma, J., Simeon, N. C., Jin, X., Liu, X., Ngabire, D., Kim, N.-H., Tarte, N. H., and Kim, G.-D. 2017b. *Sasa borealis* extract-mediated green synthesis of silver-silver chloride nanoparticles and their antibacterial and anticancer activities. *New J Chem* 41(3):1363–1371. doi:10.1039/c6nj03454c.

Patil, M. P., Rokade, A. A., Ngabire, D., and Kim, G.-D. 2016. Green synthesis of silver nanoparticles using water extract from galls of *Rhus chinensis* and its antibacterial activity. *J Clust Sci* 27(5):1737–1750. doi:10.1007/s10876-016-1037-4.

Patil, M. P., Seo, Y. B., and Kim, G.-D. 2018c. Morphological changes of bacterial cells upon exposure of silver-silver chloride nanoparticles synthesized using *Agrimonia pilosa*. *Microb Pathog* 116:84–90. doi:10.1016/j.micpath.2018.01.018.

Patil, M. P., Singh, R. D., Koli, P. B., Patil, K. T., Jagdale, B. S., Tipare, A. R., and Kim, G.-D. 2018b. Antibacterial potential of silver nanoparticles synthesized using *Madhuca longifolia* flower extract as a green resource. *Microb Pathog* 121:184–189. doi:10.1016/j.micpath.2018.05.040.

Pattekari, P., Zheng, Z., Zhang, X., Levchenko, T., Tochilin, V., and Lvov, Y. 2011. Top-down and bottom-up approaches in production of aqueous nanocolloids of low solubility drug paclitaxel. *Phys Chem Chem Phys* 13(19):9014–9019. doi:10.1039/c0cp02549f.

Prabhu, S., and Poulose, E. K. 2012. Silver nanoparticles: Mechanism of antimicrobial action, synthesis, medical applications, and toxicity effects. *Int Nano Lett* 2(1):32. doi:10.1186/2228-5326-2-32.

Prakasham, R. S., Kumar, B. S., Kumar, Y. S., and Kumar, K. P. 2014. Production and characterization of protein encapsulated silver nanoparticles by marine isolate *Streptomyces parvulus* SSNP11. *Indian J Microbiol* 54(3):329–336. doi:10.1007/s12088-014-0452-1.

Prema, P., Iniya, P. A., and Immanuel, G. 2016. Microbial mediated synthesis, characterization, antibacterial and synergistic effect of gold nanoparticles using *Klebsiella pneumoniae* (MTCC-4030). *RSC Adv* 6(6):4601–4607. doi:10.1039/c5ra23982f.

Priyadarshini, S., Gopinath, V., Priyadharsshini, N. M., MubarakAli, D., and Velusamy, P. 2013. Synthesis of anisotropic silver nanoparticles using novel strain, *Bacillus flexus* and its biomedical application. *Colloids Surf B Biointerfaces* 102:232–237. doi:10.1016/j.colsurfb.2012.08.018.

Rahimi, G., Alizadeh, F., and Khodavandi, A. 2016. Mycosynthesis of silver nanoparticles from *Candida albicans* and its antibacterial activity against *Escherichia coli* and *Staphylococcus aureus*. *Trop J Pharm Res* 15(2):371–375. doi:10.4314/tjpr.v15i2.21.

Ramakrishna, M., Babu, D. R., Gengan, R. M., Chandra, S., and Rao, G. N. 2016. Green synthesis of gold nanoparticles using marine algae and evaluation of their catalytic activity. *J Nanostruct Chem* 6(1):1–13. doi:10.1007/s40097-015-0173-y.

Ramanathan, R., Field, M. R., O'Mullane, A. P., Smooker, P. M., Bhargava, S. K., and Bansal, V. 2013. Aqueous phase synthesis of copper nanoparticles: A link between heavy metal resistance and nanoparticle synthesis ability in bacterial systems. *Nanoscale* 5(6):2300–2306. doi:10.1039/C2NR32887A.

Ramkumar, V. S., Pugazhendhi, A., Prakash, S., Ahila, N. K., Vinoj, G., Selvam, S., Kumar, G., Kannapiran, E., and Rajendran, R. B. 2017. Synthesis of platinum nanoparticles using seaweed *Padina gymnospora* and their catalytic activity as PVP/PtNPs nanocomposite towards biological applications. *Biomed Pharmacother* 92:479–490. doi:10.1016/j.biopha.2017.05.076.

Rautaray, D., Ahmad, A., and Sastry, M. 2004. Biological synthesis of metal carbonate minerals using fungi and actinomycetes. *J Mater Chem* 14(14):2333–2340. doi:10.1039/b401421f.

Raychoudhury, P., Ghosh, S., and Pal, R. 2016. Cyanobacteria mediated green synthesis of gold-silver nanoalloy. *J Plant Biochem Biotechnol* 25(1):73–78. doi:10.1007/s13562-015-0311-0.

Riddin, T. L., Gericke, M., and Whiteley, C. G. 2006. Analysis of the inter- and extracellular formation of platinum nanoparticles by *Fusarium oxysporum* f. sp. *lycopersici* using response surface methodology. *Nanotechnology* 17(14):3482–3489.

Roy, S., Das, T. K., Maiti, G. P., and Basu, U. 2016. Microbial biosynthesis of nontoxic gold nanoparticles. *Mater Sci Eng B* 203:41–51. doi:10.1016/j.mseb.2015.10.008.

Sadhasivam, S., Shanmugam, P., and Yun, K. 2010. Biosynthesis of silver nanoparticles by *Streptomyces hygroscopicus* and antimicrobial activity against medically important pathogenic microorganisms. *Colloids Surf B Biointerfaces* 81(1):358–362. doi:10.1016/j.colsurfb.2010.07.036.

Sadhasivam, S., Shanmugam, P., Veerapandian, M., Subbiah, R., and Yun, K. 2012. Biogenic synthesis of multidimensional gold nanoparticles assisted by *Streptomyces hygroscopicus* and its electrochemical and antibacterial properties. *BioMetals* 25(2):351–360. doi:10.1007/s10534-011-9506-6.

Salunke, B. K., Sawant, S. S., Lee, S.-I., and Kim, B. S. 2015. Comparative study of MnO_2 nanoparticles synthesis by marine bacterium *Saccharophagus degradans* and yeast *Saccharomyces cerevisiae*. *Appl Microbiol Biotechnol* 99(13):5419–5427. doi:10.1007/s00253-015-6559-4.

Sanchez-Harvas, M., Gil, J. V., Bisbal, F., Ramon, D., and Martinez-Culebras, P. V. 2008. Mycobiota and mycotoxin producing fungi from cocoa beans. *Int J Food Microbiol* 125(3):336–340. doi:10.1016/j.ijfoodmicro.2008.04.021.

Sanghi, R., and Verma, P. 2009. A facile green extracellular biosynthesis of CdS nanoparticles by immobilized fungus. *Chem Eng J* 155(3):886–891. doi:10.1016/j.cej.2009.08.006.

Sarikaya, M. 1999. Biomimetics: Materials fabrication through biology. *PNAS* 96(25):14183–14185.

Scarano, G., and Morelli, E. 2003. Properties of phytochelatin-coated CdS nanocrystallites formed in a marine phytoplanktonic alga (*Phaeodactylum tricornutum*, Bohlin) in response to Cd. *Plant Sci* 165(4):803–810. doi:10.1016/S0168-9452(03)00274-7.

Schrofel, A., Kratosova, G., Safarik, I., Safarikova, M., Raska, I., and Shor, L. M. 2014. Applications of biosynthesized metallic nanoparticles – A review. *Acta Biomater* 10(10):4023–4042. doi:10.1016/j.actbio.2014.05.022.

Senapati, S., Syed, A., Moeez, S., Kumar, A., and Ahmad, A. 2012. Intracellular synthesis of gold nanoparticles using alga *Tetraselmis kochinensis*. *Mater Lett* 79:116–118. doi:10.1016/j.matlet.2012.04.009.

Seshadri, S., Prakash, A., and Kowshik, M. 2012. Biosynthesis of silver nanoparticles by marine bacterium, *Idiomarina* sp. PR58-8. *Bull Mater Sci* 35(7):1201–1205.

Seshadri, S., Saranya, K., and Kowshik, M. 2011. Green synthesis of lead sulfide nanoparticles by the lead resistant marine yeast, *Rhodosporidium diobovatum*. *Biotechnol Prog* 27(5):1464–1469. doi:10.1002/btpr.651.

Shahverdi, A. R., Minaeian, S., Shahverdi, H. R., Jamalifar, H., and Nohi, A.-A. 2007. Rapid synthesis of silver nanoparticles using culture supernatants of Enterobacteria: A novel biological approach. *Process Biochem* 42(5):919–923. doi:10.1016/j.procbio.2007.02.005.

Shaligram, N. S., Bule, M., Bhambure, R., Singhal, R. S., Singh, S. K., Szakacs, G., and Pandey, A. 2009. Biosynthesis of silver nanoparticles using aqueous extract from the compactin producing fungal strain. *Process Biochem* 44(8):939–943. doi:10.1016/j.procbio.2009.04.009.

Shamsuzzaman, Mashrai, A., Khanam, H., and Aljawfi, R. N. 2017. Biological synthesis of ZnO nanoparticles using *C. albicans* and studying their catalytic performance in the synthesis of steroidal pyrazolines. *Arab J Chem* 10:S1530–S1536. doi:10.1016/j.arabjc.2013.05.004.

Shankar, P. D., Shobana, S., Karuppusamy, I., Pugazhendhi, A., Ramkumar, V. S., Arvindnarayan, S., and Kumar, G. 2016. A review on the biosynthesis of metallic nanoparticles (gold and silver) using bio-components of microalgae: Formation mechanism and applications. *Enzyme Microb Technol* 95:28–44. doi:10.1016/j.enzmictec.2016.10.015.

Sheikhloo, Z., and Salouti, M. 2011. Intracellular biosynthesis of gold nanoparticles by the fungus *Penicillium chrysogenum*. *Int J Nanosci Nanotechnol* 7(2):102–105.

Sheikhloo, Z., and Salouti, M. 2012. Intracellular biosynthesis of gold nanoparticles by the fungus *Phoma macrostoma*. *Synth React Inorg Met-Org NANO-Met Chem* 42(1):65–67. doi:10.1080/15533174.2011.609230.

Shetty, P. R., Kumar, B. S., Kumar, Y. S., and Shankar, G. G. 2012. Characterization of silver nanoparticles synthesized by using marine isolate *Streptomyces albidoflavus*. *J Microbiol Biotechnol* 22(5):614–621. doi:10.4014/jmb.1107.07013.

Shi, C., Zhu, N., Cao, Y., and Wu, P. 2015. Biosynthesis of gold nanoparticles assisted by the intracellular protein extract of *Pycnoporus sanguineus* and its catalysis in degradation of 4-nitroaniline. *NANO Res Lett* 10:147. doi:10.1186/s11671-015-0856-9.

Shivashankarappa, A., and Sanjay, K. R. 2015. Study on biological synthesis of cadmium sulfide nanoparticles by *Bacillus licheniformis* and its antimicrobial properties against food borne pathogens. *Nanosci Nanotechnol Res* 3(1):6–15. doi:10.12691/nnr-3-1-2.

Shukla, R. K., Kumar, A., Gurbani, D., Pandey, A. K., Singh, S., and Dhawan, A. 2013. TiO2 nanoparticles induce oxidative DNA damage and apoptosis in human liver cells. *Nanotoxicology* 7(1):48–60. doi:10.3109/17435390.2011.629747.

Silva, L. P. C., Oliveira, J. P., Keijok, W. J., Silva, A. P. D., Aguiar, A. R., Guimaraes, M. C. C., Ferraz, C. M., Araujo, J. V., Tobias, F. L., and Braga, F. R. 2017. Extracellular biosynthesis of silver nanoparticles using the cell-free filtrate of nematophagous fungus *Duddingtonia flagrans*. *Int J Nanomed* 12:6373–6381. doi:10.2147/IJN.S137703.

Singh, R., Wagh, P., Wadhwani, S., Gaidhan, S., Kumbhar, A., Bellare, J., and Chopade, B. A. 2013. Synthesis, optimization, and characterization of silver nanoparticles from *Acinetobacter calcoaceticus* and their enhanced antibacterial activity when combined with antibiotics. *Int J Nanomed* 8:4277–4290. doi:10.2147/IJN.S48913.

Sinha, S. N., Paul, D., Halder, N., Sengupta, D., and Patra, S. K. 2015. Green synthesis of silver nanoparticles using fresh water green alga *Pithophora eodogonia* (Mont.) Wittrock and evaluation of their antibacterial activity. *Appl Nanosci* 5(6):703–709. doi:10.1007/s13204-014-0366-6.

Soenen, S. J., Manshian, B., Montenegro, J. M., Amin, F., Meermann, B., Thiron, T., Cornelissen, M., Vanhaecke, F., Doak, S., Parak, W. J., Smedt, S. D., and Braeckmans, K. 2012. Cytotoxic effect of gold nanoparticles: A multiparametric study. *ACS Nano* 6(7):5767–5783. doi:10.1021/nn301714n.

Sondi, I., and Salopek-Sondi, B. 2004. Silver nanoparticles as antimicrobial agent: A case study on *E. coli* as a model for

Gram-negative bacteria. *J Colloid Interface Sci* 275(1):177–182. doi:10.1016/j.jcis.2004.02.012.

Southam, G., and Beveridge, T. J. 1996. The occurrence of sulfur and phosphorus within bacterially derived crystalline and pseudocrystalline octahedral gold formed in vitro. *Geochim Cosmochim Acta* 60(22):4369–4376.

Spadaro, J. A., Berger, T. J., Barranco, S. D., Chapin, S. E., and Becker, R. O. 1974. Antibacterial effects of silver electrodes with weak direct current. *Antimicrob Agents Chemother* 6(5):637–642. doi:10.1128/AAC.6.5.637.

Suryavanshi, P., Pandit, R., Gade, A., Derita, M., Zachino, S., and Rai, M. 2017. *Collectrichum* sp.-mediated synthesis of sulphur and aluminium oxide nanoparticles and its in vitro activity against selected food-borne pathogens. *LWT – Food Sci Technol* 81:188–194. doi:10.1016/j.lwt.2017.03.038.

Sweeney, R. Y., Mao, C., Gao, X., Burt, J. L., Belcher, A. M., Georgiou, G., and Iverson, B. L. 2004. Bacterial biosynthesis of cadmium sulfide nanocrystals. *Chem Biol* 11(11):1553–1559. doi:10.1016/j.chembiol.2004.08.022.

Tang, S. S., Apisarnthanarak, A., and Hsu, L. Y. 2014. Mechanisms of β-lactam antimicrobial resistance and epidemiology of major community-and healthcare-associated multidrug-resistant bacteria. *Adv Drug Deliv Rev* 78:3–13. doi:10.1016/j.addr.2014.08.003.

Uddandrao, P., and Raj, M. B. 2016. ZnS semiconductor quantum dots production by an endophytic fungus *Aspergillus flavus*. *Mater Sci Eng B* 207:26–32. doi:10.1016/j.mseb.2016.01.013.

Usha, R., Prabu, E., Palaniswamy, M., Venil, C. K., and Rajendran, R. 2010. Synthesis of metal oxide nano particles by *Streptomyces* sp for development of antimicrobial textiles. *Glob J Biotechnol Biochem* 5(3):153–160.

Vala, A. K., Shah, S., and Patel, R. 2014. Biogenesis of silver nanoparticles by marine-derived fungus *Aspergillus flavus* from Bhavnagar Coast, Gulf of Khambhat, India. *J Mar Biol Oceanogr* 3:1. doi:10.4172/2324-8661.1000122.

Waghmare, S. R., Mulla, M. N., Marathe, S. R., and Sonawane, K. D. 2015. Ecofriendly production of silver nanoparticles using *Candida utilis* and its mechanistic action against pathogenic microorganisms. *3 Biotech* 5(1):33–38. doi:10.1007/s13205-014-0196-y.

Wang, T., Yang, L., Zhang, B., and Liu, J. 2010. Extracellular biosynthesis and transformation of selenium nanoparticles and application in H2O2 biosensor. *Colloids Surf B Biointerfaces* 80(1):94–102. doi:10.1016/j.colsurfb.2010.05.041.

Yan, Z.-Y., Du, Q.-Q., Qian, J., Wan, D.-Y., and Wu, S.-M. 2017. Eco-friendly intracellular biosynthesis of CdS quantum dots without changing *Escherichia coli*'s antibiotic resistance. *Enzyme Microb Technol* 96:96–102. doi:10.1016/j.encmictec.2016.09.017.

Yang, W., Shen, C., Ji, Q., An, H., Wang, J., Liu, Q., and Zhang, Z. 2009. Food storage material silver nanoparticles interfere with DNA replication fidelity and bind with DNA. *Nanotechnology* 20(8):085102. doi:10.1088/0957-4484/20/8/085102.

Zhang, H., Li, Q., Lu, Y., Sun, D., Lin, X., Deng, X., and Zheng, S. 2005. Biosorption and bioreduction of diamine silver complex by *Corynebacterium*. *J Chem Technol Biotechnol* 8:285–290. doi:10.1002/jctb.1191.

Zhang, W., and Majidi, V. 1994. Monitoring the cellular response of *Stichococcus bacillaris* to exposure of several different metals using in vivo 31P NMR and other spectroscopic techniques. *Environ Sci Technol* 28(9):1577–1581. doi:10.1021/es00058a007.

Zheng, D., Hu, C., Gan, T., Dang, X., and Hu, S. 2010. Preparation and application of a novel vanillin sensor based on biosynthesis of Au–Ag alloy nanoparticles. *Sens Actuators B* 148(1):247–252. doi:10.1016/j.snb.2010.04.031.

Zhu, X., Kumari, D., Huang, M., and Achal, V. 2016. Biosynthesis of CdS nanoparticles through microbial induced calcite precipitation. *Mater Des* 98:209–214. doi:10.1016/j.matdes.2016.03.008.

19

Nanotechnology in Molecular Targeting, Drug Delivery, and Immobilization of Enzyme(s)

Abhishek Sharma, Kishore Kumar, Tanvi Sharma, Shweta Sharma, and Shamsher S. Kanwar

CONTENTS

19.1 Introduction

Nanotechnology is primarily attributed to the combined use of science and technology in the design, synthesis, and characterization of devices and materials at the nanoscale level (Emerich et al., 2003; Sahoo and Labhasetwar, 2003; Cai et al., 2017; Sharma et al., 2017a). Nanotechnology is used in the additional fields of applied science such as tissue engineering, protein detection, drug delivery, robotics, DNA engineering, cancer treatment, gene therapy, and medicine (Hossain and Alfroz, 2012; Mishra et al., 2017; Sharma et al., 2017a). The nanomedicines and devices designed by using nanotechnology can interact at the molecular level and enhance the effectiveness of drugs with few side effects (Nikalje, 2015). This type of technology will enable us to repair and analyze the human body comprehensively and effectively, the same way as we can repair any conventional machine today. Nanoscience involves research and technology development at 1 nm to 100 nm scale. One nanometer is a billionth of a meter, *i.e.*, nano indicates 10^{-9}. Nanotechnology is an integrative field developed from disciplines such as applied physics, materials science, colloidal science, molecular chemistry, and mechanical as well as electrical engineering. Due to its consideration in colloidal science, with a combination of new generation analytical tools such as atomic force microscope (AFM), X ray diffraction (XRD), and scanning tunneling microscope (STM), the advancement of nanotechnology is arising in the modern era (Tiwari et al., 2011).

The nanoparticles associated with the field of nanotechnology have a long history. Nanoparticles were used in the 4th century for the manufacturing of the famous Lycurgus Cup made of dichroic glass in Rome, and in the 9th century to create special effects on the surface of pots in Mesopotamia (Freestone et al., 2007). Michael Faraday was the first scientist who gave a scientific explanation of the optical properties of nanometer-scale metals in his research paper (1857) and specified that: "It is well known that when thin leaves of gold or silver are mounted upon glass and heated to a temperature that is well below a red heat (~500 °C), a remarkable change of properties takes place, whereby the continuity of the metallic film is destroyed. The result is that white light is now freely transmitted, reflection is correspondingly diminished, while the electrical resistivity is enormously increased" (Faraday, 1857). The American physicist Richard Feynman described a process of manipulation of individual atoms and molecules which might be developed using tools. In this process, he faced the problem of scaling which arose on modifying the magnitude of various physical phenomena. The role of gravity became less important and van der Waals attraction became more pronounced. Japanese scientist Norio Taniguchi of Tokyo University of Science first used the term nanotechnology in 1974. In 1980, K. Eric Drexler conceptually explored the concept of nanotechnology in depth, and gave an explanation about the technological significance of nanoscale through speeches and his book entitled *Engines of creation: The coming era of nanotechnology* (Wilsdon 2004). Presently, nanomedicine is the most actively growing area of nanotechnology, in which new technologies and devices are developed at the nanoscale level for the improvement of diagnosis and therapeutics (Moghimi and Farhangrazi, 2013; Sharma et al., 2014; Xu et al., 2015). The design, characterization, production, and application of new structures, devices, and systems are possible due to the control and manipulation of atoms and molecules at the nanoscale (Jain et al., 2012). Nanomedicine shows a dissimilarity in the response to light, magnetic or electronic irritation, physicochemical features such as pH, and temperature sensitivity in comparison to conventional medicine. As a result, nanomedicine provides many advantages in the fields of targeted drug delivery and therapeutics, which triumph over the boundaries of molecular imaging and gene/drug delivery methods in the biological systems (Miele et al., 2012). NPs often protect drugs against degradation and enhance drug stability, prolong the circulation and target searching time, reduce side effects, and improve the distribution and metabolic process in tissues (Peer et al., 2007). NPs are composed of nanomolecules in which the nanomedicine is entrapped or encapsulated and made to be delivered to a particular organ or cell in the body where medicine is actually required.

19.2 Different Classes of NPs

NPs can be classified on the basis of their size ranges between 1 and 2500 nm (Buzea et al., 2007; Fahlman 2007).The most common classes of NPs exploited in nano-biotechnology for social welfare in the modern era have been highlighted (Figure 19.1).

19.2.1 Liposomes

Liposomes are small spherical artificial vesicles composed of lipid bilayers surrounding an aqueous inner phase. They are usually composed of phospholipids (say, lecithin) or cholesterol, which is usually employed to encapsulate various active drugs (Akbarzadeh et al., 2013). Today, the liposome is a very useful tool in many scientific disciplines and especially in nanotechnology for the delivery of drugs (Hofheinz et al., 2005). Due to hydrophilic, hydrophobic, and biocompatibility characteristics, the liposome becomes an ideal drug carrier. The properties of the liposome depend on its lipid composition, surface charge, and method of preparation (Himanshu et al., 2011). Liposomes are classified on the basis of their size and number of bilayers: unilamellar vesicles and small multi-lamellar vesicles (Amarnath and Sharma, 1997). The active compound or drug is delivered to the concerned cell or organ on the basis of the solubility of the drug if the drug is water soluble, then it may be entrapped in the aqueous space and, if it is lipid soluble, it may be bound to the surface lipid membrane. Some liposomes of a certain diameter range have faced the problem of immune system interference. This problem is conquered by "stealth liposomes", a new generation of liposomes which can avoid the interference of the immune system with a longer half-life (Moghimi and Szebeni 2003). Stealth liposomes are spherical vesicles covered with a phospholipid bilayer used for the delivery of drugs or genetic material into a cell with controlled release (Gabizon, 2001). Polymers can also be used for drug delivery by constructing the polymer protein conjugates which reduce the immunogenicity and increase the stability of the proteins (Agnihotri et al., 2004). These polymer–protein conjugates target the tumor(s) *via* enhancing permeability and retention effect, as a consequence allowing lysosomotropic drug delivery (Lee, 2006).

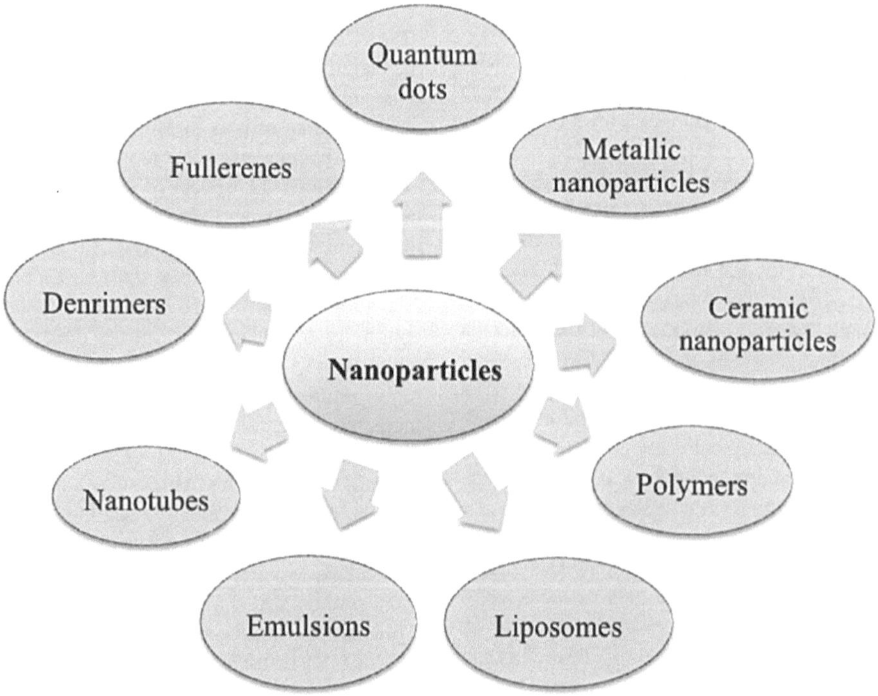

FIGURE 19.1 Various types of nanoparticles used in the field of nanotechnology.

19.2.2 Ceramic NPs

Ceramic NPs were discovered in the early 1980s. These NPs, formed by the sol-gel method, may be made up of oxides, carbides, phosphates, and carbonates of metals, or metalloids such as calcium, titanium, silicon, etc. (Thomas et al., 2015). These are inorganic units possessing porosity and are often considered as drug delivery carriers (Cherian et al., 2000). Ceramic NPs are biocompatible, *i.e.*, not very harmful to living tissues, and may be used in cancer treatment. The main drawback of these NPs is their non-biodegradability and accumulation in the body, which may produce undesirable effects.

19.2.3 Metallic NPs

Metallic NPs were first recognized by Faraday in 1857. Recently, they have been gaining great attention worldwide in biomedicine and engineering. Metallic particles include a class of supramagnetic materials that can be coated with phospholipids, dextran, etc. to obstruct aggregation. These NPs can be used as active or passive targeting agents (Gupta and Gupta, 2005). Gold NPs are the most ideal spherical particles composed of a dielectric core enclosed by a thin metallic shell. Metallic NPs have highly favorable optical and chemical properties of immense interest in medical and therapeutic applications (Hirsch et al., 2006).

19.2.4 Carbon Nanomaterials

These are allotropes of carbon with a polygonal structure, and include fullerenes and nanotubes. Fullerenes are composed of stacked graphene sheets of linked hexagonal rings. They have numerous points of attachment for tissue binding (Bosi et al., 2003). These nanomaterials are the most broadly used NPs because of their high electrical conductivity. Carbon nanotubes structurally consist of a single sheet of graphite rolled to form a cylinder which may be either single-walled or multi-walled. Multi-walled carbon nanotubes consist of single-walled nanotubes stacked one inside the other. Functional nanotubes are the ideal NPs for the delivery of nanomedicines (Pagona and Tagmatarchis, 2006).

19.2.5 Quantum Dots (QDs)

Quantum dots (QDs) are small nanocrystals of semiconductor compounds (*e.g.*, CdSe, CdTe, CdS, ZnSe, InP, and InS) which have excellent fluorescence properties. QDs were the first nanomaterial to be used in biological applications (Chan, 1998). QDs possess unique optical electronic and fluorescent properties. They are coated with other materials for their use in biological applications, so that the release of toxic heavy metals can be prevented (Weng and Ren, 2006).

19.3 Applications of Nanoparticles

Nanoparticles are used or are being evaluated for use in diverse fields (Figure 19.2) of enzyme immobilization, drug release, diagnostics, etc.

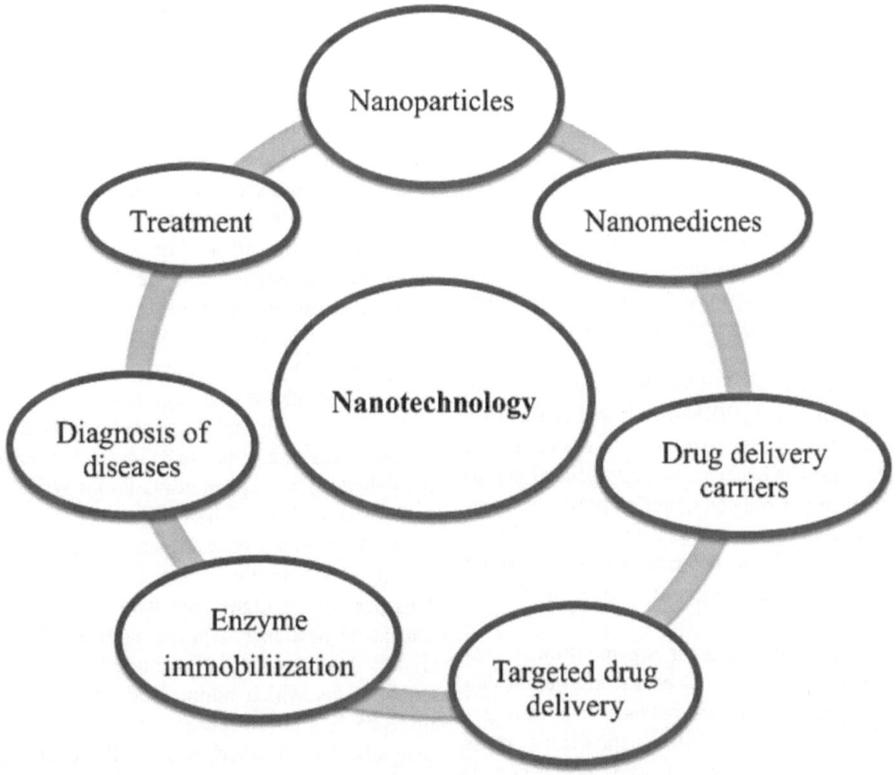

FIGURE 19.2 Most promising applications of nanotechnology for human welfare.

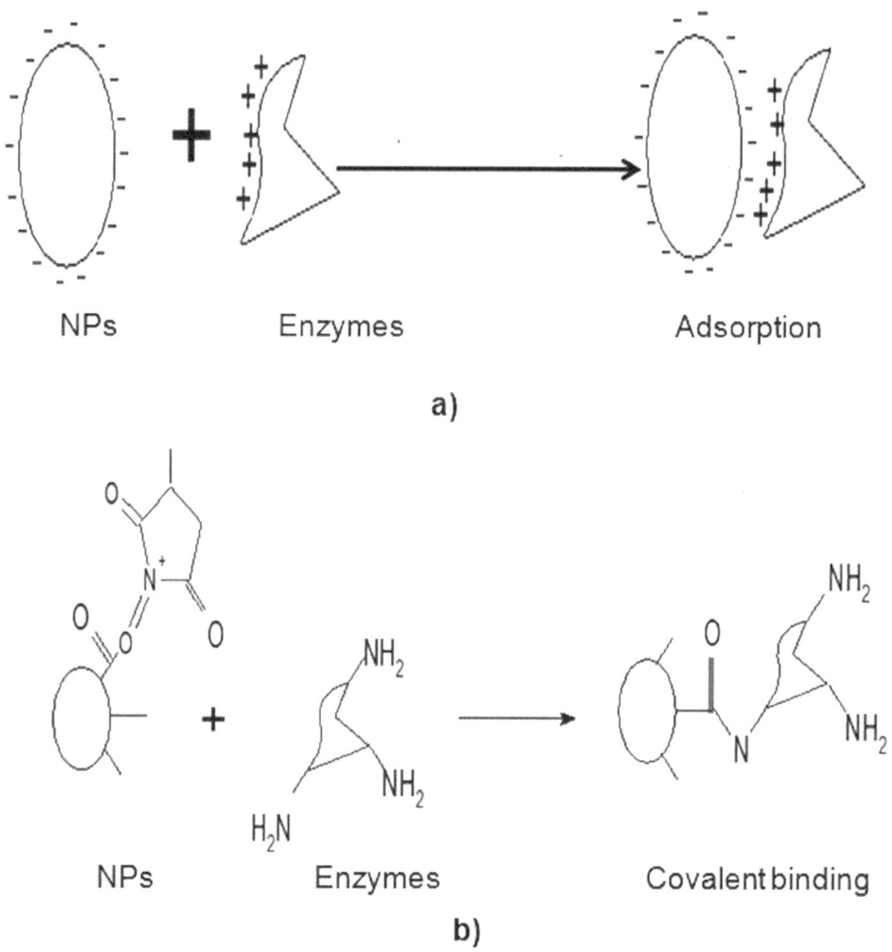

FIGURE 19.3 (a, b): Common methods of immobilization of NPs include: adsorption (a) and covalent binding (b).

19.3.1 NPs as Immobilization Matrices

The immobilization of enzyme(s) onto a variety of solid matrices remains attractive because it cuts down the expenditure of the process and often improves the robustness of the biocatalyst (Kumar et al., 2015). Enzymes, when used as biocatalysts, are superior to chemical catalysts because of their bio-catalytic nature, superior specificity, and placid reaction conditions (Sharma et al., 2016; Sood et al., 2016; Jamwal et al., 2017; Sharma et al., 2017a; Sharma et al., 2017b; Sharma et al., 2017c; Sharma et al., 2017d; Sharma et al., 2017e; Sharma et al., 2018a-b; Thakur et al., 2018; Sharma et al., 2019c). Immobilized enzymes are drawing a lot interest as marketable solutions because of their reduced operational expenses and well-organized consumption of the enzymes (Kumar et al., 2015; Sharma et al., 2017a; Sharma et al., 2017c; Sharma et al., 2019a, b). The main methods of immobilization of enzymes onto the NPs are adsorption and covalent binding (Figure 19.3).

NPs act as capable support materials for enzyme immobilization (Table 19.1) because of their perfect characteristics, which are involved in balancing the key factors that establish the effectiveness of biocatalysts, including specific surface area, mass transfer resistance, and effective enzyme loading (Feng and Ji, 2011; Gupta et al., 2011; Ansari and Husain, 2012; Verma et al., 2013a; Sharma et al., 2017a). Using NPs can solve the problem of diffusion while dealing with the macromolecular substrates. The enzyme/ biocatalyst immobilized on NPs dispersed in an aqueous medium shows Brownian movement and greater homogeneity, which often improve the enzymatic activity as compared to the unbound enzyme (Gupta et al., 2011). Magnetic NPs show the additional advantage that they can be separated easily from a reaction system by applying external magnetic fields. The immobilization of enzymes onto NPs improves the stability and performance of the enzyme, besides reduced protein folding (Ahmad and Sardar, 2015).

The NPs of Au and Ag have been used for the immobilization of lysozyme, glucose oxidase, aminopeptidase, and alcohol dehydrogenase (Lan et al., 2008; Wu et al., 2008; Keighron and Keating, 2010). The immobilization of the enzyme CALB lipase B on fumed silica NPs resulted in an increase in catalytic activity. The immobilization of acetylcholine esterase on nickel NPs provided a highly sensitive detection method for organophosphate pesticides (Ganesana et al., 2011). The immobilization of this enzyme also resulted in an improved reusability of the enzyme, which maintained specific enzymatic activity of up to 85 % after 10 reuses. The immobilization of cellulase on magnetic NPs resulted in a smaller decrease in the activity of the enzyme, but when kept at a temperature of 80 °Cit showed slightly greater activity than the free enzyme (Khoshnevisan et al., 2011). The magnetic NPs coated with silica were used for

TABLE 19.1

Examples of Enzymes Immobilized on Various Types of NPs

S. No.	Enzyme(s)	Matrix use for creating NPs	References
1.	α -Amylase	Silica NPs	Verma et al., 2013b
2.	Trypsin	Nano-diamond prepared by detonation	Wei et al., 2010
3.	Lipase	Fe_3O_4 and $MgFe_2O_4$	Huang et al., 2003 Sharma et al., 2017a Sharma et al., 2018b
4.	Glucose oxidase	Thiolated gold	Pandey et al., 2007
5.	Diastase	Silica-coated nickel	Namdeo and Bajpai, 2009
6.	α –Amylase	Silver	Mishra et al., 2013
7.	Lipase	Poly-dopamine-coated Fe_3O_4	Wang et al., 2017
8.	Trypsin	Chitosan magnetic	Sun et al., 2017
9.	Horseradish peroxidase	Fe_3O_4 magnetic nanoparticles	Mohamed et al., 2017
10.	Invertase	Magnetic diatomaceous earth	Cabrera et al., 2017
11.	Subtilisin	Polymer-coated mesoporous silica	Ozbek and Unal, 2017
12.	Cellulose	Zinc ferrite	Manasa et al., 2017
13	Esterase	Fe_3O_4	Sharma et al., 2018b

the immobilization of trypsin, and it was observed that, in each experiment in which the magnetic NPs were used for immobilization, an increased proteolytic activity was recorded compared to the use of free trypsin. This system of immobilization of trypsin was applied to a pressure-assisted digestion for proteome analysis (Lee et al., 2011). The immobilization of peroxidase, cellulose, trypsin, and α-amylase on TiO_2 nanoparticles was also reported (Ahmad et al., 2013a,b). The magnetic NPs (magnetite (Fe_3O_4)-chitosan (CS)-poly [N-benzyl-2-(methacryloxy)-N and N-dimethyl ethanaminium bromide] (PQ) nanoparticles (Fe_3O_4-CS-PQ) were used for the immobilization of lipase (Siodmiak et al., 2013) (Figure 3b)).

19.3.2 Role of NPs in Medicine

NPs can be used in the medicine field as they can target the specific cells of an organ (Sharma et al., 2014). Nanoscience contributes to the management of various diseases or disease-based pathogens, and it also counters multiple drug resistances in leukemia by blocking the drug outflow from the cancer cells as well as blocking apoptosis in sepsis (Prabhu et al., 2011). NP-based thrombocytic agents might be more indicated for use in clot exclusion, are also used in nano-dentistry, for treatments such as dentition renaturalization, alleviation of permanent hypersensitivity, etc. (Buxton, 2009). The rapid emergence of multidrug-resistant strains of pathogens has generated an urgent search for new drugs or antimicrobial agents (Meena et al., 2016; Meena et al., 2017; Meena et al., 2018; Jyoti et al., 2018;). Medical communities are now trying to resolve these problems through the use of NPs as nanomedicines. Silver NP-based wound dressing materials have proved successful in the market thanks to their positive wound healing effects. Silver NPs are water soluble and diffuse into the body after delivering the drug. Thus, nano silver has shown a strong antibacterial activity in wound and burn healing (Freitas, 2000). Nanocrystalline silver has the distinctive characteristic of inhibiting antibiotic resistance and antiseptic resistance microbes. Dendrimers are an innovative polymer with distinct construction, high molecular homogeneity, and low polydispersity, which makes them ideal to produce improvements in nanomedicine (Lyczak and Schechter, 2005). Dendrimers are used for the transport of drug(s) across the cellular barriers. Mesoporous silica particles (MSPs) and layered double hydroxides (LDHs) are used for efficient drug delivery (Samad et al., 2009; Pasqua et al., 2009). The polymeric NPs employing polylactide co-glycolide display effective anti-tubercular activity (Pandey and Khullar, 2006).

19.3.3 Nanoparticles in Drug Delivery Systems

The improvements in the field of nanotechnology have made achievable the delivery of a particular drug to specific cell(s) in a suitable amount for the treatment. The side effects of the drugs can be lowered significantly by selectively depositing the active agents to the target sites. Therefore, targeted drug delivery reduces the drug consumption and the treatment expenses, too. The delivery of nanomedicine can be achieved by modifying NPs in accordance with the target sites using nanoengineered devices. NP-based drug delivery systems (liposomes, metallic NPs, micelles, polymers, etc.) have a valuable function in drug delivery with reduced side effects (Sinha et al., 2006; Cho et al., 2008). Recent developments have shown the beneficial aspects of NPs for drug delivery, sensing of anti-cancer agents, small interference RNA (si RNA) delivery, etc. (Jiang et al., 2010; Castanotto and Rossi, 2009). Nanotechnology promises construction of artificial cells, enzymes and genes. This may work as a replacement therapy for many disorders which are due to deficiency of enzymes, mutation of genes or altered synthesis of proteins. Currently, nanodevices like respirocytes, microbivores and probes encapsulated by biologically localized embedding have greater applications in treatment of anemia and infections. (Sandhiya et al., 2009). Drug release investigation is a major issue for the successful evaluation and screening of nanomedicines. Researchers often employ chromatographic techniques for the estimation of the content(s) of drugs in cells or animals to investigate the effect of released drugs. The chromatographic technique method used for the investigation of drug release faces some problems. First, the test subject must be subjected to pretreatment, hence it is difficult to find real-time drug release information in living systems. The second problem is that the chromatographic method cannot differentiate between the drug-free and the nanocarrier-encapsulated drug. The recently developed QD-based fluorescence resonance energy technique (FRET) method resolves these problems. The QD donors and chromophore acceptors are in close vicinity so that their emission spectrum and absorption spectrum overlap with each other. The FRET efficiency is closely related to the distance between QDs and acceptors; a closer distance will lead to a stronger FRET. Therefore, the distance regulating the FRET technique allows the successful application of QDs in immuno-analysis and fabrication of nanosensors (Gill et al., 2008). Nowadays QDs and the FRET technique have made great progress on the intercellular drug release systems of many nanomedicines.

19.3.4 NPs-Targeting Tumor Sites

NP targeted drug deliver may be potentially used to deliver anti-cancer agents to the tumor site, which provides considerable benefits to cancer patients (Byrne et al., 2008; Sharma et al., 2014). The major advantages of nanomedicine in cancer treatment are that the nanostructures exploit the physiological aspects of tumors and the surrounding tissue inflammation response. Lyp-1 NPs (PEG-PLGA) is a target-specific drug delivery to lymphatic metastasis tumors (Ruoslaht 2017). Silver NPs can act as anti-angiogenic molecules by targeting the activation of the P13K/AKT signaling pathway (Gurunathan et al., 2009). The NP-mediated targeting of phosphatidylinositol-3-kinase signaling inhibits the process of angiogenesis (Harfouche et al., 2009). NPs enable the targeting of the p13K pathway, which resulted in the inhibition of endothelial cell proliferation and tumor angiogenesis (Harfouche *et al.,* 2009). Canine parvo virus (CPV) has a natural affinity for transferrin receptors (TIRs) of canine and human origin, and this property could be harnessed and overexpressed by a variety of human tumor cells (Singh, 2009). Labeling nanocrystals with immune cells acts as a platform technology for nanoimmunotherapy (Conniot et al., 2014). The combination of plasmid DNA encoding a multimeric soluble form of CD40L (Psp-d-CD40L) reduced tumor growth, which could be established through B16F-10 melanoma tumor (Stone et al., 2009). The combination of toll-like receptor (TLR) agonists, C-phosphate-G- (CpG), and poly (I:C) reduces the tumor growth and increases the rate of survival (Kaczanowska et al., 2013). It is also associated with a reduction in intra-tumoral CD11c+ dendritic cells and an influx of CD8 T cells.

Conclusion

The application of nanotechnology for the invention of new molecules and to manipulate the materials available in nature at the molecular level could lead to exciting experimental work to improve healthcare. The development in the field of nanotechnology results in advancements in the fields of genomics and proteomics, too. The current state of the art of NP-based therapeutics that are highly stable and have greater accuracy in targeted drug delivery as well as enhanced capacity, incorporation, etc. results in a reduction in drug doses and treatment expenses. Multi-functional NPs have the ability to carry active therapeutic agents, imaging contrast agent(s), diamonds, and targeting moieties, which can be used as anti-cancer agents. Nanotechnology is used by the pharmaceutical industry to reduce toxicity and improve the quality of the drugs, which obviously results in a better effectiveness of the drugs. The targeted drug delivery of NPs created the possibility of delivery of costly drugs with little side effects. Extra efforts are needed in the research about NP-based drug delivery systems for the effective treatment of cancer patients. Although the full potential of nanotechnology may be a couple of years or decades away, we are quite sure that nanotechnology shall create a major revolution in medicine.

Acknowledgments

The financial support in the form of DST-Junior Research Fellowship [DST/INSPIRE Fellowship/2013/1036; IF 131077] to one of the authors (AS) by the Department of Science and Technology, New Delhi (India) is thankfully acknowledged.

Conflict of Interest

The authors declare no conflict of interest in publishing this article in this journal.

REFERENCES

Agnihotri SA, Mallikarjuna N, Aminabhavi TM (2004). Recent advances on chitosan-based micro- and nanoparticles in drug delivery. *Journal of Controlled Release.* 100(1), 5–28.

Ahmad R, Khatoon N, Sardar M (2013a). Biosynthesis, characterization and application of TiO_2 nanoparticles in biocatalysis and protein folding. *Journal of Proteins and Proteomics.* 4, 115–121.

Ahmad R, Mishra A, Sardar M (2013b). Peroxidase-TiO_2 nanobioconjugates for the removal of phenols and dyes from aqueous solutions. 5, 1020–1025.

Ahmad R, Mishra A, Sardar M (2014). Simultaneous immobilization and refolding of heat treated enzymes on TiO_2 nanoparticles. Advanced Science, Engineering and Medicine. 6(12), 1264–1268.

Ahmad R, Sardar M (2015). Enzyme immobilization: An overview on nanoparticles as immobilization matrix. *Biochemistry and Analytical Biochemistry.* 4, 178. doi:10.4172/2161-1009.1000178.

Akbarzadeh A, Rezaei-Sadabady R, Davaran S, Joo SW, Zarghami N, Hanifehpour Y, Samiei M, Kouhi M, Nejati-Koshki K (2013). Liposome: Classification preparation and applications. *Nanoscale Research Letters.* 8(1), 102. doi:10.1186/1556-276X-8-102.

Amarnath S, Sharma US (1997). Liposomes in drug delivery: Progress and limitations. *International Journal of Pharmaceutics.* 154(2), 123–140. doi:10.1016/S0378-5173(97)00135-X.

Ansari SA, Husain Q (2012). Potential applications of enzymes immobilized on/in nano materials: A review. *Biotechnology Advances.* 30(3), 512–523.

Bosi S, Da Ros T, Spalluto G, Prato M (2003). Fullerene derivatives: An attractive tool for biological applications. *European Journal of Medicinal Chemistry.* 38(11–12), 913–923. doi:10.1016/j.ejmech.2003.09.005.

Buxton DB (2009). Nanomedicine for the management of lung and blood diseases. *Nanomedicine.* 4(3), 331–339. doi:10.2217/nnm.09.8.

Buzea C, Pacheco II, Robbie K (2007). Nanomaterials and nanoparticles: Sources and toxicity. *Biointerphases.* 2(4), 17–71.

Byrne JD, Betancourt T, Brannon-Peppas L (2008). Active targeting schemes for nanoparticle systems in cancer therapeutics. *Advanced Drug Delivery Reviews.* 60(15), 1615–1626. doi:10.1016/j.addr.2008.08.005.

Cabrera MP, Asis CRD, Neri DFM, Pereira CF, Soria F, Jr, Carvalho LB (2017). High sucrolytic activity by invertase immobilized onto magnetic diatomaceous earth nanoparticles LBC. *Biotechnology Reports.* 14, 38–46. doi:10.1016/j.btre.2017.03.001.

Cai W, Jin M, Zhao Z, Lei Z, Zhang Z, Adachi Y, Lee DJ (2017). Influence of ferrous iron dosing strategy on aerobic granulation of activated sludge and bioavailability of phosphorus accumulated in granules. *Biteb.* doi:10.1016/j.biteb.2018.03.004.

Castanotto D, Rossi JJ (2009). The promises and pitfalls of RNA-interference-based therapeutics. *Nature.* 457(7228), 426–433.

Chan WC, Nie S (1998). Quantum dot bioconjugates for ultrasensitive nonisotopic detection. *Science.* 281(5385), 1–8.

Cherian AK, Rana AC, Jain SK (2000). Self-assembled carbohydrate stabilized ceramic nanoparticles for the parenteral delivery of insulin. *Drug Development and Industrial Pharmacy.* 26(4), 459–463. doi:10.1081/DDC-100101255.

Cho KJ, Wang X, Nie SM, Shin DH (2008). Therapeutic nanoparticles for drug delivery in cancer. *Clinical Cancer Research.* 14(5), 1310–1316. doi:10.1158/1078-0432.CCR-07-1441.

Conniot J, Silva JM, Fernandes JG, Silva LC, Gaspar R, Brocchini S, Florindo HF, Barata TS (2014). Cancer immunotherapy: Nanodelivery approaches for immune cell targeting and tracking. *Frontiers in Chemistry.* 2, 105. doi:10.3389/fchem.2014.00105.

Emerich DF, Thanos CG (2003). Nanotechnology and medicine. *Expert Opinion on Biological Therapy.* 3(4), 655–663.

Fahlman BD (2007). *Materials Chemistry.* Springer, Berlin, 282–283. doi:10.1007/978-1-4020-6120-2.

Faraday M (1857). Experimental relations of gold (and other metals) to light. *Philosophical Transaction of the Royal Society of London.* 147, 145–181. doi:10.1098/rstl.1857.0011.

Feng W, Ji P (2011). Enzymes immobilized on carbon nanotubes. *Biotechnology Advances.* 29(6), 889–895.

Freestone I, Meeks N, Sax M, Higgitt C (2007). The Lycurgus cup – A roman nanotechnology. *Gold Bulletin.* 40(4), 270–277.

Freitas RA (2000). Nanodentistry. *Journal of American Dental Association.* 131(11), 1559–1565.

Ganesana M, Istarnboulie G, Marty JL, Noguer T, Andreescu S (2011). Site-specific immobilization of a (His) 6-tagged acetylcholinesterase on nickel nanoparticles for highly sensitive toxicity biosensors. *Biosensors and Bioelectronics.* 30(1), 43–48.

Gabizon AA (2001). Stealth liposomes and tumor targeting: One step further in the quest for the magic bullet. *Clinical Cancer Research.* 7(2), 223–227.

Gill R, Zayats M, Willner I (2008). Semiconductor quantum dots for bioanalysis. *Angewandtechemie International Edition.* 47(40), 7602–7625.

Gupta AK, Gupta M (2005). Synthesis and surface engineering of iron oxide nanoparticles for biomedical applications. *Biomaterials.* 26(18), 3995–4021.

Gupta MN, Kaloti M, Kapoor M, Solanki K (2011). Nanomaterials as matrices for enzyme immobilization. *Artificial Cells, Blood Substitutes and Biotechnology.* 39(2), 98–109.

Gurunathan S, Lee KJ, Kalishwaralal K, Sheikpranbabu S, Vaidyanathan R, Eom SH (2009). Antiangiogenic properties of silver nanoparticles. *Biomaterials.* 30(31), 6341–6350. doi:10.1016/j.biomaterials.2009.08.008.

Harfouche R, Basu S, Soni S, Hentschel DM, Mashelkar RA, Sengupta S (2009). Nanoparticles mediated targeting of phosphatidylinositol-3-kinase signaling inhibits angiogenesis. *Angiogenesis.* 12(4), 325–338. doi:10.1007/s10456-009-9154-4.

Himanshu A, Sitasharan P, Singhai AK. 2011. Liposomes as drug carriers. *International Journal of Pharmacy and Life Sciences.* 2, 945–951.

Hirsch LR, Gobin AM, Lowery AR, Tam F, Drezek RA, Halas NJ, West JL (2006). Metal nanoshells. *Annals of Biomedical Engineering.* 34(1), 15–22. doi:10.1007/s10439-005-9001-8.

Hofheinz RD, Gnad-Vogt SU, Beyer U, Hochhaus A (2005). Liposomal encapsulated anti-cancer drugs. *Anti-Cancer Drugs.* 16(7), 691–707. doi:10.1097/01.cad.0000167902.53039.5a.

Hossain Z, Afroz H (2012). Prospects and applications of nanobiotechnology: A medical perspective. *Journal of Nanobiotechnology.* 10, 1–8.

Huang SH, Liao MH, Chen DH (2003). Direct binding and characterization of lipase onto magnetic nanoparticles. *Biotechnology Progress.* 19(3), 1095–1100.

Jain S, Hirst DG, O'Sullivan JM (2012). Gold nanoparticles as novel agents for cancer therapy. *Brazilian Journal of Radiology.* 85(1010), 101–113.

Jamwal S, Kumar R, Sharma A, Kanwar SS (2017). Response surface methodology (RSM) approach for improved extracellular RNase production *by a Bacillus* sp. *Journal of Advances in Microbiology.* 3, 131–144.

Jiang S, Ganesammandhan MK, Zhang Y (2010). Optical imaging guided cancer therapy with fluorescent nanoparticles. *Journal of the Royal Society Interface.* 7(42), 3–18. doi:10.1098/rsif.2009.0243.

Jyoti P, Meena KR, Sharma A, Kapoor R, Sharma T, Kanwar SS (2018). Anti-microbial and anti-oxidant properties of *Cunninghamialanceolata. Insight Pharmaceutical Sciences.* 8, 21–27.

Kaczanowska S, Joseph AM, Davila E (2013). TLR agonists: Our best frenemy in cancer immunotherapy. *Journal of Leukocytic Biology.* 93(6), 847–863. doi:10.1189/jlb.1012501.

Keighron JD, Keating CD (2010). Enzyme: Nanoparticle bioconjugates with two sequential enzymes: Stoichiometry and activity of malate dehydrogenase and citrate synthase on Au nanoparticles. *Langmuir.* 26(24), 18992–19000.

Khoshnevisan K, Bordbar A-K, Zare D, Davoodi D, Noruzi M (2011). Immobilization of cellulase enzyme on superparamagnetic nanoparticles and determination of its activity and stability. *Chemical Engineering Journal.* 171(2), 669–673.

Kumar A, Sharma A, Kanwar SS (2014). Nanobiotechnology: Role in biocatalysis and bioprocesses. In: BS. Bhoop (Ed.), *Nanomedicines.*

Kumar A, Zhang S, Wu G, Wu CC, Chen JP, Baskaran R, Liu Z (2015). Cellulose binding domain assisted immobilization of lipase (GSlip–CBD) onto cellulosic nanogel: Characterization and application in organic medium. *Colloids and Surfaces, Part B: Biointerfaces.* 136, 1042–1050.

Lan D, Li B, Zhang Z (2008). Chemiluminescence flow biosensor for glucose based on gold nanoparticle-enhanced activities of glucose oxidase and horseradish peroxidase. *Biosensors and Bioelectronics.* 24(4), 940–944.

Lee B, Lopez-Ferrer D, Kim BC, Na HB, Park YI, et al. (2011). Rapid and efficient protein digestion using trypsin-coated magnetic nanoparticles under pressure cycles. *Proteomics.* 11(2), 309–318.

Lee LJ (2006). Polymer nano-engineering for biomedical applications. *Annals of Biomedical Engineering.* 34(1), 75–88. doi:10.1007/s10439-005-9011-6.

Lyczak JB, Schechter PJ. Nanocrystalline silver inhibits antibiotic, antiseptic-resistant bacteria (2005). *Clinical Pharmacology and Therapeutics.* 77(2), 60. doi:10.1016/j.clpt.2004.12.119.

Manasa P, Saroj P, Korrapati N (2017). Immobilization of cellulase enzyme on zinc ferrite nanoparticles in increasing enzymatic hydrolysis on ultrasound-assisted alkaline pretreated *Crotalaria* Juncea biomass. *Indian Journal of Science and Technology*. 10(24), 1–7. doi:10.17485/ijst/2017/v10i24/112798.

Meena KR, Dhiman R, Sharma A, Kanwar SS (2016). Applications of lipopeptide(s) from a bacillus sp: An overview. *Research Journal of Recent Sciences*. 5, 50–54.

Meena KR, Sharma A, Kanwar SS (2017). Microbial lipopeptides and their medical Applications. *Annals of Pharmacology and Pharmaceutics*. 2, 1–5.

Meena KR, Tandon T, Sharma A, Kanwar SS (2018). Lipopeptide antibiotic production by *Bacillus velezensis* KLP2016. *Journal of Applied Pharmaceutical Science*. 8, 91–98.

Miele E, Spinelli GP, Miele E, Di Fabrizio E, Ferretti E, Tomao S, Gulino A (2012). Nanoparticle-based delivery of small interfering RNA: Challenges for cancer therapy. *International Journal of Nanomedicine*. 7, 3637–3657.

Mishra A, Ahmad R, Singh V, Gupta MN, Sardar M (2013). Preparation, characterization and biocatalytic activity of a nanoconjugate of alpha amylase and silver nanoparticles. *Journal of Nanoscience and Nanotechnology*. 13(7), 5028–5033.

Mishra P, Singh L, Islam MA, Nasrullah M, Sakinah AM, Wahid ZA (2017). NiO and CoO nanoparticles mediated biological hydrogen production: Effect of Ni/Co oxide NPs-ratio. *Biteb*. doi:10.1016/j.biteb.2018.02.004.

Moghimi SM, Farhangrazi ZS (2013). Nanomedicine and the complement paradigm. *Nanomedicine: Nanotechnology, Biology and Medicine*. 9(4), 458–460.

Moghimi SM, Szebeni J (2003). Stealth liposomes and long circulating nanoparticles: Critical issues in pharmacokinetics, opsonization and protein-binding properties. *Progress in Lipid Research*. 42(6), 463–478. doi:10.1016/S0163-7827(03)00033-X.

Mohamed SA, Al-Harbi MH, Almulaiky YQ, Ibrahim IH, El-Shishtawy RM (2017). Immobilization of horseradish peroxidase on Fe_3O_4 magnetic nanoparticles. *Electronic Journal of Biotechnology*. 27, 84–90.

Namdeo M, Bajpai SK (2009). Immobilization of a-amylase onto cellulose-coated magnetite (CCM) nanoparticles and preliminary starch degradation study. *Journal of Molecular Catalysis B: Enzymatic*. 59(1–3), 134–139.

Nikalje AP (2015). Nanotechnology and its applications in medicine. *Medicinal Chemistry*. 5(2), 081–089.

Ozbek B, Unal S (2017). Preparation and characterization of polymer-coated mesoporous silica nanoparticles and their application in subtilisin immobilization. *Korean Journal of Chemical Engineering*. 34(7), 1992–2001.

Pagona G, Tagmatarchis N (2006). Carbon nanotubes: Materials for medicinal chemistry and biotechnological applications. *European Journal of Medicinal Chemistry*. 13(15), 1789–1798. doi:10.2174/092986706777452524.

Pandey P, Singh SP, Arya SK, Gupta V, Datta M, Singh S, Malhotra BD (2007). Application of thiolated gold nanoparticles for the enhancement of glucose oxidase activity. *Langmuir*. 23(6), 3333–3337.

Pandey R, Khullar GK (2006). Nanotechnology based drug delivery system for the treatment of tuberculosis. *Indian Journal of Experimental Biology*. 44(5), 357–366.

Pasqua L, Cundari S, Ceresa C, Cavaletti G (2009). Recent development, applications, and perspectives of mesoporous silica particles in medicine and biotechnology. *Current Medical Chemistry*. 16(23), 3054–3063. doi:10.2174/092986709788803079.

Peer D, Karp JM, Hong S, Farokhzad OC, Margalit R, Langer R (2007). Nanocarriers as an emerging platform for cancer therapy. *Nature Nanotechnology*. 2(12), 751–760.

Prabhu V, Uzzaman S, Viswanathan Guruvayoorappan C (2011). Nanoparticles in drug delivery and cancer therapy: The giant rats tail. *Journal of Cancer Therapy*. 2, 1–10.

Ruoslaht E (2017). Peptides penetrating for improved drug delivery. *Advanced Drug Delivery Reviews*. 110–111, 3–12. doi:10.1016/j.addr.2016.03.008.

Sahoo SK, Labhasetwar V (2003). Nanotech approaches to drug delivery and imaging. *Drug Discovery Today*. 8(24), 1112–1120.

Samad A, Alam MI, Saxena K (2009). Dendrimers: A class of polymers in the nanotechnology for the delivery of active pharmaceuticals. *Current Pharmaceutical Design*. 15(25), 2958–2969. doi:10.2174/138161209789058200.

Sandhiya S, Dkhar SA, Surendiran A (2009). Emerging trends of nanomedicines-an overview. *Fundamental and Clinical Pharmacology*. 23(3), 263–269. doi:10.1111/j.1472-8206.2009.00692.

Sharma A, Kumar A, Kanwar SS (2014). Nanobiotechnology role in human health and medicine. In: BS Bhoop (Ed.) *Nanomedicines*. 1, 231–251.

Sharma A, Kumar A, Meena K, Rana S, Singh M, Kanwar SS (2017a). Fabrication and functionalization of magnesium nanoparticle for lipase immobilization in n-propyl gallate synthesis. *Journal of King Saud University – Science*. 29(4), 536–546.

Sharma A, Meena KR, Kanwar SS (2017b). Molecular characterization and bioinformatics studies of a lipase from *Bacillus thermoamylovorans* BHK67. *International Journal of Biological Macromolecules*. 107(B), 2131–2140.

Sharma A, Meena KR, Kanwar SS (2018a). Microbial thermo-tolerant and solvent-tolerant lipases and their applications. *Current Research in Microbiology*. 2, 1–15.

Sharma A, Sharma T, Gupta R, Kanwar SS (2017d). Correlation of enzyme activity and stabilities with extreme environments: Complexities and flexibilities in adaptations. In: Vandana Rai (Ed.), *Recent Advances in Biotechnology*. Shree Publishers, New Delhi. 1, 17–31.

Sharma A, Sharma T, Meena K, Kanwar SS (2017c). Physical adsorption of lipase onto mesoporous silica. *International Journal of Current Advanced Research*. 6(5), 3837–3841. doi:10.24327/ijcar.2017.3841.0378.

Sharma A, Sharma T, Meena KR, Kumar A, Kanwar SS (2018b). High throughput synthesis of ethyl pyruvate by employing superparamagnetic iron nanoparticles-bound esterase. *Process Biochemistry*. doi:10.1016/j.procbio.2018.05.004.

Sharma A, Sharma T, Meena KR, Kumar R, Kanwar SS (2019a). Biodiesel and the potential role of microbial lipases in its production. In: PK Arora (Ed.), *Microbial Technology for the Welfare of Society*. Springer, 83–99.

Sharma A, Sharma T, Sharma T, Sharma S, Kanwar SS (2019b). Role of microbial hydrolases in bioremediation. In: A. Kumar, S. Sharma (Eds.), *Microbes and Enzymes in Soil Health and Bioremediation, Microorganisms for Sustainability*. Springer, 149–164.

Sharma T, Sharma A, Kanwar SS (2016). Purification and characterization of an extracellular mass esterase from *Bacillus pumilus*.

Journal of Advance Biotechnology and Bioengineering. 4, 9–16.

Sharma T, Sharma A, Sharma A, Kanwar SS (2017e). An overview on esterases: Structure, classification, sources and their application. In: Vandana Rai (Ed.), *Recent Advances in Biotechnology.* Shree Publishers, New Delhi. 2, 216–229.

Sharma T, Sharma S, Kamyab H, Kumar A (2019c). Energizing the CO utilization by Chemo-enzymatic approaches and potentiality of carbonic 2 anhydrases: A review. *Journal of Cleaner Production.* doi:10.1016/j.jclepro. 2019.119138.

Singh P (2009). Tumour targeting using canine parvovirus nanoparticles. *Current Topics in Microbiology and Immunology.* 327, 123–141.

Sinha R, Kim GJ, Nie S, Shin DM (2006). Nanotechnology in cancer therapeutics: Bioconjugated nanoparticles for drug delivery. *Molecular Cancer Therapeutics.* 5(8), 1909–1917. doi:10.1158/1535-7163.MCT-06-0141.

Siodmiak T, Borowska MZ, Marszałł MP (2013). Lipase-immobilized magnetic chitosan nanoparticles for kinetic resolution of (R, S)-ibuprofen. *Journal of Molecular Catalysis B: Enzymatic.* 94, 7–14.

Sood S, Sharma A, Sharma N, Kanwar SS (2016). Carboxylesterases: Sources, characterization and broader applications. *Insight Enzymatic Research.* 1, 1–11.

Stone GW, Barzee S, Snarsky V, Santucci C, Tran B, Langer R, Zugates GT, Anderson DG, Kornbluth RS (2009). Nanoparticle delivered multimeric soluble CD40 LDNA combined with like receptor agonists as a treatment for melanoma. *PLOS ONE.* 4, 7334. doi:10.1371/journal.pone.0007334.

Sun J, Xu B, Shi Y, Yang L, Ma HL (2017). Activity and stability of trypsin immobilized onto chitosan magnetic nanoparticles. *Advances in Materials Science and Engineering.* 2017, 1–10. doi:10.1155/2017/1457072.

Thakur N, Kumar A, Sharma A, Bhalla TC, Kumar D (2018). Purification and characterization of alkaline, thermostable and organic solvent stable protease from a mutant of *Bacillus* sp. *Biocatalysis and Agricultural Biotechnology.* 16, 217–224.

Thomas SC, Harshita Mishra PK, Talegoanker S (2015). Ceramic nanoparticles: Fabrication methods and applications in drug delivery. *Current Pharmaceutical Design.* 21(42), 6165–6188.

Tiwari B, Prasad H, Singh P, Sailesh SK (2011). Synthesis, characterization and thin film deposition of gold nanoparticles. *International Journal of Synthesis and Characterization.* 4, 23–26.

Verma ML, Barrow CJ, Puri M (2013a). Nanobiotechnology as a novel paradigm for enzyme immobilisation and stabilisation with potential applications in biodiesel production. *Applied Microbiology and Biotechnology.* 97(1), 23–39.

Verma ML, Chaudhary R, Tsuzuki T, Barrow CJ, Puri M (2013b). Immobilization of ß-glucosidase on a magnetic nanoparticle improves thermostability: Application in cellobiose hydrolysis. *Bioresource Technology.* 135, 2–6.

Wang C, Han H, Jiang W, Ding X, Li Q, Wang Y (2017). Immobilization of thermostable lipase QLM on core-shell structured polydopamine-coated Fe_3O_4 nanoparticles. *Catalysts.* 7(12), 49. doi:10.3390/catal7020049.

Wei L, Zhang W, Lu H, Yang P (2010). Immobilization of enzyme on detonation nanodiamond for highly efficient proteolysis. *Talanta.* 80(3), 1298–1304. doi:10.1016/j.talanta.2009.09.029.

Wilsdon J (2004). The politics of small things: Nanotechnology, risk, and uncertainty. *IEEE Technology and Society Magazine.* 23(4), 16–21.

Wu CL, Chen YP, Yang JC, Lo HF, Lin LL (2008). Characterization of lysine-tagged Bacillus stearothermophilus leucine aminopeptidase II immobilized onto carboxylated gold nanoparticles. *Journal of Molecular Catalysis B: Enzymatic.* 54(3–4), 83–89.

Xu X, Ho W, Zhang X, Bertrand N, Farokhzad O (2015). Cancer nanomedicine: From targeted delivery to combination therapy. *Trends in Molecular Medicine.* 21(4), 223–232.

Index